We Will Show Them!

Essays in Honour of
Dov Gabbay
on his 60th Birthday

Volume 1

We Will Show Them!

Essays in Honour of
Dov Gabbay
on his 60th Birthday
Volume 1

edited by

Sergei Artemov, Howard Barringer,
Artur d'Avila Garcez, Luis C. Lamb and John Woods

ISBN 1-904987-11-7
Published by College Publications
Scientific Directors: Dov Gabbay, Vincent F. Hendricks and John Symons
Managing Director: Jane Spurr
Department of Computer Science
King's College London
Strand, London WC2R 2LS, UK

http://www.collegepublications.co.uk

Cover design by Lydia Rivlin
Printed by Lightning Source, Milton Keynes, UK

Photograph Gallery

Dov Gabbay, today — actually 21st July 2005!

This picture was taken during a Logic Conference at Bar Ilan University in 1979 or 1980. You can see Dov, who is 5th from the right.

Dov, in his office at Imperial College in 1991.

This is a picture of Dov, in his office with his "muse", 7of9, and colleagues Odinaldo Rodrigues, Artur Garcez, Nicola Olivetti and Luis Lamb.

"Consistency is futile!"

CONTENTS

Preface

Visitors to Dov Gabbay's suite of offices at King's can hardly miss noticing the upper right quadrant of the white-board, directly opposite the room's sofa. In the board's other sectors, one sees the usual jottings of a logician at work, the detritus of definitions and proofs about modal and temporal logic, or labelled deduction and fibred semantics, or agenda relevance and the *GW*-schema for abduction, or neural networks and logic for learning. But the entries in the upper right aren't logic at all. In the centre is a single short sentence in English. In the rest of the space is a sprawl of different translations of it in German, Portuguese, Greek, Russian, Swedish, Hebrew, Dutch, Turkish, Latin, Italian and Hungarian and many more. In time, this babel grew to overflowing. To relieve the pressure, it was transferred by Lydia Rivlin, to an attractively backgrounded, large wooden panel affixed to the wall facing the entrance to Dov's rooms. Everyone knows what a rabbits' warren King's is, and it will come as no surprise that while the multi-lingual panel can hardly be seen by the entering visitor, it is squarely in the exitor's field of vision. What does the departing visitor see? In no fewer than 25 of the world's languages, one sees the sentence, "We will show them", having been contributed, one by one, by Dov's visitors streaming into King's from the world over.

For the most part, visitors react to "We will show them" with amusement. It is, they think, a funny thing for someone as well-known and esteemed as Dov Gabbay to be saying. For it is, is it not, a defiant saying, expressing a determination to persist until the outer world takes some notice and accords one a measure of respect? Clearly, then, an ironic joke? It is a joke of sorts, but not that one. It light-heartedly exploits an ambiguity, but it is not ironic.

When Dov was a first year student at the Hebrew University in 1963, he enrolled in the course in Mathematics and Physics. Before the year was out, he had lost his affection for physics, and his heart to pure mathematics and logic. As Dov would later say, "Physics was a speculative mess, scarcely different from romantic fiction! Only logic could deliver the certitude demanded by genuine knowledge." When he made that decision Dov was 18 years old. How could he know that logic was in process of the largest transformation since taking the mathematical turn in the middle and late nineteenth century? How could he know that he himself would be one of the driving forces of this change, and of how far it would take him from the austere dogmatism of his youth? Dov has called this transition "the practical turn in logic". Whereas the earlier, mathematical, turn had concentrated on the analysis of logically interesting properties in a highly decontextual way, the practical turn would reverse that abstraction. It would acknowledge the influence of time and place, of belief and belief change, of relevance and analogy, of consequence relations weaker than strict deduction, and the central place of agents. It would be a transformation that sought to recover something of logic's ancient and original remit, which was the exposure of the structure of human reasoning as transacted under the conditions of real-life. It is forty-two years since Dov repudiated physics for its speculativeness and its messiness. In the intervening years, he has learned a marked tolerance for these traits, and has only very recently observed,

> I believe that the computational point of view of logic, arising from computer science needs, which includes resource considerations and various other mechanisms such as abduction, revisions, actions, change, etc., will become increasingly predominant in logic. This point of view will eventually be extensively applied in traditional areas like analytic philosophy, law, economics, theology, language, and political theory among others. Logic will continue to evolve as a result of these applications to become an independent discipline akin to mathematics and physics, chemistry and biology. Mathematical logic will no longer be an important part of logic outside mathematics departments.

Dov is no enemy of mainstream mathematical logic, which is where he himself has done some of his most important work. No one admires more than he the foundational achievements of set theory, model theory, proof theory and recursion theory, not to overlook work of like importance in theoretical computer science and logic programming. Still, not every mainstream logician is happy with the turn toward the practical. It is a turn toward a messiness and a speculativeness that makes physics look rather neat and tidy. Some see the new logic as a return to the discredited Laws of Thought tradition in logic, and to a widely detested psychologism. Perhaps the shortest way of capturing the present cut of Dov Gabbay's jib as a logician is this. When Dov was a boy, he thought that logic was a purely abstract science. He now thinks, as did Aristotle, that logic is a humanities subject.

One sees, then, that there is in this view something indeed to be defiant about. Notwithstanding logic's inexorable turning toward the contextual, there are legions of practitioners of the mainstream, including some of its most eminent, for whom the last thing that a proper logic could be is one of the humanities. Dov has acquired more of his large reputation as an insider rather than an outsider. His insider reputation is wholly secure. His outsider reputation is a work in progress. Accordingly, a defiant "We will show them."

"We will show them" has another meaning that captures Dov's intellectual personality, and is not in the least defiant. People who know him will be aware that one good way to send Dov Gabbay into a quite undisguised coma is to discuss a problem rather than solve it. Colleagues who want to hold Dov's attention had better show him something, and had better not talk about showing it. Dov's hankering for demonstration wins him the status of honorary Missourian. Perhaps "hankering" is not the right word for it; it is better to think of it as a drive. Like most academics of high importance, Dov maintains a large network of contacts around the world. It would be interesting to know how many of these contacts arose from a "cold-call" from Dov, asking to be shown something. One sees this same drive at conferences. Dov is incapable of small talk. He isn't interested in tanking up at the bar or visiting baroque churches. Dov wants people to show him things. And, of course, the favour is returned as often as it is asked for. "Here, let me show you" could well be Dov's epitaph. Dov never lectures with notes. He has no affection for powerpoint presentations. These constrain spontaneity. But more importantly, they are not the best ways of showing things. The best way of showing things is with markers on a whiteboard or on overhead transparencies. The best way of showing something to others is to show it to oneself in their presence. So, again, "We will show them."

His evolving fidelity to logic's humanistic character is but one of the respects in which Dov Gabbay stands apart from the mainstream. Another is his extraordinary range. The chapters of this large book run to 60 and reflect several times that number of issues and problems. It is very nearly the literal truth to say that there isn't a single matter covered here that lies beyond Dov's own demonstrated reach. There is an amusing (and true) story about Dov as a youthful associate professor at Stanford in the 1970s. Georg Kreisel tapped on Dov's door. Would he please come to Kreisel's office to take a call from Gödel. Yes, of course. As it happens, Gödel wanted to say his piece in German; so Kreisel stayed on the phone and translated for Dov. What, Gödel demanded to know, did Dr. Gabbay think he was doing? Was it modal logic, was it temporal logic, was it intuitionistic logic, was it tense logic, or was it the formal semantics of natural languages, or was it statistics, or what? Dov allowed that it was all these, and more. Gödel was fit to be tied, fearful that an accursed dilettantism would soon strike at the brilliant young man's emerging profile. Dov did what he always does when under fire. He *negotiated*. After a good deal of to-ing and fro-ing, mediated by a growingly bemused Kreisel, Dov promised Gödel that he would cut back his involvement with statistics.

The mainstream's distrust of range is one of its defining features. Crossing the line of a sub-discipline invites rejection from both sides. Crossing over into a wholly different discipline risks opprobrium from both sides. This hostility to cross-disciplinary work is deeply entrenched. There are exceptions, of course. (One thinks of Weiner and von Neumann, to name just two.) But there are costs. To keep the mocking judgement of amateurism at bay, the discipline-crosser is expected to meet two targets. He must produce in accordance with each set of local standards. And he should develop results that couldn't (or most likely wouldn't) have been produced in either of the single disciplines alone. Part of what enables Dov to meet these expectations is that he is endlessly learning new things. Again, he never stops getting people to show him things. The things he most wants to be shown are the most important, the most central things — the essence of them. Dov is also a veritable pan-logician. Dov thinks that anything that interests him has a logic. Since his interests range all over the place, this helps explain why he is so receptive to a large pluralism in logic. It also helps explain why his multi-disciplinary efforts have been met with such success. Dov is an inveterate builder of formal models. Modelling comes to him as naturally as breathing. Dov will model anything that strikes his fancy. He will do so at the drop of a hat, especially if he has an audience. ("Here, let me show you"). What makes this so is that Dov is dominantly a visual thinker. He sees structure where others see nothing. It is also this architectural sense that mitigates Dov's drift toward the psychologically real. For all its messiness, Dov does not think that a psychologically real account of reasoning is unmodellable. This is important. While Dov wholly embraces the research targets of a practical logic of cognitive systems, his *methods* are still largely those of the logical mainstream. Any fool can make a bad model of just about anything. It takes something special to model the real in ways that subdue imprecision yet preserve complexity. Doubtless, it is Dov's successes in this regard that maintains him as a fully paid-up member of the mathematical mainstream. This is exactly as it should be for someone who thinks like Dov. Not only does the turn towards the practical re-orient logic's subject matter, it also re-sets the limits of a mathematical methodology. Thus logic's recovery of its humanistic focus is, in Dov Gabbay's hands, no cause to abandon the applicable insights of mathematical re-expression.

It is less a fact than an expository convenience that logic's turn toward the practical is a discrete and recent event. It is, in fact, not a single event and it is not all that recent. During the period in which first order classical logic was achieving its hegemony, non-classical rivals were also gaining strength. Not least of these were intuitionistic logic and the modal logics of strict implication, each of which arose in the early years of the century just past, and each of which had an applicational cast. Intuitionistic logic sought better to serve a philosophically correct appreciation of mathematical reasoning as actually practised, just as modal logic aimed for a more realistic account of entailment. In their turn, relevant, epistemic, deontic, temporal, dynamic and linear logics made for a closer tie between logic's formalisms and the structure of facts in the real world. Similar engagements with similar results arose from developments in situation and belief-change logics, and, from computer science, the logics of default, defeasible and non-monotonic reasoning, as well as logic programming. The pace of change was abated by the research programmes of cognitive science, argumentation theory and informal logic. There is these days more logic than you can shake a stick at. It is clearly an *embarras de richesse*. It is not an altogether good state for logic to be in. It threatens with prospects that anything goes. Anything-goes logic is Cole Porter logic. It embodies the conceit of pluralism without bounds. One of the attractions of the conservative mainstream is its insistence that *hardly anything* goes. It should be easy for all to see that Cole Porter logic is not a live option, that unconstrained pluralism will test the mettle of intellectual integrity as sorely as the paradoxical proofs of Presocratic conjurers. Somewhere lines have to be drawn. Limits must be placed on the theorists' freedom to say that human reasoning is anything he sees fit to stick into one or other of his models. Models require mature data as inputs, and the distortions to which they inevitably give rise must be reined in. The task of pulling together the best of logic's present pluralism into comprehensive and stable theories of the empirically real and normatively sound still lies before us, as unperformed as it is urgent. Logicism was the last great synthesis in logic (and a failed one). Logic is long overdue for another, and we can only hope that it will not fail. Perhaps, in the end, it is the new synthesis that best animates Dov's motto.

"We will show them": How likely is this to be true? Let us say simply that its prospects are good with Dov Gabbay at the helm, and significantly better than were he not.

The Editors, August 2005

Acknowledgements

We are grateful to Jane Spurr for her very effective assistance during all the phases of this project. Without her support, we would not have been able to meet the tight schedule — starting with invitations being sent out to authors in March 2005 — for the edition, production, and publication of this book by October 2005.

We would also like to thank the authors for their overwhelming response to our invitation, which has forced us to have a two-volume book, and for producing such high-quality, interesting contributions in such a short period of time.

We also thank Lydia Rivlin for producing an eye-catching, unique cover for the book, and the International Federation for Computational Logic (IF-CoLog) for their financial support, which will promote the free distribution of the book to key libraries and universities across the world.

The Editors, October 2005

List of Contributors

Samson Abramsky. *Oxford University Computing Laboratory, UK*
Email: samson.abramsky@comlab.ox.ac.uk

Name. *City University of New York, USA.*
Email: sartemov@gc.cuny.edu

David Ahn. *University of Amsterdam, The Netherlands.*
Email: ahn@science.uva.nl

Sisay Fissaha Adafre. *University of Amsterdam, The Netherlands.*
Email: sfissaha@science.uva.nl

Atocha Aliseda. *Universidad Nacional Autónoma de México*
Email: atocha@filosoficas.unam.mx

Amilhood Amir. *Bar-Ilan University, Israel, and Georgia Instiute of Technology, USA.*
Email: amir@cs.biu.ac.il

Name. Carlos Areces *INRIA Lorraine, France.*
Email: Carlos.Areces@loria.fr

Wouter van Atteveldt. *Vrije Universtiteit Amsterdam, The Netherlands.*
Email: wh.van.attevedlt@fs2.vu.nl

Arnon Avron. *Tel Aviv University, Israel.*
Email: aa@math.tau.ac.il

Matthias Baaz. *Technische Universität, Austria.*
Email: baaz@logic.at

Sebastian Bader. *Dresden University of Technology, Germany.*
Email: sebastian.bader@inf.tu-dresden.de

Howard Barringer. *Univeristy of Manchester, UK.*
Email: Howard.Barringer@manchester.ac.uk

Johan van Benthem. *Amsterdam, The Netherlands, and Stanford, USA.*
Email: johan@science.uva.nl

Patrick Blackburn. *INRIA Lorraine, France.*
Email: Patrick.Blackburn@loria.fr

Alexander Bochman. *Holon Academic Institute of Technology, Israel.*
Email: bochmana@hait.ac.il

Krysia Broda. *Imperial College London, UK*
Email: kb@doc.ic.ac.uk

Peter D. Bruza. *University of Queensland, Australia.*
Email: bruza@dstc.edu.au

Richard J. Cole. *University of Queensland, Australia.*
Email: rcole@itee.uq.edu.au

Carlos Caleiro. *Technical University of Lisbon, Portugal*
Email: ccal@math.ist.utl.pt

Walter Carnielli. *State University of Campinas, Brazil.*
Email: carniell@cle.unicamp.br

Marcelo E. Coniglio. *State University of Campinas, Brazil*
Email: coniglio@cle.unicamp.br

Ariel Cohen. *Technion - Israel Institute of Technology, Israel.*
Email: arikc@bgumail.bgu.ac.il

Marcello D'Agostino. *Dipartimento di Scienze Umane, Università di Ferrara, Italy.*
Email: dgm@unife.it

Artur S. d'Avila Garcez. *City University, London, UK.*
Email: aag@soi.city.ac.uk

Anuj Dawar. *University of Cambridge Computer Laboratory, UK.*
Email: Anuj.Dawar@cl.cam.ac.uk

Jürgen Dix. *Clausthal University of Technology, Germany.*
Email: dix@informatik.tu-clausthal.de

Kosta Došen. *Mathematical Institute, SANU, Belgrade, Serbia.*
Email: kosta@turing.mi.sanu.ac.yu

Luis Fariñas del Cerro. *Université Paul Sabatier, Toulouse, France.*
Email: farinas@irit.fr

Paolo Ferraris. *University of Texas, USA.*
Email: otto@cs.utexas.edu

Melvin Fitting. *City University of New York, USA.*
Email: melvin.fitting@lehman.cuny.edu

Marcelo Finger. *Universidade de São Paulo, Brazil.*
Email: mfinger@ime.usp.br

Chris Fox. *University of Essex, UK.*
Email: foxcj@essex.ac.uk

Michael Gabbay. *King's College London, UK.*
Email: michael.gabbay@kcl.ac.uk

Murdoch Gabbay. *King's College London, UK.*
Email: jamie@dcs.kcl.ac.uk

Olivier Gasquet. *Université Paul Sabatier, Toulouse, France.*
Email: gasquet@irit.fr

Joseph Goguen. *University of California at San Diego, USA.*
Email: goguen@cs.ucsd.edu

John Grant. *Towson University, USA.*
Email: jgrant@towson.edu

Andreas Herzig. *Université Paul Sabatier, Toulouse, France.*
Email: herzig@irit.fr

Pascal Hitzler. *University of Karlsruhe, Germany.*
Email: phitzler@aifb.uni-karlsruhe.de

Wilfrid Hodges. *Queen Mary, University of London, UK.*
Email: hodges@qml.ac.uk

Ian Hodkinson. *Imperial College London, UK.*
Email: imh@doc.ic.ac.uk

Rosalie Iemhoff. *Technische Universität, Austria.*
Email: iemhoff@logic.at

Dale Jacquette. *The Pennsylvania State University, USA.*
Email: dlj4@email.psu.edu

Michael Kaminski. *Technion - Israel Institute of Technology, Israel.*
Email: kaminski@cs.technion.ac.il

Ruth Kempson. *King's College, London, UK.*
Email: ruth.kempson@kcl.ac.uk

Sarit Kraus. *University of Maryland, USA and Bar Ilan University, Israel.*
Email: sarit@macs.biu.ac.il

Jan Willem Klop. *Vrije Universiteit, The Netherlands*
Email: jwk@cs.vu.nl

Agi Kurucz. *King's College London, UK.*
Email: kuag@dcs.kcl.ac.uk

Ugur Kuter. *University of Maryland, USA.*
Email: ukuter@cs.umd.edu

Luís C. Lamb. *Federal University of Rio Grande do Sul (UFRGS), Brazil.*
Email: luislamb@acm.org

Shalom Lappin. *King's College London, UK.*
Email: shalom.lappin@kcl.ac.uk

Daniel Leivant. *Indiana University, USA.*
Email: leivant@cs.indiana.edu

Vladimir Lifschitz. *University of Texas, USA.*
Email: vl@cs.utexas.edu

Kai Lin. *University of California at San Diego, USA.*
Email: klin@cs.ucsd.edu

Lorenzo Magnani. *University of Pavia, Italy and Sun Yat-sen University, China.*
Email: lmagnani@unipv.it

David Makinson. *King's College London, UK.*
Email: david.makinson@kcl.ac.uk

Larisa Maksimova. *Siberian Branch of Russian Acad. Sci., Russia*
Email: lmaksi@math.nsc.ru

Erica Melis. *German Research Institute for Artificial Intelligence (DFKI), Germany.*
Email: melis@dfki.de

George Metcalfe. *Technische Universität, Austria.*
Email: metcalfe@logic.at

Alice G. B. ter Meulen. *Univesity of Groningen, The Netherlands.*
Email: atm@let.rug.nl

Wilfried Meyer-Viol. *King's College, London, UK.*
Email: meyervio@dcs.kcl.ac.uk

Johann A. Makowsky. *Technion - Israel Institute of Technology, Israel.*
Email: janos@CS.Technion.AC.IL

Ben Moszkowski. *De Montfort University, UK.*
Email: benm@dmu.ac.uk

Dana Nau. *University of Maryland, USA.*
Email: nau@cs.umd.edu

Rolf Nossum. *University of Kristiansand, Norway.*
Email: Rolf.Nossum@hia.no

Hans-Jürgen Ohlbach. *Universität München, Germany*
Email: ohlbach@lmu.de

Anjolina G. de Oliveira. *Universidade Federal de Pernambuco (UFPE), Brazil.*
Email: ago@di.ufpe.br

Nicola Olivetti. *Unversity of Torino, Italy.*
Email: olivetti@di.unito.it

Don Perlis. *University of Maryland, USA.*
Email: perlis@cs.umd.edu

Zoran Petrić. *Mathematical Institute, SANU, Belgrade, Serbia.*
Email: zpetric@mi.sanu.ac.yu

Gabriella Pigozzi. *King's College London, UK.*
Email: pigozzi@kcl.ac.uk

Amir Pnueli. *Weizmann Institute of Science, Israel and Courant Institute of Mathematical Sciences New York University, USA.*
Email: amir@cs.nyu.edu; amir@wisdom.weizmann.ac.il

Ruy J. G. B. de Queiroz. *Universidade Federal de Pernambuco (UFPE), Brazil.*
Email: ruy@di.ufpe.br

Mark Reynolds. *University of Western Australia, Australia.*
Email: mark@csse.uwa.edu.au

Maarten de Rijke. *University of Amsterdam, The Netherlands.*
Email: mdr@wins.uva.nl

Odinaldo Rodrigues. *King's College London, UK.*
Email: odinaldo.rodrigues@kcl.ac.uk

Alessandra Russo. *Imperial College London, UK*
Email: ar3@doc.ic.ac.uk

Vladimir V. Rybakov. *Manchester Mentropolitan University, UK.*
Email: v.rybakov@mmu.ac.uk

Mohamad Sahade. *Université Paul Sabatier, Toulouse, France.*
Email: sahade@irit.fr

Stefan Schlobach. *Vrije Universtiteit Amsterdam, The Netherlands.*
Email: schlobac@few.vu.nl

David E. Rydeheard. *University of Manchester, UK.*
Email: david@cs.man.ac.uk

Amílcar Sernadas. *Technical University of Lisbon, Portugal*
Email: acs@math.ist.utl.pt

Cristina Sernadas. *Technical University of Lisbon, Portugal*
Email: css@math.ist.utl.pt

Valentin Shehtman. *Institute for Information Transmission Problems, Russia*
Email: shehtman@lpcs.math.msu.su

Jörg Siekmann. *German Research Institute for Artificial Intelligence (DFKI), Germany.*
Email: siekmann@dfki.uni-sb.de

Patrick Suppes. *Stanford University, USA.*
Email: suppes@csli.stanford.edu

Roel de Vrijer. *Vrije Universiteit, The Netherlands*
Email: rdv@cs.vu.nl

Jon Williamson. *University of Kent, UK.*
Email: j.williamson@kent.ac.uk

Frank Wolter. *University of Liverpool, UK.*
Email: frank@csc.liv.ac.uk

John Woods. *University of Britsh Columbia, Canada, and King's College London, UK.*
Email: jhwoods@interchange.ubc.ca
Michael Zakharyaschev. *King's College London, UK.*
Email: mz@dcs.kcl.ac.uk

A Cook's Tour of the Finitary Non-Well-Founded Sets

SAMSON ABRAMSKY

1 Some reminiscences, and an explanation

It is a great pleasure to contribute this paper to a birthday volume for Dov. Dov and I arrived at Imperial College at around the same time, and soon he, Tom Maibaum and I were embarked on a joint project, the Handbook of Logic in Computer Science. We obtained a generous advance from Oxford University Press, and a grant from the Alvey Programme, which allowed us to develop the Handbook in a rather unique, interactive way. We held regular meetings at Cosener's House in Abingdon (a facility run by what was then the U.K. Science and Engineering Research Council), at which contributors would present their ideas and draft material for their chapters for discussion and criticism. Ideas for new chapters and the balance of the volumes were also discussed. Those were a remarkable series of meetings — a veritable education in themselves. I must confess that during this long process, I did occasionally wonder if it would ever terminate .... But the record shows that five handsome volumes were produced [Abramsky et al., 1992]. Moreover, I believe that the Handbook has proved to be a really valuable resource for students and researchers. It has been used as the basis for a number of summer schools. Many of the chapters have become standard references for their topics. In a field with rapidly changing fashions, most of the material has stood the test of time — thus far at least!

A large part of this success is due to Dov. Even though this particular Handbook series (among the many he has edited) is not the closest to his own interests, he not only originally inspired the project and got it going, but he stayed with it, and his energy and enthusiasm were essential to carrying it through. The ideas he had learned from his previous experience with the Handbook of Philosophical Logic [Gabbay and Guenthner, 1983] proved important. For example, every Chapter had an official Second Reader, a friendly critic and conscience; in many cases, these Second Readers worked above and beyond the call of duty, and helped to materially improve the Chapters. Another of Dov's ideas was that each Chapter should have a broad division into three parts. A first part should be genuinely introductory, and provide a helpful overview to the browser — who might then return for a more detailed look. The second part should be the technical core of the topic — and the contents of this core should be agreed by a consensus, in the discussions at the Handbook meetings. Finally, in the last part the author was free to ride their own hobby-horses, and pursue those topics they were particularly keen on in greater depth. The wisdom behind this is that the freedom offered by this third part made accepting a wider consensus on the core of the Chapter much more palatable to authors.

Beyond these organizational ideas, Dov's presence at the Handbook meetings was crucial to establishing their distinctive, intellectually engaged but friendly and relaxed atmosphere. He infused these occasions with his inimitable sense of humour, and his vision of the great possibilities of cooperation in Science.

So, thank you Dov, for this and much else — but above all for drawing me into what proved to be such a worthwhile project.

An Explanation

The scientific paper which follows requires a few words of explanation. Recalling those days when we were engaged with the Handbook, roughly the period 1985–90, the thought occurred to me that writing up a lecture which I had given then, but never published, might be rather appropriate. For one thing, the lecture has something to say about modal logic, and hence is closer to Dov's interests than much of my work.

To whet the appetite of any modal logicians who may be reading this, let me challenge them with the following questions:

> What is the Stone space of the free modal algebra?
> Which kind of set theory does it provide a model for?

(If you can answer these questions, you are excused from reading this paper.)

Lectures on versions of this material were given on several occasions in 1988-89, including: the 1988 British Colloquium on Theoretical Computer Science in Edinburgh; the Workshop on Logic from Computer Science held at MSRI Berkeley in 1989; and the International Symposium on Topology held in Oxford in July 1989. The lecture has been referenced in several subsequent publications, e.g. [Mislove *et al.*, 1991; Barwise and Moss, 1996; Alessi *et al.*, 1997; Forti and Honsell, 1996]. It has always been rather on my conscience that I had not written it up for publication. That is what I have now done. I have added a few references to later work, and tidied up one or two points of technical detail, but essentially this is a straightforward write-up of the 1988 lecture. It is an extended discussion of a single example, which is used to illustrate some wider themes. It may not be *new*, but I hope that it can still be *useful*. In that sense, it is offered in the same spirit as the Handbook, which Dov and Tom and I were working on in those years.

2 Introduction

Our topic in this paper is a single example: the space of *finitary non-well-founded sets*, which we study from many different points of view: process models; metric approximation; topology and the Vietoris construction; modal logic and Stone duality; and domain theory. We obtain five distinct characterizations of this space. We also study some basic features of its behaviour as a set-theoretic universe.

The main purpose of this study is to illustrate some general themes, in particular:

- Alternative descriptions of models:
  - relating different semantics, and semantics and logic
  - deriving one systematically from the other.

- Taking our cue from Domain theory [Scott, 1970], the right level for mathematical modelling of computation is
  - not the strictly finite
  - not the unboundedly infinite
  - but the *finitary i.e.* those objects appearing as "limits" of finite ones.

Our example will arise from a topologizing of non-well-founded set theory [Forti and Honsell, 1984; Aczel, 1988]. We will end up with something logically weaker, but computationally more meaningful — and perhaps with some logical interest in its own right. Along the way, we shall touch on numerous points in denotational semantics, concurrency theory, modal logic and set theory.

Acknowledgements We will draw on ideas from many sources, notably: Peter Aczel on non-well-founded set theory [Aczel, 1988]; Nivat and de Bakker and Zucker on denotational models of processes based on ultrametrics [Nivat, 1979; de Bakker and Zucker, 1982]; Milner, Hennessy, Park, Bergstra and Klop on process algebra and bisimulation [Hennessy and Milner, 1985; Park, 1981; Bergstra and Klop, 1984]; and Smyth's topological perspective on computation [Smyth, 1983]. We shall also draw extensively on our own work on Domain Theory in Logical Form, A Domain Equation for Bisimulation, and Total vs. Partial Objects in Denotational Semantics [Abramsky, 1985; Abramsky, 1987; Abramsky, 1991a;

Abramsky, 1991b]. The reader in search of more details is directed to these papers.

3   First approach: finitary sets as limits of finite ones

We begin with the *hereditarily finite sets* $\mathbb{H}$. These have played a role in logic as a "roomier" and more structured alternative to the natural numbers $\mathbb{N}$. They can be defined inductively by

(1) $\dfrac{x_1 \in \mathbb{H}, \ldots, x_n \in \mathbb{H}}{\{x_1, \ldots, x_n\} \in \mathbb{H}}$

(The base case is $n = 0$, which gives $\varnothing \in \mathbb{H}$.) More formally, we can write the inductive definition

$$
\begin{aligned}
V_0 &= \varnothing \\
V_{k+1} &= \mathcal{P}(V_k) \\
V_\omega &= \bigcup_{k \in \omega} V_k
\end{aligned}
$$

and define $\mathbb{H} = V_\omega$.

This definition relies implicitly on the fact that the full powerset construction, applied to a finite set, can only yield finite subsets. A more conceptually pleasing definition is to use the finite powerset constructor $\mathcal{P}_f(\cdot)$ explicitly. We can then write a fixpoint equation for $\mathbb{H}$:

(2) $\mathbb{H} = \mu X. \mathcal{P}_f(X) = \displaystyle\bigcup_{k \in \omega} \mathcal{P}_f^k(\varnothing)$

by the least fixpoint theorem, since $\mathcal{P}_f(\cdot)$, unlike $\mathcal{P}(\cdot)$, is continuous.

This starts another train of thought. The finite powerset construction $\mathcal{P}_f(\cdot)$ builds *free semilattices* over sets — *i.e.* $(\mathcal{P}_f, \{\cdot\}, \bigcup)$ is a monad on **Set** whose algebras are the semilattices [Mac Lane, 1998]. Thus Definition (2) suggest an algebraic description of $\mathbb{H}$.

3.1 $\mathbb{H}$ as the free process algebra with one action

We consider the following algebraic theory:

A fragment of CCS [Milner, 1980]

The signature comprises a binary operation $+$, a unary *prefixing operator* $e\cdot$, and a constant 0. The equations are:

$$
\begin{aligned}
x + 0 &= x & (3) \\
x + y &= y + x & (4) \\
x + (y + z) &= (x + y) + z & (5) \\
x + x &= x & (6)
\end{aligned}
$$

which say that an algebra $(S, +, 0)$ is a semilattice (equivalently, an idempotent, commutative monoid). There are no axioms for the prefixing operator. We consider the free algebra with no generators. (Almost equivalently, we can consider the algebra BPA of Bergstra and Klop [Bergstra and Klop, 1984] over one generator.)

The intention is indicated by the following semantics:

$$
\begin{aligned}
[\![0]\!] &= \varnothing \\
[\![P + Q]\!] &= [\![P]\!] \cup [\![Q]\!] \\
[\![e \cdot P]\!] &= \{[\![P]\!]\}.
\end{aligned}
$$

Thus iterations of prefixing allows arbitrary (finite) levels of nesting of sets, while within each level the semilattice axioms give the structure of a finite powerset.[1]

We can therefore define a membership relation on this free algebra:

$$
[P] \in [Q] \iff \exists R.\, Q = e \cdot P + R.
$$

Note that this is equivalently expressed as the usual *transition relation* on processes:

$$
[P] \in [Q] \iff Q \xrightarrow{e} P
$$

This leads to a more computational, process-like description of hereditarily finite sets, as *finite, rooted trees*.

EXAMPLE 1.

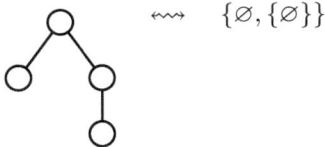

$$\rightsquigarrow \quad \{\varnothing, \{\varnothing\}\}$$

We can characterize equality on process terms or finite rooted trees directly in terms of the transition relation: this is the now-classical notion of *bisimulation* [Park, 1981; Hennessy and Milner, 1985; Milner, 1980]:

$$
\begin{aligned}
T_1 \sim T_2 \iff & T_1 \to T_1' \Rightarrow \exists T_2'.\, T_2 \to T_2' \wedge T_1' \sim T_2' \qquad (7) \\
\wedge & T_2 \to T_2' \Rightarrow \exists T_1'.\, T_1 \to T_1' \wedge T_1' \sim T_2'. \qquad (8)
\end{aligned}
$$

---

[1]Note that, if we built *complete* semilattices at each level, and iterated the construction through all ordinals, we would (disregarding set-theoretic subtleties) be building a full set-theoretic universe V as the initial algebra for this theory! Subsequently, Joyal and Moerdijk have developed a sophisticated form of "algebraic set theory" in a general categorical setting, which in essence uses this approach to the description of set-theoretic universes [Joyal and Moerdijk, 1995].

This recursively simulates extensionality:

$$x = y \iff \forall z.\, z \in x \leftrightarrow z \in y \tag{9}$$

$$x \sim y \iff \forall z.\, z \in x \Rightarrow \exists w.\, w \in y \wedge z \sim w \tag{10}$$

$$\wedge \; z \in y \Rightarrow \exists w.\, w \in x \wedge z \sim w. \tag{11}$$

This is indeed reminiscent of the way that extensionality is imposed in various constructions of models of set theory (e.g. Boolean-valued models in forcing), and pointed the way to Aczel's work on non-well-founded set theory.

The recursion in the above *coinductive definition* can, in this simple case of finite trees, be unwound inductively as follows [Hennessy and Milner, 1985]:

$$
\begin{aligned}
T_1 \sim_0 T_2 &\equiv \text{true} \\
T_1 \sim_{k+1} T_2 &\equiv T_1 \to T_1' \Rightarrow \exists T_2'.\, T_2 \to T_2' \wedge T_1' \sim_k T_2' \\
&\qquad \wedge \\
&\qquad T_2 \to T_2' \Rightarrow \exists T_1'.\, T_1 \to T_1' \wedge T_1' \sim_k T_2' \\
T_1 \sim T_2 &\equiv \forall k.\, T_1 \sim_k T_2.
\end{aligned}
$$

3.2   A metric on $\mathbb{H}$

Since every set is the directed union of its finite subsets, we need a more refined notion of limit to filter out computationally unrealistic sets. Thus topology begins to enter the picture.

The rooted tree representation of $\mathbb{H}$, and its associated bisimulation equivalence gives rise to a computationally meaningful notion of distance between hereditarily finite sets:

> The more work you have to do to distinguish between the sets,
> the closer they are.

We equate 'work' with the depth to which we have to probe the trees to discover a difference which distinguishes them as representations of sets. This leads to the following definition:

$$
d(S, T) = \begin{cases} 0, & S \sim T \\ 2^{-k}, & \text{least } k \text{ such that } S \not\sim_k T \quad \text{otherwise} \end{cases}
$$

EXAMPLE 2.

$$d(\varnothing, \{\varnothing\}) = 1/2, \qquad d(\{\varnothing\}, \{\varnothing, \{\varnothing\}\}) = 1/4.$$

We can give an inductive definition of this distance function, defined directly on hereditarily finite sets, as follows:

(1) $d(S,T) = 0$ If $S = T$

(2.1) $d(\varnothing, T) = d(S, \varnothing) = 1/2$ If $S \neq \varnothing \neq T$

(2.2) $d(S,T) = 1/2 \max(\sup_{s \in S} \inf_{t \in T} d(s,t),$ If $S \neq \varnothing \neq T$
$\sup_{t \in T} \inf_{s \in S} d(s,t))$

Note how, in the inductive case (2.2), the standard *Hausdorff metric* [Dugundji, 1966] appears as a "minimaxing" calculation of $\nsim_k$. We can also view the Hausdorff metric as the interpretation of the definition of extensional equality (9) in "real-valued logic" [Lavere, 1973].

This definition does indeed yield a metric, and in fact an ultra-metric:

$$d(S,T) \leq \max(d(S,U), d(U,T)).$$

Now that we have an (ultra-)metric space $(\mathbb{H}, d)$, we can look at Cauchy sequences to see which limits should arise.

EXAMPLE 3. The sequence $\varnothing, \{\varnothing\}, \{\{\varnothing\}\}, \{\{\{\varnothing\}\}\}, \ldots$ can be visualized as follows, with successive distances indicated:

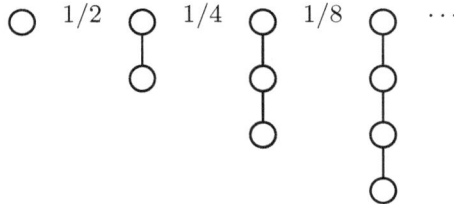

EXAMPLE 4.

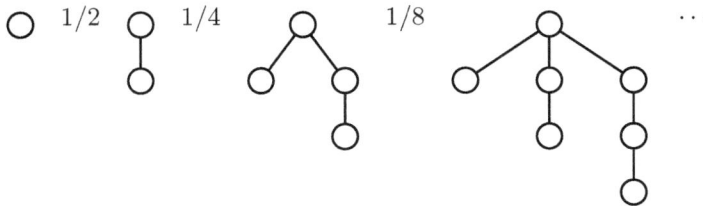

The corresponding sequence of sets is

$$\varnothing, \ \{\varnothing\}, \ \{\varnothing, \{\varnothing\}\}, \ \{\varnothing, \{\varnothing\}, \ \{\{\varnothing\}\}\}, \ldots$$

We now finally reach our first definition of the *universe of finitary sets*
$\mathbb{F}$:

DEFINITION 5. We define $\mathbb{F}$ to be the metric completion of $(\mathbb{H}, d)$; *i.e.*
(equivalence classes of) Cauchy sequences in $\mathbb{H}$. Membership is defined by:

$$[S_n] \in [T_n] \quad \equiv \quad \forall n.\, \exists m.\, S_n \in T_m.$$

Note that in Example (3), these definitions yield $S \in S$! Thus the process
of metric completion automatically gives rise to non-well-founded sets.

4 Interlude: Domain Equations

We consider domain equations $X \cong F(X)$. An *interpretation* of such
an equation is provided by specifying a category $\mathcal{C}$, and an endofunctor
$F : \mathcal{C} \longrightarrow \mathcal{C}$. We are generally interested in an *extremal solution* of such
an equation: either an *initial algebra* $\alpha : FA \to A$, or a *final coalgebra*
$\beta : A \to FA$. (The Lambek lemma [Lambek, 1970] guarantees that ini-
tiality or finality does indeed imply that the arrow is an isomorphism).
These concepts generalize the lattice-theoretic notions of least and greatest
fixpoint. In most cases of interest, initial algebras can be constructed as
colimits:

$$\varinjlim (\mathbf{0} \to F\mathbf{0} \to F^2\mathbf{0} \to \cdots)$$

generalizing the construction of the least fixpoint as $\bigvee_k F^k \bot$, while final
coalgebras can be constructed as limits:

$$\varprojlim (\mathbf{1} \leftarrow F\mathbf{1} \leftarrow F^2\mathbf{1} \leftarrow \cdots)$$

generalizing the construction of the greatest fixpoint as $\bigwedge_k F^k \top$. (In the
domain theoretic case, the *limit-colimit coincidence* [Scott, 1970] means that
the two constructions coincide, and we obtain *both* an initial algebra *and* a
final coalgebra.) For the finitary case we are considering, the functors will
be ω-continuous in the appropriate sense, and the limit or colimit can be
taken with respect to the ω-chain of finite iterations.

In [Abramsky, 1985], the view was taken that

> Domain equations can, and should, be viewed schematologically.

This means that a given language for describing functors can be interpreted
in different categories, and the solutions compared. A general result of this
kind is presented in [Abramsky, 1985]. We shall follow this point of view in
our treatment of $\mathbb{F}$.

5   Second description of $\mathbb{F}$

We will now characterize the space of finitary sets $\mathbb{F}$ as the solution of a domain equation. We take as our ambient category **CUMet**, the category of compact ultrametric spaces and continuous maps. Our functor $F : \mathbf{CUMet} \longrightarrow \mathbf{CUMet}$ is

$$F(X) = \mathbf{1} + \mathcal{P}_H(X)$$

where $\mathcal{P}_H(X)$ is the set of all non-empty closed subsets of X (note that 'closed' and 'compact' are equivalent in this context), with the Hausdorff metric:

$$d(S, T) = \max(\sup_{s \in S} \inf_{t \in T} d(s, t), \ \sup_{t \in T} \inf_{s \in S} d(s, t)).$$

The $\mathbf{1} + \cdots$ term is to code the empty set. We apply a contraction factor (for convenience, $1/2$) in taking the disjoint union $X + Y$, so that the empty set is a distance of at least $1/2$ from any non-empty set, and we shrink the Hausdorff metric by $1/2$ in each recursive iteration.

We now provide our second characterization of $\mathbb{F}$:

PROPOSITION 6. $\mathbb{F}$ *is the final coalgebra of the functor F. Moreover, $\mathbb{F}$ is cocontinuous, so $\mathbb{F}$ is constructed as the limit of the ω^{op}-chain*

$$\lim_{\leftarrow}(\mathbf{0} \leftarrow F\mathbf{0} \leftarrow F^2\mathbf{0} \leftarrow \cdots)$$

We give a picture of the first few terms of the construction:

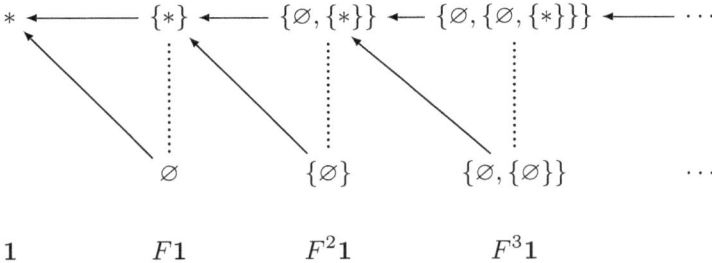

This characterization analyzes our ad hoc inductive definition into the iterated application of a general construction; similarly, the metric completion of the finite levels arises systematically from the general notion of limit used to construct the final coalgebra.

We therefore articulate the following principle:

A domain equation yields more information than an ad hoc construction.

The metric structure on $\mathbb{F}$ has some significance, e.g. we can apply the Banach fixpoint theorem to deduce that the equation $x = \{x\}$ has a unique solution, since the map $x \mapsto \{x\}$ is contractive, and similarly for

$$x = \{\varnothing\} \cup \{\{y\} \mid y \in x\}$$

(yielding our previous examples (3) and (4)). More generally, we get the existence of unique solutions for *guarded equations* (those in which the recursion variables appear under the scope of set-forming braces) [Aczel, 1988; Barwise and Moss, 1996]. However, for many purposes the topological structure suffices — and in any event, the precise definition of the metric is irrelevant to the structure. This leads naturally to the following

Question: Which metric topologies arise in **CUMet**?

PROPOSITION 7. *The category* **CUMet** *is equivalent to the category* **Stone** *of second-countable Stone spaces.*

Proof. In one direction, note that an open ball in an ultrametric is a closed set. Indeed, if $z \in B(x; \epsilon)$ and $d(x, y) \geq \epsilon$, then

$$\epsilon \leq d(x, y) \leq \max(d(x, z), d(z, y))$$

so $d(z, y) \geq \epsilon$, and $z \notin B(y; \epsilon)$. Thus $y \notin \overline{B(x; \epsilon)}$. This implies that the open balls form a clopen base in the metric topology. Since M is compact, it is second-countable.

For the converse, if a Stone space S is second countable, with dual Boolean algebra $B = \{b_n \mid n \in \mathbb{N}\}$, define

$$d(x, y) = \begin{cases} 0 & x = y \\ 2^{-n}, & \text{least } n \text{ such that } x \in b_n \Leftrightarrow y \notin b_n \quad \text{otherwise.} \end{cases}$$

■

This equivalence leads us to our next characterization of $\mathbb{F}$.

6   Third description of $\mathbb{F}$

We now characterize $\mathbb{F}$, *qua* topological space, as the solution of a domain equation in **Stone**. To do this, we need to answer the following question:

> What is the topological construction analogous to the Hausdorff metric powerspace $\mathcal{P}_H$?

The answer is provided by the *Vietoris construction* $\mathcal{P}_V$ [Johnstone, 1982]. Although it can be defined much more generally, we shall view $\mathcal{P}_V$ as a functor on **Stone**. Given a Stone space S, $\mathcal{P}_V(S)$ is the set of all compact (which since S is compact Hausdorff, is equivalent to closed) subsets of S, with topology generated by

$$\square U \;=\; \{C \mid C \subseteq U\} \tag{12}$$

$$\lozenge U \;=\; \{C \mid C \cap U \neq \varnothing\} \tag{13}$$

where U ranges over the open sets of S. We can read $\square U$ as the set of all C such that C *must* satisfy U, and $\lozenge U$ as the set of C such that C *may* satisfy U. The allusion to modal logic notation is thus deliberate, and we shall shortly see a connection to standard modal notions. Note that in our definition of the Vietoris powerspace, the empty set *is* included.

We now compare the Vietoris topology, and the metric topology arising from the Hausdorff powerspace metric. Given a metric space M, we write $\mathsf{MT}(M)$ for the topological space arising by taking the metric topology on M.

PROPOSITION 8. *For any M in* **CUMet***:* $\mathsf{MT}(1+\mathcal{P}_H(M)) = \mathcal{P}_V(\mathsf{MT}(M))$.

This result is in fact true in much greater generality, but the above is sufficient for our purposes.

Now we obtain the following description of $\mathbb{F}$.

PROPOSITION 9. *The space of finitary sets in its metric topology,* $\mathsf{MT}(\mathbb{F})$, *is the final coalgebra of the Vietoris functor on* **Stone***.*

This is an instance of a general result comparing the solutions of a class of domain equations in **CUMet** with the corresponding solutions in **Stone** [Abramsky, 1985]. (The class covers many of the ultrametric process models.) This description will be very useful in exploring the structure of $\mathbb{F}$ as a set-theoretic universe.

Since $\mathbb{F}$ is a Stone space, it has a dual Boolean algebra. This can be derived systematically — as a logic — from the domain equation used to describe $\mathbb{F}$. This illustrates the general programme for relating denotational semantics and program logics developed under the heading of 'Domain theory in logical form' [Abramsky, 1987; Abramsky, 1991a]. It will also lead us to our next description of $\mathbb{F}$.

7 Fourth description of $\mathbb{F}$

We now describe $\mathbb{F}$ as *the Stone space of the free modal algebra (on no generators)*. Here we take a modal algebra to be a Boolean algebra B equipped with a unary operator $\lozenge$ satisfying the axioms

$$(\mathrm{MA}) \quad \lozenge(a \vee b) = \lozenge a \vee \lozenge b \qquad\qquad \lozenge 0 = 0.$$

This is the algebraic variety corresponding to the minimal normal modal logic **K** [Blackburn *et al.*, 2001]. The Boolean algebra is equipped with a constant 0, so the free algebra over no generators can be non-trivial. We shall show that it is indeed non-trivial.

To derive this characterization systematically, we recall that the Vietoris construction can be described logically (or localically [Johnstone, 1982]) as an operation on *theories*. For the coherent case [Johnstone, 1982], $V(L)$, for a distributive lattice L, is the distributive lattice generated by $\Box a$, $\Diamond a$, $(a \in L)$, subject to the axioms:

$$\Box(a \wedge b) = \Box a \wedge \Box b \qquad \Diamond(a \vee b) = \Diamond a \vee \Diamond b \qquad (14)$$

$$\Box 1 = 1 \qquad \Diamond 0 = 0 \qquad (15)$$

$$\Box(a \vee b) \leq \Box a \vee \Diamond b \qquad \Diamond(a \wedge b) \geq \Diamond a \wedge \Box b. \qquad (16)$$

In the boolean case, where we have a classical negation, $\Box$ and $\Diamond$ are interdefinable (e.g. $\Box a = \neg \Diamond \neg a$), and the axiomatization simplifies to (MA). To see this, note firstly that (MA) implies that $\Diamond$ is monotone. Now from Boolean algebra we derive $a \leq \neg b \vee (a \wedge b)$, from which by monotonicity we obtain

$$\Diamond a \quad \leq \quad \Diamond(\neg b \vee (a \wedge b)) \quad = \quad \Diamond \neg b \vee \Diamond(a \wedge b),$$

and hence $\Diamond a \wedge \neg \Diamond \neg b \leq \Diamond(a \wedge b)$. The rest of (14)–(16) follows by duality. Thus we obtain a construction $\mathsf{MA}(B)$, which when applied to a Boolean algebra B constructs a new Boolean algebra with generators $\Diamond a$, $a \in B$, subject to the Boolean algebra axioms plus (MA).

Now we can iterate this construction to get the initial solution of $\mathbb{B} = \mathsf{MA}(\mathbb{B})$ in **Bool**, the category of Boolean algebras. This is constructed by taking a colimit of the finite iterates $\mathsf{MA}^k(\mathbf{2})$, starting from the 2-element Boolean algebra $\mathbf{2}$. This colimit can be constructed concretely as a union, essentially by taking the Lindenbaum algebra of the propositional theory which is inductively generated by these iterates. This propositional theory is the standard modal system **K**—but with no propositional atoms. (There *are* constants for true and/or false.) Thus another role for domain equations is revealed, as systematizing the inductive definition of the formulas and inference rules of a logic. A detailed account of how this works in a directly analogous (and rather more complex) case is given in [Abramsky, 1987; Abramsky, 1991a].

To see how hereditarily finite sets can be completely characterized by modal formulas (the "master formula" of the set; *cf.* [Hennessy and Milner,

1985]), we define:

$$\mathcal{F}(\varnothing) \;\;=\;\; \Box 0 \quad (=\neg\Diamond 1) \tag{17}$$

$$\mathcal{F}(\{x_1,\ldots,x_n\}) \;\;=\;\; \Box \bigvee_{i=1}^{n} \mathcal{F}(x_i) \,\wedge\, \bigwedge_{i=1}^{n} \Diamond\mathcal{F}(x_i). \tag{18}$$

The link between $\mathbb{B}$ and $\mathbb{F}$ is given by *Stone duality* [Johnstone, 1982]:

PROPOSITION 10.

1. $\mathbb{B}$ *is isomorphic to the Boolean algebra of clopen subsets of* $\mathbb{F}$, *with the modal operator* $\Diamond$ *defined by*

$$\Diamond U = \{S \in \mathbb{F} \mid S \subseteq U\}$$

   (recall that $\mathbb{F} \cong \mathcal{P}_V(\mathbb{F})$*).*

2. $\mathbb{F}$ *is isomorphic to Spec* $\mathbb{B}$, *the space of ultrafilters over* $\mathbb{B}$, *i.e.* models *of* $\mathbb{B}$: $f^{-1}(1)$, *for Boolean homomorphisms* $f : \mathbb{B} \longrightarrow \mathbf{2}$. *This space is topologised by:*

$$U_b = \{x \in Spec\ \mathbb{B} \mid b \in x\} \qquad (b \in \mathbb{B}).$$

Again, this is an instance of very general results [Abramsky, 1991a].

8   Fifth description of $\mathbb{F}$

Our final characterization of $\mathbb{F}$ is as the subspace of maximal elements of a *domain* [Scott, 1970; Abramsky and Jung, 1994]. Once again, this characterization will arise systematically, by comparing our previous description of $\mathbb{F}$ as the solution of a domain equation in **Stone** with the solution of a corresponding equation in a category of domains **Dom**. A convenient choice for **Dom** for our purposes is **SFP** [Plotkin, 1976], the countably-based bifinite domains [Abramsky and Jung, 1994].

We define a domain D as the solution (both initial algebra and final coalgebra) of the equation

$$(19) \quad D \;\;=\;\; \mathcal{P}_P^0(D) = \mathbf{1}_\perp \oplus \mathcal{P}_P(D).$$

Here $\mathcal{P}_P(\cdot)$ is the Plotkin or convex powerdomain [Plotkin, 1976; Abramsky and Jung, 1994]. The additional term in our domain equation codes in the empty set, by a "semi-coalesced sum":

$$\mathbf{1}_\perp \oplus D \; = \;$$

A similar but more general domain equation, which allows for an arbitrary set of possible actions, is used in [Abramsky, 1987; Abramsky, 1991b] to give a denotational semantics for process calculi, which is fully abstract with respect to strong *partial* bisimulation.

The Plotkin powerdomain on **SFP** (or more generally, Lawson-compact) domains, in fact coincides with the Vietoris construction on those domains seen as topological spaces with the Scott topology [Smyth, 1983; Abramsky, 1987; Abramsky and Jung, 1994].

Given a domain D, we write $\mathsf{Max}(D)$ for the subspace of maximal elements, viewed as a topological space. We can now state our final characterization of $\mathbb{F}$.

PROPOSITION 11. $\mathbb{F} \cong \mathsf{Max}(D)$, *where D is the solution of the domain equation (19).*

Again, this is an instance of a general result, relating solutions of a class of domain equations in **Dom** with those in **Stone** [Abramsky, 1985].

We note that D has "partial sets".

EXAMPLE 12.

$$\rightsquigarrow \quad \{\perp, \varnothing, \{\perp, \varnothing\}\}$$

The ordering on D is the Egli-Milner ordering [Plotkin, 1976]:

$$S \sqsubseteq T \;\; \equiv \;\; (\forall x \in S.\, \exists y \in T.\, x \sqsubseteq y) \;\wedge\; (\forall y \in T.\, \exists x \in S.\, x \sqsubseteq y).$$

This can be described in terms of "partial bisimulation"; see e.g. [Abramsky, 1991b].

The space of partial sets has been explored further in [Mislove*et al.*, 1991; Alessi *et al.*, 1997].

9 Set Theory in $\mathbb{F}$

We now turn to considering $\mathbb{F}$ as a set-theoretic universe, *i.e.* as a structure $(\mathbb{F}, \in, =)$. Since we have the isomorphism

$$\mathbb{F} \xrightleftharpoons[\text{fold}]{\overset{\text{unfold}}{\cong}} \mathcal{P}_V(\mathbb{F})$$

we can define

$$S \in T \;\equiv\; S \in \text{unfold}(T).$$

(We shall often elide uses of fold and unfold.)

How much of ordinary (ZFC) set theory does $\mathbb{F}$ satisfy? It certainly cannot be a model of full ZFC, since $\mathbb{F}$ satisfies the ultimate anti-foundation axiom:

$$\boxed{V \text{ is a set}}$$

where $V = \{x \mid \text{true}\} = \mathbb{F}$. So something has to give, or Russell's paradox would apply.

We begin by discussing the axioms which do straightforwardly apply. Our knowledge of the structure of the domain equation used to construct $\mathbb{F}$ makes these properties fall out very simply.

- Extensionality:

$$[\forall z. z \in x \leftrightarrow z \in y] \;\rightarrow\; x = y.$$

  This follows immediately from the injectivity of unfold.

- The following closure conditions:

  - $\varnothing \in \mathbb{F}$
  - $x \in \mathbb{F} \Rightarrow \{x\} \in \mathbb{F}$
  - $x, y \in \mathbb{F} \Rightarrow x \cup y \in \mathbb{F}$
  - $x \in \mathbb{F} \Rightarrow \bigcup x = \{z \mid \exists y \in x. z \in y\} \in \mathbb{F}$

  all hold, since $(\mathcal{P}_V(\mathbb{F}), \cup, \varnothing)$ is a semilattice (the free topological semilattice generated by $\mathbb{F}$), and $(\mathcal{P}_V, \{\cdot\}, \bigcup)$ is a monad [Johnstone, 1982].

- A suitable axiom of Infinity holds, since e.g.

$$x = \{\varnothing\} \cup \{\{y\} \mid y \in x\}$$

  has a (unique) solution in $\mathbb{F}$.

- The Powerset axiom holds, in the form:

$$x \in \mathbb{F} \;\Rightarrow\; \mathcal{P}_V(x) \in \mathbb{F}.$$

Indeed $\mathsf{unfold}(x) \subseteq \mathbb{F}$; applying $\mathcal{P}_V$ functorially to the inclusion map we get

$$\mathcal{P}_V(\mathsf{unfold}(x)) \hookrightarrow \mathcal{P}_V(\mathbb{F}).$$

The image of a compact space is compact, and hence $\mathcal{P}_V(\mathsf{unfold}(x)) \in \mathcal{P}_V(\mathcal{P}_V(\mathbb{F}))$, and $\mathsf{fold}^2(\mathcal{P}_V(\mathsf{unfold}(x))) \in \mathbb{F}$.

Clearly, we cannot have full (classical) separation, since otherwise we could derive an inconsistency from $V \in V$. We *do*, however, have *continuous* versions of Separation, Replacement, and Choice.

Separation Separation says that $\{y \in x \mid \phi(y)\}$ is a set. Consider the characteristic function of the predicate ϕ, $f_\phi : \mathbb{F} \to \mathbf{2}$. If f_ϕ is *continuous* ($\mathbf{2}$ taken as a discrete space), then

$$\{y \in x \mid f_{\phi(y)} = \mathsf{true}\} \in \mathbb{F},$$

since this is the inverse image of a closed set by a continuous map, hence closed, hence compact, hence in $\mathbb{F}$.

Replacement If $f : x \to \mathbb{F}$ is continuous, for $x \in \mathbb{F}$:

$$\{f(y) \mid y \in x\} \in \mathbb{F}.$$

Since x is compact, the image $f(x) \subseteq \mathbb{F}$ is compact, hence $f(x) \in \mathcal{P}_V(\mathbb{F})$, and $\mathsf{fold}(f(x)) \in \mathbb{F}$.

Choice If $f : x \longrightarrow \mathcal{P}_V(\mathbb{F}) \setminus \{\varnothing\}$ is continuous, for $x \in \mathbb{F}$, then for some $y \in \mathbb{F}$:

$$\forall z \in x.\, \exists w \in f(z).\, w \in y \;\wedge\; \forall w \in y.\, \exists z \in x.\, w \in f(z).$$

This arises from a standard topological result about selection functions [Michael, 1951]: given continuous $f : x \longrightarrow \mathcal{P}_V(\mathbb{F}) \setminus \{\varnothing\}$, there is a continuous selection function $g : x \to \mathbb{F}$ with $g(z) \in f(z)$, for all $z \in x$. We can then take

$$y = \{g(z) \mid z \in x\}.$$

In order to formulate a logical theory for this set-theoretic universe, we need to impose syntactic conditions on formulas to ensure that they give rise to continuous functions. These conditions will, inevitably, involve restrictions on the use of negation. They can then be used to formulate appropriate versions of the Separation, Replacement and Choice axioms.

For more extensive investigations in this area, see e.g. [Forti and Honsell, 1996].

BIBLIOGRAPHY

[Abramsky, 1985] S. Abramsky. *Total vs. Partial Objects in Denotational Semantics*. Unpublished lecture, given at Workshop on Category Theory and Computer Programming, Guildford, 1985.

[Abramsky, 1987] S. Abramsky. *Domain Theory and the Logic of Observable Properties*. Ph.D. thesis, University of London, 1987. Available at http://web.comlab.ox.ac.uk/oucl/work/samson.abramsky/.

[Abramsky, 1991a] S. Abramsky. Domain theory in logical form. *Annals of Pure and Applied Logic* 51, 1–77, 1991.

[Abramsky, 1991b] S. Abramsky. A Domain Equation for Bisimulation. *Information and Computation*, 92(2), 161–218, 1991.

[Abramsky and Jung, 1994] S. Abramsky and A. Jung. Domain Theory. *Handbook of Logic in Computer Science*, edited by S. Abramsky, D. M. Gabbay and T. S. E. Maibaum, Oxford University Press, Vol. 3, 1–168, 1994.

[Abramsky et al., 1992] *The Handbook of Logic in Computer Science*, edited by S. Abramsky, D. M. Gabbay and T. S. E. Maibaum. Oxford University Press. Volume 1: *Background: Mathematical Structures* and Volume 2: *Background: Computational Structures*, published in 1992. Volume 3: *Semantic Structures* and Volume 4: *Semantic Modelling*, published in 1995. Volume 5: *Logic and Algebraic Methods*, published in 2000.

[Aczel, 1988] P. Aczel. *Non-Well-Founded Sets*. CSLI Lecture Notes Vol. 14. Stanford University. 1988.

[Alessi et al., 1997] F. Alessi, P. Baldan and F. Honsell. Partializing Stone Spaces Using SFP Domains. *Proceedings of CAAP '97*, Springer Lecture Notes in Computer Science Vol. 1158:478–489, 1997.

[de Bakker and Zucker, 1982] J. de Bakker and J. Zucker. Processes and the denotational semantics of concurrency. *Information and Control* 54:70–120, 1982.

[Barwise and Moss, 1996] J. Barwise and L. Moss. *Vicious Circles*. CSLI Publications, 1996.

[Bergstra and Klop, 1984] J. Bergstra and J.-W. Klop. Process algebra for synchronous communication. *Information and Control* 60:109–137, 1984.

[Blackburn et al., 2001] P. Blackburn, M. de Rijke and Y. Venema. *Modal Logic*. Cambridge University Press 2001.

[Dugundji, 1966] J. Dugundji. *Topology*. Allyn and Bacon 1966.

[Forti and Honsell, 1984] M. Forti and F. Honsell. Set Theory with Free Construction Principles. *Annali Scuola Normale Superiore — Pisa Classe de Scienza 10:493–522. Serie IV*. 1983.

[Forti and Honsell, 1996] M. Forti and F. Honsell. A General Construction of Hyperuniverses. *Theoretical Computer Science* 156:203–215, 1996.

[Gabbay and Guenthner, 1983] D. Gabbay and F.Guenthner. *Handbook of Philosophical Logic Volumes I–IV*. First Edition, D. Reidel, 1983–89.

[Hennessy and Milner, 1985] M. Hennessy and R. Milner. Algebraic laws for nondeterminism and concurrency. *Journal of the ACM* 32(1):137–161, 1985.

[Johnstone, 1982] P. T. Johnstone. *Stone Spaces*. Cambridge University Press 1982.

[Joyal and Moerdijk, 1995] A. Joyal and I. Moerdijk. *Algebraic Set Theory*. Cambridge University Press, 1995.

[Lambek, 1970] J. Lambek. Subequalisers. *Canad. Math. Bull.* 13:337–349, 1970.

[Lavere, 1973] F. W. Lawvere. Metric spaces, generalized logic and closed categories. *Rendiconti Seminario Matematico e Fisico di Milano XLIII*, Pavia 1973.

[Mac Lane, 1998] S. Mac Lane. *Categories for the Working Mathematician*. Second Edition. Springer-Verlag 1998.

[Michael, 1951] E. Michael. Topologies on spaces of subsets. *Trans. American Math. Soc.* 71, 152–182, 1951.

[Milner, 1980] R. Milner. *A Calculus of Communicating Systems*. Springer Lecture Notes
    in Computer Science Vol. 92, 1980.

[Mislove*et al.*, 1991] M. Mislove, F. Oles and L. Moss. Non-well-founded sets modelled
    as ideal fixed points. *Information and Computation* 93(1):16–54, 1991.

[Nivat, 1979] M. Nivat. Infinite words, infinite trees, infinite computations. *Foundations
    of Computer Science III*. Mathematics Centrum Tracts, 1979.

[Park, 1981] D. Park. Concurrency and Automata on Infinite Sequences. *Proceedings
    of the 5th GI Conference*. Springer Lecture Notes in Computer Science, Vol. 104,
    167–183, 1981.

[Plotkin, 1976] G. D. Plotkin. A Powerdomain Construction. *SIAM Journal on Com-
    puting*, 5:452–487, 1976.

[Scott, 1970] D. S. Scott. *Outline of a Mathematical Theory of Computation*. Technical
    Monograph PRG-2 OUCL, 1970.

[Smyth, 1983] M. B. Smyth. Powerdomains and predicate transformers: a topological
    view. *Automata, Languages and Programming*, edited by J. Diaz, Springer Lecture
    Notes in Computer Science Vol. 154, 662–675, 1983.

Existential Semantics for Modal Logic

SERGEI ARTEMOV

1 Introduction

Gödel's paper [Gödel, 1933] offered, perhaps, the first rigorous approach to
the semantics of modal logic. In that paper, Gödel interpreted the modality
$\Box F$ as "F is provable," or in other words, "there exists a proof for F," and
offered the modal logic S4 as the propositional logic of provability. A similar,
existential-style semantics for modality is intrinsic to epistemic logic. Plato's
celebrated definition of knowledge as "justified true belief" (cf. *Meno* and
Theatatus) requires an *existence of a justification* for claiming $\Box F$, i.e., "F
is known."

However, Kripke semantics, which since the 1960s has been a dominant
semantics for modal logic is of a clear *universal* character, since it reads
the modality $\Box F$ as "in all possible situations, F holds." Such a reading of
modality is typical in dynamic modal logics, temporal logics, etc., where a
graph of possible runs of a computational process, or a timeline constitutes
a natural Kripke frame. Hintikka's modal logic approach to knowledge
([Hintikka, 1962]) also relies on universal-style semantics of knowledge of F
as "F holds in all possible situations."[1]

The clarity and efficiency of Kripke semantics in studying modal logics
and its numerous applications led to a situation in which modal logicians
sometimes routinely identify Kripke semantics with *the* semantics for modal
logic. Such a bias toward universal-style semantics for modal logic proved
to be restrictive and sometimes counterproductive. Some questions involving modal logic and applications required the development and use of an
existential-style formal semantics for modal logic.

Here, we will survey two such examples: formalizing Gödel's intended
provability interpretation of S4, and incorporating justification into formal
epistemic logic to capture the classical "justified true belief" definition of
knowledge.

---

[1]Though Hintikka's approach ignores the "justification" component Plato's tripartite
definition of knowledge, it captures the "true belief" components.

The existential realizability semantics we offer may be regarded as belonging to the framework of Labelled Deductive Systems described by Dov Gabbay (cf., for example, [Gabbay, 1994])[2].

2   Realizability semantics for modal logic

Gödel in [Gödel, 1933] used the modal logic S4 consisting of classical propositional calculus augmented by modal principles

1. $\Box(F \to G) \to (\Box F \to \Box G)$,
2. $\Box F \to \Box\Box F$,
3. $\Box F \to F$,
4. If $\vdash F$ then $\vdash \Box F$ *(Rule of Necessitation)*,

as the logic describing plausible laws of provability to provide the provability semantics for intuitionistic propositional logic IPC. Gödel immediately noticed, however, that the straightforward reading of S4-modality as "provable in a given formal system" contradicts his Second Incompleteness Theorem. Indeed, from 3. by Necessitation, one gets $\Box(\Box F \to F)$, which is a modal way of saying that consistency of the first-order arithmetic PA is provable in PA, given $\Box$ stands for a formal provability in PA and F is set to $0 = 1$.

Later in 1938 (first published in [Gödel, 1995]) Gödel sketched an idea of the logic of proofs and hinted that such a logic will provide a desired provability semantics for his provability calculus S4 and intuitionistic logic IPC. Unfortunately, that paper remained unpublished and virtually unknown until Kurt Gödel's Collected Work, volume III, which appeared in 1995. The format of the logic of proofs was similar to the one used in the Brouwer-Heyting-Kolmogorov provability semantics for intuitionistic logic, hence, since S4 emulates IPC via the well-known Gödel translation (cf. [Troelstra and Schwichtenberg, 1996; Chagrov and Zakharyaschev, 1997]), the logic of proofs should provide a formalization of the latter semantics as well.

In an equivalent formulation, the logic of proofs instead of $\Box F$ ("there is a proof for F") used new atoms of the form $t{:}F$, understood as "t is a proof of F." The modal postulates of S4 above have a clear reading in terms of operations on proofs. So 1. corresponds to the operation *application* "·" such that

$$s{:}(F \to G) \to (t{:}F \to (s{\cdot}t){:}G) \ .$$

Axiom 2 encodes a *proof checking* operation "!" such that

$$t{:}F \to !t{:}(t{:}F) \ .$$

Axiom 3 does not seem to correspond to any operations on proofs, but its explicit counterpart

$$t{:}F \to F$$

---

[2]The author takes this opportunity to sincerely congratulate Dov Gabbay on the occasion of his 60th birthday.

is internally provable, unlike $\Box F \to F$ itself. Indeed, for any specific t, if t is a proof of F, then F is provable and hence $t\!:\!F \to F$ is provable. Otherwise, if t is not a proof of F, then $\neg t\!:\!F$ holds and is provable as a true (primitive) recursive fact. In this case $t\!:\!F \to F$ is again provable. The Necessitation Rule 4. signifies a meta-operation of *internalization*, which given a derivation of F, returns its formalized version as a proof term p such that $p\!:\!F$ is derivable.

Gödel's papers [Gödel, 1933; Gödel, 1995] gave neither a complete description of operations on proofs needed, nor a complete system of postulates for the logic of proofs. These questions, along with the proof of the realization theorem, were settled in the 1990s (first seminar/conference talks in 1994, first publication in [Artemov, 1995], a master paper [Artemov, 2001a]). It turned out that in addition to *application* and *proof checker* one more operation on proofs is needed for building a robust logic of proofs capable of realizing the whole of modal logic S4. This new operation is *sum* "+", obeying the postulates

$$s\!:\!F \to (s+t)\!:\!F, \quad t\!:\!F \to (s+t)\!:\!F.$$

In the common representation of proofs as finite lists of formulas, the operation "+" can be interpreted as the concatenation of proofs. Naturally, we assume that a given proof proves all formulas appearing in the list; we call such proof systems *multi-conclusion*. Note that *sum* yields polymorphism of the resulting logic: if $s\!:\!F$ and $t\!:\!G$ hold, then both $(s+t)\!:\!F$ and $(s+t)\!:\!G$ hold as well, hence $s+t$ has two types, F and G.

The resulting system of proof objects, called *proof polynomials*, is defined as a system of formal terms built from variables $x, y, z, \ldots$ and constants $a, b, c, \ldots$ by means of two binary operations, *application* "·" and *sum* "+", and one unary operation *proof checker* "!". The language of the Logic of Proofs is that of classical propositional logic augmented by a new formula construction rule:

if t is a proof polynomial and F a formula, then $t\!:\!F$ is a formula.

Note that unlike typed combinatory logic or lambda calculus, any $t\!:\!F$ is a well-defined formula for any t and F. Once again, the language of the Logic of Proofs may be regarded as a special instance of Gabbay's Labelled Deductive Systems [Gabbay, 1994].

The Logic of Proofs LP has the postulates

A0. classical axioms
A1. $t{:}(F \to G) \; \to (s{:}F \to (t \cdot s){:}G)$ *(application)*
A2. $t{:}F \to F$ *(reflexivity)*
A3. $t{:}F \; \to {!}t{:}(t{:}F)$ *(proof checker)*
A4. $s{:}F \to (s+t){:}F, \quad t{:}F \to (s+t){:}F$ *(sum)*
R1. modus ponens
R2. $\vdash c{:}A,$ *where* $A \in A0\text{-}A4,$ *c is a proof constant.*   *(axiom necessitation)*

Perhaps *R2* requires a short comment. This rule is responsible for assigning proof constants to specific axioms of the system to signify given atomic proofs which the system does not analyze any further. As is, the axiom necessitation rule *R2* admits assigning any constant to any formula, thus allowing the reuse of constants as proofs of different axioms. On the other hand, one could use *R2* in the "injective" mode by picking a fresh constant for each instant of an axiom.

A key component of any semantics of LP is a *constant specification*, which is an assignment of sets of axioms to proof constants. Any derivation in LP specifies a constant specification $\{c_1{:}A_1, c_2{:}A_2, c_3{:}A_3, \ldots\}$ obtained by the rule *R2* in this derivation.

The standard provability semantics for LP is provided by a multi-conclusion proof predicate $Proof(x, y)$ in PA. We assume that such a predicate is supplied with computable operations for *application, proof checker,* and *sum.* As is usual for the provability semantics of logical languages, sentence letters of LP are interpreted as arbitrary arithmetical sentences, and proof polynomials are interpreted as proofs (their Gödel numbers, to be precise) with given operations *application, proof checker,* and *sum.* A sentence of the form $t{:}F$ is interpreted as a formal statement that t is a proof of F:

$$[t{:}F]^* \;\; = \;\; Proof(t^*, F^*) \,,$$

where t^* and F^* are interpretations of t and F, respectively. An interpretation $*$ *respects a given constant specification CS* if all formulas $c{:}A$ from *CS* are true under the interpretation $*$. The soundness theorem claims that any formula F derivable in LP with a given constant specification *CS* is true under any arithmetical interpretation respecting *CS*. In [Artemov, 1995; Artemov, 2001a] one could find arithmetical completeness theorems stating that LP derives all logical principles in its langauge that are true in every provability interpretation of the above type.

The goal of this note is to show how proof polynomials and their natural provability interpretation above provide an existential semantics of realizability for the modal logic S4 and, with corresponding modifications, to other major modal logics. All relevant technical work has been done in the papers mentioned below, we will just be using these results here.

How is the Logic of Proofs LP related to the original modal logic S4? Trivially, the forgetful projection of LP when each occurrence of $t{:}F$ is replaced by $\Box F$ is S4-compliant. The main question here was whether LP was indeed a constructive version of the whole S4, i.e., whether each S4-theorem F could be given an LP-reading by recovering proof polynomials in modalities of F. This question may be regarded as a question about skolemization of informal existential quantifiers on proofs built into modalities.

> Consider, for example, a formula $\Box F \rightarrow \Box G$. Its provability reading is *if there is x such that x is a proof of F, then there is y such that y is a proof of G*. An informal skolemization of the latter would introduce a new function $y = f(x)$ and convert the statement into *if x is a proof of F, then $f(x)$ is a proof of G*. In LP notation, this would be a formula $x{:}F \rightarrow f(x){:}G$, i.e., two displayed modalities of the original formula are realized by x and $f(x)$ respectively.

An easy first observation demonstrates that LP validates Gödel's internalization rule. The following holds in LP (cf. [Artemov, 1995; Artemov, 2001a]):

$$if \vdash F, \ then \vdash p{:}F \ for \ some \ proof \ polynomial \ p \ ,$$

which is a constructive version of the usual Necessitation Rule in modal logic $\vdash F \implies \vdash \Box F$, where the displayed occurrence of $\Box$ has been realized by a corresponding proof polynomial p. This observation creates an illusion that the general realization theorem of S4 in LP is now a cheap exercise in induction on S4-derivations. Indeed, the axioms of S4 are trivially realizable; the Necessitation Rule also has its natural counterpart in LP. However, the straightforward induction fails in the *modus ponens* step: by the induction hypothesis, realizations of F and $F \rightarrow G$ exist but they could be wildly different on F, which does not allow us to conclude that G is also realizable. The last observation also shows that a possible realization procedure of a formula in a given derivation tree cannot be local, i.e., depend only on a subtree that derives this formula.

Naturally, the solution came from an analysis of cut-free derivations in S4. The following realization theorem was first established in [Artemov, 1995], cf. also [Artemov, 2001a]:

THEOREM 1. S4 *is the forgetful projection of* LP.

In particular, there is an algorithm that constructs a realization r, which substitutes proof polynomials for all modalities in a given normal Gentzen-style S4-derivation of formula F, and thereby produces a formula F^r derivable in LP, such that

1) all negative occurrences of modalities in F are realized by variables;

2) all positive occurrences of modalities in F are realized by proof polynomials depending only on variables from (1).

The realization procedures from [Artemov, 1995] and [Artemov, 2001a] were easy to describe but not computationally efficient. In particular, they work in exponential time and produce realization terms exponential in the size of the original (cut-free) S4-derivation. An improved polynomial algorithm was found in [Brezhnev and Kuznets, 2005]. In particular, it produces realizing proof polynomials of length quadratic in the size of the original cut-free S4-derivation. A surprising semantical proof of the realization theorem for S4 was found in [Fitting, 2005a]. Fitting's proof does not depend on cut-elimination and hence could be generalized to other modal logics which do not necessarily have the cut-elimination property.

We do not present realization algorithms here, and refer the reader to the above mentioned papers. Here is an example of an S4-derivation converted into an LP-derivation which, it is hoped, gives some flavor of how the realization works.

	Derivation in S4	Derivation in LP
1.	$\Box A \to \Box A \vee B$	$x{:}A \to x{:}A \vee B$
2.	$\Box(\Box A \to \Box A \vee B)$	$a{:}(x{:}A \to x{:}A \vee B)$
3.	$\Box\Box A \to \Box(\Box A \vee B)$	$!x{:}x{:}A \to (a{\cdot}!x){:}(x{:}A\vee B)$
4.	$\Box A \to \Box\Box A$	$x{:}A \to !x{:}x{:}A$
5.	$\Box A \to \Box(\Box A \vee B)$	$x{:}A \to (a{\cdot}!x){:}(x{:}A\vee B)$
5a.		$(a{\cdot}!x){:}(x{:}A\vee B) \to (a{\cdot}!x+b{\cdot}y){:}(x{:}A\vee B)$
5b.		$x{:}A \to (a{\cdot}!x+b{\cdot}y){:}(x{:}A\vee B)$
6.	$B \to \Box A \vee B$	$B \to x{:}A \vee B$
7.	$\Box(B \to \Box A \vee B)$	$b{:}(B \to x{:}A \vee B)$
8.	$\Box B \to \Box(\Box A \vee B)$	$y{:}B \to (b{\cdot}y){:}(x{:}A \vee B)$
8a.		$(b{\cdot}y){:}(x{:}A \vee B) \to (a{\cdot}!x+b{\cdot}y){:}(x{:}A\vee B)$
8b.		$y{:}B \to (a{\cdot}!x+b{\cdot}y){:}(x{:}A\vee B)$
9.	$\Box A \vee \Box B \to \Box(\Box A\vee B)$	$x{:}A\vee y{:}B \to (a{\cdot}!x+b{\cdot}y){:}(x{:}A\vee B)$

Steps 1–4, 6, 7, and 9 admit straightforward translations from S4 to LP. Steps 5 and 8 need adjustments in LP since similar occurrences of modality there are realized with different proof terms. Operation *sum* allows us to consolidate two different proof terms for the same formula

$$(a{\cdot}!x){:}(x{:}A\vee B) \text{ and } (b{\cdot}y){:}(x{:}A \vee B)$$

into one

$$(a{\cdot}!x+b{\cdot}y){:}(x{:}A\vee B)$$

needed to conclude the derivation.

Systems of proof polynomials for the modal logics K, K4, D, D4, and T were described in [Brezhnev, 2000; Brezhnev, 2001].

Artemov, Kazakov, and Shapiro in [Artemov $et\ al.$, 1999] found a system of proof terms capable of realizing modal logic

$$\mathsf{S5} = \mathsf{S4} + (\neg \Box F \rightarrow \Box \neg \Box F) \ .$$

What differentiates this case from S4 is the presence of negative information about proofs. The paper [Artemov $et\ al.$, 1999] introduces a logic of proofs for S5 and establishes the realizability theorem for S5.

The following fundamental self-referential property of the provability semantics for modal logics was established recently by R. Kuznets ([Brezhnev and Kuznets, 2005]): there are formulas provable in S4, each realization of which requires a self-referential constant specification of the sort $c\!:\!A(c)$, where $A(c)$ contains c. An example of such a formula is

$$\neg \Box \neg (S \rightarrow \Box S) \ ,$$

where S is a sentence variable. This result shows that the provability semantics for modal logic is intrinsically self-referential.

These advances demonstrate that realizability of modalities by corresponding proof terms provides an existential-style semantics for a wide range of major modal logics. Such semantics was indispensable for finding the intended provability interpretation of Gödel's provability logic S4 and formalizing the Brouwer-Heyting-Kolmogorov semantics for the intuitionistic logic IPC ([Artemov, 2001a]).

In the next chapter, we will discuss another major application of the realizability semantics for modal logic in the logic of knowledge.

3 Introducing justification into epistemic logic

Proof polynomials constitute a robust structure. One can show that the system of proof polynomials enjoys some sort of functional completeness property with respect to the standard provability semantics: every operation on proofs which is invariant with respect to the choice of a proof system and which can be specified by a propositional formula is realized by a proof polynomial ([Artemov, 1995; Artemov, 2001b]).

In [Artemov $et\ al.$, 1999; Artemov, 2001a; Fitting, 2004; Fitting, 2005b] the Logic of Proofs is viewed as a general purpose calculus of justification. $Application$, $proof\ checker$, and sum are meaningful not only as operations on proofs: these operations are so basic that they may be regarded as operations on pieces of evidence in general.

The principle *A1* specifies the basic operation of *application*: a justification of an implication $F \to G$ applied to any justification of the premise F returns a justification of the conclusion G. *A2* is the verifiability property of evidence: for any evidence t of F, the result of applying a checker to t, $!t$, provides a justification of $t{:}F$. *A3* reflects the monotonicity principle: a justification for F remains a justification after adding any additional evidence. Finally, LP4 is the reflexivity property.

Completeness properties of the Logic of Proofs with respect to modal logic show a certain degree of sufficiency of proof polynomials as the system of justification for classical propositional logic.

A principal contribution to generalizing the Logic of Proofs to a general calculus of justification has been made by M. Fitting, who developed a Kripke-style semantics for LP ([Fitting, 2004; Fitting, 2005b]). One of the key ingredients of Fitting's semantics, the evidence function, has been studied earlier in [Mkrtychev, 1997].

In [Kuznets, 2005] it was shown that the Logic of Proofs as a system of knowledge does not suffer the logical omniscience sickness: one cannot claim in LP a knowledge assertion of a formula without actually building an evidence for this formula. An exact result by R. Kuznets states that if LP proves $t{:}F$, there is a proof of F in LP of a length less than the length of t.

A step incorporating proof polynomials as a justification component into epistemic logic has been made in [Artemov and Nogina, 2004; Artemov, 2004; Artemov and Nogina, 2005]. These papers presented a framework long wished for in epistemology — the introduction of justification for a knowledge operator. The problem of epistemic justification for knowledge assertions has been a pivotal concern in mainstream epistemology.

In [Artemov, 2004], the following three basic series of evidence-based knowledge systems were introduced. Consider n copies of modal logics representing knowledge operators of n agents, $K_1, \ldots, K_n$, along with a system of evidence assertions taken from the Logic of Proofs LP. Here we consider some representative examples.

$T_n LP$ = *all postulates of* T_n *and* LP *plus* $t{:}\varphi \to K_i \varphi$.

$S4_n LP$ = *all postulates of* $S4_n$ *and* LP *plus* $t{:}\varphi \to K_i \varphi$.

$S5_n LP$ = *all postulates of* $S5_n$ *and* LP *plus* $t{:}\varphi \to K_i \varphi$.

The intended reading of these systems is as follows. In the standard n-agent setting, there is a common LP-based system of admissible evidence accepted by all the agents. Informal examples of such situations are provided by a criminal trial (evidence admitted by a judge is accepted by all the participants), mathematics (proven facts are accepted as knowledge by all mathematicians), etc.

It is easy to notice that *internalization* holds for each of $\mathsf{T}_n\mathsf{LP}$, $\mathsf{S4}_n\mathsf{LP}$, and $\mathsf{S5}_n\mathsf{LP}$:

$$\text{If } \vdash F, \text{ then } \vdash p{:}F \text{ for some proof polynomial } p \ .$$

A new epistemic notion, that of a *Justified Knowledge*, can be obtained by taking the forgetful projection of the justification assertions in the evidence-based knowledge systems ([Artemov, 2004]).

We first introduce a new modal operator J ($\mathsf{J}\varphi$ reads φ *is justified*) axiomatically:

$\mathsf{T}_n^{\mathsf{J}} = \mathsf{T}_n + \mathsf{S4}$ (with modality J) $+ \ \mathsf{J}\varphi \rightarrow \mathsf{K}_i\varphi$, $i = 1, 2, \ldots, n$.
$\mathsf{S4}_n^{\mathsf{J}} = \mathsf{S4}_n + \mathsf{S4}$ (with modality J) $+ \ \mathsf{J}\varphi \rightarrow \mathsf{K}_i\varphi$, $i = 1, 2, \ldots, n$.
$\mathsf{S5}_n^{\mathsf{J}} = \mathsf{S5}_n + \mathsf{S4}$ (with modality J) $+ \ \mathsf{J}\varphi \rightarrow \mathsf{K}_i\varphi$, $i = 1, 2, \ldots, n$.

Then we establish that $\mathsf{J}\varphi$'s are indeed forgetful projections of $t{:}\varphi$'s.

THEOREM 2 ([Artemov, 2004]). $\mathsf{T}_n^{\mathsf{J}}$, $\mathsf{S4}_n^{\mathsf{J}}$, *and* $\mathsf{S5}_n^{\mathsf{J}}$ *are exactly the forgetful projections of* $\mathsf{T}_n\mathsf{LP}$, $\mathsf{S4}_n\mathsf{LP}$, *and* $\mathsf{S5}_n\mathsf{LP}$, *respectively*.

In particular, given any principle φ provable in any of $\mathsf{T}_n^{\mathsf{J}}$, $\mathsf{S4}_n^{\mathsf{J}}$, or $\mathsf{S5}_n^{\mathsf{J}}$, one could find a realization of all occurrences of the Justified Knowledge modality J in φ such that the resulting formula φ^r is derivable in $\mathsf{T}_n\mathsf{LP}$, $\mathsf{S4}_n\mathsf{LP}$, or $\mathsf{S5}_n\mathsf{LP}$, respectively. In view of the Realization Theorem above, $\mathsf{J}\varphi$ can be understood as

there is an access to an evidence for φ .

In $\mathsf{S4}_n^{\mathsf{J}}$ the $(n+1)$st modality J is exactly McCarthy's "any fool knows" modality ([McCarthy *et al.*, 1979]), which now receives an exact evidence-based semantics. J plays the role of a dummy sceptical agent who accepts facts only if they are supplied with checkable evidence. On the other hand, this agent is trusted by all other agents and is capable of internalizing and inspecting proofs in the system.

A simple but fundamental observation shows that the Justified Knowledge J satisfies the *fixed point axiom* for Common Knowledge

$$\mathsf{J}\varphi \ \rightarrow \ \mathsf{E}(\varphi \wedge \mathsf{J}\varphi) \ ,$$

and hence may be regarded as a constructive common knowledge epistemic operator, which can be used instead of or together with the usual Common Knowledge.

To compare Justified Knowledge J and Common Knowledge C, consider the system $\mathsf{S4}_n$ with operators J and C. It is easy to see that Justified Knowledge is a stronger modality than Common Knowledge, i.e.,

$$\mathsf{J}\varphi \rightarrow \mathsf{C}\varphi \ .$$

On the other hand, the Justified Knowledge system $S4_n^J$ is leaner then the Common Knowledge system $S4_n^C$:

$$(S4_n^J)^* \subset S4_n^C \,,$$

where $*$ is renaming C to J. In [Antonakos, 2005] the exact difference between $S4_n^J$ and $S4_n^C$ has been measured:

$$S4_n^C = (S4_n^J)^* + (\varphi \wedge C(\varphi \to E\varphi) \to C\varphi) \,,$$

where $E\varphi = K_1\varphi \wedge K_2\varphi \wedge \ldots \wedge K_n\varphi$.

Evidence-based and Justified Knowledge have certain advantages compare to the usual Common Knowledge.

1. Justified Knowledge is more practical and automation-friendly. Justified Knowledge systems are easier to work with, since they are the standard modal logics supported by a well-developed machinery. In particular, $S4_n^J$ enjoys cut-elimination, which opens a way to automated proof search methods.

2. The question of whether a given real system has a Justified Knowledge can be reduced to checking a manageable set of model-independent conditions.

3. Evidence-based and Justified Knowledge provide an additional degree of flexibility, since an evidence part can be chosen independently of knowledge systems of individual agents. The Common Knowledge operator is a derivative of the agent knowledge operators and carries the features of the latter.

4. The justification part of the evidence-based knowledge systems is free of logical omniscience. An agent cannot claim to have evidence-based knowledge without having actually built a supporting evidence term.

BIBLIOGRAPHY

[Antonakos, 2005] E. Antonakos. Comparing justified and common knowledge. Logic Colloquium, Athens, August 2005.

[Artemov and Nogina, 2004] S. Artemov and E. Nogina. Logic of knowledge with justifications from the provability perspective. Technical Report TR-2004011, CUNY Ph.D. Program in Computer Science, 2004.

[Artemov and Nogina, 2005] S. Artemov and E. Nogina. Basic epistemic logics with justifications. Technical Report TR-2005004, CUNY Ph.D. Program in Computer Science, 2005.

[Artemov et al., 1999] S. Artemov, E. Kazakov, and D. Shapiro. On logic of knowledge with justifications. Technical Report CFIS 99-12, Cornell University, 1999.

[Artemov, 1995] S. Artemov. Operational modal logic. Technical Report MSI 95-29, Cornell University, 1995.

[Artemov, 2001a] S. Artemov. Explicit provability and constructive semantics. Bulletin of Symbolic Logic, 7(1):1–36, 2001.

[Artemov, 2001b] S. Artemov. Operations on proofs that can be specified by means of modal logic. In *Advances in Modal Logic. Volume 2*, pages 59–72. CSLI Publications, Stanford University, 2001.

[Artemov, 2004] S. Artemov. Evidence-based common knowledge. Technical Report TR-2004018, CUNY Ph.D. Program in Computer Science, 2004.

[Brezhnev and Kuznets, 2005] V. Brezhnev and R. Kuznets. Making knowledge explicit: How hard it is. Technical Report TR-2005003, CUNY Ph.D. Program in Computer Science, 2005.

[Brezhnev, 2000] V. Brezhnev. On explicit counterparts of modal logics. Technical Report CFIS 2000-05, Cornell University, 2000.

[Brezhnev, 2001] V. Brezhnev. On the logic of proofs. In *Proceedings of the Sixth ESSLLI Student Session, Helsinki*, pages 35–46, 2001. http://www.helsinki.fi/esslli/.

[Chagrov and Zakharyaschev, 1997] A. Chagrov and M. Zakharyaschev. *Modal Logic*. Oxford Science Publications, 1997.

[Fitting, 2004] M. Fitting. A logic of explicit knowledge. To appear in *Logica 2004 Proceedings*. Available on web page http://comet.lehman.cuny.edu/fitting., 2004.

[Fitting, 2005a] M. Fitting. The logic of proofs, semantically. *Annals of Pure and Applied Logic*, 132(1):1–25, 2005.

[Fitting, 2005b] M. Fitting. LP - the family (invited talk). The Third New York City Logic Conference., May 2005.

[Gabbay, 1994] D.M. Gabbay. *Labelled Deductive Systems*. Oxford University Press, 1994.

[Gödel, 1933] K. Gödel. Eine Interpretation des intuitionistischen Aussagenkalkuls. *Ergebnisse Math. Kolloq.*, 4:39–40, 1933. English translation in: S. Feferman et al., editors, *Kurt Gödel Collected Works, Vol. 1*, pages 301–303. Oxford University Press, Oxford, Clarendon Press, New York, 1986.

[Gödel, 1995] K. Gödel. Vortrag bei Zilsel, 1938. In S. Feferman, editor, *Kurt Gödel Collected Works. Volume III*, pages 86–113. Oxford University Press, 1995.

[Hintikka, 1962] J. Hintikka. *Knowledge and Belief*. Cornell University Press, Ithaca, 1962.

[Kuznets, 2005] R. Kuznets. Logic of proofs as a measure of Hilbert style complexity. Logic Colloquium 2005, Athens, August 2005.

[McCarthy et al., 1979] J. McCarthy, M. Sato, T. Hayashi, and S. Igarishi. On the model theory of knowledge. Technical Report STAN-CS-79-725, Stanford University, 1979.

[Mkrtychev, 1997] A. Mkrtychev. Models for the logic of proofs. In S. Adian and A. Nerode, editors, *Logical Foundations of Computer Science' 97, Yaroslavl'*, volume 1234 of *Lecture Notes in Computer Science*, pages 266–275. Springer, 1997.

[Troelstra and Schwichtenberg, 1996] A. Troelstra and H. Schwichtenberg. *Basic Proof Theory*. Cambridge University Press, Amsterdam, 1996.

Recognizing and Interpreting Temporal Expressions in Open Domain Texts

David Ahn, Sisay Fissaha Adafre, and Maarten de Rijke

Foreword

Since the early 1990s, when the third author was a visiting PhD student at Imperial College, he and Dov Gabbay have had many interactions. Usually over the phone, they talked about science, publishing and other initiatives, and invariably their conversations ended on a strategic note. Dov and the third author of this paper share a fundamental vision on the status and future of logic. To have a bright future, the discipline needs to be strongly embedded in external uses and needs. Much of the innovation in logic over the past decades has come from computer science, with new questions, new modeling needs, new reasoning mechanisms, etc. To continue to strive, we believe that logic should be embedded alongside its application areas, with feedback back and forth through measurable evaluations, either theoretical or experimental. It is clear that in many research institutes around the world this is actually how logic and computer science have come to interact. But we believe that logic can and should play a similar role vis-à-vis other scientific areas, both traditional and non-traditional, such as analytic philosophy, law, cognitive science, economics, information science, theology, language, and political theory.

In more recent years, the interactions between Dov and the third author have become less frequent, mainly because Dov has been looking at one set of areas for inspiration and research questions (including law and philosophy), while the third author of this paper has turned to another, organized around information access. This paper deals with identifying, and assigning meaning to, temporal information. It touches on at least two of Dov's long-standing interests, *temporal logic* and *natural languages semantics*. There is very little symbolic reasoning in this paper, and the semantics we pursue is shallow at best, but the paper does contribute to our understanding of how to process natural language texts if we are ever going to have systems that deal with the information contained in, for instance, news papers or web pages, in an intelligent way.

1   Introduction

Current information retrieval (IR) systems allow us to locate documents that might contain pertinent information, but most of them leave it to the user to extract useful information from a ranked list. This leaves the user with a large amount of text to consume. *Information extraction* [Appelt and Israel, 1999] (IE) is a core technology to help reduce the amount of text that has to be read to obtain the desired information. Indeed, recognizing entities and meaningful relations between them is key to providing focused information access. *Temporal* IE provides a particularly interesting task in this respect. Temporal expressions (*timexes*) are natural language phrases that refer directly to time points or intervals. They convey temporal information on their own and also serve as anchors for events referred to in text. From a user's perspective, temporal aspects of events and entities, and of text snippets, provide a natural mechanism for organizing information. An example of an area in which accurate analysis of temporal expressions plays an important role is Question Answering (QA). For instance, answering questions like *"When was Van Gogh born?"* requires accurate identification of the date of birth of the person under consideration (*recognition*) and rendering of the answer in some standard format (*normalization*). Recognizing temporal expressions is now a "do-able" task, even without tremendous knowledge engineering efforts. Moreover, in recent years, the task of automatically interpreting (or normalizing) temporal expressions has begun to receive attention [Mani and Wilson, 2000; Schilder and Habel, 2001].

The importance of processing timexes is reflected by the large number of NLP evaluation efforts where they figure. Recognizing timexes is an integral part of many IE tasks (e.g., MUC-6 and 7 Named Entity Recognition tasks, ACE-2004 Event Recognition task). There are various annotation guidelines for timexes [Ferro *et al.*, 2004; Setzer and Gaizauskas, 2000; Pustejovsky *et al.*, 2003]. And a timex annotated corpus has been released with the aim of improving the processing of timexes [TimeBank, 2004].

The type of timexes considered in a typical IE task are limited to date and time values [MUC-6, 1995; Chinchor, 1997]. In contrast, at the 2004 Temporal Expression Recognition and Normalization [TERN, 2004] evaluation, a wide variety of timexes are considered—which makes the task more interesting and much more challenging. The participants in the 2004 TERN evaluation evinced a notable split in their approaches to temporal

IE: systems performing only recognition were all machine-learning-based, while systems performing the full recognition and normalization task were purely rule-based. The message from the recognition task was clear: given a tagged corpus, machine learning provides excellent recognition results with minimal human intervention. Furthermore, such a data-driven approach may be preferable because of its portability and robustness.

For the normalization task, it is not so obvious how to directly apply machine-learning methods based on sequence labeling. The classes involved (temporal values) are potentially unbounded, and a significant proportion of timexes require non-local context for interpretation. Additionally, many timexes require temporal computation with respect to contextually given information—the connection between form and content is mediated by both context and world knowledge.

What are the opportunities for using robust, "shrink-wrapped" machine-learning tools for temporal IE? Ultimately, we want to build a portable, maintainable system that can serve us well as a component for more high-level tasks. To this end, we want to make as much use of off-the-shelf machine-learning packages as possible. We would also like to limit the scope of rule application, in order to simplify rule-writing and maintenance. In this paper, we describe two sets of experiments that bring us closer to this goal. In the first set, we demonstrate that decoupling recognition from normalization—feeding a rule-based normalizer with an automatically learned recognizer—can improve overall performance. In the second set, we decompose the normalization task, allowing us to find opportunities for applying data-driven methods for disambiguation within normalization.

2   Background

2.1   The TERN Setting

For the experiments we report on in this paper, we adopt the tasks, data, and evaluation methodology of the 2004 TERN evaluation [TERN, 2004]. The TERN evaluation is organized under the auspices of the Automatic Content Extraction program (ACE, http://www.nist.gov/speech/tests/ace/), whose objective is to develop natural language processing technology to support automatic understanding of textual data. TERN consists of two tasks: recognition and normalization. Timex *recognition* involves correctly detecting and delimiting timexes in text. *Normalization* involves assigning recognized timexes a fully qualified temporal value. Both tasks are defined, for human annotators, in the TIDES TIMEX2 annotation guidelines [Ferro *et al.*, 2004]. These introduce an SGML element, TIMEX2, to mark timexes. TIMEX2 elements may contain a number of attributes; we focus on the VAL attribute, which indicates the actual reference of the TIMEX2. Its range of values are an extension of the ISO 8601 standard for representing time [ISO 8601, 1997].

The recognition and normalization tasks are performed with respect to corpora of transcribed broadcast news speech and news wire texts from ACE 2002–4, marked up in SGML format and hand-annotated for TIMEX2s. The training and test sets for the TERN evaluation and for all of the experiments we describe in this paper consist of 511 and 192 documents with 5326 and 1828 TIMEX2s, respectively.

We use the TERN official scorer to evaluate our performance but add several metrics. The official scorer computes precision, recall, and F-measure for identification of TIMEX2s (overlap between a gold standard and a system TIMEX2) and exact-match of TIMEX2s (exact overlap) and for the attributes; since we focus on VALs, we report only VAL scores. For VALs, the scorer computes precision as the ratio of correct VALs to attempted VALs and recall as the ratio of correct VALs to possible VALs *in the TIMEX2s recognized by the system*—not all TIMEX2s in the gold standard. Since we are interested in the end-to-end task, we report results with recall (and F-measure) computed with respect to *all* possible TIMEX2s. We refer to the TERN versions as *relative recall (RR)* and *F (RF)*, and to our versions as *absolute recall (AR)* and *F (AF)*.

2.2   Recognition

The recognition task is to identify phrases that refer to time points. The TIDES guidelines limit the set of markable timexes (indicated with the TIMEX2 tag) to those phrases headed by a temporal trigger word. The latter seem to fall into several categories. Some refer to time units of definite duration (minute, afternoon, day, night, weekend, month, summer, season, quarter, year, decade, century, millennium, era, semester). Others refer to definite points in time (January, Monday, New Year's Eve, Washington's Birthday, yesterday, today, tomorrow, midnight). Still others indicate repetition with respect to a definite period (daily, monthly, biannual, semiannual, hourly, daily, monthly, ago). And some refer to temporal concepts that can at least be oriented on a timeline with respect to some definite time point (future, past, time, period, point, recent, former, current, ago, currently, lately).

Syntactically, TIMEX2s must be one of the following: noun, noun phrase, adjective, adverb, adjective phrase, or adverb phrase. All premodifiers and postmodifiers of the timex must be included in the extent of the TIMEX2 tag, e.g.,

- Premodifiers: *8* winters, *the past* week, *four bad* years, *about 15* minutes, *less than a* week

- Postmodifiers: *Nearly* three years *later*, the week *before last*, two years *ago*, three years *in prison*, Only days *after his father was assassinated*, months *of Israeli-Palestinian bloodshed and Israeli blockades*

Either rule-based systems or machine learning can be used for recognition, but as we saw in the 2004 TERN evaluation, all the recognition-only systems (which generally achieved better recognition scores than the full-task systems) were machine-learning-based. The reasons for this are clear: recognition can easily be cast as a sequence-labeling task, for which good machine-learning systems exist. Unsurprisingly, the recognition components of the full-task systems were all rule-based. Normalization is most straightforwardly conceived of as interpreting recognition rules. Thus, it is natural to develop a single set of rules to be used for recognition and paired with interpretation functions for normalization. However, the result is a monolithic rule-based system, which requires significant engineering efforts to build and maintain.

In this paper we explore the possibility of breaking down the task into smaller pieces and using the best methods possible for each sub-task in an effort to develop a more robust solution to the temporal IE task. The first step, described in §3, is to decouple recognition from normalization. Experimental results indicate that this decoupling (which both improves recognition and allows a liberalization of the normalization rules) improves overall, end-to-end performance.

2.3 Normalization

Timex normalization is the problem of assigning an ISO 8601 value to a recognized timex. The TIDES guidelines distinguish several kinds of values; our normalization system handles time points, durations, sets, and the past, present, and future tokens. Time points are expressed by three kinds of timexes: fully qualified, deictic, and anaphoric. Fully qualified timexes, such as *March 15, 2001*, can be normalized without reference to any other temporal entities. Deictic and anaphoric timexes, on the other hand, must be interpreted relative to another temporal entity. Deictic timexes, such as *today, yesterday, three weeks ago, last Thursday, next month*, are interpreted with respect to the time of utterance—for our corpus, the document time stamp. Anaphoric timexes, such as *March 15, the next week, Saturday*, are interpreted with respect to a reference time—a salient time point previously evoked in the text that may shift as the text progresses. Some anaphoric timexes (those without an explicit direction indicator such as *next* or *previous*) depend on other factors, such as the tense and aspect of the verb they modify, to determine whether the time point they refer to is before or after the reference time.

Durations are generally expressed by timexes headed by a unit (*day*, *months*, etc.). However, even fully qualified timexes expressing durations, i.e., those in which both the quantity and the unit are specified, such as *six months*, *800 years*, *three long days*, are systematically ambiguous between a duration and an anaphoric point reading. For example, in the sentence *The Texas Seven hid out there for three weeks*, the timex *three weeks* refers to a duration, whereas in the sentence *California may run out of cash in three weeks*, the same timex refers to a point three weeks after the reference point.

Work on normalization with respect to TIMEX2-like guidelines goes back to [Mani and Wilson, 2000], who use a rule-based system to identify and normalize timexes. Like our work in §4 (but unlike more recent work on normalization), they also use an automatically learned classifier within their system, but only for one task—distinguishing specific and generic uses of *today*. Unlike our work, however, they restrict normalization to date-valued expressions; the question of distinguishing points and durations, e.g., does not arise. Other rule-based normalization systems include [Saquete *et al.*, 2002; Schilder, 2004], as well as the TERN full-task systems. Building on the approach in [Mani and Wilson, 2000], we present in §4 a modular timex normalization system architecture that allows us to separate context-independent interpretation, for which we continue to use a rule-based approach, from context-dependent processing.

3   Decoupling Recognition from Normalization

Here, we describe a set of experiments (reported on in greater detail in [Ahn *et al.*, 2005]) in which we run the same rule-based timex normalizer on the output of several different timex recognizers. Our basic monolithic rule-based system is a two-pass system. In the first pass, a document is tokenized, POS tagged and chunked using TreeTagger [TreeTagger, 2004], and then a series of regular expressions is used to find timexes. In the second pass, the document time stamp and all sentences containing TIMEX2s are extracted; the tense of each sentence (based on the first verb chunk) is also determined. Interpretation rules paired with the regular expressions for recognition are used to generate normalized values for each TIMEX2. Anaphoric and deictic expressions are evaluated with respect to the timestamp and tense information.

Recognition with Conditional Random Fields

A recently-introduced machine learning technique for labeling and segmenting sequence data is Conditional Random Fields (CRFs, [Lafferty *et al.*, 2001]). Unlike Hidden Markov Models, CRFs are based on exponential

	Correct	Incorrect	P	RR	RF	AR	AF
Rule-based	782	143	0.845	0.699	0.765	0.449	0.586
CRF	885	231	0.793	0.583	0.672	0.501	0.614
Gold standard	955	219	0.812	0.549	0.655	0.549	0.655

Table 1. Normalization results

models in which probabilities are computed based on the values of a set of features induced from both the observation and label sequences. They have been used in POS tagging, shallow parsing [Sha and Pereira, 2003], and named entity recognition [McCallum and Li, 2003]. We use the minorThird implementation of CRFs for extracting timexes from text [Cohen, 2004]. Initial recognition results with the default features are: 98%/84%/91% (P/R/F) for identification and 80%/66%/74% (P/R/F) for exact-match. An error analysis suggested several additional features which resulted in substantial performance improvements: 98%/86%/91% (P/R/F) for identification and 86%/75%/80% (P/R/F) for exact-match.

Decoupling Experiments

We ran our normalizer on the output of two different recognition systems—the rule-based recognizer of our monolithic system and the optimized CRF recognizer—and on the gold standard, in order to see the effects of recognition performance on normalization performance. Our expectation was that since the normalization rules are basically identical to the rule-based recognizer, any additional recognized timexes would not be normalized anyhow.

Table 1 lists the results of these three runs on the TERN test corpus. For our purposes, AR and AF are more important than RR and RF. The results indicate that better recognition does help normalization. We analyzed the gold standard and rule-based recognizer runs to determine why. The former obviously presents more timexes to the normalizer than the latter; the question is why, given that the patterns for the normalizer and the rule-based recognizer are the same, the normalizer attempts these extra timexes. The two main reasons are the reliance of the rule-based recognizer on unreliable upstream components (tokenizer, tagger, and chunker) and our liberalization of the normalizer rules.

4   Zooming in on Normalization

From the perspective of the end-to-end task of temporal IE, our experiments so far suggest that it is worthwhile to optimize recognition and normalization independently. By decoupling normalization from recognition, not only do we allow for independent optimization of recognition, but we also give ourselves the opportunity to conceive of normalization as an independent task—rules do not need to serve double-duty for both recognition and normalization. We now delve into normalization, exploring one way of decomposing normalization both to simplify the rule set and to bring in data-driven methods.

In the rest of this section, we lay out our analysis of the task and explain how we tackle the various sub-tasks. In §5, we focus on the machine-learning experiments we perform as part of normalization. We devote §6 to describing our end-to-end normalization experiments.

4.1 Decomposing the Normalization Task

We seek a robust, maintainable approach to the normalization task, one that allows for the use of data-driven techniques, where appropriate, and circumscribes the scope of rule application to simplify rule development. To that end, we decompose the task as described above into five discrete stages:

1. Lexical lookup: mapping names to numbers, units to ISO values, etc.

2. Context-independent composition: combining the values of the lexical tokens within a timex to produce a context-independent semantic representation.

3. Context-dependent classification: determining whether a timex is a point or duration, looks forward or backward, makes specific or generic reference, etc.

4. Reference time, or temporal focus, tracking: for anaphoric timexes, whose values must be computed with respect to a reference time.

5. Final computation: combining the results of all of these steps to produce a final normalized value.

This architecture clearly separates context-independent processing (stages 1 and 2), for which finite-state rules can be relatively easily developed, from context-dependent processing (stages 3 and 4), for which finite-state rule sets can quickly become unwieldy. It also distinguishes context-dependent classification tasks, which rely on primarily local context, from reference time tracking, which requires more global information. The final stage makes use only of meta-data produced by the earlier stages and does no linguistic processing.

In an error analysis of a purely rule-based normalization system, we have observed observed that classification of timexes (stage 3) is a significant source of error [Ahn et al., 2005]. Now, classification of expressions into a limited set of classes using local context is exactly the kind of task at which machine-learning-based classifiers excel. Thus, this separation of tasks provides us with an opportunity to deploy an off-the-shelf machine learning package within the normalization task.

4.2   Addressing the Interpretation Stages

We use a rule-based system to handle stages 1 and 2 (which we refer to as *pre-normalization*). Timex lexicons and composition mechanisms may be learned, but our lexicon and composition rule set are relatively small and unambiguous, so writing and maintaining them is more or less straightforward. In addition to generating a context-independent representation for a timex, the composition mechanism also determines whether the timex is ambiguous in a way that can be resolved by one of the context-dependent classifiers.

Stage 3 is where we study the potential contribution of data-driven methods to timex normalization. We make use of a maximum entropy classifier to perform context-dependent classification. Based on the error analysis of [Ahn *et al.*, 2005], we have isolated three classification tasks that contribute to the errors their rule-based system makes and seem amenable to machine learning from relatively local surface features:

- The first task is distinguishing whether an ambiguous unit phrase refers to a point or a duration; we refer to this as the *point-duration problem*.

- The second task (*direction problem*) is determining whether an ambiguous anaphoric point-referring phrase refers to a point before, after, or the same as the reference time. Some of the instances of this class are generated by the first classifier.

- The third task (the *today problem*) is determining whether an occurrence of the word *today* refers specifically to the day of the article or broadcast or generically to the present.

Section 5 is devoted to a detailed description of the methods used to tackle these tasks.

As to stage 4, we experiment with two very simple models of temporal focus tracking. In the first, the system uses the document time stamp as the reference time for all anaphoric expressions. This is an oversimplification of the problem of temporal focus, but it seems to be reasonable at least for day names in short news items. In the second temporal focus model, the system uses the most recent previous point-referring timex of suitable granularity as the reference time for an anaphoric expression. This, too, is a simplifying assumption, since it ignores the effects of discourse structure on focus tracking, among other things. (In both cases, deictic expressions, such as *tomorrow* and *three years ago*, are still computed with respect to the document time stamp).

Finally, we take a rule-based approach to stage 5. Temporal arithmetic is performed to derive a fully qualified temporal value from the context-independent value and the reference time of a timex, together with information from the context-dependent classifiers. Of the five stages of normalization, this stage is least obviously amenable to machine learning.

5   Developing the Classification Experiments

We now describe our machine-learning approach to the three classification tasks that make up stage 3 of our strategy for the overall normalization task. We use a maximum entropy classifier for our experiments [Berger *et al.*, 1996]. Specifically, we use the minorthird implementation of maximum entropy classifiers, which uses the same underlying model as CRFs applied to sequence labeling [Cohen, 2004].

5.1   Generating the Training Data

The training material is a tagged corpus consisting of plain text in which the timexes to be classified are marked by XML tags that encode the classes, e.g., `...<forward><dir_unknown>Tuesday</dir_unknown></forward>....` The inner XML tag marks a timex as an instance to be used for training while the outer XML tag assigns a class to the instance. The task is to learn from these example timex instances in the training corpus rules or patterns to classify new instances.

To generate the training data, we use the output of our pre-normalization stages, which tags ambiguous time unit phrases (for the point-duration task), ambiguous anaphoric timexes (for the direction task), and occurrences of *today* (for the *today* task). Not all unit phrases are ambiguous (e.g., *two years ago* is always point-referring), nor are all anaphoric timexes ambiguous with regards to direction (e.g., *a month later* is always forward-looking), so what counts as an instance for these problems is dependent on our pre-normalization.

Given that, generating training data for the point-duration and *today* tasks is straightforward, since the classes are reflected directly in the final normalized values. The final normalized value of a duration always begins with a P followed by a quantity indication, whereas that of a point begins with a digit. Similarly, occurrences of *today* that make generic reference are normalized as PRESENT_REF, while specifically referring occurrences have full date values.

Generating training data for the direction task, is non-trivial as it assumes a model of temporal focus tracking. We produced two training sets for the task, based on the two focus models our normalization system can use: timestamp-based and recency-based (see §4.2). As both are simplistic models, the generated training data, in either case, is noisy.

5.2 The Point-Duration Problem

For the *point-duration task*, ambiguous time unit phrases need to be classified into three classes: point, duration and other. Using only lexical features, both in the timex itself and in the left and right context (window of 3 words), the system achieves an accuracy of 0.73 (frequency baseline: 0.40).

5.3 The Direction Problem

Confronted with the *direction problem* for day names in their date normalization system, Mani and Wilson [Mani and Wilson, 2000] use hand-crafted rules which look at the tense of the closest verb in the same clause as a timex to determine the direction. The rule-based direction classifier we use in our experiments in §6 is based on this method, but here, we describe our machine learning approach. Tense alone cannot be used to identify direction accurately (see our error analysis in §6), so additional features, such as lexical items such as *last* and *earlier*, need to be used for the learning algorithm to produce a reasonable result. Using only such lexical features as these, with a context window of 3 words, the system, using the timestamp-based dataset, has an accuracy of 0.59 (frequency baseline: 0.44). Adding tense (derived from the POS tags of the closest finite verbs) as a feature improves the result by 0.02 (to 0.61). Using the same features with the recency-based data yields an accuracy of 0.57 (we discuss the difference in §6).

A closer look at the confusion matrix indicates that the classifier has problems in distinguishing between the *same* and *backward* classes. More than 30% of each class is assigned to the other class. These are the main sources of errors since the two classes constitute 89% of the total test instances.

5.4 The Generic-Specific Problem

Mani and Wilson [Mani and Wilson, 2000] also report that ambiguity resulting from generic vs. specific meaning of timexes are a main source of error. They single out the timex *today*, which is most subject to this ambiguity, and automatically acquire a classifier for it. Their best system achieves an accuracy of 0.80. We have developed a similar classifier for our system using the features they propose. The resulting classifier achieves an accuracy of 0.85 in a 70/30 split experiment on the training data. Unfortunately, our data for this classifier is not only sparse, but heavily skewed (90% of the instances are specific). Thus, our system still underperforms the frequency baseline 0.89. In the experiments we describe in §6, we use the baseline classifier for this task.

5.5   Reflections

In general, all of these classification tasks are more difficult than the recognition task, where machine-learning approaches achieve very good results. Although two of the systems are better than the frequency baseline, there is a lot of room for improvement. Simple lexical features, which are successful in the recognition task, do not have the same impact on these classification tasks. This suggests a potential gain by providing the learner with more semantically motivated features. For now, though, we want to see how the results from relatively straightforward application of off-the-shelf machine learning affects our normalization task.

6   Experiments

We now describe the experiments we performed in trying to answer our main research question: whether data-driven methods can be successfully applied to the timex normalization task. We describe the approaches we compared, then the metrics used, the results, and, finally, an error analysis.

6.1   Experimental Setup

The system we used for our experiments consists of the following components. For the pre-normaliation stages (1 and 2), there are regular-expression grammars written in the JAPE formalism that run within the GATE system [Cunningham *et al.*, 2002]. For stage 3, there are three kinds of classifiers for the point-duration and direction tasks: baseline classifiers (which assign the majority class), MaxEnt classifiers, described in §5 and rule-based classifiers, briefly described below. Also, there is the baseline classifier for the *today* task. For stages 4 and 5, there is a perl script (using the Time::Piece module) that traverses a document, tracking reference times and computing final normalized values.

The JAPE grammars include both simple (stage 1) grammars to normalize number expressions, month names, day names, and unit names (and identify likely years) and a larger (stage 2) grammar that takes these annotations as input and generates context-independent values for TIMEX2s. Many rules generate final normalized values, since many expressions (e.g., full dates, month/year expressions) do not require contextualizing. The remaining rules generate a context-independent value and, for ambiguous unit phrases, anaphoric points, and occurrences of *today*, a classification problem.

The rule-based classifiers are simple. The point-duration classifier has rules that match expressions that are likely point or duration indicators either within a unit phrase or immediately to the left of a unit phrase. The direction classifier uses the same heuristic as [Mani and Wilson, 2000], relying on TreeTagger for part-of-speech tagging and chunking: it looks at the closest preceding verb chunk—if it is past or perfect tense, it labels the instance as "backward"; if it is present-progressive or a present tense copula, it labels the instance as "same"; and if it is any other present-tense (including modals, such as *will, shall, may*), it labels the timex as "forward."

We used the classifier module of the minorthird package for training a maximum entropy-based classifier. The tasks of transforming the instances into feature vectors and estimating parameters are done automatically by the system.

6.2 Ten Approaches

We ran the system over the TERN test corpus (192 documents, 1828 TIMEX2s, of which 1741 have a non-null VAL attribute), varying the classifiers and the reference tracking model used. Note that for all configurations, we used the baseline classifier for the today task, which assigns *specific* to every instance.

1. Baseline classifiers (assign majority class—point-duration: duration, direction: backward); timestamp-based

2. Baseline classifier for point-duration, rule-based classifier for direction; timestamp-based

3. Rule-based classifiers for both tasks; timestamp-based

4. Baseline classifier for p-d, MaxEnt classifier for dir; timestamp-based

5. MaxEnt classifiers for both tasks; timestamp-based

6. "Perfect" classifiers; timestamp-based

7. Baseline classifiers; recency-based

8. Baseline classifier for p-d, rule-based classifier for dir; recency-based

9. Rule-based classifiers for both tasks; recency-based

10. Baseline classifier for p-d, MaxEnt classifier for dir; recency-based

11. MaxEnt classifiers for both tasks; recency-based

12. "Perfect" classifiers; recency-based.

We include two "perfect" runs (6 and 12) to set a ceiling on the overall normalization performance of the classifiers.

6.3   Metrics

We use the official TERN scorer to evaluate our normalization performance. As we mentioned in §2.1, the scorer computes precision, recall, and F-measure for each normalization attribute; since we are focusing on the core VAL normalization, we report only VAL scores. For the TERN scorer, precision is the ratio of correctly normalized VALs to attempted VALs; recall, the ratio of correct VALs to possible VALs *in the recognized TIMEX2s*; and F-measure, $(2 * P * R)/(P + R)$. Recall (and, F-measure) is computed not with respect to all possible TIMEX2s in the gold standard but only with respect to the TIMEX2s recognized by the system. Because we are interested in the end-to-end task, we also report results with recall (and F-measure) computed with respect to *all* possible TIMEX2s. We refer to the TERN versions as *relative recall* (RR) and *F-measure* (RF), and to our versions as *absolute recall* (AR) and *F-measure* (AF).

Since we use the same recognizer output for all of our runs, the difference between the relative and absolute measures is immaterial in comparing these results, but the absolute measures make comparison with other systems clearer.

6.4   Results

Our results are given in Table 2, with the results from the best system at TERN for comparison (row 13). While our results are not yet up to that level, they are competitive with other systems; AF-scores for the TERN systems ranged from 0.439 to 0.830, with an average of 0.657.

Overall, the best (non-perfect) runs for each reference tracking model use the baseline p-d classifier and the MaxEnt dir classifier (AF-scores of 71.3% and 70.1% for timestamps (run 4) and recency-based (run 10, respectively)). In each case, adding the MaxEnt p-d classifier reduces performance slightly (70.4% and 69.4% for runs 5 and 11), but in any case, the performance remains several points above that of any of the rule-based runs (F-scores between 66.5% and 66.9%). We comment on this fall-off in performance when adding point-based classifiers below. Comparing the two reference tracking models, it is clear that the timestamp-based model outperforms the recency-based model. We discuss the reasons for this below, as well.

There is clearly still room for improvement, both in the classifiers (as can be seen in comparisons with the "perfect" runs) and for the other components of the system. In addition to looking at classifier performance independently, we have performed an error analysis of the "perfect" runs to see where the other components go wrong. Again, see below.

Returning to our research questions, how can data-driven techniques be brought to bear on the normalization task? Our experiments show that

	Correct	P	RR	RF	AR	AF
1	1063	0.733	0.694	0.713	0.611	0.666
2	1066	0.736	0.696	0.716	0.612	0.669
3	1063	0.734	0.694	0.714	0.611	0.667
4	1138	0.784	0.743	0.763	0.654	0.713
5	1124	0.775	0.734	0.754	0.646	0.704
6	1230	0.848	0.803	0.825	0.706	0.771
7	1060	0.734	0.692	0.713	0.609	0.666
8	1059	0.735	0.692	0.713	0.608	0.666
9	1058	0.734	0.691	0.712	0.608	0.665
10	1117	0.774	0.730	0.751	0.643	0.701
11	1105	0.765	0.722	0.743	0.635	0.694
12	1214	0.841	0.793	0.816	0.697	0.762
13	1389	0.866	0.837	0.851	0.798	0.830

Table 2. Normalization experiments; row numbers 1–12 refer to the list of approaches in §6.2.

data-driven methods outperform rule-based methods for crucial sub-tasks within the overall normalization task. Can we use "shrink-wrapped" data-driven language-processing solutions to make the semantic interpretation problem of timex normalization more robust, modular, and easily maintainable? The answer is clear. The relative ease of generating these runs by swapping out different classifiers and temporal reference tracking modules suggests a clear "yes." Our staged architecture has done two things for us. (1) It has limited the set of absolutely required rules to both context-independent pre-normalization, where no rule has to make reference to any information outside of a given timex, and final computation, where rules make reference only to meta-data (the preliminary value generated by the pre-normalization, the classification values generated by the classifiers, and the reference time generated by the reference model). (2) It has identified a potential place for data-driven methods within the normalization process—namely, incorporating contextual information—and made it straightforward to experiment with both data-driven and rule-based methods to tackle this sub-task.

6.5 Where the Errors Originate

Our analysis of the final normalization performance of the perfect runs (6 and 12) allows us to pinpoint errors arising from the non-classifier stages of our system and allows us to limit the error analysis of the other runs to the classification results. Of the 216 incorrectly normalized TIMEX2s in run 6,

	Baseline	Rule-based	MaxEnt
Point-duration	40%	54%	73%
Direction (timestamp)	46%	39%	62%
Direction (recency)	43%	38%	57%

Table 3. Classification accuracy results.

63 result from problems in the pre-normalization stages, including a single rule bug involving time zones that resulted in 36 errors. 26 errors arise from the recognizer extent errors; 18 errors, from the temporal reference model contributes 18 errors; and 6 errors, from temporal computation in stage 5. 46 errors are the result of a bug in the generation of the gold standard for the p-d classifier: durations of unspecified length are erroneously labeled as points. 46 more errors result from classes of timexes (sets and non-specific timexes) whose instances are ambiguous with point- and duration-referring timexes. For the 85 timexes that are recognized but not normalized, the failure is attributable to a recognition extent error or an omission in the pre-normalization rule base.

To see the effect of varying temporal reference models, we turn to the other "perfect" run, run 12, and look at anaphoric timexes. Even though at least 18 errors in the timestamp-based perfect run result from poor temporal reference tracking, the recency-based model does no better. Why? In one document, 4 of the timestamp-based errors occur in a chain, and because of a recognition extent error, the timex immediately preceding the chain that establishes the anchor point is not normalized at all, so the recency-based model also fails on all four timexes. For several other timexes, the recency-based model chooses the correct reference time, but because of errors in stage 5, the wrong final value is computed. Finally, in the remaining cases, neither the timestamp-based nor the recency-based model is sufficient to account for phenomena that are known to be hard, such as anchoring to an event, discourse effects, or shifts in granularity that allow for non-specific reference. Additionally, the recency-based model introduces several errors of its own.

Turning to the ambiguous point-referring timexes that are sent by stage 2 to the p-d classifier, we see why there is an across-the-board fall-off in end-to-end performance when adding a rule-based or MaxEnt classifier even though their classification accuracy is much better than the baseline (see Table 3). 73 of the 83 ambiguous point-referring timexes are ones that even the perfect runs normalize incorrectly (for various reasons, all among those outlined above). Furthermore, none of the 102 "other" timexes (with respect to the p-d task) are correctly normalized by the perfect runs (they refer to non-specific or quantified entities, which have limited support in the final normalizer). Meanwhile, almost all of the 86 durations are normalized correctly. Since the baseline labels all instances as durations, it only misses 10 points that might have been correctly normalized in the end. The other classifiers have to have perfect accuracy on durations and recognize some of those 10 points as points to beat that, and even then, the difference is not very large.

Where the MaxEnt classifiers do make a difference is in the dir task. Here, even though the improvement in classification accuracy over the baseline is smaller than for the p-d task (see Table 3), using the MaxEnt dir classifier with either temporal reference model results in F-score improvements of 3 to 4 percentage points over both the baseline and the rule-based dir classifier. Of course, in terms of classification accuracy, the rule-based classifier, while doing better than chance, actually does worse than the baseline. While tense may be an important feature in deciding direction, it is far from the only feature.

Comparing the two MaxEnt classifiers, we see that the classifier trained and tested on data generated using the timestamp reference model outperforms the one with the recency-based model. The classifiers have a broadly similar distribution of misclassified instances: they have difficulty discriminating between the "backward" and "same" classes; the main difference in accuracy is due to recall of the "same" class. One possible explanation for the difference between the classifiers is that the stage 2 process is mistakenly labeling some deictic timexes as anaphoric ones, and a cursory examination of the data seems to bear this out: while deictic timexes such as *today* and *this week* are correctly flagged as deictic in stage 2, others, such as *this afternoon* are not.

7   Conclusions

In performing IE tasks, we would like to make as much use as possible of off-the-shelf, shrink-wrapped machine learning packages rather than going through the time-consuming process of manually developing rule-based systems. To that end, we have presented a staged temporal IE architecture that allows for experimentation with different approaches to different parts of the normalization task. Within this architecture, we have identified sub-tasks that seem amenable to a machine-learning approach, and we have performed several experiments comparing machine-learning and rule-based approaches to these sub-tasks.

Overall, we find that data-driven methods can be applied at several points within a temporal IE system. In fact, by dividing the task into stages, we find that the tasks that require the most complex rule systems—

incorporating a variety of information to make classification decisions—are precisely the ones that are well-suited for machine learning. Thus, by confining rule-based components to context-independent stages, we can vastly simplify their development and, at the same time, explore the use of data-driven methods in semantic normalization.

A major obstacle in using machine learning to acquire classifiers is generating reliable training data. We are reconsidering the ordering of the stages in our architecture—rather than have the rule-based component of stage 2 pick out, for instance, the ambiguous unit phrases for stage 3 classification as points, durations, sets, etc., perhaps all unit phrases could be classified prior to stage 2. This would result in much more training data for learning this classifier and would further simplify the rules required for context-independent interpretation.

We plan to develop more accurate and robust of temporal reference models. The main problem here is the sparsity of explicit timexes and the interaction of temporal reference with event reference. There is a lot of theoretical work on temporal reference, and the interaction between events and times; see e.g., [Mani et al., 2005]. Recent computational work [Mani and Schiffman, to appear] using machine learning to assign reference times to sequences of clauses is also relevant, as are the TimeML guidelines and TimeBank corpus, which mark up not just timexes but also events and links [Pustejovsky et al., 2002; TimeBank, 2004].

Ultimately, we want our temporal IE work to end up as a component within a broader end-user task, such as temporal question answering or text mining, both to motivate and inform future annotation and architectural choices, and because we believe this will uncover new aspects of temporal IE.

Acknowledgements

David Ahn was supported by the Netherlands Organization for Scientific Research (NWO) under project number 612.066.302. Sisay Fissaha Adafre was supported by NWO under project number 220-80-001. Maarten de Rijke was supported by grants from NWO, under project numbers 365-20-005, 612.069.006, 220-80-001, 612.000.106, 612.000.207, 612.066.302, 264-70-050, and 017.001.190.

BIBLIOGRAPHY

[Ahn et al., 2005] D. Ahn, S. Fissaha Adafre, and M. de Rijke. Extracting Temporal Information from Open Domain Text: A Comparative Exploration. *Journal of Digital Information Management*, 3(1):14–20, 2005.
[Appelt and Israel, 1999] D. Appelt and D. Israel. Introduction to information extrac-

tion technology: IJCAI-99 tutorial, 1999. URL: http://www.ai.sri.com/~appelt/ie-tutorial/.

[Berger et al., 1996] A. Berger, S. Della Pietra, and V. Della Pietra. A maximum entropy approach to natural language processing. *Computational Linguistics*, 22(1):39–71, 1996.

[Chinchor, 1997] N. Chinchor. MUC-7 named entity task definition, September 1997. URL: http://www.itl.nist.gov/iaui/894.02/related_projects/muc/proceedings/ne%_task.html.

[Cohen, 2004] W. Cohen. Methods for identifying names and ontological relations in text using heuristics for inducing regularities from data, 2004. URL: http://minorthird.sourceforge.net.

[Cunningham et al., 2002] H. Cunningham, D. Maynard, K. Bontcheva, and V. Tablan. GATE: A framework and graphical development environment for robust NLP tools and applications. In *Proceedings of the 40th Anniversary Meeting of the Association for Computational Linguistics*, 2002.

[Ferro et al., 2004] L. Ferro, L. Gerber, I. Mani, and G. Wilson. *TIDES 2003 Standard for the Annotation of Temporal Expressions*. MITRE, April 2004.

[ISO 8601, 1997] ISO 8601: Information interchange – representation of dates and times, 1997.

[Lafferty et al., 2001] J. Lafferty, F. Pereira, and A. McCallum. Conditional random fields: Probabilistic models for segmenting and labeling sequence data. In *Proceedings of the International Conference on Machine Learning*, 2001.

[Mani and Schiffman, to appear] I. Mani and B. Schiffman. Temporally anchoring and ordering events in news. In J. Pustejovsky and R. Gaizauskas, editors, *Time and Event Recognition in Natural Language*. John Benjamins, to appear.

[Mani and Wilson, 2000] I. Mani and G. Wilson. Robust temporal processing of news. In *Proceedings of the 38th ACL*, 2000.

[Mani et al., 2005] I. Mani, J. Pustejovsky, and R. Gaizauskas, editors. *The Language of Time: A Reader*. Oxford University Press, 2005.

[McCallum and Li, 2003] A. McCallum and W. Li. Early results for Named Entity Recognition with conditional random fields, feature induction and web-enhanced lexicons. In *Proceedings of the 7th CoNLL*, 2003.

[MUC-6, 1995] MUC-6. Named entity task definition, May 1995. URL: http://www.cs.nyu.edu/cs/faculty/grishman/NEtask20.book_1.html.

[Pustejovsky et al., 2002] J. Pustejovsky, R. Sauri, A. Setzer, R. Gaizauskas, and B. Ingria. *TimeML Annotation Guidelines*, 2002.

[Pustejovsky et al., 2003] J. Pustejovsky, J. Castaño, R. Ingria, R. Saurí, R. Gaizauskas, A. Setzer, and G. Katz. TimeML: Robust specification of event and temporal expressions in text. In *Proceedings of the AAAI Spring Symposium*, 2003.

[Saquete et al., 2002] E. Saquete, P. Martinez-Barco, and R. Munoz. Recognizing and tagging temporal expressions in Spanish. In *Workshop on Annotation Standards for Temporal Information in Natural Language, LREC 2002 (Third International Conference on Language Resources and Evaluation)*, pages 44–51, 2002.

[Schilder and Habel, 2001] F. Schilder and C. Habel. From temporal expressions to temporal information: Semantic tagging of news messages. In *Proceedings of the ACL-2001 Workshop on Temporal and Spatial Information Processing*, 2001.

[Schilder, 2004] F. Schilder. Extracting meaning from temporal nouns and temporal prepositions. *ACM Transactions on Asian Language and Information Processing*, 2004.

[Setzer and Gaizauskas, 2000] A. Setzer and R. Gaizauskas. Annotating events and temporal information in newswire texts. In *Proceedings of LREC2000*, 2000.

[Sha and Pereira, 2003] F. Sha and F. Pereira. Shallow parsing with conditional random fields. In *Proceedings of Human Language Technology-NAACL*, 2003.

[TERN, 2004] TERN. Temporal Expression Recognition and Normalization, 2004. URL: http://timex2.mitre.org/tern.html.

[TimeBank, 2004] TimeBank.   Annotated corpus, 2004.   URL: `http://www.cs.`
`brandeis.edu/~jamesp/arda/time/timebank.html`.

[TreeTagger, 2004] TreeTagger. A language independent part-of-speech tagger, 2004.
URL: `http://www.ims.uni-stuttgart.de/projekte/corplex/TreeTagger/`.

What is a Logical System?
A Commentary[1]

ATOCHA ALISEDA

1 Introduction

When I think of Dov's Gabbay legacy to logic, in so far as what has been the greatest influence on my own work, I cannot help to think of his marvelous contribution to the axiomatic theory of consequence relations. This type of analysis started with Danna Scott and was inspired in the works of logical consequence by Tarski and those of natural deduction by Gentzen. As widely known, it describes a style of inference at a very abstract structural level, giving its pure combinatorics. It has proved very successful in artificial intelligence for studying different types of plausible reasoning [Kraus *et al.*, 1990], and indeed as a general framework for non-monotonic consequence relations [Gabbay, 1985]. Another area where it has proved itself is dynamic semantics, where not one but many new notions of dynamic consequences are to be analyzed [van Benthem, 1996].

The contribution of Dov Gabbay to this field, as I see it, was on the one hand, to provide a general framework for analysis and comparison to the proliferation of new proposed logics in artificial intelligence, and on the other hand, is his attempt to provide a set of minimal properties for a consequence relation to be considered a logic. As we shall see, however, there seems to be no specific minimal set of properties for that purpose. Still, the question remains in regard to what extent we may use this framework as an attempt to provide a logical criterion of demarcation, and I believe this is fertile area to explore.

In occasion of the workshop on *Abduction in Science* held in Groningen in

---

[1] I was kindly invited by John Woods to take part in this publication because of my work in the field of abduction, where Dov has made a significant contribution. In particular, I was very fortunate to learn about the research wich led to '*The Reach of Abduction: Insight and Trial*' [Gabbay and Woods, 2005] in its early development. However, I know the recently published version is substantially changed and moreover, I'd like to take this opportunity to continue a brief conversation I started with Dov three years ago.

2002, I had the opportunity to talk briefly with Dov about my ideas of how the axiomatic theory of consequence relations may be used in connection to the problem of demarcation in logic, but these ideas were then neither yet published nor presented as such[1]. Therefore, I'd like to take this opportunity to present Dov's development on this issue, and my response in connection to his work.

2   In search of a common logical framework

As clearly stated in [Ohlbach and Reyle, 1999]:

> *'In 1984 there was in the AI literature a multitude of proposed non-monotonic logical sustems defined for a variety of reasons for a large number of applicactions. In an attempt to put some order in what was then a chaotic field, Gabbay asked himself what minimal properties do we require of a consequence relation $A_1, \ldots, A_n \vdash B$ in order for it to be considered as a logic. In his seminal paper [Gabbay, 1985] he proposed the following.*

- **Reflexivity:** $\Delta, A \vdash A$
- **Restricted Monotonicity:**

$$\frac{\Delta \vdash A \quad \Delta \vdash B}{\Delta, A \vdash B}$$

- **Cut:**

$$\frac{\Delta, A \vdash B \quad \Delta \vdash A}{\Delta \vdash B}$$

> *The idea is to clasify non-monotonic systems by properties of their consequence relations. Kraus–Lehman–Magidor developed preferential semantics corresponding to various additional conditions on $\vdash$ and this has started the area now known as the axiomatic approach to non–monotonic logics.'*

This new framework served to analyze and compare many proposed logical systems. One important contribution is that it goes beyond the view of classifying a set of logical systems for what they dot not validate –not surprinsingly were labelled *non–monotonic logics*– and rather looks in a positive way for the properties that they do observe.

---

[1]In [Aliseda, 2005] I depart from Haack's classification of logics amongst extensions, deviations and inductive ones; and then use the axiomatic consequence relations approach to characterize abduction and argue that it is indeed a logic.

However, the question of whether these three specific rules were indeed validated by every system was refuted.

Ten years or so later on, it was acknowledged by Gabbay himself that '*although some classification was obtained and semantical results were proved, the approach does not seem to be strong enough. Many systems do not satisfy restricted monotonicty. Other systems such as relevance logic, do not satisfy even reflexivity. Others have richness of their own which is lost in a simple presentation as an axiomatic consequence relation. Obviously, a different approach is needed, one which would be more sensitive to the variety of features of the systems in the field*' [Gabbay, 1994]. As is well-known, Gabbay then moved to propose his well-known Labelled Deductive Systems, certainly a much more robust framework for logical systems.

In any case, the question of what structural properties does a formal system should satisfy in order to be considered as logical was left unanswered. What I'd like to suggest is that rather than aiming at an specific set of minimal rules, we should be asking for a minimal schema set of structural properties a system should satisfy to be considered a logical one. Many logics deviced for AI applicactions, althoug they are non-monotonic by nature, that does dot mean they do not satisfy any particular form of monotonicity, be this one cautious or any other restricted form of monotonicity. The same applies for properties like reflexivity or cut.

3 A Criterion of Logical Demarcaction

Providing a set of minimal schema rules a formal logical system should observe is closely connected to the problem of demarcation in logic, that is, to the question of providing a proper division in order to distinguish between those formal logical systems from those which are not. Although we may decide in the end that the demarcation in logic is an open question without a unique answer (c.f. [Aliseda, 2005]), we may put forward a criterion of demarcation based on a minimal list of structural schema rules which characterize logical formal systems.

Following Gabbay's already mentioned first attempt, my particular suggestion is that for a formal system to be considered as a logical one, it must have a safe way to preserve inference validity when we insert additional premisses, it must somehow allow to safely chain inferences and it must also have the capacity of autoreflexion. That is to say, it must have some form of monotonicity, transivity or cut, and of reflexivity. And these forms need not be the same ones as those for classical logic.

In what follows, I present the case of what may be called Bolzano's consequence.

3.1   Bolzano's Consequence

We may define a consequence relation $\Rightarrow$ as follows[2]:

$\Delta, \alpha \Rightarrow \varphi$ iff

(i) $\Delta, \alpha \models \varphi$

(ii) Δ, α are consistent

The following are proposed structural rules of reflexivity, monotonicity and cut for this type of consequence relation:

- Conditional Reflexivity:

$$\frac{\Delta, A \Rightarrow B}{\Delta, A \Rightarrow A}$$

- Modified Monotonicity:

$$\frac{\Delta \Rightarrow A \qquad \Delta, B \Rightarrow C}{\Delta, B \Rightarrow A}$$

- Simultaneous Cut:

$$\frac{\Delta \Rightarrow A \qquad \Delta, A \Rightarrow B}{\Delta \Rightarrow B}$$

These rules state the following. Conditional Reflexivity requires that the sequence Δ, A derive something else (B), as this ensures consistency. Modified Monotonicity requires that the sequence Δ, B derives something else (C) in order to ensure consistency when a premise B is added. Simultaneous Cut is a combination of Cut and Contraction in which the sequent A may be omitted in the conclusion when it is consistently derived by Δ and it consistently derives B. Note that the above list is neither exhaustive of what rules may be validated by this consequence relation nor it is the core list for an structural characterization. Modified Monotonicity is a derived rule of the other two. (Cf.[Aliseda, 2005] for a complete set of rules and a representation theorem).

---

[2]This type of consequence is characterized a such in [van Benthem, 1985]. It is also characterized as consistent abduction by me in [Aliseda, 2005] and previous publications, for the consistency requirement is the minimum taken for abductive inference.

BIBLIOGRAPHY

[Aliseda, 2005] A. Aliseda. *Abductive Reasoning: Logical Investigations into Discovery and Explanation*. Synthese Library. Kluwer–Springer Academic Press. To appear in 2005.

[van Benthem, 1985] J. van Benthem. 'Lessons from Bolzano'. Center for the Study of Language and Information. Technical Report CSLI-84-6. Stanford University. 1984. Later published as 'The variety of consequence, according to Bolzano'. *Studia Logica* 44, pp. 389–403. 1985.

[van Benthem, 1996] J. van Benthem. *Exploring Logical Dynamics*. CSLI Publications, Stanford University. 1996.

[Gabbay, 1985] D.M. Gabbay. 'Theoretical foundations for non-monotonic reasoning', in K. Apt (ed), *Expert Systems, Logics and Models of Concurrent Systems*, pp. 439–459. Springer–Verlag. Berlin, 1985.

[Gabbay, 1994] D.M. Gabbay (ed). *What is a Logical System?*. Clarendon Press. Oxford. 1994.

[Gabbay, 1996] D.M. Gabbay. *Labelled Deductive Systems*. Oxford University Press. 1996.

[Gabbay and Kempson, 1994] D.M. Gabbay, R. Kempson. *Labeled Abduction and Relevance Reasoning*. Manuscript. Department of Computing, Imperial College and Department of Linguistics, School of Oriental and African Studies. London. 1994.

[Gabbay and Woods, 2005] D.M. Gabbay and J. Woods. *A Practical Logic of Cognitive Systems. The Reach of Abduction: Insight and Trial (volume 2)*. Amsterdam: North-Holland. 2005.

[Kraus et al., 1990] S. Kraus, D. Lehmann, M. Magidor. 'Nonmonotonic Reasoning, Preferential Models and Cumulative Logics'. *Artificial Intelligence*, 44: 167–207. 1990.

[Ohlbach and Reyle, 1999] H.J.Ohlbach and U. Reyle (eds.). 'Research Themes of Dov Gabbay', in *Logic, Language and Reasoning*, pp. 13–30. Kluwer Academic Publishers. 1999.

Two Glass Balls and a Tower

Amihood Amir

1 Introduction

For many years, one of the favorite questions that interviewers for high-tech companies in Israel asked prospective job applicants was the following.

Suppose a company developed a very resilient glass. They created identical glass balls and desire to find the height from which a fall of the glass ball will cause it to shatter. The tests are conducted by dropping glass balls from various floors of an n-story building. We require the floor number k such that the glass breaks if dropped from it, but does not break from floor $k - 1$.

We would like to know what is the minimum number of tests (ball drops) necessary to find the number k, in the worst case.

Three situations are considered:

1. **One ball:** The new applicant is presented with a single ball for the tests (due to the high cost of the new material).

   Clearly, the applicant has no choice but to start from floor 1 and drop the ball, if the ball breaks then the answer is 1. As long as the ball does not break, the tester needs to sequentially move up floor by floor, and drop the ball from each floor. When eventually the ball breaks upon the drop from floor k, that is the result. The number of tests is then $O(n)$.

2. **Unbounded number of balls:** The tester gets as many balls as her requires.

   Clearly, this is a case for the *divide-and-conquer* approach. Binary search over the n floors of the building can guarantee the result in time $O(\log n)$.

3. **Two balls:** The tester has exactly two balls.

   This is the intriguing case. Superficially, it seems like another ball can not improve the situation over the one-ball case by more than a constant. A second look at the problem shows that divide-and-conquer can still be used, but only to depth 2. This is a *bounded depth divide-and-conquer*. Simply divide the n floors to groups of $\sqrt{n}$

sequential floors. This creates $\sqrt{n}$ such groups. Use the first ball to find the *group* where the ball breaks, in $O(\sqrt{n})$ tests. Then use the second ball to "fine tune" the result and find what floor within the group is the critical one. Since the group size is $\sqrt{n}$, the number of tests for this level is also $O(\sqrt{n})$, for a total of $O(\sqrt{n})$ tests.

The above problem demonstrates in a very clear way the bounded-depth divide-and-conquer technique. This technique has been playing a role in pattern matching for over two decades.

String matching, the problem of finding all occurrences of a given pattern in a given text, is a classical problem in computer science. The problem has pleasing theoretical features and a number of direct applications to "real world" problems. The Boyer-Moore [Boyer and Moore, 1977] algorithm is directly implemented in the *emacs* "s" and *UNIX* "grep" commands.

Advances in Multimedia, Digital Libraries and Computational Biology have shown that a much more generalized theoretical basis of string matching could be of tremendous benefit [Pentland, 1992; Olson, 1995]. To this end, string matching has had to adapt itself to increasingly broader definitions of "matching". Two types of problems need to be addressed – *generalized matching* and *approximate matching*. In generalized matching, one still seeks all exact occurrences of the pattern in the text, but the "matching" relation is defined differently. The output is all locations in the text where the pattern "matches" under the new definition of match. The different applications define the matching relation. Examples can be seen in Amir and Farach's *less-than matching* ([Amir and Farach, 1995]) or Amir, Iliopoulos, Kapah and Porat's *weighted matching* ([Amir *et al.*, 2005]). The second model is that of approximate matching. In approximate matching, one defines a distance metric between the objects (e.g. strings, matrices) and seeks all text location where the pattern matches the text by a pre-specified "small" distance.

One of the earliest and most natural metrics is the *Hamming distance*, where the distance between two strings is the number of mismatching characters. Let n be the text length and m the pattern length. Abrahamson [Abrahamson, 1987] showed that the Hamming distance problem, also known as the *string matching with mismatches* problem can be solved in time $O(n\sqrt{m \log m})$, i.e. within these time bounds one can find the hamming distance of the pattern at *every* text location. This is an asymptotic improvement over the $O(nm)$ bound even in the worst case.

All problems mentioned above, i.e. the *less-than matching, weighted matching,* and *Hamming distance*, were all solved by the bounded-depth divide-and-conquer method. We will review these problems and point out the application of this method.

2   Hamming Distance

2.1   Problem Definition and Preliminaries

DEFINITION 1.

1. Let $a, b \in \Sigma$. Define

$$neq(a, b) =^{def} \begin{cases} 1 & \text{if } a \neq b; \\ 0 & \text{if } a = b. \end{cases}$$

2. Let $X = x_0 x_1 ... x_{n-1}$ and $Y = y_0 y_1 ... y_{n-1}$ be two strings over alphabet Σ. Then the *Hamming distance* between X and Y $(ham(X, Y))$ is defined as

$$ham(X, Y) =^{def} \sum_{i=0}^{n-1} neq(x_i, y_i).$$

3. The *The Hamming Distance* Problem is defined as follows:

Input: Text $T = t_0 ... t_{n-1}$, pattern $P = p_0 ... p_{m-1}$, where $t_i, p_j \in \Sigma$, $i = 0, ... n - 1$; $j = 0, ..., m - 1$.

Output: For every text location i, output $ham(P, T^{(i)})$, where $T^{(i)} = t_i t_{i+1} ... t_{i+m-1}$.

Abrahamson [Abrahamson, 1987] developed a seminal algorithm that finds $ham(P, T^{(i)})$ $\forall i$ in total time $O(n\sqrt{m \log m})$, using bounded-depth divide-and-conquer. We present a simplified version of that algorithm.

The two glass balls symbolize two different techniques to solve the problem. One easily handles the case of a small alphabet, and the other handles the case of every alphabet symbol appearing a small number of times. We will make a slight change in the problem requirements, though. We will count *matches* rather than *mismatches*. The Hamming distance can be easily calculated from the number of matches since it is simply the difference between the length of the pattern (m) and the number of matches.

2.2   Small alphabet

We consider the case where P has a *small* alphabet, e.g. less than $\sqrt{m}$ different alphabet symbols. We will use convolutions, as introduced by Fischer and Paterson [Fischer and Paterson, 1974]. Fischer and Paterson observed that string matching is a special case of a generalized convolution.

DEFINITION 2. Let $X = \langle x_0, ..., x_m \rangle$, $Y = \langle y_0, ..., y_n \rangle$ be two given vectors, $x_i, y_i \in D$. Let $\otimes$ and $\oplus$ be two given functions where

$$\otimes : D \times D \to E,$$

$$\oplus : E \times E \to E, \quad \oplus \text{ associative}.$$

Then the *convolution of X and Y with respect to $\otimes$ and $\oplus$* is:

$$X \langle \otimes, \oplus \rangle Y = \langle z_0, ..., z_{n+m} \rangle$$

where

$$z_k = \bigoplus_{i+j=k} x_i \otimes y_j \qquad \text{for} \quad k = 0, ..., m+n.$$

EXAMPLE 3.
  Boolean Product: $\otimes$ is $\wedge$ and $\oplus$ is $\vee$.
  Polynomial product: $\otimes$ is $\times$ and $\oplus$ is $+$.
  Exact string matching: $\otimes$ is $=$ and $\oplus$ is $\wedge$ but the pattern is transposed.
For all matches of pattern $b\,a\,a$ in text $b\,a\,a\,b\,a$, do $b\,a\,a\,b\,a \langle =, \wedge \rangle b\,a\,a^R$.

		b	a	a	b	a
			a	a	b	
—	—	—	—	—	—	—
		1	0	0	1	0
	0	1	1	0	1	
0	1	1	0	1		
—	—	—	—	—	—	—
0	0	1	0	0	1	0
		—	—	—		

Note that $(X \langle =, \wedge \rangle Y)_k = 1$ iff $\langle x_{k-n}, ..., x_k \rangle = \langle y_n, ..., y_0 \rangle$ for $n \leq k \leq m$. We conclude that there is a match in position 2, i.e.

$$\begin{array}{ccccc} b & a & a & b & a \\ - & - & - & & \\ \langle & a & a & b & \rangle^R \end{array}$$

The problem is that most such convolutions require time $O(nm)$ to compute. An exception is polynomial multiplication that can be achieved in time $O(n \log m)$ using the Fast Fourier Transform (FFT). Thus it is necessary to reduce the desired convolution to polynomial multiplication in the complex field in order to take advantage of the FFT algorithm. We show how this is done in the counting matches problem.

We need some definitions first.

DEFINITION 4. Define

$$\chi_\sigma(x) = \begin{cases} 1 & \text{if } x = \sigma \\ 0 & \text{if } x \neq \sigma \end{cases} \qquad\qquad \chi_{\bar{\sigma}}(x) = \begin{cases} 1 & \text{if } x \neq \sigma \\ 0 & \text{if } x = \sigma \end{cases}$$

If $X = x_0 \ldots x_{n-1}$ then $\chi_\sigma(X) = \chi_\sigma(x_0) \ldots \chi_\sigma(x_{n-1})$. Similarly define $\chi_{\bar{\sigma}}(X)$.

For string $S = s_0 \ldots s_{n-1}$, S^R is the reversal of the string, i.e. $s_{n-1} \ldots s_0$.

We return to the match problem for small alphabets. The product $\chi_\sigma(T)$ by $\chi_\sigma(P^R)$ is an array where the number in each location is the number of matches of a σ text element with a σ in the pattern. If we multiply $\chi_\sigma(T)$ by $\chi_\sigma(P)^R$, for every $\sigma \in \Sigma$, and add the results, we get the total number of matches. Since polynomial multiplication can be done in time $O(n \log m)$ using FFT, and we do $|\Sigma|$ multiplications, the total time for finding all matches using this scheme is $O(|\Sigma| n \log m)$.

Time: Our alphabet size is $O(\sqrt{m})$, so the problem can be solved in time $O(n\sqrt{m} \log m)$.

Notice that this technique knows how to count matches for "groups" of symbol occurrences, all those who are equal to the symbol being considered.

2.3 Every Symbols occurs only a Few Times in Pattern

Consider now the case where the alphabet is large, but also where every alphabet symbol appears in the pattern only a small number of times, e.g. no more than $\sqrt{m}$ times.

The most naive algorithm for counting matches is, for every text location i, to count all matches of the pattern and text, and record it in $M[i]$, where M is a *number of matches* array. This can be implemented by the following simple algorithm.

Naive Algorithm version A
{ Initialize M to 0 }
$M \leftarrow 0$
{ Main Loop }
For $i = 0$ to $n - 1$ do:
   For $j = 0$ to $m - 1$ do:
   If $t_{i+j} = p_j$ then $M[i] \leftarrow M[i] + 1$
   enfFor j
endFor i
end Algorithm

It is clear that version A is equivalent to version B below, where we simply check all pattern element versus the text element scanned at the moment and update the appropriate counters.

Naive Algorithm version B
{ Initialize M to 0 }
$M \leftarrow 0$
{ Main Loop }
For $i = 0$ to $n - 1$ do:
   For $j = 0$ to $m - 1$ do:
   If $t_i = p_j$ then $M[i - j] \leftarrow M[i - j] + 1$
   enfFor j
endFor i
end Algorithm

The time for both above algorithms is, naturally, $O(nm)$. However, note that in version B, we may make a slight improvement. We can pre-process the pattern and create, for every alphabet symbol σ, a list $L_\sigma[0], ..., L_\sigma[\ell_\sigma]$ of locations where σ occurs in the pattern, i.e. all i such that $p_i = \sigma$.

EXAMPLE 5. Let
$$P = \begin{matrix} 0 & 1 & 2 & 3 & 4 & 5 & 6 & 7 & 8 & 9 \\ A & A & B & A & C & B & A & B & C & B \end{matrix}$$
Then A's list is: $L_A = 0, 1, 3, 6$, B's list is: $L_B = 2, 5, 7, 9$, and C's list is: $L_C = 4, 8$.

We can modify version B to only consider t_i's list in the comparison, rather than go through the entire pattern. This achieves the following algorithm:

Not-So-Naive Algorithm C
{ Initialize M to 0 }
$M \leftarrow 0$
{ Main Loop }
For $i = 0$ to $n - 1$ do:
Let $\sigma \leftarrow t_i$
   For all $j = 0$ to ℓ_σ do:
   $M[i - L_\sigma[j]] \leftarrow M[i - L_\sigma[j]] + 1$
   enfFor j
endFor i
end Algorithm

In the worst case, Algorithm C has exactly the same time as version B. However, in the fortunate case we are considering, where every symbol occurs at most $\sqrt{m}$ times in the pattern, the lengths of the L lists never exceeds $\sqrt{m}$ and algorithm C's running time is then $O(n\sqrt{m})$.

Notice that this second glass ball counts matches *within* groups of equal symbol occurrences. We sequentially go through the entire list of indices of every symbol.

2.4 General Alphabets

We are now ready to present the general algorithm.

DEFINITION 6. A symbol that appears in the pattern at least $\sqrt{m \log m}$ times is called *frequent*. Otherwise, it is called *rare*.

It is clear from the definition that there are at most $O(\sqrt{m}/\sqrt{\log m})$ frequent symbols, thus we can count matches of all frequent symbols in time $O(n\sqrt{m/\log m}\log m) = O(n\sqrt{m \log m})$ as shown in Subsection 2.2.

The matches of rare symbols are counted as in Algorithm C of Subsection 2.3. Since a rare symbol appears at most $\sqrt{m \log m}$ times, this stage is done in time $O(n\sqrt{m \log m})$.

3 Less-Than Matching

The Less-Than Matching problem was introduced by Amir and Farach [Amir and Farach, 1995] as a tool for solving the approximate matching problem in non-rectangular two dimensional images. The problem is defined as follows.

The *less-than matching problem* is:

Input: Text string $T = t_0, ..., t_{n-1}$ and pattern string $P = p_0, ..., p_{m-1}$ where $t_i, p_i \in \mathbb{N}$ (the set of natural numbers).

Output: All locations i in T where $t_{i+k-1} \geq p_k$, $k = 1, ..., m$.

In words, every matched element of the pattern is not greater than the corresponding text element. If the text and pattern are drawn schematically, we are interested in all position where the pattern lies completely below the text. See Figure 1.

The less-than matching problem was also solved by the bounded-depth divide-and conquer technique [Amir and Farach, 1995]. As in the Hamming distance case, we also have two algorithms. The first handles small alphabets (groups of floors) and is solved by the FFT. The second handles the situation within the groups. This process is trickier than in the mismatch case that we saw in Section 2.3.

3.1 Small Alphabet

Assume that there are g different elements in the pattern.

NOTATION 7. For $\sigma, x \in \mathbb{N}$, let

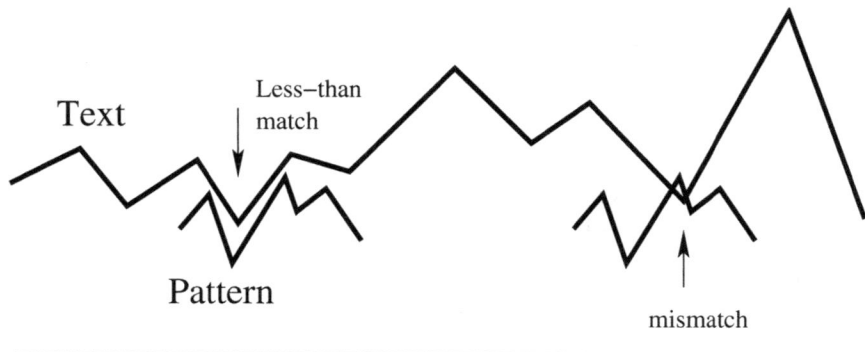

Figure 1. The first appearance of the pattern lies completely below the text, thus there is a match. The second is a mismatch.

$$\chi_\sigma(x) = \begin{cases} 1 & \text{if } x = \sigma \\ 0 & \text{if } x \neq \sigma \end{cases}$$

$$\chi_{<\sigma}(x) = \begin{cases} 1 & \text{if } x < \sigma \\ 0 & \text{if } x \geq \sigma \text{ or } x = \phi \end{cases}$$

If $X = x_1, \ldots, x_n$ then $\chi_\sigma(X) = \chi_\sigma(x_1), \ldots, \chi_\sigma(x_n)$. Similarly define $\chi_{<\sigma}(X)$.

We would like to know for each element of the pattern, where it is lined up with something less than it. We can achieve this by computing, for each σ in P, $\chi_{<\sigma}(T) \otimes \chi_\sigma(P^R)$ (where $\otimes$ is polynomial multiplication), and considering all non-zero locations.

Let $\Sigma = \{\sigma_1, \sigma_2, \ldots, \sigma_g\}$ be the set of all different numbers appearing in P. Let $M_i = \chi_{<\sigma_i}(T) \otimes \chi_{\sigma_i}(P^R)$ (where $\otimes$ is polynomial multiplication). Then M_i is non-zero at position t iff there is a σ_i in the pattern matched with something smaller than σ_i when the pattern is lined up at t. These cases are exactly when we get a mismatch. If we let M be the sum of all the M_i's we get a non-zero if there was a mismatch caused by any $\sigma \in \Sigma$. By using FFT we can calculate each of the polynomial multiplications in time $O(n \log m)$, for a total of $O(g\, n \log m)$.

3.2 The General Case

As we had mentioned, dealing with the elements within groups is not immediate. We need to define the groups in a somewhat different manner.

We therefore provide here the general framework and then show how to efficiently "fine tune" within groups.

Our input is text $T = t_0, \ldots, t_{n-1}$ and pattern $P = p_0, \ldots, p_{m-1}$. Without loss of generality we may assume that the text alphabet is the same as the pattern alphabet. If this is not the case, replace every text number by the largest pattern number that does not exceed it.

In this case also frequent numbers are handled differently from rare numbers. A number is *frequent* if it appears at least $\sqrt{m \log m}$ times, otherwise it is *rare*. There are at most $\sqrt{m}/\sqrt{\log m}$ frequent numbers.

We can use the FFT in the manner described in Subsection 3.1 to disqualify all text locations where there is a text element smaller than a frequent element. This can be done in time $O(n\sqrt{m \log m})$. Now in all remaining locations we just need to make sure that the text location is at least as large as the pattern element corresponding to it. This is done by splitting into groups.

Dividing into Groups:

Let $p_{j_0}, \ldots, p_{j_{g_1}}$ be the non-frequent elements of P.

Consider every pattern element as a pair $\langle s, d \rangle$ where s is a number and d is the location of the number in the array P. We get a list $L = \langle p_{j_0}, j_0 \rangle$, $\langle p_{j_1}, j_1 \rangle, \ldots, \langle p_{j_{g_1}}, j_{g_1} \rangle$.

Sort L lexicographically. (There are no more than m elements in L.) Call the sorted array L'.

Divide L' into $g = O(\sqrt{m/\log m})$ blocks, each containing no more than $2\sqrt{m \log m}$ elements, in a manner that no number appears in more than one block. We are assured that such a division is possible because all remaining numbers are non-frequent.

Possible Implementation: Put the first $\sqrt{m}$ occurrences in block 1. Let σ be the symbol of the last occurrence in block 1. Make sure all occurrences of σ are added to block 1 (since σ is rare it will not add more than $\sqrt{m \log m}$ elements. Continue in a similar fashion to create all blocks.

EXAMPLE 8. Let the pattern P be:

0	1	2	3	4	5	6	7	8	9	10	11	12	13	14	15
G	A	H	A	C	B	D	E	C	B	F	F	E	G	A	B

$\sqrt{16} = 4$.

Sorting the occurrences gives: $\langle A, 1 \rangle, \langle A, 3 \rangle, \langle A, 14 \rangle, \langle B, 5 \rangle, \langle B, 9 \rangle, \langle B, 15 \rangle$, $\langle C, 4 \rangle, \langle C, 8 \rangle, \langle D, 6 \rangle, \langle E, 7 \rangle, \langle E, 12 \rangle, \langle F, 10 \rangle, \langle F, 11 \rangle, \langle G, 0 \rangle, \langle G, 13 \rangle, \langle H, 2 \rangle$.

The suggested implementation produces the blocks:

block 1: $\langle A, 1 \rangle, \langle A, 3 \rangle, \langle A, 14 \rangle, \langle B, 5 \rangle, \langle B, 9 \rangle, \langle B, 15 \rangle$.
block 2: $\langle C, 4 \rangle, \langle C, 8 \rangle, \langle D, 6 \rangle, \langle E, 7 \rangle, \langle E, 12 \rangle$.
block 3: $\langle F, 10 \rangle, \langle F, 11 \rangle, \langle G, 0 \rangle, \langle G, 13 \rangle$.
block 4: $\langle H, 2 \rangle$.

Notice that we have 4 blocks, none larger than 6 elements.

Construct the Text and Pattern of Representatives:

For each block B_i, $i = 0, \ldots, g \leq \sqrt{m/\log m}$ let b_i be the smallest (leftmost) element in the block; call b_i the *representative* of block B_i.

Let T' and P' be T and P such that every t_i and p_i is replaced by the representative of the block it is in. This step can be implemented by a sequential scan of L'.

We now find all less-than matches of P' in T', by the FFT. P' and T' can be considered "flattened out" versions of P an T. When we seek all less-than matches of P' in T' we only detect the "large" mismatches, i.e., those between elements that are so different that they are in different blocks. However, mismatches between elements of the same block are undetected. At this stage we must "fine tune" our approximate solution. We now come to the second level of our bounded divide-and-conquer. The one that needs to adjust for the mismatches that may exist but were not recorded. These are precisely the cases where a text number is smaller than a pattern number that is **in the same block** as the text number.

3.3   Checking Elements within Blocks

Scan T and for every element t_i of T only compare it to the $O(\sqrt{m \log m})$ elements of P that are in t_i's block. The subroutine is very similar to algorithm C in Section 2.3.

Fine Tuning Algorithm

For $i = 0$ to $n - 1$ do:
    Let B_{t_i} be the block of L' that t_i is in,
    Let $P^{B_{t_i}} \leftarrow \{\langle s, d \rangle | \langle s, d \rangle \in B_{t_i}\}$
    For every element $\langle s, d \rangle$ in $P^{B_{t_i}}$ { at most $2\sqrt{m \log m}$ elements }
        if $t_i < s$ then $M[i - d] \leftarrow M[i - d] + 1$
    endFor
endFor i

end Algorithm

The vector M is now correct since the first part of the algorithm included all the errors between blocks and the last part found all the errors within a block.

Time: $O(m \log m)$ for sorting and $O(n \log m)$ for reducing to alphabet
       of pattern.
      $O(n\sqrt{m \log m})$ for less-than matching of the representatives
      $O(n\sqrt{m \log m})$ for correcting mismatches within blocks

Total: $O(n\sqrt{m}\log m)$

4   Approximate Matching in Weighted Sequences

Weighted sequences have been recently introduced by Iliopoulos et al. [Iliopoulos *et al.*, 2003] as a tool to handle a set of sequences that are not identical but have many local similarities. The weighted sequence is a "statistical image" of this set, where the probability of every symbol's occurrence at every text location is given.

DEFINITION 9. A *weighted sequence* $T = t_0, ..., t_n$ over alphabet Σ is a sequence of sets t_i, $i = 0, ..., n$. Every t_i is a set of pairs $(s_j, \pi_i(s_j))$, where $s_j \in \Sigma$ and $\pi_i(s_j)$ is the probability of having symbol s_j in location i. Formally,

$$t_i = \{(s_i, \pi_i(s_j)) \mid s_j \neq s_\ell \text{ for } j \neq \ell, \text{ and } \sum_j \pi_i(s_j) = 1\}.$$

For a finite alphabet $\Sigma = \{a_1, ..., a_{|\Sigma|}\}$ we can view a weighted sequence as a $|\Sigma| \times n$ matrix T of numbers in $[0, 1]$, where $T[j, i] = \pi_i(a_j)$. For the rest of this paper we assume a finite fixed alphabet Σ.

EXAMPLE 10. Let $\Sigma = \{A, B, C, D\}$, then an example of a text of length 7 is:

A	1/2	1/3	1/4	1	1/6	1/4	0
B	1/4	1/3	1/4	0	1/6	0	1/5
C	1/4	0	1/4	0	1/6	1/2	3/5
D	0	1/3	1/4	0	1/2	1/4	1/5

DEFINITION 11. $P = p_0, ..., p_m$ is a *solid sequence* over alphabet Σ if $p_i \in \Sigma$, $i = 0, ..., m$.

We say that *solid pattern* P (or simply *pattern* P) *occurs in location* i of weighted text T with probability at least α if $\prod_{j=0}^{m} \pi_j(p_j) \geq \alpha$.

EXAMPLE 12. Let T be the weighted text in the previous example, $P = ADCC$, and $\alpha = 0.1$. Then P occurs at location 3 with probability at least α, since the probability of P at location 3 is $\frac{3}{2 \cdot 2 \cdot 5} = \frac{3}{20} = 0.15 > 0.1$, but P does not appear in locations $0, 1$ and 2 since the probability at each of these locations is 0.

DEFINITION 13. The *exact weighted matching problem* is defined as follows:

Input: Weighted text T over alphabet Σ, solid pattern P over alphabet Σ, and probability $\alpha \in [0, 1]$.
Output: All locations i in T where pattern P occurs with probability at least α.

The following is a straightforward efficient algorithm for exact weighted matching.

Algorithm
1. Convert all values of T to their logarithms.
2. For each $\sigma \in \Sigma$ do:
   { Denote by T_σ the σ-th row of T, i.e. the list of probabilities of σ in all locations. }
   $\quad S_\sigma \leftarrow T_\sigma \otimes \chi_\sigma(P)$.
   endFor
3. $Sum \leftarrow \sum_{\sigma \in \Sigma} T_\sigma$
4. For $i = 0$ to $n - m$ do: if $Sum[i] \geq \log \alpha$ then there is a match at location i
end Algorithm

Algorithm Time: Steps 1., 3. and 4. are trivially done in time $O(|\Sigma|n) = O(n)$ (since Σ is a fixed finite alphabet). Step 2. is done in time $O(|\Sigma|n \log m) = O(n \log m)$. Total time: $O(n \log m)$.

What interests us in the context of our bounded-depth divide-and-conquer method is the Hamming distance in weighted sequences problem.

Computing the Hamming distance between two (solid) strings assumes that a number of symbols were replaced. The Hamming distance is the number of these replaced symbols. In the case of weighted subsequences it makes a difference where these symbols were replaced. The simpler case, which we consider in this section, assumes replacement in the text. The assumption is that some text symbols are erroneous and, in fact, there should have been a probability 1 for the symbol that happens to match the pattern, rather than the probabilities that appear in the text.

EXAMPLE 14. For the text in example 10, consider pattern $P = BADC$ and $\alpha = 0.25$. There is no exact match. However, if we allow one mismatch, there is a match in location 2. Simply assume that the probability of having a B in location 2 is 1. At that point, the total probability at location 2 is $1 \cdot 1 \cdot 1 \cdot \frac{1}{2} \cdot \frac{1}{2} = \frac{1}{4}$. In location 0 even with one mismatch the probability is still $\frac{1}{48}$. In location 1 and one mismatch the probability is $\frac{1}{72}$, and in location 3 with one mismatch the probability is $\frac{1}{40}$.

Note that by this definition, allowing enough mismatches can guarantee a match at every location, no matter how close to 1 we choose α.

DEFINITION 15. The *Weighted Hamming Distance with Mismatches in the Text problem* is the following:

Input: Weighted text T over alphabet Σ, solid pattern P over alphabet Σ, and probability $\alpha \in [0,1]$.

Output: For every location i in T, the minimum k such that if k text probabilities were changed to 1 then pattern P would occur at location i with probability at least α.

There does not seem to be a natural way to use the powerful constraint that the numbers in the weighted text are probabilities. However, it seems like we can solve the problem without it. We reduce the weighted Hamming distance with mismatches in the text problem to the *minimum ignored mask bits problem*. The idea is to consider a text whose elements are non-positive numbers, and a pattern which is a mask, i.e. its symbols are 0's and 1's. Suppose we are interested in finding out, for each text location i, the sum of the text numbers that are aligned with 1's in the pattern.

EXAMPLE 16. $T = -1, -3, -17, 0, -5, -6, -1$, $P = 1001$. Then the result at location 0 is -1, at location 1 is -8, at location 2 is -23 and at location 3 is -1.

Clearly this is a simple convolution of the pattern and text. However, we add a complication, we also have a non-positive integer α and for every text location i we seek the smallest number of mask bits that, if set to 0, would make the sum of text numbers that are aligned with (the remaining) 1's in the pattern, be no less than α.

EXAMPLE 17. In the example above, if $\alpha = -5$ then the result at location 0 is -1, with 0 dropped 1 bits from the mask, at location 1 it is -3 if the last mask bit is 0-ed, i.e. the mask becomes 1000, or -5 if the first mask is 0-ed, i.e. the mask becomes 0001. At location 2 we need to 0 two 1 bits in the mask to make the mask 0000 and the result 0. At location 3 the result is -1 with 0 mask bits altered.

We formally define the problem.

DEFINITION 18. The *Minimum Ignored Mask Bits problem* is the following:

Input: Solid text T of length $n + 1$ whose elements are non-positive integers, solid pattern P of length $m + 1$ over alphabet $\{0, 1\}$, and integer $\alpha \le 0$.

Output: For every location i in T, the minimum k such that if k pattern bits are changed from 1 to 0, and M' is the pattern resulting from those k changes, then $\sum_{j=0}^{m} T[i + j]M[j] \ge \alpha$.

CLAIM 19. The weighted Hamming distance with mismatches in the text problem is linearly reducible to the minimum ignored mask bits problem.

Proof. Given weighted text T in matrix format, where the value in $T[i, j]$ is $\log \pi_j(s_i)$, let solid text T' be a linear listing of matrix T in column-major order, i.e. $T' = T[1, 0], T[2, 0], T[3, 0], ..., T[|\Sigma|, 0],$
$T[1, 1], T[2, 1], T[3, 1], ..., T[|\Sigma|, 1], ...,$
$T[1, n], T[2, n], T[3, n], ..., T[|\Sigma|, n]$. Let M be a string of length $|\Sigma|(m + 1)$ over $\{0, 1\}$ where M is the concatenation of strings $B(p_0), B(p_1), ..., B(p_m)$. $B(a)$ is defines as follows. Let $a = s_\ell$, where $\Sigma = \{s_1, s_2, ..., s_{|\Sigma|}\}$. Then $B(a)$ is a bit string of length Σ, where the ℓ-th element is 1 and all other elements are 0.

EXAMPLE: If $\Sigma = \{A, B, C, D\}$ and $P = BBAD$,
then $M = 0100\ 0100\ 1000\ 0001$.

Clearly, the reduction is linear. It is also clear that turning a 1 bit in the mask M to 0, is equivalent to changing the probability in the text position corresponding to it to 1. Thus a solution to the minimum ignored mask bits problem will provide the solution to the weighted Hamming distance with mismatches in the text problem. ∎

Algorithm's Idea: We consider the two cases and show an easy efficient solution for each of them. Subsequently, we use the bounded-depth divide-and-conquer strategy, that splits a general input into the two straightforward cases, and thus solves each separately.

4.1 Bounded Alphabet

The first special case is one where the domain of numbers appearing in the text is bounded, i.e. there are only r different numbers that can appear as text elements. Formally, let $R = \{n_1, ..., n_r\} \subset \mathbb{Z}^- \cup \{0\}$, and let T be a text over R. Assume that $n_1, ..., n_r$ are sorted in descending order. Once we know that the only possibilities for text values are from R, we can calculate, for every location i the r sums $S_{i,j}$, $j = 1, ..., r$, where $S_{i,j}$ is the sum of n_j's that are matched to 1's in the mask at location i.

Since we are interested in finding the smallest number k of mask 1 bits that, when turned to 0 will make the sum greater than α, and since **all numbers are non-positive**, the following observation is crucial to the algorithm:

OBSERVATION 20. For any location i where $\sum_{j=1}^{r} S_{i,j} < \alpha$, the solution to the minimum ignored mask bits problem can be found by sequentially adding numbers that participate in the sum starting from the ones that contribute least to decreasing it, i.e. the largest (n_1). Stop adding them when the remaining sum is no longer less than α.

This elimination would normally require $O(m)$ work per location. However, since there are only r different values, and we know how many instances of each value participate in the sum at location i ($S_{i,j}/n_j$), we can do this in time $O(r)$ per location.

This gives rise to the following algorithm:

Algorithm Bounded Alphabet

1. For $j = 1$ to r do:

    { Denote by S_j the array whose elements are $S_{0,j}, S_{1,j}, ..., S_{n,j}$,
    i.e. $S_j[i] = S_{i,j}$.}

    $S_j \leftarrow \chi_{n_j}(T) \otimes M$

    endFor

2. For $i = 0$ to $n - m$ do:

    $S, j \leftarrow 0$

    While $S \geq \alpha$ do:

        $j \leftarrow j + 1$

        $S \leftarrow S + S_j[i]$

    endWhile

    { The situation is $S < \alpha$ but $S - S_j[i] \geq \alpha$.}

$$k_i \leftarrow \left(\sum_{\ell=1}^{j-1} \frac{S_\ell[i]}{n_\ell} \right) + \left\lfloor \frac{(S-\alpha)}{n_j} \right\rfloor$$

    endFor

end Algorithm

Algorithm's Time: $O(rn \log m)$ for step 1. and $O(rn)$ for step 2. for a total of $O(rn \log m)$. Note that step 2. could be done faster by binary search, but since it is dominated by the time of step 1., we wrote the simpler pseudocode.

4.2 Bounded Number of Large Numbers

A second special case we consider is when there is no bound on the number of different text elements, but we do know that for every text substring of length m there are at most r elements greater than α. This means that for location i, there is no point in even considering all elements except those r.

Here the algorithm is much simpler. Assume the r elements are sorted in non-increasing order. For every text location i, it suffices to consider the r elements from largest to smallest. For each element check whether it correspond to a 1 bit in the mask. If so, add it to the sum and check if it is still above α. When all elements are considered, or when the sum drops below α, if ℓ elements were added to the sum and if the mask has s 1 bits, then k is $s - \ell$.

Formally, let $B_i = \{\langle b_{i,1}, \ell_{i,1} \rangle, \langle b_{i,2}, \ell_{i,2} \rangle, ..., \langle b_{i,r}, \ell_{i,r} \rangle\}$ be the set of pairs such that $b_{i,1}, b_{i,2}, ..., b_{i,r}$ are the values of $\{T[i], T[i+1], ..., T[i+m]\}$ that are greater than α, sorted in non-increasing order, and $\ell_{i,j}$ is the index of $b_{i,j}$ in T, $j = 1, ..., r$, i.e. $T[\ell_{i,j}] = b_{i,j}$.

The algorithm is as follows:

Algorithm Bounded Relevant Numbers
  For $i = 0$ to $n - m$ do:
    $S \leftarrow 0$ { S is the masked sum at location i. }
    $j \leftarrow 0$ { j is a counter of the relevant text elements. }
    $x \leftarrow 0$ { x is a counter of the number of elements in the masked
sum. }
    While $S \geq \alpha$ and $j < r$ do:
      $j \leftarrow j + 1$
      If $M[\ell_{i,j}] = 1$ then $S \leftarrow S + b_{i,j}$
                                $x \leftarrow x + 1$
    endWhile
    $k_i \leftarrow s - x + 1$
  endFor
end Algorithm

Algorithm's Time: $O(nr)$.

4.3   Divide and Conquer

We are now ready to present our divide-and-conquer algorithm. Assume first, that the text length is at most $2m$. This is a standard assumption and can be made without loss of generality because of the following lemma.

LEMMA 21. *Assume that there exists an algorithm that solves the less-than matching problem in time $O(mf(m))$ for $n \leq 2m$, then there exists an algorithm that solves the less-than matching problem in time $O(nf(m))$ for any n-length text.*

Proof. Simply divide the text into $2\frac{n}{2m}$ overlapping $2m$-length segments (see Fig. 2) and solve the matching problem separately for each. Clearly, if there is a match in any location of T it will appear in one of the segments. ∎

We now have a situation where the text is of size $2m$, the pattern of size m. Sort all text elements and split them into r blocks of size at most $2|\Sigma|\frac{m}{r}$ each.

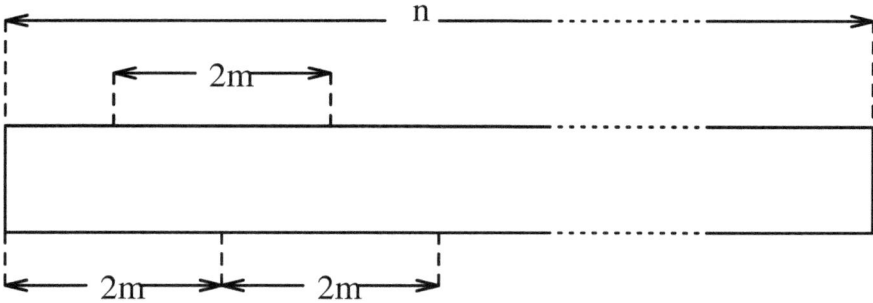

Figure 2. Slicing problem to smaller problems.

The idea is to use *Algorithm Bounded Alphabet* on the blocks, and *Algorithm Bounded Relevant Numbers* to find the border of the numbers participating in the sum within the block that tips under α.

Since the greatest contribution to the masked sum lies with the larger numbers, it is clear that for the mismatches, our strategy should be replacing the smaller numbers where necessary. Therefore, we go block by block, from the largest down, and compute the following two sets of values for every location:

1. How many numbers from each block participate in the masked sum for that location.

2. What is the sum of each block for that location.

Given the above two values, we can compute in time $O(r)$ for every location, within the granularity of blocks, what is the Hamming distance. Note that there will be one block at the "seam" where some values could possibly participate in the product and yet still not plunge below α. We need to adjust for these values. This will be done by *Algorithm Bounded Relevant Numbers* and take time $O(\frac{m}{r})$ per text location.

Algorithm's Time: The time for this algorithm is $O(rf(m))+O(m\frac{m}{r})$, where $f(m)$ is the time it takes to compute the block information. We do it by convolutions, as in *Algorithm Bounded Alphabet* so $f(m) = m \log m$. The optimal r is then the one where

$$rm \log m = m\frac{m}{r}$$

$$r^2 = \frac{m}{\log m}$$

$$r = \sqrt{\frac{m}{\log m}}.$$

Thus the algorithm's time is $O(n\sqrt{m \log m})$.

5   Conclusion

We considered three problems, the Hamming distance problem, the less-than matching problem and the Hamming distance problem in weighted sequences. Muthukrishnan [Muthukrishnan and Palem, 1994] showed that those problems can not be solved by convolutions alone in time faster than $O(nm)$. Nevertheless the bounded-depth divide-and-conquer technique allows them all to be solved in time $O(n\sqrt{m \log m})$.

6   Acknowledgements

Professor Dov Gabbay was my instructor in undergraduate logic and then in a number of advanced logic courses and seminars. He was also my Ph.D. advisor and my thesis was on functional completeness in temporal logics. Since then I have taken some other turns, to complexity, computational biology, and algorithms. This paper is a purely algorithmic paper, but Professor Gabbay's influence on it is apparent to all who know him.

One of the traits that made taking courses with Professor Gabbay fun was the anecdotal examples accompanying the mathematics. This certainly helped garner a better intuitive understanding of the material and this is what I try to employ in my classes, and in this "glass balls and tower" exposition. I would like to point out several examples from Professor Gabbay's undergraduate logic class that illustrate this concept.

When explaining to the class the difference between *inclusive or* and *exclusive or* Professor Gabbay gave the following two examples:

"Suppose there was a sign on a movie theater saying that the entrance costs 2 Liras **or** show a soldier's ID. If a simple soldier went and bought a ticket, he would certainly not be denied entrance.

On the other hand, if you walk in the street and see a mother followed by a wailing child who then threatens the child: "stop crying **or** I will smack you". If the terrified child shuts up we will look unkindly at the mother if she will hit the child and triumphantly declare: "**inclusive or!!!**. Therefore," concluded Gabbay, "if you ever find yourself in a bank when a character in a sky mask and brandishing an *Uzi* barges in and shouts: "your money or your life!", be sure that you ascertain whether he means *inclusive* or *exclusive or.*"

Another monumental statement of Professor Gabbay was the following. When he defined *tautology*, Professor Gabbay explained that it is a statement that is always true. "Therefore," he said, "if you are a politician, you should attempt to always say tautologies."

The next day (and you will soon find out how dated this story is) the following headline appeared in the newspaper:

Kissinger: There may or may not be important developments.

I prepared a sign with both above statements and it hung on my walls for many years:

Gabbay: If you are a politician, you should attempt to always say tautologies.

Kissinger: There may or may not be important developments.

For teaching me to do research, and for showing me that it should be fun - Thanks.

More Acknowledgements

Partially supported by NSF grant CCR-01-04494 and ISF grant 35/05.

BIBLIOGRAPHY

[Abrahamson, 1987] K. Abrahamson. Generalized string matching. *SIAM J. Comp.*, 16(6):1039–1051, 1987.

[Amir and Farach, 1995] A. Amir and M. Farach. Efficient 2-dimensional approximate matching of half-rectangular figures. *Information and Computation*, 118(1):1–11, April 1995.

[Amir et al., 2005] A. Amir, C. Iliopoulos, O. Kapah, and E. Porat. Approximate matching in weighted sequences. submitted for publication, 2005.

[Boyer and Moore, 1977] R.S. Boyer and J.S. Moore. A fast string searching algorithm. *Comm. ACM*, 20:762–772, 1977.

[Fischer and Paterson, 1974] M.J. Fischer and M.S. Paterson. String matching and other products. *Complexity of Computation, R.M. Karp (editor), SIAM-AMS Proceedings*, 7:113–125, 1974.

[Iliopoulos et al., 2003] C. S. Iliopoulos, L. Mouchard, K. Perdikuri, and A. Tsakalidis. Computing the repetitions in a weighted sequence. In *Proceeding of the Prague Stringology Conference*, pages 91–98, 2003.

[Muthukrishnan and Palem, 1994] S. Muthukrishnan and K. Palem. Non-standard stringology: Algorithms and complexity. In *Proc. 26th Annual Symposium on the Theory of Computing*, pages 770–779, 1994.

[Olson, 1995] M. V. Olson. A time to sequence. *Science*, 270:394–396, 1995.

[Pentland, 1992] A. Pentland. Invited talk. NSF Institutional Infrastructure Workshop, 1992.

Reichenbach, Prior and Montague: A Semantic Get-together

CARLOS ARECES AND PATRICK BLACKBURN

1 Introduction

What do you write on for the logician who has written (or edited a Handbook) on just about everything? Finding something new isn't an option. So in the hope of reviving pleasant memories, we've decided instead to present some material related to work that Dov did quite some time ago, in the 1970s.

Dov Gabbay is a pioneer in the development of multi-dimensional modal and tense logic for natural language semantics; a record of his work in this area can be found in his book *Investigations into Modal and Tense Logics, with Applications to Problems in Linguistics and Philosophy* [Gabbay, 1976]. Multidimensional tense logics can improve on ordinary tense logic for natural language semantics in a number of ways; the most important is that they (or some versions of them at least) make it possible to 'store' times so that they can later be referred back to. This is important because, as is explained below, temporal reference is ubiquitous in natural language.

There is another style of modal and tense logic which makes reference to times possible, namely hybrid logic. And as the paper [Blackburn, 1994] pointed out, even a relatively modest hybrid logic (namely *nominal tense logic*) makes it possible to combine the ideas of Hans Reichenbach (see [Reichenbach, 1947]), who emphasized the importance of temporal reference, with those of Arthur Prior (see [Prior, 1967]), who emphasized the importance of the internal perspective on temporal structure provided by modal and tense logics. So Reichenbach and Prior have already met. But the paper that brought them together used a propositional version of nominal tense logic. To give semantic definitions that can be used with real grammars, more powerful systems are needed.

The pioneer of the use of richer logics in natural language semantics was Richard Montague who (most famously in his paper "The Proper Treatment of Quantification in Ordinary English" [Montague, 1973]) showed that higher-order logic was a superb tool for semantic construction. The higher-order logic that Montague developed for this purpose was called $\mathcal{IL}$ (Intensional Logic) and it made use of Prior's tense operators. Hence Prior and Montague have met up too.

So why not bring all three together to Dov's birthday party? That's what this paper is about. We are going to hybridize Montague's $\mathcal{IL}$ in the simplest possible way (we'll simply add nominals) to form $\mathcal{NIL}$ (Nominal Intensional Logic). We'll then show that the resulting system is capable of assigning nominal tense logical representations, which capture the insights of both Reichenbach and Prior, in a compositional way.

2   Prior and Reichenbach

Tense logic is a simple form of modal logic used for reasoning about time. It was invented by Arthur Prior, who introduced the F and P modalities (meaning "at some Future time", and "at some Past time" respectively) and their respective duals G and H ("it is always Going to be the case", and "it always Has been the case"). Why did he do this? Because Prior viewed tensed talk as fundamental: we exist in time, and deal with temporal information from the inside. He felt that the internal perspective offered by modal logics — where we evaluate formulas *inside* models, at some particular point — made it an ideal tool for capturing the situated nature of our experience and the way we talk about it.

For example, suppose we represent the meaning of the present-tensed English sentence "Dov smiles" by the propositional symbol *dov-smile*. If we prefix this with the P operator we obtain P*dov-smile*, and this is true at a time t if and only if Dov does indeed smile at some time t' previous to t. This captures (part of) the meaning of the past-tensed English sentence "Dov smiled". Moreover, the syntactic relationship between "Dov smiles" and "Dov smiled" (which differ only in their tense inflection) is reminiscent of the syntactic relationship between *dov-smile* and P*dov-smile*.

This is an interesting observation, and Prior's insistence on the importance of the internal perspective offered by tense logic deserves to be taken seriously. However, Prior's insight, though useful, fails to take into account the importance of *temporal reference* (that is, reference to specific times) in the semantics of tense (and indeed, in other natural language constructions).

Let's return to our example. The sentence "Dov smiled" does *not* mean that at some completely unspecified past time Dov did in fact smile (which is the meaning the tense logical representation P*dov-smile* gives it). Rather, it means that at some *particular*, contextually determined, past time Dov did in fact smile. Prior's tense logical representations are interesting, and correct as far as they go, but they do not go far enough. Moreover, in orthodox tense logic there is no mechanism to enable them to go further.

Structure	Name	English example
E–R–S	Pluperfect	I had seen
E,R–S	Past	I saw
R–E–S	Future-in-the-past	I would see
R–S,E	Future-in-the-past	I would see
R–S–E	Future-in-the-past	I would see
E–S,R	Perfect	I have seen
S,R,E	Present	I see
S,R–E	Prospective	I am going to see
S–E–R	Future perfect	I will have seen
S,E–R	Future perfect	I will have seen
E–S–R	Future perfect	I will have seen
S–R,E	Future	I will see
S–R–E	Future-in-the-future	(Latin: abiturus ero)

Figure 1. Reichenbach's referential analysis of tense

Hans Reichenbach, on the other hand, viewed temporal reference as central to the semantics of tense in natural language. He distinguished tenses in terms of the reference they make to three temporal markers, namely what he called the *point of speech* (S), the *point of event* (E), and the *point of reference* (R). Now, much of what he says is compatible with Prior's views. For a start, point of speech is the time at which the sentence is uttered, and this concept is fundamental to the internal perspective of tense logic: it's simply the particular time at which we chose to evaluate a formula in a given model. The point of event is the time at which the eventuality the sentence is talking about takes place. This might be the same time as the point of speech, or to its past, or to its future. This concept also fits naturally with Prior's tense logic. If φ is the representation of some eventuality, then evaluating φ at some time amounts to identifying point of event with point of speech. Prefixing P to form $P\varphi$ locates the point of event to the past of the point of speech. Prefixing F to form $F\varphi$ locates the point of event to the future of the point of speech.

However, Reichenbach's key innovation was the point of reference, and here we encounter something that orthodox tense logic cannot handle. Figure 1 tabulates Reichenbach's analyses of the tense forms of English. What do these analyses say?

Consider Reichenbach's account of the pluperfect. When we utter "I had seen", there is a clear intuition that we refer to some past time (this is the point of reference) and assert that the seeing event took place *before that*. Accordingly, Reichenbach analyzes the pluperfect form as E–R–S, which means that the point of event lies to the past of the point of reference, which in turn lies to the past of the point of speech.

What about the simple past? When we discussed "Dov smiled" we said that the function of the simple past in English was to locate an event at some particular (contextually determined) past time. In effect, this is what Reichenbach's treatment of the simple past gives us. He analyzes this tense as E,R–S. That is, the point of event and the point of reference coincide and lie to the past of the point of speech. The point of reference is the contextually determined past time, and by co-locating it with the point of event, we account for the way the simple past tense works.

Reichenbach's analyses are open to criticism. Some linguists have objected, for example, to the use of three distinct diagrams (namely R–E–S, R–E,S, and R–S–E) to account for the future-in-the-past tense (and indeed, the future perfect): as there is only a single natural language form, they demand a single representation. Moreover, many linguists would feel that his analysis of the present perfect (which amounts to saying that point of reference corresponds to point of speech) does not get to grips with the subtleties of this construction. Nonetheless, in spite of their shortcomings, Reichenbach's views on temporal semantics are highly influential in contemporary natural language semantics, and Reichenbach-inspired ideas lie at the heart of much recent work. This is because temporal reference in natural language is ubiquitous, and without some way of capturing its effects, we cannot adequately analyze many temporal constructions. Orthodox tense logic has fallen into disuse in natural language semantics largely because it offers no such mechanism.

3   Nominal Tense Logic

Prior didn't really like Reichenbach's ideas, but if you read the little he has to say on the subject (see pages 12–15 of [Prior, 1967]) you'll see that he doesn't offer any really solid criticisms. Rather, his dislike seems to stem from his conviction that what Reichenbach was saying was incompatible with a *logical* analysis of tense. This is ironic, since in the same book that he criticizes Reichenbach on these grounds, Prior also introduced a tool that allows his views to be integrated with Reichenbach's in a very smooth way! What was this tool? The key idea underlying modern hybrid logic: *sort* the propositional symbols, and use *terms as formulas.*

Let's see what this involves. Take a language of tense logic (with propositional symbols p, q, r, and so on) and add a second sort of propositional symbol. The new symbols are called *nominals*, and are typically written i, j, k, and l. Both types of propositional symbol can be freely combined to form complex formulas in the usual way. But the key ingredient is the following: *we insist that each nominal be true at exactly one time in any model.* Formally, the set of times used to interpret a nominal must be a *singleton* set. A nominal 'names' a time by being true at that single time and nowhere else.

This is a simple change, but it has important consequences. It immediately yields a more expressive logic. Consider the following (orthodox) tense logical formula:

$$\mathsf{F}(r \wedge p) \wedge \mathsf{F}(r \wedge q) \rightarrow \mathsf{F}(p \wedge q).$$

This can be falsified. The first conjunct in the antecedent says that in the future there is time where both r and p are true together, and the second asserts that in the future there is a time where r and q are true together. The conclusion then asserts that in the future there is a time where p and q are true together. But this is obviously unjustified: the future times that witness p and q may be distinct.

Now consider the following formula of nominal tense logic:

$$\mathsf{F}(i \wedge p) \wedge \mathsf{F}(i \wedge q) \rightarrow \mathsf{F}(p \wedge q).$$

This is identical to the preceding formula, except that we have replaced the propositional symbol r by the nominal i, but the resulting formula is impossible to falsify. We now have some extra information: the p-witnessing and q-witnessing future times both make i true, and there is only one time which does this, for i is a nominal. Hence these future times must be identical, and the conclusion follows.

However, what is important for present purposes is that in nominal tense logic it is possible to merge Prior and Reichenbach's ideas on tense. Prior never made the connection, but nominals are precisely the missing component needed to handle Reichenbach's points of reference! Figure 2 shows the table given earlier, but with nominal tense logical representations added in the final column.

Consider the representation $\mathsf{P}(i \wedge \mathsf{P}\varphi)$ of the pluperfect. This says that there is some time in the past labelled i and that the event φ happened *before that*. This representation combines Reichenbach's insight into the role played by temporal reference with Prior's insistence on the privileged role of tensed talk.

Note that in some cases the nominal tense logical representations improve on Reichenbach. In particular, note that the future-in-the-past (and the future perfect) now has a single representation. The formula $\mathsf{P}(i \wedge \mathsf{F}\varphi)$ asserts that there is a reference time i in the past, and that the point of event occurs to the future of i, which is what is wanted. We're not forced (as Reichenbach was) to spell out the irrelevant relationships that can hold between the point of event and the point of speech.

Structure	Name	English example	Representation
E–R–S	Pluperfect	I had seen	$P(i \wedge P\varphi)$
E,R–S	Past	I saw	$P(i \wedge \varphi)$
R–E–S	Future-in-the-past	I would see	$P(i \wedge F\varphi)$
R–S,E	Future-in-the-past	I would see	$P(i \wedge F\varphi)$
R–S–E	Future-in-the-past	I would see	$P(i \wedge F\varphi)$
E–S,R	Perfect	I have seen	$P\varphi$
S,R,E	Present	I see	φ
S,R–E	Prospective	I am going to see	$F\varphi$
S–E–R	Future perfect	I will have seen	$F(i \wedge P\varphi)$
S,E–R	Future perfect	I will have seen	$F(i \wedge P\varphi)$
E–S–R	Future perfect	I will have seen	$F(i \wedge P\varphi)$
S–R,E	Future	I will see	$F(i \wedge \varphi)$
S–R–E	Future-in-the-future	(Latin: abiturus ero)	$F(i \wedge F\varphi)$

Figure 2. Reichenbach's analysis in nominal tense logic

This is neat. However, as we pointed out at the start of the paper, the discussion has been conducted within the confines of *propositional* nominal tense logic. If we want to apply these ideas to real grammars for natural language, that's not good enough. And this leads us to the main topic of the paper: adding nominals to Montague's $\mathcal{IL}$.

4 Nominal Intensional Logic

In this section we take Montague's $\mathcal{IL}$ (which makes use of Prior's F and P operators) and add nominals to it. The result is called $\mathcal{NIL}$ (Nominal Intensional Logic). The following definitions follow Montague's treatment faithfully; the only deviations from his original work are the clauses we have added for handling nominals.

DEFINITION 1 (Syntax of $\mathcal{NIL}$). Let t, e, and s be any fixed objects. Then the set TYPES of types of $\mathcal{NIL}$ is defined recursively as follows:

$$\text{TYPES} ::= t \mid e \mid \langle a, b \rangle \mid \langle s, a \rangle$$

where $a, b \in$ TYPES.

Basic Expressions: For each type a, $\mathcal{NIL}$ contains a denumerably infinite set of *non-logical constants* $c_{n,a}$, for each natural number n. The set of all non-logical constants of type a is called CON_a. For each type a, $\mathcal{NIL}$ contains a denumerably infinite set of *variables* $v_{n,a}$ for each natural number n. The set of all variables of type a is called VAR_a. Moreover, $\mathcal{NIL}$ contains a denumerably infinite set of *nominals* i_n for each natural number n. Nominals are of type t. The set of all nominals is called NOM.

Meaningful Expressions: The set ME_a of *meaningful expressions of type a* is defined recursively as:

1. Every variable of type a and every constant of type a are in ME_a.

2. Every nominal is in ME_t.

3. If $\alpha \in \mathsf{ME}_a$ and u is a variable of type b, then $\lambda u a \in \mathsf{ME}_{\langle b, a \rangle}$.

4. If $\alpha \in \mathsf{ME}_{\langle a, b \rangle}$ and $\beta \in \mathsf{ME}_a$ then $\alpha(\beta) \in \mathsf{ME}_b$.

5. If α and β are both in ME_a, then $\alpha = \beta \in \mathsf{ME}_t$.

6. If φ and ψ are in ME_t and u is a variable of any type, then the following are also in ME_t: $\neg\psi$, $(\varphi \wedge \psi)$ and $\exists u \varphi$.

7. If φ in ME_t, then $\Diamond\varphi$, $\mathsf{F}\varphi$ and $\mathsf{P}\varphi$ are in ME_t.

8. If $\alpha \in \mathsf{ME}_a$, then $^{\wedge}\alpha \in \mathsf{ME}_{\langle s, a \rangle}$.

9. If $\alpha \in \mathsf{ME}_{\langle s, a \rangle}$, then $^{\vee}\alpha \in \mathsf{ME}_a$.

DEFINITION 2 (Semantics of $\mathcal{NIL}$). A (standard) *model for $\mathcal{NIL}$* is a 5-tuple $\langle A, W, T, <, F \rangle$ such that A, W and T are non-empty sets, $<$ is a linear ordering on the set T, and F is a function whose domain is the set of all non-logical constants of $\mathcal{NIL}$ together with the set of nominals NOM, F assigns to each non-logical constant a *sense* (as defined below). Moreover, for any nominal i, $F(i)$ must be a function with domain $W \times T$ and range $\{0, 1\}$ such that for some unique $t \in T$, we have $[F(i)](\langle w, t \rangle) = 1$ (for all $w \in W$), and for all $t' \neq t$ (and all $w \in W$) we have $[F(i)](\langle w, t' \rangle) = 0$. That is, nominals denote functions that return 1 on one unique value of their temporal argument (the value of the world argument is irrelevant).

The set D_a of *possible denotations* of type a in a model $\langle A, W, T, <, F \rangle$ is defined as follows (where a and b are types):

$$
\begin{aligned}
\mathsf{D}_e &= A. \\
\mathsf{D}_t &= \{0, 1\}. \\
\mathsf{D}_{\langle a, b \rangle} &= \mathsf{D}_b^{\mathsf{D}_a}, \text{for } a \neq s \\
\mathsf{D}_{\langle s, a \rangle} &= \mathsf{D}_a^{W \times T}
\end{aligned}
$$

The set S_a of *senses of type a* is defined as $\mathsf{D}_{\langle s, a \rangle}$. We can now complete the definition of the range of F in a model. The function F will assign to each non-logical constant of $\mathcal{NIL}$ of type a a member of S_a (we've already specified what it does with the nominals, and we can notice now that nominals just receive a special kind of sense in S_t).

An *assignment of values to variables g* is a function having as domain the set of all variables such that for any variable $v_{n,a}$, $g(v_{n,a}) \in \mathsf{D}_a$. We say that an assignment g' is a v-variant of an assignment g if it coincides with g in all values except perhaps in the value assigned to v.

Given a model $M = \langle A, W, T, <, F \rangle$, an assignment g, a world $w \in W$, and a time $t \in T$ we define, for any expression α the *extension of α with respect to model M, w, t, g*, denoted $[\![\alpha]\!]^{M,w,t,g}$, recursively as indicated in Figure 3.

5 Reichenbach and Prior via Montague

With the formalities out of the way, let's see how we can put $\mathcal{NIL}$ to work. The key tools we need are the following macros. These encapsulate in $\mathcal{NIL}$ the basic patterns needed for building nominal tense logical representations compositionally:

$$
\begin{array}{rcl}
\mathsf{PAST} & =_{def} & \lambda V \lambda x^{\wedge}\mathsf{P}(i \wedge V(x)) \\
\mathsf{PLUPERF} & =_{def} & \lambda V \lambda x^{\wedge}\mathsf{P}(i \wedge \mathsf{P}V(x)) \\
\mathsf{PERF} & =_{def} & \lambda V \lambda x^{\wedge}(i \wedge \mathsf{P}V(x)) \\
\mathsf{FUTPAST} & =_{def} & \lambda V \lambda x^{\wedge}\mathsf{P}(i \wedge \mathsf{F}V(x))
\end{array}
$$

Let's look at some examples. In what follows, we make use of standard Montague semantic representations. For example, the proper name "Dov" will have the representation

$$\mathsf{DOV} \quad =_{def} \quad \lambda P^{\vee}(P\,dov),$$

and the semantic representation of "smile" will simply be SMILE. That is, what follows is classical Montague semantics, except that we make use of our Reichenbach-meets-Prior macros.

Here's a first example. Let's build a representation for "Dov smiled". We'll assume that this can be grammatically analyzed as (Dov(past(smile))). Hence we build its representation as follows:

$$
\begin{array}{rcl}
\text{Dov} & \hookrightarrow & \mathsf{DOV} \\
& \hookrightarrow & \lambda P^{\vee}(P\,dov) \\
\text{smiled} & \hookrightarrow & \mathsf{PAST}(\mathsf{SMILE}) \\
& \hookrightarrow & \lambda V \lambda x^{\wedge}\mathsf{P}(i \wedge V(x))\,smile \\
& \hookrightarrow & \lambda x^{\wedge}\mathsf{P}(i \wedge smile(x)) \\
\text{Dov smiled} & \hookrightarrow & \mathsf{DOV}(\mathsf{PAST}(\mathsf{SMILE})) \\
& \hookrightarrow & \lambda P^{\vee}(P\,dov)\lambda x^{\wedge}\mathsf{P}(i \wedge smile(x)) \\
& \hookrightarrow & ^{\vee}(\lambda x^{\wedge}\mathsf{P}(i \wedge smile(x))\,dov) \\
& \hookrightarrow & ^{\vee\wedge}\mathsf{P}(i \wedge smile(dov)) \\
& \hookrightarrow & \mathsf{P}(i \wedge smile(dov))
\end{array}
$$

$$\llbracket c_{n,a} \rrbracket^{M,w,t,g} \;=\; [F(c_{n,a})](\langle w,t\rangle),$$
$$\text{for } c_{n,a} \in \mathsf{CON}_a$$

$$\llbracket i_n \rrbracket^{M,w,t,g} \;=\; [F(i_n)](\langle w,t\rangle),$$
$$\text{for } i_n \in \mathsf{NOM}$$

$$\llbracket v_{n,a} \rrbracket^{M,w,t,g} \;=\; g(\alpha),$$
$$\text{for } v_{n,a} \in \mathsf{VAR}_a$$

$$\llbracket \lambda u\alpha \rrbracket^{M,w,t,g} \;=\; h, \text{where } h \text{ is the function with range } \mathsf{D}_b, \text{ s.t.}$$
$$h(o) = \llbracket \alpha \rrbracket^{M,w,t,g'} \text{ for } g' = g[u \mapsto o]$$
$$\text{for } u \text{ a variable of type } \mathsf{D}_b \text{ and } \alpha \in \mathsf{ME}_a$$

$$\llbracket \alpha(\beta) \rrbracket^{M,w,t,g} \;=\; \llbracket \alpha \rrbracket^{M,w,t,g}(\llbracket \beta \rrbracket^{M,w,t,g}),$$
$$\text{for } \alpha \in \mathsf{ME}_{\langle a,b\rangle} \text{ and } \beta \in \mathsf{ME}_a$$

$$\llbracket \alpha = \beta \rrbracket^{M,w,t,g} \;=\; 1 \text{ iff } \llbracket \alpha \rrbracket^{M,w,t,g} = \llbracket \beta \rrbracket^{M,w,t,g},$$
$$\text{for } \alpha, \beta \in \mathsf{ME}_t$$

$$\llbracket \neg\varphi \rrbracket^{M,w,t,g} \;=\; 1 \text{ iff } \llbracket \varphi \rrbracket^{M,w,t,g} = 0,$$
$$\text{for } \alpha \in \mathsf{ME}_t$$

$$\llbracket (\varphi \wedge \psi) \rrbracket^{M,w,t,g} \;=\; 1 \text{ iff } \llbracket \varphi \rrbracket^{M,w,t,g} = 1 \text{ and } \llbracket \psi \rrbracket^{M,w,t,g} = 1,$$
$$\text{for } \varphi, \psi \in \mathsf{ME}_t$$

$$\llbracket \exists u\varphi \rrbracket^{M,w,t,g} \;=\; 1 \text{ iff } \llbracket \varphi \rrbracket^{M,w,t,g'} = 1$$
$$\text{for some } u\text{-variant } g' \text{ of } g,$$
$$\text{for } u \text{ a variable and } \varphi \in \mathsf{ME}_t$$

$$\llbracket \Diamond\varphi \rrbracket^{M,w,t,g} \;=\; 1 \text{ iff } \llbracket \varphi \rrbracket^{M,w',t',g} = 1$$
$$\text{for some } w' \in W \text{ and } t' \in T,$$
$$\text{for } \varphi \in \mathsf{ME}_t$$

$$\llbracket \mathsf{F}\varphi \rrbracket^{M,w,t,g} \;=\; 1 \text{ iff } \llbracket \varphi \rrbracket^{M,w,t',g} = 1$$
$$\text{for some } t' \in T, \text{ s.t., } t < t',$$
$$\text{for } \varphi \in \mathsf{ME}_t$$

$$\llbracket \mathsf{P}\varphi \rrbracket^{M,w,t,g} \;=\; 1 \text{ iff } \llbracket \varphi \rrbracket^{M,w,t',g} = 1$$
$$\text{for some } t' \in T, \text{ s.t., } t' < t,$$
$$\text{for } \varphi \in \mathsf{ME}_t$$

$$\llbracket {}^\wedge\alpha \rrbracket^{M,w,t,g} \;=\; h, \text{where } h \text{ is the function with range } W \times T, \text{ s.t.}$$
$$h(\langle w',t'\rangle) = \llbracket \alpha \rrbracket^{M,w',t',g} \text{ for any } \langle w',t'\rangle \in W \times T,$$
$$\text{for } \alpha \in \mathsf{ME}_a$$

$$\llbracket {}^\vee\alpha \rrbracket^{M,w,t,g} \;=\; \llbracket \alpha \rrbracket^{M,w,t,g}(\langle w,t\rangle),$$
$$\text{for } \alpha \in \mathsf{ME}_{\langle s,a\rangle}$$

Figure 3. Semantics of $\mathcal{NIL}$

This is the semantic representation given in our tables. Let's now build a representation for "Dov had smiled". We assume that this can be grammatically analyzed as (Dov(had(smile))). Hence we build its representation as follows:

$$
\begin{array}{rcl}
\text{Dov} & \hookrightarrow & \mathsf{DOV} \\
& \hookrightarrow & \lambda P^{\vee}(P\,dov) \\
\text{had} & \hookrightarrow & \mathsf{PLUPERF} \\
& \hookrightarrow & \lambda V \lambda x^{\wedge}\mathsf{P}(i \wedge \mathsf{P}V(x)) \\
\text{had smiled} & \hookrightarrow & \mathsf{PLUPERF(SMILE)} \\
& \hookrightarrow & \lambda V \lambda x^{\wedge}\mathsf{P}(i \wedge \mathsf{P}V(x))\,smile \\
& \hookrightarrow & \lambda x^{\wedge}\mathsf{P}(i \wedge \mathsf{P}\,smile(x)) \\
\text{Dov had smiled} & \hookrightarrow & \mathsf{DOV(PLUPERF(SMILE))} \\
& \hookrightarrow & \lambda P^{\vee}(P\,dov)\lambda x^{\wedge}\mathsf{P}(i \wedge \mathsf{P}\,smile(x)) \\
& \hookrightarrow & {}^{\vee}(\lambda x^{\wedge}\mathsf{P}(i \wedge \mathsf{P}\,smile(x))\,dov) \\
& \hookrightarrow & {}^{\vee\wedge}\mathsf{P}(i \wedge \mathsf{P}\,smile(dov)) \\
& \hookrightarrow & \mathsf{P}(i \wedge \mathsf{P}\,smile(dov))
\end{array}
$$

Again, this is the nominal tense logical representation we would expect from our earlier table. We leave the reader to experiment with other examples. The most important example to consider is, of course, "Dov will have smiled", but this is bound to work out right given that the birthday books are coming his way!

6   Conclusion

In this paper we have taken a first (short) step towards applying ideas from hybrid logic to natural language semantics. There's a lot left to do. For a start, we made use of only the basic tool of hybrid logics (namely nominals) but other tools (such as the @ operator) are clearly relevant to temporal semantics too (for example, @ can be used to help build up temporal representations for entire discourse, not just sentences). For further information on richer hybrid logics of time, see [Areces *et al.*, 2000]. Moreover, Montague's models are point based (that is, time is conceived of as a succession of instants), but for more detailed work it is useful to have access to temporal interval structure and to be able to evaluate formulas with respect to extended periods of time. It would be interesting to incorporate an interval based semantics to this system and add new modalities (such as a subinterval modality) to exploit it. Furthermore, it would be natural (following [Blackburn, 1994]) to shift to a two-dimensional pattern of evaluation to cover *temporal indexicals* such as "now", "yesterday", "today" and "tomorrow". The hybrid approach to these words makes use of the semantic machinery developed by Hans Kamp (see [Kamp, 1971]),

Frank Vlach (see [Vlach, 1973]) and David Kaplan (see [Kaplan, 1977; Kaplan, 1989]), but exploits it using nominal-like propositional symbols *now, yesterday, today* and *tomorrow* rather than by adding new modalities. The move to a two-dimensional pattern of evaluation would also bring us one step nearer Dov's pioneering work.

But to conclude we want to note that there is a deeper sense in which this paper is linked to Dov's work: the simple fact that it uses hybrid logic. As far we are aware, Dov has never explicitly written on hybrid logic (though [Gabbay and Malod, 2002] comes close) but his work has influenced its development. In particular, his paper "An irreflexivity lemma" [Gabbay, 1981] and the development of labeled deduction (see [Gabbay, 1996]) provided ideas and insights that proved crucial to the development of hybrid deduction. More generally, via his writing Dov Gabbay has helped inspire key developments in modal (and many other kinds of) logic, and via his editorial work he has made it possible for logic to mature and come to grips with it relations with other fields. We are very happy to have this opportunity to pay our respects to Dov in this volume. Happy Birthday!

BIBLIOGRAPHY

[Areces *et al.*, 2000] Carlos Areces, Patrick Blackburn, and Maarten Marx. The computational complexity of hybrid temporal logics. *Logic Journal of the IGPL*, 8(5):653–679, 2000.

[Blackburn, 1994] Patrick Blackburn. Tense, temporal reference and tense logic. *Journal of Semantics*, 11:83–101, 1994.

[Gabbay and Malod, 2002] Dov Gabbay and Guillaume Malod. Naming worlds in modal and temporal logic. *Journal of Logic, Language and Information*, 13(1):3–22, 2002.

[Gabbay, 1976] Dov Gabbay. *Investigations into Modal and Tense Logics, with Applications to Problems in Linguistics and Philosophy*. Reidel, 1976.

[Gabbay, 1981] Dov Gabbay. An irreflexivity lemma. In U. Mönnich, editor, *Aspects of Philosophical Logic*, pages 67–89. Reidel, 1981.

[Gabbay, 1996] Dov Gabbay. *Labelled Deductive Systems. Vol. 1*. The Clarendon Press Oxford University Press, New York, 1996. Oxford Science Publications.

[Kamp, 1971] Hans Kamp. Formal properties of "now". *Theoria*, 37:227–273, 1971.

[Kaplan, 1977] David Kaplan. Demonstratives: An essay on the semantics, logic, metaphysics and epistemology of demonstratives and other indexicals, 1977. Draft No 2, UCLA xerox.

[Kaplan, 1989] David Kaplan. Demonstratives: An essay on the semantics, logic, metaphysics and epistemology of demonstratives and other indexicals. In J. Almog, J. Perry, and H. Wettstein, editors, *Themes from Kaplan*, pages 481–563. Oxford University Press, Oxford, 1989.

[Montague, 1973] Richard Montague. The Proper Treatment of Quantification in Ordinary English. In J. Hintikka, J. Moravcsik, and P. Suppes, editors, *Approaches to Natural Language*, pages 221–242. Reidel, Dordrecht, 1973.

[Prior, 1967] Arthur Prior. *Past, Present and Future*. Clarendon Press, Oxford, 1967.

[Reichenbach, 1947] Hans Reichenbach. *Elements of Symbolic Logic*. Random House, New York, 1947.

[Vlach, 1973] Frank Vlach. *Now and Then: a formal study in the logic of tense anaphora*. PhD thesis, UCLA, 1973.

A Modal View on Polder Politics[1]

WOUTER VAN ATTEVELDT AND STEFAN SCHLOBACH

1 Introduction

The media plays an increasing role in the functioning of our society, often being the main informant of citizens and policy-makers. Controversies easily arise when the media are accused of inaccurate or biased information provision, such as the affair around the BBC and the suicide of David Kelly in the UK and the alleged demonisation of Pim Fortuyn before his assassination in the Netherlands. The messages of politicians reach most citizens through the media rather than directly, making the image of a party or candidate and their reported positions on key issues an important factor in election campaigns.

Therefore, it is important to analyse the content of the media and its effects on media consumers, such as it is often done in election studies, agenda setting, and framing research [Bryant and Zillman, 2002; Goffman, 1974; McCombs and Shaw, 1972]. These studies generally annotate news items using Content Analysis which has resulted in a large number of annotated articles over the past decades. A uniform way to store and query this media data can lead to easier and more accurate (meta-)research. Moreover, this can allow general purpose analysis and visualisation tools to be developed.

In this paper we provide such a uniform representation by formalising media data as Kripke structures. Not only do we get unified representations with formal semantics, we also inherit well studied query languages in the form of modal logics. This allows us to use off-the-shelf model checking software to query newspaper articles, or at least the formalisation of these articles. In order to reproduce recent results of a media study about the Dutch elections in 2002 [Kleinnijenhuis *et al.*, 2003], we show that quite expressive modal languages (such as Hybrid Logic $H(@, \downarrow)$) are needed to capture important political terms, such as internal debate or criticism.

In our experiments, we take almost 7500 newspaper articles, and transform them into Kripke structures with around 35.000 edges. Using an exist-

---

[1]The Polder model is a phenomenon in politics in the Netherlands to reach an agreement despite differences, however great these might be.[Wikipedia, 2005]

ing model-checker for hybrid logics [Dragone, 2005], we query for a number of politically relevant patterns, and statistically evaluate the number of matches. As our experiments were designed to reproduce the results of the original study by other means, it is good news, but does not come as a surprise, that our results are in line with the original study of [Kleinnijenhuis *et al.*, 2003]. Nevertheless, the investigation shows the practical feasibility of our approach on realistic data.

There are three significant advantages of our approach: it offers an explicit semantics to the formal representation of newspaper documents, which makes the study much more transparent and therefore reproducible. Secondly, by using existing tools, the amount of purpose-written and ad-hoc software is greatly reduced, and, finally, it becomes straightforward to integrate formally represented background knowledge.

2   Analysing the News

Quantitative Content Analysis is a field of Communication Science that analyses media messages by categorising units of text into scientifically relevant concepts [Krippendorff, 2004; Holsti, 1969]. Network or Relational Content Analysis methods, such as the NET method described below, additionally construct a network from these texts where the nodes are the categories mentioned above and the vertices determine certain relations between these nodes as expressed in the text [Roberts, 1997; Carley, 1990]. These relations are often of a fairly subjective or opinionative nature, such as (dis)like between actors and issue positions, but can also include causal relations or actions, such as passing a law or going on strike.

In this section, a simplified version of the Network Content Analysis method called the NET method (Network analysis for Evaluative Texts [Kleinnijenhuis *et al.*, 2003; van Cuilenburg *et al.*, 1988]) will be used. This method, an extension of Evaluative Assertion Analysis [Osgood *et al.*, 1956], assumes that all relevant information in a sentence can be expressed as ⟨subject,relation, object⟩ triples. The subject and object of these statements are drawn from a (generally fixed) list of relevant concepts. In the original method, the relation consists of a relationship type (such as causal or associative) and quantifications of the ambivalence, direction, and strength of the relationship. To keep matters simple, in this initial study, we only consider binary relations of association versus dissociation.

The resulting network can be queried in different ways depending on the research question. In general, the researcher defines relevant concepts as patterns over these networks. In election studies, such as the study that provided the data underlying this section, it is interesting to know what the general tone and type of the news is, who is portrayed positively or negatively, and what issue positions the different actors have according to the media. Moreover, more specific concepts such as internal dissent, unclear issue positions and criticism from political actors can also be important [Kleinnijenhuis *et al.*, 2003].

2.1 NET by example: the Rise and Fall of Fortuyn

The examples in this section are all drawn from the media coverage of the 2002 parliamentary elections in the Netherlands as it was described in [Kleinnijenhuis *et al.*, 2003]. Initially, these elections appeared to become a duplicate of the 1998 polarisation between the ruling parties, the PvdA (labour) and VVD (conservative), with both criticising each other but generally intending to continue the coalition. The PvdA, however, was hindered by internal dispute, as exemplified in an article published in the Dutch newspaper "Trouw" on 27 November 2001. Figure (1) shows two sentences of this article which are indicative of internal strife, and which are coded as ⟨Melkert,−, vanGijzel⟩ and ⟨PvdA,−, vanGijzel⟩.

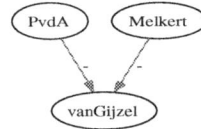

Figure 1. 'Rob van Gijzel fights back'
(Trouw, 27 November 2001)

PvdA-leader Melkert took away the spokesmanship on the construction fraud affair from Rob van Gijzel. [..] This shows that he can no longer assume support within his party

Not all went as expected by the ruling parties. On 9 February 2002, the newcomer Pim Fortuyn gave an interview in the Volkskrant which caused him to be seen as extreme right, and forced him to establish his own party to continue in Dutch politics (Figure 2). The displayed sentences all reflect issue positions: ⟨Fortuyn,−, Islam⟩, ⟨Fortuyn,−, Immigrants⟩, and ⟨Fortuyn,+, Discrimination⟩.

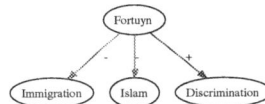

Figure 2. 'Fortuyn: Islamites out of the country'
(de Volkskrant, 9 February 2002)

If it is up to Pim Fortuyn, no more asylum seekers will enter the Netherlands. [..] Fortuyn also wants to abolish the Article 1 of the Constitution, that forbids discrimination.

These radical issue positions caused him to receive heavy criticism from the established parties. The sentences quoted in figure 3 are all direct criticism of Fortuyn by the other party leaders ⟨Dijkstal,–, Fortuyn⟩, ⟨Melkert,–, Fortuyn⟩ and ⟨De Graaf,–, Fortuyn⟩. Note that any meaningful interpretation of this graph requires some background knowledge, for example, that Dijkstal, Melkert and De Graaf were the party leaders of the VVD, PvdA and D66 at the time.

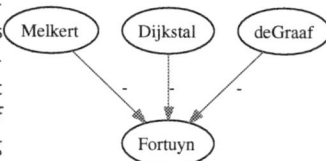

Figure 3. "Public urinator"
(Trouw, 13 February 2002)

[Before the interview] people had some second thoughts about Fortuyn, but now he is said to 'encroach on the heart of our civilisation' (Dijkstal), 'have crossed a line, that you cannot cross' (Melkert) [..]. De Graaf could not help citing Anne Frank and calling him a fascist and a racist.

Fortuyn did surprisingly well in the local elections in Rotterdam later that month, which caused the leader of the PvdA, Melkert, to come under heavy fire. Moreover, the PvdA was still plagued with internal disagreement, for example about the participation in the Joint Strike Fighter project (Figure 4). The first sentence is coded as ⟨Timmermans,–, JSF⟩, ⟨PvdA,+, JSF⟩, the second as ⟨Timmermans,–, Melkert⟩ and ⟨Melkert,+, JSF⟩.

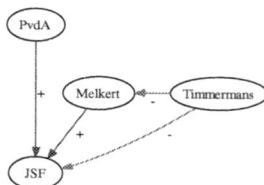

Figure 4. 'Difficult, but I never threatened to resign'
(de Volkskrant, 10 April 2002)

PvdA parliamentary Frans Timmermans remains skeptical about the JSF decision of his party [..] after the PvdA-fraction finally said 'Yes' to the JSF. Timmermans played the main part and diametrically oppose party leader Melkert on this issue

However, as the elections approached, Fortuyn tempered with his positions, leaving behind opponents scrambling to adopt radical positions in an atmosphere of crisis (figure 5). The first two sentences are coded as opposing issue positions on immigration. The last part of the second sentence and the last sentence are encoded as direct criticism from the media.

The tragic end of the campaign is known: Pim Fortuyn was assassinated nine days prior to the elections. His party (the List Pim Fortuyn) made an unprecedented debut and was made part of the government, but it was plagued by internal dissent over the succession and was diminished in the next election after the cabinet fell less than three months after its formation.

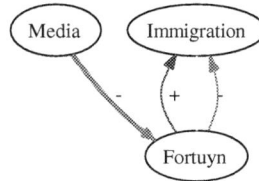

Figure 5. 'Fortuyn's Ambitions'
(Trouw, 8 March 2002)

With his plea for a general pardon for asylum seekers, Pim Fortuyn again causes great confusion. Retaining his ideas of strict admission policy and restricting family reunification, this new proposal surprised friend and enemy alike. This is hardly credible.

2.2   Healthy Conflict or Internal Dissent

In campaign time, conflict and criticism are the rule rather than the exception. However, there is an important distinction between different types of conflict: conflicting or disputing issues with opponents can be beneficial as it allows a party to create a distinct profile. Internal criticism, however, is generally harmful, as is criticism from societal actors. Also, if different members of a party adopt opposing viewpoints on an issue, the position of the party is unclear to the voter and can be seen as indicating internal dissent or lack of leadership. Fortuyn's being able to frame the election as him against the incumbent coalition, rather than Labour against Conservatives, might be a large factor explaining his success, while the PvdA's lack of leadership and internal troubles can explain their strong decline.

Thus, it is important to identify whether the criticiser is a member of the same party or a different party, and whether opposing viewpoints are adopted by members of different parties or of the same party. Thus, we need to know the party membership of the actors that are under investigation. Using this background knowledge, we can see that figure 1 indicates *internal criticism*, figure 4 indicates *internal disagreement*, while figure 3 indicates *criticism by opponents* of Fortuyn. These concepts are thus defined as patterns over a combination of media data and background knowledge.

3   Querying opinions through model checking

In the previous section we introduced a prototypical problem exemplifying the way social scientists use information about opinions to analyse political situations. We now show that NET graphs can be mapped to particular Kripke structures, and that hybrid modal languages are useful query languages over these structures. In particular, we show that we can reformulate the case study of the previous section as a hybrid model checking problem.

Not only can we reproduce the published results with our approach, as will be shown in the following section, but the expressiveness of the querying languages allows one to extend the complexity of possible sociological analysis of the underlying newspaper articles. Furthermore, it provides a formal interpretation to the experiments. Moreover, it makes it easier to do meta-research over different communication scientific studies by formalising the semantics and hence allows for improved theory building.

3.1   NET graphs and background knowledge

In section 2 we loosely introduced NET graphs as a result of a specific method of content analysis. These graphs can be characterised more formally as follows: let $\mathcal{N} = \{n_1, \ldots, n_l\}$ be a finite set of names, which usually represent political actors, groups and issues. A *NET graph* is then defined as a finite set of triples $\langle n_i, *, n_j \rangle$, where $n_i, n_j \in \mathcal{N}$ are names, and $* \in \{+, -\}$ is an opinionated relation, denoting consent and dissent, respectively. As an example, consider the NET graph NET_6 which was introduced to exemplify internal disagreement between two members of the same party, and which is repeated on the left-hand side of figure 6. Here, we have four names, Melkert, Timmermans, PvdA and JSF, with triples $\langle$Timmermans,-, JSF$\rangle$, $\langle$PvdA,+, JSF$\rangle$, $\langle$Timmermans,-, Melkert$\rangle$ and $\langle$Melkert,+, JSF$\rangle$.

Background knowledge is usually given as a simple ontology over political actors and concepts. Formally, let $\mathcal{N}$ again denote the set of names, $\{R_1, \ldots, R_n\}$ a finite set of binary relations, and $\mathcal{P} = \{P_1, \ldots, P_m\}$ a finite set of propositional variables. The *domain ontology* $\mathcal{O}$ is then defined as a finite set of axioms of the form $(n_i : P), P \sqsubseteq Q$ or $R(n_i, n_j)$, where $n_i, n_j \in \mathcal{N}, \{P, Q\} \subseteq \mathcal{P}$ and $R \in \{R_1, \ldots, R_n\}$. In our example, both Timmermans and Melkert are politicians, and members of the PvdA, which is a party, while JSF is an issue. This can be represented formally in the ontology $\mathcal{O}_6 = \{$Timmermans:PolActor, Timmermans:memberOf PvdA, Melkert:PolActor, Melkert:memberOf PvdA, JSF:Issue, PvdA:Party$\}$.

The next step in our formalisation process is to integrate the background knowledge and NET graphs into one joint formalism. For this purpose, we define a valuation function V, from the set $\mathcal{P}$ of propositional variables to the power set of $\mathcal{N}$, the names of the individuals as the smallest set $V(P)$ such that $\{n \mid n : P \in \mathcal{O}\} \subseteq V(P)$ and $\bigcup_{(P \sqsubseteq Q) \in \mathcal{O}} V(Q) \subseteq V(P)$ for each $P \in \mathcal{P}$. With this definition we get to the Kripke structure $\mathcal{K}_6$ on the right-hand side of figure 6 for our ongoing example.

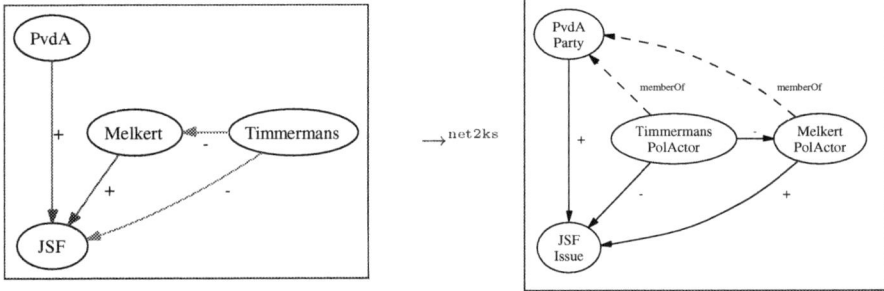

Figure 6. A Kripke structure $\mathcal{K}_6$ for the NET graph of figure 4

For the sake of a simplified presentation of our results in this paper, we omitted a number of steps that should lead us from NET graphs and background knowledge to Kripke structures. First, we avoided specifying a formal semantics for the combination of NET graphs and ontologies, and provided a syntactic rather than semantic characterisation of the information in the newspaper articles under consideration. Furthermore, we need a strong Closed World Assumption, because our resulting semantics assumes that we have complete knowledge about our actors and issues w.r.t. the propositional variables at hand. But in our application these are acceptable assumptions because of the restricted domain.

3.2 Modal logics as query languages over Kripke structures

Kripke structures over a set of propositional variables *PROP*, are triples of the form $\langle M, \mathcal{R}, V \rangle$, where M is a set of states, $\mathcal{R}$ a family of binary accessibility relations on M, and V is a valuation function from PROP to the power-set of M. In our case the states M correspond to the names $\mathcal{N}$ and the family $\mathcal{R}$ of relations to both the domain relations and the opinionated relations $\{R_1, \ldots, R_n\} \cup \{+, -\}$.

Interpreting NET graphs and their domain ontologies as Kripke structures has two big advantages: first, it provides a unified representation about all the knowledge that is represented about a certain political statement. Secondly, Kripke structures come with a number of very powerful and well studied query languages, namely modal logics [Blackburn *et al.*, 2001]. In our case, multi-modal logics seem to be the obvious choice for querying multi-relation structures. They extend propositional logic by adding a family of modal operators $\{\Diamond_1, \ldots, \Diamond_d\}$. Formally, let PROP $= \{p, q, \ldots\}$ be a set of propositional variables. The syntax of multi-modal logic MML is as usual: $\varphi := \top \mid p \mid \neg \varphi \mid \varphi \wedge \psi \mid \Diamond_i$. To simplify matters, $\varphi \vee \psi$ and $\Box_i \varphi$

are defined as abbreviations for $\neg(\neg\varphi \wedge \neg\psi)$ and $\neg\Diamond_i\neg\varphi$. Let $\mathcal{K}$ be a Kripke structure and $m \in M$. The semantics of multi-modal logic is defined as usual:

$$\mathcal{K}, m \models \top$$
$$\mathcal{K}, m \models p \quad\quad \text{iff} \quad m \in V(p),\, p \in \mathsf{PROP}$$
$$\mathcal{K}, m \models \neg\varphi \quad\quad \text{iff} \quad \mathcal{K}, m \not\models \varphi$$
$$\mathcal{K}, m \models \varphi \wedge \psi \quad\quad \text{iff} \quad \mathcal{K}, m \models \varphi \text{ and } \mathcal{K}, m \models \psi$$
$$\mathcal{K}, m \models \Diamond_i\varphi \quad\quad \text{iff} \quad \exists m'\, (R_i mm' \wedge \mathcal{K}, m' \models \varphi)$$

To see the potential of modal logic as a query language let us study some examples, e.g., *give me politicians criticising other politicians*. Asking this query over the Kripke structure $\mathcal{K}_6$ should return the name Timmermans, because $\langle\mathtt{Timmermans}, -, \mathtt{Melkert}\rangle \in NET_6$, as well as Melkert:PolActor$\in \mathcal{O}_6$ and Timmermans:PolActor$\in \mathcal{O}_6$. Formally, the set of answers to this query is the set of all worlds n for which $\mathcal{K}_6, n \models \mathsf{PolActor} \wedge \Diamond_-\mathsf{PolActor}$, where $\Diamond_-$ is interpreted as the relation $-$. More generally, given any Kripke structure $\mathcal{K}$, and an arbitrary formula φ in the basic language, it is well known that it can be checked in linear time whether φ is true at a state in $\mathcal{K}$ or not.[]

Finally, we have what we need for querying newspaper articles: a unified formal representation of opinions, combined with ontological information about political actors and issues, and an expressive querying language. Two simple examples of what we can do:

- **Conflict news:** Politician $\wedge$ ($\Diamond_+$ Politician $\vee \Diamond_-$ Politician)
- **Issue news:** Politician $\wedge (\Diamond_+ \mathsf{Issue} \vee \Diamond_- \mathsf{Issue})$

Unfortunately, the logic is not yet expressive enough for some important queries which are needed in our case study. First, we are unable to ask for adversaries of particular political actors, or opinions of politicians w.r.t. specific political issues. This is because modal logics usually do not allow to address worlds by their names, i.e. they usually do not have support for nominals. So, if we were to ask about those politicians opposing immigration, we would need the name of the issue immigration in our language. Then the query (Politician$\wedge\Diamond_-$immigration) would capture our intuition.

But we need even more. A type of query that is required for the political investigations is about internal disagreement, that also cannot be expressed, even with nominals. In the following example we want to query for a political party, which has suffered criticism on a particular issue it supports from one of its members. Basically the query should include the proposition Party, i.e. we search for a party, and an issue this party supports, i.e. $\Diamond_m\Diamond_+\mathsf{Issue}$. Finally, we need to find a member of the party, which is critical of the same issue. But here we fail, as we cannot state equality of two issues or two parties, respectively. What is needed is a further extension of our logic with weak quantification, or variable binding.

3.3   Hybrid Logics

Hybrid Logic (HL, for short) extends multi-modal logic with devices for naming states and accessing states by names. Let $\mathsf{NOM} = \{i, j, \ldots\}$ and $\mathsf{WVAR} = \{x, y, \ldots\}$ be sets of nominals and state variables, respectively. HL's syntax is: $\varphi := \mathsf{MML} \mid i \mid x \mid @_t\varphi \mid {\downarrow}x.\varphi \mid \exists x.\varphi$, with $i \in \mathsf{NOM}$, $x \in \mathsf{WVAR}$, $t \in \mathsf{NOM} \cup \mathsf{WVAR}$. The operators in $\{@, {\downarrow}, \exists\}$ are called *hybrid operators*. We call $\mathsf{WSYM} = \mathsf{NOM} \cup \mathsf{WVAR}$ the set of *state symbols*, and $\mathsf{ATOM} = \mathsf{PROP} \cup \mathsf{NOM} \cup \mathsf{WVAR}$ the set of *atoms*. For simplicity, we omit parenthesis after the ${\downarrow}$. For instance, ${\downarrow}x.p \wedge @_x q$ reads as ${\downarrow}x.(p \wedge @_x q)$.

Hybrid Logic is interpreted over *hybrid Kripke structures*, i.e., Kripke structures $\langle M, R, V \rangle$ where the valuation function V assigns singleton subsets of M to nominals $i \in \mathsf{NOM}$. To give meaning to the formulas, we also need the notion of *assignment*. An assignment g is a mapping $g : \mathsf{WVAR} \to M$. Given an assignment g, we define g_m^x by $g_m^x(x) = m$ and $g_m^x(y) = g(y)$ for $x \neq y$. For any atom a, let $[V, g](a) = \{g(a)\}$ if a is a state variable, and $V(a)$ otherwise.

The semantics of hybrid logic is defined over a hybrid Kripke structure $\mathcal{M} = \langle M, R, V \rangle$, where $m \in M$, and g is an assignment; the semantics for Boolean and modal operators is as before.

$$
\begin{array}{lll}
\mathcal{M}, g, m \models a & \text{iff} & m \in [V, g](a), \ a \in \mathsf{ATOM} \\
\mathcal{M}, g, m \models @_t\varphi & \text{iff} & \mathcal{M}, g, m' \models \varphi, \text{ where } [V, g](t) = \{m'\}, \ t \in \mathsf{WSYM} \\
\mathcal{M}, g, m \models {\downarrow}x.\varphi & \text{iff} & \mathcal{M}, g_m^x, m \models \varphi \\
\mathcal{M}, g, m \models \exists x.\varphi & \text{iff} & \text{there is } m' \in M \text{ such that } \mathcal{M}, g_{m'}^x, m \models \varphi
\end{array}
$$

This means that $@_t$ shifts evaluation to the state named by t, where t is a nominal or a variable. The down-arrow ${\downarrow}x$ binds the state variable x to the *current* state, while the existential binder $\exists x$ binds to *some* state in the model; ${\downarrow}$ and $\exists$ do not shift evaluation away from the current state. The definition of basic hybrid logic almost literally stems from the first part of [Franceschet and de Rijke, 2005]. For an introduction to hybrid logics we also refer to [Blackburn and Seligman, 1998]

Let us study how we can use this expressiveness to query our Kripke structures with the example we discussed before. Now, given the ${\downarrow}x.$ binder, we can formulate *internal disagreement* with the following query: ${\downarrow}x.\mathsf{Party} \wedge (\Diamond_m \Diamond_+ (\mathsf{Issue} \wedge {\downarrow}y.(@_x \Diamond_m \Diamond_- y)))$. The outer binder binds the current state, i.e. the party we query for to a variable. Inside the formula, we then enforce a criticism relation between a party member and the issue in question.

Hybrid Logics as query languages over Kripke structures stemming from NET graphs with background knowledge offer sufficient expressiveness to emulate realistic investigations in content analysis. Moreover, as there is a fully-fledged model checker publicly available, these results almost come for free, as we will show in the following section. Unfortunately, the model checking problem for the hybrid logics we consider in this section is SPACE-complete [Franceschet and de Rijke, 2005]. In our case, however, both queries and models remain small enough to ensure that there will be no problems in practise.

4   Experiments and results

The experimental validation of our proposed method is a secondary analysis based on articles coded for the investigation of the Dutch 2002 parliamentary campaign by [Kleinnijenhuis *et al.*, 2003]. This data set consists of 7,478 newspaper articles and television items from 20 November 2001 to 14 May 2002 of which the headline and leading paragraph were coded, yielding 35,031 triples. These triples cover the period from the rise of Pim Fortuyn to shortly after his assassination, and most of the study focusses on the events described in section 2.

Each article was translated into a Kripke structure as described in section 3.1, using a background ontology consisting of the party membership of all politicians. Finally, propositions were added that identify politicians, parties, political actors (comprising politicians and parties), actors, and issues. We then formalised the main political frames as modal or hybrid logic formulas. Using the model checker described in [Dragone, 2005, see also [Franceschet and de Rijke, 2005]], these formal HL queries were then checked in each of these Kripke structures, resulting in a list of parties, politicians or issues for which these formulas were true.

These outcomes were summarised per period and compared to the results presented in the original study. Small discrepancies occur because in the original study the number of edges were counted rather than the number of worlds. Thus, if three actors dispute an issue, [Kleinnijenhuis *et al.*, 2003] count three edges whereas we count one world.

4.1   Modal logic: General political frames

To measure the general tone of the campaign news, three modal logic queries were used to discover the three general news frames used for the original study. Table 1 shows the relative attention for these frames, i.e. the number of Kripke structures in which there was a world in which the corresponding formula was true. The total percentages are compared to the percentages published in [Kleinnijenhuis *et al.*, 2003, table 2.2, p. 23], which we restate in the last column of the table for convenience.

4.2 $H(@)$: Issue positions of politicians and parties

To discover the average position of politicians on one of the key issues, immigration, two $H(@)$ queries were used. Note that this requires inter-

Horserace news: reality $\wedge$ ($\Diamond_+$PolActor $\vee$ $\Diamond_-$PolActor)
Conflict news: Actor $\wedge$ ($\Diamond_+$PolActor $\vee$ $\Diamond_-$PolActor)
Issue news: PolActor $\wedge$ ($\Diamond_+$Issue $\vee$ $\Diamond_-$Issue)

Concept	until interview	until loc. elections	until murder	after murder	total	n	*Orig.*
Conflict	40%	44%	46%	57%	45%	4,613	*49%*
Horserace	14%	13%	19%	24%	17%	1,721	*20%*
Issue	47%	42%	35%	19%	39%	3,968	*31%*

Table 1. Relative attention to news types

preting the hybrid Kripke structures as we want to know the names of the worlds (i.e. politicians) in which the formula holds[1]. Table 2 reports the average direction of the position on immigration per party per period, i.e. $(anti - pro)/(anti + pro)$, compared to table 4.3 from [Kleinnijenhuis *et al.*, 2003, p. 54]. Interesting to see is the tempering of Fortuyn's position in the period between the local elections and his assassination, also reported in figure 4.1 of the original study. Note also the 'jerk to the right' by both incumbent parties (VVD and PvdA) after the interview while the CDA actually becomes more positive about immigration.

Pro-immigration: PolActor $\wedge$ $\Diamond_+$immigration
Anti-immigration: PolActor $\wedge$ $\Diamond_-$immigration

Party	until interview	until loc. elections	until elections	total	(n)	*Original*
VVD	-0.42	-0.50	-0.76	-0.53	(60)	*-0.66*
PvdA	0.29	-0.45	-0.11	-0.07	(97)	*-0.11*
CDA	-0.71	-0.56	0.00	-0.57	(41)	*-0.66*
Fortuyn	-1.00	-1.00	-0.48	-0.62	(104)	*-0.52*

Table 2. Average position on immigration

4.3 $H(@, \downarrow)$: Disputed issues

To discover which issues were hotly debated, a formula was used with inverse relations (defined as $\Diamond_i^{inv}\varphi \equiv \exists x.(\Diamond_i x \wedge \varphi)$). For the four most disputed issues, table 3 reports the total number of articles in which that issue was debated per period, along with the percentage of total issue debate this issue received. The results from [Kleinnijenhuis *et al.*, 2003] are the relative total amount of attention per issue, as they did not consider the specific

---

[1]Note that to bind to the parties rather than the politicians, we would have to use the inverse membership relation, which requires $H(@, \downarrow)$; see also section 4.3

concept of disputed issues, but the general percentages do agree: the only interesting *issue news* seem to be those involving disputed issues. Interesting to see is the debate on crime and right-wing issues after the local elections in Rotterdam, suggesting Fortuyn managed to focus the debate on 'his' issues. Also note the sudden discussion on safety and the campaign itself after the assassination, and the fact that the sharpest debates on immigrants immediately followed the interview of 9 February.

Disputed Issue: Issue $\wedge \Diamond_+^{inv}$ Actor $\wedge \Diamond_-^{inv}$ Actor

Issue	until interview		until loc. elections		until murder		after murder		total		Orig.
Crime	73	23%	9	10%	51	18%	9	26%	142	19%	*18%*
Right	54	17%	9	10%	56	20%	3	9%	122	17%	*18%*
Immigration	21	7%	21	23%	34	12%	1	3%	77	11%	*12%*
Left	26	8%	18	20%	21	7%	3	9%	68	9%	*8%*
Campaign	6	2%	2	2%	10	4%	11	31%	29	4%	*5%*

Table 3. Hotly debated issues

4.4 $H(@, \downarrow)$ with background knowledge: Internal dissent

To discover different forms of internal strife, three $H(@, \downarrow)$ formulas were used on the combined media-data and background knowledge. In these formula $\Diamond_m$ represents the memberOf relation between a party and a person. Table 4 gives the total number of occurrences of these three frames, as summarised in figure 5.1 from [Kleinnijenhuis *et al.*, 2003], the *support − criticism* per week is also given and compared to the original number. The difference in scale is a result of the counting of nodes (parties) rather than edges as explained above. An interesting result is that no party seems to be able to have internal discussion without internal (personal) criticism, although the PvdA shows in the last period that the internal discussion need not keep step with the criticism.

5 Conclusion

In this paper we have describe first attempts at using modal logics, with their well studied formalisms and tools in media studies. As a proof of concept, we reproduce relevant results of a case study about the Dutch election campaign of 2002. We do this by formalising newspaper articles as Kripke structures, and by using an existing hybrid model checker to search for instances of politically interesting patterns. This approach has some very nice properties: we can use an existing logic with well-studied semantics and out-of-the-box tools. We can do this, because of the close relation of the

Internal criticism: $\downarrow x.\mathsf{Party} \wedge (\lozenge_m^{inv} \lozenge_-(x \vee \lozenge_m x) \vee \lozenge_-(x \vee \lozenge_m x))$

Internal support: $\downarrow x.\mathsf{Party} \wedge (\lozenge_m^{inv} \lozenge_+(x \vee \lozenge_m x) \vee \lozenge_+(x \vee \lozenge_m x))$

Internal disagreement: $\downarrow x.\mathsf{Party} \wedge (\lozenge_m \lozenge_+(\mathsf{Issue} \wedge \downarrow y.(@_x \lozenge_m \lozenge_- y)))$

Party	until interview					until elections					until murder				
	disa	crit	supp	diff	*orig*	disa	crit	supp	diff	*orig*	disa	crit	supp	diff	*orig*
PvdA	59	64	42	-1.9	*-3*	9	12	14	0.6	*1*	46	110	76	-4.0	*-8*
VVD	39	55	66	1.0	*3*	11	6	16	2.8	*5*	55	66	70	0.5	*6*
CDA	7	6	7	0.1	*0*	0	8	15	2.0	*4*	5	7	9	0.2	*0*
LPF	3	1	3	0.2	*0*	7	4	11	2.0	*4*	34	28	40	1.4	*1*

Table 4. Internal Criticism and Debate

social scientist's intermediate NET representation with Kripke structures. As a result, we can offer explicit formal semantics, easy integration of background knowledge, and an expressive query language, as well as, with the hybrid model checker, a generic tool for querying.

Relevance to Communication Science

Using this method has three advantages to communication sciences. First, it allows for an easy way to formally describe the concepts used in a study. This can make a study more transparent and hence more repeatable.

Second, by making the investigated concepts and domain knowledge formal and explicit, the amount of specific code or SPSS syntax that has to be written to conduct an analysis is greatly reduced. In fact, apart from the general code to translate NET articles to Kripke structures and to invoke the model checker on those structures with the queries, no code or syntax was written at all for this study.

Finally, it allows the separation of the domain specific knowledge (in the ontology) from the definition of the concepts/frames. This makes it possible to define and maintain the domain ontology without knowing the specific queries that will be asked and, more importantly, to formulate the concepts in general enough terms to compare concepts in different domains. The current queries could also be used on the Dutch election campaigns of 1994 and 1998 or even the American presidential campaigns.

Of course, whether the findings of one study, or even the concepts used in those findings, are useful in different circumstances is another question, one that is crucial for many studies or theoretical works: How far can the proposed concepts and theories be generalised? The method described here can be a tool to help investigate this question, thus allowing for more systematic meta-research and theory-building.

Shortcomings, limitations and challenges

Obviously, the streamlining into this particular modal logic formalism comes at a price. The reduction of querying NET graphs with background knowledge from complex reasoning to model checking is only possible because of two relatively strong assumptions: first, we assume complete knowledge about the actors and issues (at least w.r.t. the propositions in our domain), and we restrict our background knowledge to very simple "unfoldable" axioms. There has recently been a surge in interest in simple terminologies to reduce the complexity of reasoning (for example [Calvanese *et al.*, 2005]), and we would claim that it should be reasonably simple to extend our approach to more expressive ontologies.

To simplify this presentation, we did not combine the semantics of the NET relations into the modal reasoning, such as transitivity or combination of arrows. Instead we applied explicit constraint propagation before translating to Kripke structures. This, however, is somehow inconsistent with our unifying approach taken otherwise. Another limitation is our failure to capture the quantitative aspects of NET representations. In part, we avoid the issue by considering individual documents rather than aggregated NET graphs. Thus, the statistical variance is expressed in the results of the analysis, rather than explicitly in the representation itself. Nevertheless, a certain level of approximation and fuzziness is lost in the process.

Another strong restriction of our current approach is the lack of any temporal or epistemic dimension: queries for *changes of mind over time* or *reactions on someone*, let alone *differences between newspapers* cannot be expressed. Nevertheless, we take this shortcoming as a challenge for future work, rather than a limitation, as there is extensive work on multi-dimensional modal logics to start with (just to mention [Gabbay *et al.*, 2003]).

Acknowledgements

We would like to thank Massimo Franceschet and Luigi Dragone for their help on hybrid logics and the HL model checker. Also many thanks to Nel Ruigrok for her help with the script.

BIBLIOGRAPHY

[Blackburn and Seligman, 1998] P. Blackburn and J. Seligman. What are hybrid languages? In M. Kracht, M. de Rijke, H. Wansing, and M. Zakharyaschev, editors, *Advances in Modal Logic*, volume 1, pages 41–62. CSLI Publications, Stanford University, 1998.

[Blackburn *et al.*, 2001] P. Blackburn, M. de Rijke, and Y Venema. *Modal Logic*. Cambridge University Press, 2001.

[Bryant and Zillman, 2002] J. Bryant and D. Zillman, editors. *Media Effects: Advances in Theory and Research*. Lawrence Erlbaum, Mahwah, NJ, 2002.

[Calvanese et al., 2005] D. Calvanese, G. De Giacomo, D. Lembo, M. Lenzerini, and R. Rosati. Dl-lite: Tractable description logics for ontologies. In *Proc. of the 20th Nat. Conf. on Artificial Intelligence (AAAI 2005)*, 2005.

[Carley, 1990] Kathleen Carley. Content analysis. In R.E. Asher, editor, *The Encyclopedia of Language and Linguistics*. Pergamon, Edinburgh, 1990.

[Dragone, 2005] L. Dragone. Hybrid logics model checker. C implementation published on http://www.luigidragone.com/hlmc/ (2005-07-06), 2005.

[Franceschet and de Rijke, 2005] Massimo Franceschet and Maarten de Rijke. Model checking for hybrid logics (with an application to semistructured data). *Journal of Applied Logic*, 2005.

[Gabbay et al., 2003] D. M. Gabbay, A. Kurucz, F. Wolter, and M. Zakharyaschev. *Many-Dimensional Modal Logics: Theory and Applications*. Elsevier, 2003.

[Goffman, 1974] E. Goffman. *Frame Analysis: An Essay on the Organization of Experience*. Harper and Row, London, 1974.

[Holsti, 1969] O. Holsti. *Content Analysis for the Social Sciences and Humanities*. Addison-Wesley, Reading MA, 1969.

[Kleinnijenhuis et al., 2003] J.. Kleinnijenhuis, D. Oegema, J. de Ridder, A. van Hoof, and R. Vliegenthart. *De puinopen in het nieuws*, volume 22 of *Communicatie Dossier*. Kluwer, Alpen aan de Rijn (Netherlands), 2003.

[Krippendorff, 2004] K. Krippendorff. *Content Analysis: An Introduction to Its Methodology (second edition)*. Sage Publications, 2004.

[McCombs and Shaw, 1972] M.E. McCombs and d.l. Shaw. The agenda-setting function of mass media. *Public Opinion Quarterly*, 36:176–187, 1972.

[Osgood et al., 1956] C.E. Osgood, S. Saporta, and J.C. Nunnally. Evaluative assertion analysis. *Litera*, 3:47–102, 1956.

[Roberts, 1997] C.W. Roberts, editor. *Text Analysis for the Social Sciences; Methods for Drawing Statistical Inferences from Texts and Transcripts*. Lawrence Erlbaum Associate, Mahwah, New Jersey, 1997.

[van Cuilenburg et al., 1988] J.J. van Cuilenburg, J. Kleinnijenhuis, and J.A. de Ridder. *Tekst en Betoog: naar een Computergestuurde Inhoudsanalyse van Betogende Teksten*. Coutinho, Muiderberg (Netherlands), 1988.

[Wikipedia, 2005] URL://www.wikipedia.org, 2005.

Logical Non-determinism as a Tool for Logical Modularity: An Introduction

ARNON AVRON

1 Introduction

It is well known that every propositional logic which satisfies certain very natural conditions can be characterized semantically using a multi-valued matrix ([Łos and Suszko, 1958; Wójcicki, 1988; Urquhart, 2001]). However, there are many important decidable logics whose characteristic matrices necessarily consist of an infinite number of truth values. In such a case it might be quite difficult to find any of these matrices, or to use one when it is found. Even in case a logic does have a finite characteristic matrix it might be difficult to discover this fact, or to find such a matrix. The deep reason for these difficulties is that in an ordinary multi-valued semantics the rules and axioms of a system should be considered as a whole, and there is no method for separately determining the semantic effects of each rule alone.

In this introductory paper (which is based on several previous papers of the author) it is shown that by allowing the use of non-deterministic operations, one can provide in a lot of cases simple modular semantics of rules of inference, so that the semantics of a system is obtained by joining the semantics of its rules in the most straightforward way. Our main tool for this task is the use of finite Nmatrices ([Avron and Lev, 2001; Avron and Lev, 2004; Avron and Lev, 2005]. Nmatrices are multi-valued structures in which the value assigned by a valuation to a complex formula can be chosen non-deterministically out of a certain nonempty set of options. The use of finite structures of this sort has the benefit of preserving all the advantages of logics with ordinary finite-valued semantics (in particular: decidability and compactness), while it is applicable to a much larger family of logics. The central idea in using Nmatrices for providing semantics for rules is that the main effect of a "normal" rule is to reduce the degree of non-determinism of operations, by forbidding some options (in non-deterministic computations of truth values) which we could have had otherwise. Many examples of applying this idea are provided below.

2 Basic Concepts

2.1 Consequence Relations, Logics, and Pure Rules

DEFINITION 1.

1. A *Scott consequence relation* (*scr* for short) for a language $\mathcal{L}$ is a binary relation $\vdash$ between sets of formulas of $\mathcal{L}$ that satisfies the following three conditions:

   strong reflexivity: if $\Gamma \cap \Delta \neq \emptyset$ then $\Gamma \vdash \Delta$.
   monotonicity: if $\Gamma \vdash \Delta$ and $\Gamma \subseteq \Gamma'$, $\Delta \subseteq \Delta'$ then $\Gamma' \vdash \Delta'$.
   Transitivity (cut):   if $\Gamma \vdash \psi, \Delta$ and $\Gamma', \psi \vdash \Delta'$ then $\Gamma, \Gamma' \vdash \Delta, \Delta'$.

2. An scr $\vdash$ for $\mathcal{L}$ is *structural* if for every uniform $\mathcal{L}$-substitution σ and every Γ and Δ, if $\Gamma \vdash \Delta$ then $\sigma(\Gamma) \vdash \sigma(\Delta)$. $\vdash$ is *finitary* if the following condition holds for all $\Gamma, \Delta \subseteq \mathcal{W}$: if $\Gamma \vdash \Delta$ then there exist finite $\Gamma' \subseteq \Gamma$ and $\Delta' \subseteq \Delta$ such that $\Gamma' \vdash \Delta'$. $\vdash$ is *consistent* (or *non-trivial*) if there exist non-empty Γ and Δ s.t. $\Gamma \nvdash \Delta$. [1]

3. A propositional *logic* is a pair $\langle \mathcal{L}, \vdash \rangle$, where $\mathcal{L}$ is a propositional language and $\vdash$ is an scr for $\mathcal{L}$ which is structural and consistent. The logic $\langle \mathcal{L}, \vdash \rangle$ is finitary if $\vdash$ is finitary.

DEFINITION 2.

1. A *pure rule* in a propositional language $\mathcal{L}$ is any ordered pair $\langle \Gamma, \Delta \rangle$, where Γ and Δ are finite sets of formulas in $\mathcal{L}$ (We shall usually denote such a rule by $\Gamma \Rightarrow \Delta$ rather than by $\langle \Gamma, \Delta \rangle$).

2. Let $\langle \mathcal{L}, \vdash_1 \rangle$ be a propositional logic, and let S be a set of rules in a propositional language $\mathcal{L}'$. By the *extension* of $\langle \mathcal{L}, \vdash_1 \rangle$ by S we mean the logic $\langle \mathcal{L}^*, \vdash^* \rangle$, where $\mathcal{L}^* = \mathcal{L} \cup \mathcal{L}'$, and $\vdash^*$ is the least *structural* scr $\vdash$ such that $\Gamma \vdash \Delta$ whenever $\Gamma \vdash_1 \Delta$ or $\langle \Gamma, \Delta \rangle \in S$.

REMARK 3. Obviously, the extension of $\langle \mathcal{L}, \vdash_1 \rangle$ by S is well-defined (i.e. a logic) only if $\vdash^*$ is consistent. In all the cases we consider below this will easily be guaranteed by the semantics we provide (and so we shall not even mention it).

REMARK 4. It is easy to see that $\vdash^*$ is the closure under cuts and weakenings of the set of all pairs $\langle \sigma(\Gamma), \sigma(\Delta) \rangle$, where σ is a uniform substitution in

---

[1] See [Avron and Lev, 2001; Avron and Lev, 2005] for the importance of the consistency property.

$\mathcal{L}^*$, and either $\Gamma \vdash_1 \Delta$ or $\langle \Gamma, \Delta \rangle \in S$. This in turn implies that an extension of a finitary logic by a set of pure rules is again finitary.

CONVENTION: To emphasize the fact that the presence of a rule in a system means the presence of all its instances, we shall usually describe a rule using the metavariables φ, ψ, θ rather than the atomic formulas $p_1, p_2, \ldots$. Thus although formally the rule $(\supset\Rightarrow)$ is the rule $p_1, p_1 \supset p_2 \Rightarrow p_2$, we shall write it as $\varphi, \varphi \supset \psi \Rightarrow \psi$.

REMARK 5. Suppose that the formula θ occurs in a pure rule of a logic $\mathcal{L}$, and we decide to select θ as the "principal formula" of that rule. Assume e.g. that the rule is of the form $\varphi_1, \ldots, \varphi_n \Rightarrow \psi_1, \ldots, \psi_k, \theta$ (the consideration in the other case is similar). Suppose further that $\Gamma_i \vdash \Delta_i, \varphi_i$ for $i = 1, \ldots, n$ and $\psi_j, \Gamma_j \vdash \Delta_j$ for $j = 1, \ldots, k$. Then $\Gamma_1, \ldots, \Gamma_n \vdash \Delta_1, \ldots, \Delta_k, \theta$ (by n+k cuts). It follows that $\mathcal{L}$ is closed in this case under the Gentzen-type rule:

$$\frac{\Gamma_i \Rightarrow \Delta_i, \varphi_i \ (i = 1, \ldots, n) \quad \psi_j, \Gamma_j \Rightarrow \Delta_j \ (j = 1, \ldots, k)}{\Gamma_1, \ldots, \Gamma_n \Rightarrow \Delta_1, \ldots, \Delta_k, \theta}$$

Conversely, if $\mathcal{L}$ is closed under this Gentzen-type rule then by applying it to the reflexivity axioms $\varphi_i \vdash \varphi_i \ (i = 1, \ldots, n)$ and $\psi_j \vdash \psi_j \ (j = 1, \ldots, k)$ we get $\varphi_1, \ldots, \varphi_n \vdash \psi_1, \ldots, \psi_k, \theta$. It follows that every pure rule in the sense of Definition 2 is equivalent to some *multiplicative* (in the terminology of [Girard, 1987]) or *pure* (in the terminology of [Avron, 1991]) Gentzen-type rule. Moreover: it is easy to see that most standard rules used in Gentzen-type systems are equivalent to finite sets of pure rules in the sense of Definition 2. For example: the usual $(\supset\Rightarrow)$ rule of classical logic is equivalent by what we have just shown to the pure rule $\varphi, \varphi \supset \psi \Rightarrow \psi$. The classical $(\Rightarrow\supset)$, in turn, can be split into the following two rules:

$$\frac{\Gamma, \varphi \Rightarrow \Delta}{\Gamma \Rightarrow \Delta, \varphi \supset \psi} \qquad \frac{\Gamma \Rightarrow \Delta, \psi}{\Gamma \Rightarrow \Delta, \varphi \supset \psi}$$

Hence $(\Rightarrow\supset)$ is equivalent to the set $\{\psi \Rightarrow \varphi \supset \psi, \ \Rightarrow \varphi, \varphi \supset \psi\}$. [2]

2.2 Non-deterministic Matrices

Our main semantic tool in what follows will be the following generalization from [Avron and Lev, 2001; Avron and Lev, 2004; Avron and Lev, 2005] of the concept of a matrix: [3]

---

[2]Recall that formally we should have written here $\{p_2 \Rightarrow p_1 \supset p_2, \ \Rightarrow p_1, p_1 \supset p_2\}$.

[3]A special two-valued case of this definition was essentially introduced in [Batens *et al.*, 1999]. Another particular case of the same idea, using a similar name, was used in [Crawford and Etherington, 1998]. It should also be noted that Carnielli's "possible-translations semantics" (see [Carnielli and Marcos, 2002]) was originally called "non-deterministic semantics", but later the name was changed to the present one.

DEFINITION 6.

1. A *non-deterministic matrix* (*Nmatrix* for short) for a propositional language $\mathcal{L}$ is a tuple $\mathcal{M} = \langle \mathcal{V}, \mathcal{D}, \mathcal{O} \rangle$, where:

   (a) $\mathcal{V}$ is a non-empty set of *truth values*.

   (b) $\mathcal{D}$ is a non-empty proper subset of $\mathcal{V}$.

   (c) For every n-ary connective $\diamond$ of $\mathcal{L}$, $\mathcal{O}$ includes a corresponding n-ary function $\widetilde{\diamond}$ from $\mathcal{V}^n$ to $2^{\mathcal{V}} - \{\emptyset\}$.

   We say that $\mathcal{M}$ is *(in)finite* if so is $\mathcal{V}$.

2. Let $\mathcal{W}$ be the set of formulas of $\mathcal{L}$. A *(legal) valuation* in an Nmatrix $\mathcal{M}$ is a function $v : \mathcal{W} \to \mathcal{V}$ that satisfies the following condition for every n-ary connective $\diamond$ of $\mathcal{L}$ and $\psi_1, \dots, \psi_n \in \mathcal{W}$:

$$v(\diamond(\psi_1, \dots, \psi_n)) \in \widetilde{\diamond}(v(\psi_1), \dots, v(\psi_n))$$

3. A valuation v in an Nmatrix $\mathcal{M}$ is a *model* of (or *satisfies*) a formula ψ in $\mathcal{M}$ (notation: $v \models^{\mathcal{M}} \psi$) if $v(\psi) \in \mathcal{D}$. v is a *model* in $\mathcal{M}$ of a set Γ of formulas (notation: $v \models^{\mathcal{M}} \Gamma$) if it satisfies every formula in Γ.

4. $\vdash_{\mathcal{M}}$, the consequence relation induced by the Nmatrix $\mathcal{M}$, is defined by: $\Gamma \vdash_{\mathcal{M}} \Delta$ if for every v such that $v \models^{\mathcal{M}} \Gamma$, there is $\varphi \in \Delta$ such that $v \models^{\mathcal{M}} \varphi$.

5. A logic $\mathbf{L} = \langle \mathcal{L}, \vdash_{\mathbf{L}} \rangle$ is *sound* for an Nmatrix $\mathcal{M}$ (where $\mathcal{L}$ is the language of $\mathcal{M}$) if $\vdash_{\mathbf{L}} \subseteq \vdash_{\mathcal{M}}$. $\mathbf{L}$ is *complete* for $\mathcal{M}$ if $\vdash_{\mathbf{L}} \supseteq \vdash_{\mathcal{M}}$. $\mathcal{M}$ is *characteristic* for $\mathbf{L}$ if $\mathbf{L}$ is both sound and complete for it (i.e.: if $\vdash_{\mathbf{L}} = \vdash_{\mathcal{M}}$). $\mathcal{M}$ is *weakly-characteristic* for $\mathbf{L}$ if for every formula φ of $\mathcal{L}$, $\vdash_{\mathbf{L}} \varphi$ iff $\vdash_{\mathcal{M}} \varphi$.

REMARK 7. We shall identify an ordinary (deterministic) matrix with an Nmatrix whose functions in $\mathcal{O}$ always return singletons.

THEOREM 8 ([Avron and Lev, 2005; Avron and Lev, 2004]). *A logic which has a finite characteristic Nmatrix is finitary and decidable.*

The following Definition and Theorem are from [Avron, 2005a]:

DEFINITION 9. Let $\mathcal{M}_1 = \langle \mathcal{V}_1, \mathcal{D}_1, \mathcal{O}_1 \rangle$ and $\mathcal{M}_2 = \langle \mathcal{V}_2, \mathcal{D}_2, \mathcal{O}_2 \rangle$ be Nmatrices for a language $\mathcal{L}$.

1. A reduction of $\mathcal{M}_1$ to $\mathcal{M}_2$ is a function $F : \mathcal{V}_1 \to \mathcal{V}_2$ such that:

   (a) For every $x \in \mathcal{V}_1$, $x \in \mathcal{D}_1$ iff $F(x) \in \mathcal{D}_2$.

(b) $F(y) \in \tilde{\diamond}_{\mathcal{M}_2}(F(x_1), \ldots, F(x_n))$ for every n-ary connective $\diamond$ of $\mathcal{L}$ and every $x_1, \ldots, x_n, y \in \mathcal{V}_1$ such that $y \in \tilde{\diamond}_{\mathcal{M}_1}(x_1, \ldots, x_n)$.

2. $\mathcal{M}_1$ is a *refinement* of $\mathcal{M}_2$ if there exists a reduction of $\mathcal{M}_1$ to $\mathcal{M}_2$.

THEOREM 10. *If $\mathcal{M}_1$ is a refinement of $\mathcal{M}_2$ then $\vdash_{\mathcal{M}_2} \subseteq \vdash_{\mathcal{M}_1}$.*

Proof. Assume that F is a reduction of $\mathcal{M}_1$ to $\mathcal{M}_2$. We first show that if v is a legal valuation in $\mathcal{M}_1$ then $v' = F \circ v$ (the composition of F and v) is a legal valuation in $\mathcal{M}_2$. Indeed, let $\diamond$ be an n-ary connective of $\mathcal{L}$, and let $\varphi_1, \ldots, \varphi_n$ be n formulas of $\mathcal{L}$. We show that $v'(\diamond(\varphi_1, \ldots, \varphi_n)) \in \tilde{\diamond}_{\mathcal{M}_2}(v'(\varphi_1), \ldots, v'(\varphi_n))$. Let $y = v(\diamond(\varphi_1, \ldots, \varphi_n))$, and $x_i = v(\varphi_i)$ ($i = 1, \ldots, n$). Then $y \in \tilde{\diamond}_{\mathcal{M}_1}(x_1, \ldots, x_n)$, and so $F(y) \in \tilde{\diamond}_{\mathcal{M}_2}(F(x_1), \ldots, F(x_n))$. Since $v'(\diamond(\varphi_1, \ldots, \varphi_n)) = F(y)$ and $v'(\varphi_i) = F(x_i)$ ($i = 1, \ldots, n$), our claim follows.

Now assume that $\Gamma \vdash_{\mathcal{M}_2} \Delta$. We show that $\Gamma \vdash_{\mathcal{M}_1} \Delta$ as well. So let v be a model of Γ in $\mathcal{M}_1$. Then $v(\varphi) \in \mathcal{D}_1$ for every $\varphi \in \Gamma$. Hence $F(v(\varphi)) \in \mathcal{D}_2$ for every $\varphi \in \Gamma$. Since $F \circ v$ is a legal valuation in $\mathcal{M}_2$, this means that $F \circ v$ is a model of Γ in $\mathcal{M}_2$, and so $F(v(\psi)) = (F \circ v)(\psi) \in \mathcal{D}_2$ for some $\psi \in \Delta$. Since F is a reduction function, this implies that $v(\psi) \in \mathcal{D}_1$ for some $\psi \in \Delta$, as required. ∎

REMARK 11. An important case in which $\mathcal{M}_1 = \langle \mathcal{V}_1, \mathcal{D}_1, \mathcal{O}_1 \rangle$ is a refinement of $\mathcal{M}_2 = \langle \mathcal{V}_2, \mathcal{D}_2, \mathcal{O}_2 \rangle$ is when $\mathcal{V}_1 \subseteq \mathcal{V}_2$, $\mathcal{D}_1 = \mathcal{D}_2 \cap \mathcal{V}_1$, and $\tilde{\diamond}_{\mathcal{M}_1}(\vec{x}) \subseteq \tilde{\diamond}_{\mathcal{M}_2}(\vec{x})$ for every n-ary connective $\diamond$ of $\mathcal{L}$ and every $\vec{x} \in \mathcal{V}_1^n$. It is easy to see that the identity function on $\mathcal{V}_1$ is in this case a reduction of $\mathcal{M}_1$ to $\mathcal{M}_2$. A refinement of this sort will be called *simple*. [4]

3 Canonical Rules and 2-Valued Nmatrices

In this section we establish a strong connection between the simplest type of pure rules, and the simplest type of Nmatrices.

DEFINITION 12. A *2-Nmatrix* is an Nmatrix in which $\mathcal{V} = \{t, f\}$ and $\mathcal{D} = \{t\}$ (Obviously, any two-valued Nmatrix is isomorphic to some 2-Nmatrix).

DEFINITION 13. A *canonical* (propositional) rule is a pure rule of the form $\diamond(p_1, \ldots, p_n), \Gamma \Rightarrow \Delta$ or $\Gamma \Rightarrow \Delta, \diamond(p_1, \ldots, p_n)$, where $\diamond$ is an n-ary connective, $p_1, \ldots, p_n$ are n distinct atomic formulas, $\Gamma \subseteq \{p_1, \ldots, p_n\}$, $\Delta \subseteq \{p_1, \ldots, p_n\}$, and $\Gamma \cap \Delta = \emptyset$.

---

[4]What we call here "a simple refinement" is what was called "a refinement" in [Avron, 2004]. The present definition of "a refinement" is a refinement of the definition given to that concept there.

EXAMPLE 14. The three pure rules for implication described in Remark 5 ($\varphi, \varphi \supset \psi \Rightarrow \psi$, $\psi \Rightarrow \varphi \supset \psi$, and $\Rightarrow \varphi, \varphi \supset \psi$) are all canonical.

DEFINITION 15. A set S of canonical rules is *coherent* if whenever both $\diamond(p_1, \ldots, p_n), \Gamma_1 \Rightarrow \Delta_1$ and $\Gamma_2 \Rightarrow \Delta_2, \diamond(p_1, \ldots, p_n)$ are rules of S, we have that $(\Gamma_1 \cup \Gamma_2) \cap (\Delta_1 \cup \Delta_2) \neq \emptyset$.

EXAMPLE 16. The three pure rules for implication from Example 14 form a coherent set of rules. In contrast, the following two pure rules for the famous connective "tonk" of Prior ([Prior, 1960]) form an incoherent set: $\{p \Rightarrow pTq, \quad pTq \Rightarrow q\}$.

REMARK 17. It is easy to see that if we extend a logic by an incoherent set of canonical rules we get an inconsistent scr. Hence coherence is a minimal constraint on sets of canonical rules.

DEFINITION 18. A propositional logic $\langle \mathcal{L}, \vdash \rangle$ is called *canonical* if it is an extension of the trivial logic $\langle \mathcal{L}, \vdash_{triv} \rangle$ (where $\Gamma \vdash_{triv} \Delta$ iff $\Gamma \cap \Delta \neq \emptyset$) by a coherent set of canonical rules.

THEOREM 19. *A propositional logic is canonical iff it has a characteristic 2-Nmatrix.*

Proof. Suppose first that $\langle \mathcal{L}, \vdash \rangle$ is an extension of $\langle \mathcal{L}, \vdash_{triv} \rangle$ by a coherent set S of canonical rules. We define a corresponding 2-Nmatrix $\mathcal{M}_S$ as follows. Given an n-ary connective $\diamond$ of $\mathcal{L}$ and $x_1, \ldots, x_n \in \{t, f\}$, we let:

- $\widetilde{\diamond}(x_1, \ldots, x_n) = \{f\}$ if there is a rule $\diamond(p_1, \ldots, p_n), \Gamma \Rightarrow \Delta$ in S such that $x_i = t$ if $p_i \in \Gamma$, and $x_i = f$ if $p_i \in \Delta$.

- $\widetilde{\diamond}(x_1, \ldots, x_n) = \{t\}$ if there is a rule $\Gamma \Rightarrow \Delta, \diamond(p_1, \ldots, p_n)$ in S such that $x_i = t$ if $p_i \in \Gamma$, and $x_i = f$ if $p_i \in \Delta$.

- $\widetilde{\diamond}(x_1, \ldots, x_n) = \{t, f\}$ otherwise.

Since S is coherent, $\widetilde{\diamond}$ is well-defined. It is also easy to see that $\Gamma \vdash_{\mathcal{M}_S} \Delta$ whenever $\Gamma \Rightarrow \Delta$ is an instance of a rule in S. Hence $\langle \mathcal{L}, \vdash \rangle$ is sound with respect to $\mathcal{M}_S$. To prove completeness, suppose $\Gamma_0 \not\vdash \Delta_0$. We construct a model of Γ_0 in $\mathcal{M}_S$ which is not a model of any formula in Δ_0. For this extend Γ_0 to a maximal set T of formulas such that $T \not\vdash \Delta_0$. Obviously, $\varphi \notin T$ iff $T, \varphi \vdash \Delta_0$. This entails that T is $\vdash$-prime, in the sense that for every finite set Δ, if $T \vdash \Delta$ then $T \vdash \varphi$ for some $\varphi \in \Delta$ (otherwise we could have derived $T \vdash \Delta_0$ using n cuts, where n is the number of formulas in Δ). Define now a valuation v_T by $v_T(\varphi) = t$ iff $\varphi \in T$. The facts that T is $\vdash$-prime, and that $\Gamma \vdash \Delta$ whenever $\Gamma \Rightarrow \Delta$ is (an instance of) a rule of S, together imply that v_T is a legal valuation in $\mathcal{M}_S$. Since obviously $T \cap \Delta_0 = \emptyset$, v_T is the required model of Γ_0 which is not a model of Δ_0.

For the converse, let $\mathcal{M}$ be a 2-Nmatrix. We show that $\vdash_{\mathcal{M}}$ is canonical. Let $S(\mathcal{M})$ be the set of all canonical rules having one of the following forms:

- $\diamond(p_1, \ldots, p_n), \Gamma \Rightarrow \Delta$, where $\Gamma \cap \Delta = \emptyset$, $\Gamma \cup \Delta = \{p_1, \ldots, p_n\}$, and $\diamond$ is an n-ary connective of the language of $\mathcal{M}$ such that $\widetilde{\diamond}(x_1, \ldots, x_n) = \{f\}$, where $\langle x_1, \ldots, x_n \rangle$ is defined by: $x_i = t$ if $p_i \in \Gamma$, $x_i = f$ if $p_i \in \Delta$.

- $\Gamma \Rightarrow \Delta, \diamond(p_1, \ldots, p_n)$, where $\Gamma \cap \Delta = \emptyset$, $\Gamma \cup \Delta = \{p_1, \ldots, p_n\}$, and $\diamond$ is an n-ary connective of the language of $\mathcal{M}$ such that $\widetilde{\diamond}(x_1, \ldots, x_n) = \{t\}$, where $\langle x_1, \ldots, x_n \rangle$ is defined by: $x_i = t$ if $p_i \in \Gamma$, $x_i = f$ if $p_i \in \Delta$.

It is easy to see that $\mathcal{M}_{S(\mathcal{M})} = \mathcal{M}$. Therefore it follows from the first part of this proof that $\vdash_{\mathcal{M}}$ is the extension of $\vdash_{triv}$ by the set $S(\mathcal{M})$. ∎

REMARK 20. Using the procedure described in Remark 5 (where the principal formula is taken to be $\diamond(p_1, \ldots, p_n)$), we can see that every canonical rule in the sense used here is equivalent to what was called in [Avron and Lev, 2001] a *separated canonical* Gentzen-type rule. A much more extensive class of canonical Gentzen-type rules and systems was investigated in [Avron and Lev, 2001; Avron and Lev, 2005]. Accordingly, a more complicated coherence criterion was used there (which applies to all type of sets of canonical rules, and is easily seen to be equivalent to the one used here in the case of a set of separated canonical rules). Since it was proved in [Avron and Lev, 2001] that every canonical Gentzen-type rule is equivalent to a finite set of separated canonical Gentzen-type rules, Theorem 19 is actually equivalent to the characterization given in [Avron and Lev, 2001; Avron and Lev, 2005] to the whole class of canonical Gentzen-type systems.

REMARK 21. In [Avron and Lev, 2005] it is shown that any logic which has a characteristic 2-Nmatrix $\mathcal{M}$ can have a finite characteristic (ordinary) matrix if and only if $\mathcal{M}$ itself is deterministic (in which case it is unique). It follows that a canonical propositional logic is either a fragment of classical logic, or else it has no finite characteristic deterministic matrix (although it has of course a characteristic 2-valued non-deterministic matrix).

4 Negation Rules and 4-Valued Nmatrices

The most common type of non-canonical rules used in proof systems for the classical connectives, are rules which handle combinations of negation with other connectives. In this section we show how to use Nmatrices with four values (or less) in order to provide modular semantics for many rules of this sort. For simplicity of presentation, we investigate only rules involving the unary connective $\neg$, and the binary connective $\supset$ (where $\neg$ has some features of classical negation, while $\supset$ has some features of classical implication). Extending the methods for similar rules involving other connectives (like disjunction or conjunction) is a straightforward matter. [5]

DEFINITION 22.

1. Let NIR be the the following set of rules:

$$(\neg \Rightarrow) \qquad \neg\varphi, \varphi \Rightarrow$$
$$(\Rightarrow \neg) \qquad \Rightarrow \neg\varphi, \varphi$$
$$(\supset\Rightarrow) \qquad \varphi, \varphi \supset \psi \Rightarrow \psi$$
$$(\Rightarrow\supset)_1 \qquad \psi \Rightarrow \varphi \supset \psi$$
$$(\Rightarrow\supset)_2 \qquad \Rightarrow \varphi \supset \psi, \varphi$$
$$(\Rightarrow \neg\neg) \qquad \varphi \Rightarrow \neg\neg\varphi$$
$$(\neg\neg \Rightarrow) \qquad \neg\neg\varphi \Rightarrow \varphi$$
$$(\neg \supset\Rightarrow)_1 \qquad \neg(\varphi \supset \psi) \Rightarrow \varphi$$
$$(\neg \supset\Rightarrow)_2 \qquad \neg(\varphi \supset \psi) \Rightarrow \neg\psi$$
$$(\Rightarrow \neg \supset) \qquad \varphi, \neg\psi \Rightarrow \neg(\varphi \supset \psi)$$

2. For $S \subseteq NIR$ let $\mathbf{L}[S]$ be the extension by S of the trivial logic in the language $\{\neg, \supset\}$ (i.e. $\mathbf{L}[S]$ is the minimal logic for which all the rules in S are valid).

The basic idea in providing semantics for the logics $\mathbf{L}[S]$ ($S \subseteq NIR$) is to let the value assigned to a sentence φ provide information not only about the truth/falsity of φ, but also about the truth/falsity of its negation. This leads to the use of elements from $\{0, 1\}^2$ as our truth-values, where the intended intuitive meaning of $v(\varphi) = \langle x, y \rangle$ is the following:

- $x = 1$ iff φ is "true" (i.e. $v(\varphi) \in \mathcal{D}$).

- $y = 1$ iff $\neg\varphi$ is "true" (i.e. $v(\neg\varphi) \in \mathcal{D}$).

This interpretation of the components of the truth-values dictates the following constraint on any valuation v (where $P_1(\langle x_1, \ldots, x_k \rangle) = x_1$ and $P_2(\langle x_1, \ldots, x_k \rangle) = x_2$:

$$P_1(v(\neg\varphi)) = P_2(v(\varphi))$$

---

[5]The material in this section extends and systematizes results and methods from [Avron, 2004; Avron, 2005a].

In terms of Nmatrices this constraint translates into the condition:

$$(\text{NEG}) \quad \tilde{\neg}a \subseteq \{y \mid P_1(y) = P_2(a)\}$$

We start our semantic investigation of NIR with the weakest Nmatrix which satisfies Condition (NEG):

DEFINITION 23. The Nmatrix $\mathcal{M}_4^B = \langle \mathcal{V}_4, \mathcal{D}_4, \mathcal{O}_4^B \rangle$ is defined as follows:

- $\mathcal{V}_4 = \{t, \top, \bot, f\}$ where:

$$
\begin{aligned}
t &= \langle 1, 0 \rangle \\
\top &= \langle 1, 1 \rangle \\
\bot &= \langle 0, 0 \rangle \\
f &= \langle 0, 1 \rangle
\end{aligned}
$$

- $\mathcal{D}_4 = \{a \in \mathcal{V}_4 \mid P_1(a) = 1\} = \{t, \top\}$
- Let $\mathcal{V} = \mathcal{V}_4$, $\mathcal{D} = \mathcal{D}_4$, $\mathcal{F} = \mathcal{V}_4 - \mathcal{D}$. The operations in $\mathcal{O}_4^B$ are:

$$
\tilde{\neg}a = \begin{cases} \mathcal{D} & \text{if } P_2(a) = 1 \quad (\text{i.e. } a \in \{f, \top\}) \\ \mathcal{F} & \text{if } P_2(a) = 0 \quad (\text{i.e. } a \in \{t, \bot\}) \end{cases}
$$

$$a \tilde{\supset} b = \mathcal{V}$$

Below we shall see that every logic which is defined by some subset of NIR is characterized by some simple refinement of $\mathcal{M}_4^B$. In fact, it is quite easy to compute for each rule in NIR the semantic condition on such refinement which corresponds to that rule:

DEFINITION 24.

1. The general refining conditions induced by the conditions in NIR are:

$C(\neg \Rightarrow)$: If $P_1(a) = 1$ then $P_2(a) = 0$

$C(\Rightarrow \neg)$: If $P_1(a) = 0$ then $P_2(a) = 1$

$C(\supset \Rightarrow)$: If $P_1(a) = 1$ and $P_1(b) = 0$ then $a \tilde{\supset} b \subseteq \{x \mid P_1(x) = 0\}$
  (Equivalently: if $a \in \mathcal{D}$ and $b \in \mathcal{F}$ then $a \tilde{\supset} b \subseteq \mathcal{F}$)

$C(\Rightarrow \supset)_1$: If $P_1(b) = 1$ then $a \tilde{\supset} b \subseteq \{x \mid P_1(x) = 1\}$
  (Equivalently: if $b \in \mathcal{D}$ then $a \tilde{\supset} b \subseteq \mathcal{D}$)

$C(\Rightarrow \supset)_2$: If $P_1(a) = 0$ then $a \tilde{\supset} b \subseteq \{x \mid P_1(x) = 1\}$
  (Equivalently: if $a \in \mathcal{F}$ then $a \tilde{\supset} b \subseteq \mathcal{D}$)

$C(\Rightarrow \neg\neg)$: If $P_1(a) = 1$ then $\tilde{\neg}a \subseteq \{x \mid P_2(x) = 1\}$

$\mathrm{C}(\neg\neg\Rightarrow)$: If $P_1(a) = 0$ then $\widetilde{\neg}a \subseteq \{x \mid P_2(x) = 0\}$

$\mathrm{C}(\neg\supset\Rightarrow)_1$: If $P_1(a) = 0$ then $a\widetilde{\supset}b \subseteq \{x \mid P_2(x) = 0\}$

$\mathrm{C}(\neg\supset\Rightarrow)_2$: If $P_2(b) = 0$ then $a\widetilde{\supset}b \subseteq \{x \mid P_2(x) = 0\}$

$\mathrm{C}(\Rightarrow\neg\supset)$: If $P_1(a) = 1$ and $P_2(b) = 1$ then $a\widetilde{\supset}b \subseteq \{x \mid P_2(x) = 1\}$

2. For $S \subseteq NIR$, let $C(S) = \{Cr \mid r \in S\}$

The conditions in $C(NIR)$ were formulated in a way which can be applied whenever we employ finite sequences of 0's and 1's as the truth-values, the designated elements are those for which the first component is 1, and condition (NEG) is satisfied (We present another, more complicated example of this type of semantics in the next section). In the case of simple refinements of $\mathcal{M}_4^B$ these conditions can easily be translated into the following more specific ones:

$\mathrm{C}(\neg\Rightarrow)$: Use only t, f and $\bot$

$\mathrm{C}(\Rightarrow\neg)$: Use only t, f and $\top$

$\mathrm{C}(\supset\Rightarrow)$: If $a \in \{t, \top\}$ and $b \in \{f, \bot\}$ then $a\widetilde{\supset}b \subseteq \{f, \bot\}$

$\mathrm{C}(\Rightarrow\supset)_1$: If $b \in \{t, \top\}$ then $a\widetilde{\supset}b \subseteq \{t, \top\}$

$\mathrm{C}(\Rightarrow\supset)_2$: If $a \in \{f, \bot\}$ then $a\widetilde{\supset}b \subseteq \{t, \top\}$

$\mathrm{C}(\Rightarrow\neg\neg)$: $\widetilde{\neg}t = \{f\}$, $\widetilde{\neg}\top = \{\top\}$

$\mathrm{C}(\neg\neg\Rightarrow)$: $\widetilde{\neg}f = \{t\}$, $\widetilde{\neg}\bot = \{\bot\}$

$\mathrm{C}(\neg\supset\Rightarrow)_1$: If $a \in \{f, \bot\}$ then $a\widetilde{\supset}b \subseteq \{t, \bot\}$

$\mathrm{C}(\neg\supset\Rightarrow)_2$: If $b \in \{t, \bot\}$ then $a\widetilde{\supset}b \subseteq \{t, \bot\}$

$\mathrm{C}(\Rightarrow\neg\supset)$: If $a \in \{t, \top\}$ and $b \in \{\top, f\}$ then $a\widetilde{\supset}b \subseteq \{\top, f\}$

Here are some examples of how the conditions have been derived:

$\mathrm{C}(\neg\Rightarrow)$: The rule means that if $v(\varphi)$ is designated (i.e. $P_1(v(\varphi)) = 1$) then $v(\neg\varphi)$ is not (i.e, $P_1(v(\neg\varphi)) = P_2(v(\varphi)) = 0$). This is the case iff there is no $a \in \mathcal{V}$ such that $P_1(a) = P_2(a) = 1$, i.e.: if $\top$ is not available.

$C(\Rightarrow \neg\neg)$: The rule is valid iff $\neg\neg a \in \mathcal{D}$ whenever $a \in \mathcal{D}$ (where $\neg x$ denotes some element in $\tilde{\neg}x$). This means that if $P_1(a) = 1$ then $P_1(\neg\neg a) = P_2(\neg a) = 1$. It follows, e.g., that $\tilde{\neg}t \subseteq \{f, \top\}$. On the other hand, condition (NEG) above implies that $\tilde{\neg}t \subseteq \{f, \bot\}$. Hence $\tilde{\neg}t = \{f\}$. The considerations in the case of $\tilde{\neg}\top$ are similar.

$C(\neg \supset\Rightarrow)_2$: The rule is valid if $\neg(a \supset b)$ is in $\mathcal{F}$ whenever $\neg b$ is in $\mathcal{F}$ (where $x \supset y$ denotes some element in $x\tilde{\supset}y$). This is equivalent to: if $P_2(b) = 0$ then $P_2(a \supset b) = 0$. This is exactly the condition $C(\neg \supset\Rightarrow)_2$.

DEFINITION 25. For $S \subseteq NIR$, let $\mathcal{M}_S$ be the weakest simple refinement (Remark 11) of $\mathcal{M}_4^B$ in which the conditions in $C(S)$ are all satisfied. In other words: $\mathcal{M}_S = \langle \mathcal{V}_S, \mathcal{D}_S, \mathcal{O}_S \rangle$, where $\mathcal{V}_S$ is the set of values from $\mathcal{V}_4$ which are not rejected by any condition in S, $\mathcal{D}_S = \mathcal{D}_4 \cap \mathcal{V}_S$, and for any connective $\diamond \in \mathcal{O}$ and any $\vec{x} \in \mathcal{V}_S^n$ (where n is the arity of $\diamond$), the interpretation in $\mathcal{O}_S$ of $\diamond$ assigns to $\vec{x}$ the set of all the values in $\tilde{\diamond}_B(\vec{x})$ which are not forbidden by any condition in $C(S)$ (it is straightforward to check that for $S \subseteq NIR$ this set is never empty. The same is true for $\mathcal{D}_S$).

EXAMPLES 26.

1. Let $C_{min} = \{(\supset\Rightarrow), (\Rightarrow\supset)_1, (\Rightarrow\supset)_2, (\Rightarrow \neg), (\neg\neg \Rightarrow)\}$. Then $\mathcal{M}_{C_{min}}$ is the three-valued Nmatrix[6] $\langle \mathcal{V}, \mathcal{D}, \mathcal{O} \rangle$, where:

   - $\mathcal{V} = \{t, \top, f\}$
   - $\mathcal{D} = \{t, \top\}$
   - The operations in $\mathcal{O}$ are:

$$\tilde{\neg}a = \begin{cases} \mathcal{D} & \text{if } a = \top \\ \{f\} & \text{if } a = t \\ \{t\} & \text{if } a = f \end{cases}$$

$$a\tilde{\supset}b = \begin{cases} \mathcal{D} & \text{if } a = f \text{ or } b \in \mathcal{D} \\ \{f\} & \text{otherwise} \end{cases}$$

2. Let $\mathcal{FOUR} = NIR - \{(\neg \Rightarrow), (\Rightarrow \neg)\}$. Then $\mathcal{M}_{\mathcal{FOUR}}$ is the 4-valued (deterministic) *matrix* $\langle \mathcal{V}_4, \mathcal{D}_4, \mathcal{O}_4 \rangle$, where the operations in $\mathcal{O}_4$ are:[7]

$$\tilde{\neg}t = \{f\} \quad \tilde{\neg}f = \{t\} \quad \tilde{\neg}\top = \{\top\} \quad \tilde{\neg}\bot = \{\bot\}$$

---

[6]What is denoted here by $L(C_{min})$ is the negation-implication fragment of the paraconsistent logic C_{min} studied in [Carnielli and Marcos, 1999]. The 3-valued Nmatrix for this logic described here was first introduced in [Avron and Lev, 2004; Avron and Lev, 2005].

[7]The connective $\tilde{\supset}$ of $\mathcal{O}_4$ was introduced in [Arieli and Avron, 1996]. The soundness and completeness of the logic $\mathcal{FOUR}$ (augmented with disjunction and conjunction) for $\mathcal{M}_{\mathcal{FOUR}}$ was also first stated and proved there.

$$a \tilde{\supset} b = \begin{cases} t & \text{if } a \notin \mathcal{D} \\ b & \text{otherwise} \end{cases}$$

THEOREM 27. *For $S \subseteq NIR$, $\mathcal{M}_S$ is a characteristic Nmatrix for $\mathbf{L}[S]$.*

Proof. It is easy to verify that for any $r \in NIR$, the satisfaction of $C(r)$ in some simple refinement of $\mathcal{M}_4^B$ guarantees the validity of r in that refinement. This entails the soundness of $\mathbf{L}[S]$ with respect to $\mathcal{M}_S$.

For the converse, assume $\Gamma \nvdash_{\mathbf{L}[S]} \Delta$. We construct a model of Γ in $\mathcal{M}_S$ which is not a model of any formula in Δ. For this extend Γ to a maximal set Γ^* of formulas such that $\Gamma^* \nvdash_{\mathbf{L}[S]} \Delta$. Then Γ^* is closed under the rules in S, and $\varphi \notin \Gamma^*$ iff $\Gamma^*, \varphi \vdash_{\mathbf{L}[S]} \Delta$. Define now a valuation v by $v(\varphi) = \langle x(\varphi), y(\varphi) \rangle$, where:

$$x(\varphi) = \begin{cases} 1 & \varphi \in \Gamma^* \\ 0 & \varphi \notin \Gamma^* \end{cases} \qquad y(\varphi) = \begin{cases} 1 & \neg\varphi \in \Gamma^* \\ 0 & \neg\varphi \notin \Gamma^* \end{cases}$$

It is trivial that v is a legal valuation in $\mathcal{M}_4^B$. To show that it is also a legal valuation in $\mathcal{M}_S$, we need to check that it respects the conditions in $C(S)$. We do half of the cases, leaving the other half for the reader:

$C(\neg \Rightarrow)$: Assume $(\neg \Rightarrow) \in S$. Then there can be no sentence φ such that $\{\varphi, \neg\varphi\} \subseteq \Gamma^*$. Hence $v(\varphi) \neq \top$ for all φ.

$C(\Rightarrow \neg)$: Assume $(\Rightarrow \neg) \in S$. Suppose that there is φ such that $v(\varphi) = \bot$. Then $\varphi \notin \Gamma^*$ and $\neg\varphi \notin \Gamma^*$. It follows that $\Gamma^*, \varphi \vdash_{\mathbf{L}[S]} \Delta$ and $\Gamma^*, \neg\varphi \vdash_{\mathbf{L}[S]} \Delta$. Since also $\Gamma^* \vdash_{\mathbf{L}[S]} \varphi, \neg\varphi$ in this case, we get that $\Gamma^* \vdash_{\mathbf{L}[S]} \Delta$. A contradiction. Hence $v(\varphi) \neq \bot$ for all φ.

$C(\Rightarrow \neg\neg)$: Assume $(\Rightarrow \neg\neg) \in S$.

- Suppose $v(\varphi) = t$. Then $\varphi \in \Gamma^*$ and $\neg\varphi \notin \Gamma^*$. By $(\Rightarrow \neg\neg)$, also $\neg\neg\varphi \in \Gamma^*$. Hence $v(\neg\varphi) = f$ by definition of v.
- Suppose $v(\varphi) = \top$. Then $\varphi \in \Gamma^*$ and $\neg\varphi \in \Gamma^*$. By $(\Rightarrow \neg\neg)$, also $\neg\neg\varphi \in \Gamma^*$. Hence $v(\neg\varphi) = \top$ by definition of v.

$C(\supset\Rightarrow)$ Assume $(\supset\Rightarrow) \in S$. Suppose that $v(\varphi) \in \mathcal{D}$ and $v(\psi) \notin \mathcal{D}$. Then $\varphi \in \Gamma^*$ and $\psi \notin \Gamma^*$. It follows that $\varphi \supset \psi \notin \Gamma^*$ (since Γ^* is closed under $(\supset\Rightarrow)$ in this case), and so $v(\varphi \supset \psi) \notin \mathcal{D}$.

$C(\neg \supset\Rightarrow)_1$: Assume $(\neg \supset\Rightarrow)_1 \in S$. Suppose that $v(\varphi) \notin \mathcal{D}$. Then $\varphi \notin \Gamma^*$, and so also $\neg(\varphi \supset \psi) \notin \Gamma^*$. It follows that $v(\varphi \supset \psi) \in \{t, \bot\}$.

Obviously, v is a model of Γ in $\mathcal{M}_S$ which is not a model of Δ. ■

COROLLARY 28. $\mathbf{L}[S]$ *is decidable for every $S \subseteq NIR$.*

5 Logics of Formal (In)consistency

In this section we present an example of a family of logics for which the modular semantic analysis provided by Nmatrices has proved to be particularly useful: the paraconsistent logics of da Costa's school (see [da Costa., 1974; Carnielli and Marcos, 2002; Carnielli *et al.*, 2006]). The main ideas of this school is to limit the applicability of the $(\neg \Rightarrow)$ rule (which amounts to "a single contradiction entails everything") to the case where φ is "consistent", and to express the assumption of this consistency of φ within the language. The easiest way to implement these ideas is to include in the language a special connective $\circ$, with the intended meaning of $\circ\varphi$ being "φ is consistent". Then one can explicitly add the assumption of the consistency of φ to the problematic (from a paraconsistent point of view) classical (and intuitionistic) rule $(\neg \Rightarrow)$, getting the rule called (**b**) below. Various other rules concerning $\neg$, $\circ$ and other connectives can then be added, leading to a large family of logics known as "Logics of Formal Inconsistency" (LFIs - see [Carnielli and Marcos, 2002; Carnielli *et al.*, 2006]). For simplicity of presentation, we again consider only the binary connective $\supset$ in addition to the unary connectives $\neg$ and $\circ$, and investigate some of the most common rules connected with these 3 connectives which have been investigated in the literature concerning LFIs. Extending the methods for similar rules involving other connectives (like disjunction or conjunction) is again a straightforward matter. [8]

DEFINITION 29.

1. Let FCR be the the following set of rules:

   (**b**) $\circ\varphi, \neg\varphi, \varphi \Rightarrow$

   (**d1**) $\Rightarrow \circ\varphi, \varphi$

   (**d2**) $\Rightarrow \circ\varphi, \neg\varphi$

   (**i1**) $\neg\circ\varphi \Rightarrow \varphi$

   (**i2**) $\neg\circ\varphi \Rightarrow \neg\varphi$

   (**a$_\neg$**) $\circ\varphi \Rightarrow \circ\neg\varphi$

   (**a$_\supset$**) $\circ\varphi, \circ\psi \Rightarrow \circ(\varphi \supset \psi)$

   (**o$_\supset^1$**) $\circ\varphi \Rightarrow \circ(\varphi \supset \psi)$

   (**o$_\supset^2$**) $\circ\psi \Rightarrow \circ(\varphi \supset \psi)$

   (**4**) $\circ\varphi \Rightarrow \circ\circ\varphi$

---

[8]The material in this section extends and systematizes results and methods from [Avron, 2005a].

2. Let $LFIR = NIR \cup FCR$.

3. For $S \subseteq LFIR$ let $\mathbf{L}[S]$ be the extension by S of the trivial logic in the language $\{\neg, \circ, \supset\}$ (i.e. $\mathbf{L}[S]$ is the minimal logic for which all the rules in S are valid).

This time the basic idea in providing semantics for the logics of the form $\mathbf{L}[S]$ ($S \subseteq LFIR$) is to let the value assigned to a sentence φ provide information not only about the truth/falsity of φ and $\neg\varphi$, but also about the truth/falsity of $\circ\varphi$. This leads to the use of elements from $\{0,1\}^3$ as our truth-values, where now the intended intuitive meaning of $v(\varphi) = \langle x, y, z \rangle$ is the following:

- $x = 1$ iff φ is "true" (i.e. $v(\varphi) \in \mathcal{D}$).

- $y = 1$ iff $\neg\varphi$ is "true" (i.e. $v(\neg\varphi) \in \mathcal{D}$).

- $z = 1$ iff $\circ\varphi$ is "true" (i.e. $v(\circ\varphi) \in \mathcal{D}$).

In addition to (NEG), which remains unchanged, this interpretation dictates this time also the following condition:

$$(\text{CON}) \quad \tilde{\circ}a \subseteq \{y \mid P_1(y) = P_3(a)\}$$

Accordingly, this time we start our semantic investigation of $LFIR$ with the weakest Nmatrix which satisfies both (NEG) and (CON). Then we show that every logic which is defined by some subset of $LFIR$ is characterized by some (easily computable) simple refinement of that Nmatrix.

DEFINITION 30.

1. The Nmatrix $\mathcal{M}_8^B = \langle \mathcal{V}_8, \mathcal{D}_8, \mathcal{O}_8^B \rangle$ is defined as follows:

- $\mathcal{V}_8 = \{0,1\}^3$
- $\mathcal{D}_8 = \{a \in \mathcal{V}_8 \mid P_1(a) = 1\}$
- Let $\mathcal{V} = \mathcal{V}_8$, $\mathcal{D} = \mathcal{D}_8$, $\mathcal{F} = \mathcal{V}_8 - \mathcal{D}$. The operations in $\mathcal{O}_8^B$ are:

$$\tilde{\neg}a = \begin{cases} \mathcal{D} & \text{if } P_2(a) = 1 \\ \mathcal{F} & \text{if } P_2(a) = 0 \end{cases}$$

$$\tilde{\circ}a = \begin{cases} \mathcal{D} & \text{if } P_3(a) = 1 \\ \mathcal{F} & \text{if } P_3(a) = 0 \end{cases}$$

$$a\tilde{\supset}b = \mathcal{V}$$

2. The general refining conditions induced by the conditions in NIR are identical to those given in Definition 24.

3. The general refining conditions induced by the conditions in FCR are:

   C(b): If $P_1(a) = 1$ and $P_2(a) = 1$ then $P_3(a) = 0$

   C(d1): If $P_1(a) = 0$ then $P_3(a) = 1$

   C(d2): If $P_2(a) = 0$ then $P_3(a) = 1$

   C(i1): If $P_1(a) = 0$ then $\tilde{o}a \subseteq \{x \mid P_2(x) = 0\}$

   C(i2): If $P_2(a) = 0$ then $\tilde{o}a \subseteq \{x \mid P_2(x) = 0\}$

   C(a$_\neg$): If $P_3(a) = 1$ then $\tilde{\neg}a \subseteq \{x \mid P_3(x) = 1\}$

   C(a$_\supset$): If $P_3(a) = 1$ and $P_3(b) = 1$ then $a\tilde{\supset}b \subseteq \{x \mid P_3(x) = 1\}$

   C(o$_\supset^1$): If $P_3(a) = 1$ then $a\tilde{\supset}b \subseteq \{x \mid P_3(x) = 1\}$

   C(o$_\supset^2$): If $P_3(b) = 1$ then $a\tilde{\supset}b \subseteq \{x \mid P_3(x) = 1\}$

   C(4): If $P_3(a) = 1$ then $\tilde{o}a \subseteq \{\langle 1,0,1\rangle, \langle 1,1,1\rangle\}$

4. For $S \subseteq LFIR$, let $C(S) = \{Cr \mid r \in S\}$, and let $\mathcal{M}_S$ be the weakest simple refinement of $\mathcal{M}_8^B$ in which the conditions in $C(S)$ are all satisfied (again it is not difficult to check that this is well-defined for every $S \subseteq LFIR$).

THEOREM 31. *For $S \subseteq LFIR$, $\mathcal{M}_S$ is a characteristic Nmatrix for $\mathbf{L}[S]$.*

Proof. The proof is similar to that of Theorem 27, using the intended interpretation of a triple $\langle x, y, z\rangle$. ∎

COROLLARY 32. $\mathbf{L}[S]$ *is decidable for every $S \subseteq LFIR$.*

EXAMPLES 33.

1. Let $\mathbf{B} = \{(\supset\Rightarrow), (\Rightarrow\supset)_1, (\Rightarrow\supset)_2, (\Rightarrow \neg), (b)\}$. Then $\mathbf{L}[\mathbf{B}]$ is the basic logic of formal inconsistency mbC from [Carnielli *et al.*, 2006] [9]. By Theorem 31, the following Nmatrix $\mathcal{M}_5^B = \langle \mathcal{V}_5, \mathcal{D}_5, \mathcal{O}_5^B\rangle$ is a characteristic Nmatrix for this logic:

   - $\mathcal{V}_5 = \{t, t_I, I, f_I, f\}$ where:

$$
\begin{aligned}
t &= \langle 1,0,1\rangle \\
t_I &= \langle 1,0,0\rangle \\
I &= \langle 1,1,0\rangle \\
f &= \langle 0,1,1\rangle \\
f_I &= \langle 0,1,0\rangle
\end{aligned}
$$

---

[9]More precisely: it is the $\{\supset, \neg, \circ\}$-fragment of mbC. mbC itself has in addition conjunction and disjunction, together with their usual classical rules.

- $\mathcal{D}_5 = \{t, I, t_i\}$ $(= \{\langle x, y, z \rangle \in \mathcal{V}_5 \mid x = 1\})$.

- Let $\mathcal{D} = \mathcal{D}_5$, $\mathcal{F} = \mathcal{V}_5 - \mathcal{D}$. The operations in $\mathcal{O}_5^B$ are defined by:

$$a \widetilde{\supset} b = \begin{cases} \mathcal{D} & \text{if either } a \in \mathcal{F} \text{ or } b \in \mathcal{D} \\ \mathcal{F} & \text{if } a \in \mathcal{D} \text{ and } b \in \mathcal{F} \end{cases}$$

$$\widetilde{\neg} a = \begin{cases} \mathcal{D} & \text{if } a \in \{I, f, f_I\} \\ \mathcal{F} & \text{if } a \in \{t, t_I\} \end{cases}$$

$$\widetilde{\circ} a = \begin{cases} \mathcal{D} & \text{if } a \in \{t, f\} \\ \mathcal{F} & \text{if } a \in \{I, t_I, f_I\} \end{cases}$$

2. Let $S = \mathbf{B} \cup \{(\Rightarrow \neg \supset), (\mathbf{i1}), (\mathbf{a}_\neg)\}$. Then $\mathcal{M}_S = \langle \mathcal{V}_S, \mathcal{D}_S, \mathcal{O}_S \rangle$, where:

- $\mathcal{V}_S = \{t, t_I, I, f\}$

- $\mathcal{D}_S = \{t, I, t_i\}$

- $a \widetilde{\supset} b = \begin{cases} \mathcal{D}_S & \text{if either } a = f \text{ or } b \in \{t, t_I\} \\ \{I\} & \text{if } a \in \mathcal{D}_S \text{ and } b = I \\ \{f\} & \text{if } a \in \mathcal{D}_S \text{ and } b = f \end{cases}$

- $\widetilde{\neg} t = \widetilde{\neg} t_I = \{f\}$ $\widetilde{\neg} I = \mathcal{D}_S$ $\widetilde{\neg} f = \{t\}$

- $\widetilde{\circ} t = \mathcal{D}_S$ $\widetilde{\circ} t_I = \widetilde{\circ} I = \{f\}$ $\widetilde{\circ} f = \{t, t_I\}$

3. Let $\mathbf{Cia} = \{(\supset \Rightarrow), (\Rightarrow \supset)_1, (\Rightarrow \supset)_2, (\Rightarrow \neg), (\neg\neg \Rightarrow), (\mathbf{b}), (\mathbf{i1}), (\mathbf{i2}), ((\mathbf{a}_\supset)\}$.[10] Then $\mathcal{M}_{Cia} = \langle \mathcal{V}_{Cia}, \mathcal{D}_{Cia}, \mathcal{O}_{Cia} \rangle$, where:

- $\mathcal{V}_{Cia} = \{t, I, f\}$

- $\mathcal{D}_{Cia} = \{t, I\}$

- $a \widetilde{\supset} b = \begin{cases} \{f\} & \text{if } a \in \{t, I\} \text{ and } b = f \\ \{t\} & \text{if either } a = f, \ b \in \{f, t\} \text{ or } a = t, \ b = t \\ \{t, I\} & \text{otherwise} \end{cases}$

- $\widetilde{\neg} t = \{f\}$ $\widetilde{\neg} I = \{I\}$ $\widetilde{\neg} f = \{t\}$

- $\widetilde{\circ} t = \widetilde{\circ} f = \{t\}$ $\widetilde{\circ} I = \{f\}$

---

[10]$\mathbf{L[Cia]}$ is the $\{\supset, \neg, \circ\}$-fragment of the logic Cia from [Carnielli and Marcos, 2002; Carnielli *et al.*, 2006].

6 Further Research

We have demonstrated the potential of using non-deterministic matrices in providing modular semantics for logics. This leaves us with two main research problems. The first is to develop a general theory determining the type of systems for which the methods exemplified here work, and under what conditions (such a theory has been developed so far only for the special case of systems defined by a set of canonical rules - see Section 3). The second is to extend the methods for other types of logical systems. Here are some specific directions requiring further research:

1. Every set S of rules dealt with in Sections 4 and 5 has the property that the corresponding Nmatrix $\mathcal{M}_S$ actually exists. This should not always be the case. One obvious case in which this may fail is when the conditions in $C(S)$ leave no designated element or no non-designated one. In this case the resulting system should be trivial. A simple example in which this happens is when we add to $LFIR$ the rule $\circ\varphi \Rightarrow \varphi$ (corresponding to the modal axiom usually denoted by (T)). The corresponding condition is: if $P_1(a) = 0$ then $P_3(a) = 0$ (thus rejecting $\langle 0, 1, 1\rangle$ and $\langle 0, 0, 1\rangle$). Together with C($\mathbf{d1}$), (T) leaves no non-designated element, and indeed it is easy to verify that a system with both (T) and ($\mathbf{d1}$) is trivial. A generalization of the criterion of coherence for canonical systems is therefore needed here. A more subtle possible problem is that the conditions in $C(S)$ may leave in some cases only an empty set of options for applications of operations. Here is an example: suppose we add to NIR (see Section 4) the rule (MT): $\neg\psi, \varphi \supset \psi \Rightarrow \neg\varphi$. The corresponding condition, $C(MT)$, is: if $P_2(a) = 0$ and $P_2(b) = 1$ then $a\tilde{\supset}b \subseteq \{x \mid P_1(x) = 0\}$. Thus $\langle 0, 1\rangle\tilde{\supset}\langle 1, 1\rangle$ is not defined in a system which contains both $(\Rightarrow\supset)_1$ and (MT), although neither $\langle 0, 1\rangle$ nor $\langle 1, 1\rangle$ is rejected by any of the two rules. However, in this case the resulting system is not trivial, and can in fact be characterized by a finite Nmatrix (obtained as the product of simpler Nmatrices of the type used in Section 4). Hence an even more sophisticated coherence theory is needed to deal with this phenomenon, and a more general theory for constructing characteristic Nmatrices is needed to overcome the difficulty it causes.

2. Not every logic defined by a finite system of rules has a finite characteristic Nmatrix. Let, for example, (1) be the following rule from [Carnielli and Marcos, 2002; Carnielli et al., 2006]:

(1) $\neg(\varphi \wedge \neg\varphi) \vdash \circ\varphi$

In [Avron, 2005b] it was shown that $\mathbf{L}[\{(\Rightarrow \neg), (\neg\neg \Rightarrow), (\mathbf{b}), (\mathbf{l})\}]$ (which is called there Cl) has no finite characteristic Nmatrix. However, it does have an infinite characteristic Nmatrix $\mathcal{M}$ with is "semifinite" in the sense that determining the validity in this logic of any formula requires only a finite sub-Nmatrix of $\mathcal{M}$ (which can be determined in advance by the formula). Accordingly, a general theory is needed when a system has a finite characteristic Nmatrix, and when it has a "semifinite" characteristic Nmatrix.

3. This paper deals only with Scott consequence relations defined by a set of *pure* rules like in Definition 2. It is not clear how in general to extend it to deal with rules that are not treated as pure (like the necessitation rule in modal logics). An important case in which this can be done (using both possible worlds semantics and Nmatrices) is described in [Avron, 2004], where extensions of positive intuitionistic logic by subsets of NIR are investigated.

4. The idea of modular non-deterministic semantics has been applied so far mainly to propositional logics. It is important to extend it also to logics with quantifiers (some steps in this direction have been made in [Avron and Zamansky, 2005; Avron and Zamansky, 2004]).

Acknowledgement

This research was supported by THE ISRAEL SCIENCE FOUNDATION founded by The Israel Academy of Sciences and Humanities.

BIBLIOGRAPHY

[Arieli and Avron, 1996] Ofer Arieli and Arnon Avron. Reasoning with logical bilattices. *Journal of Logic, Language and Information*, 5(1):25–63, 1996.
[Avron and Lev, 2001] Arnon Avron and Iddo Lev. Canonical propositional Gentzen-type systems. In R. Goré, A Leitsch, and T. Nipkow, editors, *Proc. of the 1st International Joint Conference on Automated Reasoning (IJCAR 2001)*, number 2083 in Lecture Notes in AI, pages 529–544. Springer Verlag, 2001.
[Avron and Lev, 2004] Arnon Avron and Iddo Lev. Non-deterministic matrices. In *Proc. of the Thirty-fourth International Symposium on Multiple-valued Logic (ISMVL 2004)*, pages 282–287. IEEE Computer Society, 2004.
[Avron and Lev, 2005] Arnon Avron and Iddo Lev. Non-deterministic multiple-valued structures. *Journal of Logic and Computation*, 15:241–261, 2005.
[Avron and Zamansky, 2004] Arnon Avron and Anna Zamansky. Cut-elimination and quantification. (submitted), 2004.
[Avron and Zamansky, 2005] Arnon Avron and Anna Zamansky. Quantification in non-deterministic multi-valued structures. In *Proceedings of the 35th IEEE International Symposium on Multiple-Valued Logic (ISMVL 2005)*, pages 296–301. IEEE Computer Society Press, 2005.
[Avron, 1991] Arnon Avron. Simple consequence relations. *Information and Computation*, 92(1):105–139, 1991.

[Avron, 2004] Arnon Avron. A non-deterministic view on non-classical negations. *Studia Logica* (to appear), 2004.

[Avron, 2005a] Arnon Avron. Non-deterministic matrices and modular semantics of rules. In J.-Y. Beziau, editor, *Logica Universalis*, pages 149–167. Birkhüser Verlag, 2005.

[Avron, 2005b] Arnon Avron. Non-deterministic semantics for paraconsistent c-systems. In *Proceedings of ECSQARU 2005*, 2005.

[Batens et al., 1999] D. Batens, K. De Clercq, and N. Kurtonina. Embedding and interpolation for some paralogics. The propositional case. *Reports on Mathematical Logic*, 33:29–44, 1999.

[Carnielli and Marcos, 1999] W. A. Carnielli and J. Marcos. Limits for paraconsistent calculi. *Notre Dame Journal of Formal Logic*, 40:375–390, 1999.

[Carnielli and Marcos, 2002] W. A. Carnielli and J. Marcos. A taxonomy of C-systems. In W. A. Carnielli, M. E. Coniglio, and I. L. M. D'Ottaviano, editors, *Paraconsistency — the logical way to the inconsistent*, Lecture notes in pure and applied Mathematics, pages 1–94. Marcell Dekker, 2002.

[Carnielli et al., 2006] W. A. Carnielli, M. E. Coniglio, and J. Marcos. Logics of formal inconsistency. In D. Gabbay and F. Guenthner, editors, *Handbook of Philosophical Logic*. 2006. To appear.

[Crawford and Etherington, 1998] J. M. Crawford and D. W. Etherington. A non-deterministic semantics for tractable inference. In *Proc. of the 15th International Conference on Artificial Intelligence and the 10th Conference on Innovative Applications of Artificial Intelligence*, pages 286–291. MIT Press, Cambridge, 1998.

[da Costa., 1974] Newton C. A. da Costa. On the theory of inconsistent formal systems. *Notre Dame Journal of Formal Logic*, 15:497–510, 1974.

[Girard, 1987] J. Y. Girard. Linear logic. *Theoretical Computer Science*, 50:1–101, 1987.

[Los and Suszko, 1958] J. Los and R. Suszko. Remarks on sentential logics. *Indagationes Mathematicae*, 20:177–183, 1958.

[Prior, 1960] A. N. Prior. The runabout inference ticket. *Analysis*, 21:38–9, 1960.

[Urquhart, 2001] Alasdair Urquhart. Many-valued logic. In Dov M. Gabbay and Franz Guenthner, editors, *Handbook of Philosophical Logic*, volume 2, pages 249–295. Kluwer Academic Publishers, second edition, 2001.

[Wójcicki, 1988] R. Wójcicki. *Theory of Logical Calculi: Basic Theory of Consequence Operations*. Kluwer, 1988.

On the Proof Theory of the Existence Predicate

MATTHIAS BAAZ AND ROSALIE IEMHOFF

1 Introduction

In this paper we have brought together several results on the proof theory of the existence predicate in intuitionistic logic. This predicate, E, denotes whether a term exists or not: Et is read as t *exists*. Such a predicate was first introduced by Dana Scott in [Scott, 1979] in 1979. In this paper the author, in his one words, advocates in a mild way an extension of intuitionistic logic allowing reference to partial terms. One example from the paper pointing out, if not the necessity, at least the usefulness of having an existence predicate available is the following. In the context of rings the statement

$$\forall x \, \varphi(x) \;\Rightarrow\; \varphi(0)$$

is unconditionally true, since 0 is an element of all rings. In contrast to this, however,

$$\forall x \, \varphi(x) \;\Rightarrow\; \varphi(1/x)$$

is not generally true as not every element in a ring has an inverse. One could of course turn the latter into a conditional statement by formulating it as

$$\forall x \, \varphi(x) \;\Rightarrow\; \forall y \big(x \cdot y = 1 \;\Rightarrow\; \varphi(y) \big).$$

Now, still following Scott in [Scott, 1979], using the existence predicate one could express this more succinctly and directly as

$$\forall x \, \varphi(x) \wedge E(1/x) \;\Rightarrow\; \varphi(1/x).$$

In classical logic there is less need for this as one can always split definitions, theorems and proof in cases: either the object exists and then ...., or it does not exist. It is shown in [Scott, 1979] that in the presence of equality one can define the existence predicate via

$$Et \;\Leftrightarrow\; \exists y \, (t = y),$$

for y not free in t. But still, not only allows the existence predicate for more elegant and direct formulations, also when equality is not there it can be interpreted in a meaningful way. The latter is in accordance with the fact that existence comes for equality. For example, in term rewriting systems equality presupposes existence:

$$t = s \;\Rightarrow\; t{\downarrow} \wedge s{\downarrow}\,.$$

In this paper we study the existence predicate on this basic level. We consider intuitionistic logic IQC without equality and extended with the existence predicate, i.e. we add E to the language of predicate logic and extend IQC by certain axioms capturing the notion of existence.

What caused the renewed interest in the existence predicate is that recently the predicate was put to use in intuitionistic logic by providing satisfying answers to various problems that did not seem solvable in the setting of intuitionistic logic pure. The two main examples of this, one in the context of Skolemization, the other in the context of truth-value logics, will be discussed below.

This paper is meant as a survey of recent results on the proof theory of E. No results included here are new. The survey contains the introduction of various sequent calculi capturing the notion of existence, and the proofs that they all satisfy a form of cut-elimination, interpolation and the Beth definability property. Furthermore it introduces a semantics based on Kripke models but with a slightly different forcing relation, together with the proofs that this semantics is sound and complete for the calculi.

Our systems diverge slightly from other approaches in the literature in that we mostly consider mixed systems. By this we mean that we consider languages $\mathcal{L} \subseteq \mathcal{L}'$ and for terms in $\mathcal{L}$ assume that they exist, but for terms in $\mathcal{L}'\backslash\mathcal{L}$ we do not assume this. The systems consist of a basic Gentzen calculus LJE to which we add axioms of the form Et for all t in $\mathcal{L}$. This allows us to consider various other systems in the literature as subsystems of our systems, depending on how we choose $\mathcal{L}$. Also the applications discussed at the end of the paper show why allowing such a mixture is desirable.

1.1 Brouwer and Kant

On a more philosophical level there might be a reason that the existence predicate has not played an important role in the context of intuitionistic logic up till now, relating to Kant's view on existence.

We quote below from *Kritik der reinen Vernunft*, B626–B628. The precise passage was pointed out to us by Mark van Atten. The context of the quotation are the proofs of God's existence, according to which the concept God includes all perfections of realities, and that it is more perfect of more real to have the existence property than to have it not. In contrast to this, Kant claims that existence is no predicate and therefore cannot be part of the concept God. Kant writes:

"Sein ist offenbar kein reales Prädicat, d. i. ein Begriff von irgend etwas, was zu dem Begriffe eines Dinges hinzukommen könne. Es ist bloß die Position eines Dinges oder gewisser Bestimmungen an sich selbst.

Im logischen Gebrauche ist es lediglich die Copula eines Urtheils. Der Satz: Gott ist allmächtig, enthält zwei Begriffe, die ihre Objecte haben: Gott und Allmacht; das Wörtchen: ist, ist noch nicht ein Prädicat obenein, sondern nur das, was das Prädicat beziehungsweise aufs Subject setzt. Nehme ich nun das Subject (Gott) mit allen seinen Prädicaten (worunter auch die Allmacht gehört) zusammen und sage: Gott ist, oder es ist ein Gott, so setze ich kein neues Prädicat zum Begriffe von Gott, sondern nur das Subject an sich selbst mit allen seinen Prädicaten und zwar den Gegenstand in Beziehung auf meinen Begriff. Beide müssen genau einerlei enthalten, und es kann daher zu dem Begriffe, der bloß die Möglichkeit ausdrückt, darum daß ich dessen Gegenstand als schlechthin gegeben (durch den Ausdruck: er ist) denke, nichts weiter hinzukommen.

Und so enthält das Wirkliche nichts mehr als das bloß Mögliche. Hundert wirkliche Thaler enthalten nicht das Mindeste mehr, als hundert mögliche. Denn da diese den Begriff, jene aber den Gegenstand und dessen Position an sich selbst bedeuten, so würde, im Fall dieser mehr enthielte als jener, mein Begriff nicht den ganzen Gegenstand ausdrücken und also auch nicht der angemessene Begriff von ihm sein. Aber in meinem Vermögenszustande ist mehr bei hundert wirklichen Thalern, als bei dem bloßen Begriffe derselben (d. i. ihrer Möglichkeit). Denn der Gegenstand ist bei der Wirklichkeit nicht bloß in meinem Begriffe analytisch enthalten, sondern kommt zu meinem Begriffe (der eine Bestimmung meines Zustandes ist) synthetisch hinzu, ohne daß durch dieses Sein außerhalb meinem Begriffe diese gedachte hundert Thaler selbst im mindesten vermehrt werden.

Wenn ich also ein Ding, durch welche und wie viel Prädicate ich will, (selbst in der durchgängigen Bestimmung) denke, so kommt dadurch, da ich noch hinzusetze: dieses Ding ist, nicht das mindeste zu dem Dinge hinzu."

In the english translation[1]: "Being is evidently not a real predicate, that is, a conception of something which is added to the conception of some other thing. It is merely the positing of a thing, or of certain determinations in it. Logically, it is merely the copula of a judgement. The proposition, God is omnipotent, contains two conceptions, which have a certain object or content; the word is, is no additional predicate — it merely indicates the relation of the predicate to the subject. Now, if I take the subject (God)

---

[1] Translation by J.M.D. Meicklejohn 1969.

with all its predicates (omnipotence being one), and say: God is, or, There is a God, I add no new predicate to the conception of God, I merely posit or affirm the existence of the subject with all its predicates — I posit the object in relation to my conception. The content of both is the same; and there is no addition made to the conception, which expresses merely the possibility of the object, by my cogitating the object - in the expression, it is — as absolutely given or existing. Thus the real contains no more than the possible. A hundred real dollars contain no more than a hundred possible dollars. For, as the latter indicate the conception, and the former the object, on the supposition that the content of the former was greater than that of the latter, my conception would not be an expression of the whole object, and would consequently be an inadequate conception of it. But in reckoning my wealth there may be said to be more in a hundred real dollars than in a hundred possible dollars — that is, in the mere conception of them. For the real object — the dollars — is not analytically contained in my conception, but forms a synthetical addition to my conception (which is merely a determination of my mental state), although this objective reality — this existence — apart from my conceptions, does not in the least degree increase the aforesaid hundred dollars.

By whatever and by whatever number of predicates — even to the complete determination of it — I may cogitate a thing, I do not in the least augment the object of my conception by the addition of the statement: This thing exists."

As was pointed out to us by Mark van Atten, Brouwer certainly knew the passage above, as in a letter[2] to his PhD supervisor he mentions having read Kritik der reinen Vernunft very thoroughly.

The above expressed viewpoint however overlooks the fact, that (conditions of) existence can be described in various sometimes surprising ways and the corresponding concepts can be interrelated via intuitionistic logic. It is the same story as with termination, another manifestation of existence: termination of a computable function with respect to a given input is a mere fact independent of the logical viewpoint. To analyze however the notion of termination itself logical frameworks are needed to represent (conditions of) termination in a general setting: this step is necessary even to state the totality of a program, i.e. the termination with respect to all potential input values.

1.2 Section contents

The paper is build up as follows. In Section 2 the main proof system is introduced. In Section 3 it is proved that to have a form of cut-elimination, in

---

[2]A letter from 5.11.1906 to Diederik Korteweg

Section 5 it is shown that it satisfies interpolation and the Beth definability property. Section 4 discusses the relation of our Gentzen calculi to systems capturing E that have been considered before. In Section 6 semantics are introduced which in Section 7 are proved to be sound and complete with respect to the calculi. Section 8 contains the applications to Skolemization and truth-value logics mentioned above.

2 A Gentzen calculus

In this section we define the Gentzen calculus LJE, an extension of LJ for intuitionistic predicate logic extended by the existence predicate, that covers the intuition that Et means t *exists*. Hilbert type systems for the existence predicate were first introduced by D. Scott in [Scott, 1979]. Natural deduction formulations were given by M. Unterhalt in [Unterhalt, 1986]. The Gentzen calculi given below were first introduced by the authors in [Baaz and Iemhoff, 2005b]. The relation between these systems will be discussed in Section 4.

2.1 Preliminaries

Given an existence predicate, terms, including variables, typically range over existing as well as non-existing objects, while the quantifiers range over existing objects only. Proofs are assumed to be trees.

We consider languages $\mathcal{L} \subseteq \mathcal{L}'$ for intuitionistic predicate logic plus the existence predicate E, without equality. $E \in \mathcal{L}$ and $\mathcal{L}'_-$ denotes $\mathcal{L}'$ without E. For convenience we assume that $\mathcal{L}$ contains at least one constant and no variables, and that $\mathcal{L}'$ contains infinitely many variables. The reason for requiring that $\mathcal{L}$ contains at least one constant is given in Remark 2. We assume that the variables belong only to the bigger language because in our main system we will assume terms in $\mathcal{L}$ to exist, and assuming variables to exist would block the free substitution of terms for variables. Moreover, we think it the more natural approach to let free variables be as free as possible, no restrictions put on them. More on this in Section 4.

The languages contain $\bot$, and $\neg A$ is defined as $A \to \bot$. $A, B, C, D, E, ..$ range over formulas in $\mathcal{L}'$, $s, t, ..$ over terms in $\mathcal{L}'$. Γ, Δ, Π range over multisets of formulas in $\mathcal{L}'$. Sequents are expressions of the form $\Gamma \Rightarrow C$, where Γ is a finite multiset. A sequent is in $\mathcal{L}$ if all its formulas are in $\mathcal{L}$. And similarly for $\mathcal{L}'$. A formula is *closed* when it does not contain free variables. A sequent $\Gamma \Rightarrow C$ is closed if C and all formulas in Γ are closed. We often write Ax for $A(x)$.

In the final proof system ($\Rightarrow Et$) will hold for the terms in $\mathcal{L}$, but not necessarily for the terms in $\mathcal{L}'\backslash\mathcal{L}$. $\mathcal{T}_\mathcal{L}$ denotes the set of terms in $\mathcal{L}$, $\mathcal{F}_\mathcal{L}$ denotes the set of formulas in $\mathcal{L}$, $\mathcal{S}_\mathcal{L}$ denotes the set of sequents in $\mathcal{L}$, and similarly for $\mathcal{L}'$.

2.2 The system LJE

Ax $\Gamma, P \Rightarrow P$ (P atomic) $L\bot$ $\Gamma, \bot \Rightarrow C$

$$L\wedge \frac{\Gamma, A, B \Rightarrow C}{\Gamma, A \wedge B \Rightarrow C} \qquad\qquad R\wedge \frac{\Gamma \Rightarrow A \qquad \Gamma \Rightarrow B}{\Gamma \Rightarrow A \wedge B}$$

$$L\vee \frac{\Gamma, A \Rightarrow C \qquad \Gamma, B \Rightarrow C}{\Gamma, A \vee B \Rightarrow C} \qquad\qquad R\vee \frac{\Gamma \Rightarrow A_i}{\Gamma \Rightarrow A_0 \vee A_1} \; i = 0, 1$$

$$L\to \frac{\Gamma, A \to B \Rightarrow A \qquad \Gamma, B \Rightarrow C}{\Gamma, A \to B \Rightarrow C} \qquad\qquad R\to \frac{\Gamma, A \Rightarrow B}{\Gamma \Rightarrow A \to B}$$

$$L\forall \frac{\Gamma, \forall x Ax, At \Rightarrow C \qquad \Gamma, \forall x Ax \Rightarrow Et}{\Gamma, \forall x Ax \Rightarrow C} \qquad\qquad R\forall \frac{\Gamma, Ey \Rightarrow Ay}{\Gamma \Rightarrow \forall x A[x/y]} \; *$$

$$L\exists \frac{\Gamma, Ay, Ey \Rightarrow C}{\Gamma, \exists x Ax \Rightarrow C} \; * \qquad\qquad R\exists \frac{\Gamma \Rightarrow At \qquad \Gamma \Rightarrow Et}{\Gamma \Rightarrow \exists x A[x/t]}$$

$$\mathrm{Cut} \; \frac{\Gamma \Rightarrow A \qquad \Gamma, A \Rightarrow C}{\Gamma \Rightarrow C}$$

Where (∗) denotes the condition that y does not occur free in Γ and C.

We write LJE $\vdash$ S if the sequent S is derivable in LJE. For a set of sequents $\mathcal{S}$ and a sequent S, we say that S *is derivable from $\mathcal{S}$ in* LJE, and write $\mathcal{S} \vdash_{\mathsf{LJE}} S$, if S is derivable in LJE extended by axioms $\mathcal{S}$. We define

$$\mathsf{LJE}(\mathcal{S}) \equiv_{def} \{S \in \mathcal{S}_{\mathcal{L}'} \mid \mathcal{S} \vdash_{\mathsf{LJE}} S\}.$$

In the system LJE no existence of any term is assumed. This implies e.g. that we cannot derive $\Rightarrow \exists x Ex$. And we cannot derive $\forall x Px \Rightarrow Pt$ either, but only $\forall x Px, Et \Rightarrow Pt$. Note however that the former is derivable in LJE from ($\Rightarrow Et$). This is the reason why we consider derivations from extra axioms, especially axioms of the form ($\Rightarrow Et$). Therefore, we define the following sets of sequents

$$\Sigma_{\mathcal{L}} \equiv_{def} \{\Gamma \Rightarrow Et \mid t \in \mathcal{T}_{\mathcal{L}}, \Gamma \text{ a multiset}\}.$$

Note that for all sequents $\Gamma \Rightarrow Et$ in $\Sigma_{\mathcal{L}}$, t is a closed term, and that because of the assumptions on $\mathcal{L}$, $\Sigma_{\mathcal{L}}$ contains at least one sequent. Given two languages $\mathcal{L} \subseteq \mathcal{L}'$, we write

$$\mathsf{LJE}(\Sigma_{\mathcal{L}}) \equiv_{def} \{S \in \mathcal{S}_{\mathcal{L}'} \mid \Sigma_{\mathcal{L}} \vdash_{\mathsf{LJE}} \Rightarrow S\}.$$

The $\mathcal{L}'$ is not denoted in $\mathsf{LJE}(\Sigma_{\mathcal{L}})$, but most of the time it is clear what 'bigger' language $\mathcal{L}'$ of which $\mathcal{L}$ is a subset is.

We often write $\vdash_{\mathcal{L}}$ for $\vdash_{\mathsf{LJE}(\Sigma_{\mathcal{L}})}$.

EXAMPLE 1.

$$\nvdash_{\mathsf{LJE}} \Rightarrow \exists x Ex \qquad \vdash_{\mathsf{LJE}} \Rightarrow \forall x Ex.$$

$$\vdash_{\mathsf{LJE}(\Sigma_{\mathcal{L}})} \Rightarrow \exists x Ex \wedge \forall x Ex.$$

In Proposition 8 the relation between LJ and LJE is explained.

REMARK 2. The requirement that $\mathcal{L}$ contains at least one constant is needed to make the construction of the reduction trees work: see Definition 18, the remark at the case $R\forall$.

2.3 Uniqueness

Observe that given another predicate E' that satisfies the same rules of LJE as E', it follows that

$$\vdash_{\mathcal{L}} Et \Rightarrow E't \quad \text{and} \quad \vdash_{\mathcal{L}} E't \Rightarrow Et.$$

We namely have that $\vdash_{\mathcal{L}} (\Rightarrow \forall x Ex \wedge \forall x E'x)$, and also $\vdash_{\mathcal{L}} (\forall x Ex, E't \Rightarrow Et)$ and $\vdash_{\mathcal{L}} (\forall x E'x, Et \Rightarrow E't)$. Finally, two cuts do the trick. This shows that the existence predicate E is unique up to provable equivalence.

3 Cut elimination

We assume eigenvariables, free and bound variables to be three distinct sets of variables. The variable y in $L\exists$ and $R\forall$ is called an eigenvariable. The depth of a sequent in a proof is inductively defined as the sum of the depths of its upper sequents plus 1. Thus axioms have depth 1. The complexity $|C|$ of a formula is the number of occurrences of connectives and quantifiers in C. The rank of a cut is $1 +$ the complexity of the cut formula. The level of a cut is the sum of the depths of its two hypotheses. The cutrank $cr(P)$ of a proof P is the maximal rank of cuts in P. The depth of a proof, $dp(P)$, is the depth of its endsequent. We write $\mathsf{LJE} \vdash_d S$ when S has a proof of depth $\leq d$ in LJE, We write $\mathsf{LJE} \vdash^c S$ when S has a proof of cutrank $\leq c$. Similarly for $\mathsf{LJE}(\Sigma_{\mathcal{L}})$. For a proof P, $P[t/y]$ denotes the result of substituting t for y everywhere in P.

3.1   Substitution, Weakening and Contraction

We start with the substitution lemma.

LEMMA 3. *For* $\mathsf{L} \in \{\mathsf{LJE}(\Sigma_{\mathcal{L}}), \mathsf{LJE}\}$:
If P is a proof in L of a sequent S in $\mathcal{L}'$ in which y occurs free, and if t is a term in $\mathcal{L}'$ that does not contain eigenvariables or bound variables of P, then $P[t/y]$ is a proof of $S[t/y]$ in L. Moreover, $cr(P[t/y]) \leq cr(P)$ and $dp(P[t/y]) \leq dp(P)$.

Proof. We treat the case $\mathsf{L} = \mathsf{LJE}(\Sigma_{\mathcal{L}})$. We use induction to the depth d of P. Let $P' = P[t/y]$, $S' = S[t/y]$. First $d = 1$, the case that P is an instance of an axiom. The axioms Ax, $\mathsf{L}\bot$ in P are replaced by instances of the same axioms in P', so these will not be violated under the transformation. For axioms $\Pi \Rightarrow Es$ in $\Sigma_{\mathcal{L}}$ it follows that s is a closed term in $\mathcal{L}$. Hence the sequent that results from the substitution, $(\Pi[t/y] \Rightarrow Es)$, belongs to $\Sigma_{\mathcal{L}}$ too. This completes the case $d = 1$.

Suppose $d > 1$. First note that because eigenvariables are distinct from free variables in a proof, y cannot be an eigenvariable in P. We distinguish by cases according to the last rule in P. The connective rules and cuts in P are replaced by instances of the same rules in P', so these will not be violated under the transformation. Thus the quantifier rules remain.

Suppose the last inference in P is a quantifier rule. In the case of $\mathsf{L}\forall$ and $\mathsf{R}\exists$ there are no side conditions, whence these rules will not be violated in going from P to P'. We treat $\mathsf{R}\forall$, the case $\mathsf{L}\exists$ is similar. Consider an application of $\mathsf{R}\forall$ in P:

$$\frac{\begin{array}{c} P_1 \\ \Pi, Ez \Rightarrow Bz \end{array}}{\Pi \Rightarrow \forall u Bu}$$

Thus z is not free in Π, and $z \neq y$ and $u \neq y$, since y is no eigenvariable or bound variable. By assumption on t, u does not occur in t. Under the transformation this will become

$$\frac{\begin{array}{c} P_1[t/y] \\ \Pi[t/y], Ez \Rightarrow Bz[t/y] \end{array}}{\Pi[t/y] \Rightarrow \forall u Bu[t/y]}$$

To see that this a valid application of $\mathsf{R}\forall$, it suffices to see that z is not free in $\Pi[t/y]$, which is clear from the assumption on t.

To check that $cr(P') \leq cr(P)$ and $dp(P') \leq dp(P)$ is left to the reader. ∎

LEMMA 4. *For* $\mathsf{L} \in \{\mathsf{LJE}(\Sigma_{\mathcal{L}}), \mathsf{LJE}\}$: $\mathsf{L} \vdash^c_d \Gamma \Rightarrow C$ *implies* $\mathsf{L} \vdash^c_d \Gamma, A \Rightarrow C$.

Proof. Left to the reader. For the quantifier rules, use Lemma 3 to repair variable clashes. ∎

LEMMA 5. *For* $\mathsf{L} \in \{\mathsf{LJE}(\Sigma_{\mathcal{L}}), \mathsf{LJE}\}$: L *has contraction. In fact:*

$$\mathsf{L} \vdash^c_d \Gamma, A, A \Rightarrow C \quad \text{implies} \quad \mathsf{L} \vdash^c_d \Gamma, A \Rightarrow C \tag{1}$$

Proof. To show that the system has contraction we need the following claim.

CLAIM. For $d > 0$, it holds that

$$
\begin{aligned}
\mathsf{L} \vdash^c_d \Gamma, A \wedge B \Rightarrow C &\quad \text{implies} \quad \mathsf{L} \vdash^c_d \Gamma, A, B \Rightarrow C \\
\mathsf{L} \vdash^c_d \Gamma, A \vee B \Rightarrow C &\quad \text{implies} \quad \mathsf{L} \vdash^c_d \Gamma, A \Rightarrow C \text{ and } \mathsf{L} \vdash_d \Gamma, B \Rightarrow C \\
\mathsf{L} \vdash^c_d \Gamma \Rightarrow A \to B &\quad \text{implies} \quad \mathsf{L} \vdash^c_d \Gamma, A \Rightarrow B \\
\mathsf{L} \vdash^c_d \Gamma, \exists x A x \Rightarrow C &\quad \text{implies} \quad \mathsf{L} \vdash^c_d \Gamma, Ey, Ay \Rightarrow C, \text{ for all } y.
\end{aligned}
$$

Proof of Claim. The only detail here is the possibility of variable clashes. We only treat the case of the existential quantifier, with induction to d. If $\Gamma, \exists x A x \Rightarrow C$ is an axiom, then so is $\Gamma, Ey, Ay \Rightarrow C$. Suppose it it not an axiom. If in the last inference in the proof of $\Gamma, \exists x A x \Rightarrow C$, $\exists x A x$ is not principal, then the induction hypothesis applies: for the rules without eigenvariables this is immediate. For the rules with eigenvariables, if the eigenvariable is y, we just replace it by a fresh eigenvariable not occurring in the proof, and then using the induction hypothesis we obtain a proof of $\Gamma, Ey, Ay \Rightarrow C$ of same rank and depth. If $\exists x A x$ is principal in the last rule, the result follows immediately. This proofs the claim.

Using this claim we prove (1) with induction to the depth d of the proof of $\Gamma, A, A \Rightarrow C$ in L. If $d = 1$, the sequent is an axiom, and so $\Gamma, A \Rightarrow C$ clearly is an axiom too (also in the case of $\Sigma_{\mathcal{L}}$). Consider the case $d + 1$. If the last rule in the proof is a right rule or the principal formula is in Γ, then the induction hypothesis applies. Therefore, suppose it is a left rule and the principal formula is not in Γ. We distinguish by cases. We treat $\mathsf{L} \wedge$ and leave the other cases to the reader. In this case the last part of the proof then looks as follows.

$$
\begin{array}{c}
\vdots \\
\hline
\Gamma, A \wedge B, A, B \Rightarrow C \\
\hline
\Gamma, A \wedge B, A \wedge B \Rightarrow C
\end{array}
$$

Assume the cutrank of the proof is n. Let P be the proof of $\Gamma, A \wedge B, A, B \Rightarrow C$. Note that P has depth d. Thus we can apply the claim and obtain a proof of $\Gamma, A, B, A, B \Rightarrow C$ of depth $\leq d$ and cutrank $\leq n$. Then we apply the induction hypothesis, first to A and then to B, and obtain a proof of $\Gamma, A, B \Rightarrow C$ of depth $\leq d$ and cutrank $\leq n$. An application of L$\wedge$ provides a proof of $\Gamma, A \wedge B \Rightarrow C$ of depth $\leq d+1$ and cutrank $\leq n$, as desired. ∎

3.2   Restriction to Ecuts

THEOREM 6.   *For* $\mathsf{L} \in \{\mathsf{LJE}(\Sigma_\mathcal{L}), \mathsf{LJE}\}$:
Every sequent in $\mathcal{L}'$ *provable in* L *has a proof in* L *in which the only cuts are instances of the ECut rule:*

$$ECut: \quad \frac{\Gamma \Rightarrow Et \in \Sigma_\mathcal{L} \qquad \Gamma, Et \Rightarrow C}{\Gamma \Rightarrow C}$$

In particular, LJE *has cut-elimination.*

Proof. For a smooth induction it is convenient to replace the Cut rule in LJE by the following generalization of it, the so-called *Mix rule*:

$$\text{Mix} \quad \frac{\Gamma \Rightarrow A \qquad \Gamma', A \Rightarrow C}{\Gamma\Gamma' \Rightarrow C}$$

In the Mix rule A is called the cutformula. When we speak about cuts in a proof, we refer to instances of the Cut or the Mix rule. The notions of cutrank are extended to proofs with the Mix rule in the obvious way. To prove the theorem we then show that applications of Mix can be removed from a proof, unless they are instances of EMix, which is

$$\text{EMix:} \quad \frac{\Gamma \Rightarrow Et \in \Sigma_\mathcal{L} \qquad \Gamma', Et \Rightarrow C}{\Gamma\Gamma' \Rightarrow C}$$

Note that this indeed implies that all provable sequents have a proof in which the only cuts are instances of ECut: $\Gamma \Rightarrow Et \in \Sigma_\mathcal{L}$ implies $\Gamma' \Rightarrow Et \in \Sigma_\mathcal{L}$ for all Γ', and thus the conclusion of the EMix as above can be obtained also via the ECut

$$\frac{\Gamma\Gamma' \Rightarrow Et \in \Sigma_\mathcal{L} \qquad \Gamma\Gamma', Et \Rightarrow C}{\Gamma\Gamma' \Rightarrow C}$$

For now, we call a proof *ecutfree* if all applications of Mix are instances of EMix, and we call it *cutfree* when it contains no cuts at all. Recall that the cutrank $cr(P)$ of a proof P is $1 +$ the maximal complexity of cutformulas in P.

The proof of the theorem consists of two claims. The first shows how to remove cuts of rank > 1 from a proof, and the second shows how cuts of rank 1 that are not instances of EMix can be removed from a proof. These two claims together imply the theorem.

CLAIM. For $\mathsf{L} \in \{\mathsf{LJE}(\Sigma_{\mathcal{L}}), \mathsf{LJE}\}$: Every sequent in $\mathcal{L}'$ provable in L has a proof in L in which all cuts have rank 1.

Proof of Claim. We treat the case $\mathsf{LJE}(\Sigma_{\mathcal{L}})$, the case LJE is similar. It suffices to show that a proof P ending in a cut

$$
\frac{\begin{array}{cc} P_1 & P_2 \\ \Gamma \Rightarrow A & \Gamma', A \Rightarrow C \end{array}}{\Gamma\Gamma' \Rightarrow C}
$$

with $|A| > 0$ and with $cr(P_1), cr(P_2) \leq |A|$, can be transformed into a proof P' of $\Gamma\Gamma' \Rightarrow C$ such that $cr(P') < cr(P)$. Note that $cr(P) = |A| + 1 > 1$. We prove this by induction on the cutrank of P with a sub induction to the level of the lowest cut of maximal rank in P (the level of a cut is the sum of the depths of its two hypotheses). We call $\Gamma \Rightarrow A$ and $\Gamma', A \Rightarrow C$ the hypotheses of the cut and $\Gamma\Gamma' \Rightarrow C$ the conclusion. Since $|A| > 0$, A cannot be principal in an axiom, including $\Sigma_{\mathcal{L}}$. Note also that A cannot be of the form Et. Therefore, we only have to distinguish the following two cases:
(a) the cutformula is not principal in one of the hypotheses,
(b) cutformula is principal in both hypotheses, which are not axioms.

(a) Suppose the cut formula is not principal in one of the hypotheses. If this hypothesis is an instance of axioms Ax or $\mathsf{L}\bot$, then so is the conclusion of the cut, and whence we have a cutfree proof of it. If this hypothesis is an instance of an axiom $\Gamma \Rightarrow Et$ in $\Sigma_{\mathcal{L}}$, then since $|A| > 0$ it has to be the right hypothesis. Observe that $(\Gamma \Rightarrow Et) \in \Sigma_{\mathcal{L}}$, implies that $(\Pi \Rightarrow Et) \in \Sigma_{\mathcal{L}}$ for all Π. Hence the conclusion of the cut is a sequent in $\Sigma_{\mathcal{L}}$, in which case we have a cutfree proof of it.

Next suppose that the hypothesis in which A is not principal is the lower sequent of an application of one of the rules. In this case we can cut higher up. That is, suppose the cutformula is not principal in the left hypothesis, and assume this is a two hypotheses rule R, say $R\vee$. Then P looks as follows.

$$\text{R}\lor \quad \frac{\dfrac{P_1 \qquad\quad P_2}{\dfrac{\Gamma_1 \Rightarrow A \qquad \Gamma_2 \Rightarrow A}{\Gamma \Rightarrow A}} \qquad \dfrac{P_3}{\Gamma', A \Rightarrow C}}{\Gamma\Gamma' \Rightarrow C}$$

Note that by assumption $cr(P_i) < cr(P)$ for $i = 1, 2, 3$. Then we transform the proof into a proof P' as follows.

$$\text{R} \quad \frac{\dfrac{P_1 \qquad\qquad P_3}{\dfrac{\Gamma_1 \Rightarrow A \qquad \Gamma', A \Rightarrow C}{\Gamma_1\Gamma' \Rightarrow C}} \qquad \dfrac{P_2 \qquad\qquad P_3}{\dfrac{\Gamma_2 \Rightarrow A \qquad \Gamma', A \Rightarrow C}{\Gamma_2\Gamma' \Rightarrow C}}}{\Gamma\Gamma' \Rightarrow C}$$

Now we have two cuts on A, but the level of the lowest cut of maximal rank in P' is one of these cuts. Thus $cr(P') = cr(P)$, but the level of the lowest cut of maximal rank in P' is smaller than the level of the lowest cut of maximal rank in P. Therefore, we can apply the induction hypothesis and are done. The other cases are similar. Note that in the case that R is a cut, it is by assumption a cut of rank $< |A| + 1$. Hence also in this case the induction hypothesis applies to P'.

(b) In this case the cut is principal in both hypotheses, and both hypotheses are not axioms. We distinguish by cases according to the outermost logical symbol in A: the cases $\land$, $\lor$, $\rightarrow$ are treated in the same way as in the case of LJ, see e.g. [Troelstra and Schwichtenberg, 1996]. We treat the quantifiers.

$\forall$: then P looks as follows:

$$\frac{\dfrac{P_1}{\dfrac{\Gamma, Ey \Rightarrow Ay}{\Gamma \Rightarrow \forall xAx}}\,d_1 \qquad \dfrac{\dfrac{P_2}{\Gamma', \forall xAx, At \Rightarrow C} \qquad \dfrac{P_3}{\Gamma', \forall xAx \Rightarrow Et}}{\Gamma', \forall xAx \Rightarrow C}\,d_2}{\Gamma\Gamma' \Rightarrow C}$$

Note that y is not free in Γ because of the conditions on $R\forall$, and y is not free in Γ', C and t because of the conditions on eigenvariables in a proof. By assumptions on variables, t does not contain eigenvariables or bound variables in P.

We can transform the above proof into the following proof P':

$$\frac{\dfrac{\dfrac{\dfrac{P_1}{\dfrac{\Gamma, Ey \Rightarrow Ay}{\Gamma \Rightarrow \forall xAx}} \qquad \dfrac{P_3}{\Gamma', \forall xAx \Rightarrow Et}}{\Gamma\Gamma' \Rightarrow Et} \qquad \dfrac{P_1[t/y]}{\Gamma, Et \Rightarrow At}}{\Gamma\Gamma\Gamma' \Rightarrow At} \qquad \dfrac{\dfrac{\dfrac{P_1}{\Gamma, Ey \Rightarrow Ay}}{\Gamma \Rightarrow \forall xAx} \qquad \dfrac{P_2}{\Gamma', \forall xAx, At \Rightarrow C}}{\Gamma\Gamma', At \Rightarrow C}}{\Gamma\Gamma\Gamma'\Gamma' \Rightarrow C}$$

Note that the endsequent of $P_1[t/y]$ indeed is $\Gamma, Et \Rightarrow At$ as y is not free Γ. By Lemma 3, $P_1[t/y]$ is a proof of $(\Gamma, Et \Rightarrow At)$ in $\mathsf{LJE}(\Sigma_{\mathcal{L}})$ such that $cr(P_1[t/y]) \leq cr(P_1) < cr(P)$. The cuts on $\forall x Ax$ both have a lower level and the same rank as in P. Therefore, we can apply the induction hypothesis and obtain proofs of their conclusions of cutrank $< cr(P)$. Whence there is a proof of $\Gamma\Gamma\Gamma'\Gamma' \Rightarrow C$ of cutrank $< cr(P)$. Application of some contractions, Lemma 5, gives a proof of $\Gamma\Gamma' \Rightarrow C$ of cutrank $< cr(P)$. This proves the case $\forall$.

$\exists$: Similar. Here P looks as follows:

$$
\frac{
\dfrac{\begin{array}{cc} P_1 & P_2 \\ \Gamma \Rightarrow At & \Gamma \Rightarrow Et \end{array}}{\Gamma \Rightarrow \exists x Ax}
\qquad
\dfrac{\begin{array}{c} P_3 \\ \Gamma', Ey, Ay \Rightarrow C \end{array}}{\Gamma', \exists x Ax \Rightarrow C}
}{\Gamma\Gamma' \Rightarrow C}
$$

Because of the side condition that y is not free in Γ' and C we can transform this proof into the following proof P':

$$
\frac{
\dfrac{\begin{array}{c} P_1 \\ \Gamma \Rightarrow At \end{array}}{}
\qquad
\dfrac{\begin{array}{cc} P_2 & P_3[t/y] \\ \Gamma \Rightarrow Et & \Gamma', Et, At \Rightarrow C \end{array}}{\Gamma\Gamma', At \Rightarrow C}
}{\Gamma\Gamma\Gamma' \Rightarrow C}
$$

By Lemma 3, $cr(P_3[t/y]) \leq cr(P_3)$. Thus $cr(P') < cr(P)$. This completes (b) and thereby the proof of the claim.

CLAIM. For $\mathsf{L} \in \{\mathsf{LJE}(\Sigma_{\mathcal{L}}), \mathsf{LJE}\}$: Every sequent in $\mathcal{L}'$ that has a proof in L of cutrank 1, has a proof in L in which all cuts are instances of EMix.

Proof of Claim. We treat the case $\mathsf{LJE}(\Sigma_{\mathcal{L}})$. We use induction to the depth d of a proof P of cutrank ≤ 1 of a sequent S. The case $d = 1$ is trivial, as then P consists of an axiom only. Suppose $d > 1$. If the last inference in P is not a cut or it is an application of EMix, we can apply the induction hypothesis and are done. Therefore, suppose P ends in a cut that is not an instance of EMix:

$$
\frac{\begin{array}{cc} P_1 & P_2 \\ \Gamma \Rightarrow A & \Gamma', A \Rightarrow C \end{array}}{\Gamma\Gamma' \Rightarrow C} \; d
$$

Thus by the induction hypothesis P_1 and P_2 are ecutfree, i.e. all cuts they contain are instances of EMix. And as P has cutrank ≤ 1, A is atomic or $\bot$ or of the form Et. Denote $\Gamma\Gamma' \Rightarrow C$ by S. We distinguish the following cases:

(c) the cutformula is principal in the rigth hypothesis,

(d) the cutformula is not principal in the right hypothesis.

(c) Assume the cutformula is principal in the right hypothesis. The form of A implies that whence the right hypothesis $\Gamma', A \Rightarrow C$ has to be an axiom. Since A is principal in it, $C = A$ or $A = \bot$. In the former case we can obtain a ecutfree proof of S by weakening the sequent $\Gamma \Rightarrow A$. If $A = \bot$, then it follows that either $\bot \in \Gamma$ or A is not principal in the left hypothesis. In the former case S is an instance of $L\bot$ and we are done. In the latter case, since A is not principal in it, $\Gamma \Rightarrow \bot$ is the conclusion of a rule R in which $\bot$ is not principal. In this case one can cut higher up, like in case (b) in the proof of the first claim: we treat the case that R is an EMix, and leave the other cases to the reader. In this case P looks as follows.

$$\frac{\dfrac{\Gamma \Rightarrow Et \in \Sigma_{\mathcal{L}} \qquad \overset{\displaystyle P_1}{\Gamma', Et \Rightarrow \bot}}{\Gamma\Gamma' \Rightarrow \bot} \qquad \Gamma'', \bot \Rightarrow C}{\Gamma\Gamma'\Gamma'' \Rightarrow C}$$

We transform this proof into the proof P':

$$\frac{\Gamma \Rightarrow Et \in \Sigma_{\mathcal{L}} \qquad \dfrac{\overset{\displaystyle P_1}{\Gamma', Et \Rightarrow \bot} \qquad \Gamma'', \bot \Rightarrow C}{\Gamma'\Gamma'', Et \Rightarrow \bot}}{\Gamma\Gamma'\Gamma'' \Rightarrow \bot}$$

We apply the induction hypothesis to P' and are done.

(d) Assume the cutformula is not principal in the right hypothesis. If $\Gamma', A \Rightarrow C$ is an axiom, then $\bot \in \Gamma'$, $C \in \Gamma'$ or $C = Et$ for some $t \in \mathcal{T}_{\mathcal{L}}$. In all cases S is an instance of the same axiom. If the right hypothesis is an application of a rule R we proceed as follows. We treat the cases that R is a two hypothesis rule that is not a cut, and the case that it is a cut, and leave the other cases to the reader. First, suppose R is not a cut. Then P looks as follows.

$$\frac{\overset{\displaystyle P_1}{\Gamma \Rightarrow A} \qquad \dfrac{\overset{\displaystyle P_2}{\Gamma_1, A \Rightarrow C_1} \qquad \overset{\displaystyle P_3}{\Gamma_2, A \Rightarrow C_2}}{\Gamma', A \Rightarrow C}\ R}{\Gamma\Gamma' \Rightarrow C}$$

Note that by the induction hypothesis the P_i are ecutfree. Then we transform the proof into a proof P' as follows.

$$
\frac{
\begin{array}{cc}
P_1 & P_2 \\
\Gamma \Rightarrow A \quad \Gamma_1, A \Rightarrow C_1 \\
\hline
\Gamma\Gamma_1 \Rightarrow C_1
\end{array}
\qquad
\begin{array}{cc}
P_1 & P_3 \\
\Gamma \Rightarrow A \quad \Gamma_2, A \Rightarrow C_2 \\
\hline
\Gamma, \Gamma_2 \Rightarrow C_2
\end{array}
}{\Gamma\Gamma' \Rightarrow C} \text{R}
$$

Since R is not a cut we can apply the induction hypothesis to P' and are done.

Finally, we treat the case that R is a cut. By the induction hypothesis it is an instance of EMix. Hence P looks like this:

$$
\frac{
\begin{array}{c}
P_1 \\
\Gamma \Rightarrow A
\end{array}
\qquad
\frac{
\Gamma', A \Rightarrow Et \ \in \ \Sigma_{\mathcal{L}} \qquad
\begin{array}{c}
P_2 \\
\Gamma'', Et, A \Rightarrow C
\end{array}
}{\Gamma'\Gamma'', A \Rightarrow C}
}{\Gamma\Gamma'\Gamma'' \Rightarrow C}
$$

Then we transform the proof into a proof P' as follows:

$$
\frac{
\Gamma' \Rightarrow Et \ \in \ \Sigma_{\mathcal{L}} \qquad
\frac{
\begin{array}{cc}
P_1 & P_2 \\
\Gamma \Rightarrow A \quad \Gamma'', Et, A \Rightarrow C
\end{array}
}{\Gamma\Gamma'', Et \Rightarrow C}
}{\Gamma\Gamma'\Gamma'' \Rightarrow C}
$$

To see that this is indeed a proof, note that $(\Gamma', A \Rightarrow Et) \in \Sigma_{\mathcal{L}}$ implies $t \in \mathcal{T}_{\mathcal{L}}$, which implies $(\Gamma' \Rightarrow Et) \in \Sigma_{\mathcal{L}}$. Now the induction hypothesis applies to P', and we are done. This proves the second claim.

As explained above, the two claims imply the theorem. ∎

COROLLARY 7. $\mathsf{LJE}(\Sigma_{\mathcal{L}})$ *is consistent.*

The cut elimination theorem allows us to proof the following correspondence between LJ and $\mathsf{LJE}(\Sigma_{\mathcal{L}})$, one direction of which has already been proved above.

PROPOSITION 8. *For every sequent S in $\mathcal{L}$ not containing E:*

$$\mathsf{LJ} \vdash S \text{ if and only if } \mathsf{LJE}(\Sigma_{\mathcal{L}}) \vdash S.$$

Proof. The direction from right to left: show with induction to the depth of the proof that for Γ and A not containing E, if $Et_1, \ldots, Et_n, \Gamma \Rightarrow A$ is derivable in $\mathsf{LJE}(\Sigma_{\mathcal{L}})$ by a proof in which all cuts are instances of ECut, then $\Gamma \Rightarrow A$ is derivable in LJ. We leave the other direction to the reader. ∎

PROPOSITION 9. *For quantifier free closed sequents the relations* $\vdash_{\mathsf{LJE}}$ *and* $\vdash_{\mathcal{L}}$ *are decidable.*

Proof. Show, using the theorem on ECuts above, that when t is a term that does not occur in a quantifier free sequent $\Gamma \Rightarrow C$, not even as a subterm, then

$$\vdash_{\mathcal{L}} \Gamma, Et \Rightarrow C \text{ implies } \vdash_{\mathcal{L}} \Gamma \Rightarrow C,$$

and similarly for LJE.   ∎

4   IQCE and IQCE$^+$

As remarked above, given an existence predicate, closed terms typically range over existing as well as non-existing elements, while quantifiers range over existing objects only. As to the choice of the domain for the variables, there have been different approaches. Scott in [Scott, 1979] introduces a system IQCE for the predicate language with the distinguished predicate E, in which variables range over all objects, like in LJE and LJE$(\Sigma_{\mathcal{L}})$. On the other hand, Beeson in [Beeson, 1985] discusses a system in which variables range over existing objects only.

The formulation of the system IQCE in [Scott, 1979], where logic with an existence predicate was first introduced was in Hilbert style, where the axioms and rules for the quantifiers were the following:

$$\forall x Ax \wedge Et \rightarrow At \qquad \frac{\vdots \atop B \wedge Ey \rightarrow Ay}{B \rightarrow \forall x Ax} *$$

$$\frac{\vdots \atop Ay \wedge Ey \rightarrow B}{\exists x Ax \rightarrow B} * \qquad At \wedge Et \rightarrow \exists x Ax$$

Here $*$ are the usual side conditions on the eigenvariable y.

The following formulation of IQCE in natural deduction style was first given in [Unterhalt, 1986]. We recall the system as given in [Troelstra and van Dalen, 1988]. We call the system NDE (Natural Deduction Existence). It consists of the axioms and quantifier rules of the standard natural deduction formulation of IQC (as e.g. given in [Troelstra and van Dalen, 1988]), where the quantifier rules are replaced by the following rules:

$$\forall I \ \frac{\begin{array}{c}[Ey]\\ \vdots\\ Ay\end{array}}{\forall x A x}* \qquad \forall E \ \frac{\begin{array}{cc}\vdots & \vdots\\ \forall x A x & Et\end{array}}{At}$$

$$\exists I \ \frac{\begin{array}{cc}At & Et\end{array}}{\exists x A x} \qquad \exists E \ \frac{\begin{array}{cc}\vdots & [Ay][Ey]\\ & \vdots\\ \exists x A x & C\end{array}}{C}*$$

Again, the $*$ are the usual side conditions on the eigenvariable y. It is easy to see that the following holds.

FACT 10. $\forall A \in \mathcal{F}_{\mathcal{L}'}$: $\vdash_{\mathsf{IQCE}} A$ if and only if $\vdash_{\mathsf{NDE}} A$ if and only if $\vdash_{\mathsf{LJE}} \Rightarrow A$.

Existence logic in which terms range over all object while quantifiers and variables only range over existing objects is denoted by IQCE^+ and has e.g. been used by M. Beeson in [Beeson, 1985]. The logic is the result of leaving out Ey in the two rules for the quantifiers in IQCE given above and adding Ex as axioms for all variables x. A formulation in natural deduction style is obtained from NDE by replacing the $\forall I$ and $\exists E$ by their standard formulations for IQC and adding Ex as axioms for all variables x. We call the system NDE^+. There are some details concerning substitutions for these systems, but we will not discuss them here, but only remark in how far these systems are equivalent:

FACT 11. $\forall A \in \mathcal{F}_{\mathcal{L}}$:
$\vdash_{\mathsf{IQCE}^+} A$ iff $\vdash_{\mathsf{NDE}^+} A$ iff $\{\Gamma \Rightarrow Ex \mid x$ a variable, Γ a multiset$\} \vdash_{\mathsf{LJE}(\Sigma_{\mathcal{L}})} \Rightarrow A$.

M. Unterhalt in [Unterhalt, 1986] thoroughly studied the Kripke semantics of these logics and proved respectively completeness and strong completeness for the systems IQCE and IQCE^+. Section 6 discusses his and our completeness results.

5 Interpolation

Recall that we say that a single conclusion Gentzen calculus L has *interpolation* if whenever $\mathsf{L} \vdash \Gamma_1, \Gamma_2 \Rightarrow C$, there exists an I in the common language of Γ_1 and $\Gamma_2 \cup \{C\}$ such that

$$\Gamma_1 \vdash_{\mathsf{L}} I \text{ and } I, \Gamma_2 \vdash_{\mathsf{L}} C.$$

In the context of existence logics, the *common language* of two multisets Γ_1 and Γ_2, denoted by $\mathcal{L}(\Gamma_1, \Gamma_2)$, consists of all variables, $\top$, $\bot$ and E, and all predicates and non-variable terms that occur both in Γ_1 and Γ_2.

We say that a Gentzen calculus L satisfies the *Beth definability property* if whenever $A(R)$ is a formula with R an n-ary relation symbol in a language $\mathcal{L}$, and R', R'' are two relation symbols not in $\mathcal{L}$ such that

$$\mathsf{L} \vdash A(R') \wedge A(R'') \Rightarrow \forall \bar{x}(R'\bar{x} \ \leftrightarrow \ R''\bar{x}),$$

then there is a formula S in $\mathcal{L}$ such that

$$\mathsf{L} \vdash \Rightarrow \forall \bar{x}(S\bar{x} \ \leftrightarrow \ R'\bar{x}).$$

In this section we prove that the calculus LJE and $\mathsf{LJE}(\Sigma_{\mathcal{L}})$ have interpolation. To this end we use a calculus LJE' that is equivalent to LJE but in which the structural rules are not hidden.

The system LJE'

Ax $\Gamma, P \Rightarrow P$ (P atomic) $L\bot$ $\Gamma, \bot \Rightarrow C$

$$\mathrm{LW} \ \frac{\Gamma \Rightarrow C}{\Gamma, A \Rightarrow C} \qquad\qquad \mathrm{LC} \ \frac{\Gamma, A, A \Rightarrow C}{\Gamma, A \Rightarrow C}$$

$$\mathrm{L}\wedge \ \frac{\Gamma, A, B \Rightarrow C}{\Gamma, A \wedge B \Rightarrow C} \qquad\qquad \mathrm{R}\wedge \ \frac{\Gamma \Rightarrow A \qquad \Gamma \Rightarrow B}{\Gamma \Rightarrow A \wedge B}$$

$$\mathrm{L}\vee \ \frac{\Gamma, A \Rightarrow C \qquad \Gamma, B \Rightarrow C}{\Gamma, A \vee B \Rightarrow C} \qquad\qquad \mathrm{R}\vee \ \frac{\Gamma \Rightarrow A_i}{\Gamma \Rightarrow A_0 \vee A_1} \ i = 0, 1$$

$$\mathrm{L}{\rightarrow} \ \frac{\Gamma \Rightarrow A \qquad \Gamma, B \Rightarrow C}{\Gamma, A \rightarrow B \Rightarrow C} \qquad\qquad \mathrm{R}{\rightarrow} \ \frac{\Gamma, A \Rightarrow B}{\Gamma \Rightarrow A \rightarrow B}$$

$$\mathrm{L}\forall \ \frac{\Gamma, At \Rightarrow C \qquad \Gamma \Rightarrow Et}{\Gamma, \forall x Ax \Rightarrow C} \qquad\qquad \mathrm{R}\forall \ \frac{\Gamma, Ey \Rightarrow Ay}{\Gamma \Rightarrow \forall x A[x/y]} \ *$$

$$\mathrm{L}\exists \ \frac{\Gamma, Ay, Ey \Rightarrow C}{\Gamma, \exists x A[x/y] \Rightarrow C} \ * \qquad\qquad \mathrm{R}\exists \ \frac{\Gamma \Rightarrow At \qquad \Gamma \Rightarrow Et}{\Gamma \Rightarrow \exists x Ax}$$

$$\mathrm{ECut:} \ \frac{\Gamma \Rightarrow Et \in \Sigma_{\mathcal{L}} \qquad \Gamma, Et \Rightarrow C}{\Gamma \Rightarrow C}$$

The calculus $\mathsf{LJE}'(\Sigma_{\mathcal{L}})$ is the system LJE' extended by the axioms $\Sigma_{\mathcal{L}}$ (Section 2).

LEMMA 12. *For al formulas A in $\mathcal{L}'$:*

$$\mathsf{LJE} \vdash A \;\Leftrightarrow\; \mathsf{LJE}' \vdash A \qquad \mathsf{LJE}(\Sigma_{\mathcal{L}}) \vdash A \;\Leftrightarrow\; \mathsf{LJE}'(\Sigma_{\mathcal{L}}) \vdash A.$$

Proof. Use Theorem 6 and Lemma's 4 and 5. ■

Recall that we write $\mathcal{L}(\Gamma_1, \Gamma_2)$ for the common language of Γ_1 and Γ_2, i.e. the language consisting of the predicates and non-variable terms that occur both in Γ_1 and Γ_2, plus $\top$, $\bot$ and E and the variables.

THEOREM 13. LJE' *and* $\mathsf{LJE}'(\Sigma_{\mathcal{L}})$ *have interpolation.*

Proof. We first prove the theorem for LJE' and then for $\mathsf{LJE}'(\Sigma_{\mathcal{L}})$ by showing how this case can be reduced to the LJE' case. We write $\vdash$ for $\vdash_{\mathsf{LJE}'}$ in this proof. Assume $\vdash \Gamma_1, \Gamma_2 \Rightarrow C$. We look for a formula I in the common language $\mathcal{L}(\Gamma_1, \Gamma_2 \cup \{C\})$ of Γ_1 and $\Gamma_2 \cup \{C\}$ such that

$$(2) \quad \vdash \Gamma_1 \Rightarrow I \quad \vdash I, \Gamma_2 \Rightarrow C.$$

We prove the theorem with induction to the depth d of P. Recall that the depth of a sequent in a proof is inductively defined as the sum of the depths of its upper sequents plus 1. Thus axioms have depth 1. The depth of a proof is the depth of its endsequent.

$d = 1$: P is an instance of an axiom. When the axiom is Ax we have $\Gamma_1 \Gamma_2, Q \Rightarrow Q$, where Q is an atomic formula. There are two cases: we look for interpolants I and J such that

$$\vdash \Gamma_1, Q \Rightarrow I \quad \vdash I, \Gamma_2 \Rightarrow Q \quad \text{and} \quad \vdash \Gamma_1 \Rightarrow J \quad \vdash J, Q, \Gamma_2 \Rightarrow Q.$$

This case is trivial: take $I = Q$ and $J = \top$. The case that P is an instance of $L\bot$ is equally simple: again there are two possibilities, like above, and the interpolants are $\top$ and $\bot$.

$d > 1$. We distinguish by cases according to the last rule applied in P. If it is a LC, P looks as follows.

$$\vdots$$
$$\frac{\Gamma_1 \Gamma_2, A, A \Rightarrow C}{\Gamma_1 \Gamma_2, A \Rightarrow C}$$

Again there are several cases: we look for interpolants

$$\vdash \Gamma_1, A \Rightarrow I \quad \vdash I, \Gamma_2 \Rightarrow C \quad \text{and} \quad \vdash \Gamma_1 \Rightarrow J \quad \vdash J, A, \Gamma_2 \Rightarrow C.$$

By the induction hypothesis there are interpolants I' and J' such that the sequents $\Gamma_1, A, A \Rightarrow I'$ and $I', \Gamma_2 \Rightarrow C$, and $\Gamma_1 \Rightarrow J'$ and $J', A, A, \Gamma_2 \Rightarrow C$ are derivable. Moreover, I' is in $\mathcal{L}(\Gamma_1 \cup \{A\}, \Gamma_2 \cup \{C\})$, and J' is in $\mathcal{L}(\Gamma_1, \Gamma_2 \cup \{A, C\})$. Hence taking $I = I'$ and $J = J'$ and applying contraction gives the desired result. The case LW is equally trivial.

The connective cases are completely straightforward. For completeness sake we treat $\rightarrow$. Suppose the last rule is L$\rightarrow$:

$$\frac{\vdots \qquad\qquad \vdots}{\Gamma_1\Gamma_2 \Rightarrow A \qquad \Gamma_1\Gamma_2, B \Rightarrow C}{\Gamma_1\Gamma_2, A \rightarrow B \Rightarrow C}$$

We have to find $I \in \mathcal{L}(\Gamma_1 \cup \{A \rightarrow B\}, \Gamma_2 \cup \{C\})$ and $J \in \mathcal{L}(\Gamma_1, \Gamma_2 \cup \{A \rightarrow B, C\})$ such that

$$\vdash \Gamma_1, A \rightarrow B \Rightarrow I \quad \vdash I, \Gamma_2 \Rightarrow C \quad \text{and} \quad \vdash \Gamma_1 \Rightarrow J \quad \vdash J, A \rightarrow B, \Gamma_2 \Rightarrow C.$$

For I, note that by the induction hypothesis there are $I' \in \mathcal{L}(\Gamma_2, \Gamma_1 \cup \{A\})$ and $J' \in \mathcal{L}(\Gamma_1 \cup \{B\}, \Gamma_2 \cup \{C\})$ such that

$$(3) \quad \vdash \Gamma_2 \Rightarrow I' \quad \vdash I', \Gamma_1 \Rightarrow A \quad \text{and} \quad \vdash \Gamma_1, B \Rightarrow J' \quad \vdash J', \Gamma_2 \Rightarrow C.$$

By applying LW with I' to the 3rd sequent in (3), then applying L$\rightarrow$ to this and the 2nd sequent, and then applying R$\rightarrow$, gives

$$\vdash \Gamma_1, A \rightarrow B \Rightarrow I' \rightarrow J'.$$

Applying L$\rightarrow$ to the 1st and 4th sequents in (3) gives $I' \rightarrow J', \Gamma_2 \Rightarrow C$. Hence we can take $I = I' \rightarrow J'$. Note that I indeed is in the common language $\mathcal{L}(\Gamma_1 \cup \{A \rightarrow B\}, \Gamma_2 \cup \{C\})$.

For J, observe that by the induction hypothesis there are $I'' \in \mathcal{L}(\Gamma_1, \Gamma_2 \cup \{A\})$ and $J'' \in \mathcal{L}(\Gamma_1, \Gamma_2 \cup \{B, C\})$ such that

$$(4) \quad \vdash \Gamma_1 \Rightarrow I'' \quad \vdash I'', \Gamma_2 \Rightarrow A \quad \text{and} \quad \vdash \Gamma_1 \Rightarrow J'' \quad \vdash J'', B, \Gamma_2 \Rightarrow C.$$

From this it follows that

$$\vdash \Gamma_1 \Rightarrow I'' \wedge J'' \quad \vdash I'' \wedge J'', A \rightarrow B, \Gamma_2 \Rightarrow C.$$

Hence we can take $J = I'' \wedge J''$ in this case. Note that J indeed is in the common language $\mathcal{L}(\Gamma_1, \Gamma_2 \cup \{A \rightarrow B, C\})$.

Suppose the last rule is R$\rightarrow$:

$$\frac{\vdots}{\Gamma_1\Gamma_2, A \Rightarrow B}{\Gamma_1\Gamma_2 \Rightarrow A \rightarrow B}$$

By the induction hypothesis there is a interpolant $I \in \mathcal{L}(\Gamma_1, \Gamma_2 \cup \{A, B\})$ for the upper sequent: $\vdash \Gamma_1 \Rightarrow I$ and $\vdash I, A, \Gamma_2 \Rightarrow B$. I is an interpolant for the lower sequent too.

We treat the universal quantifier and leave the existential quantifier to the reader. Suppose the last rule is $R\forall$:

$$\vdots$$
$$\frac{\Gamma_1 \Gamma_2, Ey \Rightarrow A(y)}{\Gamma_1 \Gamma_2 \Rightarrow \forall x A[x/y]}$$

By the induction hypothesis there is a interpolant $I \in \mathcal{L}(\Gamma_1, \Gamma_2 \cup \{Ey, A(y)\})$ for the upper sequent: $\vdash \Gamma_1 \Rightarrow I$ and $\vdash I, Ey, \Gamma_2 \Rightarrow A(y)$. In case y is not free in I the sequent $I, \Gamma_2 \Rightarrow \forall x A[x/y]$ is derivable too. Hence we can take I as an interpolant of the lower sequent and are done. Therefore, suppose y occurs free in I. By the side conditions y is not free in $\Gamma_1 \Gamma_2$. Hence we have the following derivation:

$$\vdots$$
$$\frac{\Gamma_1, Ey \Rightarrow I}{\Gamma_1 \Rightarrow \forall z I[z/y]}$$

Thus the following derivation shows that $\forall z I[z/y]$ is an interpolant for the lower sequent:

$$\vdots$$
$$\frac{\dfrac{I, Ey, \Gamma_2 \Rightarrow A(y) \qquad Ey, \Gamma_2 \Rightarrow Ey}{\forall z I[z/y], Ey, \Gamma_2 \Rightarrow A(y)}}{\forall z I[z/y], \Gamma_2 \Rightarrow \forall x A[x/y]}$$

Finally, we treat $L\forall$:

$$\vdots \qquad\qquad \vdots$$
$$\frac{\Gamma_1 \Gamma_2, A(t) \Rightarrow C \qquad \Gamma_1 \Gamma_2 \Rightarrow Et}{\Gamma_1 \Gamma_2, \forall x A(x) \Rightarrow C}$$

We have to find $I \in \mathcal{L}(\Gamma_1 \cup \{\forall x A(x)\}, \Gamma_2 \cup \{C\})$ and $J \in \mathcal{L}(\Gamma_1, \Gamma_2 \cup \{\forall x A(x), C\})$ such that

$$\vdash \Gamma_1, \forall x A(x) \Rightarrow I \quad \vdash I, \Gamma_2 \Rightarrow C \quad \text{and} \quad \vdash \Gamma_1 \Rightarrow J \quad \vdash J, \forall x A(x), \Gamma_2 \Rightarrow C.$$

First we treat the case J. Note that by the induction hypothesis there are three formulas $I' \in \mathcal{L}(\Gamma_1, \Gamma_2 \cup \{A(t), C\})$, $J' \in \mathcal{L}(\Gamma_1, \Gamma_2 \cup \{Et\})$ and $H' \in \mathcal{L}(\Gamma_2, \Gamma_1 \cup \{Et\})$ such that

(5) $\vdash \Gamma_1 \Rightarrow I' \ \vdash I', A(t), \Gamma_2 \Rightarrow C$ and $\vdash \Gamma_1 \Rightarrow J' \ \vdash J', \Gamma_2 \Rightarrow Et$

$$\vdash \Gamma_2 \Rightarrow H' \ \vdash H', \Gamma_1 \Rightarrow Et.$$

Note that I', J' and H' may contain t. If t does not occur in I' and J' or it occurs in $\mathcal{L}(\Gamma_1, \Gamma_2 \cup \{\forall x A(x), C\})$, then $I', J' \in \mathcal{L}(\Gamma_1, \Gamma_2 \cup \{\forall x A(x), C\})$. Moreover, (5) implies

$$\vdash \Gamma_1 \Rightarrow I' \wedge J' \ \vdash I' \wedge J', \forall x A(x), \Gamma_2 \Rightarrow C.$$

Thus in this case we can take $J = I' \wedge J'$.

On the other hand, if t does occur in I' or J' and not in $\mathcal{L}(\Gamma_1, \Gamma_2 \cup \{\forall x A(x), C\})$ we proceed as follows. Either t occurs not in Γ_1 or t does not occur in $\Gamma_2 \cup \{\forall x A(x), C\}$. In the first case, it follows that t does not occur in I' and not in J', contradicting our assumptions. Thus t occurs in Γ_1 but not in $\Gamma_2 \cup \{\forall x A(x), C\}$. Hence t does not occur in H'. Note that we have a derivation

$$
\begin{array}{cc}
\vdots & \vdots \\
\dfrac{H', \Gamma_1 \Rightarrow I' \wedge J' \qquad H', \Gamma_1 \Rightarrow Et}{\dfrac{H', \Gamma_1 \Rightarrow \exists x(I' \wedge J')[x/t]}{\Gamma_1 \Rightarrow \big(H' \rightarrow \exists x(I' \wedge J')[x/t]\big)}}
\end{array}
$$

Now note something important: because t does not occur in $\forall x A(x)$, this implies that $\forall x A(x) = \forall x A[x/t]$ (for the difference between $A(x)$ and $A[x/t]$ see the preliminaries, Section 2.1). Thus also $\forall x(A[y/t])[x/y] = \forall x A(x)$. And because t does not occur in Γ_2 or C, by the substitution lemma, Lemma 3, we also have a derivation for a variable y not occurring in P of

$$
\begin{array}{ll}
 & \vdots \qquad\qquad\qquad\qquad\qquad \vdots \\
\vdots & \dfrac{(I' \wedge J')[y/t], Ey, A[y/t], \Gamma_2 \Rightarrow C \qquad Ey, (I' \wedge J')[y/t], \Gamma_2 \Rightarrow Ey}{Ey, (I' \wedge J')[y/t], \forall x A(x), \Gamma_2 \Rightarrow C} \\
\Gamma_2 \Rightarrow H' & \dfrac{}{\dfrac{\exists x(I' \wedge J')[x/t], \forall x A(x), \Gamma_2 \Rightarrow C}{\big(H' \rightarrow \exists x(I' \wedge J')[x/t]\big), \forall x A(x), \Gamma_2 \Rightarrow C}}
\end{array}
$$

Hence we can take $J = \big(H' \rightarrow \exists x(I' \wedge J')[x/t]\big)$ and are done.

The last case we have to treat is the one where we look for the interpolant $I \in \mathcal{L}(\Gamma_1 \cup \{\forall x A(x)\}, \Gamma_2 \cup \{C\})$ such that

(6) $\vdash \Gamma_1, \forall x A(x) \Rightarrow I \ \vdash I, \Gamma_2 \Rightarrow C.$

Note that by the induction hypothesis there are $I' \in \mathcal{L}(\Gamma_1 \cup \{A(t)\}, \Gamma_2 \cup \{C\})$, $J' \in \mathcal{L}(\Gamma_2, \Gamma_1 \cup \{Et\})$ and $H' \in \mathcal{L}(\Gamma_2, \Gamma_1 \cup \{Et\})$ such that

$$\vdash \Gamma_1, A(t) \Rightarrow I' \ \vdash I', \Gamma_2 \Rightarrow C \quad \text{and} \quad \vdash \Gamma_2 \Rightarrow J' \ \vdash J', \Gamma_1 \Rightarrow Et$$

$$\vdash \Gamma_1 \Rightarrow H' \quad \vdash H', \Gamma_2 \Rightarrow Et.$$

Observe that whence we have $\vdash (J' \to I'), \Gamma_2 \Rightarrow C$. Furthermore, we have a derivation

$$\frac{\dfrac{\vdots \qquad\qquad \vdots}{J', A(t), \Gamma_1 \Rightarrow I' \qquad J', \Gamma_1 \Rightarrow Et}}{\dfrac{J', \forall x A(x), \Gamma_1 \Rightarrow I'}{\forall x A(x), \Gamma_1 \Rightarrow J' \to I'}}$$

Thus, in case t belongs to the common language $\mathcal{L}(\Gamma_1 \cup \{\forall x A(x)\}, \Gamma_2 \cup \{C\}))$ we can take $I = (J' \to I')$ and are done. Therefore, assume t does not belong to the common language. In case it does not belong to $\Gamma_2 \cup \{C\}$, it follows that both I' and J' cannot contain t and we can again take $I = (J' \to I')$. Therefore, assume t does not belong to $\Gamma_1 \cup \{\forall x A(x)\}$. Hence H' does not contain t. But then we can infer, by Lemma 3, for a fresh variable y, from $\vdash \forall x A(x), \Gamma_1 \Rightarrow J' \to I'$ above, that we have the following derivation

$$\frac{\dfrac{\vdots}{\forall x A(x), Ey, \Gamma_1 \Rightarrow (J' \to I')[y/t]}}{\dfrac{\forall x A(x), \Gamma_1 \Rightarrow \forall z (J' \to I')[z/t] \qquad \dfrac{\vdots}{\Gamma_1 \Rightarrow H'}}{\forall x A(x), \Gamma_1 \Rightarrow \forall z (J' \to I')[z/t] \wedge H'}}$$

On the other hand we also have

$$\frac{\dfrac{\vdots \qquad\qquad \vdots}{H', J' \to I', \Gamma_2 \Rightarrow C \qquad H', J' \to I', \Gamma_2 \Rightarrow Et}}{\dfrac{H', \forall z (J' \to I')[z/t], \Gamma_2 \Rightarrow C}{H' \wedge \forall z (J' \to I')[z/t], \Gamma_2 \Rightarrow C}}$$

Hence we take $I = H' \wedge \forall z (J' \to I')[z/t]$ as the interpolant.

It is interesting to note that (6) also holds for $I = (Et \to I') \wedge H'$. But in this case I belongs in general not to the common language.

Finally, we show that $\mathsf{LJE}'(\Sigma_\mathcal{L})$ has interpolation too, by reducing this case to the case LJE' in the following way. Given a proof P of $\Gamma_1 \Gamma_2 \Rightarrow C$ in $\mathsf{LJE}'(\Sigma_\mathcal{L})$ we consider all axioms of the form $\Pi \Rightarrow Et \in \Sigma_\mathcal{L}$ that occur in P. Suppose there are n of them: $\Pi_1 \Rightarrow Et_1, \ldots, \Pi_n \Rightarrow Et_n$. Note that all t_i have to be closed. Clearly, there is a proof of $Et_1, \ldots, Et_n, \Gamma_1 \Gamma_2 \Rightarrow C$ in LJE' by replacing the axioms $\Pi_i \Rightarrow Et_i$ by the logical axioms $\Pi_i, Et_i \Rightarrow Et_i$. Now we consider the following partition $\Gamma_1' \Gamma_2' \Rightarrow C$ of $Et_1, \ldots, Et_n, \Gamma_1 \Gamma_2 \Rightarrow C$:

$$\Gamma_1' = \Gamma_1 \cup \{Et_j \mid j \leq n, \ t_j \text{ occurs in } \Gamma_1 \text{ or not in } \Gamma_1 \cup \Gamma_2\}.$$

$$\Gamma'_2 = \Gamma_2 \cup \{Et_j \mid j \leq n, \ t_j \text{ occurs in } \Gamma_2\}.$$

By the interpolation theorem for LJE$'$ there exists an interpolant I such that $\vdash_{\mathsf{LJE}'} \Gamma'_1 \Rightarrow I$ and $\vdash_{\mathsf{LJE}'} I, \Gamma'_2 \Rightarrow C$ where I is in the common language of Γ'_1 and $\Gamma'_2 \cup \{C\}$. It is not difficult to see that whence I is in the common language of Γ_1 and $\Gamma_2 \cup \{C\}$ too. By cutting on the Et_i's we obtain

$$\vdash_{\mathsf{LJE}'(\Sigma_{\mathcal{L}})} \Gamma_1 \Rightarrow I \qquad \vdash_{\mathsf{LJE}'(\Sigma_{\mathcal{L}})} I, \Gamma_2 \Rightarrow C.$$

This proves that LJE$'(\Sigma_{\mathcal{L}})$ has interpolation too. ∎

COROLLARY 14. LJE *and* LJE$(\Sigma_{\mathcal{L}})$ *have interpolation.*

5.1 Beth's theorem

Following standard proofs for the Beth definability property of LJ, it is easy to prove the following theorem.

THEOREM 15. LJE *and* LJE$(\Sigma_{\mathcal{L}})$ *satisfy the Beth definability property.*

6 Two notions of forcing

In [Unterhalt, 1986] Unterhalt proved the completeness of IQCE$^+$ with respect to a certain Kripke semantics that is similar to the semantics defined below. Here we define a semantics via Kripke models equipped with a slightly different notion of forcing, called *eforcing* and denoted by $\Vdash^e$. We then show that LJE is sound and complete with respect to $\Vdash^e$.

As we will define various kinds of Kripke models and various kinds of forcing, let us start with listing the notions we are going to define:

Kripke models, *models* for short: standard Kripke models,

Kripke existence models, *emodels* for short: Kripke models with constant domains in which the existence predicate plays a special role as it is assumed to be nonempty,

total emodels: emodels in which the image of a function on arguments that exist always exist,

forcing: the standard notion of forcing,

existence forcing, *eforcing* for short: a notion of forcing in which predicates and connectives are forced in the usual way, but for which the quantifiers range over existing objects only. This notion of forcing is only defined for emodels.

6.1   Kripke models and Kripke existence models

A *classical structure for* $\mathcal{L}'$ is a pair (D_w, I_w) such that D_w is a nonempty set and I_w is a map from $\mathcal{L}'_{D_w}$ such that

for every n-ary predicate P in $\mathcal{L}'$, $I_w(P)$ is an n-ary predicate on D_w,

for every n-ary function f in $\mathcal{L}'_D$, $I_w(f)$ is an n-ary function on D_w (constants are considered as 0-ary functions),

$I_w(a) = a$ for every constant $a \in D_w$.

A *classical existence structure for* $\mathcal{L}'$ is a classical structure for $\mathcal{L}'$ satisfying the extra requirement

$I_w(E)$ is a *nonempty* unary predicate on D_w.

Note that in a classical structure the existence predicate plays no special role, while in a classical existence structure.

For any closed $\mathcal{L}'_D$-term t, $I_w(t)$ denotes the interpretation of t under I_w in D, which is defined as usual. $I_w(t_1, \ldots, t_n)$ is short for $I_w(t_1), \ldots, I_w(t_n)$. For $\mathcal{L}'_D$-sentences A, let $(D, I_w) \models A$ denote that A holds in the structure (D, I_w), which is defined as usual for classical structures. Note that the interpretation of any closed term in $\mathcal{L}'_D$ is an element of and the same in all domains.

A *frame* is a pair $(W, \preccurlyeq)$ where W is a nonempty set and $\preccurlyeq$ is a partial order on W with a root. A *Kripke model*, *model* for short, on a frame $F = (W, \preccurlyeq)$ is a triple $K = (F, D, I)$, where $D = \{D_w \mid w \in W\}$ is a collection of nonempty sets and I is a collection $\{I_w \mid w \in W\}$, such that the (D_w, I_w) are classical structures for $\mathcal{L}'$ that satisfy the persistency requirements:

$$w \preccurlyeq v \;\Rightarrow\; D_w \subseteq D_v,$$

and for all predicates $P(\bar{x})$ in $\mathcal{L}$ and for all closed $\mathcal{L}'_D$-terms $\bar{t}$,

$$w \preccurlyeq v \;\Rightarrow\; \big((D, I_w) \models P(\bar{t}) \;\Rightarrow\; (D, I_v) \models P(\bar{t})\big),$$

$$w \preccurlyeq v \;\Rightarrow\; I_w(\bar{t}) = I_v(\bar{t}).$$

In particular, $I_w(t) = I_v(t)$ for all closed terms in $\mathcal{L}'_{D_w}$. Therefore, we sometimes write $I(t)$ instead of $I_w(t)$. A *Kripke existence model*, *emodel* for short, is a Kripke model in which the (D_w, I_w) are classical existence structures and in which for all nodes w and v: $D_w = D_v$, i.e. Kripke existence model have constant domains. Therefore, we denote emodels often by $K = (F, D, I)$, where D is now a non empty set, and not a collection of sets as in the case of models. We call an emodel *total* when for all its nodes k and for all functions $f(x_1, \ldots, x_n)$ in $\mathcal{L}'$

$$\forall\, a_1, \ldots, a_n \in D : k \Vdash \bigwedge_{i=1}^{n} E a_i \to E f(a_1, \ldots, a_n).$$

6.2   Forcing and existence forcing

Given a Kripke model the notion of *forcing*, $\Vdash$, is defined as usual. Given a Kripke existence model $K = (D, \preccurlyeq, I)$, the *existence forcing relation* $\Vdash^e$, *eforcing* for short, is defined as follows, and denoted by $\Vdash^e$ to distinguish it from the standard forcing relation. For our purposes it suffices to define the eforcing relation $K, w \Vdash^e A$ at node w inductively only for *sentences* in $\mathcal{L}'_D$. For predicates $P(\bar{x})$ in $\mathcal{L}'$ (including E) and closed $\mathcal{L}'_D$-terms t, we put

$$K, w \Vdash^e P(\bar{t}) \equiv_{def} (D, I_w) \models P(\bar{t}),$$

and extend $K, w \Vdash^e A$ to all sentences in $\mathcal{L}'_D$ in the usual way for connectives, but differently for the quantifiers:

$$
\begin{aligned}
&k \nVdash^e \bot \\
&k \Vdash^e A \wedge B &&\text{iff}&& k \Vdash^e A \text{ and } k \Vdash^e B \\
&k \Vdash^e A \vee B &&\text{iff}&& k \Vdash^e A \text{ or } k \Vdash^e B \\
&k \Vdash^e A \to B &&\text{iff}&& \forall k' \succcurlyeq k : k' \Vdash^e A \Rightarrow k' \Vdash^e B \\
&k \Vdash^e \exists x A(x) &&\text{iff}&& \exists d \in D\; k \Vdash^e Ed \wedge A(d) \\
&k \Vdash^e \forall x A(x) &&\text{iff}&& \forall d \in D : k \Vdash^e Ed \to A(d).
\end{aligned}
$$

Note that the upwards persistency requirement

$$k \preccurlyeq l \,\wedge\, k \Vdash^e A \;\Rightarrow\; l \Vdash^e A.$$

is fulfilled. Moreover, note that

$$k \Vdash^e \forall x A(x) \;\Leftrightarrow\; \forall l \succcurlyeq k \forall d \in D\, l \Vdash^e Ed \to Ad.$$

When K is clear from the context we write $k \Vdash^e A$ instead of $K, k \Vdash^e A$. We call an emodel K an $\mathcal{L}$-emodel when

$$\forall k \forall t \in \mathcal{T}_{\mathcal{L}} : \; k \Vdash^e Et.$$

A *valuation on K* is a map α from variables to the domain at the root. Thus in the case of existence models it is a map from variables to D. For a formula A in $\mathcal{L}'_D$ we write $A[\alpha]$ for the formula that is the result of substituting $\alpha(x)$ for x, for every variable x in A. For formulas A we write $K \Vdash^e A$ and say that *A is eforced in K* if for all nodes k, for all valuations

α on K, $K, k \Vdash^e A[\alpha]$. We say that A is $\mathcal{L}$-*eforced*, written $\Vdash^e_\mathcal{L} A$, when $K \Vdash^e A$ for all $\mathcal{L}$-emodels K. We define similar notions for sequents $\Gamma \Rightarrow C$, considering them as formulas $\bigwedge \Gamma \to C$. We say that a collection of sequents $\mathcal{S}$ ($\mathcal{L}$-)eforces S when for all ($\mathcal{L}$-)models K, if $K \Vdash^e_\mathcal{L} S'$ for all $S' \in \mathcal{S}$, then $K \Vdash^e_\mathcal{L} S$. Similar notions are defined for forcing, reading forcing everywhere for eforcing and model for emodel.

7   Soundness and completeness

For the soundness and completeness proof of LJE with respect to $\Vdash^e$ to come, it will be convenient to work in Gentzen calculus LJE^∞ that is equivalent to LJE for finite sequents but that can deal with sequents of the form $\Gamma \Rightarrow \Delta$, where Γ and Δ may be infinite and Δ may contain more than one formula. It is similar to LJE, and in the case of R$\forall$ and R$\to$ the antecedent may still contain only one formula, like in LJE. Furthermore, it has structural rules weakening and contraction. Because of this, in L$\to$ and L$\forall$ the principal formula does not have to occur in the hypotheses.

Important: From now on Γ, Π, Δ and Λ range over countably infinite multisets of formulas, and sequents may be infinite from now on. Except in the setting of LJE or $\vdash_\mathcal{L}$ that only apply to finite sequents: in these cases we tacitly assume the sequents to be finite.

The system LJE^∞

$$Ax \;\; \Gamma, P \Rightarrow P, \Delta \;\; (P \text{ atomic}) \qquad L\bot \;\; \Gamma, \bot \Rightarrow \Delta$$

$$LW \; \frac{\Gamma \Rightarrow \Delta}{\Gamma, A \Rightarrow \Delta} \qquad\qquad RW \; \frac{\Gamma \Rightarrow \Delta}{\Gamma \Rightarrow A, \Delta}$$

$$LC \; \frac{\Gamma, A, A \Rightarrow \Delta}{\Gamma, A \Rightarrow \Delta} \qquad\qquad RC \; \frac{\Gamma \Rightarrow A, A, \Delta}{\Gamma \Rightarrow A, \Delta}$$

$$L\wedge \; \frac{\Gamma, A, B \Rightarrow \Delta}{\Gamma, A \wedge B \Rightarrow \Delta} \qquad R\wedge \; \frac{\Gamma \Rightarrow A, \Delta \quad \Gamma \Rightarrow B, \Delta}{\Gamma \Rightarrow A \wedge B, \Delta}$$

$$L\vee \; \frac{\Gamma, A \Rightarrow \Delta \quad \Gamma, B \Rightarrow \Delta}{\Gamma, A \vee B \Rightarrow \Delta} \qquad R\vee \; \frac{\Gamma \Rightarrow A, B, \Delta}{\Gamma \Rightarrow A \vee B, \Delta}$$

$$L\rightarrow \frac{\Gamma \Rightarrow A, \Delta \qquad \Gamma, B \Rightarrow \Delta}{\Gamma, A \rightarrow B \Rightarrow \Delta} \qquad\qquad R\rightarrow \frac{\Gamma, A \Rightarrow B}{\Gamma \Rightarrow A \rightarrow B}$$

$$L\forall \ \frac{\Gamma, At \Rightarrow \Delta \qquad \Gamma \Rightarrow Et, \Delta}{\Gamma, \forall x Ax \Rightarrow \Delta} \qquad\qquad R\forall \ \frac{\Gamma, Ey \Rightarrow Ay}{\Gamma \Rightarrow \forall x A[x/y]} \ *$$

$$L\exists \ \frac{\Gamma, Ay, Ey \Rightarrow \Delta}{\Gamma, \exists x A[x/y] \Rightarrow \Delta} \ * \qquad\qquad R\exists \ \frac{\Gamma \Rightarrow At, \Delta \qquad \Gamma \Rightarrow Et, \Delta}{\Gamma \Rightarrow \exists x Ax, \Delta}$$

$$\text{Cut} \ \frac{\Gamma \Rightarrow A \qquad \Gamma, A \Rightarrow \Delta}{\Gamma \Rightarrow \Delta}$$

We write $\mathsf{LJE}^\infty(\Sigma_\mathcal{L})$ for the system obtained from LJE^∞ by adding the sequents $\Sigma_\mathcal{L}$ as axioms. We say that LJE^∞ derives $\Gamma \Rightarrow \Delta$, $\vdash_{\mathsf{LJE}^\infty} \Gamma \Rightarrow \Delta$, when there are finite $\Gamma' \subseteq \Gamma$ and $\Delta' \subseteq \Delta$ such that $\vdash_{\mathsf{LJE}^\infty} (\Gamma' \Rightarrow \Delta')$. We say that a set of sequents $\mathcal{S}$ derives a sequent S in LJE^∞, when there are finite $\Gamma' \subseteq \Gamma$ and $\Delta' \subseteq \Delta$ such that $\Gamma' \Rightarrow \Delta'$ is derivable in the system LJE^∞ to which the sequents in $\mathcal{S}$ are added as axioms. We have similar notions for $\mathsf{LJE}^\infty(\Sigma_\mathcal{L})$. We often write $\vdash_\mathcal{L}^\infty$ for $\mathsf{LJE}^\infty(\Sigma_\mathcal{L}) \vdash$.

We leave it to the reader to verify that the following holds, using the fact that LJE has weakening and contraction, Lemma's 4 and 5:

LEMMA 16. *For finite Γ and Δ: $\vdash_\mathcal{L}^\infty \Gamma \Rightarrow \Delta$ if and only if $\vdash_\mathcal{L} \Gamma \Rightarrow \bigvee \Delta$.*

7.1 Soundness

THEOREM 17. *For all sets of closed sequents $\mathcal{S}$ and all closed sequents S in $\mathcal{L}'$:*

$$\mathcal{S} \vdash_\mathcal{L}^\infty S \text{ implies } \mathcal{S} \Vdash_\mathcal{L}^e S.$$

Proof. We only consider the case that $\mathcal{S}$ is empty and that S is a sequent with at most one formula in the succedent and leave the other cases to the reader. For a smooth induction we prove that all axioms of $\mathsf{LJE}^\infty(\Sigma_\mathcal{L})$ are $\mathcal{L}$-forced, and that for all its rules, if the hypotheses of the rule are $\mathcal{L}$-forced, then so is the conclusion. The case of the axioms is simple and so are most of the rules. We treat the axiom $\Sigma_\mathcal{L}$ and the rules $R\forall$ and $R\exists$. Let K be an $\mathcal{L}$-model. Recall that we write $k \Vdash^e \Gamma$ meaning that $k \Vdash^e A$ for all $A \in \Gamma$.

Consider a sequent $\Gamma \Rightarrow Et$ in $\Sigma_{\mathcal{L}}$. Hence t is a closed term in $\mathcal{T}_{\mathcal{L}}$. Let $\bar{x}$ be all the free variables that occur in Γ. By assumption on $\mathcal{L}$-models it follows that $K \Vdash^e_{\mathcal{L}} (\Gamma \Rightarrow Et)[\bar{a}/\bar{x}]$ for all $\bar{a} \in D$.

For R$\forall$ suppose $\Vdash^e \Pi, Ey \Rightarrow Ay$ and y not free in Π. Consider k in K, suppose that the free variables in Π and Ay are among $\bar{x}y$, let $\bar{a} \in D$, and assume $k \Vdash^e \Pi[\bar{a}/\bar{x}]$. We have to show that

$$\forall d \in D : \ k \Vdash^e (Ed \rightarrow Ad)[\bar{a}/\bar{x}].$$

Therefore, consider $l \succcurlyeq k$ and $d \in D$ such that $l \Vdash^e Ed$. We have to show that $l \Vdash^e Ad[\bar{a}/\bar{x}]$. As the side condition on R$\forall$ implies that y does no occur free in Π, we have $l \Vdash^e (\Pi \wedge Ey)[\bar{a}d/\bar{x}y]$. As $\Vdash^e \Pi, Ey \Rightarrow Ay$, this implies $l \Vdash^e Ay[\bar{a}d/\bar{x}y]$, that is, $l \Vdash^e Ad[\bar{a}/\bar{x}]$.

For $R\exists$ suppose $\Vdash^e \Pi \Rightarrow At$, $\Vdash^e \Pi \Rightarrow Et$, and let all free variables of Π and At be among $\bar{x}$, pick $\bar{a} \in D$ and assume $k \Vdash^e \Pi[\bar{a}/\bar{x}]$. We have to show that

$$\exists d \in D : \ k \Vdash^e (Ed \wedge Ad)[\bar{a}/\bar{x}].$$

Since $\Vdash^e \Pi \Rightarrow Et$ and $k \Vdash^e \Pi[\bar{a}/\bar{x}]$, this gives $k \Vdash^e Et[\bar{a}/\bar{x}]$. Similarly, $k \Vdash^e At[\bar{a}/\bar{x}]$. Let $d = t[\bar{a}/\bar{x}]$. Then we have $k \Vdash^e (Ed \wedge Ad)[\bar{a}/\bar{x}]$, as desired. ∎

7.2   Completeness

As mentioned above, the completeness proof given follows the pattern of the completeness proof for LJ as given in [Takeuti, 1987]. The idea is that if a sequent is underivable we apply the inference rules in the reversed order as long as possible, resulting is a so-called reduction tree with at least one branch along which all sequents are underivable. This branch will be a node in the Kripke model that we obtain by repeating this process, and that will refute the sequent we started with. Therefore, we first have to introduce the notion of a reduction tree, a notion similar to that of a Beth tableau.

DEFINITION 18.   Given a (possibly infinite) sequent S, the *reduction tree for S* is inductively defined as follows. Recall that we assumed that $\mathcal{L}$ contains at least one constant and no variables, and that $\mathcal{L}'$ has an infinite set of variables. Furthermore, we assume that at every stage of the construction we have infinetely many fresh variables of $\mathcal{L}'$ available, i.e. variables that do not occur in the sequents constructed so far.

The construction of the reduction tree for $S = (\Gamma \Rightarrow \Delta)$ consists of repeated application of steps 0,1, 2, ..., 7, which correspond to inference rules of LJE without the structural rules, R$\forall$ and R$\rightarrow$. We leave it to the reader to check that at every stage of the construction we deal with countably infinite sequents only, i.e. with sequents for which the antecedent and succedent contain countably inifnite many formulas only.

Step $n = 0$: write S at the bottom of the tree.

Step $n > 0$: if every leaf is an axiom of LJE or a sequent in $\Sigma_{\mathcal{L}}$, then stop. If this is not the case, then this stage is defined according to $n \equiv 0, 1, \ldots, 8 \bmod 9$. Let $\Pi \Rightarrow \Lambda$ be any leaf of the tree defined at stage $n-1$.

$n \equiv 0$: L$\wedge$ reduction. Let α be a set such that $\{A_{i0} \wedge A_{i1} \mid i \in \alpha\}$ consists exactly of all formulas in Π with outermost logical symbol $\wedge$ to which no reduction has yet been applied. Then above S write the sequent

$$\Pi, \{A_{i0}, A_{i1} \mid i \in \alpha\} \Rightarrow \Lambda.$$

$n \equiv 1$: R$\wedge$ reduction. Let α be a set such that $\{A_{i0} \wedge A_{i1} \mid i \in \alpha\}$ consists exactly of all formulas in Λ with outermost logical symbol $\wedge$ to which no reduction has yet been applied. Then above S write all sequents of the form

$$\Pi \Rightarrow \{A_{if(i)} \mid i \in \alpha\}, \Lambda$$

for any map $f : \alpha \to \{0, 1\}$.

$n \equiv 2$: L$\vee$ reduction. Defined in a similar way as R$\wedge$ reduction.

$n \equiv 3$: R$\vee$ reduction. Defined in a similar way as L$\wedge$ reduction.

$n \equiv 4$: L$\to$ reduction. Let α be a set such that $\{A_i \to B_i \mid i \in \alpha\}$ consists exactly of all formulas in Π with outermost logical symbol $\to$ to which no reduction has yet been applied. Then for all $f : \alpha \to \{0, 1\}$, write above S the sequent

$$\Pi, \{B_i \mid f(i) = 1\} \Rightarrow \{A_i \mid f(i) = 0\}, \Lambda.$$

$n \equiv 5$: L$\forall$ reduction. Let α be a set such that $\{\forall x_i A_i(x_i) \mid i \in \alpha\}$ consists exactly of all formulas in Π with outermost logical symbol $\forall$. Let $\mathcal{T}$ consists of all terms t for which Et occurs in Π. Above S write the sequent

$$\Pi, \{A_i(t) \mid i \in \alpha, t \in \mathcal{T}\} \Rightarrow \Lambda.$$

Note that if $\{Et \mid t \in \mathcal{T}_{\mathcal{L}}\} \subseteq \Pi$ we can always carry out this step, since there is at least one constant in $\mathcal{L}$, which implies there is at least one expression of the form Et in $\{Et \mid t \in \mathcal{T}_{\mathcal{L}}\}$, and thus in Π.

$n \equiv 6$: L$\exists$ reduction. Let α be a set such that $\{\exists x_i A_i(x_i) \mid i \in \alpha\}$ consists exactly of all formulas in Π with outermost logical symbol $\exists$ to which no reduction has yet been applied. Introduce fresh variables $\{y_i \mid i \in \alpha\}$ of $\mathcal{L}'$, and above S write the sequent

$$\Pi, \{A_i(y_i), Ey_i \mid i \in \alpha\} \Rightarrow \Lambda.$$

$n \equiv 7$: R$\exists$ reduction. Defined in a similar way as L$\forall$ reduction.

$n \equiv 8$: if $\Pi \Rightarrow \Lambda$ is an axiom of LJE or a sequent in $\Sigma_\mathcal{L}$, then stop. If this is not the case write the same sequent $\Pi \Rightarrow \Lambda$ above it.

This completes the definition of reduction trees.

The follwing is straightforward.

LEMMA 19. *If all leaves of the reduction tree of a sequent S are axioms of* LJE($\Sigma_\mathcal{L}$), *then S is provable in* LJE($\Sigma_\mathcal{L}$).

The following lemma, Lemma 21, is non-trivial and crucial in the completeness proof. It is an analogue of a lemma in [Takeuti, 1987] for LJ, and its main ingredient is the following generalization of König's Lemma.

PROPOSITION 20. *(A generalized König's Lemma, [Takeuti, 1987]) Let X be any set. Let $*(\cdot)$ be a property on partial functions $f : X \to \{0,1\}$. If*

1. $*(f)$ *holds if and only if there is a finite subset $Z \subseteq X$ such that $*(f \uparrow Z)$ (here $f \uparrow Z$ is the restriction of f to Z), and*

2. $*(f)$ *holds for all total functions f on X,*

then there exists a finite set $X' \subseteq X$ such that $*(f)$ for any f with $X' \subseteq dom(f)$ ($dom(f)$ is the domain of f).

Proof. For completeness sake we repeat Takeuti's proof from [Takeuti, 1987]. Let Y be the product of $|X|$ times $\{0,1\}$. Give $\{0,1\}$ the discrete topology and Y the product topology. Since $\{0,1\}$ is compact, so is Y by Tychonoff's theorem. For maps f and g call g an *extension* of f, when $dom(f) \subseteq dom(g)$ and f and g are equal on $dom(f)$. For every f with finite domain, let

$$\mathcal{N}_f \equiv_{def} \{g \mid g \text{ is total and an extension of } f\}.$$

Furthermore, let

$$\mathcal{C} \equiv_{def} \{\mathcal{N}_f \mid dom(f) \text{ is finite and } *(f)\}.$$

$\mathcal{C}$ is an open cover of Y. Therefore, $\mathcal{C}$ has a finite subcover, say $\mathcal{N}_{f_1}, \ldots, \mathcal{N}_{f_n}$. Let

$$X' = dom(f_1) \cup \ldots \cup dom(f_n).$$

Then X' satisfies the conditions of the theorem: assume $Z \subseteq dom(f)$. Let f' be a total extension of f. Then $*(f')$ by 2. and $f \in \mathcal{N}_{f_1} \cup \ldots \cup \mathcal{N}_{f_n}$, say $f \in \mathcal{N}_{f_i}$. Whence f is an extension of f_i and $*(f_i)$. Therefore, $*(f)$ by 1. ∎

LEMMA 21. *If a sequent S is not provable in* LJE($\Sigma_\mathcal{L}$), *then its reduction tree has a branch along which all sequents are underivable in* LJE($\Sigma_\mathcal{L}$).

Proof. In this proof provable will always mean provable in $\mathsf{LJE}(\Sigma_{\mathcal{L}})$, $\vdash$ stands for $\mathsf{LJE}(\Sigma_{\mathcal{L}}) \vdash$. We prove the lemma by proving the following: if in a reduction tree, for some set α, $\Gamma_\beta \Rightarrow \Delta_\beta$ ($\beta = 1, 2 \ldots, \alpha$) are all the immediate successors of $\Gamma \Rightarrow \Delta$, then if all these successors are provable, then so is $\Gamma \Rightarrow \Delta$. Recall that a sequent $\Pi \Rightarrow \Lambda$ is provable when there are finite $\Pi' \subseteq \Pi$ and $\Lambda' \subseteq \Lambda$ such that $\Pi' \Rightarrow \Lambda'$ is provable.

We distinguish by cases acceding to the rule that is applied to $\Gamma \Rightarrow \Delta$ resulting in the immediate successors $\Gamma_\beta \Rightarrow \Delta_\beta$.

$\mathsf{L}\wedge$ reduction: then $\Gamma \Rightarrow \Delta$ has one upper sequent, which is of the form $\Gamma, \{A_{i0}, A_{i1} \mid i \in \alpha\} \Rightarrow \Delta$, where $A_{i0} \wedge A_{i1}$ are all the sequents in Γ with outermost logical symbol $\wedge$. By assumption there are finite $\Gamma' \subseteq \Gamma$ and $\Delta' \subseteq \Delta$ and $B_i \in \{A_{i0}, A_{i1}\}$ for $i \leq n$ such that $\vdash \Gamma, B_1, \ldots, B_n \Rightarrow \Delta'$. Hence $\vdash \Gamma', \{A_{i0}, A_{i1} \mid i \leq n\} \Rightarrow \Delta'$, which again implies $\vdash \Gamma', \{A_{i0} \wedge A_{i1} \mid i \leq n\} \Rightarrow \Delta'$. Thus $\vdash \Gamma \Rightarrow \Delta$.

$\mathsf{R}\wedge$ reduction: then $\Gamma \Rightarrow \Delta$ has immediate successors $\Gamma \Rightarrow \{A_{if(i)} \mid i \in \alpha\}, \Delta$ for any map $f : \alpha \to \{0, 1\}$, where $\{A_{i0} \wedge A_{i1} \mid i \in \alpha\}$ consists exactly of all formulas in Δ with outermost logical symbol $\wedge$. By assumption for all $f : \alpha \to \{0, 1\}$ there are finite $\Gamma' \subseteq \Gamma$, $\Delta' \subseteq \Delta$ and $n_f \in \omega$ such that $\vdash \Gamma' \Rightarrow \{A_{if(i)} \mid i \leq n_f\}, \Delta'$. Now we are going to use the generalized König's Lemma. We define a property $*(\cdot)$ on the partial functions $f : \alpha \to \{0, 1\}$ as follows ($dom(f)$ denotes the domain of f):

$$*(f) \equiv \exists m \exists a_1 \ldots a_m \in dom(f) \exists \text{ finite } \Gamma' \subseteq \Gamma, \Delta' \subseteq \Delta :$$
$$\vdash \Gamma' \Rightarrow \{A_{\alpha_i f(a_i)} \mid i \leq m\}, \Delta'.$$

Then conditions 1. and 2. of the generalized König's Lemma 20 are satisfied. Hence there is a finite subset $\beta \subseteq \alpha$ such that $*(f)$ whenever $\beta \subseteq dom(f)$. Let $\mathcal{F}$ be the collection of f for which $dom(f) = \beta$. Thus for all $f \in \mathcal{F}$ there are finite $\Gamma^f \subseteq \Gamma$ and $\Delta^f \subseteq \Delta$ such that

$$\vdash \Gamma^f \Rightarrow \{A_{if(i)} \mid i \in \beta\}, \Delta^f.$$

Hence by weakening and repeated application of $\mathsf{R}\wedge$, one obtains

$$\vdash \{\Gamma^f \mid f \in \mathcal{F}\} \Rightarrow \{A_{i0} \wedge A_{i1} \mid i \in \beta\}, \{\Delta^f \mid f \in \mathcal{F}\}.$$

This implies that $\vdash \Gamma \Rightarrow \Delta$.

The case $\mathsf{R}\vee$ is similar to $\mathsf{R}\wedge$, and $\mathsf{L}\vee$ and $\mathsf{L}\to$ are similar to $\mathsf{R}\wedge$.

$\mathsf{L}\exists$ reduction: then $\Gamma \Rightarrow \Delta$ has immediate successor

$$\Gamma, \{A_i(y_i), Ey_i \mid i \in \alpha\} \Rightarrow \Delta,$$

where $\{\exists x_i A_i(x_i) \mid i \in \alpha\}$ consists exactly of all formulas in Δ with outermost logical symbol $\exists$. By assumption there are finite $\Gamma' \subseteq \Gamma$ and $\Delta' \subseteq \Delta$

and $n \in \omega$ such that

$$\vdash \Gamma', \{A_i(y_i), Ey_i \mid i \leq n\} \Rightarrow \Delta'.$$

Applications of L$\exists$ imply that then $\Gamma \Rightarrow \Delta$ is provable too.

The cases R$\exists$ and L$\forall$ are similar. This proves the lemma. ■

THEOREM 22. *For all sets of closed sequents $\mathcal{S}$ and all closed sequents S in $\mathcal{L}'$:*

$$\mathcal{S} \Vdash_{\mathcal{L}}^e S \ implies \ \mathcal{S} \vdash_{\mathcal{L}}^{\infty} S.$$

Proof. We treat the case that $\mathcal{S}$ is empty and leave the other case to the reader. The proof we give is similar to the elegant completeness proof for LJ in [Takeuti, 1987]. Recall that $\vdash_{\mathcal{L}}^{\infty}$ stands for $\vdash_{\mathsf{LJE}(\Sigma_{\mathcal{L}})}$. In the proof we will write $\vdash$ for $\vdash_{\mathsf{LJE}(\Sigma_{\mathcal{L}})}$. Let $S = (\Gamma \Rightarrow \Delta)$ be a closed sequent and assume that $\nvdash S$. We will construct a $\mathcal{L}$-model K such that $K \nVdash^e S$ in the following way.

K will be defined in ω many steps using reduction trees, which will be the nodes of K. We assume that $\mathcal{L}'$ contains infinitely many variables that do not occur in S.

Step 0: Let T_0 be the reduction tree for $\Gamma, \{Et \mid t \in \mathcal{T}_{\mathcal{L}}\} \Rightarrow \Delta$. Call this node 0. Since $\nvdash \Gamma \Rightarrow \Delta$, also

$$\nvdash \Gamma, \{Et \mid t \in \mathcal{T}_{\mathcal{L}}\} \Rightarrow \Delta.$$

By Lemma 21 there is a branch b_0 in T_0 containing only unprovable sequents. We proceed with b_0 and T_0 to the next step 1.

Step $i+1$. For any reduction tree T with branch b along which all sequents are unprovable constructed at step i, we consider Π and Λ, which are the respective unions of the formulas in the antecedents and succedents along b. Note that thus $\nvdash \Pi \Rightarrow \Lambda$. Let k range over all formulas in Λ with outermost logical symbol $\rightarrow$ or $\forall$. We proceed in the following way.

If k is a formula of the form $A \rightarrow B$. Then construct the reduction tree T_k for $\Pi, A \Rightarrow B$. This tree will be an immediate successor of T. Note that $\nvdash \Pi, A \Rightarrow B$. Thus by Lemma 21 there is a branch b in T containing only unprovable sequents. We proceed with b and T to the next step $i + 2$.

If k is a formula $\forall x A(x)$. Then construct the reduction tree T_k for $\Pi, Ey \Rightarrow A(y)$, where y is a variable in $\mathcal{L}$ that has not yet occurred in the construction of K. This tree will be an immediate successor of T. Observe that if $\Pi, Ey \Rightarrow A(y)$ is derivable, then so is $\Pi \Rightarrow \forall x Ax$, since y does not occur in Π. Thus $\nvdash \Pi, Ey \Rightarrow A(y)$, and whence by Lemma 21 there is a branch b in T containing only unprovable sequents. We proceed with b and T to the next step $i + 2$.

This process is continued ω times. Let W be the union of 0 and all k's in the construction and let $\preccurlyeq$ be the reflexive transitive closure of the immediate successor relation constructed at the stages. Define D to be the set of all terms appearing in the construction. Given a k in the construction, let T_k be the reduction tree at k and let Π_k and Λ_k be the respective unions of the formulas in the antecedents and succedents along the chosen infinite branch b_k in T_k. Then define an interpretation I as follows:

$$I_k(R) \equiv_{def} \{\bar{d} \in D \mid R(\bar{d}) \in \Pi_k\},$$

and I_k is the identity on function symbols: $I_k(f)(\bar{a}) = f(\bar{a}) \in D$. Since $\mathcal{L}$ contains at least one constant c, it also implies that $I_k(E)$ is nonempty. Note that $K = ((W, \preccurlyeq), D, I)$ indeed is a Kripke existence model. The fact that we started with the sequent $\Gamma, \{Et \mid t \in \mathcal{T}_\mathcal{L}\} \Rightarrow \Delta$ implies that $Et \in \Pi_k$ for all k and all terms t in $\mathcal{L}$. Hence $K \Vdash^e Et$ for all terms $t \in \mathcal{T}_\mathcal{L}$, and thus K is an $\mathcal{L}$-model. It is not difficult to show with formula induction that we have

$$A \in \Pi_k \;\Rightarrow\; k \Vdash^e A$$
$$A \in \Lambda_k \;\Rightarrow\; k \nVdash^e A$$

we treat the case $A = B \to C$ and leave the other cases to the reader. This will complete the theorem.

First assume $B \to C \in \Pi_k$. We have to show that $k \Vdash^e B \to C$. Therefore, consider $l \succcurlyeq k$ such that $l \Vdash^e B$. Thus by the induction hypothesis $B \in \Pi_l$. By the construction of the reduction tree, $C \in \Pi_l$ or $B \in \Lambda_l$. Since $B \notin \Lambda_l$, otherwise the branch would be derivable, it follows that $C \in \Pi_l$, and thus $l \Vdash^e C$.

Second, assume $B \to C \in \Lambda_k$. By the construction of the model, there is a node $l \succcurlyeq k$ such that $B \in \Pi_l$ and $C \in \Lambda_l$. This implies that $l \Vdash^e B$ and $l \nVdash^e C$. Hence $k \nVdash^e B \to C$. ∎

By Lemma 16 it follows that :

COROLLARY 23. *For all sets of finite closed sequents $\mathcal{S}$ and all finite closed sequents S in $\mathcal{L}'$:*

$$\mathcal{S} \vdash_\mathcal{L} S \text{ if and only if } \mathcal{S} \Vdash^e_\mathcal{L} S.$$

COROLLARY 24. *For all sets of closed sequents $\mathcal{S}$ and all closed sequents S in $\mathcal{L}'$:*
$\mathcal{S} \vdash_\mathcal{L} S$ *if and only if $K \Vdash^e S$ for all $\mathcal{L}$-models K based on frames that are conversely well-founded trees that force $\mathcal{S}$.*

Proof. Immediate from Lemma 16 and the proof of Theorem 22. ∎

8 Applications

8.1 Skolemization

One use of the existence predicate is in the setting of Skolemization. Recall that the Skolemization of a formula is the result of replacing strong quantifiers, i.e. positive universal and negative existential quantifiers, by fresh function symbols, thus obtaining a formula without strong quantifiers that is equiconsistent with the original formula. As is well-known, Skolemization is not complete with respect to IQC. That is, there are formulas that are underivable, but for which their Skolemized version is derivable in IQC. For example,

$$\text{IQC} \nvdash \forall x (Ax \vee B) \rightarrow (\forall x Ax \vee B) \qquad \text{IQC} \vdash \forall x (Ax \vee B) \rightarrow (Ac \vee B).$$

In [Baaz and Iemhoff, 2005] an alternative Skolemization method called *eSkolemization* is introduced and is shown to be sound and complete with respect to IQC for a large class of formulas, including all formulas in which every strong quantifier is existential or of the form $\forall x \neg \neg Ax$. This class is much larger than the class of formulas for which the standard Skolemization method is sound and complete. This eSkolemization method makes use of the existence predicate. It replaces negative occurrences of existential quantifiers $\exists x Bx$ by $(Ef(\bar{y}) \wedge Bf(\bar{y}))$, and positive occurrences of universal quantifiers $\forall x Bx$ by $(Ef(\bar{y}) \rightarrow Bf(\bar{y}))$. For example, the eSkolemization of the displayed formula above is

$$\text{IQCE} \nvdash \forall x (Ax \vee B) \rightarrow ((Ec \rightarrow Ac) \vee B).$$

The eSkolemization method is extended to sequents $\Gamma \Rightarrow C$ by considering them as formulas $\bigwedge \Gamma \rightarrow C$. The eSkolemization of a sequent S is denoted by S^s. Clearly,

$$\text{LJE} \vdash A \Rightarrow A^s.$$

Then it is shown in [Baaz and Iemhoff, 2005] that

THEOREM 25. *[Baaz and Iemhoff, 2005] For each closed sequent S in $\mathcal{L}'$ in which all strong quantifiers are existential: $\vdash_{\mathcal{L}} S$ if and only if $\vdash_{\mathcal{L}} S^s$.*

By Lemma 8 this implies that

COROLLARY 26. *[Baaz and Iemhoff, 2005] For each closed sequent S in $\mathcal{L} \backslash E$ in which all strong quantifiers are existential: $\vdash_{\text{LJ}} S$ if and only if $\vdash_{\mathcal{L}} S^s$.*

There is an extension of the main result to a larger class of formulas than the one occurring in the theorem above. This class of formulas is not syntactically defined and therefore less useful. However, it contains a syntactically defined class of formulas strictly larger than the class of formulas in which all strong quantifiers are existential: the class of formulas in which all strong universal quantifiers are of the form $\forall x \neg \neg Ax$. Hence the result, Theorem 29, that eSkolemization is sound and complete for this class of formulas is a genuine extension of Theorem 25.

DEFINITION 27. For a formula A that occurs in a sequent S, $S[B/A]^p$ (p for positive) denotes the result of replacing every positive occurrence of A in S by B. Note that we do not put restrictions on the possible occurrences of free variables in A or S. We say that *all strong quantifiers in S are almost existential* if for every subformula $\forall x Ax$ of S, it holds that

$$S[\neg \exists x \neg Ax / \forall x Ax]^p \vdash_{\mathcal{L}} S.$$

Note that we always have

$$S \vdash_{\mathcal{L}} S[\neg \exists x \neg Ax / \forall x Ax]^p.$$

Thus almost existential sequents are sequents that, as a formula, are equivalent to a formula in which all strong quantifiers are existential.

REMARK 28. Clearly, all strong quantifiers in S are almost existential, if no strong universal quantifiers occur in S. But the class of sequents in which all strong quantifiers are almost existential also contains the formulas in which all strong universal quantifiers are of the form $\forall x \neg \neg Ax$. But the class contains more: for example, $\bot \Rightarrow \forall x Ax$ does not belong to the mentioned classes but every quantifier in this formula is almost existential.

THEOREM 29. *[Baaz and Iemhoff, 2005] For each closed sequent S in $\mathcal{L}'$ in which all strong quantifiers are almost existential:*

$$\vdash_{\mathcal{L}} S \text{ if and only if } \vdash_{\mathcal{L}} S^s.$$

COROLLARY 30. *[Baaz and Iemhoff, 2005] For all closed sequents S in $\mathcal{L} \backslash E$ in which all strong quantifiers are almost existential:*

$$\vdash_{\mathsf{LJ}} S \text{ if and only if } \vdash_{\mathcal{L}} S^s.$$

As was first proved by Mints in [Mints, 1966] we have the following corollary when using Proposition 9.

COROLLARY 31. *[Baaz and Iemhoff, 2005] For the fragment of sentences without weak quantifiers and in which all strong quantifiers are almost existential, derivability on* IQC *is decidable.*

In [Baaz and Iemhoff, 2005] also an analogue of Herbrand's theorem is provided, which together with eSkolemization links derivability in intuitionistic predicate logic to derivability in intuitionistic propositional logic, at least for formulas in which all strong quantifiers are almost existential.

8.2 Truth-value logics

Another application of the existence predicate is in the context of truth-value logics. These are logics based on truth-value sets V, i.e. closed subsets of the unit interval $[0, 1]$, also called *Gödel sets*. One can, for a given Gödel set V, interpret formulas by mapping them to elements of V. The logical symbols receive a meaning via restrictions on these interpretations, e.g. by stipulating that the interpretation of $\wedge$ is the infimum of the interpretations of the respective conjuncts, or that the interpretation of $\exists\, x Ax$ is the supremum of the values of Aa for all elements a in the domain. Given these interpretations, one can associate a logic with such a Gödel set V: the logic of all sentences that are mapped to 1 under any interpretation on V. Here we work only with the languages $\mathcal{L}'$ and $\mathcal{L}'_- = \backslash E$.

We define the following frame logics:

$$
\begin{aligned}
L_F \Vdash A \;\; &\equiv_{def} \;\; K \Vdash A \text{ for all models } K \text{ on } F \\
L_F \;\; &\equiv_{def} \;\; \{A \mid A \text{ a sentence in } \mathcal{L}'_-, F \Vdash A\} \\
L_F^{cd} \;\; &\equiv_{def} \;\; \{A \mid A \text{ a sentence in } \mathcal{L}'_-, \\
&\qquad\qquad K \Vdash A \text{ for all models } K \text{ on } F \text{ with constant domain}\} \\
L_F^{cdte} \;\; &\equiv_{def} \;\; \{A \mid A \text{ a sentence in } \mathcal{L}'_-, \\
&\qquad\qquad K \Vdash A \text{ for all total emodels } K \text{ on } F \text{ with constant domain}\}
\end{aligned}
$$

Gödel logics are a famous example of truth value logics. Given a Gödel set V and a nonempty set D, a *Gödel logic interpretation* I is defined as follows.

$$
\begin{aligned}
I(P\bar{t}) \;\; &= \;\; I(P)(I(\bar{t})) \\
I(A \wedge B) \;\; &= \;\; \inf(I(A), I(B)) \\
I(A \vee B) \;\; &= \;\; \sup(I(A), I(B)) \\
I(A \to B) \;\; &= \;\; \begin{cases} 1 & \text{if } I(A) \leq I(B) \\ I(B) & \text{otherwise,} \end{cases} \\
I(\exists\, x Ax) \;\; &= \;\; \sup\{I(Aa) \mid a \in D\} \\
I(\forall\, x Ax) \;\; &= \;\; \inf\{I(Aa) \mid a \in D\}.
\end{aligned}
$$

The Gödel logic G_V consists of those sentences in $\mathcal{L}'_-$ that receive value 1 under all such Gödel logic interpretations I, for all possible domains D. A. Beckmann and N. Preining in [Beckmann and Preining, 2005] proved that Gödel logics correspond to the logics of linear frames with constant domain in the following way.

THEOREM 32. *(A. Beckmann and N. Preining [Beckmann and Preining, 2005]) For any countable linear frame F there exists a Gödel set V such that*

(7) $G_V = L_F^{cd}$,

and vice versa: for every Gödel set V there exists a countable linear frame F such that (7).

Based on these ideas, in [Iemhoff, 2005] an analogue was found for the case of linear frames without the extra restriction to constant domains. So-called *Scott logics* S_V were defined, where S_V consists of all sentences A in $\mathcal{L}'_-$ that receive the value 1 for any domain assignment and any Scott logic interpretation on V. Here a domain assignment is a pair (D, e) where D is a nonempty set and e is a function $e : D \to V$ satisfying

$$\exists\, a \in D\; e(a) = 1.$$

Given a domain assignment (D, e), a *Scott logic interpretation* I interprets terms and predicate symbols on D, satisfies

$$\inf_i e(a_i) \le e(I(f)(\bar{a}))$$

for all n-ary function symbols f in the language and all sequences $\bar{a} = a_1, \dots, a_n$ in D^n, and extends to all formulas as follows:

$$
\begin{array}{rcl}
I(P\bar{t}) & = & I(P)(I(\bar{t})) \\
I(A \wedge B) & = & \inf(I(A), I(B)) \\
I(A \vee B) & = & \sup(I(A), I(B)) \\
I(A \to B) & = & \begin{cases} 1 & \text{if } I(A) \le I(B) \\ I(B) & \text{otherwise,} \end{cases} \\
I(\exists\, xAx) & = & \sup\{e(a) \wedge I(Aa) \mid a \in D\} \\
I(\forall\, xAx) & = & \inf\{e(a) \to I(Aa) \mid a \in D\}.
\end{array}
$$

Note that a Gödel logic interpretation is a Scott logic interpretation where e is the constant 1 function. Then one can show that Scott logics correspond to the logics of linear frames:

THEOREM 33. *[Iemhoff, 2005] For every countable linear frame F there exists a Gödel set V such that*

(8) $S_V = L_F$,

and vice versa: for every countable Gödel set V there exists a countable linear frame F such that (8).

Note that this correspondence is not quite as strong as in the case of Gödel logics where every V is linked to a frame. In the case of Scott logics we could establish this only for countable V. We do not know whether the stronger form also holds, but conjecture it to be the case.

In the same paper [Iemhoff, 2005] it has been shown that there is a natural and faithful translation $(\cdot)^e$ from Scott logics into Gödel logics that makes use of the existence predicate. Roughly, we extend the notion of Gödel logics to the language $\mathcal{L}'$, and then we let the E in the Gödel logic of V correspond to the e in the Scott logic of V. Thus in this setting we first have to extend the notion of a Gödel logics to the language $\mathcal{L}'$, i.e. to E.

A *Gödel existence logic interpretation* I on (V, D) is a Gödel logic interpretation on (V, D) that satisfies the extra requirements that

$$\exists\, a \in D\ I(Ea) = 1,$$

and for all functions h in $\mathcal{L}$, for all $\bar{a} = a_1, \ldots, a_n \in D$,

$$I(\bigwedge_{i \leq n} Ea_i) \leq I(Eh(\bar{a})).$$

The Gödel existence logic G_V^e of a Gödel set V consists of all $\mathcal{L}'$-sentences A such that receive value 1 under all Gödel existence logic interpretations on all domain assignments.

Given these definitions, $(\cdot)^e$ is defined as follows.

$\bigl(P(\bar{t})\bigr)^e = P(\bar{t})$ for atomic P and terms $\bar{t}$,

$(\cdot)^e$ commutes with the connectives,

$(\exists\, xA(x))^e = \exists\, x\bigl(Ex \wedge (A(x))^e\bigr),$

$(\forall\, xA(x))^e = \forall\, x\bigl(Ex \rightarrow (A(x))^e\bigr).$

Given this translation we then have the following theorem.

THEOREM 34. *[Iemhoff, 2005] For any Gödel set V, $(\cdot)^e$ is a faithful translation of S_V into G_V, i.e. for all sentences A in $\mathcal{L}'_-$:*

$$S_V \models A \ \Leftrightarrow\ G_V^e \models A^e.$$

Furthermore, we have: (in [Iemhoff, 2005] L_F^{cdte} is denoted L_F^{cde})

PROPOSITION 35. *[Iemhoff, 2005] For any frame F, $(\cdot)^e$ is faithful translation of L_F into L_F^{cdte}, i.e. for all sentences A in $\mathcal{L}'_-$:*

$$L_F \Vdash A \ \Leftrightarrow\ L_F^{cdte} \Vdash A^e.$$

PROPOSITION 36. *[Iemhoff, 2005] For every frame F, $L_F^{cdte} = G_V^e$.*

Note the similarity between the different applications of the existence predicate: the translation $(\cdot)^e$ does a similar thing to quantifiers as eSkolemization does.

Acknowledgement

We thank Mark van Atten for pointing out to us the particular passage of Kritik der reinen Vernunft quoted in Section 1.2 and for the remarks on Brouwer.

BIBLIOGRAPHY

[Baaz and Iemhoff, 2005] Baaz, M. and Iemhoff, R., Skolemization in intuitionistic logic, *Submitted*, 2005.

[Baaz and Iemhoff, 2005b] Baaz, M. and Iemhoff, R., Gentzen calculi for the existence predicate, *Studia Logica*, to appear.

[Baaz and Iemhoff, 2005c] Baaz, M. and Iemhoff, R., On interpolation in existence logics, *Submitted*, 2005.

[Baaz and Leitsch, 1994] Baaz, M. and Leitsch, A., On Skolemization and proof complexity, *Fundamenta Informaticae* 20, pp. 353-379, 1994.

[Beckmann and Preining, 2005] Beckmann, A. and Preining, N., Linear Kripke Frames and Gödel Logics, *Submitted*, 2005.

[Beeson, 1985] Beeson, M., Foundations of Constructive Mathematics, Springer, Berlin, 1985.

[Corsi, 1989] Corsi, G., A cut-free calculus for Dummett's LC quantified, *Zeitschrift für Mathematische Logik und Grundlagen der Mathematik* 35, pp. 289-301, 1989.

[Corsi, 1989b] Corsi, G., A logic characterized by the class of connected models with nested domains, *Studia Logica* 48(1), pp. 15-22, 1989.

[Corsi, 1992] Corsi, G., Completeness theorem for Dummett's LC quantified and some of its extensions, *Studia Logica* 51(2), pp. 317-335, 1992.

[Dyckhoff and Pinto, 1999] Dyckhoff, R. and Pinto, L., Proof search in constructive logics, *Sets and proofs*, Lond. Math. Soc. Lect. vol. 258, Cambridge University Press (1999), pp. 53-65.

[Fitting, 1999] Fitting, M., Herbrand's theorem for a modal logic, *Logic and foundations of mathematics. Selected contributed papers of the 10th international congress of logic, methodology and philosophy of science*, Kluwer (1999), pp. 219-225.

[Herbrand, 1930] Herbrand, J., *Recherches sur la théorie de la demonstration*, PhD thesis University of Paris (1930).

[Iemhoff, 2005] Iemhoff, R., A note on linear Kripke models, *Journal of Logic and Computation*, 2005, to appear.

[Mints, 1962] Mints, G.E., An analogue of Hebrand's theorem for the constructive predicate calculus, *Sov. Math. Dokl.* 3 (1962), pp. 1712-1715.

[Mints, 1966] Mints, G.E., Hebrand's theorem for the predicate calculus with equality and function symbols, *Sov. Math.,Dokl.* 7 (1966), pp. 911-914

[Mints, 1972] Mints, G.E., The Skolem method in intuitionistic calculi, *Proc. Steklov Inst. Math.* 121 (1972), pp. 73-109

[Mints, 1994] Mints, G.E., Resolution strategies for the intuitionistic predicate logic, *Constraint Programming. Proceedings of the NATO Advanced Study Institute*, Comput. Syst. Sci. 131, Springer (1994), pp. 289-311.

[Mints, 2000] Mints, G.E., Axiomatization of a Skolem function in intuitionistic logic, *Formalizing the dynamics of information*, Faller, M. (ed.) et al., CSLI Lect. Notes 91, pp. 105-114, 2000.

[Preining, 2003] Preining, N., *Complete Recursive Axiomatizability of Gödel Logics*, PhD-thesis, Technical University Vienna, 2003.

[Scott, 1979] Scott, D.S., Identity and existence in intuitionistic logic, *Applications of sheaves, Proc. Res. Symp. Durham 1977*, Fourman (ed.) et al., Lect. Notes Math. 753, pp. 660-696, 1979.

[Shankar, 1992] Shankar, N., Proof search in the Intuitionistic Sequent Calculus, *Automated deduction-CADE-11*, Lect. Notes Comput. Science 607 (1992), pp. 522-536.

[Skolem, 1920] Skolem, T., Logisch-kombinatorische Untersuchungen über die Erfüllbarkeit oder Beweisbarkeit mathematischer Sätze nebst einem Theorem über dichte Mengen , *Skrifter utgitt av Videnskapsselskapet i Kristiania, I, Mat. Naturv. Kl.* 4 (1920), pp. 1993-2002.

[Smoryński, 1973] Smoryński, C., Elementary intuitionistic theories, *Journal of Symbolic Logic* 38 (1973), pp. 102-134.

[Smoryński, 1977] Smoryński, C., On axiomatizing fragments, *Journal of Symbolic Logic* 42 (1977), pp. 530-544.

[Takano, 1987] Takano, M., Another proof of the strong completeness of the intuitionistic fuzzy logic, *Tsukuba J. Math.* 11(1), pp. 101-105, 1987.

[Takeuti, 1987] Takeuti, G., *Proof Theory*, Studies in Logic and the Foundations of Mathematics 81, North-Holland Publishing Company, 1987.

[Takeuti and Titani, 1984] Takeuti, G. and Titani, M., Intuitionistic fuzzy logic and intuitionistic fuzzy set theory, *Journal of Symbolic Logic* 49, pp. 851-866, 1984.

[Troelstra and van Dalen, 1988] Troelstra, A.S., and van Dalen, D., *Constructivism in Mathematics*, vol. I North-Holland, 1988.

[Troelstra and Schwichtenberg, 1996] Troelstra, A.S., and Schwichtenberg, H.,, *Basic Proof Theory*, Cambridge Tracts in Theoretical Computer Science 43, Cambridge University Press, 1996.

[Unterhalt, 1986] Unterhalt, M., *Kripke-Semantik mit partieller Existenz*, PhD-thesis, University of Münster, 1986.

Dimensions of Neural-symbolic Integration — A Structured Survey

Sebastian Bader[1] and Pascal Hitzler[2]

1 Introduction

Research on integrated neural-symbolic systems has made significant progress in the recent past. In particular the understanding of ways to deal with symbolic knowledge within connectionist systems (also called artificial neural networks) has reached a critical mass which enables the community to strive for applicable implementations and use cases. Recent work has covered a great variety of logics used in artificial intelligence and provides a multitude of techniques for dealing with them within the context of artificial neural networks.

Already in the pioneering days of computational models of neural cognition, the question was raised how symbolic knowledge can be represented and dealt with within neural networks. The landmark paper [McCulloch and Pitts, 1943] provides fundamental insights how propositional logic can be processed using simple artificial neural networks. Within the following decades, however, the topic did not receive much attention as research in artificial intelligence initially focused on purely symbolic approaches. The power of machine learning using artificial neural networking was not recognized until the 80s, when in particular the backpropagation algorithm [Rumelhart *et al.*, 1986] made connectionist learning feasible and applicable in practice.

These advances indicated a breakthrough in machine learning which quickly led to industrial-strength applications in areas such as image analysis, speech and pattern recognition, investment analysis, engine monitoring, fault diagnosis, etc. During a training process from raw data, artificial neural networks acquire expert knowledge about the problem domain, and the ability to generalize this knowledge to similar but previously unencountered situations in a way which often surpasses the abilities of human experts. The

[1] Supported by the German Research Foundation (DFG) under GK334.
[2] Supported by the German Federal Ministry of Education and Research (BMBF) under the SwartWeb project.

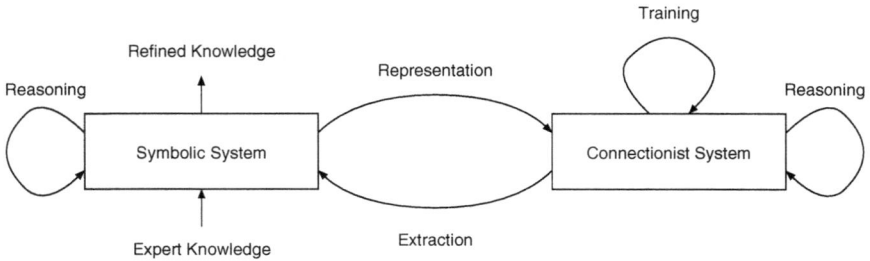

Figure 1. Neural-symbolic learning cycle

knowledge obtained during the training process, however, is hidden within the acquired network architecture and connection weights, and not directly accessible for analysis, reuse, or improvement, thus limiting the range of applicability of the neural networks technology. For these purposes, the knowledge would be required to be available in structured symbolic form, most preferably expressed using some logical framework.

Likewise, in situations where partial knowledge about an application domain is available before the training, it would be desirable to have the means to guide connectionist learning algorithms using this knowledge. This is the case in particular for learning tasks which traditionally fall into the realm of symbolic artificial intelligence, and which are characterized by complex and often recursive interdependencies between symbolically represented pieces of knowledge.

The arguments just given indicate that an integration of connectionist and symbolic approaches in artificial intelligence provides the means to address machine learning bottlenecks encountered when the paradigms are used in isolation. Research relating the paradigms came into focus when the limitations of purely connectionist approaches became apparent. The corresponding research turned out to be very challenging and produced a multitude of very diverse approaches to the problem. Integrated systems in the sense of this survey are those where symbolic processing functionalities emerge from neural structures and processes.

Most of the work in integrated neural-symbolic systems addresses the neural-symbolic learning cycle depicted in Figure 1. A front-end (symbolic system) is used to feed symbolic (partial) expert knowledge to a neural or connectionist system which can be trained on raw data, possibly taking the internally represented symbolic knowledge into account. Knowledge acquired through the learning process can then be extracted back to the symbolic system (which now also acts as a back-end), and made available for further processing in symbolic form. Studies often address only parts of the neural-symbolic learning cycle (like the representation or extraction of knowledge), but can be considered to be part of the overall investigations concerning the cycle.

We assume that the reader has a basic familiarity with artificial neural networks and symbolic artificial intelligence, as conveyed by any introductory courses or textbooks on the topic, e.g. in [Russell and Norvig, 2003]. However, we will refrain from going into technical detail at any point, but rather provide ample references which can be followed up at ease. The selection of research results which we will discuss in the process is naturally subjective and driven by our own specific research interests. Nevertheless, we hope that this survey also provides a helpful and comprehensive albeit unusual literature overview to neural-symbolic integration.

This chapter is structured as follows. In Section 2, we introduce some of those integrated neural-symbolic systems, which we consider to be foundational for the majority of the work undertaken within the last decade. In Section 3, we will explain our proposal for a classification scheme. In Section 4, we will survey recent literature by means of our classification. Finally, in Section 5, we will give an outlook on possible further developments.

2 Neural-Symbolic Systems

As a reference for later sections, we will review some well-known systems here. We will start with the landmark results by McCulloch and Pitts, which relate finite automata and neural networks [McCulloch and Pitts, 1943]. Then we will discuss a method for representing structured terms in a connectionist systems, namely the recursive autoassociative memories (RAAM) [Pollack, 1990]. The SHRUTI System, proposed in [Shastri and Ajjanagadde, 1993], is discussed next. Finally, *Connectionist Model Generation using the Core Method* is introduced as proposed in [Hölldobler and Kalinke, 1994]. These approaches lay the foundations for most of the more recent work on neural-symbolic integration which we will discuss in this chapter.

2.1 Neural Networks and Finite Automata

The advent of automata theory and of artificial neural networks, marked also the advent of neural-symbolic integration. In their seminal paper [McCulloch and Pitts, 1943] Warren Sturgis McCulloch and Walter Pitts showed that there is a strong relation between symbolic systems and artificial neural networks. In particular, they showed that for each finite state machine there is a network constructed from binary threshold units – and vice versa – such that the input-output behaviour of both systems coincide. This is

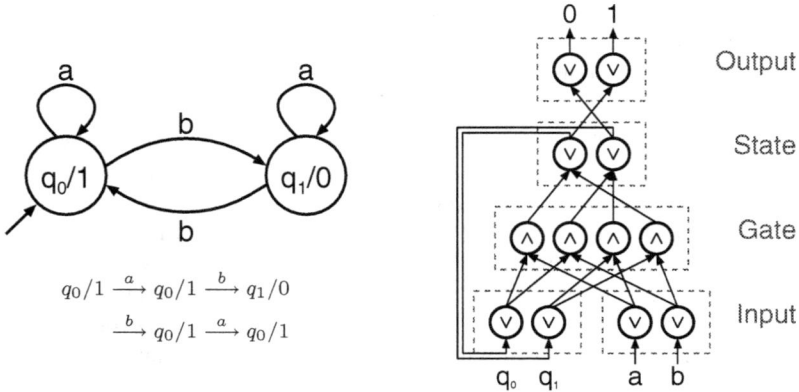

Figure 2. A simple Moore-machine, the processing of the input word *abba* (left) and a corresponding McCulloch-Pitts-network (right).

due to the fact that simple logical connectives such as conjunction, disjunction and negation can easily be encoded using binary threshold units, with weights and thresholds set appropriately. To illustrate the ideas, we will discuss a simple example in the sequel.

EXAMPLE 1. Figure 2 on the left shows a simple Moore-machine, which is a finite state machine with outputs attached to the states [Hopcroft and Ullman, 1989]. The corresponding network is shown on the right. The network consists of four layers. For each output-symbol $(0, 1)$ there is a unit in the output-layer, and for each input-symbol (a, b) a unit in the right part of the input-layer. Furthermore, for each state (q_0, q_1) of the automaton, there is a unit in the state-layer and in the left part of the input layer. In our example, there are two ways to reach the state q_1, namely by being in state q_1 and receiving an 'a', or by being in state q_0 and receiving a 'b'. This is implemented by using a disjunctive neuron in the state-layer receiving inputs from two conjunctive units in the gate layer, which are connected to the corresponding conditions, as e.g. being in state q_0 and reading a 'b'.

A network of n binary threshold units can be in 2^n different states only, and the change of state depends on the current input to the network only. These states and transitions can easily be encoded as a finite automaton, using a straightforward translation [McCulloch and Pitts, 1943; Kleene, 1956]. An extension to the class of weighted automata is given in [Bader et al., 2004a].

Input		Hidden		Output
$(A\ B)$	$\rightarrow$	$R_1(t)$	$\rightarrow$	$(A'(t)\ B'(t))$
$(C\ D)$	$\rightarrow$	$R_2(t)$	$\rightarrow$	$(C'(t)\ D'(t))$
$(R_1(t)\ R_2(t))$	$\rightarrow$	$R_3(t)$	$\rightarrow$	$(R_1'(t)\ R_2'(t))$

Table 1. Extracted training samples from the tree shown in Figure 3.

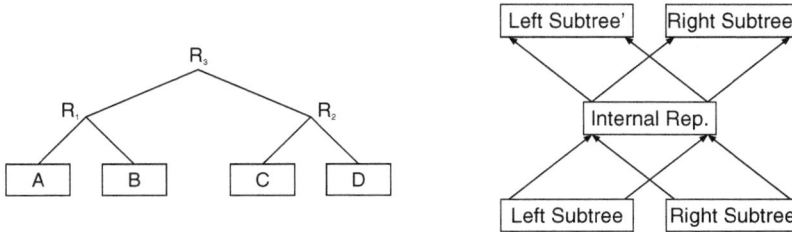

Figure 3. Example tree and a RAAM for binary trees.

2.2 Connectionist Term Representation

The representation of possibly infinite structures in a finite network is one of the major obstacles on the way to neural-symbolic integration [Bader *et al.*, 2004b]. One attempt to solve this will be discussed in this section, namely the idea of *recursive autoassociative memories (RAAMs)* as introduced in [Pollack, 1990], where a fixed length representation of variable sized data is obtained by training an artificial neural network using backpropagation. Again, we will try to illustrate the ideas by discussing a simple example.

EXAMPLE 2. Figure 3 shows a small binary tree which shall be encoded in a fixed-length real vector. The resulting RAAM-network is depicted in Figure 3, where each box depicts a layer of 4 units. The network is trained as an encoder-decoder network, i.e. it reproduces the input activations in the output layer [Bishop, 1995]. In order to do this, it must create a compressed representation in the hidden layer. Table 1 shows the activations of the layers during the training of the network. As the training converges we shall have $A = A'$, $B = B'$, etc. To encode the terminal symbols A, B, C and D we use the vectors $(1,0,0,0)$, $(0,1,0,0)$, $(0,0,1,0)$ and $(0,0,0,1)$ respectively. The representations of R_1, R_2 and R_3 are obtained during training. After training the network, it is sufficient to keep the internal representation R_3, since it contains all necessary information for recreating the full tree. This is done by plugging it into the hidden layer and recursively using the output activations, until binary vectors, hence terminal symbols, are reached.

Rules	Facts
$Owns(y, z) \leftarrow Gives(x, y, z)$	$Gives(john, josephine, book)$
$Owns(x, y) \leftarrow Buys(x, y)$	$Buys(carl, x)$
$Can-sell(x, y) \leftarrow Owns(x, y)$	$Owns(josephine, ball)$

Table 2. A knowledge base for Own and $Can-sell$

While recreating the tree from its compressed representation, it is necessary to distinguish terminal and non-terminal vectors, i.e. those which represent leafs of the trees from those representing nodes. Due to noise or inaccuracy, it can be very hard to recognise the "1-of-n"-vectors representing terminal symbols. In order to circumvent this problem different solutions were proposed, which can be found in [Stolcke and Wu, 1992; Sperduti, 1994a; Sperduti, 1994b]. The ideas described above for binary vectors apply also for trees with larger, but fixed, branching factors, by simply using bigger input and output layers. In order to store sequences of data, a version called S-RAAM (for sequential RAAM) can be used [Pollack, 1990]. In [Blair, 1997] modifications were proposed to allow the storage of deeper and more complex data structures than before, but their applicability remains to be shown [Kalinke, 1997]. Other recent approaches for enhancement have been studied e.g. in [Sperduti *et al.*, 1995; Kwasny and Kalman, 1995; Sperduti *et al.*, 1997; Hammerton, 1998; Adamson and Damper, 1999], which also include some applications. A recent survey which includes RAAM architectures and addresses structured processing can be found in [Frasconi *et al.*, 2001]. The related approach on *Holographic reduced representations (HRRs)* [Plate, 1991; Plate, 1995] also uses fixed-length representations of variable-sized data, but using different methods.

2.3 Reflexive Connectionist Reasoning

A wide variety of tasks can be solved by humans very fast and efficiently. This type of reasoning is sometimes referred to as *reflexive reasoning*. The SHRUTI system [Shastri and Ajjanagadde, 1993] provides a connectionist architecture performing this type of reasoning. Relational knowledge is encoded by clusters of cells and inferences by means of rhythmic activity over the cell clusters. It allows to encode a (function-free) fragment of first-order predicate logic analyzed in [Hölldobler *et al.*, 1999b]. Binding of variables – a particularly difficult aspect of neural-symbolic integration – is obtained by time-synchronization of activities of neurons.

EXAMPLE 3. Table 2 shows a knowledge base describing what it means to own something and to be able to sell it. Furthermore it states some facts. The resulting SHRUTI network is shown in Figure 4.

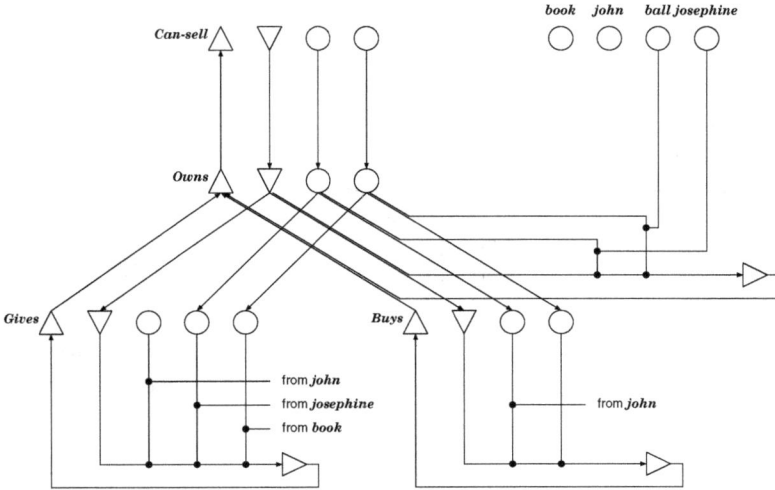

Figure 4. A SHRUTI network for the knowledge base from Table 2. Each predicate is represented by two *relais units* ($\triangle, \triangledown$) and a set of *argument units* ($\bigcirc$). Constants are represented as $\bigcirc$-units in the upper right. Facts are implemented using $\triangleright$-units.

Recent enhancements, as reported in [Shastri, 1999] and [Shastri and Wendelken, 1999], allow e.g. the support of negation and inconsistency. [Wendelken and Shastri, 2003] adds very basic learning capabilities to the system, while [Wendelken and Shastri, 2004] addresses the problem of multiple reuse of knowledge rules, an aspect which limits the capabilities of SHRUTI.

2.4   Connectionist Model Generation using the Core Method

In 1994, Hölldobler and Kalinke proposed a method to translate a propositional logic program into a neural network [Hölldobler and Kalinke, 1994] (a revised treatment is contained in [Hitzler *et al.*, 2004]), such that the network will settle down in a state corresponding to a model of the program. To achieve this goal, not the program itself, but rather the associated consequence operator was implemented using a connectionist system. The realization is close in spirit to [McCulloch and Pitts, 1943], and Figure 5 shows a propositional logic program and the corresponding network.

EXAMPLE 4.   The simple logic program in Figure 5 states that a is a fact, b follows from a, etc. This "follows-from" is usually captured by the associated consequence operator T_P [Lloyd, 1988]. The figure shows also the

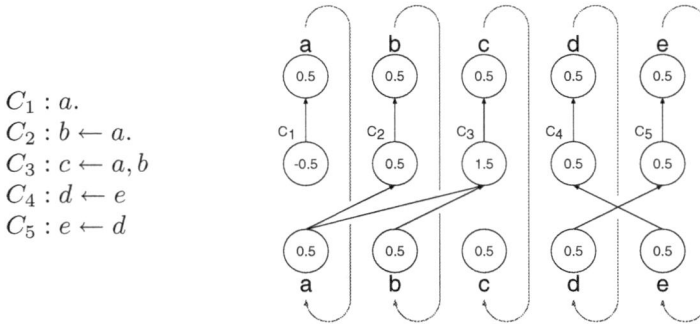

$C_1 : a.$
$C_2 : b \leftarrow a.$
$C_3 : c \leftarrow a, b$
$C_4 : d \leftarrow e$
$C_5 : e \leftarrow d$

Figure 5. A simple propositional logic program and the corresponding network. Numbers within the units denote the thresholds. All weights are set to 1.

corresponding network, obtained by the algorithm given in [Hölldobler and Kalinke, 1994]. For each atom (a, b, c, d, e) there is a unit in the input- and output layer, whose activation represents the truth value of the corresponding atom. Furthermore, for each rule in the program there is a unit in the hidden layer, acting as a conjunction. If all requirements are met, this unit becomes active and propagates its activation to the consequence-unit in the output layer.

It can be shown that every logic program can be implemented using a 3-layer network of binary threshold units, and that 2-layer networks do not suffice. It was also shown that under some syntactic restrictions on the programs, their semantics could be recovered by recurrently connecting the output- and the input layer of the network (as indicated in Figure 5) and propagating activation exhaustively through the resulting recurrent network. Key idea to [Hölldobler and Kalinke, 1994] was to represent logic programs by means of their associated *semantic operators*, i.e. by connectionist encoding of an operator which captures the meaning of the program, instead of encoding the program directly. More precisely, the functional input-output behaviour of a semantic operator T_P associated with a given program P is encoded by means of a feedforward neural network N_P which, when presented an encoding of some I to its input nodes, produces $T_P(I)$ at its output nodes. Output nodes can also be connected recurrently back to the input nodes, resulting in a connectionist computation of iterates of I under T_P, as used e.g. in the computation of the semantics or meaning of P [Lloyd, 1988]. I, in this case, is a (Herbrand-)interpretation for P, and T_P is a mapping on the set I_P of all (Herbrand-)interpretations for P.

This idea for the representation of logic programs spawned several investigations in different directions. As [Hölldobler and Kalinke, 1994] employed binary threshold units as activation functions of the network nodes, the results were lifted to sigmoidal and hence differentiable activation functions in [Garcez et al., 1997; Garcez and Zaverucha, 1999]. This way, the connectionist representation of logic programs resulted in a network architecture which could be trained using standard backpropagation algorithms. The resulting connectionist inductive learning and reasoning system CILP was completed by providing corresponding knowledge extraction algorithms [Garcez et al., 2001]. Further extensions to this include modal [Garcez et al., 2002b] and intuitionistic logics [Garcez et al., 2003]. Metalevel priories between rules were introduced in [Garcez et al., 2000]. An in-depth treatment of the whole approach can be found in [Garcez et al., 2002a]. The *knowledge based artificial neural networks (KBANN)* [Towell and Shavlik, 1994] are closely related to this approach, by using similar techniques to implement propositional logic formulae within neural networks, but with a focus on learning.

Another work following up on [Hölldobler and Kalinke, 1994] concerns the connectionist treatment of first-order logic programming. [Seda, 2005] and [Seda and Lane, 2005] approach this by approximating given first-order programs P by finite subprograms of the grounding of P. These subprograms can be viewed as propositional ones and encoded using the original algorithm from [Hölldobler and Kalinke, 1994]. [Seda, 2005] and [Seda and Lane, 2005] show that arbitrarily accurate encodings are possible for certain programs including definite ones (i.e. programs not containing negation as failure). They also lift their results to logic programming under certain multi-valued logics.

A more direct approach to the representation of first-order logic programs based on [Hölldobler and Kalinke, 1994] was pursued in [Hölldobler et al., 1999a; Hitzler and Seda, 2000; Hitzler et al., 2004; Hitzler, 2004; Bader et al., 2005a; Bader et al., 2005b]. The basic idea again is to represent semantic operators $T_P : I_P \rightarrow I_P$ instead of the program P directly. In [Hölldobler and Kalinke, 1994] this was achieved by assigning propositional variables to nodes, whose activations indicate whether the nodes are true or false within the currently represented interpretation. In the propositional setting this is possible because for any given program only a finite number of truth values of propositional variables plays a role – and hence the finite network can encode finitely many propositional variables in the way indicated. For first-order programs, infinite interpretations have to be taken into account, thus an encoding of ground atoms by one neuron each is impossible as it would result in an infinite network, which is not computationally feasible to work with.

The solution put forward in [Hölldobler *et al.*, 1999a] is to employ the capability of standard feedforward networks to propagate real numbers. The problem is thus reduced to encoding I_P as a set of real numbers in a computationally feasible way, and to provide means to actually construct the networks starting from their input-output behaviour. Since sigmoidal units can be used, the resulting networks are trainable by backpropagation. [Hölldobler *et al.*, 1999a] spelled out these ideas in a limited setting for a small class of programs, and was lifted in [Hitzler and Seda, 2000; Hitzler *et al.*, 2004] to a more general setting, including the treatment of multi-valued logics. [Hitzler, 2004] related the results to logic programming under non-monotonic semantics. In these reports, it was shown that approximation of logic programs by means of standard feedforward networks is possible up to any desired degree of accuracy, and for fairly general classes of programs. However, no algorithms for practical generation of approximating networks from given programs could be presented. This was finally done in [Bader *et al.*, 2005b], and implementations of the approach are currently under way, and shall yield a first-order integrated neural-symbolic system with similar capabilities as the propositional system CILP.

There exist two alternative approaches to the representation of first-order logic programs via their semantic operators, which have not been studied in more detail yet. The first approach, reported in [Bader and Hitzler, 2004], uses insights from fractal geometry as in [Barnsley, 1993] to construct iterated function systems whose attractors correspond to fixed points of the semantic operators. The second approach builds on Gabbay's *Fibring logics* [Gabbay, 1999], and the corresponding Fibring Neural Networks [Garcez and Gabbay, 2004]. The resulting system, presented in [Bader *et al.*, 2005a], employs the fibring idea to control the firing of nodes such that it corresponds to term matching within a logic programming system. It is shown that certain limited kinds of first-order logic programs can be encoded this way, such that their models can be computed using the network.

3   A New Classification Scheme

In this section we will introduce a classification scheme for neural-symbolic systems. This way, we intend to bring some order to the heterogeneous field of research, whose individual approaches are often largely incomparable. We suggest to use a scheme consisting of three main axes as depicted in Figure 6, namely *Interrelation*, *Language* and *Usage*.

For the interrelation-axis, depicted in Figure 7, we roughly follow the scheme introduced and discussed in [Hilario, 1995; Hatzilygeroudis and Prentzas, 2004], but adapted to the particular focus which we will put

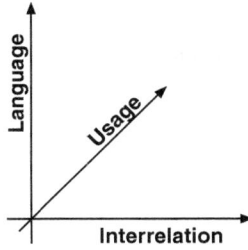

Figure 6. Main Axes

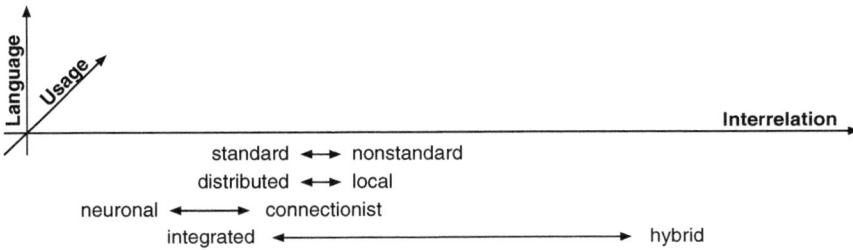

Figure 7. Interrelation

forward. In particular, the classifications presented in [Hilario, 1995; Hatzi-lygeroudis and Prentzas, 2004] strive to depict each system at exactly one point in a taxonomic tree. From our perspective, certain properties or design decisions of systems are rather independent, and should be understood as different *dimensions*. From this perspective approaches can first be divided into two main classes, namely into *integrated* (called *unified* or *translational* in [Hilario, 1995; Hatzilygeroudis and Prentzas, 2004]) and *hybrid* systems. Integrated are those, where full symbolic processing functionalities emerge from neural structures and processes – further details will be discussed in Section 4.1. Integrated systems can be further subdivided into neuronal and connectionist approaches, as discussed in Section 4.1. Neuronal indicates the usage of neurons which are very closely related to biological neurons. In connectionist approaches there is no claim to neurobiological plausibility, instead general artificial neural network architectures are used. Depending on their architecture, they can be split into standard and non-standard networks. Furthermore, we can distinguish local and distributed representation of the knowledge which will also be discussed in more detail in Section 4.1.

Note that the subdivisions belonging to the interrelation axis are again independent of each other. They should be understood as independent sub-dimensions, and could also be depicted this way by using further coordinate axes. We hope that our simplified visualisation makes it easier to maintain an overview. But to be pedantic, for our presentation we actually understand the neural-connectionist dimension as a subdivision of integrated systems, and the distributed-local and standard-nonstandard dimensions as independent subdivisions of connectionist systems – simply because this currently suffices for classification.

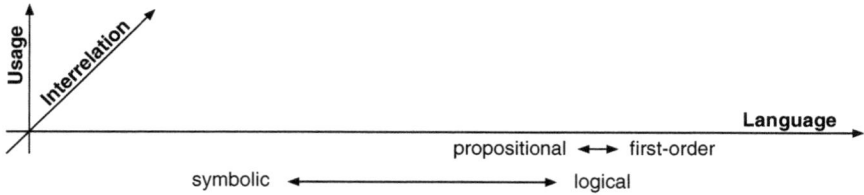

Figure 8. Language

Figure 8 depicts the second axis in our scheme. Here, the systems are divided according to the language used in their symbolic part. We distinguish between *symbolic* and *logical* languages. Symbolic approaches include the relation to automata as in [McCulloch and Pitts, 1943], to grammars [Elman, 1990; Fletcher, 2001] or to the storage and retrieval of terms [Pollack, 1990], whereas the logical approaches require either propositional or first order logic systems, as e.g. in [Hölldobler and Kalinke, 1994] and discussed in Section 2.4. The language axis will be discussed in more detail in Section 4.2.

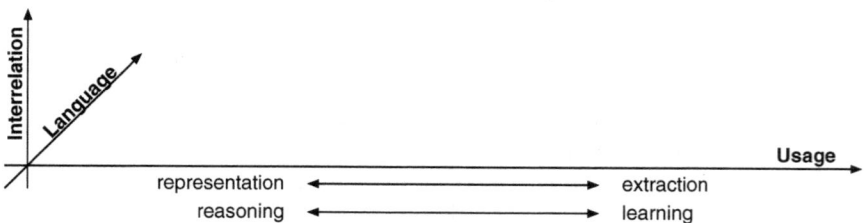

Figure 9. Usage

Most systems focus on one or only a few aspects of the neural-symbolic learning cycle depicted in Figure 1, i.e. either the representation of symbolic knowledge within a connectionist setting, or the training of preinitialized networks, or the extraction of symbolic systems from a network. Depending on this main focus we can distinguish the systems as shown in Figure 9. The issues of *extraction* vs. *representation* on the one hand and *learning* vs. *reasoning* on the other hand, are discussed in Section 4.3. Systems may certainly cover several or all of these aspects, i.e. they may span whole subdimensions.

4 Dimensions of Neural Symbolic Integration

In this section, we will survey main research results in this area by classifying them according to eight dimensions, marked by the arrows in Figures 7-9.

- Interrelation

    1. Integrated versus hybrid

    2. Neuronal versus connectionist

    3. Local versus distributed

    4. Standard versus nonstandard

- Language

    5. Symbolic versus logical

    6. Propositional versus first-order

- Usage

    7. Extraction versus representation

    8. Learning versus reasoning

As discussed above, we believe that these dimensions mark the main points of distinction between different integrated neural-symbolic systems. The chapter is structured accordingly, examining each of the dimensions in turn.

4.1 Interrelation

Integrated versus Hybrid

This section serves to further clarify what we understand by *neural-symbolic integration*. Following the rationale laid out in the introduction, we understand why it is desirable to combine symbolic and connectionist approaches, and there are obviously several ways how this can be done. From a bird's eye view, we can distinguish two main paradigms, which we call *hybrid* and *integrated* (or following [Hilario, 1995], *unified*) systems, and this survey is concerned with the latter.

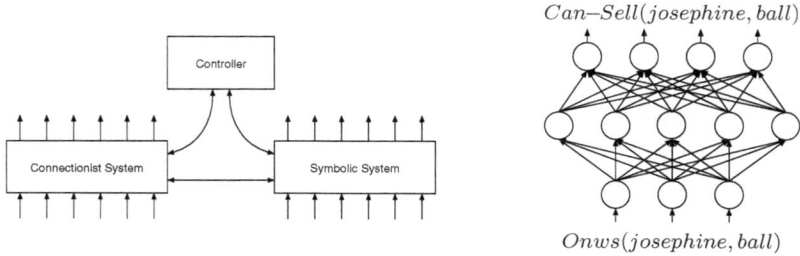

Figure 10. Hybrid (left) versus integrated (right) architecture.

Hybrid systems are characterized by the fact that they combine two or more problem-solving techniques in order to address a problem, which run in parallel, as depicted in Figure 10.

An *integrated* neural-symbolic system differs from a hybrid one in that it consists of one connectionist main component in which symbolic knowledge is processed, see Figure 10 (right). Integrated systems are sometimes also referred to as *embedded* or *monolithic* hybrid systems, cf. [Sun, 2001]. Examples for integrated systems are e.g. those presented in Sections 2.2-2.4.

For either architecture, one of the central issues is the representation of symbolic data in connectionist form [Bader *et al.*, 2004b]. For the hybrid system, these transformations are required for passing information between the components. The integrated architecture must implicitly or explicitly deal with symbolic data by connectionist means, i.e. must also be capable of similar transformations.

This survey covers integrated systems only, the study of which appears to be particularly challenging. For recent selective overview literature see e.g. [Browne and Sun, 2001; Garcez *et al.*, 2002a; Bader *et al.*, 2004b]. The first, [Browne and Sun, 2001], focuses on reasoning systems. The field of propositional logic is thoroughly covered in [Garcez *et al.*, 2002a], where the authors revisit the approach of [Hölldobler and Kalinke, 1994] and explain their extensions including applications to real world problems, like fault diagnosis. In [Bader *et al.*, 2004b] the emphasis is on the challenge problems arising from first-order neural-symbolic integration.

Neuronal versus Connectionist

There are two driving forces behind the field of neural-symbolic integration: On the one hand it is the striving for an understanding of human cognition, and on the other it is the vision of combining connectionist and symbolic artificial intelligence technology in order to arrive at more powerful reasoning and learning systems for computer science applications.

In [McCulloch and Pitts, 1943] the motivation for the study was to understand human cognition, i.e. to pursue the question how higher cognitive – logical – processes can be performed by artificial neural networks. In this line of research, the question of *biological feasability* of a network architecture is prominent, and inspiration is often taken from biological counterparts.

The SHRUTI system [Shastri and Ajjanagadde, 1993] as described in Section 2.3, for example, addresses the question how it is possible that biological networks perform certain reasoning tasks very quickly. Indeed, for some complex recognition tasks which involve reasoning capabilities, human responses occur sometimes at *reflexive* speed, particularly within a time span which allows processing through very few neuron layers only. As mentioned above, time-synchronization was used for the encoding of variable binding in SHRUTI.

The recently developed spiking neurons networks [Maass, 2002] take an even more realistic approach to the modelling of temporal aspects of neural activity. Neurons, in this context, are considered to be firing so-called *spike trains*, which consist of patterns of firing impulses over certain time intervals. The complex propagation patterns within a network are usually analysed by statistical methods. The encoding of symbolic knowledge using such temporal aspects has hardly been studied so far, an exception being [Sougne, 2001]. We perceive it as an important research challenge to relate the neurally plausible spiking neurons approach to neural-symbolic integration research. To date, however, only a few preliminary results on computational aspects of spiking neurons have been obtained [Natschläger and Maass, 2002; Maass and Markram, 2004; Maass *et al.*, 2005].

Another recent publication, [van der Velde and de Kamps, 2005], shows how natural language could be encoded using biologically plausible models of neural networks. The results appear to be suitable for the study of neural-symbolic integration, but it remains to be investigated to which extent the provided approach can be transfered to symbolic reasoning. Similarly inspiring might be the recent book [Hawkins and Blakeslee, 2004] and accompanying work, though it discusses neural-symbolic relationships on a very abstract level only.

The lines of research just reviewed take their major motivation from the goal to achieve biologically plausible behaviour or architectures. As already mentioned, neural-symbolic integration can also be pursued from a more technically motivated perspective, driven by the goal to combine the advantages of symbolic and connectionist approaches by studying their interrelationships. The work on the Core Method, discussed in Section 2.4, can be subsumed under this technologically inspired perspective.

Local versus Distributed Representation of Knowledge

For integrated neural-symbolic systems, the question is crucial how symbolic knowledge is represented within the connectionist system. If standard networks are being trained using backpropagation, the knowledge acquired during the learning process is spread over the network in diffuse ways, i.e. it is in general not easy or even possible to identify one or a small number of nodes whose activations contain and process a certain symbolic piece of knowledge.

The RAAM architecture and their variants as discussed in Section 2.2 are clearly based on distributed representations. Technically, this stems from the fact that the representation is initially learned, and no explicit algorithm for translating symbolic knowledge into the connectionist setting is being used.

Most other approaches to neural-symbolic integration, however, represent data locally. SHRUTI (Section 2.3) associates a defined node assembly to each logical predicate, and the architecure does not allow for distributed representation. The approaches for propositional connectionist model generation using the Core Method (Section 2.4) encode propositional variables as single nodes in the input resp. output layer, and logical formulae (rules) by single nodes in the hidden layer of the network.

The design of distributed encodings of symbolic data appears to be particular challenging. It also appears to be one of the major bottlenecks in producing applicable integrated neural-symbolic systems with learning and reasoning abilities [Bader et al., 2004b]. This becomes apparent e.g. in the difficulties faced by the first-order logic programming approaches discussed in Section 2.4. Therein, symbolic entities are not represented directly. Instead, interpretations (i.e. valuations) of the logic are being represented, which contain truth value assignments to language constructs. Concrete representations, as developed in [Bader et al., 2005b], distribute the encoding of the interpretations over several nodes, but in a diffuse way. The encoding thus results in a distributed representation. Similar considerations apply to the recent proposal [Gust and Kühnberger, 2005], where first-order logic is first converted into variable-free form (using topoi from category theory), and then fed to a neural network for training.

Standard versus Non/standard Network Architecture

Even though neural networks are a widely accepted paradigm in AI it is hard to make out a standard architecture. But, all so called standard-architecture systems agree at least on the following:

- only real numbers are propagated along the connections

- units compute very simple functions only

- all units behave similarly (i.e. they use similar simple functions and the activation values are always within a small range)

- only simple recursive structures are used (e.g. connecting only the output back to the input layer, or use selfrecursive units only)

When adhering to these standard design principles, powerful learning techniques as e.g. backpropagation [Rumelhart *et al.*, 1986] or Hebbian Learning [Hebb, 1949] can be used to train the networks, which makes them applicable to real world problems.

However, these standard architectures do not easily lend themselves to neural-symbolic integration. In general, it is easier to use non-standard architectures in order to represent and work with structured knowledge, with the drawback that powerful learning abilities are often lost.

Neural-symbolic approaches using standard networks are e.g. the CILP system [Garcez and Zaverucha, 1999], KBANN [Towell and Shavlik, 1994], RAAM (Section 2.2) and [Seda and Lane, 2005] (Section 2.4). Usually, they consist of a layered network, consisting of three or in case of KBANN more layers, and sigmoidal units are being used. For these systems experimental results are available showing their learning capabilities. As discussed above, these systems are able to handle propositional knowledge (or first order with a finite domain). Similar observations can be made about the standard architectures used in [Hölldobler *et al.*, 1999a; Hitzler *et al.*, 2004; Bader *et al.*, 2005b] for first-order neural-symbolic integration.

Non-standard networks were used e.g. in the SHRUTI system [Shastri and Ajjanagadde, 1993], and in the approaches described in [Bader and Hitzler, 2004] and [Bader *et al.*, 2005a]. In all these implementations non-standard units and non-standard architectures were used, and hence none of the usual learning techniques are applicable. However, for the SHRUTI system limited learning techniques based on Hebbian Learning [Hebb, 1949] were developed [Shastri, 2002; Shastri and Wendelken, 2003; Wendelken and Shastri, 2003].

4.2 Language

Symbolic versus Logical

One of the motivations for studying neural-symbolic integration is to combine connectionist learning capabilities with symbolic knowledge processing, as already mentioned. While our main interest is in pursuing logical aspects of symbolic knowledge, this is not necessarily always the main focus of investigations.

Work on representing automata or weighted automata [Kleene, 1956; McCulloch and Pitts, 1943; Bader *et al.*, 2004a] (Section 2.1) using artificial neural networks, for example focuses on computationally relevant structures, such as automata, and not directly on logically encoded knowledge. Nevertheless, such investigations show how to deal with structural knowledge within a connectionist setting, and can serve as inspiration for corresponding research on logical knowledge.

Recursive autoassociative memory, RAAM, and their variants as discussed in Section 2.2, deals with terms only, and not directly with logical content. RAAM allows connectionist encodings of first-order terms, where the underlying idea is to present terms or term trees sequentially to a connectionist system which is trained to produce a compressed encoding characterized by the activation pattern of a small collection of nodes. To date, storage capacity is very limited, and connectionist processing of the stored knowledge has not yet been investigated in detail.

A considerable body of work exists on the connectionist processing and learning of structured data using recurrent networks [Sperduti *et al.*, 1995; Sperduti *et al.*, 1997; Frasconi *et al.*, 2001; Hammer, 2002; Hammer, 2003; Hammer *et al.*, 2004a; Hammer *et al.*, 2004b]. The focus is on tree representations and manipulation of the data.

[Hölldobler *et al.*, 1997; Kalinke and Lehmann, 1998] study the representation of counters using recurrent networks, and connectionist unification algorithms as studied in [Hölldobler, 1990; Hölldobler and Kurfess, 1992; Hölldobler, 1993] are designed for manipulating terms, but already in a clearly logical context. The representation of grammars [Giles *et al.*, 1991] or more generally of natural language constructs [van der Velde and de Kamps, 2005] also has a clearly symbolic (as opposed to logical) focus.

It remains to be seen, however, to what extent the work on connectionist processing of structured data can be reused in logical contexts for creating integrated neural-symbolic systems with reasoning capabilities. Integrated reasoning systems like the ones presented in Sections 2.3 and 2.4 currently lack the capabilities of the term-based systems, so that a merging of these efforts appears to be a promising albeit challenging goal.

Propositional versus First-Order

Logic-based integrated neural-symbolic systems differ as to the knowledge representation language they are able to represent. Concerning the capabilities of the systems, a major distinction needs to be made between those which deal with propositional logics, and those based on first-order predicate (and related) logics.

What we mean by propositional logics in this context includes propositional modal, temporal, non-monotonic, and other non-classical logics. One of their characteristic feature which distinguishes them from first-order logics for neural-symbolic integration is the fact that they are of a finitary nature: propositional theories in practice involve only a finite number of propositional variables, and corresponding models are also finite. Also, sophisticated symbol processing as needed for nested terms in the form of substitutions or unification is not required.

Due to their finiteness it is thus fairly easy to implement propositional logic programs using neural networks [Hölldobler and Kalinke, 1994] (Section 2.4). A considerable body of work deals with the extension of this approach to non-classical logics [Garcez *et al.*, 2005; Garcez *et al.*, 2000; Garcez *et al.*, 2002b; Garcez *et al.*, 2003; Garcez *et al.*, 2004a; Garcez *et al.*, 2004b; Garcez and Lamb, 200x]. This includes modal, intuitionistic, and argumentation-theoretic approaches, amongst others. Earlier work on representing propositional logics is based on Hopfield networks [Pinkas, 1991b; Pinkas, 1991a] but has not been followed up on recently.

In contrast to this, predicate logics – which for us also include modal, non-monotonic, etc. extensions – in general allow to use function symbols as language primitives. Consequently, it is possible to use terms of arbitrary depth, and models necessarily assign truth values to an infinite number of ground atoms. The difficulty in dealing with this in a connectionist setting lies in the finiteness of neural networks, necessitating to capture the infinitary aspects of predicate logics by finite means. The first-order approaches presented in [Hölldobler *et al.*, 1999a; Hitzler and Seda, 2000; Bader and Hitzler, 2004; Hitzler *et al.*, 2004; Bader *et al.*, 2005a; Bader *et al.*, 2005b] (Section 2.4) solve this problem by using encodings of infinite sets by real numbers, and representing them in an approximate manner. They can also be carried over to non-monotonic logics [Hitzler, 2004].

[Bader *et al.*, 2005a], which builds on [Garcez and Gabbay, 2004] and [Gabbay, 1999] uses an alternative mechanism in which unification of terms is controlled via fibrings. More precisely, certain network constructs encode the matching of terms and act as gates to the firing of neurons whenever corresponding symbolic matching is achieved.

A prominent subproblem in first-order neural-symbolic integration is that of variable binding. It refers to the fact that the same variable may occur in several places in a formula, or that during a reasoning process variables may be bound to instantiate certain terms. In a connectionist setting, different parts of formulae and different individuals or terms are usually represented independently of each other within the system. The neural network paradigm, however, forces subnets to be blind with respect to detailed activation patterns in other subnets, and thus does not lend itself easily to the processing of variable bindings.

Research on first-order neural-symbolic integration has led to different means of dealing with the variable binding problem. One of them is to use temporal synchrony to achieve the binding. This is encoded in the SHRUTI system (Section 2.3), where the synchronous firing of variable nodes with constant nodes encodes a corresponding binding. Other approaches, as discussed in [Browne and Sun, 1999], encode binding by relating the propagated activations, i.e. real numbers.

Other systems avoid the variable binding problem by converting predicate logical formulae into variable-free representations. The approaches in [Hölldobler et al., 1999a; Hitzler and Seda, 2000; Hitzler et al., 2004; Hitzler, 2004; Seda, 2005; Seda and Lane, 2005; Bader et al., 2005a; Bader et al., 2005b] (Section 2.4) make conversions to (infinite) propositional theories, which are then approximated. [Gust and Kühnberger, 2005] use topos theory instead.

It shall be noted here that SHRUTI (Section 2.3) addresses the variable binding problem, but allows to encode only a very limited fragment of first-order predicate logic [Hölldobler et al., 1999b]. In particular, it does not allow to deal with function symbols, and thus could still be understood as a finitary fragment of predicate logic.

4.3   Usage

Extraction versus Representation

The representation of symbolic knowledge is necessary even for classical applications of connectionist learning. As an example, consider the neural-networks-based Backgammon playing program TD-Gammon [Tesauro, 1995], which achieves professional players' strength by temporal difference learning on data created by playing against itself. TD-Gammon represents the Backgammon board in a straightforward way, by encoding the squares and placement of pieces via assemblies of nodes, thus representing the structured knowledge of a board situation directly by a certain activation pattern of the input nodes.

In this and other classical application cases the represented symbolic knowledge is not of a complex logical nature. Neural-symbolic integration, however, attempts to achieve connectionist processing of complex logical knowledge, learning, and inferences, and thus the question how to represent logical knowledge bases in suitable form becomes dominant. Different forms of representation have already been discussed in the context of local versus distributed representations.

Returning to the TD-Gammon example, we would also be interested in the complex knowledge as acquired by TD-gammon during the learning process, encoding the strategies with which this program beats human players. If such knowledge could be extracted in symbolic form, it could be used for further symbolic processing using inference engines or other knowledge based systems.

It is apparent, that both the representation and the extraction of knowledge are of importance for integrated neural-symbolic systems. They are needed for closing the neural-symbolic learning cycle (Figure 1). However, they are also of independent interest, and are often studied separately.

As for the representation of knowledge, this component is present in all systems presented so far. The choice how representation is done often determines whether standard architectures are used, if a local or distributed approach is taken, and whether standard learning algorithms can be employed.

A large body of work exists on extracting knowledge from trained networks, usually focusing on the extraction of rules. [Jacobsson, 2005] gives a recent overview over extraction methods. A method from 1992 [Giles et al., 1991] is still up to date, where a method is given to extract a grammar represented as a finite state machine from a trained recurrent neural network. [McGarry et al., 1999] show how to extract rules from radial basis function networks by identifying minimal and maximal activation values. Some of the other efforts are reported in [Towell and Shavlik, 1993; Andrews et al., 1995; Bologna, 2000; Garcez et al., 2001; Lehmann et al., 2005]

It shall be noted that only a few systems have been proposed to date which include representation, learning, and extraction capabilities in a meaningful way, one of them being CILP [Garcez et al., 1997; Garcez and Zaverucha, 1999; Garcez et al., 2001]. It is to date a difficult research challenge to provide similar functionalities in a first-order setting.

Learning versus Reasoning

Ultimately, our goal should be to produce an effective AI system with added reasoning and learning capabilities, as recently pointed out by Valiant [Valiant, 2003] as a key challenge for computer science. It turns out that most current systems have either learning capabilities or reasoning capabilities, but rarely both. SHRUTI (Section 2.3), for example, is a reasoning system with very limited learning support.

In order to advance the state of the art in the sense of Valiant's vision mentioned above, it will be necessary to install systems with combined capabilities. In particular, learning should not be independent of reasoning, i.e. initial knowledge and logical consequences thereof should help guiding the learning process. There is no system to-date which realizes this in any way, and new ideas will be needed to attack this problem.

5 Conclusions and Further Work

Intelligent systems based on symbolic knowledge processing, on the one hand, and on artificial neural networks, on the other, differ substantially. Nevertheless, these are both standard approaches to artificial intelligence and it would be very desirable to combine the robustness of neural networks with the expressivity of symbolic knowledge representation. This is the reason why the importance of the efforts to bridge the gap between the connectionist and symbolic paradigms of Artificial Intelligence has been widely recognised. As the amount of hybrid data containing symbolic and statistical elements as well as noise increases in diverse areas such as bioinformatics or text and web mining, neural-symbolic learning and reasoning becomes of particular practical importance. Notwithstanding, this is not an easy task, as illustrated in the survey.

The merging of theory (background knowledge) and data learning (learning from examples) in neural networks has been indicated to provide learning systems that are more effective than e.g. purely symbolic and purely connectionist systems, especially when data are noisy [Garcez and Zaverucha, 1999]. This has contributed decisively to the growing interest in developing neural-symbolic systems, i.e. hybrid systems based on neural networks that are capable of learning from examples and background knowledge, and of performing reasoning tasks in a massively parallel fashion.

However, while symbolic knowledge representation is highly recursive and well understood from a declarative point of view, neural networks encode knowledge implicitly in their weights as a result of learning and generalisation from raw data, which are usually characterized by simple feature vectors. While significant theoretical progress has recently been made on knowledge representation and reasoning using neural networks, and on direct processing of symbolic and structured data using neural methods, the integration of neural computation and expressive logics such as first order logic is still in its early stages of methodological development.

Concerning knowledge extraction, we know that neural networks have been applied to a variety of real-world problems (e.g. in bioinformatics, engineering, robotics), and they were particularly successful when data are noisy. But entirely satisfactory methods for extracting symbolic knowledge from such trained networks in terms of accuracy, efficiency, rule comprehensibility, and soundness are still to be found. And problems on the stability and learnability of recursive models currently impose further restrictions on connectionist systems.

In order to advance the state of the art, we believe that it is necessary to look at the biological inspiration for neural-symbolic integration, to use more formal approaches for translating between the connectionist and symbolic paradigms, and to pay more attention to potential application scenarios.

The general motivation for research in the field of neural-symbolic integration (just given) arises from conceptual observations on the complementary nature of symbolic and neural network based artificial intelligence described above. This conceptual perspective is sufficient for justifying the mainly foundations-driven lines of research being undertaken in this area so far. However, it appears that this conceptual approach to the study of neural-symbolic integration has now reached an impasse which requires the identification of use cases and application scenarios in order to drive future research.

Indeed, the theory of integrated neural-symbolic systems has reached a quite mature state but has not been tested extensively so far on real application data. The current systems have been developed for the study of general principles, and are in general not suitable for real data or application scenarios that go beyond propositional logic. Nevertheless, these studies provide methods which can be exploited for the development of tools for use cases, and significant progress can now only be expected as a continuation of the fundamental research undertaken in the past.

In particular, first-order neural-symbolic integration still remains a widely open issue, where advances are very difficult, and it is very hard to judge to date to what extent the theoretical approaches can work in practice. We believe that the development of use cases with varying levels of expressive complexity is, as a result, needed to drive the development of methods for neural-symbolic integration beyond propositional logic [Hitzler et al., 2005].

BIBLIOGRAPHY

[Adamson and Damper, 1999] M. J. Adamson and R. I. Damper. B-RAAM: A connectionist model which develops holistic internal representations of symbolic structures. *Connection Science*, 11(1):41–71, 1999.

[Andrews et al., 1995] R. Andrews, J. Diederich and A. Tickle. A survey and critique of techniques for extracting rules from trained artificial neural networks. *Knowledge–Based Systems*, 8(6), 1995.

[Bader and Hitzler, 2004] S. Bader and P. Hitzler. Logic programs, iterated function systems, and recurrent radial basis function networks. *Journal of Applied Logic*, 2(3):273–300, 2004.

[Bader et al., 2004a] S. Bader, S. Hölldobler and A. Scalzitti. Semiring artificial neural networks and weighted automata. In G. Palm S. Biundo, T. Frühwirth, editor, *KI 2004: Advances in Artificial Intelligence. Proceedings of the 27th Annual German Conference on Artificial Intelligence, Ulm, Germany, September 2004*, volume 3238 of *Lecture Notes in Artificial Intelligence*, pages 281–294. Springer, 2004.

[Bader et al., 2004b] S. Bader, P. Hitzler and S. Hölldobler. The integration of connectionism and knowledge representation and reasoning as a challenge for artificial intelligence. In L. Li and K. K. Yen, editors, *Proceedings of the Third International Conference on Information, Tokyo, Japan*, pages 22–33. International Information Institute, 2004. ISBN 4-901329-02-2.

[Bader et al., 2005a] S. Bader, A. S. d'Avila Garcez and P. Hitzler. Computing first-order logic programs by fibring artificial neural network. In I. Russell and Z. Markov, editors, *Proceedings of the 18th International Florida Artificial Intelligence Research Symposium Conference, FLAIRS05, Clearwater Beach, Florida, May 2005*, pages 314–319. AAAI Press, 2005.

[Bader et al., 2005b] S. Bader, P. Hitzler and A. Witzel. Integrating first-order logic programs and connectionist systems — a constructive approach. In A. S. d'Avila Garcez, J. Elman and P. Hitzler, editors, *Proceedings of the IJCAI-05 Workshop on Neural-Symbolic Learning and Reasoning, NeSy'05, Edinburgh, UK*, 2005.

[Barnden, 1995] J. A. Barnden. High-level reasoning, computational challenges for connectionism, and the conposit solution. *Applied Intelligence*, 5(2):103–135, 1995.

[Barnsley, 1993] M. Barnsley. *Fractals Everywhere*. Academic Press, San Diego, CA, USA, 1993.

[Bishop, 1995] C. M. Bishop. *Neural Networks for Pattern Recognition*. Oxford University Press, 1995.

[Blair, 1997] A. Blair. Scaling-up RAAMs. Technical Report CS-97-192, University of Brandeis, 1997.

[Bologna, 2000] G. Bologna. Rule extraction from a multi layer perceptron with staircase activation functions. In *IJCNN (3)*, pages 419–424, 2000.

[Browne and Sun, 1999] A. Browne and R. Sun. Connectionist variable binding. *Expert Systems*, 16(3):189–207, 1999.

[Browne and Sun, 2001] A. Browne and R. Sun. Connectionist inference models. *Neural Networks*, 14(10):1331–1355, 2001.

[Elman, 1990] J. L. Elman. Finding structure in time. *Cognitive Science*, 14:179–211, 1990.

[Fletcher, 2001] P. Fletcher. Connectionist learning of regular graph grammars. *Connection Science*, 13(2):127–188, 2001.

[Frasconi et al., 2001] P. Frasconi, M. Gori, A. Kuchler and A. Sperduti. From sequences to data structures: Theory and applications. In J. Kolen and S. C. Kremer, editors, *Dynamical Recurrent Networks*, pages 351–374. IEEE Press, 2001.

[Gabbay, 1999] D. M. Gabbay. *Fibring Logics*. Oxford Univesity Press, 1999.

[Gallant, 1993] S. I. Gallant. *Neural network learning and expert systems*. MIT Press, Cambridge, MA, 1993.

[Garcez and Gabbay, 2004] A. S. d'Avila Garcez and D. M. Gabbay. Fibring neural networks. In *In Proceedings of the 19th National Conference on Artificial Intelligence (AAAI 04). San Jose, California, USA, July 2004*. AAAI Press, 2004.

[Garcez and Lamb, 200x] D. M. Gabbay, A. S. d'Avila Garcez and L. C. Lamb. Value-based argumentation frameworks as neural-symbolic learning systems. *Journal of Logic and Computation*, 200x. To appear.

[Garcez and Zaverucha, 1999] A. S. d'Avila Garcez and G. Zaverucha. The connectionist inductive lerarning and logic programming system. *Applied Intelligence, Special Issue on Neural networks and Structured Knowledge*, 11(1):59–77, 1999.

[Garcez et al., 1997] A. S. d'Avila Garcez, G. Zaverucha and L. A. V. de Carvalho. Logical inference and inductive learning in artificial neural networks. In C. Hermann, F. Reine and A. Strohmaier, editors, *Knowledge Representation in Neural networks*, pages 33–46. Logos Verlag, Berlin, 1997.

[Garcez et al., 2000] A. S. d'Avila Garcez, K. Broda and D. M. Gabbay. Metalevel priorities and neural networks. In *Proceedings of the Workshop on the Foundations of Connectionist-Symbolic Integration, ECAI'2000, Berlin*, August 2000.

[Garcez et al., 2001] A. S. d'Avila Garcez, K. Broda and D. M. Gabbay. Symbolic knowledge extraction from trained neural networks: A sound approach. *Artificial Intelligence*, 125:155–207, 2001.

[Garcez et al., 2002a] A. S. d'Avila Garcez, K. B. Broda and D. M. Gabbay. *Neural-Symbolic Learning Systems — Foundations and Applications*. Perspectives in Neural Computing. Springer, Berlin, 2002.

[Garcez *et al.*, 2002b] A. S. d'Avila Garcez, L. C. Lamb and D. M. Gabbay. A connectionist inductive learning system for modal logic programming. In *Proceedings of
the IEEE International Conference on Neural Information Processing ICONIP'02,
Singapore*, 2002.
[Garcez *et al.*, 2003] A. S. d'Avila Garcez, L. C. Lamb and D. M. Gabbay. Neural-
symbolic intuitionistic reasoning. In M. Koppen A. Abraham and K. Franke, editors,
Frontiers in Artificial Intelligence and Applications, Melbourne, Australia, Decem-
ber 2003. IOS Press. Proceedings of the Third International Conference on Hybrid
Intelligent Systems (HIS'03).
[Garcez *et al.*, 2004a] A. S. d'Avila Garcez, D. M. Gabbay and L. C. Lamb. Argumen-
tation neural networks. In *Proceedings of 11th International Conference on Neural
Information Processing (ICONIP'04)*, Lecture Notes in Computer Science LNCS,
Calcutta, November 2004. Springer-Verlag.
[Garcez *et al.*, 2004b] A. S. d'Avila Garcez, L. C. Lamb, K. Broda and D. M. Gabbay.
Applying connectionist modal logics to distributed knowledge representation prob-
lems. *International Journal of Artificial Intelligence Tools*, 2004.
[Garcez *et al.*, 2005] A. S. d'Avila Garcez, D. M. Gabbay and L. C. Lamb. *Connectionist
Non-Classical Logics*. Springer-Verlag, 2005. To appear.
[Giles *et al.*, 1991] C. Giles, D. Chen, C. Miller, H. Chen, G. Sun and Y. Lee. Second-
order recurrent neural networks for grammatical inference. In *Proceedings of the
International Joint Conference on Neural Networks 1991*, volume 2, pages 273–281,
New York, 1991. IEEE.
[Gust and Kühnberger, 2005] H. Gust and K.-U. Kühnberger. Learning symbolic infer-
ences with neural networks. In *CogSci 2005*, 2005. to appear.
[Hammer *et al.*, 2004a] B. Hammer, A. Micheli, A. Sperduti and M. Strickert. Recursive
self-organizing network models. *Neural Networks*, 17(8–9):1061–1085, 2004. Special
issue on New Developments in Self-Organizing Systems.
[Hammer *et al.*, 2004b] B. Hammer, A. Micheli, M. Strickert and A. Sperduti. A general
framework for unsupervised processing of structured data. *Neurocomputing*, 57:3–35,
2004.
[Hammer, 2002] B. Hammer. Recurrent networks for structured data — a unifying
approach and its properties. *Cognitive Systems Research*, 3(2):145–165, 2002.
[Hammer, 2003] B. Hammer. Perspectives on learning symbolic data with connectionis-
tic systems. In R. Kühn, R. Menzel, W. Menzel, U. Ratsch, M. M. Richter and I.-O.
Stamatescu, editors, *Adaptivity and Learning*, pages 141–160. Springer, 2003.
[Hammerton, 1998] J. A. Hammerton. *Exploiting Holistic Computation: An evaluation
of the Sequential RAAM*. PhD thesis, University of Birmingham, 1998.
[Hatzilygeroudis and Prentzas, 2000] I. Hatzilygeroudis and J. Prentzas. Neurules: In-
tegrating symbolic rules and neurocomputing. In D. Fotiades and S. Nikolopoulos,
editors, *Advances in Informatics*, pages 122–133. World Scientific, 2000.
[Hatzilygeroudis and Prentzas, 2004] I. Hatzilygeroudis and J. Prentzas. Neuro-
symbolic approaches for knowledge representation in expert systems. *International
Journal of Hybrid Intelligent Systems*, 1(3-4):111–126, 2004.
[Hawkins and Blakeslee, 2004] J. Hawkins and S. Blakeslee. *On Intelligence*. Henry Holt
and Company, LLC, New York, 2004.
[Healy, 1999] M. J. Healy. A topological semantics for rule extraction with neural net-
works. *Connection Science*, 11(1):91–113, 1999.
[Hebb, 1949] D. O. Hebb. *The Organization of Behavior*. Wiley, New York, 1949.
[Hilario, 1995] M. Hilario. An overview of strategies for neurosymbolic integration. In
R. Sun and F. Alexandre, editors, *Proceedings of the Workshop on Connectionist-
Symbolic Integration: From Unied to Hybrid Approaches*, Montreal, 1995.
[Hitzler and Seda, 2000] P. Hitzler and A. K. Seda. A note on relationships between
logic programs and neural networks. In Paul Gibson and David Sinclair, editors,
Proceedings of the Fourth Irish Workshop on Formal Methods, IWFM'00, Electronic
Workshops in Comupting (eWiC). British Computer Society, 2000.

[Hitzler et al., 2004] P. Hitzler, S. Hölldobler and A. K. Seda. Logic programs and connectionist networks. *Journal of Applied Logic*, 3(2):245–272, 2004.

[Hitzler et al., 2005] P. Hitzler, S. Bader and A. S. d'Avila Garcez. Ontology learning as a use case for artificial intelligence. In A. S. d'Avila Garcez, J. Elman and P. Hitzler, editors, *Proceedings of the IJCAI-05 workshop on Neural-Symbolic Learning and Reasonin, NeSy'05, Edinburgh, UK, August 2005*, 2005.

[Hitzler, 2004] P. Hitzler. Corollaries on the fixpoint completion: studying the stable semantics by means of the clark completion. In D. Seipel, M. Hanus, U. Geske and O. Bartenstein, editors, *Proceedings of the 15th International Conference on Applications of Declarative Programming and Knowledge Management and the 18th Workshop on Logic Programming, Potsdam, Germany, March 4-6, 2004*, volume 327 of *Technichal Report*, pages 13–27. Bayerische Julius-Maximilians-Universität Würzburg, Institut für Informatik, 2004.

[Hölldobler and Kalinke, 1994] S. Hölldobler and Y. Kalinke. Towards a massively parallel computational model for logic programming. In *Proceedings ECAI94 Workshop on Combining Symbolic and Connectionist Processing*, pages 68–77. ECCAI, 1994.

[Hölldobler and Kurfess, 1992] S. Hölldobler and F. Kurfess. CHCL – A connectionist inference system. In B. Fronhöfer and G. Wrightson, editors, *Parallelization in Inference Systems*, pages 318 – 342. Springer, LNAI *590*, 1992.

[Hölldobler et al., 1997] S. Hölldobler, Y. Kalinke and H. Lehmann. Designing a counter: Another case study of dynamics and activation landscapes in recurrent networks. In *Proceedings of the KI97: Advances in Artificial Intelligence*, volume 1303 of *LNAI*, pages 313–324. Springer, 1997.

[Hölldobler et al., 1999a] S. Hölldobler, Y. Kalinke and H.-P. Störr. Approximating the semantics of logic programs by recurrent neural networks. *Applied Intelligence*, 11:45–58, 1999.

[Hölldobler et al., 1999b] S. Hölldobler, Y. Kalinke and J. Wunderlich. A recursive neural network for reflexive reasoning. In S. Wermter and R. Sun, editors, *Hybrid Neural Systems*. Springer, Berlin, 1999.

[Hölldobler, 1990] S. Hölldobler. A structured connectionist unification algorithm. In *Proceedings of AAAI*, pages 587–593, 1990.

[Hölldobler, 1993] S. Hölldobler. *Automated Inferencing and Connectionist Models*. Fakultät Informatik, Technische Hochschule Darmstadt, 1993. Habilitationsschrift.

[Hopcroft and Ullman, 1989] J. E. Hopcroft and J. D. Ullman. *Introduction to Automata Theory, Languages and Computation*. Addison Wesley, 1989.

[Jacobsson, 2005] H. Jacobsson. Rule extraction from recurrent neural networks: A taxonomy and review. *Neural Computation*, 17(6):1223–1263, 2005.

[Kalinke and Lehmann, 1998] Y. Kalinke and H. Lehmann. Computations in recurrent neural networks: From counters to iterated function systems. In G. Antoniou and J. Slaney, editors, *Advanced Topics in Artificial Intelligence*, volume 1502 of *LNAI*, Berlin/Heidelberg, 1998. Proceedings of the 11th Australian Joint Conference on Artificial Intelligence (AI'98), Springer–Verlag.

[Kalinke, 1997] Y. Kalinke. Using connectionist term representation for first–order deduction – a critical view. In F. Maire, R. Hayward and J. Diederich, editors, *Connectionist Systems for Knowledge Representation Deduction*. Queensland University of Technology, 1997. CADE–14 Workshop, Townsville, Australia.

[Kleene, 1956] S. C. Kleene. Representation of events in nerve nets and finite automata. In C. E. Shannon and J. McCarthy, editors, *Automata Studies*, volume 34 of *Annals of Mathematics Studies*, pages 3–41. Princeton University Press, Princeton, NJ, 1956.

[Kwasny and Kalman, 1995] S. Kwasny and B. Kalman. Tail-recursive distributed representations and simple recurrent networks. *Connection Science*, 7:61–80, 1995.

[Lehmann et al., 2005] J. Lehmann, S. Bader and P. Hitzler. Extracting reduced logic programs from artificial neural networks. In A. S. d'Avila Garcez, J. Elman and P. Hitzler, editors, *Proceedings of the IJCAI-05 Workshop on Neural-Symbolic Learning and Reasoning, NeSy'05, Edinburgh, UK*, 2005.

[Lloyd, 1988] J. W. Lloyd. *Foundations of Logic Programming*. Springer, Berlin, 1988.

[Maass and Markram, 2004] W. Maass and H. Markram. On the computational power of recurrent circuits of spiking neurons. *Journal of Computer and System Sciences*, 69(4):593–616, 2004.

[Maass et al., 2005] W. Maass, T. Natschläger and H. Markram. On the computational power of circuits of spiking neurons. *J. of Physiology (Paris)*, 2005. in press.

[Maass, 2002] W. Maass. Paradigms for computing with spiking neurons. In J. L. van Hemmen, J. D. Cowan and E. Domany, editors, *Models of Neural Networks*, volume 4 of *Early Vision and Attention*, chapter 9, pages 373–402. Springer, 2002.

[McCulloch and Pitts, 1943] W. S. McCulloch and W. Pitts. A logical calculus of the ideas immanent in nervous activity. *Bulletin of Mathematical Biophysics*, 5:115–133, 1943.

[McGarry et al., 1999] K. J. McGarry, J. Tait, S. Wermter and J. MacIntyre. Rule-extraction from radial basis function networks. In *Ninth International Conference on Artificial Neural Networks (ICANN'99)*, volume 2, pages 613–618, Edinburgh, UK, 1999.

[Natschläger and Maass, 2002] T. Natschläger and W. Maass. Spiking neurons and the induction of finite state machines. *Theoretical Computer Science: Special Issue on Natural Computing*, 287:251–265, 2002.

[Niklasson and Linåker, 2000] Lars Niklasson and Fredrik Linåker. Distributed representations for extended syntactic transformation. *Connection Science*, 12(3–4):299–314, 2000.

[Pinkas, 1991a] G. Pinkas. Propositional non-monotonic reasoning and inconsistency in symmetrical neural networks. In *IJCAI*, pages 525–530, 1991.

[Pinkas, 1991b] G. Pinkas. Symmetric neural networks and logic satisfiability. *Neural Computation*, 3:282–291, 1991.

[Plate, 1991] T. A. Plate. Holographic Reduced Representations: Convolution algebra for compositional distributed representations. In J. Mylopoulos and R. Reiter, editors, *Proceedings of the 12th International Joint Conference on Artificial Intelligence, Sydney, Australia, August 1991*, pages 30–35, San Mateo, CA, 1991. Morgan Kauffman.

[Plate, 1995] T. A. Plate. Holographic reduced representations. *IEEE Transactions on Neural Networks*, 6(3):623–641, May 1995.

[Pollack, 1990] J. B. Pollack. Recursive distributed representations. *AIJ*, 46:77–105, 1990.

[Rodriguez, 1999] P. Rodriguez. A recurrent neural network that learns to count. *Connection Science*, 11(1):5–40, 1999.

[Rumelhart et al., 1986] D. E. Rumelhart, G. E. Hinton and R. J. Williams. Learning internal representations by error propagation. In D. E. Rumelhart, J. L. McClelland and the PDP Research Group, editors, *Parallel Distributed Processing, vol. 1: Foundations*, pages 318–362. MIT Press, 1986.

[Russell and Norvig, 2003] S. Russell and P. Norvig. *Artificial Intelligence: A Modern Approach*. Prentice Hall, 2 edition, 2003.

[Seda and Lane, 2005] A. K. Seda and M. Lane. On approximation in the integration of connectionist and logic-based systems. In *Proceedings of the Third International Conference on Information (Information'04)*, pages 297–300, Tokyo, November 2005. International Information Institute.

[Seda, 2005] A. K. Seda. On the integration of connectionist and logic-based systems. In T. Hurley, M. Mac an Airchinnigh, M. Schellekens, A. K. Seda and G. Strong, editors, *Proceedings of MFCSIT2004, Trinity College Dublin, July 2004*, Electronic Notes in Theoretical Computer Science, pages 1–24. Elsevier, 2005.

[Shastri and Ajjanagadde, 1993] L. Shastri and V. Ajjanagadde. From associations to systematic reasoning: A connectionist representation of rules, variables and dynamic bindings using temporal synchrony. *Behavioural and Brain Sciences*, 16(3):417–494, September 1993.

[Shastri and Wendelken, 1999] L. Shastri and C. Wendelken. Soft computing in SHRUTI: — A neurally plausible model of reflexive reasoning and relational information processing. In *Proceedings of the Third International Symposium on Soft Computing, Genova, Italy*, pages 741–747, June 1999.

[Shastri and Wendelken, 2003] L. Shastri and C. Wendelken. Learning structured representations. *Neurocomputing*, 52–54:363–370, 2003.

[Shastri, 1999] L. Shastri. Advances in Shruti — A neurally motivated model of relational knowledge representation and rapid inference using temporal synchrony. *Applied Intelligence*, 11:78–108, 1999.

[Shastri, 2002] L. Shastri. Episodic memory and cortico-hippocampal interactions. *Trends in Cognitive Sciences*, 6:162–168, 2002.

[Sougne, 2001] J. P. Sougne. Binding and multiple instantiation in a distributed network of spiking nodes. *Connection Science*, 13(1):99–126, 2001.

[Sperduti et al., 1995] A. Sperduti, A. Starita and C. Goller. Learning distributed representations for the classifications of terms. In *Proceedings of the 14th International Joint Conference on AI, IIJCAI-95*, pages 509–517. Morgan Kaufmann, 1995.

[Sperduti et al., 1997] A. Sperduti, A. Starita and C. Goller. Distributed representations for terms in hybrid reasoning systems. In Ron Sun and Frédéric Alexandre, editors, *Connectionist Symbolic Integration*, chapter 18, pages 329–344. Lawrence Erlbaum Associates, 1997.

[Sperduti, 1994a] A. Sperduti. Labeling RAAM. *Connection Science*, 6(4):429–459, 1994.

[Sperduti, 1994b] A. Sperduti. Encoding labeled graphs by labeling RAAM. In Jack D. Cowan, G. Tesauro and J. Alspector, editors, *Advances in Neural Information Processing Systems 6, [7th NIPS Conference, Denver, Colorado, USA, 1993]*, pages 1125–1132. Morgan Kaufmann, 1994.

[Stolcke and Wu, 1992] A. Stolcke and D. Wu. Tree matching with recursive distributed representations. Technical Report tr-92-025, ICSI, Berkeley, 1992.

[Sun, 2001] R. Sun. Hybrid systems and connectionist implementationalism. In *Encyclopedia of Cognitive Science*. MacMillan Publishing Company, 2001.

[Tesauro, 1995] G. Tesauro. Temporal difference learning and TD-Gammon. *Communications of the ACM*, 38(3), March 1995.

[Towell and Shavlik, 1993] G. G. Towell and J. W. Shavlik. Extracting refined rules from knowledge-based neural networks. *Machine Learning*, 13:71–101, 1993.

[Towell and Shavlik, 1994] G. G. Towell and J. W. Shavlik. Knowledge-based artificial neural networks. *Artificial Intelligence*, 70(1–2):119–165, 1994.

[Valiant, 2003] L. G. Valiant. Three problems in computer science. *Journal of the ACM*, 50(1):96–99, 2003.

[van der Velde and de Kamps, 2005] F. van der Velde and M. de Kamps. Neural blackboard architectures of combinatorial structures in cognition. *Behavioral and Brain Sciences*, 2005. to appear.

[Wendelken and Shastri, 2003] C. Wendelken and L. Shastri. Acquisition of concepts and causal rules in shruti. In *Proceedings of Cognitive Science, Boston, MA, August 2003*, 2003.

[Wendelken and Shastri, 2004] C. Wendelken and L. Shastri. Multiple instantiation and rule mediation in shruti. *Connection Science*, 16:211–217, 2004.

Modelling Evolvable Systems:
A Temporal Logic View

HOWARD BARRINGER AND DAVID E. RYDEHEARD

1 Motivation

There is a growing trend to develop and deploy computing systems that adapt or evolve themselves dynamically according to use and observed changes in their environment. Evolution may be triggered either from within the system or be forced by some external, and not necessarily human, agent. The traditional view of system specification being invariant for the duration of a run will no longer hold for such an evolvable system; it will be rare to be able to predict all possible ways in which a system may evolve.

In this contribution, we consider one approach to how temporal logic, and specifications based on temporal logic, can be used to support formal reasoning about evolvable systems in a natural way. Our approach builds on our past work in temporal logic-based specification, the imperative view of temporal logic as in our work on executable temporal logic, METATEM, developed jointly with Dov Gabbay in the late 1980s [Barringer *et al.*, 1996], [Barringer *et al.*, 1995], and the more recent work on the EAGLE logic for run-time verification [Barringer *et al.*, 2004], [Artho *et al.*, 2005].

A rather crude view of formal systems development has basic steps comprising:

1. formalisation of requirements as a set of system properties,

2. definition/construction of a formal specification,

3. revision of the specification,

4. development of an implementation that satisfies the specification,

   - via refinement and/or decomposition of specification, together with theorem proving, or

   - via checking individual properties against a purported model,

5. revision of the specification and/or implementation,

6. test analysis,

7. deployment of system.

Although this is recognized as being overly simplistic and generally unrealizable, a key point is that once the system is deployed, its specification remains invariant during its deployed lifetime. Of course, later releases of the system, including bug fixes, new features and general updates, are quite usual. However, one again regards the specifications of these new releases as invariant during their (often too short a) lifetime.

This static view of specification and system behaviour is unsatisfactory from many points of view. In some highly specialized application areas, for example deep space mission software, control software for future unmanned exploration of planets, e.g. NASA's Martian rovers, a certain degree of adaptive software has already been prototyped. In Earth-based deployment, software systems are beginning to be constructed that feature in limited ways *autonomy*, *adaptation* and *evolution*. Of course, there are many applications which include algorithms that adapt, or attempt to optimize, their performance to the current situation, e.g. network routing, dynamic load-balancing of work for processors, etc. In these latter scenarios, however, the overall system specification remains invariant, e.g. correct delivery of packets or the timely execution of work. At a quite different organisational level, the evolutionary behaviour of businesses and business processes, and the formal descriptions of such behaviour, has attracted considerable interest (see, for example, [Greenwood *et al.*, 1998], [Greenwood *et al.*, 1999] or [Cunin *et al.*, 2001]).

The question then is, as one changes the nature of systems so that adaptation or evolution becomes a dominant feature and to be present at significantly higher levels of system organisation, how does one specify and reason about systems with such features?

1.1 Behavioural specification

The possible worlds model, which is usually used to provide meaning for temporal and modal logics might at first sight appear appropriate for capturing the behaviour of evolving systems. The accessibility relation captures the evolution between the system in one world and another. But in most uses the notion of evolution of the system is at a very low computational level. For example, typically, in temporal logic-based specification, the world represents the state of a system, providing, for example, values to variables, and the accessibility relation captures the next-state relation of the "computation" of the system being specified. The dynamic behaviour of a system, the specification of which remains invariant, can thus be modelled.

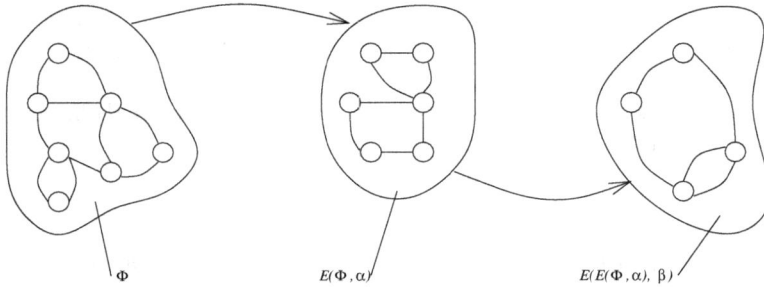

Φ $E(\Phi,\alpha)$ $E(E(\Phi,\alpha),\ \beta)$

Figure 1. Evolving Worlds

Now consider the following scenario. A system is running according to some specified behaviour, say Φ. At some point, the system is interrupted (the apparatus for doing that is not important here), it is then modified and re-started but now running according to a modified specification, say $E(\Phi,\alpha)$. Let us view the entirety of the interrupt, modify and re-start process as an evolutionary step of the original system. Indeed, the transformer $E : Spec \times Modification \rightarrow Spec$ performs the evolution.

It appears sensible to separate the system level evolution from the computational state evolution of the system. Thus, in a possible worlds setting, at the system evolution level, let the individual worlds be associated with the possible evolved system specifications; indeed, one might take the worlds as Kripke structures capturing the computational evolution of the specified system. The relation between the Kripke structures is then that of evolutionary steps. We thus have a two-level model description of the overall system.

As presented above in pictorial form, the set-up seems rather straightforward! Perhaps one reason for being fooled is that we are reading the diagram operationally: execute the system according to the specification Φ, then evolve at some point with modification α, and then continue executing $E(\Phi,\alpha)$, etc.. But what is α? Where did it come from? If we know what α is at the time the original system Φ is developed then indeed one can reason about the future evolution. This, however, is rather unlikely. Indeed, in the absence of any constraints on what the modifications α and β may be, one would need to reason about the system Φ and all its possible evolutions — clearly a challenging if not impossible task.

As an aside, it is worth noting that one could construct a temporal logic capable of expressing the situation outlined above. It would be a multi-modal temporal logic containing in addition to the basic next-step temporal

operator, a collection of system evolution operators (each characterising the
respective underlying evolution relation). However, the logic would also
require a means to embed meta-level action, i.e. changing behaviour, at the
object level. Rather simplistically, the following (heavily extended) fixpoint
interval temporal logic could represent an evolving system:

$$\nu B(F).(F; \exists \alpha.(input(\alpha) \wedge \bigcirc B(E(F, \alpha))))$$

Read the above formula as follows: F is a specification of the "current"
system. At some point F stops and is followed by an input of the modification α which is then applied to the specification F to yield a new specification for execution. This is a very powerful logic, undoubtedly far too
powerful, and one has to question its virtue here, over and above just formal
description. We will later propose a different approach to the use of temporal logic which appears to provide a restricted but flexible and sufficiently
general framework to modelling and reasoning about evolvable systems.

1.2   Modifying a specification

Let us briefly comment on some crucial and difficult issues associated with
modifying specifications, in particular, behavioural specifications given
purely logically. Such a specification is essentially a logical theory given
as a set of constraints over a set of observables which are expressed as predicates and functions. As an example, imagine the behaviour of a buffer is
specified in terms of observations of input to and output from the buffer;
indeed we assume predicates $in(v)$ and $out(v)$ hold if value v passes in to,
respectively out from, the buffer. Without going into the details of what the
specification is (just imagine it!), name it Φ and let us assume that at some
stage during the buffer's operation something causes the system to evolve
through the ADDITION of a new constraint α on the buffer. If we further
assume that this new constraint is not in conflict with, i.e. is consistent
with, the old specification Φ, then the new specification can be given by
$\Phi \wedge \alpha$. Further restricting behaviour is, in effect, relatively straightforward
and can be embedded easily as a logical operation. If, however, the required
evolution required REMOVAL (or replacement) of some behaviour, characterised by a logical constraint, we are in deep water. Deletion is a very
difficult issue and has been stretching some computer scientists and logicians for many years. Of course, physical deletion of some given constraints
from a known specification is clearly possible - this is a meta-level operation.
However, physical deletion doesn't usually yield the desired effect of removing unwanted behaviours. The deletion we most probably want is a logical
deletion, i.e. to ensure that α is not amongst the consequences of the modified specification. How this is achieved, or whether this can be achieved,

appears to be a fundamental issue for giving a logical characterisation of an evolving system. There are interesting ideas based on the notion of Gabbay's anti-formulae [Gabbay *et al.*, 2002] - which effectively annihilate the use of a particular formula during proof, thus disabling its use. However, at present, such ideas won't actually achieve what is required here, for some other formula is sure to characterise the undesirable behaviour, which then requires addition of all such anti-formulae! This is clearly an avenue to be explored but it is one that is certainly strewn with numerous pitfalls.

1.3   Checking and monitoring behaviour

Assume that we have a solution to creating formal logic-based specification for an evolving or adaptive software system. What can one do with it? The ultimate challenge is to prove, preferably in an automated fashion, that certain properties are invariant over evolution or adaptation. The properties include both safety and liveness. How meaningful the properties are depends crucially on what constraints are placed on how a system evolves. As indicated above, if one allows complete freedom in evolution then there is probably very little that can be achieved. On the other hand, one can turn this around to try to answer synthesis questions such as what constraints must be placed on the evolution of this system in order to preserve a desired property.

The automated checking and/or synthesis of properties is an intractable, if not impossible, problem and even with today's best technology it is likely to be of limited capability and unable to handle fully the behaviour of large-scale evolvable systems. This argument should not be used to rule out research into the development and application of techniques to tackle these problems for specialized situations. We're just pointing out there'll be no silver bullets for full-scale formal verification.

An alternative to static verification, as achieved by model-checking, automated theorem proving and their combination, is dynamic formal analysis, or run-time verification. In run-time analysis, the system under scrutiny is observed during execution and properties checked either in an on-line (potentially interacting with the system) or off-line (a post hoc analysis of an execution). Clearly this can be significantly less costly than attempting to analyse the complete behaviour of a system, which is what static analyses, such as model-checking, will attempt to achieve. On-line analyses have the benefit of being able to take action on the system as soon as it is discovered that some property fails. Integration of observer and analyst with the system under scrutiny must ensure, however, that there is no unacceptable degradation in overall system performance. Furthermore, integration of an observer shouldn't change the desired observational behaviour of the system.

Most interesting and challenging of all is the integration of 'action' into the logical specification, analysis and execution frameworks. Is adaptation and evolution treated at a meta-level or embedded with the logical specification itself? To return to the rather expressive fixpoint formula given above, one can certainly see how it might be used in a run-time analysis. However, apart from the semantic cleanliness, it is not clear what else may be gained.

1.4 Towards a logic for evolvable systems

In this article, we present our initial steps at using linear-time temporal logic for the specification, and the static and dynamic logical analysis, of evolvable systems. We use as an illustrative example a model of an evolvable bounded buffer. A process, which we refer to as an *evolver* process, monitors the dynamic behaviour of a bounded buffer. If the evolver finds that the buffer needs to store more items than the buffer's current capacity, the evolver triggers an evolution of the buffer to increase its size. On the other hand, if the evolver finds that the buffer is nearly empty most of the time, the evolver triggers an evolution that contracts the capacity of the buffer. Of course, there are different ways in which a buffer may expand or indeed contract. For example, it may expand by first creating a new empty buffer, of the same size and type, and then re-directing the output of the original buffer to the input of the new one and then taking the new output and directing at the target of the original buffer. To put this another way, the two buffers are serially piped together. Alternatively, an expansion could have been achieved by putting two buffers in parallel. More simply, the buffer may have a specific capacity that can directly be modified. To make this all concrete, we present an implementation of this example in Java. A property that we'd like to establish of the subsequent *evolvable* buffer – the buffer and its evolver – is that the composite system has effectively unbounded capacity.

Rather than collapse the reasoning all down to the object level, we present a two-level temporal logic that can be applied, quite elegantly, to capture both the base-level behaviour and meta-level evolutionary behaviour of the buffer. We require a model structure, which we call an *E-structure*, that captures the two-level structure suggested by Figure 1. That particular figure presents too simplistic a view. Indeed, the set-up has to be more flexible and a little more complex. For instance, in order to able to reason about the (meta-level) evolution of a system, predicate abstraction is used to abstract away from lower-level detail. Indeed, it is precisely this abstraction process that simplifies the reasoning over different Kripke-like structures enabling us to extract features common to these structures and examine their behaviour with respect to evolutionary steps. Sections 3 and 4 present this model-theoretic framework for a two-level approach to evolvable systems.

We then introduce an appropriate fixpoint temporal logic over E-structures that supports system specification and reasoning. We introduce three types of *next* modalities: (i) the familiar $\bigcirc$ modality for moving to the next state of a local system; (ii) a family of action modalities $\mathbf{X}_e$ for evolutionary transition e from one local system to another; and (iii) the anonymous modality $\mathbf{X}$ denoting that an (un-named) evolutionary step occurs.

Sufficient technical machinery is then in place for us to present some examples of evolvable buffering systems.

2   An Evolvable Bounded Buffer

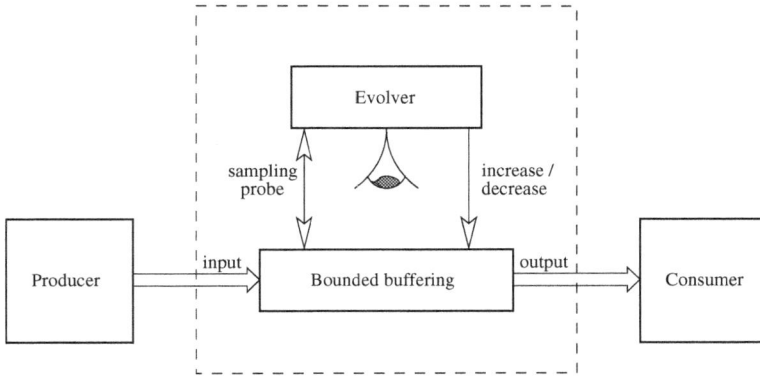

Figure 2. Communication via an Evolvable Buffer

Figure 2 depicts a typical scenario for an application of a bounded buffer. A producer process passes items to a consumer process. The producer and consumer processes are assumed to be independent, asynchronous processes and hence typically have different rates of production and consumption of items. A bounded buffer is used to smooth the transmission between the producer and consumer and avoid unnecessary blocking that may occur through the buffer being full and thus blocking further input and slowing the producer process. If we further assume that the rates of production and consumption of items are highly variable, it is then difficult to predict an optimal size for the bounded buffer. The buffering system therefore needs to be adaptive. The buffer capacity must evolve dynamically to suit the varying rates of input from the producer process and output demanded by

the consumer process. An evolver process samples the performance of the
bounded buffer via probes and determines when to increase, or decrease,
the buffer's capacity.

We provide a simple model of this example in Java. The class definition
below describes a bounded buffer (that holds strings) implemented as a
passive object, indeed in concurrency terminology as a monitor. The class
has two synchronized methods for input to and output from the buffer. The
methods are synchronized to ensure exclusive access to the shared instance
variables of class. The constructor of the class creates buffer objects of fixed
size. If an input method is called from a thread t when the buffer is full,
the calling thread t is suspended on a wait list until there is notification
that buffer has space for more items. Similarly, if a thread attempts to
obtain an item from the buffer when there are none, the calling thread is
also suspended until the buffer has some actual content to pass out.

```
public class Buffer {
  protected int max;
  protected String [] content;
  protected int head;

  Buffer(int rqdMax){
    max = rqdMax;
    head = 0;
    content = new String [max];
  }

  public synchronized void input(String item)
          throws InterruptedException
  {
    while (!(head < max) ) { wait();}
    content[head++] = item;
    notifyAll();
  }

  public synchronized String output()
          throws InterruptedException
  {
    while ((head == 0) ) { wait(); }
    String value = content[--head];
    notifyAll();
    return value;
  }
} //class Buffer
```

The next class definition extends the Buffer class with features to support

the notion of an evolvable buffer. The possible evolutions of the buffer are
kept very simple. A method is provided to increase the size of the buffer and
a method to decrease its size. Naturally, both methods are synchronized
methods to ensure exclusive access to the shared information. The class also
provides two probe methods, shouldExpand and shouldContract, which
can be called to determine whether the buffer should evolve, i.e. its capacity
expanded or contracted.

```
public class EvolvableBuffer extends Buffer
{
  private final int initialSize;
  private double limit;
  private int factor;

  EvolvableBuffer(int size, double rqdLimit, int rqdFactor){
    super(size);
    initialSize = size;
    limit = rqdLimit;
    factor = rqdFactor;
  }

  public boolean shouldExpand(){
    return (head >= limit*max);
  }

  public boolean shouldContract(){
    return (max > initialSize && head <= (1-limit)*max);
  }

  public synchronized void expand()
  {
    reSize(max*factor);
    max = max*factor;
    limit += (1-limit)/factor*(factor-1);
    notifyAll();
  }

  public synchronized void contract()
  {
    reSize(max/factor);
    max = max/factor;
    limit = 1-((1-limit)/(factor-1)*factor);
    notifyAll();
  }
```

```
  private void reSize(int newSize)
  {
    String [] copy = new String [newSize];
    for (int i = 0; i < head; i++){
      copy[i] = content[i];
    }
    content = copy;
  }

} //class EvolvableBuffer
```

Next, in the code below, we provide a very simple evolver process. An evolver is an active object that samples the state of a bounded buffer using the probe methods shouldExpand and shouldContract of the Buffer object with which it is associated. If either method returns true, the evolver invokes the appropriate evolutionary action via expand or contract. We also give a suitable class definition for the example main program that sets up a producer, consumer, evolvable buffer and evolver process, as depicted in Figure 2.

```
public class Evolver extends Thread {
  EvolvableBuffer x;

  Monitor(EvolvableBuffer givenBuffer){
    this.x = givenBuffer;
  }

  public void run()
  {
    while (true) {
      try {
        if (x.shouldExpand()) { x.expand(); }
        else if (x.shouldContract()) { x.contract(); }
        Thread.sleep(500);
      }
      catch (InterruptedException e) {}
    }
  }

} //class Evolver

public class Example
{
  public static void main (String [] args) {
```

```
EvolvableBuffer x = new EvolvableBuffer(8, 0.666667, 2);
Producer p = new Producer("A", x);
Consumer c = new Consumer("B", x);
Evolver o = new Evolver(x);
p.start();
c.start();
o.start();
}
} //class Example
```

When this program is run, the producer and consumer processes are always eventually able to input or output. In fact, this would also be the case for a producer, consumer and basic, non-evolvable, buffer. The key difference, however, is that the evolvable buffer will not block on input nearly as often as the basic buffer. Indeed, the evolver process will (crudely) adjust the size of the evolvable buffer to match the varying rates of producer and consumer. If, on the other hand, the consumer process were to stop, the evolver process would ensure that the evolvable buffer kept increasing in size and never permanently blocking the producer. This is a typical property that we'd like to establish of the system.

3   Models for Evolvable Systems

We now begin a mathematical analysis of the structure and behaviour of evolvable systems such as the system of buffers given in the previous section.

Here are some of the key ideas introduced in this analysis:

- The computational structure of evolvable systems is modelled using transition systems. This is one of several possibilities for computational models - one which gives us sufficient generality for the purposes of this paper.

  We introduce the notion of an *E-structure*, a two-level structure consisting of a Kripke-like structure whose states (or worlds) are themselves Kripke structures. Evolutionary steps are labelled and provide a transition relation connecting Kripke structures. In fact, they link states in one structure with states in a possibly different structure.

- In the example of the buffers above, we model the individual buffers, and the evolutionary steps between buffers as an E-structure. However, the evolver and its interaction with the buffer system is not included in the model. To specify the behaviour of the evolver and reason about, and verify, properties of the overall evolvable system we introduce a temporal logic interpreted in E-structures. To do so we need to adapt the usual presentation of temporal logic expanding the range of modal operators so that we can describe both the behaviour of local systems and the behaviour of the evolutionary steps. Similar forms of two-level modal and temporal logics have been proposed and investigated elsewhere, see for example [Schobbens *et al.*, 2002] and [Sernadas *et al.*, 1995].

A range of temporal logics is available and most could be adapted to this purpose. In fact, we present a Linear Time Logic where propositions are interpreted along computational paths in the model. Paths include both steps within local systems and evolutionary steps. The logic we present is based on EAGLE-like primitives as one motivation for this work lies in the automated verification and run-time monitoring of evolvable systems, an application for which EAGLE was designed [Barringer *et al.*, 2004], [Artho *et al.*, 2005].

- In this description of models, evolutionary steps may take place between arbitrary and unrelated source and target systems. However, inherent in the notion of 'evolution' is that the source and target systems share common structure, so that the target is indeed a modification or evolution of the source. To reason about the overall behaviour of an evolvable system we need to identify certain common features occurring across all the local systems. Note that, in general, such common structure may be represented in very different ways in different local systems. Indeed one reason for an evolutionary step might be to re-represent some internal structure for reasons of capacity (as in the buffer example), or efficiency, or otherwise.

Again, it appears that there are several possible approaches to identifying and reasoning about common features of the local systems. One possibility, adopted here, is to use *predicate abstraction*. Thus properties of states in local systems, which are common to all systems, are abstracted to form a new transition system on which the evolutionary steps act at an abstracted level. We may then reason about the underlying evolvable system using the structure of the abstracted system. Later, we will define the relevant form of predicate abstraction and show how it can be used to specify and reason about various systems of buffers.

We begin with the definition of a model as a two-level Kripke structure. DEFINITION 1. An *E-structure*, E, is a 5-tuple

$$E = (K, K_0, \Lambda, R, I)$$

where

- K is a set of labels of Kripke structures, i.e. for each $k \in K$ there is a Kripke structure $(S(k), S_0(k) \subseteq S(k), R(k) \subseteq S(k) \times S(k))$, where $S(k)$ is the set of states, $S_0(k) \subseteq S(k)$ the set of initial states, and $R(k) \subseteq S(k) \times S(k)$ the transition relation,

- $K_0 \subseteq K$ is the set of initial structures,

- Λ is a set of labels (of evolutionary steps), and

- $R \subseteq (k : K)S(k) \times \Lambda \times (k : K)S(k)$, where (borrowing notation from type theory) $(k : K)S(K) = \{(k, s) | k \in K, s \in S(k)\}$ is the set of all (dependent) pairs of k in K and states s in $S(k)$.

- I is an interpretation of propositional symbols. Propositional symbols are of two kinds, the global symbols in a set P describing properties of members of K, and a K-indexed collection of local symbols $P_k, k \in K$, where $p \in P_k$ is a propositional symbol describing properties of states in the Kripke structure k.

  An *interpretation* of propositional symbols $(P, \{P_k | k \in K\})$ consists of

  - a function $I : K \to \mathcal{P}(P)$, where $\mathcal{P}(P)$ is the powerset of P, and
  - for each $k \in K$ a function $I_k : S(k) \to \mathcal{P}(P_k)$.

  This definition extends to an interpretation of n-place predicate symbols as n-place functions from states to truthvalues.

As defined, E-structures have just two levels of structure, but we envisage systems where more levels are natural. For example, we may observe not only the behaviour of the local systems but also the computational history of the evolutionary steps. An evolver process may then change the entire system if the pattern of evolutionary steps is not as desired - a circumstance that we may model as evolutionary steps between E-structures of the above kind.

3.1   Example: Modelling an Evolvable Buffer

As an example of an E-structure, consider a case of the evolvable buffer described in the previous section, for simplicity expanding only when full and contracting only when empty.

Firstly, we model a buffer of capacity $N > 1$ as a vector of values from set V of length at most N. We denote the collection of these vectors as $V^{[N]}$. For each $N > 1$ this defines a Kripke structure $B_N = (S_N, S_N^0, R_N)$, where

- $S_N = V^{[N]}$,

- the initial states $S_N^0 = \{[\]\}$ consist of the empty vector $[\]$ only, and

- R_N is the transition relation defined by $\forall s \in S_N, v \in V. \, |s| < N \implies (s, s : v) \in R_N$ and $\forall s \in S_N, v \in V. \, (s : v, s) \in R_N$. Here $s : v$ is the vector s extended with value v. The first clause corresponds to input of v to the buffer with contents s, the second clause to output value v from buffer with contents $s : v$.

The only predicate symbol on this buffer is $contents_N : V^{[N]}$, interpreted in state $s \in V^{[N]}$ as $contents_N(s')$ iff $s = s'$.

We now introduce evolutionary steps of **expand** and **contract** as follows, for each $N > 1$:

$$\text{expand} : (B_N, s) \to (B_{2N}, s) \text{ where } |s| = N$$
$$\text{contract} : (B_{2N}, [\]) \to (B_N, [\]).$$

Thus there is an evolutionary step of expansion when a full state is encountered and then the buffer capacity is doubled but the contents left unchanged. Likewise, when the empty state is encountered, there is a contraction step to a buffer of half the capacity.

With these evolutionary steps, the system of buffers B_N for $N = 2^n, n = 1, 2, \ldots$ defines an E-structure B.

We illustrate part of this system in Figure 3. The left-hand structure k_0 represents a Kripke structure corresponding to the behaviour of a buffer of size $N = 2$, the right-hand structure is a buffer of size $N = 4$, and some evolutionary steps have been illustrated.

This example is too simple to display all aspects of evolvable systems. In particular, the same notion of contents applies across all the local systems. In general, internal structure like this will appear differently in different local systems and evolutionary steps will allow such modification. As a simple example, consider Figure 4 which illustrates an evolutionary step in a buffer system in which instead of an expansion of a sequential buffer, a parallel buffer is introduced. Then the notion of contents in the source system is spread over two structures, each of which has contents, in the target system.

As mentioned previously, we handle this changing representation of internal structure using predicate abstraction. We now present the relevant notion of predicate abstraction and show how to define it for evolvable systems.

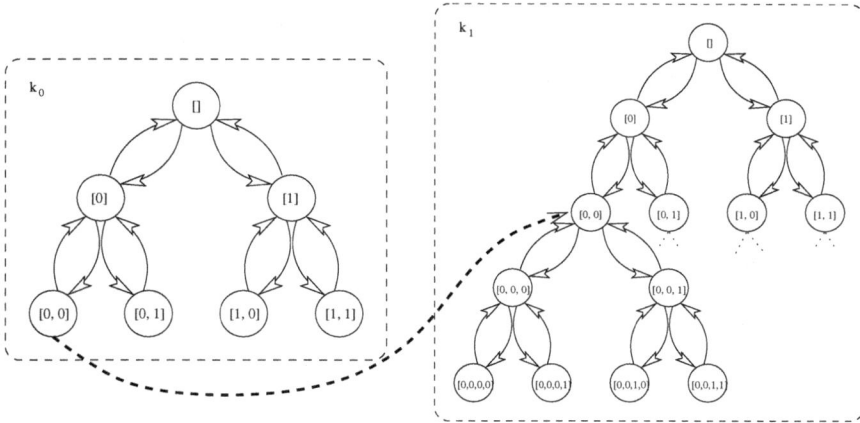

Figure 3. Part of an E-structure for an Evolvable Buffer model

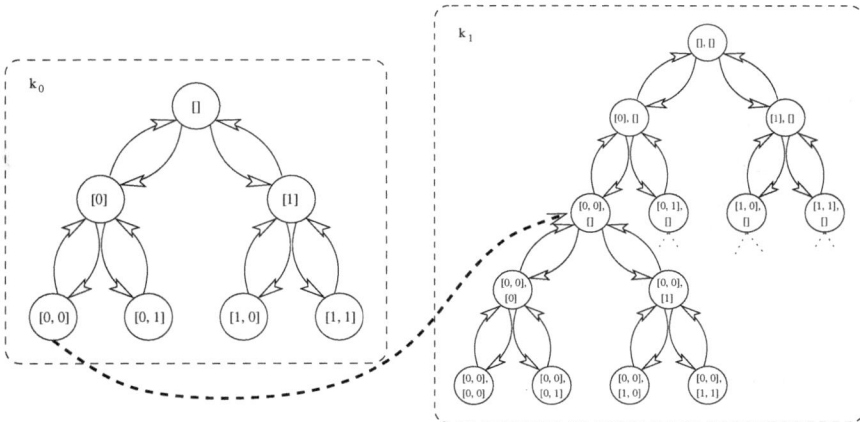

Figure 4. Part of an E-structure for a modified Evolvable Buffer

4 Predicate abstraction

We first consider individual Kripke structures and describe the relation of predicate abstraction between two structures.

Consider two Kripke structures, the 'abstract' structure $k_A = (S_A, S_A^o, R_A)$ over predicate symbols P_A and interpretation I_A, and the 'concrete' structure $k_C = (S_C, S_C^o, R_C)$ over predicate symbols P_C and interpretation I_C. Predicate abstraction is defined in terms of the extension of each of the interpretations to satisfiability relations of the form:

$k_A, s \models_{I_A} \phi$, where $s \in S_A$, and ϕ is a formula in predicate symbols P_A,

$k_C, s \models_{I_C} \phi$, where $s \in S_C$, and ϕ is a formula in predicate symbols P_C.

The relationship of predicate abstraction arises from a collection Δ of definitions of the form $p \triangleq \phi$, one for each $p \in P_A$ with ϕ a formula built from the predicate symbols in P_C:

DEFINITION 2. We say k_A is a *predicate abstraction* of k_C via Δ iff

$$\forall s, t \in S_A. \ R_A(s, t) \text{ iff}$$
$$\exists s', t' \in S_C. \ R_C(s', t') \text{ and for each } p \triangleq \phi \in \Delta$$
$$k_A, s \models_{I_A} p \text{ iff } k_C, s' \models_{I_C} \phi \text{ and}$$
$$k_A, t \models_{I_A} p \text{ iff } k_C, t' \models_{I_C} \phi.$$

The idea is that the definitions in Δ allow us to consider states in k_C as equivalent according to whether or not they are distinguished under the predicate symbols in P_A. Two states are related in the abstraction if any of the corresponding concrete states are related.

Given an abstraction k_A of k_C via Δ, we say k_A is a minimal abstraction iff there is no k_A' which is an abstraction of k_C via Δ such that $|S_A'| \leqslant |S_A|$, where S_A' is the set of states of k_A' and S_A of k_A.

Note that a minimal abstraction over predicates P_A has no more than $2^{|P_A|}$ states.

4.1   Predicate abstraction and E-structures

We now describe how predicate abstraction interacts with evolutionary structures.

Consider the case where there is a correspondence between the abstract and concrete Kripke structures: Each concrete structure k_C having a corresponding abstract structure which we denote k_A, with predicate symbols P_C and P_A and interpretations I_C and I_A. For each initial concrete structure, the corresponding abstract structure is initial.

To define the notion of predicate abstraction over models we need an indexed collection of definitions Δ_{k_A, k_C}, one set of definitions for each k_C and corresponding k_A, which define the predicate symbols in P_A, in terms of formulas ϕ built from the predicate symbols in P_C.

Given two corresponding E-structures and an indexed collection of definitions, we now define when the E-structures are related by predicate abstraction:

DEFINITION 3. Consider E-structures $E_A = (K_A, K_A^0, \Lambda, R_A, I_A)$ and $E_C = (K_C, K_C^0, \Lambda, R_C, I_C)$ with, for each concrete structure $k_C \in K_C$ a corresponding abstract structure $k_A \in K_A$, and predicate symbols P_A and P_C and the interpretations I_A and I_C extending to satisfiability relations $\models_{I_A}$ and $\models_{I_C}$.

Let Δ_{k_A, k_C} be a collection of definitions for each k_C and corresponding k_A.

We say E_A is a *predicate abstraction* of E_C iff

1. k_A is a predicate abstraction of k_C via Δ_{k_A, k_C} for all corresponding $k_A \in K_A$ and $k_C \in K_C$,

2. We say $s \in k_C$ represents $s_A \in k_A$ iff for all $p \triangleq \phi$ in Δ_{k_A, k_C}

$$k_C, s \models_{I_C} \phi \text{ iff } k_A, s_A \models_{I_A} p.$$

Then we require that there is an evolutionary step $(k_A, s_A, e, k'_A, s'_A) \in R_A$ iff for all k_C and k'_C corresponding to k_A and k'_A respectively, and for all $s \in S(k_C)$ representing s_A, there is an evolutionary step $(k_C, s, e, k'_C, s') \in R_C$ with $s' \in S(k'_C)$ representing s'_A, and all such s' represent s'_A.

4.2 Example: Evolvable buffers and predicate abstraction

As a simple example of predicate abstraction, we consider the above system B of evolvable buffers $B_N = (S_N, S_N^0, R_N)$, $N = 2^n$, $n = 1, 2, \ldots$, where S_N is defined as the collection of vectors $V^{[N]}$. The evolutionary steps between these systems consist of those of expand and contract.

A predicate abstraction from buffers in this form is illustrated in Figure 5.

Formally this E-structure is defined as follows:

$$E = (\{k_0, k_1\}, \{k_0\}, \{\text{expand}, \text{contract}\},$$
$$\{ \quad ((k_0, s_2), \text{expand}, (k_1, s_1)),$$
$$((k_1, s_0), \text{contract}, (k_0, s_0)),$$
$$((k_1, s_0), \text{contract}, (k_1, s_0)),$$
$$((k_1, s_2), \text{expand}, (k_1, s_1)) \},$$
$$I \,)$$

where

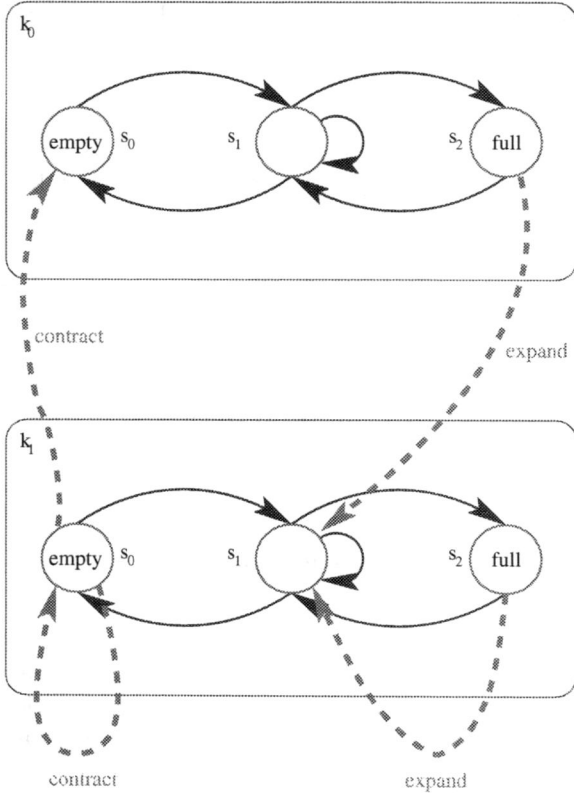

Figure 5. Abstraction of an Evolvable Buffer

$$k_0 = (\{s_0, s_1, s_2\}, \{s_0\}, \quad \{ \quad (s_0, s_1),$$
$$(s_1, s_0), (s_1, s_1), (s_1, s_2),$$
$$(s_2, s_1)\})$$
$$k_1 = (\{s_0, s_1, s_2\}, \{s_0, s_1\}, \quad \{ \quad (s_0, s_1),$$
$$(s_1, s_0), (s_1, s_1), (s_1, s_2),$$
$$(s_2, s_1)\})$$

There are two local propositions that denote whether the buffer
is full or empty. The interpretation I of these propositions
is given by the following mapping,

$$I_{k_i} = \{s_0 \mapsto \{empty\}, s_1 \mapsto \{\}, s_2 \mapsto \{full\}\} \quad \text{for } i = 0, 1.$$

We now show that this system is in fact a predicate abstraction of the system of buffers B.

Linking k_0 and k_1 with the buffers are the collections of definitions of the predicates, which in this case (for both k_0 and k_1) are:

$$empty \triangleq contents_N([\,])$$
$$full \triangleq \exists s \in V^{[N]}.\ contents_N(s) \wedge |s| = N.$$

To verify that k_0 is a predicate abstraction of B_2, and k_1 is a predicate abstraction of B_N for each $N = 2^n, n = 2, 3, \ldots$, we need to show that for each transition in k_0 and k_1 there is a transition in the corresponding buffers B_N such that, at both the source and the target states of these transitions, the interpretation of the left-hand sides of each of the definitions coincides with the interpretations of the right-hand sides.

For example, for the transition from 'neither' to 'full' in k_0, there is a transition in B_2 from $s \in S_2$ with $|s| = 1$ to $s : v$ for any $v \in V$. For the definition of $empty$, its interpretation at 'neither' is false as is the interpretation of the right-hand side at any s with $|s| = 1$. The same interpretation applies to the definition of $full$ at state 'neither' and the right-hand side at each s with $|s| = 1$. For the target states, again the interpretations of the definitions of $empty$ coincide. For the case of the definition of $full$ this is interpreted as true in k_0 and also $s : v$ is full in B_2 (i.e. satisfies the right-hand side of the definition), as required.

A similar verification for each transition in k_0 and k_1 establishes that they are predicate abstractions of the corresponding buffers B_N.

Now let us extend this abstraction to the E-structure B of evolvable buffers. That is, we show that the E-structure in Figure 5 is a predicate abstraction of this E-structure B. We have already verified that k_0 and k_1 are predicate abstractions of the corresponding buffers. All that remains to show is that this abstraction respects the evolutionary steps.

It may help to have an illustration of this situation. In Figure 6 we depict an evolutionary step of **expand** and how it acts both on the buffers and on the abstracted systems.

The evolutionary step of **expand** (both from k_0 to k_1, and from k_1 to k_1) is a step from 'full' (i.e. state s_2) to 'neither' (i.e. state s_1). Now, full states s in buffer B_N are those that satisfy $|s| = N$ and for each such state the evolutionary step **expand** maps s to s as defined above, and s is neither empty nor full in B_{2N} hence represents this abstract state. A similar argument applies to the **contract** step.

This establishes that the E-structure in Figure 5 is indeed a predicate abstraction of the simple system of evolvable buffers.

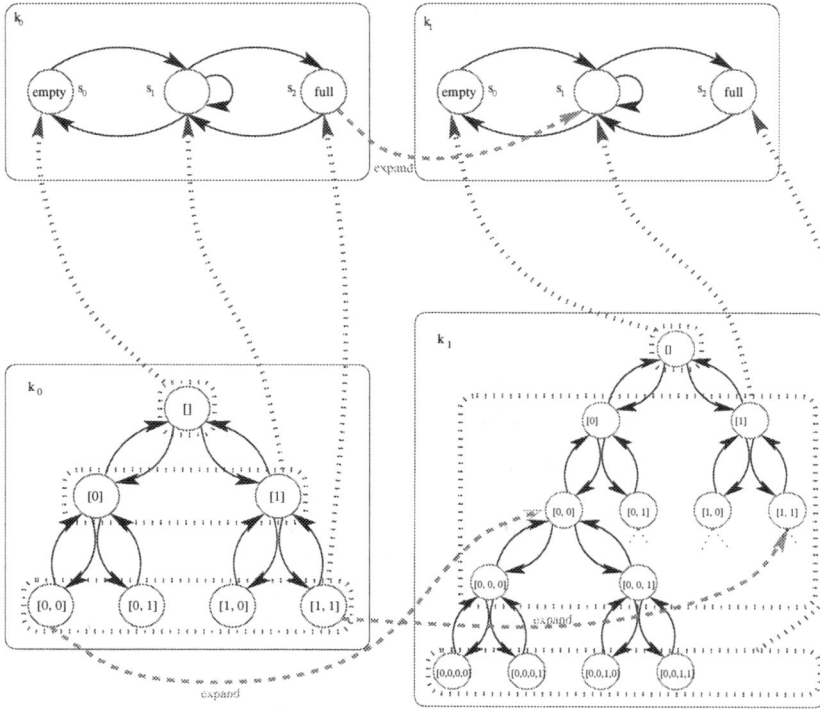

Figure 6. An evolutionary step and its predicate abstraction

5   The temporal logic $\mathcal{E}$-tl

We introduce a two-level temporal logic, $\mathcal{E}$-tl, that will be interpreted over E-structures. $\mathcal{E}$-tl is essentially a linear time temporal logic, although it has a branching capability, and possesses temporal modalities for reasoning (i) purely locally within a system, (ii) locally and globally within and across systems and (iii) purely globally considering just the evolutionary steps being made. To reason directly about possible evolutionary choice, the logic is equipped with a (branching) modality $<e>$ (for each label e of an evolutionary step) such that $<e>\ \phi$ holds when there is a possible evolution e enabled after which ϕ would hold.

We have based the logic $\mathcal{E}$-tl on the style and ideas behind the rather expressive temporal logic EAGLE. EAGLE was designed specifically to support the run-time monitoring of complex temporal patterns of observable system behaviour. In essence, it can be characterized as a fixpoint temporal logic over finite (linear) traces, where, additionally, the fixpoints can be parameterized by data values. However, rather than have the user write formulas in fixpoint form, EAGLE supports user-defined recursive rule schemata and, formally, has only a primitive next, previous, concatenation and chop temporal operators. We chose EAGLE as a basis for $\mathcal{E}$-tl because of its richness and intimate connection with run-time monitoring, which is central to the notion of an evolver monitoring the behaviour of system components. On the other hand, $\mathcal{E}$-tl differs from EAGLE in that we require the logic to reason about infinite traces (and/or trees) in addition to finite ones.

5.1 Syntax

Figure 7 provides a concise syntactic definition of $\mathcal{E}$-tl.

$$
\begin{array}{lll}
D & ::= & R^* \\
R & ::= & \{\mathbf{max} \mid \mathbf{min}\ \} N(x_1, \ldots, x_n) = F \\
P & ::= & atomic\ proposition \\
F & ::= & \mathbf{true} \mid \mathbf{false} \mid P \mid x_i \mid \\
& & \neg F \mid F_1 \wedge F_2 \mid F_1 \vee F_2 \mid F_1 \to F_2 \mid F_1 \leftrightarrow F_2 \mid \\
& & \bigcirc F \mid \odot F \mid \\
& & \mathbf{X}_e F \mid \mathbf{Y}_e F \mid \mathbf{X} F \mid \mathbf{Y} F \mid \\
& & <e> F \mid \\
& & N(F_1, \ldots, F_n)
\end{array}
$$

Figure 7. Syntax of $\mathcal{E}$-tl.

Formulas of $\mathcal{E}$-tl are constructed in the context of declarations in the form of rule schemata. These provide a syntactic mechanism for user-defined temporal modalities, allowing both maximal and minimal fixpoints of recursive definitions. The body of a rule is an $\mathcal{E}$-tl formula and rules can take formulas as arguments. For example, one might define the rule schema

$$\mathbf{max}\ \Box(\varphi) = \varphi \wedge \bigcirc \Box(\varphi)$$

and then, informally speaking, for proposition p, the formula $\Box(p)$ will hold if and only if p holds now and in every next local computation step.

$\mathcal{E}$-tl also allows mutually recursive sets of rule schemata definitions, e.g.

$$\mathbf{max}\ \mathsf{A}(f, g) = f \wedge \bigcirc \mathsf{A}(f, g) \vee g \wedge \bigcirc \mathsf{B}(f, g)$$
$$\mathbf{min}\ \mathsf{B}(f, g) = g \wedge \bigcirc \mathsf{B}(f, g) \vee f \wedge \bigcirc \mathsf{A}(f, g)$$

The arguments of each rule in the set of mutually recursive definitions must be identical.

An atomic formula is either one of the propositional constants TRUE and FALSE, or it is an *atomic proposition*, i.e. a propositional variable p, q, r, etc., or a predicate formula, $P(d_1, \ldots, d_n)$, over expressions d_1 to d_n, or a rule argument x_i.

Propositional combinations of formulas are constructed in the usual way using the standard propositional connectives.

Formulas can be built using the local temporal modality, $\bigcirc f$, informally meaning that f holds in the next local state along a computation path. The past time mirror of next, i.e. $\odot f$, means that in the immediately preceding local state of the computation f held. EAGLE also has primitive modalities for the "concatenation" and "sequential composition" (overlapping concatenation) of two formulas. For the present, we omit such interval-oriented primitives from $\mathcal{E}$-tl more for simplicity of presentation than for any other reason.

Using the global temporal modality, the formula $\mathbf{X}_e f$ holds if the system evolves under event e to a new system for which f holds. The logic is equipped with the corresponding past-time operator, i.e. $\mathbf{Y}_e f$ holds if the system has just evolved under e and in the previous state f held.

Formulas can also be built using an 'enablement' modality. For example, the formula $<e>$ TRUE signifies that, in the current state, there is a possible evolution step labelled by event e.

Finally, formulas can be built using the anonymous evolution modality, $\mathbf{X}f$, meaning that the system performs an evolution step (not specified as to which event) after which the formula f holds. Similarly, we have the past-time mirror formula $\mathbf{Y}f$ meaning that the system has just evolved, with the formula f holding immediately prior to the evolution step. Note that for known, finite, sets of possible types of evolution, $\mathbf{X}$ and $\mathbf{Y}$ are definable using $\mathbf{X}_e$ and $\mathbf{Y}_e$, respectively.

5.2   Semantics

To simplify the presentation of the semantics of $\mathcal{E}$-tl, we define a core language, $\mathcal{E}$-tl$^{\mathtt{min}}$, from which the full logic can defined. The principal simplification is the removal of rule schemata and replacement of freely occurring applications of user-defined rule schema by the more compact fixpoint form, $\mu x.\chi(x)$ denoting the minimal fixpoint solution to the equation $x = \chi(x)$ for temporal formula χ free in x. The maximal solution is denoted by $\nu x.\chi(x)$. However, we will rely on the duality between the minimal and maximal forms, namely,

$$\nu x.\chi(x) = \neg\mu x.\neg\chi(\neg x)$$

and use only the minimal form in $\mathcal{E}$-tl$^{\mathtt{min}}$.

As an example, in the context of the two rule schemata:

min Eventually$(f) = f \vee \bigcirc$Eventually$(f) \vee \mathbf{X}$Eventually(f)
max Always$(f) = f \wedge \bigcirc$Always$(f) \wedge \mathbf{X}$Always(f)

a free application of Eventually(p) would be replaced by the instantiated fixpoint representation

$$\mu y.p \vee \bigcirc y \vee \mathbf{X} y$$

whereas a free application of Always(p) would be replaced by the instantiated fixpoint representation

$$\nu x.p \wedge \bigcirc x \wedge \mathbf{X} x$$

which, when written using the dual (minimal) fixpoint form, becomes

$$\neg \mu x. \neg p \vee \neg \bigcirc \neg x \vee \neg \mathbf{X} \neg x$$

Again, for compactness in the presentation of fixpoint formulas, we will allow both maximal and minimal forms, even though the $\mathcal{E}$-tl^{min} has only the minimal form. Also remember that when interpreting next-time modalities over both finite traces, the usual equivalence

$$\bigcirc \phi \Leftrightarrow \neg \bigcirc \neg \phi$$

is no longer valid, i.e. it fails to hold in the last element of the trace. The corresponding situation holds for the previous modality, $\bigodot$.

Now consider a pair of mutually recursive rule definitions, such as the ones given earlier, namely:

max A$(f, g) = f \wedge \bigcirc$A$(f, g) \vee g \wedge \bigcirc$B$(f, g)$
min B$(f, g) = g \wedge \bigcirc$B$(f, g) \vee f \wedge \bigcirc$A$(f, g)$.

A free application of A(p, q) would be replaced by the fixpoint form

$$\nu x.p \wedge \bigcirc x \vee q \wedge \bigcirc(\mu y.q \wedge \bigcirc y \vee p \wedge \bigcirc x)$$

A free application of B(p, q) would be replaced by the fixpoint form

$$\mu y.q \wedge \bigcirc y \vee p \wedge \bigcirc(\nu x.p \wedge \bigcirc x \vee q \wedge \bigcirc y)$$

which, when written using just minimal fixpoint forms, becomes

$$\mu y.q \wedge \bigcirc y \vee p \wedge \bigcirc \neg(\mu x.(\neg p \vee \neg \bigcirc \neg x) \wedge (\neg q \vee \neg \bigcirc y)).$$

DEFINITION 4. An *anonymous* $\mathcal{E}$-tl formula φ' of φ is the formula obtained from φ by replacing all free applications to rule definitions by their equivalent anonymous fixpoint representations.

Next, we provide a few necessary preliminary definitions on finite and infinite sequences, define the linear paths on which $\mathcal{E}$-tl$^{\min}$ formula are interpreted, then provide an inductively defined semantic interpretation.

DEFINITION 5. For any set H, let $H^\dagger$ denote the set of finite and infinite sequences of elements of H, i.e. $H^\dagger = H^* \cup H^\omega$. Indexing of sequences starts from 0. Given a sequence ρ, $\rho(i)$ denotes the i-th element of ρ, $\rho(0..i)$ denotes the initial prefix of ρ up to and including the element at index i. For finite sequences ρ, i.e. $\rho \in H^*$ for some H, $|\rho|$ denotes the length of the sequence.

DEFINITION 6. Assume $\sigma : S^\dagger$. A *local path* σ of a Kripke structure $k = (S, S_0, T)$ is a (finite or infinite) sequence of states,

$$\sigma = s_0, s_1, \ldots, s_i, s_{i+1}, \ldots$$

such that $(s_i, s_{i+1}) \in T$ with $i \in \mathbb{N}$ for infinite sequences and $0 \leqslant i < |\sigma| - 1$ for finite sequences.

DEFINITION 7. A local path σ of a Kripke structure $k = (S, S_0, T)$ is said to be *rooted* in k if $\sigma(0) \in S_0$.

DEFINITION 8. Let $\pi \in (K \times S^\dagger \times \Lambda)^\dagger$. A *global path* in an E-structure $E = (K, K_0, \Lambda, R, I)$ from $k \in K$ is a (finite or infinite) sequence

$$\pi = (k_0, \sigma_0, e_0), (k_1, \sigma_1, e_1), \ldots, (k_i, \sigma_i, e_i), \ldots$$

such that: (i) $k_i \in K$ and $e_i \in \Lambda$, for all indices i, (ii) $k = k_0$; (iii) for each index i of π, σ_i is a rooted local path of k_i; and (iv) for all pairs of indices, i and $i+1$, $((k_i, \sigma_i(|\sigma_i| - 1)), e_i, (k_{i+1}, \sigma_{i+1}(0))) \in R$.

If π is of infinite length then each σ_i is an element of S^*, i.e. of finite length. If π is of finite length then for each i, $0 \leqslant i < |\pi| - 1$, $\sigma_i \in S^*$, i.e. is of finite length, but we have $\sigma_{|\pi|-1} \in S^\dagger$ (i.e. may be finite or infinite), with $e_{|\pi|-1}$ not present.

For notational convenience, we assume that the index i of a global path π can be applied componentwise.

DEFINITION 9. A global path π of E-structure $E = (K, K_0, \Lambda, R, I)$, which starts with (k_0, σ_0, e_0), is said to be *rooted* in E if and only if $k_0 \in K_0$.

DEFINITION 10. Given an E-structure E and a rooted global path π, we say that an $\mathcal{E}$-tl^{min} formula ϕ holds on π if and only if $E, \pi, 0, 0 \models \phi$, where, for global and local path indices i and j, $E, \pi, i, j \models \phi$ is defined inductively in Figure 8.

Given E-structure $E = (K, K_0, \Lambda, R, I)$ and global path,

$$\pi = (k_0, \sigma_0, e_0), (k_1, \sigma_1, e_1), \ldots, (k_i, \sigma_i, e_i), \ldots$$

indices $i \in ind(\pi)$ and $j \in ind(\sigma_i)$, and an $\mathcal{E}$-tl^{min} formula ϕ we define:-

$E, \pi, i, j \models \mathrm{TRUE}$

$E, \pi, i, j \not\models \mathrm{FALSE}$

$E, \pi, i, j \models P$ iff $P \in I(k_i)$ — for global symbol P

$E, \pi, i, j \models p$ iff $p \in I_{k_i}(\sigma_i(j))$ — for local symbol p

$E, \pi, i, j \models \neg\phi$ iff $E, \pi, i, j \not\models \phi$

$E, \pi, i, j \models \phi \wedge \psi$ iff $E, \pi, i, j \models \phi$ and $E, \pi, i, j \models \psi$

$E, \pi, i, j \models \phi \vee \psi$ iff $E, \pi, i, j \models \phi$ or $E, \pi, i, j \models \psi$

$E, \pi, i, j \models \bigcirc\phi$ iff $j + 1 \in ind(\sigma_i)$ and $E, \pi, i, j + 1 \models \phi$

$E, \pi, i, j \models \mathbf{X}_e\phi$ iff $e = e_i$ and $|\sigma_i| = j + 1$ and $E, \pi, i + 1, 0 \models \phi$

$E, \pi, i, j \models <e>\phi$ iff for some path π' s.t. $((k_i, \sigma_i(j)), e, (k_0', \sigma_0'(0))) \in R$

                             and $E, \pi(0..i) \frown \pi', i + 1, 0 \models \phi$

$E, \pi, i, j \models \mathbf{X}\phi$ iff $E, \pi, i, j \models \mathbf{X}_e$ for some e

$E, \pi, i, j \models \odot\phi$ iff $j - 1 \in ind(\sigma_i)$ and $E, \pi, i, j - 1 \models \phi$

$E, \pi, i, j \models \mathbf{Y}_e\phi$ iff $i > 0$ and $e = e_{i-1}$ and $j = 0$ and $E, \pi, i - 1, |\sigma_{i-1}| \models \phi$

$E, \pi, i, j \models \mathbf{Y}\phi$ iff $E, \pi, i, j \models \mathbf{Y}_e$ for some e

$E, \pi, i, j \models \mu x.\chi(x)$ iff $E, \pi, i, j \models \chi_\vee^\alpha(\mathrm{FALSE})$ for some ordinal α

$E, \pi, i, j \models \chi_\vee^\alpha(x)$ iff $E, \pi, i, j \models \chi_\vee^\beta(x)$ for some $\beta < \alpha$, for limiting ordinal α

$E, \pi, i, j \models \chi_\vee^\alpha(x)$ iff $E, \pi, i, j \models \chi(\chi_\vee^{\alpha-1}(x))$, for non-limiting ordinal α

Figure 8. $\mathcal{E}$-tl^{min} interpretation on a global path

DEFINITION 11. An $\mathcal{E}$-tl^{min} formula ϕ is said to be *true* in an E-structure

E if and only ϕ holds for all rooted global paths of E.

5.3 Examples

We now provide a few examples of specifications using $\mathcal{E}$-tl. We continue with the abstracted evolvable buffer presented in Section 4.2. For convenience, we repeat Figure 5 as Figure 9 below, presenting a graphical representation of an E-structure and an interpretation of abstract propositions. As explained previously, this system has been abstracted from the follow-

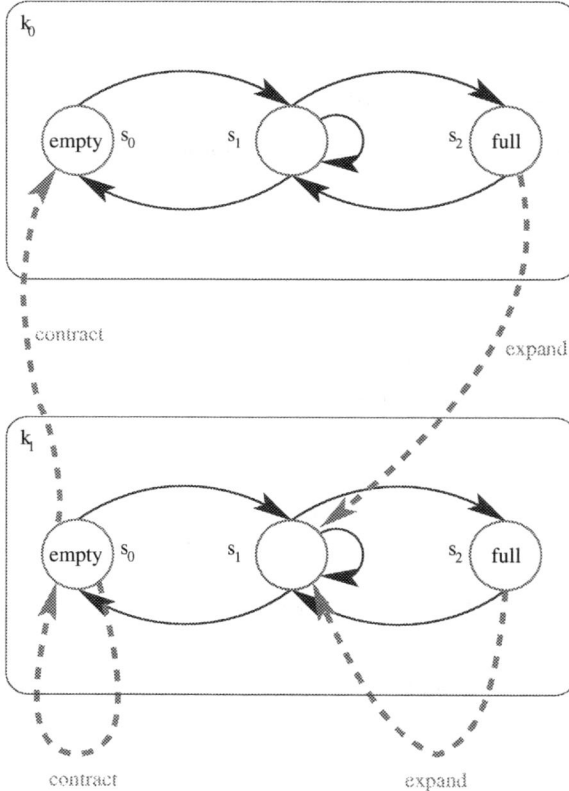

Figure 9. Abstraction of an Evolvable Buffer

ing evolvable system: We assume there is an initial buffer of fixed, finite, size. A evolver process observes the buffer behaviour and if it sees it full it triggers an evolution. The evolved buffer is also finite, however, it may be

triggered either to evolve to a buffer of larger capacity, or, when empty, to
contract in capacity. The initial buffer and all its evolutions are abstracted
by the predicates *full* and *empty*. The buffer is abstracted to be in one
of three states: empty; neither empty nor full; full. The initial buffer is
represented by the E-structure element k_0. All evolved buffers of greater
fixed capacity are then represented by k_1. The thick dashed directed edges,
labelled by either *contract* or *expand* denote evolution steps of the system.
The thinner solid edges denote local transitions of the buffer between the
three abstracted states.

There are two local system propositions that denote whether the buffer is
full or whether it is empty. The interpretation of these propositions is given
formally by the following mapping.

$$I_{k_i} = \{s_0 \mapsto \{empty\}, s_1 \mapsto \{\}, s_2 \mapsto \{full\}\} \qquad \text{for } i = 0, 1$$

An example of a finite local path of the Kripke structure k_0 is

$$\sigma = s_0, s_1, s_1, s_1, s_0, s_2, s_1, s_2,$$

but since the Kripke structure k_1 has a different initial state set, the following is a finite rooted local path of k_1 but not for k_0

$$\sigma = s_1, s_1, s_1, s_0, s_2, s_1, s_2.$$

An example of a finite global path of the E-structure E is

$$\pi_1 = ((k_0, (s_0, s_1, s_2), \text{expand}), (k_1, (s_1, s_1, s_1, s_1, s_2, s_1, s_1, s_1, s_1), \text{contract})),$$

whereas an infinite global path of E might be

$$\pi_2 = ((k_0, (s_0, s_1, s_2), \text{expand}), ((k_1, (s_1, s_2), \text{expand}), ((k_1, (s_1, s_2), \text{expand}), \dots$$

Let us now consider some properties of this system. First assume the
following definitions for localised linear-time temporal modalities: 'unless',
'until', 'always' and 'eventually in the future'.

$$\textbf{max } \texttt{Unless}(f, g) = g \vee (f \wedge \bigcirc \texttt{Unless}(f, g))$$
$$\textbf{min } \texttt{Until}(f, g) = g \vee (f \wedge \bigcirc \texttt{Until}(f, g))$$
$$\textbf{max } \Box(f) = f \wedge \bigcirc \Box(f)$$
$$\textbf{min } \Diamond(f) = f \vee \bigcirc \Diamond(f)$$

We also give the definitions for a global path 'always' and 'eventually in the
future', both for all evolution steps, and respectively a named evolution step.

max Always$(f) = f \wedge (\bigcirc$Always$(f) \vee \mathbf{X}$Always$(f))$
min Eventually$(f) = f \vee (\bigcirc$Eventually$(f) \vee \mathbf{X}$Eventually$(f))$
max AllEvolutions$(f) = f \wedge ($Unless$(\mathbf{true}, \mathbf{X}$AllEvolutions$(f)))$
min SomeEvolution$(f) = f \vee ($Until$(\mathbf{true}, \mathbf{X}$SomeEvolution$(f)))$

The formula AllEvolutions(f) thus holds at a state s when the argument formula f holds on s and at the start of all subsequent evolutions of the system.

Property Example 1: As a first, somewhat trivial, property, let us formulate that whenever the buffer is empty, its next local state, if one exists, must be neither empty nor full.

$$f_1 = \square((empty \wedge \bigcirc\text{TRUE}) \Rightarrow \bigcirc(\neg empty \wedge \neg full))$$

It is straightforward to determine that this property is true for the given example E-structure. Consider any rooted global path π. $\square(\phi)$ will be true on π if ϕ is true for each of the local states of the first trace σ_0 of π. By inspection of k_0 from which σ_0 is derived, a state satisfying the *empty* proposition (i.e. s_0) will either be locally terminal or followed by a state of k_0 in which neither *empty* nor *full* propositions hold, i.e. s_1.

Note that because local paths may be finite and the semantics of $\bigcirc$ is defined to be existential, i.e. there is a next local state, our formulation had to be guarded by the existence of a local next state. An alternative formulation can be given by noting that $\neg \bigcirc \neg\phi$ will evaluate to true at an end state of some finite local state trace. Thus f_1 could be re-written as

$$f_1 = \square(empty \Rightarrow \neg \bigcirc \neg(\neg empty \wedge \neg full))$$

If we wish to express the fact that f_1 should be true for all evolutions of the buffer system, then we could write any of the following.

$f_{1a} = $ Always$(empty \Rightarrow \neg \bigcirc \neg(\neg empty \wedge \neg full))$
$f_{1b} = $ Always$(\square(empty \Rightarrow \neg \bigcirc \neg(\neg empty \wedge \neg full)))$
$f_{1c} = $ AllEvolutions$(\square(empty \Rightarrow \neg \bigcirc \neg(\neg empty \wedge \neg full)))$

Property Example 2: Consider the property that whenever the buffer is full, an evolutionary expansion must occur and leave the buffer in a non-full state.

$$f_2 = \text{Always}(full \Rightarrow \mathbf{X}_{\text{expand}}\neg full)$$

Whilst there are global paths derivable from the given E-structure that satisfy the formula f_2, this property clearly doesn't hold for all rooted global paths. Both Kripke structures k_0 and k_1 contain local transitions from a full state to a non-full state, corresponding to the consumption of an item from the buffer.

Consider, however, the following formula.

$$f_{2a} = \texttt{Always}(full \Rightarrow \texttt{<expand>} \neg full)$$

This formula specifies that whenever the system is in a full state there is a choice of undertaking an **expand** evolution to a non-full state in the evolved system. There is no guarantee that the expansion will occur, it is just a possibility.

Property Example 3: As another straightforward constraint, the following formula characterizes the property that a buffer will always make room for further input items.

$$f_3 = \texttt{Always}(full \Rightarrow (\bigcirc \neg full \vee \mathbf{X}_{\mathbf{expand}} \neg full))$$

In other words, the buffer will never stop in a full state.

Property Example 4: As a form of fairness constraint on the buffering system, we might wish to express that a buffer can't locally cycle infinitely through its full state. Under such a constraint, a buffer will locally (i) operate only finitely and then terminate, or (ii) evolve after some finite duration, or (iii) cycle infinitely but not through a full state.

$$f_4 = \texttt{Always}(\neg \square(\bigcirc \Diamond (full)))$$

Again, note that this property is not true for the given E-structure, however, the formula might be used as a fairness constraint on the given model behaviour.

Property Example 5: Let us now try to specify logically when it is possible for a buffer to contract in size. Our model, as depicted in the E-structure of Figure 9 shows that this is possible when the buffer is empty, but only when the buffer is in some expanded state. The following formula

$$\texttt{Always}(empty \Rightarrow \texttt{<contract>} \, empty)$$

clearly doesn't hold; no contraction is possible in the initial Kripke structure k_0. The premise $empty$ is too weak. How can this be strengthened to uniquely characterize the empty state of k_0? For our particular concrete model, there are two ways ahead. The easiest is to introduce a global proposition, say *initialbuffer*, which is an abstraction of the size of the buffer at the concrete model level, and is true only in k_0.

$$\texttt{Always}(empty \wedge \neg initialbuffer \Rightarrow \texttt{<contract>} \, empty)$$

An alternative approach is to write a formula that characterises whether the numbers of expansion and contraction evolutions the system has undertaken so far are equal, which, because of the concrete interpretations of expand and contract, will mean that the buffer is in its initial size configuration. $\mathcal{E}$-tl as defined above is not expressive enough to do this, although the inclusion of the cut (interval) modality, or data parametrization of rule names, will provide sufficient expressivity, as in the EAGLE logic. However, this is not a general solution.

Property Example 6: Although we have, so far, specified properties of the evolvable buffer at an abstract level, $\mathcal{E}$-tl can equally be used for property specification at the concrete level. For example, one would wish to state that after an evolution expansion or contraction of the buffer, the contents remained the same. Clearly, we must assume a standard first order extension of $\mathcal{E}$-tl to handle predicates and quantification over data.

$$f_{6a} = \text{Always}(\forall s.contents(s) \land \mathbf{X}\text{TRUE} \Rightarrow \mathbf{X}contents(s))$$

Furthermore, if we assume a predicate $size(m)$ that is true if and only the buffer is of capacity m, then we can specify that an expand evolution will indeed expand the buffer, and similarly for contract.

$$f_{6b} = \text{Always}(\forall m.\exists n.size(m) \land \mathbf{X}_{\text{expand}}\text{TRUE}$$
$$\Rightarrow \mathbf{X}_{\text{expand}}(size(n) \land m < n))$$
$$f_{6c} = \text{Always}(\forall m.\exists n.size(m) \land \mathbf{X}_{\text{contract}}\text{TRUE}$$
$$\Rightarrow \mathbf{X}_{\text{contract}}(size(n) \land m > n))$$

5.4   Specifying an Evolver

Our illustrative Java implementation of an evolvable buffer used the idea of an "evolver" thread of control that monitored the behaviour of the buffer and forced an evolution, either an expansion or contraction, when particular conditions were fulfilled (determined via probe methods). The examples given in the above subsection are of *declarative* specifications that can be checked against complete system behaviours, for example the E-structure of Figure 9. Here we very briefly indicate how our logic $\mathcal{E}$-tl can be used directly for the executable specification of an evolver process. Our approach adopts the imperative view of temporal logic that we developed jointly with Gabbay in METATEM and, in a sense, extends the run-time monitoring approach of EAGLE with the inclusion of a notion of 'action'.

An evolver is specified by a set of temporal conditional evolution rules of the form

$$\text{Always}(condition \Rightarrow evolution)$$

where the *condition* is treated *declaratively* but the *evolution* is treated *imperatively*. The *condition* should be viewed as a property that is to be monitored on the system. Whenever it holds, the *evolution* formula determines how the system must evolve, in other words, the system under the watchful eye of the evolver is stopped, evolved in the way specified and then re-started.

Here are some simple examples of rules for specifying evolvable buffering.

Example 1: Whenever the buffer is full, it must be expanded to allow for more input. The following formula achieves this.

$$\texttt{Always}(full \Rightarrow \mathbf{X}_{\texttt{expand}} \neg full)$$

Let's analyze what happens in the imperative, or executable, view of the logic. At each monitoring step the premises of the rules are checked to determine which evolution actions are to be undertaken. Thus, when the proposition $full$ is observed to be true, the execution constructs a future to satisfy the consequents of the triggered rules. In this case, an expand action must be undertaken. If one assumes that this rule is the only rule for determining when expand occurs, the imperative view ensures that expansion can only occur under these conditions; this differs, of course, from the declarative interpretation of temporal logic that would allow models where $\mathbf{X}_{\texttt{expand}} \neg full$ is true with $\neg full$ true prior to the evolution.

Example 2: Of course, the rule above is undoubtedly too strong. A weaker version would be to require (i.e. force) an expansion as soon as the buffer has been full twice.

$$\texttt{Always}(full \wedge \texttt{HasBeen}(full) \Rightarrow \mathbf{X}_{\texttt{expand}} \neg full)$$

in the context of the $\mathcal{E}$-tl rule definition

$$\textbf{min } \texttt{HasBeen}(f) = \odot f \vee \odot \texttt{HasBeen}(f)$$

Thus, as soon as the premise is observed to be true, the right hand side of the rule is fired. The premise will be first observed to be true the second time that $full$ is seen to be true and so an expansion is executed there and then. Again, this is very different from the declarative view of the temporal formula.

Example 3: The EAGLE logic allows rule schema names to be parameterized by data as well as by EAGLE formulas. Assume such an extension to the $\mathcal{E}$-tl logic. One could then define a rule for counting occurrences of formulas holding over the past.

$$\textbf{min Count}(f, n) = \begin{array}{l} (\neg \odot \text{ TRUE} \wedge n == 0 \wedge \neg f) \vee \\ (\neg \odot \text{ TRUE} \wedge n == 1 \wedge f) \vee \\ (\odot \text{ TRUE} \wedge \neg f \wedge \odot \text{Count}(f, n)) \vee \\ (\odot \text{ TRUE} \wedge f \wedge \odot \text{Count}(f, n - 1)) \end{array}$$

The formula $\text{Count}(full, n)$ will be true at some state if there have been precisely n occurrences of the buffer in the full state of its current size or configuration. This could therefore be used to trigger when an expansion occurs, for example:

$$\texttt{Always}(full \wedge \odot \text{Count}(full, 10) \Rightarrow \mathbf{X}_{\texttt{expand}} \neg full)$$

6   Conclusions

This is very much a preliminary foray into the area, opening up a range of possibilities and applications, and raising many issues relating to the description and analysis of adaptive and evolutionary computational systems.

The aspect that distinguishes evolvable systems from more general transition systems is that of 'constancy through change'. It is not sufficient that at each evolutionary step part of the structure (albeit possibly re-represented) is preserved. As in the parlour game of 'Chinese whispers', this may lead to little or nothing being preserved overall. Inherent in the notion of evolution is the idea of constancy and the more that is preserved globally the more we can say about the overall behaviour of the system. In keeping with the logical approach of this paper, we have used predicate abstraction to identify some of the common structure preserved through evolutionary steps. It is not clear that this is the final answer, and perhaps a more algebraic approach to this aspect of evolution may serve equally well or better in identifying common structure and allowing us to model the overall behaviour of evolvable systems in terms of their components.

It is the role of the evolver process, not only to monitor the behaviour of local systems, and to evoke and compute evolutionary steps, but also to maintain invariant properties. The use of temporal logic to describe evolver processes opens up the possibility, through executable temporal specifications, of not only verifying existing evolver processes or guiding the design of such processes, but also in actually generating prototype evolvers. This aspect of the treatment above requires further understanding, especially in the relationship of logical description to execution and the models we have adopted.

We have used a simple illustrative example of a buffering system in this paper. Adaptive and evolutionary systems are of widespread occurrence. For example, hybrid systems (incorporating both discrete and continuous effects) are often evolvable, where the physical environment provides a changing context which may require adaptive and evolutionary behaviour of a system. Again, businesses and business processes are naturally adaptive evolvable systems changing according to external factors or internal drivers for change. We intend to investigate to what extent the techniques of this paper capture the behaviour of these more complex systems.

Almost finally, one of the key motivations for this work is to investigate run-time monitoring and model checking for evolvable systems. To this end, we have used an EAGLE-like approach to temporal logic so that we may, in future work, implement and run temporal descriptions to both statically check evolvable systems and to provide run-time monitoring.

7   Afterword

At last, Dov, we've been able to reveal in more detail some of the ideas we've discussed earlier this year with you. We hope you enjoy. As is clear, it is far from the final word and we trust you are still keen to participate in our small, but evolving, programme of work. Indeed, let us raise our glasses to wish you a fantastic 60th birthday and series of celebrations, and to the next joint chapter on evolution! It's always been so much fun and stimulation working (and plotting) alongside you. Happy birthday, Dov.

BIBLIOGRAPHY

[Artho et al., 2005] Cyrille Artho, Howard Barringer, Allen Goldberg, Klaus Havelund, Sarfraz Khurshid, Michael R. Lowry, Corina S. Pasareanu, Grigore Rosu, Koushik Sen, Willem Visser and Richard Washington. Combining test case generation and runtime verification. *Theor. Comput. Sci.* 336(2–3): 209–234. 2005.

[Barringer et al., 1995] H. Barringer, M. Fisher, D. Gabbay, G. Gough, and R. Owens. METATEM: An introduction. *Formal Aspects of Computing*, 7(5): 533–549, 1995.

[Barringer et al., 1996] H. Barringer, M. Fisher, D. Gabbay, R. Owens and M. Reynolds. *The Imperative Future: Principles of Executable Temporal Logic*. Research Studies Press. 1996.

[Barringer et al., 2004] Howard Barringer, Allen Goldberg, Klaus Havelund, Koushik Sen. Rule-Based Runtime Verification. Proceedings of the *VMCAI'04, 5th International Conference on Verification, Model Checking and Abstract interpretation*, Venice. Volume 2937, Lecture Notes in Computer Science, Springer-Verlag. 2004.

[Caleiro et al., 2001] Carlos Caleiro, Amílcar Sernadas and Cristina Sernadas. Fibring Logics: Past, Present and Future. *This volume.*

[Cunin et al., 2001] P.Y. Cunin, R.M. Greenwood, L. Francou, I. Roberston and B.C. Warboys. The PIE Methodology - Concept and Application. *EWSPT 2001 Proceedings 8th European Software Process Technology Workshop*, LNCS 2077, 3–26. Springer Verlag. 2001.

[Gabbay, 1999] Dov. M. Gabbay. *Fibring Logics*. Oxford University Press. 1999.

[Gabbay et al., 2002] Dov. M. Gabbay, Odinaldo Rodrigues, and John Woods. Belief Contraction, Anti-formulae and Resource Overdraft: Part I Deletion in Resource Bounded Logics. *Logic Journal of the IGPL*, 10(6): 601–652. Oxford Univ. Press. 2002

[Greenwood et al., 1999] R.M. Greenwood, I. Robertson, B.C. Warboys, and B.S. Yeomans. An Evolutionary Approach to Process System Development. Proceedings of the *International Process Technology Workshop*, Villard de Lans (Grenoble). 1999.

[Greenwood *et al.*, 1998] R.M. Greenwood, B.C. Warboys, R. Harrison and P. Henderson. An Empirical Study of the Evolution of a Software System. Proceedings of the *13th IEEE Conference on Automated Software Engineering*, Honolulu. IEEE Computer Society Press. 1998.

[Schobbens *et al.*, 2002] P.Y. Schobbens, G. Saake, A. Sernadas, C. Sernadas. A two-level temporal logic for evolving specifications. *Information Processing Letters*, 83: 167–172. 2002.

[Sernadas *et al.*, 1995] Amílcar Sernadas, Cristina Sernadas and José Félix Costa. Object Specification Logic. *Journal of Logic and Computation*, 5(5): 603–630. 1995.

[Sernadas *et al.*, 1999] Amílcar Sernadas, Cristina Sernadas and Carlos Caleiro. Fibring of logics as a categorial construction. *Journal of Logic and Computation*, 9(2): 149–179. 1999.

Open Problems in Logic and Games

JOHAN VAN BENTHEM

1 The Setting, the Purpose, and a Warning

Dov Gabbay is a prolific logician just by himself. But beyond that, he is quite good at making other people investigate the many further things he cares about. As a result, King's College London has become a powerful attractor in our field worldwide. Thus, it is a great pleasure to be an organizer for one of its flagship events: the Augustus de Morgan Workshop of 2005. Benedikt Loewe and I proposed the topic of 'interactive logic' for this occasion, with an emphasis on *social software* — the logical analysis and design of social procedures — and on *games*, arguably the formal interactive setting par excellence. This choice reflects current research interests in our logic community at ILLC Amsterdam and beyond. In this broad new area of interfaces between logic, computer science, and game theory, the present paper is my own attempt at playing Dov. I am, perhaps not telling, but at least *asking* other people to find out for me what I myself cannot.

A word of historical clarification may help here. The last time the Dutch came up the Thames (in 1667), we messed up the harbour, burnt down a few buildings, and took the English flagship the Royal Charles with us as a souvenir. The Medway Raid was still commemorated as late as 1967 in a joint ceremony. This time, however, our intentions are wholly peaceful. May the Great Game of logic flourish to the benefit of all!

What follows is a sketch of some research lines on logic and games, which occur in logic itself, computer science, and game theory. What I personally find most interesting about these recent interfaces is the importance of *interaction* between several agents as a fundamental theme in logic itself, and the new ways in which mathematical logics of computation and philosophical logics of epistemic attitudes come together. The following story is just my attempt at systematizing issues and problems. There is no pretense at completeness. Comments — and solutions — are highly welcome!

A warning beforehand is in order though, to set the right expectations for the reader. The descriptions of themes and open problems in this paper are more like a light tourist guide for Places To Visit, than one for specific

Things To Do. This reflects the still tentative state of the area of Logic and Games. It is less centered around one well-established family of formal systems than dynamic epistemic logic of information update, where I have surveyed open problems earlier this year [van Benthem, 2005c].

2   Logic and Games

Logic and games meet in several different ways.

Logic Games

First, argumentation itself is a sort of game where opposing players can win or lose. And thus, in addition to its more dominant semantic or deductive underpinnings, logical validity also has a game-like aspect of *winning strategies* for players defending valid conclusions from given discourse positions. In addition to argumentation or dialogue games, modern logicians also use a host of other scenarios, usually two-player games of perfect information, for tasks of semantic evaluation in given models, model construction, comparison of two models, proof search, or even general interaction. Some well-known names in these developments are Lorenzen, Ehrenfeucht-Fraïssé, Hintikka — but one can also mention more recent authors like Hodges, Abramsky, Girard, or Hirsch & Hodkinson. For references to this literature, cf. my lecture notes *Logic in Games* [van Benthem, 1999–], still under construction, whose main line of exposition for the 'Logic & Games' interface is followed here. A compact expert survey of logic games by Wilfrid Hodges is found in the Stanford Electronic Encyclopedia of Philosophy: http://plato.stanford.edu/entries/logic-games/.

Game Logics

But indeed, general games of any sort have an obvious logical structure as multi-actor process graphs that can be described in some logical language. This invites the use of logical machinery, in addition to the standard mathematics of game theory. One stream here consist of process languages like modal or dynamic logic, fixed-point languages, temporal logic, or linear logic. This links up with research in logics of computation, and in principle, it provides all the benefits achieved there for games as well: such as better understanding of algorithms, and perhaps even better design.

But there is also another stream. From the outset, the predictions of game theory about equilibria that 'rational' players will or must choose have been a matter of intensive debate. Here, logic has entered since the 1970s as an analysis of the knowledge and beliefs of players underpinning their choices, and the deliberations that go into them. Thus, epistemic logic, conditional logic, and other high-lights of philosophical logic have entered the scene (partially discovered independently by game theorists), promising conceptual clarification of the issues involved, as well as a more systematic view of options for 'rational' agents and rational procedures.

'Game logics' are logical systems designed for the purpose of analyzing games. Modern game logics often combine the preceding two aspects, so that one could – and does – have 'epistemic dynamic logics' for analyzing the strategies that a player might consider or choose in a given game. Other current topics in this area concern more generic structure of games in general, such as the analysis of general game-forming operations and the resulting algebraic laws. Such issues often cross over into the special area of logic games – making the above distinction between logic games and game logics one of convenience, rather than of principle. E.g., [van Benthem, 2003b] shows that special predicate-logical evaluation games are complete for the algebra of sequential operations on arbitrary games. Finally, as in other newly developing areas, there is a tendency to design *new logics* and coin new terminology, rather than using the more boring expedient of using existing ones, like standard first-order and modal logic. Who wants to use old tools when the World looks fresh and new? We will also be guilty of this, though we also mention some more conservative approaches.

General Activities and Information

Material on the Logic/Games interface may be found at several places in the literature, though there is no standard source yet, let alone a textbook. But cf. [van Benthem, 1999–; Hodges, 1998; van der Hoek and Pauly, 2005] for some broader perspectives. The Amsterdam web page http://www.illc.uva.nl/~lgc is a public resource under construction, pointing at relevant papers and journals. Regular conferences serving as a forum for work in this area include *TARK* (www.tark.org), and *LOFT* (http://www.econ.ucdavis.edu/faculty/bonanno/LOFT6.html and the more ad-hoc but quite frequent *GLC* meetings (http://www.illc.uva.nl/lgc/events.html). Congenial activities at the games and computer science interface include networks such as the EC-sponsored *GAMES*, based in Aachen (http://www.games.rwth-aachen.de).

Here and elsewhere, for precise definitions of basic notions concerning games, we refer to the literature. A good compact reference is Osborne & Rubinstein 1994, while Hofbauer & Sigmund 2002 is an up-to-date one on modern evolutionary game theory.

3   Extensive Games as Processes

Extensive games of perfect information are trees whose nodes represent stages of the game, while leaves represent possible final outcomes, which players can evaluate, and compare as to their individual preferences. Players' turns are indicated at all non-final nodes, and arrows pointing to daughter nodes represent their possible moves there. Game trees can be finite or infinite. In the latter case, infinite branches may sometimes be a nuisance, such as a computer getting stuck forever in a loop, or in argumentation, a person's eternal inability to come to the point. But infinite branches can also be viewed positively as unbounded histories of successful interaction, as with unlimited computational facilities like the internet, or indeed, the functioning of social life. Whether finite or infinite, game structures are much like those used in computer science or logic for representing processes via graphs, trees, or other mathematical notions. Thus, immediate analogies spring to mind with well-known logics for describing computational processes and general action. We list a few topics here.

Caveat

The following questions are mainly concerned with comparing different approaches, and unification across different traditions. A more ambitious goal would be the study of wholly new topics that belong to neither logic nor game theory as normally conceived. A good mine for clues is the extensive body of work on games in *temporal logic*, where issues arise by mixing standard questions about games with those about computational processes (cf. van der Meyden 2005, Ramanujam 2005).

3.1   Game Equivalence and Bisimulation

Computational logics do not have one fixed level of detail for studying processes and actions. Just as in mathematical theories of space (affine or metric geometry, topology, linear algebra), there are legitimate choices of structural similarity relations, reflecting what structure of a process one finds of interest [van Benthem, 1996]. The spectrum runs from output-oriented identifications like *finite trace equivalence*, through the finer *modal bisimulation* which also records internal choice points for agents involved in the process, to the most demanding notion of *isomorphism*. The same spectrum makes sense for games [van Benthem, 2002a], from *equivalence of players' powers* for determining final outcomes, through modal game bisimulation, to stronger notions of isomorphism preserving more game structure.

Action Equivalence versus Outcome Equivalence

Consider the two games pictured in Figure 1, which represent evaluation of the two sides of the logical law of Distribution

$$p \wedge (q \vee r) \leftrightarrow (p \wedge q) \vee (p \wedge r).$$

Are these games the same? The answer depends on our level of interest:

(a) *If we focus on turns and moves, then the two games are not equivalent.*

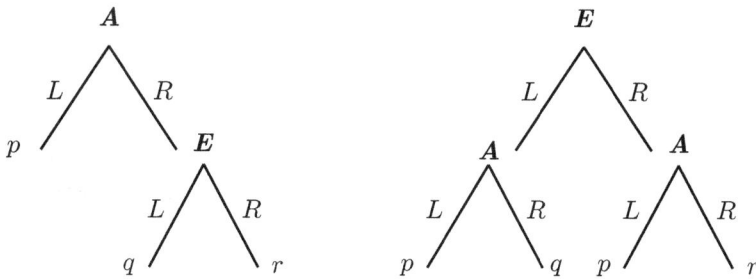

Figure 1.

For they differ in 'protocol' (who gets to play first) and in choice structure. This natural level for looking at games with local moves and choices is that of modal bisimulations. But one might also look at the achievable outcomes only. And then,

(b) *If we focus on outcomes only, then the two games are equivalent.*

The reason is that players can force the same sets of outcomes across games:

A can force the outcome to fall in the sets $\{p\}$, $\{q, r\}$,
E can force the outcome to fall in the sets $\{p, q\}$, $\{p, r\}$.

Here 'forcing' refers to sets of outcomes guaranteed by strategies for players, their 'powers' of control. A strategy forces a set X if all outcomes of the game, with arbitrary play by the others, fall inside X. In the left-hand tree, player **A** has 2 strategies, and so does **E**, yielding the listed sets. In the right-hand tree, **E** has again 2 strategies, while **A** has 4: LL, LR, RL, RR. Of these, LL yields the outcome set $\{p\}$, and RR yields $\{q, r\}$. But LR, RL guarantee only supersets $\{p, r\}$, $\{q, p\}$ of $\{p\}$: i.e., weaker powers. Thus the same control results in both games. More generally, at an input–output level, propositional distribution switches the scheduling of a game without affecting players' powers.

An appropriate bisimulation for this outcome level of game equivalence has been proposed in many areas independently [van Benthem *et al.*, 1994]:

A *power bisimulation* between game models M,N is a relation Z between game states in M,N satisfying the two conditions:

(1)   if xZy, then x, y satisfy the same proposition letters.

(2a)   for each i, if xZy and i can force U starting from x, then there is a set V which i can force starting from y, such that $\forall v \in V \exists u \in U : uZv$.

(2b)   vice versa from y to x.

Thus, game equivalences come in varieties depending on one's level of interest: from coarser to finer. But there has been no systematic logical theory so far of all natural levels.

♣ PROBLEM 1. What are natural structural equivalences for games?

So far, we have only considered 'game forms' without the preferences among outcomes that make for real games. But the same equivalence levels approach also works when extra structure is present in games, such as players' preferences (cf. Section 2.3).

Invariants and Languages

Structural similarities among processes or games induce a set of properties that are *invariant*: i.e., they hold for one structure if and only if they hold for its structural equivalents. Invariants can typically be described in a *language* – usually some logical formalism describing just those properties that are relevant at the given equivalence level. A typical result underpinning this connection says that two finite models are isomorphic iff they satisfy the same first-order sentences. There are much more sophisticated connection results in this vein, some of them involving (...) logic games of model comparison, viz. the well-known Ehrenfeucht–Fraïssé games. In process logics, one simple connection result in the same spirit says that

> There exists a bisimulation between two finite rooted process models $(\boldsymbol{M}, s)$ and $(\boldsymbol{N}, t)$ iff the roots s, t satisfy the same formulas in the modal propositional language describing available moves and atomic properties of nodes.

A similar approach works for extensive games. Choosing a description level matches choosing a particular logical language, modal, first-order, or yet other, to describe properties of nodes in game trees. We start with some obviously available candidates.

3.2   Modal and Dynamic Logics for Moves and Strategies

Modal Logic

Propositional modal logic describes process models

$$\boldsymbol{M} = (S, (R_a)_a, V)$$

with modal formulas ϕ stating properties of states $s \in S$, such as

$[a]\langle b \rangle p$:   after every R_a-step from s to any t, there is an R_b-step
from t to some state u where p holds.

This $\forall \exists$ pattern of successive modalities is typical for interaction between players: any a-move can be countered by some suitably chosen b-move leading to an outcome p. Thus, modal logic describes possible moves and choices for players in a game tree. The modal similarity type for the latter looks roughly like this:

$$\boldsymbol{M} = (NODES,\ MOVES,\ PLAYERS,\ \boldsymbol{turn},\boldsymbol{end},\ VAL)$$

Dynamic Logic

A more explicit account of players' plans and strategies requires a richer propositional dynamic logic *PDL* which also has programs π describing binary relations between states, representing the transitions corresponding to successful executions of π. One can think here of computation steps — but by now, *PDL* is also widely used as a very general logic for describing any complex actions. These are constructed

from atomic moves a and tests $?\phi$ by the three sequential operations of composition $;$, choice $\cup$, and finite iteration $*$.

This language describes game trees in more detail than basic modal logic, using e.g., iterations of single moves to describe arbitrary finite paths. But even more importantly, *PDL* can describe the fundamental game-theoretic notion of a *strategy*. For, a player's strategy is nothing but a binary relation giving her a move at each of her turns — where non-deterministic strategies may even allow more than one option. Natural descriptions of interactive strategies have precisely the sequential and conditional format of *PDL*:

"*IF* your opponent plays a, *THEN* play b, *ELSE* play c",
"*WHILE* your have not reached some goal, *DO* move a".

It is also easy to see that, at least in finite games, *PDL* can easily describe the unique outcome states of games when the players i play a profile of functional strategies σ_i.

μ-calculus and LFP(FO)

Beyond *PDL*, richer fixed-point languages such as the modal μ-calculus define arbitrary smallest and greatest fixed-point predicates in the modal language by means of recursive definitions. This genuine extension of *PDL* is needed for a faithful rendering of basic game-theoretic algorithms such as

Zermelo Colouring when showing that finite extensive two-player zero-sum games are determined. E.g., winning nodes for player i in such a game tree are defined by the following recursion:

$$WIN_i \leftrightarrow (\mathbf{end}\& WIN_i) \vee (\mathbf{turn}_i \& \langle E \rangle WIN_i) \vee (\mathbf{turn}_j \& [A] WIN_i)$$

Thus we can view the predicate WIN_i as a smallest fixed-point defined by

$$\mu p \bullet (\mathbf{end}\& WIN_i) \vee (\mathbf{turn}_i \& \langle E \rangle p) \vee (\mathbf{turn}_j \& [A]p)$$

Van Benthem [2003c] claims that game-theoretic equilibria essentially express fixed-points in a general mathematical sense. But which ones, and definable where?

Note in any case that the μ-calculus can also define behaviour of *infinite* branches, by means of modal ν-operators for *greatest fixed-points*. This reflects another strong intuition about games, viz. the infinite-stream-like behaviour of strategies. If I am ill, my strategy is to consult my doctor, an eternal resource, and extract an advice. After that, my strategy returns — intuitively — to exactly the same state as before. This suggests that game logics will either involve both types of fixed-point, and may even suggest a co-algebraic treatment of strategies — as has been proposed by Baltag, Moss, Venema and others.

I feel that existing modal fixed-point languages, or their first-order extensions such as *LFP(FO)* [Ebbinghaus and Flum, 1995; van Benthem, 2004b] provide excellent means for describing interactive game forms, i.e., the structure of moves and abstract outcomes — while preference structure can also be added later without major difficulties. One good way of testing this idea is by looking at existing notions and results in game theory:

♣ PROBLEM 2. Do a standard formalization for key theorems, proofs, and algorithms in game theory, and see which existing logics are needed.

E.g., De Bruin [2004] analyzes Backward Induction in a μ-calculus setting, with some atomic propositions added for utility values. Van Benthem [1999–] analyzes the proof of the Gale-Stewart Theorem (extending Zermelo's colouring argument to infinite games), identifying its Key Lemma as a law in a temporal logic of players' powers. Harrenstein [2004] provides further game logics (developed jointly with van der Hoek, Meijer, and others) for similar purposes. So, why don't we just use standard logical systems — and then import what we already know about their deductive apparatus and computational complexity for task like model checking or satisfiability? Why create new game logics? Part of this may just be the New World philosophy mentioned before: 'never keep old clothes when you can buy new ones'. A more respectable part of the new system design, however, is the general 'modal philosophy' in process logic: try to see what simple special-purpose languages do the job of analyzing classes of games, striking a good balance between expressive power and computational complexity.

3.3   Adding Preferences

A bare game form only becomes a genuine game with some real drama for rational, or irrational, actors when we look at pay-offs and preferences. For instance, consider the earlier two games for Distribution, but now with the following preferences for players:

$$\boldsymbol{E} \qquad p:0 \quad q:2 \quad r:1$$
$$\boldsymbol{A} \qquad p:1 \quad q:0 \quad r:2$$

Here are the pairs ($\boldsymbol{A}$-value, $\boldsymbol{E}$-value) computed by the usual BI algorithm for node values in a bottom-up manner, starting from the leaves:

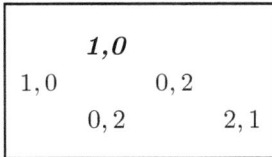

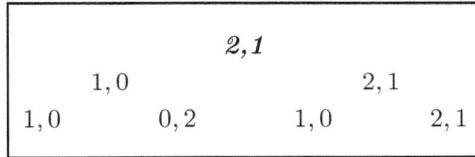

1,0			
1,0	0,2		
	0,2	2,1	

2,1			
1,0		2,1	
1,0	0,2	1,0	2,1

These numerically annotated trees have roots which predict unique outcomes for the joint behaviour of the players. And these predictions are different on the two sides!

♣ PROBLEM 3.   Define good game equivalences when preferences are present.

Further analysis of games with preferences can be done in many ways [van Benthem, 1999; JoLLI], but it does need a serious merge of modal dynamic logics with *preference logics*.

♣ PROBLEM 4.   Integrate game logics with the older preference logics.

I do not see one best formalism yet, but cf. [Bonanno, 1991; Harrenstein *et al.*, 1999; van der Hoek *et al.*, 2004; Pauly, 2001; Van Otterloo, 2005; Van Otterloo and Roy, 2005] for ongoing attempts. In particular, the latter paper has a perspicuous analysis of Backward Induction arguments with minimal means, viz. a dynamic-preference logic with simple reduction axioms relating backward induction subgames to available future moves. One limitation to all these analyses, however, is the compositional simplicity of Backward Induction, with current best actions built up in terms of those in subtrees lower down. Not all game solutions are like this.

Logical analyses of more complex game-theoretic solution concepts in a direct modal-dynamic-preference setting are scarce. This will become particularly acute once games with imperfect information are considered (Section 4). But cf. again [de Bruin, 2004] on extensive games in a 'proof-theoretic' format which achieves greater generality.

Deontic Logic

Another take on preferences involves *deontic logics* of obligations and permissions. For the standard account of this and other branches of philosophical logic, see the relevant chapter in the *Handbook of Philosophical Logic*. In particular, a deontic statement $O\phi$ says that ϕ is true in all 'best' worlds accessible to the present one, with 'best' as seen from the viewpoint of some moral authority. More generally, *conditional obligations $O\phi\psi$* say that

ψ is true in all the best worlds satisfying the antecedent condition ϕ.

Now, generalizing the original motivation of deontic logic, one can let the 'authority' vary here — including players in games themselves, who try to achieve best outcomes from their own private perspectives. Deontic logic in its standard guise does not use full-fledged binary comparisons between situations — as in "anything you can do I can do better". But it does suggest other interesting variations of relevance to games!

E.g., van der Meyden [1996] has an account of deontic preferences in dynamic logic as located, not between *worlds*, but between *available actions* in a process. This is the deontic sense in which we can command someone to do something, rather than achieve something — letting the responsability for the resulting world lie with the commander. Likewise, we might say: players' preferences could also lie between moves in a game, not just between their outcome states — and these two perspectives on preference might be complementary, rather than mutually reducible.

♣ PROBLEM 5. Integrate deontic logic with dynamic preference logic.

3.4   Rationality Assumptions

The bulk of the mathematics underlying standard game theory consists of a definition of games and Nash equilibrium between strategies, some basic existence theorems for strategic equilibrium due initially to Von Neuman and Nash – and a host of refined notions of equilibrium in the decades after the 1950s, whose proponents tried to zero in more closely on natural or useful equilibria. But there has always been controversy surrounding game-theoretic predictions, or recommendations, as the mathematical model seems too poor to make all relevant considerations explicit. In particular, whether players will play a Backward Induction (BI) solution depends on assumptions that underlie their deliberation. This is well-illustrated in the following *Centipede Game*:

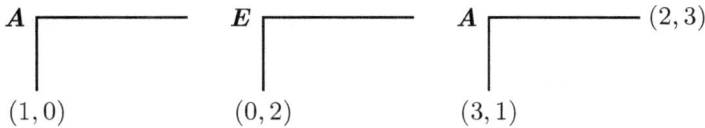

A ────────── E ────────── A ────────── $(2,3)$

$(1,0)$ $(0,2)$ $(3,1)$

Backward Induction predicts that A plays *down* here, blocking the better right region! So, which additional assumptions underwrite the *BI* prediction? A famous example is *Rationality*: the statement that every player will always opt for those available moves that make her off best in future play. Accordingly, much work on game logics is about formalisms defining Rationality and the reasoning based upon it, locking players into the *BI* solution, or whatever other notion of equilibrium may be bolstered by additional assumptions like this. In particular, Aumann has shown in a series of results (cf. de Bruin [2004] for an extensive survey and analysis) that, if players have enough common knowledge of rationality in an extensive game, then backward induction must result.

To a first approximation, these characterization arguments require just dynamic-preference logics, as above, for their formalization. And then, their structure turns out to be remarkably similar to earlier arguments found in the philosophy of action. E.g., the famous Practical Syllogism runs as follows from 'Is's to 'Ought's:

You *can do* a and b.
You *prefer* the outcome of a over that of b.
Therefore, you *will do* a.

♣ PROBLEM 6. Connect the game logic of rationality with the earlier logical tradition in the philosophy of action and practical reasoning.

3.5   Epistemic and Doxastic Logic

Even so, there is much more to rationality analysis for deriving equilibria. One crucial aspect to the whole scenario of deliberation is that players do not know yet what others will do, and what they are going to do themselves. Thus, *knowledge* becomes important, and here is where *epistemic logic* has first entered game theory (cf. [Aumann, 1976]). This requires an expansion of the notion of a game to that of a *model for a game* [Stalnaker, 1999], with a space of possible worlds containing different strategy profiles. These models encode players' knowledge about what actions are possible in the usual manner by relations of epistemic indistinguishability. Since this topic takes us into philosophical logic, we postpone further details until Section 4.

More refined models of games in the literature also bring in doxastic structure: i.e., beliefs, and their revision, as well as probabilistic expectations concerning moves and strategies. Our logical approach to this baroque setting is 'deconstructionist': we start by looking at subsystems of all this, putting the ingredients only together at the end.

3.6 Game Theory and Logic: What Good Does It Do?

Game logics look at fine-structure of games below the usual strategic forms, and they describe step-by-step behaviour over time which remains hidden in the usual strategic equilibria. But what good does this do? There have been deep mathematical existence results for equilibria in games, and on the logic side, there are deep results on meta-properties of first-order, modal, or yet other calculi. Still, the combination of good things does not automatically cook well — witness Churchill's famous response about combining his intelligence with a lady's beauty. But one is of course allowed to hope. Here are a few things that might be expected:

(a) Systematizing the theory of equilibria by their properties as programs

(b) Cleaning up complicated notions of game model, such as 'type spaces'.

(c) Finding new mathematical results of a certain depth which integrate Nash-style equilibrium existence theorems with logical meta-theorems.

Admittedly, beyond logical formalization per se, very little has happened so far in the literature on logic and games justifying these expectations.

♣ PROBLEM 7. Do any of the preceding!

3.7 Infinite Games and Temporal Logic

For infinite games in extensive form, tree models still suffice — but they suggest other languages in addition to the preceding modal and dynamic ones. In particular, the computational tradition of linear and *branching temporal logics* seems relevant, as these describe universes of branching time where infinite games can unfold. This is the perspective suggested by Fagin *et al.* [1995], Parikh and Ramanujam [2003] (both with epistemic structure added), Goranko [2001b], and many others. In particular, Halpern [2003] uses the Fagin *et al.*-style epistemic-temporal modeling of run-based systems to take a fresh look at open issues in game theory. Current expressive game logics of this sort include alternating-time temporal logic *ATL* [Alur *et al.*,

1998] and its epistemic version *ATEL* [van der Hoek and Wooldridge, 2003]. Also relevant is the temporal *STIT* formalism of Belnap *et al.* [2001].

♣ PROBLEM 8. Connect existing modal and temporal game logics.

3.8   Games in Set Theory

Infinite games have also played a role in descriptive set theory, where the Gale-Stewart Theorem culminated in 'Martin's Theorem' saying that each infinite two-player game of perfect information is determined, i.e., one of the players has a winning strategy, provided the set of winning histories for one of the players lies in the Borel Hierarchy. More foundationally, the Axiom of Determinacy even says that all such games are determined, contradicting the Axiom of Choice and thereby providing an alternative set theory. Loewe [2002; 2003] provide nice links between set theory and game theory, introducing mixed strategy game solutions into the set-theoretic tradition, as well as investigating extended notions of Backward Induction on infinite games, provided that players have suitably simple preference ranking over the possible outcomes.

♣ PROBLEM 9. Merge descriptive set theory and (infinite) game theory.

3.9   Operations on Games

Another tradition in process logics looks at processes as models identified under some equivalence relation, such as bisimulation, and then studies operations defined on equivalence classes that form new processes out of old. Examples are the operations of *process algebra* (cf. the *Handbook of Process Algebra*, or other official sources), which include Choice, Sequential Composition, as well as Parallel Merges of various kinds, There is also a logical system in this tradition, viz. *linear logic*, whose game semantics (initially started by Andreas Blass; cf. [Abramsky and Jagadeesan, 1994; Japaridze, 1997] studies natural operations on infinite game trees such as

choices $G_1 + G_2$, role switch G^d, and parallel compositions $G_1 \bullet G_2$.

Linear logic provides a sort of algebra for dealing with equivalences and implications between complex games — with 'validity' meaning that some designated player P has a winning strategy in all games of the given form. Still, these systems were designed to model some pre-given repertoire of operations whose original motivation came from logical proof theory. What happens if we open our mind a priori?

♣ PROBLEM 10. What is a natural repertoire of operations on games?

In particular, there are several parallel compositions for playing a number of games 'together'. Here, one would like to have an expressive completeness result for a natural set of game operations, on the analogy of those for process algebra in Hollenberg [1998].

4   Strategic Forms and Powers

Next consider the level where we are only interested in players' control over outcomes, without getting ensnared in details of what happens *en route* as the game unfolds. This is related to, but not quite identical with, the standard level of 'strategic forms' — as one might say that strategies, and especially, full-fledged 'game models' based on them — *pre-package* all stepwise developments of the extensive game. The difference would rather be more 'global' logical languages expressing what we find of interest now.

4.1   Powers and Game Representation

At a coarser level of identification, one looks merely at player's powers. Let's write

$\rho_G^i s, X$ player i has a strategy for playing game G from state s onward whose resulting states are always in the set X

There is some interesting structure on outcome sets. For instance, it is very easy to see that all games satisfy the following two conditions:

$C1$ if $\rho_G^i s, Y$ and $Y \subseteq Z$, then $\rho_G^i s, Z$ *Monotonicity*
$C2$ if $\rho_G^A s, Y$ and $\rho_G^E s, Z$, then Y, Z overlap   *Consistency*

Determined two-player games, like in Zermelo's Theorem, also satisfy

$C3$ if not $\rho_G^A s, Y$, then $\rho_G^E s, S - Y$; and the same
holds for E vis-á-vis A *Completeness*

Here is a simple converse representation theorem:

Fact   Any two families F_1, F_2 of subsets of a set S satisfying $C1$, $C2$, $C3$ are the root powers for the two players in some two-step game.

This result concerns games of perfect information: an analogue for games with imperfect information requires just $C1, C2$ [van Benthem, 2001]. The proofs of these representation results show one peculiarity though: it is crucial to be able to have the same outcome at different leaves of the game tree. If we require that each leaf in the game tree is a unique outcome, then further principles beyond $C1$, $C2$, $C3$ will be valid on finite games, and the restriction to two-step games is no longer appropriate. Here is a more technical question for puzzle solvers:

♣ PROBLEM 11. Find a representation theorem like the preceding for players' powers in determined games with unique outcomes.

Much richer calculi of powers for two or more players, based on earlier work in game theory and social choice, are found in the *coalition logics* of Pauly [2001], whose crucial axiom describes how powers of disjoint sets of players combine for their union. Here, too, plausible further axioms emerge when uniqueness of outcomes is assumed. In particular, then, it seems that, in the converse direction, powers of coalitions may be decomposed as conjunctions of weaker powers for their disjoint component sets.

4.2 Dynamic Game Logic

Modal Logic of Powers

At this level of game structure, one can now introduce more global modal languages that deal with players' powers directly, in the following format:

$$\boldsymbol{M}, s \vDash \{G, i\}\phi \text{ iff there is a set } X \text{ with } \rho_G^i s, X \text{ and } \forall s \in X; \boldsymbol{M}, s \vDash \phi.$$

These are related to so-called 'neighbourhood models' for modal logic (cf. [Pauly, 2001]). This more general semantics differs from the usual relational models and their minimal modal logic mainly in that the new modality {} with its $\forall\exists$ clause no longer validates distribution laws over either $\wedge$ or $\vee$.

Game Operations

This modal language becomes more expressive when we also add *operations* on games to obtain 'dynamic game logic' *DGL* (cf. [Parikh, 1985]). Models $\boldsymbol{M} = (S, \{R_g\}_g, V)$ then stand for 'game boards' with a universe S of states associated with — though not necessarily identical to — nodes in some extensive game, plus hard-wired forcing relations R_g for given atomic games. Over these structures, appropriate semantic clauses define forcing relations for compound games constructed using sequential composition $G; H$ union (choice) $G \cup H$, game dual G^d (role switch), and finite game iteration G^*. Modalities now refer to a game expression:

$$\{G, i\}\phi \quad i \text{ has a strategy making sure game } G \text{ ends}$$
$$\text{only in states satisfying } \phi.$$

Typical axioms validated by this semantics resemble those of *PDL*:

$$\{G \cup H, i\}\phi \quad \leftrightarrow \quad \{G, i\}\phi \vee \{H, i\}\phi$$
$$\{G; H, i\}\phi \quad \leftrightarrow \quad \{G, i\}\{H, i\}\phi.$$

The characteristic axiom for role switch is as follows in determined games:

$$\{G^d, i\}\phi \quad \leftrightarrow \quad \neg\{G, i\}\neg\phi.$$

In non-determined games, these principles need to refer to different players explicitly. Cf. [Parikh and Pauly, 2003] for the state of the art with *DGL*. In particular, the system is completely axiomatizable and decidable without the dual operation. Here is one long-standing issue, concerning the latter operation, which has no counterpart in the standard propositional dynamic logic of programs *PDL*:

♣ PROBLEM 12. Axiomatize *DGL* completely with game dual G^d added.

Clearly, the modality { } is an existential quantifier over strategies, which are not mentioned explicitly. One obvious question then is how *DGL* relates to more explicit *PDL*-style analysis of games. Van Benthem [1999–] suggests a two-pronged approach of game boards in tandem with families of games over them, relating the *DPL*-language over those games with a *DGL* language of the board.

♣ PROBLEM 13. Merge *DGL* and *DPL* in some natural way.

Game Algebra

DGL's main novelty is its *game algebra*. Dynamic game logic encodes the notion $G \approx H$: players have same powers in every concrete interpretation of G, H on a game board. Valid principles of game algebra include:

(a) *De Morgan Algebra* for the choice and dual:
    Boolean Algebra minus its special laws for **0**,**1**

(b) *Relation Algebra* for composition, choice and dual:
    ; is *associative, left-distributive, right-monotone*,
    and also we have that $(G; H)^d \approx G^d; H^d$.

Typically invalid for games is *right*-distributivity: $G; (H \cup K) \approx (G; H) \cup (G; K)$. Complete axiomatizations were given in [Goranko, 2001a; Venema, 2001]. Here is a connection with logic games after all [van Benthem 2003b]:

THEOREM 14. *First-order evaluation games are complete for basic game algebra of arbitrary games.*

Lacking here, however, are *parallel* game operations describing simultaneous or joint moves, which go beyond the usual sequential modal or dynamic framework. One natural stipulation for games played concurrently is the following product operation GxH, which involves taking ordered pairs of states in both component games:

$$\rho^i_{GxH}(s,t), X \text{ iff } \exists U\colon \rho^i_{G^s}, U, \exists V\colon \rho^i_H t, V\colon UxV \subseteq X.$$

♣ PROBLEM 15. Find the complete game algebra of DGL plus product.

4.3   Temporal Game Logics

Games with players' powers can also be described in a temporal language. E.g., [van Benthem, 1999–] identifies the following key lemma in the proof of the Gale-Stewart Theorem:

$$\{G, E\}\phi \vee \{G, A\}\boldsymbol{A}\neg\{G, E\}\phi$$

with $\boldsymbol{A}$ the temporal operator "always on the current branch". This *generalized determinacy* principle holds for all games, and it says that either one player has a winning strategy, or the other has a strategy preventing the first from ever reaching a position where she has a winning strategy.

♣ PROBLEM 16. Axiomatize the temporal logic of players' powers over arbitrary games.

The branching temporal logic ATL in computer science (Section 2.7) does part of this job, but it does not seem to have sufficient expressive power.

On the analogy of DGL, one would also want to add game operations in this setting. The best known system for that purpose is *linear logic*, as mentioned above.

♣ PROBLEM 17. Design and axiomatize a temporal version of linear logic which can also speak of truth on branches of infinite games.

There are many epistemic versions of temporal logics, such as the earlier-mentioned run-based systems of Fagin *et al.* [1995], or $ATEL$ [van der Hoek and Wooldridge, 2003]. These also make sense in our current setting. E.g., non-determined games are of the essence in linear logic, and they naturally involve imperfect information (see our next Section 4) and lack of knowledge by players, even in the finite case.

♣ PROBLEM 18. Find good epistemic versions of DGL and Linear Logic.

In particular, standard game semantics for linear logics has to use infinite games of perfect information as non-determined counter-examples to the validity of Excluded Middle. With games of imperfect information, finite models might suffice for completeness.

5 Knowledge, Belief and Update

As stated in Sections 2.4, 2.5, logics of knowledge, belief, and other paraphernalia of philosophical logic have entered game theory through the analysis of 'rationality' underpinning the choice of specific strategic equilibria. In particular, epistemic logic plays several roles in this connection, and it raises some new questions.

5.1 Epistemic Characterizations

First, there is an extensive literature on epistemic characterizations of various game-theoretic notions of equilibrium, including Iterated Removal of Strictly Dominated Strategies, or Perfect Rationalizability. The usual format is as follows:

> Model $M(G)$ of game G satisfies epistemic condition E iff the only strategy profiles occurring in the model are those satisfying game solution concept C.

Often, the condition E is some form of iterated or common knowledge of rationality among players. De Bruin [2004] is a thorough survey, as well as a proposal for a uniform logical format of analysis for these results. Still, the current literature consists mainly of a small bunch of such characterization results, without any obvious system.

♣ PROBLEM 19. Find a more general logical analysis of epistemic characterization results which establish a systematic equilibrium theory.

The full form of such results in game theory usually also involves notions of *belief* and *probability* — where the latter serves both to define equilibria requiring mixed strategies, and also beliefs of players in the sense of subjective probability. This makes the logical analysis more complex, and it points to further issues below.

5.2 Update About the Future: Knowledge and Belief

The main epistemic characterization results have been about games in strategic form. But knowledge and belief also come up naturally in thinking about player's moves in an extensive game, and their deliberations about the remainder to be played. In this area, one would want to move away from the excessive emphasis on Backward Induction, and consider other scenarios — cf. [van Benthem, 2004d] for some alternatives, such as 'repaying past favours'. Presumably, existing logic of update and revision can handle the game situation, but the issue is what good they can do.

♣ PROBLEM 20. Use epistemic update and belief revision to obtain a richer set of solution methods for extensive games.

5.3   Imperfect Information and Dynamic-Epistemic Logic

Knowledge also prepares the way for games of *imperfect information*, where players do not know their current situation exactly. Card games are a good example, and so is warfare. It is easy to add epistemic structure to games of imperfect information.

Dynamic-epistemic Logic for Extensive Games

The usual game-theoretic trees with information sets, or dotted lines, are already models for a combined dynamic-epistemic language combining action modalities $[\pi]\phi$ with epistemic operators. Here, knowledge assertions occur for both individuals and groups:

$M, s \vDash K_i \phi$ player i knows that ϕ is the case if ϕ is true in all states i-indistinguishable from s

$M, s \vDash C_G \phi$ ϕ is common knowledge in the group G if ϕ is true in all states reachable from s by a finite sequence of accessibility steps for any player.

This is a modal language again, now over models consisting of worlds with equivalence relations for players' indistinguishabilities, and typical interactions. Much is known about this system in the modal literature.

A combined dynamic-epistemic language *DEL* can describe many situations of interest. The following is a brief survey of van Benthem 2001, which explains the *DEL* view of imperfect information games for both extensive and strategic forms. For the sake of concrete illustration, consider the two-step game in Figure 2:

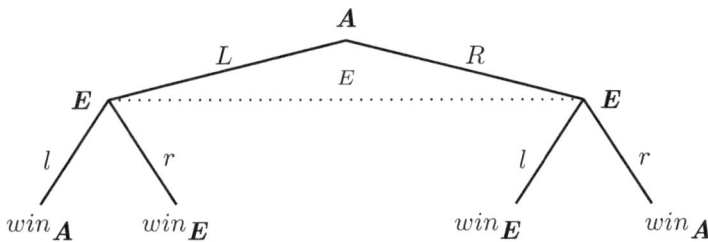

Figure 2.

Allowing only strategies that can be played despite her uncertainty, E has only two: 'left', 'right'. Note that determinacy is lost here: neither player has a winning strategy. The following formulas describe player E's plight at her right-most turn:

(a) $K_E(\langle l \rangle win_E \vee \langle r \rangle win_E)$
    E knows that some move will make her win

(b) $\neg K_E \langle i \rangle win_E \wedge \neg K_E \langle r \rangle win_E$
    there is no particular move of which E knows it will make her win.

The complete logic of this system is a well-known fusion of *PDL* plus epistemic *S5*. It has been proposed by Moore [1985] as a logic of *planning* in situations where agents do not know all the relevant information. In particular, no interaction axioms occur between knowledge and action modalities. When these occur, they express special features of agents. This is brought out by standard modal *correspondence* analysis.

THEOREM 21. *The DEL formula $K_E[a]\phi \rightarrow [a]K_E\phi$ is true in a frame iff E has Perfect Recall in the sense that, if $R_a xy$ and $y \sim_E z$, then there exists some u with $x \sim_E u$ and $R_a uz$.*

The latter condition requires this commutative diagram in the game trees:

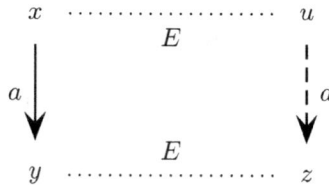

More complex versions of Perfect Recall yield to exactly the same analysis [Bonanno, 2003]. And one can also analyze other types of player, such as those having finite-state memories. One virtue of logic analysis here is the discovery that complexity of valid reasoning may differ widely for these different types of agent [Halpern and Vardi, 1989]. Here are two questions about further possible connections with game theory:

♣ PROBLEM 22. Use *DEL* to analyse the various notions of game-theoretic equilibrium that have been proposed for imperfect information games.

♣ PROBLEM 23. The 'Harsanyi trade-off' relates incomplete information about the future for players in perfect information games with games of imperfect information. Analyze this reduction in logical *DEL* terms.

Uniform Strategies

One can also add a calculus of strategies again, in a *PDL*-style extension. In game theory, imperfect information games involve *uniform strategies*, which prescribe the same move for a player across her uncertainty link. These can be correlated with programs whose only test conditions are formulas which the relevant players know. Cf. Fagin et al. [1995] on so-called 'knowledge programs' for this purpose, or the definability result for uniform strategies in van Benthem [2001].

Knowledge and Powers

DEL also has a counterpart at the *outcome level*. As we noted above, any two families of sets satisfying only the monotonicity and consistency conditions on powers can be realized by means of a two-step *imperfect* information game. Here is an example displaying the necessary tricks. Suppose we have:

$$\text{powers for } \boldsymbol{E}: \quad \{A, B, C\}, \{C, D\}$$
$$\text{powers for } \boldsymbol{A}: \quad \{B, C\}, \{A, D\}$$

An appropriate game with just these powers is given in Figure 3. Player *E* has 3 strategies, with two powers; *A* has 4 uniform strategies, with just the powers indicated, up to monotonicity:

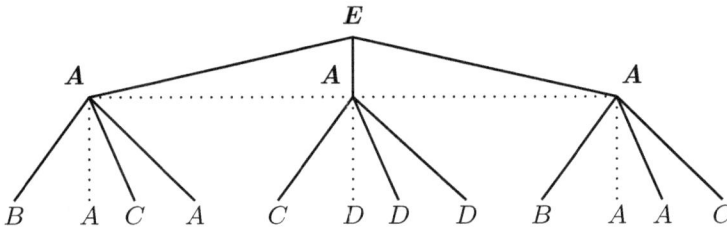

Figure 3.

But equally well, the language of powers, or even the full *DGL* of Section 3 can be easily combined with epistemic logic. In a sense [van Benthem, 2001], Hintikka's '*IF* logic' of Section 5 below is already a calculus of game operations plus implicit knowledge operators.

5.4   Information Update in Games

Concrete Epistemic Actions

'Dynamic epistemic logic' is also used in another sense these days for concrete systems that update information (cf. [Baltag *et al.*, 1998]). Van Benthem [2005b] is an extensive survey of open problems in this area. Here, we just mention a few issues specifically related to games. For the purposes of this Section, we use the letter combination *EDL* for this second sense of dynamics, which may be viewed as an instantiation of *DEL* in the above more abstract sense. One now adds *PDL* style operators referring to concrete informational actions. The basic case is *public announcement* !*A* of a proposition *A*, removing all worlds from the current model M where *A* does not hold to obtain the relativized submodel $M|A$:

$$M, s \vDash [!A]\phi \quad \text{iff} \quad M|A, S \vDash \phi.$$

This says that, after truthful announcement of A, ϕ holds. Public announcement has a complete decidable logic (cf. [van Benthem *et al.*, 2005] for an up-to-date version), whose axioms essentially compute $[!A]\phi$ by *relativizing* ϕ to *A*.

More complex informative events over an epistemic model M involve hiding of one's own actions, or partial observation by others. This happens frequently in ordinary communication, where we whisper in lecture theatres, use *bcc* in emails to make information flow in complex manners, or just cheat... A sharp focus for all this are *parlour games* where not all players can see all details of some current move — cf. [van Ditmarsch, 2000] on the logical analysis of 'knowledge games' of this kind.

Product Update via Event Models

To deal with such more sophisticated informative events, the right update mechanism is not just world elimination, but rather a product of the current epistemic model M with some new dynamic structure.

One first takes an *event model* A consisting of all relevant actions with the preconditions for their occurrence, while encoding also which agent can distinguish which events. E.g., if I read my card in your view, you know that I am performing one of a number of possible reading actions, but you do not know which one. And if I read my card secretly, you even think my action is the same as doing nothing.

The next stage then computes a *product model* $M \! x \! A$ whose worlds are pairs (s, a) of current states s in M and those events a from A which are possible at s [Baltag *et al.*, 1998]. For the special case of game trees, this product construction specializes as follows:

Take a game with current state x, and uncertainty relations $\sim_i$ among the nodes at x's tree level computed so far. Let a new move be made. The new states are the nodes at the next level of the game tree, which can be identified with ordered pairs (previous state, action last made). Now we define the uncertainties at the next tree level as follows:

$$(y, b) \sim_i (z, c) \text{ iff } y \sim_i z \text{ and } b \sim_i c$$

Thus, new uncertainty is 'old uncertainty + indistinguishable actions'.

EXAMPLE 24. Updates during play: propagating ignorance along a game.

Figure 4 depicts a game of imperfect information and the corresponding action model. Figure 5 shows how product update would compute the right uncertainty links for players in the game tree, going down from one horizontal level to the next.

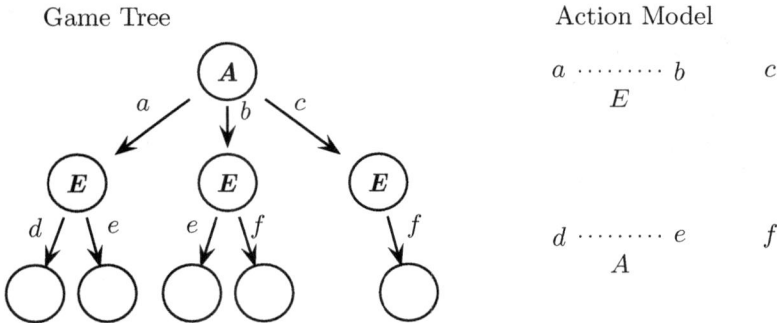

Figure 4.

One can analyze this update mechanism by its dynamic-epistemic properties. For product update, we then find two major ingredients, the earlier *Perfect Recall*, and a sort of converse property of

Uniform No Learning If two actions are ever indistinguishable, they will never make indistinguishable states distinguishable.

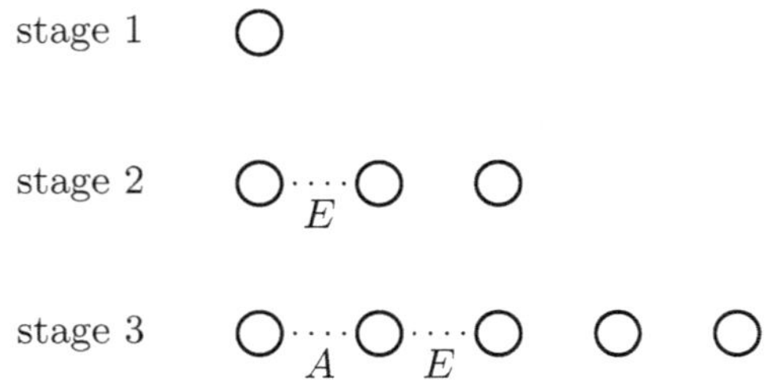

stage 1   ◯

stage 2   ◯⋯◯   ◯
               E

stage 3   ◯⋯◯⋯◯   ◯   ◯
               A E

Figure 5.

Uniform No Learning typically validates the converse direction of the earlier knowledge-action interchange law for players who obey it [van Benthem, 2001]. Note that this means that product update is not a neutral stipulation: it only works for idealized agents.

Evolution in the Update Universe

The impact of product update, in whatever formulation, may be studied in the following mathematical setting. One considers a process universe of infinite trees whose nodes are finite sequences of events (cf. [Fagin *et al.*, 1995; Parikh and Ramanujam, 2003]), and allows arbitrary uncertainty relations for agents between nodes. Such uncertainty patterns will encode what agents know, and more generically, what types of agent are present. One important special kind of such trees then arises from product update. Define the infinite tree-like structure

$Tree(\boldsymbol{M}, \boldsymbol{A})$ to consist of all successive product levels
$$\boldsymbol{M}, \boldsymbol{M}x\boldsymbol{A}, (\boldsymbol{M}x\boldsymbol{A})x\boldsymbol{A}, \ldots$$

The above game tree was an example. Now we can ask which patterns of epistemic uncertainty relations are characteristic for the latter setting. Van Benthem and Liu [2004] give the following representation result:

THEOREM 25. *An arbitrary epistemic event tree is isomorphic to some model Tree($\boldsymbol{M}, \boldsymbol{A}$) iff its uncertainty relations for all agents satisfy both Perfect Recall and Uniform No Learning.*

The same paper also discusses the possibility of a classification of agents' *strategies* in terms of restrictions on their format of definability.

General epistemic trees support epistemic temporal languages, as much as dynamic epistemic logics of information update (van Benthem [2004c]):

♣ PROBLEM 26. Can the theory of *EDL* be generalized to an epistemic temporal logic allowing general temporal pre- and postconditions for events?

Yap [2005] is a first attempt in this direction.

5.5   Diversity of Agents

The preceding characterization result can be extended to other types of agent, corresponding to other structural constraints. E.g., van Benthem [2001] characterizes the dynamic-epistemic properties corresponding to the case of *memory-free agents*:

♣ PROBLEM 27. Find a general *EDL*-style classification of types of agent based on logically expressible dynamic-epistemic constraints.

Perhaps the more interesting question, however, is what happens with groups of agents with different logical behaviour. How do we detect the 'memory-type' of an opponent, and can we take advantage of it, once we know? The recent movie "Memento" provides nice examples of this scenario, and so do many other more *SF*-like current films.

5.6   Other uses of *EDL*: Iterative Solution Algorithms for Strategic Games

Another striking aspect of games is the dynamic-epistemic character of the usual solution procedures. Notably, Iterated Removal of Strictly Dominated Strategies (SD^ω) successively removes rows and columns from game matrices to arrive at a submodel of the full set of strategy profiles where common knowledge of rationality reigns.

Van Benthem [2003c] uses repeated announcement of rationality assertions for players to analyze game solutions, not as sets of strategy profiles, but rather directly on the algorithmic procedures themselves which solve games. This works over a simple account of *game models* as sets of strategy profiles where players know their own action, but not that of the others. Standard algorithms iteratively eliminate profiles from the initial model, in a process whose steps can be viewed as announcements of some suitable epistemic statement. In any model M, any statement ϕ, repeated sufficiently often, gets to a fixed-point $\sharp(M, \phi)$. Now, either the latter submodel is empty (ϕ is then *self-defeating*), or it is non-empty, while announcing it has no further effect, so that the statement has become common knowledge in $\sharp(M, \phi)$ (ϕ is *self-fulfilling*).

THEOREM 28. *The algorithm SD^ω produces just the models $\natural(\boldsymbol{M}, WR)$ where WR is the statement that no one plays an action for which there is one that she knows to be better.*

This type of result establishes a correspondence between (a) epistemic assertions whose announcement can be iterated, and (b) game solution algorithms. Van Benthem [2003c] gives an illustration, defining a new solution algorithm stronger than SD^ω using the new statement that "no one thinks that her current action is ever a worst response".

♣ PROBLEM 29. Develop the dynamic epistemic analysis of game solution procedures more systematically.

As for the logical background of these scenarios, for iterated existential epistemic assertions ϕ the domains of the solution models $\natural(\boldsymbol{M}, \phi)$ can be defined explicitly inside the initial model by greatest fixed-point formulas in an *epistemic μ-calculus*. But the general case requires an extension with *inflationary fixed-points*. Now, the general language of epistemic inflationary fixed-point logic is ill-behaved. E.g., it is undecidable [Dawar *et al.*, 2004] and other quirks occur, too.

♣ PROBLEM 30. Find well-behaved fragments of epistemic inflationary fixed-point logic which suffice for game analysis.

5.7 Belief Revision

EDL may be viewed also as a logic for update of *belief*, rather than knowledge. Nothing much changes on a doxastic interpretation of its language, semantics, or logical laws as describing players' beliefs about moves and their effects. For instance,

♣ PROBLEM 31. Extend the analysis of game solution with iterated epistemic statements to a scenario where players' beliefs are taken into account.

But, in the context of games, we also need *belief revision*, as we encounter unexpected behaviour of our opponents contradicting our expectations so far (cf. [Gärdenfors and Rott, 1995]). Now, as explained in van Benthem [2005c], doxastic product update does not perform belief revision: contradictory new evidence leads to inconsistent belief. But its methodology can be extended to a system which does, by enriching models with *plausibility values* for worlds according to agents [Spohn, 1988]. Then, product update can be extended to plausibility update in a natural manner [Aucher, 2003], supporting the same sorts of perspicuous logics with reduction axioms which are the trade-mark of *EDL*.

Still, there is a difference in spirit. As pointed out in Liu [2004], *diversity of revision policies* (from more radical to more conservative) is a desideratum in belief revision theory, and hence no unique update rule can work for everyone. Her proposal is then to parametrize the plausibility assignment function to pairs (s, a) with weight for the past world s and the last-observed event a. This is like parametrized update rules in inductive logic or Bayesian probability.

♣ PROBLEM 32. Apply the Spohn-Aucher-Liu analysis to extensive games.

Long-term Learning

Viewed in the long run, belief revision becomes a special case of *learning strategies* over time. The time seems ripe for a merge between temporal logic (cf. Sections 3.3, 4.4), systems of update and belief revision, and the game perspective on learning developed, e.g., in Kelly [1996]. After all, a {Student, Teacher} class-room setting is very much like a two-person game. Learning agents over time have been considered in Kelly [2002]; Hendricks [2002]. The above detailed level of local moves and their effects in extensive games seems relevant here, but so does the more global analysis in terms of powers, and general topological structure of sets of runs.

♣ PROBLEM 33. Find a merge between belief revision, learning theory, and temporal logic. Are there significant completeness theorems capturing non-trivial properties of learning mechanisms in logical terms?

5.8 Preference Dynamics

While we are dynamifying various aspects of games, we might also consider doing the same with preferences. One might imagine that there are actions which change preferences. Zarnic [1999] analyses goals and planning in this way, with commands of the form *FIAT* ϕ: "make ϕ become the case". Yamada [2004] takes the same thinking to *deontic logic*. A command like "you ought to make sure that ϕ" is some authority's instruction to give possible future outcomes satisfying ϕ high(er) priority. (Related ideas are found in [Tan and van der Torre, 1998].) Thus, one can have an account of dynamic commands and their effects in the same style as *EDL*. But still, there are non-trivial issues here, having to do with the fact that *EDL* events are 'precondition oriented'. They convey information about the situation when the event *took place*, following which we just see or deduce what will hold afterwards — perhaps after a change took place. By contrast, most commands are *postcondition-oriented*. They tell us to do something that will make sure that a certain condition will hold afterwards.

♣ PROBLEM 34. Develop a complete dynamic logic of commands.

6 Logic Games

As mentioned in Section 1, many logical key tasks themselves have a game character. Most of these involve two-person zero-sum games of perfect information, some with finite extensive game trees, some allowing infinite runs. Van Benthem [1999–; 2002c] have chapters on most of the basic varieties. We run through a few basic cases.

6.1 Standard Logic Games

Evaluation Games

Semantic evaluation games exist for most logical languages. The original case are Hintikka's evaluation games between a Verifier and a Falsifier for formulas of first-order logic, whose basic feature is an equivalence between

  (a) truth of ϕ in some model M with assignment s,

  (b) the existence of a *winning strategy* for Verifier in the associated evaluation game $game(M, s, \phi)$.

  These games have a finite length measured by the quantifier depth of the formula. With this bridge, many logical and game-theoretic notions can be correlated. E.g., the game-theoretic import of Excluded Middle is that evaluation games are *determined* by Zermelo's Theorem. More complex games are needed for more expressive languages, such as *LFP(FO)* with monotone fixed-point operators. These involve infinite runs, and counting of parity of infinitely recurring μ- or ν-subformulas.

♣ PROBLEM 35. Find a systematic logical use for the surplus information in the game account, viz. the different specific winning strategies for players.

♣ PROBLEM 36. Develop an abstract model theory of logical languages based on a natural classification of game types.

Model Construction Games

Games for constructing models exist in great variety (cf. [Hodges, 1985]). One simple option is to turn *semantic tableaus* into logic games [van Benthem, 2005b]. Now there can be infinite runs, corresponding to the construction of an infinite model. The tableau rules and processing procedure can be manipulated in many different ways for this purpose. Here a model exists for a given set of statements iff Builder has a winning strategy, and

different such strategies even encode different possible models directly. This
setting also suggests introducing new vocabulary, such as an explicit instruction for dealing with a true universal quantifier more than once. Here is a
question which arises then:

♣ PROBLEM 37. Use variations on tableau games to model various substructural (categorial, linear) versions of first-order logic.

Proof Games and Argumentation

The oldest logic games are Lorenzen's for dialogue, where the main result
is that sequents are intuitionistically provable iff the Proponent of the conclusion has a winning strategy against an Opponent granting the premises.
Even more constructively, there is an effective correspondence between

  (a) *proofs*,

  (b) *winning strategies* in the dialogue game.

These dialogue systems are driven by a mixture of 'logical rules' for decomposing complex statements, and 'procedural rules' setting the schedule
and determining other points of order. More informal versions of Lorenzen
games, with both features, are used in modern argumentation theory, as a
model for rational debate.

♣ PROBLEM 38. Analyse the procedural component of dialogue games
more systematically, and find a systematic prediction what logic comes out
of what package of logical and procedural rules.

As finding proofs is the dual of looking for counter-examples, there is also

♣ PROBLEM 39. Compare dialogue games with model construction games,
and also the co-existence of strategies-as-proofs and strategies-as-models.

By the way, van Benthem 2004A has a more gloomy game analysis of
argumentation, as involving little logic, but rather a game calculus of values of arguments decreasing with familiarity, and making one's statements,
whether weak or strong, with the right timing.

Model Comparison Games

The most widely used logic games are Ehrenfeucht–Fraïssé games of model
comparison, where two models M, N satisfy the same first-order sentences
up to quantifier depth k iff the player called Duplicator has a winning strategy against the counter-player Spoiler in the game $comp(M, N, k)$ over k
rounds. Here, the concrete correspondence is between

(a) first-order formulas distinguishing the models, and

(b) winning strategies for the 'difference player' Spoiler.

Comparison games give fine-structure to more global structural equivalences such as isomorphism, potential isomorphism, or bisimulation. Thus, we could also use them to compare extensive games! Here we just mention a technical question related to the strong logical systems that we have advocated for analyzing general games:

♣ PROBLEM 40. Find a useful comparison game matching first-order logic with fixed-points *LFP(FO)*.

One intriguing thing is that model comparison games measure similarity of structures. Thus, one can play such special logic games to study similarity between general games!

Operations on Logic Games

Other logic games include more complex constructions, such as model extension [Hirsch and Hodkinson, 2002] or recursion-theoretic priority arguments [Moschovakis, 1980]. Many of these, intuitively, involve combination of subgames into larger games. Some of these combinations are the earlier-mentioned choices, compositions, and role switches. But others involve more complex forms of parallel product formation. Here is an illustration from van Benthem [1999–]:

THEOREM 41. *Ehrenfeucht-Fraïssé games are isomorphic to 'interleaved products' of Hintikka evaluation games.*

♣ PROBLEM 42. Find a typology of game operations used for logic games.

6.2   Modified Logic Games

Next, even though logic games have largely developed in isolation from game theory proper, there have been some cultural influences from game theory coming this way.

Imperfect Information

Logic games with imperfect information were proposed by Hintikka and the '*IF* school' [Hintikka and Sandu, 1997] with the aim of analysing more freely scoped quantifier languages, with motivations from linguistics, philosophy, mathematics, and these days even quantum mechanics. Here is the well-known 'slash notation' for such a new kind of logical evaluation game:

$$\forall x \; \exists y/x \; x \neq y.$$

This expresses, when played in any model M, that Verifier has a uniform strategy for winning this semantic game, even when she does not know the value of x chosen earlier by her opponent Falsifier. At the level of powers, this game is equivalent to the reverse scope order $\exists y\ \forall x/y\ x \neq y$. This observation is just one instance of general *IF* logic, for which we refer to Sandu and Tulenheimo [2005] (cf. also [Dechesne, 2005]). In particular, the full power of the system is equivalent to that of existential second-order logic. But Tulenheimo [2004] shows that some *IF* variants of modal logic remain decidable. In addition to the host of more specialized questions in this technical area, we mention

♣ PROBLEM 43. What is the complete game algebra of *IF* logic?

♣ PROBLEM 44. What are natural decidable fragments of *IF* logic?

The *IF* move from sequential to uniform strategies makes general sense for any logic game.

♣ PROBLEM 45. Develop *IF* versions for games of argumentation or model comparison.

Preferences

Finally, one can also add more finer preferences to logic games, and introduce more sophisticated equilibria than just those far, being of the form 'winning strategy versus any counter-play'. Harrenstein 2004 develops some quite original extensions of logical notions of consequence along these lines. Other possibilities abound:

(a) Measure *effort*, in making branch length to end node an additional factor in the pay-offs for players.

(b) Form new games out of existing logical tasks.

A good source for the latter additional structure is computing updates, as in Section 4. Parts of the update universe can be turned into *conversation games* with a restricted set of available assertions where the point is to be the first to satisfy some 'knowledge condition' (cf. [van Otterloo, 2005]). Nash values for these games can be hard to compute, as complexity increases quickly. Winning strategies clearly have something to do with solving the *Reachability Problem* from one epistemic model to another by a sequence of admissible actions satisfying certain epistemic assertions.

♣ PROBLEM 46. Develop a general theory of conversation games.

7   Further Topics

This survey of issues and questions, broad as it is, has still left out a number of issues that are crucial to games. We conclude by merely listing a few:

Probability

All of the above needs to be integrated with probability theory, in particular, Bayesian update: cf. [van Benthem, 2003a] for general product update in this setting. One good reason is that game solutions in general involve probabilistic *mixed strategies* in an essential way as admissible forms of behaviour. This trend is also emerging, interestingly, in *IF* logic with Hintikka's use of 'generalized Skolem functions' (cf. [van Benthem, 2004c]) to deal with quantum-mechanical particles whose positions and moments satisfy the Heisenberg Uncertainty Principle.

Infinite and Evolutionary Games

Infinitely repeated games are essential in modern game theory, starting from Axelrod's famous work on the emergence of cooperation in Prisoner's Dilemma encounters, and continuing into modern evolutionary game theory (cf. [Skyrms, 2004]). A systematic connection between such games and those found in logic seems rewarding. E.g., how does the strategy calculus of infinite linear logic games, with its emphasis on 'memory-free strategies' like Copy-Cat (i.e., Tit-for-Tat in Axelrod's account) or finite-automaton computable strategies relate to similar topics in game theory? More generally, evolutionary game theory turns around the mathematics of *dynamical systems* — and its relationship with epistemic temporal and update logics seems a potentially important short-term/long-term interface.

Joint Action

Games naturally involve collective behaviour. Just as epistemic logic has moved to considering knowledge of *groups* which need not be reducible to knowledge of individuals, it makes sense to look at group action per se. Compare also the philosophical tradition on shared agency [Bratman, 1993], which addresses congenial issues. Coalition logics do part of this job, but there is much more potential structure to collective action and collective information than what has been considered so far.

8   Conclusion

This concludes our survey of current directions and open problems at the interface of logic and game theory. If all of this comes together in a meaningful way, we would have a broad theory of interaction and information that should be of interest to logicians, game theorists, computer scientists, and philosophers.

BIBLIOGRAPHY

[Abramsky and Jagadeesan, 1994] S. Abramsky and R. Jagadeesan. Games and Full Completeness for Multiplicative Linear Logic, *Journal of Symbolic Logic*, **59**, 543–574, 1994.

[Alur et al., 1998] R. Alur, T. Henzinger and O. Kupferman. Alternating-Time Temporal Logic. In *Compositionality: the Significant Difference*, Springer Lecture Notes in Computer Science 1536, pp. 23–60, 1998.

[Aucher, 2003] G. Aucher. *A Joint System of Update Logic and Belief Revision*, Master of Logic Thesis, ILLC University of Amsterdam, 2003.

[Aumann, 1976] R. Aumann. Agreeing to Disagree, *Annals of Statistics*, **4**, 1236–1239, 1976.

[Baltag et al., 2004] A. Baltag, B. Coecke and M. Sadrzadeh. Algebra and Sequent Calculus for Epistemic Actions, ENTCS *Proceedings of Logic and Communication in Multi-Agent Systems (LCMAS) Workshop*, ESSLLI 2004, Nancy, France, 2004.

[Baltag et al., 1998] A. Baltag, L. Moss and S. Solecki. The Logic of Public Announcements, Common Knowledge and Private Suspicions. In *Proceedings TARK 1998*, pp. 43–56. Morgan Kaufmann Publishers, Los Altos, 1998. Many updated versions.

[Belnap et al., 2001] N. Belnap, M. Perloff and M. Xu. *Facing the Future*, Oxford University Press, Oxford, 2001.

[van Bemthem, 1996] J. van Benthem. *Exploring Logical Dynamics*, CSLI Publications and Chicago University Press, Stanford, 1996.

[van Benthem, 1999–] J. van Benthem. *Logic in Games*, Electronic lecture notes, ILLC University of Amsterdam and Department of Philosophy, Stanford, University, 1999–present. http://staff.science.uva.nl/~johan/Teaching

[van Benthem, 2001] J. van Benthem. Games in Dynamic Epistemic Logic, *Bulletin of Economic Research*, **53**, 219–248, 2001.

[van Benthem, 2002a] J. van Benthem. Extensive Games as Process Models, *Journal of Logic, Language and Information*, **11**, 289–313, 2002.

[van Benthem, 2002b] J. van Benthem. The Epistemic Logic of IF Games, Tech Report PP-2002-24, ILLC Amsterdam, 2002. To appear in L. Hahn, ed., *Jaakko Hintikka, Library of Living Philosophers*, Southern Illinois University/Carus Press.

[van Benthem, 2002c] J. van Benthem. What Logic Games are Trying to Tell Us, Tech Report PP-2002-25, ILLC, University of Amsterdam, 2002.

[van Benthem, 2003a] J. van Benthem. Conditional Probability Meets Update Logic, *Journal of Logic, Language and Information*, **12**, 409–421, 2003.

[van Benthem, 2003b] J. van Benthem. Logic Games are Complete for Game Logics, *Studia Logica*, **75**, 183–203, 2003.

[van Benthem, 2003c] J. van Benthem. Rational Dynamics and Epistemic Logic in Games. In S. Vannucci, ed., *Logic, Game Theory and Social Choice III*, University of Siena, department of political economy, pp. 19–23, 2003. To appear in *International Journal of Game Theory*.

[van Benthem, 2004a] J. van Benthem. De Kunst van het Vergaderen, in Wiebe van der Hoek, ed., *Liber Amicorum 'John-Jules Charles Meijer 50'*, pp. 5–7, Onderzoeksschool SIKS, Utrecht, 2004. A version of a lecture given at GLC Groningen 2003.

[van Benthem, 2004b] J. van Benthem. Minimal Predicates, Fixed Points, and Definability, Tech Report PP-2004-01, ILLC Amsterdam, 2004. To appear in *Journal of Symbolic Logic*.

[van Benthem, 2004c] J. van Benthem. Probabilistic Features in Logic Games, invited presentation, Open Court Symposium, APA Chicago. In D. Kolak and J. Symons, eds., *Quantifiers, Questions, and Quantum Physics,* Springer Verlag, New York, pp. 189–194, 2004.

[van Benthem, 2004d] J. van Benthem. Update and Revision in Games, course notes, Department of Philosophy, Stanford University, and ILLC, University of Amsterdam, 2004.

[van Benthem, 2005a] J. van Benthem. An Essay on Sabotage and Obstruction. In D. Hutter ed., *Festschrift for Jörg Siekmann*, Springer Verlag, Berlin, 2005.

[van Benthem, 2005b] J. van Benthem. Logical Construction Games, Lecture at Borjomi Autumn School 2001. To appear in A-V Pietarinen, ed., *Festschrift for Gabriel Sandu*, Philosophical Institute, University of Helsinki, 2005.

[van Benthem, 2005c] J. van Benthem. Open Problems in Logical Dynamics, ILLC Preprint:, DARE electronic archive 148382, 2005. To appear in D. Gabbay, S. Gontcharov and M. Zakharyashev, eds., *Logics for the XXIst Century: Mathematical Problems From Applied Logic*, Russian Academy of Sciences, Novosibirsk and Kluwer/Plenum, New York.

[van Benthem *et al.*, 2005] J. van Benthem, J. van Eijck and B. Kooi. A Logic of Communication and Change, ILLC Preprint:, University of Amsterdam, DARE electronic archive 148524. Shorter version published as 'Common Knowledge in Update Logics', in R. van der Meyden, ed., *Proceedings TARK 10*, Singapore, pp. 253–261, 2005.

[van Benthem *et al.*, 1994] J. van Benthem, J. van Eijck and V. Stebletsova. Modal Logic, Transition Systems and Processes, *Journal of Logic and Computation*, 4, 811–855, 1994.

[van Benthem and Liu, 2004] J. van Benthem and F. Liu. Diversity of Logical Agents in Games, *Philosophia Scientiae*, 8, 163–178, 2004.

[Bonanno, 1991] G. Bonanno. The Logic of Rational Play in Games of Perfect Information, *Economics and Philosophy* 7, 37–65, 1991.

[Bonanno, 2003] G. Bonanno. Memory of Past Beliefs and Actions, *Studia Logica*, 75, 7–30, 2003.

[Bratman, 1993] M. Bratman. Shared Intention, *Ethics*, 104, 97–113, 1993.

[de Bruin, 2004] B. de Bruin. *Explaining Games*, Dissertation 2004–03, ILLC Amsterdam, 2004.

[Dawar *et al.*, 2004] A. Dawar, E. Grädel and S. Kreutzer. Inflationary Fixed Points in Modal Logic, *ACM Transactions on Computational Logic*, 5, 282–315, 2004.

[Dechesne, 2005] F. Dechesne. *Game, Set, Maths*, Ph.D. Dissertation, Philosophical Institute, Katholieke Universiteit Brabant, Tilburg, 2005.

[van Ditmarsch, 2000] H. van Ditmarsch. *Knowledge Games*, dissertation DS-2000-06, Institute for Logic, Language and Computation, University of Amsterdam and Department of Informatics, University of Groningen, 2000.

[van Ditmarsch *et al.*, to appear] H. van Ditmarsch, W. van der Hoek and B. Kooi. *Dynamic Epistemic Logic*, Kluwer-Springer Academic Publishers, Dordrecht, to appear.

[Ebbinghaus and Flum, 1995] H.-D. Ebbinghaus and J. Flum. *Finite Model Theory*, Springer, Berlin, 1995.

[Fagin *et a.*, 1995] R. Fagin, J. Halpern, Y. Moses and M. Vardi. *Reasoning about Knowledge*, The MIT Press, Cambridge, MA, 1995.

[Gärdenfors and Rott, 1995] P. Gärdenfors and H. Rott. Belief Revision, in D. M. Gabbay, C. J. Hogger and J. A. Robinson, eds., *Handbook of Logic in Artificial Intelligence and Logic Programming* 4, Oxford University Press, Oxford 1995.

[Goranko, 2001a] V. Goranko. The Basic Algebra of Game Equivalences, in G. Sandu and M. Pauly, eds., *ESSLLI Workshop on Logic and Games*, Institute of Philosophy, University of Helsinki, 2001.

[Goranko, 2001b] V. Goranko. Coalition Games and Alternating Temporal Logics, in J. van Benthem, ed., *Proceedings TARK VIII Siena*, Morgan Kaufmann Publishers, San Francisco, 2001.

[Halpern, 2003] J. Halpern. A Computer Scientist Looks at Game Theory, *Games and Economic Behavior*, 45, 114–131, 2003.

[Halpern and Vardi, 1989] J. Halpern and M. Vardi. The Complexity of Reasoning about Knowledge and Time, *Journal of Computer and Systems Science*, 38, 195–237, 1989.

[Harrenstein et al., 1999] P. Harrenstein, W. van der Hoek, J-J. Meijer and C. Witteveen. Subgame-Perfect Nash Equilibria in Dynamic Logic, in M. Pauly and A, Baltag, eds., *Proceedings of the ILLC Workshop on Logic and Games*, Tech Report PP-1999-25, Institute for Logic, Language and Computation, University of Amsterdam, pp. 29–30, 1999.

[Harrenstein, 2004] P. Harrenstein. *Logic in Conflict*, Ph.D. thesis, Department of Computer Science, University of Utrecht, 2004.

[Hendricks, 2002] V. Hendricks. Active Agents, PHILOG Newsletter, Roskilde. In J. van Benthem and R. van Rooy, eds., special issue on Information Theories, *Journal of Logic, Language and Information*, **12**, 469–495, 2002.

[Hintikka and Sandu, 1997] J. Hintikka and G. Sandu. Game-Theoretical Semantics, in J. van Benthem and A. ter Meulen, eds., *Handbook of Logic and Language*, Elsevier, Amsterdam, pp. 361–410, 1997.

[Hirsch and Hodkinson, 2002] R. Hirsch and I. Hodkinson, 2002, *Relational Algebras by Games*, Elsevier Science Publishers, Amsterdam, 2002.

[Hodges, 1985] W. Hodges. *Building Models by Games*, Cambridge University Press, Cambridge, 1985.

[Hodges, 1998] W. Hodges. *An Invitation to Logical Games*, Department of Mathematics, Queen Mary's College, London, 1998.

[van der Hoek et al., 2004] W. van der Hoek, S. van Otterloo and M. Wooldridge. Preferences in Game Logics, *Proceedings AAMAS 2004*, New York, 2004.

[van der Hoek and Pauly, 2005] W. van der Hoek and M. Pauly. Modal Logic and Game Theory, to appear in P. Blackburn, J. van Benthem and F. Wolter, eds., *Handbook of Modal Logic*, Elsevier, Amsterdam, 2005.

[van der Hoek and Wooldridge, 2003] W. van der Hoek and M. Wooldridge. Cooperation, Knowledge, and Time: Alternating-time temporal epistemic logic and its applications, *Studia Logica*, **75**, 125–157, 2003.

[Hofbauer and Sigmund, 2002] J. Hofbauer and K. Sigmund. *Evolutionary Games and Population Dynamics*, Cambridge University Press, Cambridge, 2002.

[Hollenberg, 1998] M. Hollenberg. *Logic and Bisimulation*, Ph.D. Thesis, Philosophical Institute, University of Utrecht, 1998.

[Japaridze, 1997] Georgi Japaridze. A Constructive Game Semantics for the Language of Linear Logic, *Annals of Pure and Applied Logic*, **85**, 87–156, 1997.

[Kelly, 1996] K. Kelly. *The Logic of Reliable Enquiry*, Oxford University Press, Oxford, 1996.

[Kelly, 2002] K. Kelly. Knowledge as Reliably Inferred Stable True Belief, Department of Philosophy, Carnegie-Mellon University, Pittsburgh, 2002.

[Liu, 2004] F. Liu. *Diversity of Logical Agents*, Master's Thesis, ILLC Amsterdam, 2004.

[Löwe, 2002] B. Löwe. Playing with Mixed Strategies on Infinite Sets, *International Journal of Game Theory*, **31**, 137–150, 2002.

[Löwe, 2003] B. Löwe. Determinacy for Infinite Games with More than Two Players with Preferences, Working paper, ILLC University of Amsterdam, 2003.

[van der Meyden, 1996] R. van der Meyden. The Dynamic Logic of Permission, *Journal of Logic and Computation*, **6**, 465–479, 1996.

[van der Meyden, 1998] R. van der Meyden. Common Knowledge and Update in Finite Environments, *Information and Computation*, **140**, 115–157, 1998.

[van der Meyden, 2005] R. van der Meyden. Model Checking the Logic of Knowledge, Tutorial, First Indian Congress on Logic and its Relationship with Other Disciplines, Indian Institute of Technology, Mumbai, 2005.

[Moore, 1985] R. Moore. A Formal Theory of Knowledge and Action, Research Report, SRI International, Menlo Park, 1985.

[Moschovakis, 1980] Y. Moschovakis. *Descriptive Set Theory*, North-Holland, Amsterdam, 1980.

[Osborne and Rubinstein, 1994] M. Osborne and A. Rubinstein. *A Course in Game Theory*, The MIT Press, Cambridge, MA, 1994.

[van Otterloo, 2005] S. van Otterloo. *A Strategic Analysis of Multi-Agent Protocols*, Ph.D. Thesis, Department of Computer Science, University of Liverpool and ILLC, University of Amsterdam, 2005.

[van Otterloo and Roy, 2005] S. van Otterloo and O. Roy. Verification of Voting Protocols, Working paper, ILLC, University of Amsterdam, 2005.

[Parikh, 1985] R. Parikh. The Logic of Games and its Applications, *Annals of Discrete Mathematics*, **24**, 111–140, 1985.

[Parikh and Pauly, 2003] R. Parikh and M. Pauly. Game Logic – An Overview, *Studia Logica*, **75**, 165–182, 2003.

[Parikh and Ramanujam, 2003] R. Parikh and R. Ramanujam. A Knowledge Based Semantics of Messages, CUNY New York and Chennai, India. In J. van Benthem and R. van Rooy, eds., special issue on Information Theories, *Journal of Logic, Language and Information*, **12**, 453–467, 2003.

[Pauly, 2001] M. Pauly. *Logic for Social Software*, dissertation DS-2001-10, Institute for Logic, Language and Computation, University of Amsterdam, 2001.

[Ramanujam, 2005] R. Ramanujam. Closing Remarks, First Indian Congress on Logic and its Relationship with Other Disciplines, IIT Mumbai – IMSc Chennai, 2005.

[Sandu and Tulenheimo, 2005] G. Sandu and T. Tulenheimo. *Logics of Imperfect Information*, Book manuscript, Philosophical Institute, University of Helsinki and Sorbonne, Paris, 2005.

[Skyrms, 2004] B. Skyrms. *The Stag Hunt and the Evolution of Social Structure*, Cambridge University Press, Cambridge, 2004.

[Spohn, 1988] W. Spohn. Ordinal Conditional Functions: A Dynamic Theory of Epistemic States, in W. L. Harper et al., eds., *Causation in Decision, Belief Change and Statistics* II, Kluwer, Dordrecht, pp. 105–134, 1988.

[Stalnaker, 1999] R. Stalnaker. Extensive and Strategic Form: Games and Models for Games, *Research in Economics*, **53**, 93–291, 1999.

[Tan and van der Torre, 1998] Y. Tan and L. van der Torre. An Update Semantics for Prima Facie Obligations. In H. Prade (ed.), *Proceedings of the Thirteenth European Conference on Artificial Intelligence* (ECAI'98), pp. 38–42, 1998.

[Tulenheimo, 2004] T. Tulenheimo. *Independence-Friendly Modal Logic*, Ph.D. Thesis, Philosophical Institute, University of Helsinki, 2004.

[Venema, 2001] Y. Venema. Representation of Game Algebras, Manuscript, ILLC Amsterdam. To appear in *Studia Logica*, 2001.

[Yamada, 2004] T. Yamada. Commands and Deontic Update, class paper Philosophy 298, Stanford University, 2004.

[Yap, 2005] A. Yap. Product Update and Looking Backward. Department of Philosophy, Stanford University, 2005.

[Zarnic, 1999] B. Zarnic. Validity of Practical Inference, ILLC Preprint series PP-1999-23, University of Amsterdam, 1999.

Two Theories of Nonmonotonic Reasoning

ALEXANDER BOCHMAN

1 What is nonmonotonic reasoning?

A theory of nonmonotonic reasoning is purported to give us a more adequate description of the actual ways we think about the world. Too often we reason and act in situations where we do not or even cannot have sufficient information, so deductive reasoning, taken by itself, cannot help us in such situations. Still, we need to act in such situations in a reasonable way, and it is here that nonmonotonic reasoning finds its place. Human rational activity is not reducible to collecting facts and deriving their consequences; it embodies an active epistemic attitude that involves making assumptions and wholesale theories about the world and acting in accordance with them. We do not only perceive the world, we also give it structure in order to make it intelligible and controllable. Nonmonotonic reasoning in this respect is just an embodiment of a general scientific methodology.

The relationship between nonmonotonic reasoning and logic is part of a larger story of the relations between AI and logic (see [Thomason, 2003]). John McCarthy, one of the founders of AI, has developed a research methodology that used logical techniques for resolving reasoning problems in AI. McCarthy's objective was to formalize *commonsense reasoning* used in dealing with such problems. In a sense, nonmonotonic reasoning is an outgrowth of McCarthy's program. But though commonsense reasoning has always appeared to be an attractive standard, the study of 'artificial reasoning' need not and has not been committed to representing common sense. Traditional formalisms of nonmonotonic reasoning could hardly be called formalizations of commonsense reasoning. Still, in trying to cope with typical commonsense reasoning tasks, the suggested formalisms have succeeded in capturing important features of the latter and thereby have broken new territory for logic.

The authors of first nonmonotonic theories have tried, of course, to express their formalisms using available logical means, ranging from classical to modal logics. McCarthy, for example, has always believed that anything that can be expressed, can be expressed in first order logic. Still, from its very beginning, nonmonotonic reasoning research has created formalisms that had no counterpart in existing logical theories, and refined them to sophisticated logical systems. It is even advantageous to see nonmonotonic reasoning as a brand new approach to logical reasoning; already at this stage of its development, it is not yet another application of logic, but a relatively independent field of logical research that has a great potential in informing both future logical theory and many areas of philosophical enquiry.

The origins of nonmonotonic reasoning lie in dissatisfaction with the traditional logical methods in representing the problems posed by AI. Namely, reasoning necessary for an intelligent behavior and decision making in realistic situations has turned out to be difficult, even impossible, to represent as deductive inferences in some logical system. And the way out suggested by nonmonotonic reasoning consisted in using reasonable beliefs and assumptions that can guide us in our decisions. In this respect, nonmonotonic reasoning can be described as a theory of a rational use of assumptions.

The sentence "Birds (normally) fly" is weaker than "All birds fly"; there is a seemingly open-ended list of exceptions – ostriches, penguins, etc. etc. So, if we would try to use classical logic for representing "Birds fly", the first problem would be that it is practically impossible to enumerate all exceptions to flight with an axiom of the form

$$(\forall x).Bird(x)\&\neg Penguin(x)\&\neg Emu(x)\&\neg Dead(x)\&... \supset Fly(x)$$

The second crucial problem is that, even if we could enumerate all such exceptions, we still could not derive $Fly(Tweety)$ from $Bird(Tweety)$, since we are not given that Tweety is not a penguin, or dead, etc. The antecedent of the above implication cannot be derived, so there is no way of deriving the consequent. Nevertheless, if told only about a particular bird, say Tweety, without being told anything else about it, we would be justified in concluding that Tweety can fly. So the problem is how we can actually *make* such assumptions in the absence of information to the contrary.

Clearly, default beliefs and assumptions are *defeasible*; they might give wrong predictions in certain circumstances. This makes nonmonotonic reasoning a risky business. Still, in most cases such beliefs are useful and give desired results, and hence they are worth the risk of making an error. But what is even more important, more often than not we simply have no safe replacement for such a reasoning strategy. That is why it is worth teaching robots and computers to reason in this way.

It is this suppositional character of commonsense reasoning that conflicts with the monotonic nature of logical derivations. Monotonicity follows directly from the very notion of a proof being a sequence of steps starting with accepted axioms and proceeding by inference rules that are valid in any context of their use. Consequently, if a set of propositions entails a consequence C, then a larger set will also entail C. Commonsense reasoning is not monotonic in this sense, because adding new facts may invalidate some of the assumptions made earlier.

1.1 Two Paradigms of Nonmonotonic Reasoning

Studies in nonmonotonic reasoning have given rise to two essentially different approaches that will be called, respectively, *preferential* and *explanatory* nonmonotonic reasoning. The first approach was initiated by Dov Gabbay in [Gabbay, 1985] on the logical side, and by Shoham [Shoham, 1988] on the AI side. It encompasses nonmonotonic inference relations of [Kraus *et al.*, 1990], and a general theory of belief change [Alchourrón *et al.*, 1985]. A detailed description of this approach can be found in [Bochman, 2001; Schlechta, 2004; Makinson, 2005]. The second, explanatory approach is older, and it includes default and modal nonmonotonic logics, as well as logic programming with negation as failure. A description of this approach can be found, e.g., in [Marek and Truszczyński, 1993; Bochman, 2005].

The difference between the two approaches can be found on a number of levels. To begin with, there are two different senses in which a logical formalism may be called nonmonotonic. First, it may be *locally* nonmonotonic in that its rules do not admit addition of new premises, that is, they do not satisfy Strengthening the Antecedent. Second, it may be *globally* nonmonotonic in the sense that adding further rules to the system may possibly invalidate conclusions obtained earlier. These two kinds of nonmonotonicity are largely independent. Thus, preferential inference relations [Kraus *et al.*, 1990] are locally nonmonotonic (*Birds fly* does not imply *Penguins fly*). However, they are globally monotonic, since addition of new preferential conditionals does not invalidate previous conclusions. On the other hand, default logic [Reiter, 1980] exemplifies the combination of local monotonicity with global nonmonotonicity. Any default theory can be safely extended with default rules obtained from existing ones by strengthening their prerequisites and justifications; such additional rules will not change the set of extensions. On the other hand, adding arbitrary new rules to the default theory may result in creating new extensions, so nonmonotonic conclusions made earlier will not, in general, be preserved.

Taken by itself, however, the above distinction is a purely formal difference, and it does not necessarily imply that the two approaches are essentially different. Actually, one of the main incentives behind the preferential approach, explicitly expressed in [Shoham, 1988], was the hope that default logic and other formalisms can be subsumed by representing extensions as preferred models under some generalized notion of preference. Unfortunately, subsequent studies have raised grave doubts about this hope. Thus, the nonmonotonic semantics of default logic has turned out to violate even the most basic postulates of cumulative inference (see [Makinson, 1989]). A similar situation has been found in logic programming (see [Dix, 1991]).

In a hindsight, this outcome should have been expected, since the selection of intended models in the explanatory approach is not preferential in a usual sense. Namely, the explanatory approach determines the intended models as models satisfying certain justification conditions with respect to the rules. On a most abstract level, such models are expressible as fixed points of some operator which is not even monotonic. Accordingly, the supposed preference that singles out these models appears to be a trivial, zero-one preference that basically differentiates only right models from bad ones. In this sense, the above formal difference can be viewed only as a symptom of deeper conceptual distinctions between the two approaches.

Both preferential and explanatory nonmonotonic reasoning can be seen as theories of a reasoned use of assumptions. Now, preferential reasoning treats such assumptions as *defaults*, namely as normality assumptions we can use whenever there is no evidence to the contrary. This presumptive reading has a semantic counterpart in the notion of *normality*; default assumptions are considered as holding for normal circumstances, and the theory tells us to always assume that the world is as normal as is compatible with the known facts. This naturally creates a preferential setting, in which the normality of models is measured by the set of defaults they support (see below).

It turns out, however, that the explanatory nonmonotonic reasoning implicitly assigns a different role to assumptions. Using the name adopted by David Poole (see [Poole, 1989]) it makes such assumptions *conjectures*. Conjectures are assumptions that we make in order to explain observations. The supposition of normality (or abnormality, for that matter) is not essential for such conjectures. A certain combination of symptoms may lead to a conjecture about a rare and unusual disease, while in other cases some 'ordinary' illness will suffice for explaining the observations; it seems beside the point in this case to order diseases with respect to their 'normality'. As was rightly noted by Poole, we make conjectures only if there is evidence that requires them for explanation, in contrast to defaults that can be freely assumed, unless they contradict the facts and other assumed defaults. It was also strongly argued by Poole that the distinction between normality defaults and conjectures is closely related to the distinction between *prediction* and *explanation*: while we use defaults in order to predict facts that are yet unknown, conjectures are invoked when we have to explain known facts. Untreated syphilis can explain paresis, though the syphilis assumption does not have predictive force of deriving paresis.

A related conceptual distinction between nonmonotonic rules has been suggested by Dov Gabbay in [Gabbay, 1998a] as a difference between *expectation rules* and the *scenario rules*. Expectation rules are based on laws or conventions, and they can be used as extra deduction rules for extending our knowledge in the absence of contrary information. Reasoning with such rules obeys the basic principle of preferential inference, namely Restricted (Cumulative) Monotony. On the other hand, the scenario rules are used as arguments in constructing possible scenarios consistent with data (e.g., in diagnosis). They arise in situations of uncertainty where there are several possible scenarios, each important in its own right, so it does not have much sense to enquire about what holds for all of them.

Unfortunately, the above distinction has been obscured in the short history of nonmonotonic reasoning, because from the very beginning the main formalisms of nonmonotonic reasoning, including default logic and circumscription, have claimed their rights and responsibility on representation of normality defaults. Thus, Ray Reiter has suggested in [Reiter, 1980] that we can identify such normality assertions with a special case of default rules of the form $A{:}B/B$, appropriately called normal default rules.

On our view, the preferential approach provides a more adequate analysis of the notion of normality and, in particular, of normal defaults. The examples in the literature that reveal a discrepancy between Reiter's normal defaults $A{:}B/B$ and corresponding preferential conditionals $A \mathrel{|\!\sim} B$ (see, e.g., [Kraus *et al.*, 1990; Lehmann, 1995]) point out in favor of the latter. So the criticism raised against default logic in the preferential camp seems justified so far as we are talking about which notion better reflects our understanding of normality. Still, there is nothing wrong or nonintuitive about having both $A : B/B$ and $A : \neg B/\neg B$ in a default theory, though it is certainly counterintuitive to treat such rules simultaneously as normality defaults. In the setting of default logic, such rules say simply that, when A holds, both B and $\neg B$ are equally admissible assumptions, or conjectures. This indicates, however, that default logic has a subject of its own that should not be extrapolated to the entire field of nonmonotonic reasoning.

The above picture is of course incomplete, mainly because it stresses the differences and ignores numerous similarities and connections between the two approaches. For example, circumscription with abnormality predicates [McCarthy, 1986] can be understood in both ways, so it constitutes a borderline case covered by both approaches. The same can be said about Poole's abductive theory [Poole, 1988]. Furthermore, many reasoning tasks involve both these forms of nonmonotonic reasoning. Still, a clear understanding of both can be achieved only if they are considered first as independent reasoning paradigms with subject, principles and problems of their own.

1.2   Nonmonotonic Reasoning and Logic

Nonmonotonic reasoning changes the ways logic is used. So, before we begin,
it is essential to lay down a general conceptual structure that establishes a
proper relationship between nonmonotonic reasoning and logic.

To begin with, there are many reasons to believe that a formal representation of nonmonotonic reasoning cannot be obtained in the form of a
logical inference in some ingenious 'nonmonotonic logic'. Some of these
reasons will be detailed in what follows. The paradise of classical logic
in which the semantic notion of validity is equivalent to syntactic notion
of derivation is hardly achievable for nonmonotonic reasoning. As we will
see, logical systems provide only a more or less tight *framework* in which
nonmonotonic reasoning can be represented and performed. This pertains
not only to monotonic logical systems, but also to so-called nonmonotonic
logical formalisms and inference relations. For example, we will see that
preferential inference relations will serve only as a set of logical constraints
nonmonotonic reasoning should obey, and we should not expect that some
strengthening of such inference relations with new postulates would capture
the intended system of nonmonotonic reasoning.

Many nonmonotonic formalisms have two components. The first is a logical framework, e.g., classical logic in circumscription, a modal logic in modal
nonmonotonic logics, or cumulative inference relations in preferential nonmonotonic reasoning. The second, nonmonotonic, component determines
which of the possible models should be considered as intended ones.

In ordinary logical systems, the semantics determines the set of logical
consequences of a given theory, but also, and most importantly, it provides
a semantic interpretation for the syntax itself. Namely, it provides propositions and rules of a syntactic formalism with *meaning*, and its theories
with *informational content*. By its very design, however, the nonmonotonic
semantics is defined as a certain *subset* of the set of possible models, and
consequently it does not determine, in turn, the meaning of the propositions
and rules of the syntax. As a result, two radically different theories may
have the same nonmonotonic semantics. Furthermore, such a difference
cannot be viewed as apparent, since it may well be that by adding further
rules or facts to both these theories, we obtain new theories that already
have different nonmonotonic models.

The above situation is quite similar to the distinction between meaning
(intension) and extension of logical names, a distinction that is fundamental
for modern logic. Nonmonotonic semantics provides, in a sense, the extensional content of a theory in a particular context of its use. In order to
determine the meaning, or informational content, of a theory, we have to
consider the underlying logic. This requires a *clear separation of logical and*

nonmonotonic aspects of nonmonotonic reasoning. The separation suggests the following understanding of nonmonotonic reasoning:

Nonmonotonic Reasoning = Logic + Nonmonotonic Semantics

Logic and its associated logical semantics are responsible for providing the meaning of the rules of the formalism, while the nonmonotonic semantics provides us with nonmonotonic consequences of a theory. As the main benefit of this separation, it has been shown in [Bochman, 2005] that common kinds of explanatory nonmonotonic reasoning are reducible, in effect, to different kinds of logical reasoning with respect to a single nonmonotonic semantics. In addition, such an understanding will allow us to clarify the proper subject and tasks of preferential nonmonotonic reasoning. As we will see, however, both approaches to nonmonotonic reasoning will have to be elaborated in order to comply with the above conceptual schema.

2 Preferential Nonmonotonic Reasoning

As a theory of a rational use of assumptions, the main problem nonmonotonic reasoning deals with is that assumptions are often incompatible with one another, or with known facts. In such cases of conflict we must have a reasoned choice. The preferential approach follows the slogan *"Choice presupposes preference"*. According to this approach, the choice of assumptions should be made by forming the space of options for choice and establishing preference relations among them. This makes preferential approach a special case of a general methodology that is at least as old as the decision theory and theory of social choice. In the nonmonotonic reasoning field, this approach forms a research program that has begun with McCarthy's circumscription and continued in the works of Gabbay, Lifschitz, Shoham, Makinson, Lehmann, and many others after them.

2.1 Epistemic States

Semantic interpretation constitutes one of the main components of a viable reasoning system, monotonic or not. A formal inference engine, though important, can be effectively used for representing and solving reasoning tasks only if its basic notions have clear meaning allowing to discern them from a description of a situation at hand. It has been suggested in [Bochman, 2001] that a general representation framework for preferential nonmonotonic reasoning can be given in terms of epistemic states, defined below.

DEFINITION 1. An *epistemic state* is a triple $(\mathcal{S}, l, \prec)$, where $\mathcal{S}$ is a set of *admissible belief states*, $\prec$ a preference relation on $\mathcal{S}$, while l is a labelling function assigning a deductively closed *belief set* to every state from $\mathcal{S}$.

On the intended interpretation, admissible belief states are generated by allowable combinations of default assumptions. Such states are taken to be the options for choice. This structure implicitly reflects dependencies among beliefs. The latter, even if logically consistent, can be based on incompatible defaults and expectations, and hence cannot always be held simultaneously. Thus, potential beliefs are correlated, and some of them serve as reasons, or justification, for others. Accordingly, epistemic states embody at least some features of a foundational (or justification-based) belief acceptance.

The preference relation on admissible belief states reflects the fact that not all admissible combinations of defaults constitute equally preferred options for choice. For example, defaults are presumed to hold, so an admissible belief state generated by a larger set of defaults is normally preferred to an admissible state generated by a smaller set of defaults. In addition, not all defaults are born equal, so they may have some priority structure that imposes, in turn, additional preferences among belief states (see below).

Epistemic states guide our decisions what to believe in particular situations. They are epistemic, however, precisely because they say nothing directly about what is actually true, but only what is believed (or presumed) to hold. This makes epistemic states relatively stable entities; change in facts and situations will not necessary lead to change in epistemic states. The actual assumptions made in particular situations are obtained by choosing preferred admissible belief states that are *consistent* with the facts.

The notion of an epistemic state is broader than its initial motivation, and it can have its own life, independent of the original understanding as formed by combinations of defaults. Thus, epistemic states can also be described in terms of *expectation relations* on all propositions of the language, as has been suggested in [Gärdenfors and Makinson, 1994]. This expectation-based theory can be viewed as a special case of our general representation.

Prioritization

A formal description of epistemic states that are generated by default bases provides us with characteristic properties of epistemic states arising in particular reasoning contexts. An epistemic state is *base-generated* by a set of propositions Δ with respect to a classical consequence relation Th if

- the set of its admissible states is the set $\mathcal{P}(\Delta)$ of subsets of Δ;

- l is a function assigning each $\Gamma \subseteq \Delta$ a theory $\text{Th}(\Gamma)$;

- the preference order is monotonic on $\mathcal{P}(\Delta)$: if $\Gamma \subset \Phi$, then $\Gamma \prec \Phi$.

The preference order on admissible belief states is usually derived in some way from priorities among individual defaults. This task turns out to be a special case of a general problem of combining a set of preference relations into a single 'consensus' preference order. Let us suppose that the set of defaults Δ is ordered by some *priority relation* $\lhd$ which will be assumed to be a strict partial order: $\alpha \lhd \beta$ will mean that α *is prior to* β.

Recall that defaults are beliefs we are willing to hold insofar as it is consistent to do so. Hence any default δ determines a primary preference relation $\preccurlyeq_\delta$ on $\mathcal{P}(\Delta)$ by which admissible belief sets containing the default are preferred to belief sets that do not contain it:

$$\Gamma \preccurlyeq_\delta \Phi \equiv \text{ if } \delta \in \Gamma \text{ then } \delta \in \Phi$$

Each $\preccurlyeq_\delta$ is a weak order having just two equivalence classes, namely sets of defaults that contain δ, and sets that don't. In this setting, the problem of finding a global preference order amounts to constructing an operator that maps a set of preference relations $\{\preccurlyeq_\delta | \ \delta \in \Delta\}$ to a single preference relation $\preccurlyeq$ on $\mathcal{P}(\Delta)$. As has been shown in [Andreka *et al.*, 2002], any finitary operator of this kind satisfying the so-called Arrow's conditions is definable using a *priority graph* $(N, \lhd, v)$, where $\lhd$ is a priority order on a set of nodes N, and v is a labelling function assigning each node a preference relation. The priority graph determines a single resulting preference relation via the *lexicographic rule*, by which t is weakly preferred to s overall if it is weakly preferred for each argument preference, except possibly those for which there is a prior preference that strictly prefers t to s:

$$s \preccurlyeq t \equiv \forall i \in N(s \preccurlyeq_{v(i)} t \vee \exists j \in N(j \lhd i \wedge s \prec_{v(j)} t))$$

In our case, the prioritized base $(\Delta, \lhd)$ can be viewed as a priority graph in which every node δ is assigned a preference relation $\preccurlyeq_\delta$. Consequently, we can apply the lexicographic rule and arrive at the following definition:

$$\Gamma \preccurlyeq \Phi \equiv (\forall \alpha \in \Gamma \backslash \Phi)(\exists \beta \in \Phi \backslash \Gamma)(\beta \lhd \alpha)$$

$\Gamma \preccurlyeq \Phi$ holds when, for each default in $\Gamma \backslash \Phi$, there is a prior default in $\Phi \backslash \Gamma$. The corresponding strict preference $\Gamma \prec \Phi$ is defined as $\Gamma \preccurlyeq \Phi \wedge \Gamma \neq \Phi$.

[Lifschitz, 1985] was apparently the first to use this construction in prioritized circumscription, while [Geffner, 1992] employed it for defining preference relations among sets of defaults (see also [Grosof, 1991]).

Despite all its virtues, the above definition of preference still does not cover some applications, because it is based on absolute, unconditional priorities. Thus, inheritance hierarchies (see below) turn out to be representable in terms of priorities that are conditional upon presence of other defaults.

A conditional priority relation on Δ is a ternary relation $\alpha \lhd_\Gamma \beta$ (where $\Gamma \subseteq \Delta$) saying that α *is prior to* β *with respect to* Γ. A conditional priority relation is a *conditional priority order* if $\lhd_\Gamma$ is a strict partial order for every fixed Γ. Let $\downarrow_\Gamma \alpha$ be the set of defaults that are prior to α with respect to Γ. Then the conditional priority order is *local* if defaults that are prior to α with respect to Γ depend only on those elements of Γ that are themselves prior to α:

Locality   If $\Gamma \cap \downarrow_\Gamma \alpha = \Phi \cap \downarrow_\Gamma \alpha$, then $\downarrow_\Gamma \alpha = \downarrow_\Phi \alpha$.

Now we can use conditional priorities to define a preference relation on sets of defaults in a way quite similar to the earlier use of absolute priorities.

$$\Gamma \prec \Phi \ \equiv\ \Gamma \neq \Phi \ \& \ (\forall \alpha \in \Gamma \backslash \Phi)(\exists \beta \in \Phi \backslash \Gamma)(\beta \lhd_\Gamma \alpha)$$

It turns out that this generalized preference relation has the same nice features as in the absolute case. Let us say that a set of defaults is *well-founded* with respect to a conditional priority order if it does not contain infinite descending chains with respect to some $\lhd_\Gamma$. Then we have

THEOREM 2. *The preference relation generated by a local conditional priority order is a monotonic strict partial order on well-founded sets.*

2.2   Nonmonotonic inference and its kinds

In particular situations, we restrict our attention to admissible belief sets that are consistent with the facts, and choose preferred among them. The latter are used to support the assumptions and conclusions we make about the situation at hand. Accordingly, all kinds of nonmonotonic inference relations, described below, presuppose a two-step selection procedure: for a current evidence A, we consider admissible belief states that are consistent with A and choose preferred elements in this set.

An admissible belief state $s \in \mathcal{S}$ *supports* a proposition A, if A belongs to its belief set, that is, $A \in l(s)$. The set of all admissible states that do *not* support A will be denoted by $]A[$. Accordingly, $]\neg A[$ is the set of admissible belief states that are compatible with A.

A *skeptical inference* (or prediction) with respect to an epistemic state is obtained when we infer only what is supported by each of the preferred states. In other words, B will be a skeptical conclusion from the evidence A in an epistemic state $\mathbb{E}$ if each preferred admissible belief set in $\mathbb{E}$ that is consistent with A, taken together with A itself, implies B.

DEFINITION 3. *B is a skeptical consequence of A (notation $A \mathrel{|\!\sim} B$) in an epistemic state if $A \to B$ is supported by all preferred belief states in $]\neg A[$.*

A set of conditionals $A \mathrel{|\!\sim} B$ that are valid in an epistemic state $\mathbb{E}$ will be called a *skeptical inference relation determined by* $\mathbb{E}$.

The above definition generalizes the notion of prediction from [Poole, 1988], as well as an expectation-based inference of [Gärdenfors and Makinson, 1994]. But in fact, it is much older. While the semantics of nonmonotonic inference from [Kraus *et al.*, 1990] derives from the possible worlds semantics of Stalnaker and Lewis, the above definition can be traced back to the era before the discovery of possible worlds, namely to Frank Ramsey and John S. Mill:

> "In general we can say with Mill that 'If p then q' means that q is inferrable from p, that is, of course, from p together with certain facts and laws not stated but in some way indicated by the context." [Ramsey, 1978, page 144]

This definition has also been used in the 'premise-based' semantics for counterfactuals [Veltman, 1976; Kratzer, 1981] (see also [Lewis, 1981]).

A *credulous inference* (or explanation) with respect to an epistemic state is obtained by assuming that we can reasonably infer (or explain) conclusions that are supported by at least one preferred belief state consistent with the facts. In other words, B will be a credulous conclusion from A if at least one preferred admissible belief set in $\mathbb{E}$ that is consistent with A, taken together with A itself, implies B. In still other words,

DEFINITION 4. B is a *credulous consequence* of A in an epistemic state if $A{\rightarrow}B$ is supported by at least one preferred belief state in $]\neg A[$.

The set of conditionals that are credulously valid in an epistemic state $\mathbb{E}$ forms a *credulous inference relation determined by* $\mathbb{E}$. The above definition constitutes a generalization of the corresponding definition of explanation in Poole's abductive system [Poole, 1988]. Credulous inference is only one, though important, instance of a broad range of non-skeptical inference relations (see [Bochman, 2003]).

2.3 Syntactic characterizations

Gabbay's paper [Gabbay, 1985] was a starting point of the approach to nonmonotonic reasoning based on describing associated inference relations (see [Makinson, 1994] for an overview). This approach was primarily designed to capture the skeptical view of nonmonotonic reasoning. Many reasoning tasks in AI, such as abduction and diagnosis, are based, however, on a credulous understanding of nonmonotonic reasoning. It has turned out that a common ground for both skeptical and credulous inference is provided by a logic of conditionals suggested in [van Benthem, 1984].

The main idea behind van Benthem's approach was that a conditional can be seen as a generalized quantifier representing a relation between the respective sets of instances or situations supporting and refuting it. A situation *confirms* a conditional $A \mathrel{|\!\sim} B$ if it supports the classical implication $A{\rightarrow}B$, and *refutes* $A \mathrel{|\!\sim} B$ if it supports $A{\rightarrow}\neg B$. Then the validity of a conditional in a set of situations is determined by appropriate, 'favorable' combinations of confirming and refuting instances. Whatever they are, we can assume (together with van Benthem) that adding new confirming instances to a valid conditional, or removing refuting ones, cannot change its validity. Accordingly, we can accept the following principle:

> If all situations confirming $A \mathrel{|\!\sim} B$ confirm also $C \mathrel{|\!\sim} D$, and all situations refuting $C \mathrel{|\!\sim} D$ refute also $A \mathrel{|\!\sim} B$, then validity of $A \mathrel{|\!\sim} B$ implies validity of $C \mathrel{|\!\sim} D$.

The above principle is already sufficient for justifying the rules of the basic inference relation, given below. It supports also a representation of conditionals in terms of expectation relations. More exactly, we can say that a conditional is valid if the set of its confirming situations *is sufficiently good* compared with the set of its refuting situations. Accordingly, $A \mathrel{|\!\sim} B$ can be defined as $A{\rightarrow}\neg B < A{\rightarrow}B$ for an appropriate expectation relation $<$, while the above principle secures that this is a general expectation relation as defined in [Bochman, 2001].

A *basic inference relation* $\mathcal{B}$ satisfies the following postulates:

Reflexivity $A \mathrel{|\!\sim} A$

Left Equivalence   If $\vDash A \leftrightarrow B$ and $A \mathrel{|\!\sim} C$, then $B \mathrel{|\!\sim} C$

Right Weakening   If $A \mathrel{|\!\sim} B$ and $B \vDash C$, then $A \mathrel{|\!\sim} C$

Antecedence   If $A \mathrel{|\!\sim} B$, then $A \mathrel{|\!\sim} A \wedge B$

Deduction   If $A \wedge B \mathrel{|\!\sim} C$, then $A \mathrel{|\!\sim} B \rightarrow C$

Conjunctive Cautious Monotony   If $A \mathrel{|\!\sim} B \wedge C$, then $A \wedge B \mathrel{|\!\sim} C$

A set Γ of conditionals implies a conditional α with respect to basic inference if α belongs to the least basic inference relation containing Γ. Note, however, that all the above postulates involve at most one conditional premise. As a result, the basic entailment boils down to a derivability relation among single conditionals. The following theorem describes this derivability relation in terms of the classical entailment.

THEOREM 5. $A \mathrel{|\!\sim} B \Vdash_{\mathcal{B}} C \mathrel{|\!\sim} D$ *if and only if either* $C \vDash D$, *or* $A \rightarrow B \vDash C \rightarrow D$ *and* $C{\rightarrow}\neg D \vDash A{\rightarrow}\neg B$.

Theorem 5 justifies the principle stated earlier: $A \mathrel{|\!\sim} B$ implies $C \mathrel{|\!\sim} D$ if and only if either all situations confirm $C \mathrel{|\!\sim} D$, or else all confirming instances of the former are confirming instances of the latter and all refuting instances of the latter are refuting instances of the former.

Basic inference relation is a weak inference system, first of all because it does not allow to combine different conditionals. Nevertheless, it is in a sense complete so far as we are interested in derivability among individual conditionals. More exactly, basic derivability captures exactly the one-premise derivability of both skeptical and credulous inference relations.

As was noted already in [Gabbay, 1985], a characteristic feature of sceptical inference is the validity of the following postulate:

(And) If $A \mathrel{|\!\sim} B$ and $A \mathrel{|\!\sim} C$, then $A \mathrel{|\!\sim} B \wedge C$.

Indeed, in the framework of basic inference, And is all we need for capturing precisely the preferential inference relations from [Kraus *et al.*, 1990]. Such inference relations provide a complete axiomatization of skeptical inference with respect to epistemic states. An important special case of preferential inference, rational inference relations, are determined by linearly ordered epistemic states; they are obtained by adding further

(Rational Monotony) If $A \mathrel{|\!\sim} B$ and $A \mathrel{|\!\not\sim} \neg C$, then $A \wedge C \mathrel{|\!\sim} B$.

In contrast, credulous inference relations do not satisfy And. Still, they are axiomatized as basic inference relations satisfying Rational Monotony.

Rational Monotony is not a 'Horn' rule, so it does not allow us to derive new conditionals from given ones. In fact, credulous inference relations do not derive much more conditionals than what can be derived already by basic inference (see [Bochman, 2001]). This indicates that there should be no hope to capture credulous nonmonotonic reasoning by derivability in some nonmonotonic logic. Something else should be added to the above logical framework in order to represent the relevant form of nonmonotonic reasoning. Though less evident, the same holds for skeptical inference. Both these kinds of inference need to be augmented with an appropriate globally nonmonotonic semantics that would provide a basis for the associated systems of defeasible entailment, as described in the next section.

2.4 Defeasible entailment

A theory of reasoning about default conditionals occupies an important place in the general theory of nonmonotonic reasoning. Thus, the question whether a proposition B is derivable from an evidence A in a default base is reducible to the question whether the conditional $A \mathrel{|\!\sim} B$ is derivable from the base, so practically all reasoning problems about default conditionals are reducible to the question what conditionals can be derived from a conditional default base. The latter problem constitutes therefore the main task of a theory of default conditionals (cf. [Lehmann and Magidor, 1992]).

For a skeptical reasoning (or prediction), a most plausible understanding of default conditionals is obtained by treating them as skeptical inference rules in the framework of epistemic states. Accordingly, preferential inference relations can be considered as a *logic* behind skeptical nonmonotonic reasoning; the rules of the former should be taken for granted by the latter.

This does not mean, however, that nonmonotonic reasoning about default conditionals is reducible to preferential derivability. Preferential inference is severely sub-classical and does not allow us, for example, to infer "Red birds fly" from "Birds fly". In fact, this is precisely the reason why such inference relations have been called nonmonotonic. Clearly, there are good reasons for not accepting such a derivation as a *logical* rule for preferential inference; otherwise "Birds fly" would imply also "Birds with broken wings fly" and even "Penguins fly". Still, this should not prevent us from accepting "Red birds fly" on the basis of "Birds fly" as a reasonable *nonmonotonic* (or *defeasible*) conclusion, namely a conclusion made in the absence of information against it. By doing this, we would just follow the general strategy of non-monotonic reasoning that involves making reasonable assumptions on the basis of available information. Thus, the logical core of skeptical inference, preferential inference relations, should be augmented with a mechanism of making nonmonotonic conclusions. Actually, the choice of assumptions in this setting can be restricted to conditionals that are classically derivable from a default base. Speaking generally, we would like to keep reasoning classically about default conditionals insofar as this does not conflict with the default base and given evidence. This kind of reasoning will of course be defeasible, or *globally nonmonotonic*, since addition of new condition-als can block some of the conclusions made earlier. Recall in this respect that, though preferential inference is locally nonmonotonic with respect to premises of conditionals, it is nevertheless globally monotonic: adding new conditionals does not change the validity of previous derivations.

On the semantic side, default conditionals are constraints on epistemic states in the sense that the latter should make them skeptically valid. Still, usually there is a huge number of epistemic states that satisfy a given set of conditionals, so we have both an opportunity and necessity to choose among them. Our guiding principle in this choice can be the same basic principle of nonmonotonic reasoning, namely that the intended epistemic states should be as normal as is permitted by the current constraints. By choosing particular such states, we thereby will adopt conditionals that will not be derivable from a given base by preferential inference alone.

The above considerations lead to a seemingly inevitable conclusion that default conditionals possess a clear logical meaning and associated *logical* semantics based on epistemic states (or possible worlds models), but they still lack a globally *nonmonotonic semantics* that would provide an interpretation for the associated defeasible entailment.

Actually, the literature on nonmonotonic reasoning is abundant with theories of defeasible entailment. A history of studies on this subject could be summarized as follows. Initial formal systems, namely Pearl's system Z [Pearl, 1990] and Lehmann's rational closure [Lehmann, 1989; Lehmann and Magidor, 1992], have turned out to be equivalent. This encouraging development has followed by a realization that both theories are insufficient for representing defeasible entailment, since they do not allow to make certain intended conclusions. Hence, they have been refined in a number of ways, giving such systems as lexicographic inference [Benferhat *et al.*, 1993; Lehmann, 1995], and similar modifications of Pearl's system [Goldszmidt *et al.*, 1993; Tan and Pearl, 1995]. Unfortunately, these refined systems have encountered an opposite problem, namely, together with some desirable properties, they invariably produced some unwanted conclusions. All these systems have been based on a supposition that defeasible entailment should form a rational inference relation. A more general approach in the framework of preferential inference has been suggested in [Geffner, 1992].

A different, more syntactic, approach to defeasible entailment has been pursued in the framework of inheritance hierarchies (see [Horty, 1994]). Inheritance reasoning deals with a quite restricted class of conditionals constructed from literals. Nevertheless, in this restricted domain it has achieved a remarkably close correspondence between what is derived and what is expected intuitively. Accordingly, inheritance reasoning has emerged as an important test bed for adjudicating proposed theories.

Despite the diversity, the systems of defeasible entailment have a lot in common, and take as a starting point a few basic principles. Thus, most of them presuppose that intended models should be described, ultimately, in terms of material implications corresponding to a given set of conditionals. More exactly, these classical implications should serve as defaults in the nonmonotonic reasoning sanctioned by a default base. This idea can be made precise by requiring that the intended epistemic states for a default base $\mathfrak{B}$ should be base-generated by the corresponding set $\vec{\mathfrak{B}}$ of material implications (see Section 2.1). Namely, their admissible states should be formed by subsets of $\vec{\mathfrak{B}}$, and the preference order should be monotonic on these subsets. In addition, it should be required that all the conditionals from $\mathfrak{B}$ should be skeptically valid in the resulting epistemic state.

Already these constraints on intended epistemic states allow us to derive "Red birds fly" from "Birds fly" for all default bases that do not contain conflicting information about redness. The constraints also sanction defeasible entailment across exception classes: if penguins are birds that normally do not fly, while birds normally fly and have wings, then we are able to conclude that penguins normally have wings, despite being abnormal birds. This excludes, in effect, Pearl's system Z and rational closure that cannot make such a derivation. Still, the constraints are quite weak and do not produce problematic conclusions that plagued other systems.

Unfortunately, though the above constraints deal successfully with many examples of defeasible entailment, they are still insufficient for capturing some its important patterns. What is missing in our construction is a principled way of constructing a preference order on default sets. This problem has turned out to be far from being trivial or univocal. As for now, two most plausible solutions to this problem, suggested in the literature, are Geffner's conditional entailment and inheritance reasoning.

Conditional entailment [Geffner, 1992] determines a *prioritization* of default bases by making use of the following relation among conditionals:

DEFINITION 6. A conditional α *dominates* a set of conditionals Γ if the set of implications $\{\vec{\Gamma}, \vec{\alpha}\}$ is incompatible with the antecedent of α.

The origins of this relation can be found already in [Adams, 1975], and it has been used in practically all studies of defeasible entailment, including the notion of preemption in inheritance reasoning. A suggestive reading of dominance says that if α dominates Γ, it should have priority over at least one conditional in Γ. Accordingly, a priority order on the default base is *admissible* if it satisfies this condition. Then the intended models can be identified with epistemic states that are generated by all admissible priority orders on the default base (using the lexicographic rule - see above).

Conditional entailment has shown itself as a serious candidate on the role of a general theory of defeasible entailment. Still, it does not capture inheritance reasoning. The main difference between the two theories is that conditional entailment is based on absolute priorities among defaults, while inheritance hierarchies determine such priorities in a context-dependent way, namely in presence of other defaults that provide a (preemption) link between two defaults (see [Dung and Son, 2001]). Indeed, it has been shown in [Bochman, 2001] that inheritance reasoning is representable by epistemic states that are base-generated by default conditionals ordered by certain *conditional* priority orders (see Section 2.1). Still, the corresponding construction could hardly be called simple or natural.

A more natural representation of inheritance reasoning has been given in [Dung and Son, 2001] as an instantiation of an argumentation theory that belongs already to explanatory nonmonotonic formalisms. Furthermore, Geffner himself has shown in [Geffner, 1992] that conditional entailment still does not capture some important derivations, and it should be augmented with an explicit representation of causal reasoning. In fact, the causal generalization suggested by Geffner in the last chapters of his book has served as one of the inspirations for a causal theory of reasoning about actions and change (see [Turner, 1999]). This theory will be described later as an essential part of the explanatory approach to nonmonotonic reasoning.

Finally, a most glaring omission of the above picture of defeasible entailment is that it does not cover *credulous*, or explanatory, nonmonotonic reasoning. Furthermore, for now it is even unclear whether the above approach in terms of epistemic states is capable of representing such a reasoning, though the representation of inheritance reasoning in this framework suggests that it might. Anyway, explanatory reasoning is a well-established theory in another part of nonmonotonic reasoning field, so our next task is to single out the basic principles of the reasoning of this kind.

3   Explanatory Nonmonotonic Reasoning

An influential alternative approach to nonmonotonic reasoning includes default and modal nonmonotonic logics, as well as logic programming. We have called it an *explanatory* nonmonotonic reasoning, since explanation can be seen as its basic ingredient. Propositions and facts may be not only true or false in a model of a problem situation, but some of them are explainable (justified) by other facts and rules that are accepted. In the epistemic setting, some of the propositions are *derivable* from other propositions using rules that are admissible in the situation. In the objective setting, some of the facts are *caused* by other facts and causal rules acting in the domain. Furthermore, explanatory nonmonotonic reasoning is based on very strong principles of *Explanation Closure* or *Causal Completeness* (see [Reiter, 2001]), according to which any fact holding in a model should be explained, or caused, by the rules that describe the domain. Incidentally, it is these principles that make explanatory reasoning nonmonotonic.

By the above description, *abduction*, that is, reasoning from facts to their explanations, is an integral part of explanatory nonmonotonic reasoning. Ultimate explanations, or abducibles, correspond not to normality defaults, but to conjectures representing base causes or facts that do not require explanation; we assume the latter only for explaining some evidence.

In some domains, explanatory formalisms adopt simplifying assumptions that exempt, in effect, certain propositions from the burden of explanation. *Closed World Assumption* [Reiter, 1978] is the most important assumption of this kind. According to it, negative assertions do not require explanation. Nonmonotonic reasoning in databases and logic programming are domains for which such an assumption turns out to be most appropriate. It is important to note that the well-known *minimization principle* is actually a re-

sult of combining Explanation Closure with the Closed World Assumption
(see below). Consequently, the minimization principle need not be viewed
as a principle of scaled preference of negative information; rather, it is a
by-product of the stipulation that negated propositions can be accepted
without any further explanation, while positive assertions always require
explanation. This understanding allows us to explain why McCarthy's circumscription, that is based on the principle of minimization, is subsumed
also by explanatory formalisms.

The above principles form an ultimate basis for all formal systems of
explanatory nonmonotonic reasoning. They presuppose, however, a richer
picture of what is in the world than what is usually captured in logical
models of the latter. The traditional understanding of possible worlds in
logic stems from the Tractatus' metaphysics [Wittgenstein, 1961], where
'[e]ach item can be the case or not the case while everything else remains
the same'. Consequently, there is no way of making an inference from one
fact to another, and there is no causal nexus to justify such an inference.
The only restriction on the structure of the world is the principle of non-
contradiction. Such worlds leave no place for dependencies among facts and
related notions, in particular for causation. Wittgenstein himself concluded
that belief in such dependencies is a superstition.

An alternative picture depicts the world not as a mere assemblage of
unrelated facts, but as something that has a structure. This structure determines dependencies among occurrent facts that serve as a basis for our
explanatory and causal claims. It is this structure that makes the world
intelligible and, what is especially important, controllable. By this picture,
explanatory and causal relations form an integral part of understanding
of and acting in the world. Consequently, such relations should form an
essential part of knowledge representation, at least in Artificial Intelligence.

3.1 Explanatory nonmonotonic reasoning and logic

For a number of reasons, the rules of default logic and its relatives have
lacked a direct semantic interpretation. As a result, representation of reasoning problems in such systems has been largely a syntactic enterprize.
In order to determine acceptable combinations of defaults in default logic,
the user is required to provide explicit information about when one default
rule can 'block' another default (by specifying appropriate justifications in
premises of such rules). This strategy can be remarkably successful in resolving difficult cases of default interaction, which can be seen as the main
reason why the explanatory nonmonotonic reasoning so far has had a greater
impact on practical applications of nonmonotonic reasoning in AI than its
preferential counterpart. Still, such a methodology puts sometimes a heavy

burden on the user by requiring from him to foresee and control the results of default interactions. In other words, it does not give us a transparent and systematic way of representing empirical data, and this makes the task of knowledge representation in such systems really an art. On our diagnosis, the main reason of these shortcomings amounts to the absence of an adequate logical basis in such nonmonotonic systems.

Such important formalisms as default and autoepistemic logics, and semantics of logic programming, are usually described in a shortcut way in accordance with the following identity:

Nonmonotonic Logic = Syntax + Nonmonotonic Semantics.

The very name 'Nonmonotonic *Logic*' conveys in such cases similarity with ordinary logical systems, for which the above equality is clearly appropriate. The analogy is clear, but unfortunately misleading. As has been argued earlier, the nonmonotonic semantics does not determine the meaning of the propositions and rules of the syntax. Consequently, the above 'shortcut' definition of Nonmonotonic Logic leaves us without an exact or even clear meaning of the source syntax. Default rules do not bear on their heads information about when and how they can be applied. This is the main reason why the knowledge representation methodology in such systems is essentially an art based on accumulated experience.

All these shortcomings have become especially vivid in logic programming. From its very beginning, logic programming was based on the idea that program rules should have both a procedural and declarative (logical) meaning. This dual interpretation was purported to elevate the programming process by basing it on a transparent and systematic logical representation of real world information. A declarative meaning of a definite (Prolog) program rule was commonly taken to be the meaning of the corresponding classical implication, though the unique minimal model was singled out as determining the set of conclusions (answers to queries) of the program [van Emden and Kowalski, 1976]. Already this setting involved a discrepancy, since the *logical* meaning of a program (i.e., the set of implications) is not determined by the minimal model, but by the set of all its models.

This double understanding was further challenged with the introduction of negation as failure as a replacement of the classical negation in logic programs. Clark's completion [Clark, 1978] was commonly accepted as giving more adequate interpretations of logic programs and deductive databases. Later it has been discovered that logic programs with negation as failure allow us to represent significant parts of nonmonotonic reasoning. Moreover, general nonmonotonic formalisms inspired a new kind of semantics for logic programs, the stable and answer set semantics [Gelfond and Lifschitz, 1988;

Gelfond and Lifschitz, 1991]. These developments have advanced logic programming to the role of a general computational mechanism for knowledge representation and nonmonotonic reasoning (see [Baral, 2003]).

The idea of a dual interpretation of logic programs persisted in all these developments. However, the original question *'What is a declarative meaning of a program rule?'* has been replaced with the question *'What is a declarative meaning of a logic program?'*, and an answer to the latter question has been thought as settled by assigning logic programs some nonmonotonic semantics. Of course, there were reasons for this shift, since already the completion of a logic program does not allow us to single out a modular meaning of a program rule. Unfortunately, the association of logic programs with a meaning of this kind has turned out to be useless for most purposes we could possibly have in invoking the notion of meaning, be it knowledge representation or the problem of updating programs. The absence of a logical basis behind logic programs has resulted in an unbridled proliferation of suggested semantics. In part, this was due to the absence of clear grounds for evaluating such semantics, apart from their computational virtues, or coincidence with 'respectable' semantics in special cases.

As has been argued in the introduction, in order to determine the logical meaning, or informational content, of a nonmonotonic theory, we should consider its underlying logic. Fortunately, such a logic can often be restored on the basis of the associated nonmonotonic semantics. Namely, logical rules and postulates that preserve a given nonmonotonic semantics can be considered as a logical basis for the corresponding nonmonotonic formalism. A more precise description of such a 'reverse engineering' is as follows. Suppose we are given a nonmonotonic logic in its virgin sense, namely a syntactic formalism $\mathbb{F}$ coupled with a nonmonotonic semantics $\mathbb{S}$. The syntactic formalism determines the basic informational units that we will call default theories, for which the semantics supplies the nonmonotonic interpretation (i.e., a set of models). Default theories could be plain logical theories, as in circumscription or modal nonmonotonic logics, or sets of (default) rules, as in default logic. For a default theory Δ, let $\mathbb{S}(\Delta)$ denote the nonmonotonic semantics of Δ. Then two default theories can be called *(nonmonotonically) equivalent* if they have the same nonmonotonic interpretation. This equivalence does not determine, however, the logical meaning of default theories. Note that precisely due to the nonmonotonicity of $\mathbb{S}$, we may have $\mathbb{S}(\Delta) = \mathbb{S}(\Gamma)$, though $\mathbb{S}(\Delta \cup \Phi) \neq \mathbb{S}(\Gamma \cup \Phi)$, for some default theory Φ.

Still, the above considerations suggest a simple solution. Recall that a standard definition of meaning in logic says that two notions have the same meaning if they determine the same extension in all contexts (e.g., in all possible worlds). In our case, a context can be seen simply as a larger theory including a given one, which leads us to the following notion:

DEFINITION 7. Default theories Γ and Δ are called *strongly* $\mathbb{S}$*-equivalent*, if $\mathbb{S}(\Delta \cup \Phi) = \mathbb{S}(\Gamma \cup \Phi)$, for any default theory Φ.

This notion of strong equivalence has actually been suggested in logic programming (see [Lifschitz *et al.*, 2001]), but it turns out to have general significance. Strong equivalence is already a logical notion, since strongly equivalent theories are interchangeable in any larger theory without changing the associated nonmonotonic semantics. This suggests that there may exist a logic $\mathbb{L}$ formulated in the syntax $\mathbb{F}$ such that default theories are strongly equivalent if and only if they are logically equivalent in $\mathbb{L}$. In this case, the logic $\mathbb{L}$ can be viewed as the underlying logic of the formalism that will determine the logical meaning of default theories and, in particular, of the rules and propositions of the syntactic framework $\mathbb{F}$.

The attention to the underlying logics behind nonmonotonic reasoning is also rewarded with a better understanding of the range of such logics that are appropriate for explanatory nonmonotonic reasoning. It reveals, in particular, that the traditional map of such a reasoning is patently incomplete, and should be completed with a number of important formalisms, creating a continuous range from logic programming to modal nonmonotonic logics.

A distinctive feature of both default and modal nonmonotonic logics is that they are inherently *epistemic* formalisms. Namely, they are essentially based on such notions as belief and knowledge, so the intended semantic models represent possibly incomplete sets of beliefs, while their rules allow us to make inferences based on absence of belief, or consistency. The epistemic interpretation also strongly influenced logic programming in that the negation as failure has often been formulated as absence of knowledge (or derivation). Due to its epistemic character, default logic is a logically weak formalism that does not support many classical inferences (such as reasoning by cases). It also has other well-known shortcomings, and numerous variants of default logic have been suggested in attempts to make it more in accord with intuitions. However, relatively modest success of these attempts has shown that it is impossible to radically improve default logic without abandoning its underlying epistemic interpretation.

On the other side of the map, it has been shown that logic programs under the stable semantics can be faithfully embedded into the above epistemic formalisms, *though not vice versa*. This indicates that logic programming is a more specific nonmonotonic formalism with a richer underlying logic. A logical account of this formalism will be given below in the general framework of biconsequence relations. In this way we will also restore the lost connection between Logic and Logic Programming.

New nonmonotonic formalisms have emerged also in one of the primary application fields of nonmonotonic reasoning, a formal representation of actions and change. It was realized quite early on that classical logic cannot provide an efficient representation for such a reasoning due to the famous *frame problem*, the problem of a succinct representation of the effects of actions and their preconditions. It was only natural to expect that nonmonotonic reasoning should help in resolving this problem. After some less successful attempts to formalize temporal nonmonotonic reasoning in existing nonmonotonic logics, a dominant recent approach to solving this problem has been based on causal reasoning. The causal approach constitutes yet another instance of explanatory nonmonotonic reasoning and employs the already mentioned distinction between facts that hold in a situation versus facts that are caused (explained) by other facts and the rules. The corresponding explanation closure assumption asserts that all facts that hold in a situation should be either caused by other occurrent facts, or else preserve their truth-values in time (due to the associated *inertia assumption*). A natural formalization of these principles has been given in the framework of causal theories, introduced in [McCain and Turner, 1997].

Causal reasoning constitutes an important turning point in the development of explanatory nonmonotonic reasoning, since from its very beginning it was designed as a formalism that should provide an objective description of factual and causal (explanatory) information about action domains. In other words, it has shown that an epistemic view of explanatory nonmonotonic reasoning is not the only possibility.

3.2 Biconsequence Relations

Biconsequence relations are specialized consequence relations for reasoning with respect to a pair of contexts. On the interpretation suitable for nonmonotonic reasoning, one of these contexts is the main (objective) one, while the other context provides assumptions, or explanations, that justify inferences in the main context. This separation of inferences and their justifications creates a possibility of explanatory nonmonotonic reasoning.

The two contexts will be termed, respectively, the *context of truth* and the *context of falsity*. In the truth context the propositions are evaluated as being either true or non-true, while in the falsity context they can be false or non-false. As a benefit of this terminological decision, a bi-context reasoning can also be interpreted as a reasoning with possibly inconsistent and incomplete information. Furthermore, such a reasoning can also be viewed as a four-valued reasoning (see [Belnap, 1977]).

A *bisequent* is an inference rule of the form $a : b \Vdash c : d$, where a, b, c, d are finite sets of propositions. We will employ two informal interpretations of bisequents. According to the *four-valued interpretation*, it says

> *'If all propositions from a are true and all propositions from b are false, then either one of the propositions from c is true or one of the propositions from d is false'.*

According to the *explanatory* (or *assumption*) *interpretation*, it says

> *'If no proposition from b is assumed, and all propositions from d are assumed, then all propositions from a hold only if one of the propositions from c holds'.*

A *biconsequence relation* is a set of bisequents satisfying the rules:

Monotonicity $\dfrac{a : b \Vdash c : d}{a' : b' \Vdash c' : d'}$, if $a \subseteq a'$, $b \subseteq b'$, $c \subseteq c'$, $d \subseteq d'$;

Reflexivity $A : \Vdash A :$ and $: A \Vdash : A;$

Cut $\dfrac{a : b \Vdash A, c : d \quad A, a : b \Vdash c : d}{a : b \Vdash c : d} \qquad \dfrac{a : b \Vdash c : A, d \quad a : A, b \Vdash c : d}{a : b \Vdash c : d}.$

A biconsequence relation can be seen as a fusion, or *fibring*, of two Scott consequence relations much in the sense of [Gabbay, 1998b]. This fusion gives a syntactic expression to combining two independent contexts.

The definition of a biconsequence relation is extendable to arbitrary sets of propositions by accepting the *compactness requirement*:

Compactness $u : v \Vdash w : z$ iff $a : b \Vdash c : d,$

for some finite sets a, b, c, d such that $a \subseteq u$, $b \subseteq v$, $c \subseteq w$ and $d \subseteq z$.

For a set u of propositions, $\overline{u}$ will denote the set of propositions that do not belong to u. A pair (u, v) of sets of propositions will be called a *bitheory* of a biconsequence relation if $u : \overline{v} \nVdash \overline{u} : v$. A set u of propositions is a *(propositional) theory* of $\Vdash$, if (u, u) is a bitheory of $\Vdash$.

Bitheories can be seen as pairs of sets that are closed with respect to the bisequents of a biconsequence relation. A bitheory (u, v) of $\Vdash$ is *positively minimal*, if there is no bitheory (u', v) of $\Vdash$ such that $u' \subset u$. Such bitheories will play an important role in describing nonmonotonic semantics.

By a *bimodel* we will mean a pair of sets of propositions. A set of bimodels will be called a *binary semantics*.

DEFINITION 8. A bisequent $a : b \Vdash c : d$ is *valid* in a binary semantics $\mathcal{B}$, if, for any $(u, v) \in \mathcal{B}$, if $a \subseteq u$ and $b \subseteq \overline{v}$, then either $c \cap u \neq \emptyset$, or $d \cap \overline{v} \neq \emptyset$.

The set of bisequents that are valid in a binary semantics forms a bi-consequence relation. On the other hand, any biconsequence relation $\Vdash$ is determined in this sense by its canonical semantics defined as the set of bitheories of $\Vdash$. Consequently, the binary semantics provides an adequate interpretation of biconsequence relations.

According to Belnap's idea, the four truth-values $\{\top, \mathbf{t}, \mathbf{f}, \bot\}$ of a four-valued interpretation can be identified with the four subsets of the set $\{t, f\}$ of classical truth-values, namely $\{t, f\}$, $\{t\}$, $\{f\}$ and $\emptyset$. Thus, $\top$ means that a proposition is both true and false (i.e., contradictory), $\mathbf{t}$ means that it is 'classically' true (that is, true without being false), $\mathbf{f}$ means that it is classically false, while $\bot$ means that it is neither true nor false (undetermined). This representation allows us to see any four-valued interpretation ν as a pair of ordinary interpretations corresponding, respectively, to independent assignments of truth and falsity to propositions:

$$\nu \models A \quad \text{iff} \quad t \in \nu(A) \qquad \nu =\!\mid A \quad \text{iff} \quad f \in \nu(A).$$

Now, a bimodel (u, v) can be viewed as a four-valued interpretation, where u is the set of true propositions, while v is the set of propositions that are not false. Biconsequence relations provide in this sense a syntactic formalism for four-valued reasoning.

A *bisequent theory* is an arbitrary set of bisequents. For any bisequent theory Δ there is a least biconsequence relation $\Vdash_\Delta$ containing it that describes the logical content of Δ. This allows us to extend the notions of a bitheory and propositional theory to arbitrary bisequent theories.

Default vs. autoepistemic rules

Two kinds of bisequents are of special interest for nonmonotonic reasoning. The first are *default* bisequents $a{:}b \Vdash c{:}$ without negative conclusions that are related to rules of default logic. The second are *autoepistemic* bisequents $:b \Vdash c{:}d$ that turn out to be related to autoepistemic logic. A default bisequent says that if no proposition from b is assumed, then all propositions from a hold only if one of the propositions from c holds. Such a bisequent can be viewed as a Scott inference rule $a \vdash c$ that is conditioned by a set of negative assumptions (i.e., absence of b's). Thus, such bisequents involve full inference capabilities with respect to the main context, but permit only negative assumptions. In contrast, an autoepistemic bisequent $:b \Vdash c{:}d$ says that if no proposition from b is assumed, and all propositions from d are assumed, then one of the propositions from c holds. Such rules have rich assumption capabilities, but allow us to make only unconditional assertions.

In addition to the above distinction, we are often interested in *singular* bisequents, namely bisequents $a{:}b \Vdash C{:}d$ having a single proposition C

as a positive conclusion. Such bisequents can be seen as counterparts of Tarski rules. The formalism of default consequence relations, introduced in [Bochman, 1994], provided, in effect, a uniform description of singular default and autoepistemic bisequents as species of default rules of the form $a{:}b \vdash C$. The difference of default vs. autoepistemic rules has been reflected, however, as a difference in corresponding logics for default rules. This common representation has given a convenient basis for a comparative study of default and modal nonmonotonic reasoning (see [Bochman, 1998c]).

Structural rules

An important feature of biconsequence relations is a possibility of imposing structural constraints on the binary semantics by accepting additional structural rules. Some of them play an important role in nonmonotonic reasoning. Thus, a biconsequence relation is *consistent*, if it satisfies

Consistency $\qquad A : A \Vdash$

On the explanatory interpretation, Consistency says that no proposition can be taken to hold without assuming it. This amounts to restricting the binary semantics to *consistent* bimodels, that is, bimodels (u, v) such that $u \subseteq v$. On the four-valued representation, Consistency requires that no proposition can be both true and false, so it determines a semantic setting of *partial logic* (see, e.g., [Blamey, 1986]) that usually deals only with possible incompleteness of information.

A biconsequence relation is *regular* if it satisfies

Regularity $\qquad \dfrac{b : a \Vdash a : b}{\ : a \Vdash : b}$

Regularity is a kind of an assumption coherence constraint. It says that a coherent set of assumptions should be such that it is compatible with taking these assumptions as actually holding. A semantic counterpart of Regularity is a *quasi-reflexive* binary semantics in which, for any bimodel (u, v), (v, v) is also a bimodel.

Four-valued connectives

Any four-valued connective is definable in biconsequence relations via a pair of introduction rules and a pair of elimination rules corresponding to the two valuations of a four-valued interpretation. We are primarily interested, however, in information a bi-context reasoning can give us about ordinary, classical truth and falsity, so we restrict attention to *classical* connectives that are conservative on the subset $\{\mathbf{t}, \mathbf{f}\}$. Such connectives give classical truth-values when their arguments receive classical values $\mathbf{t}$ or $\mathbf{f}$.

There are four natural connectives that are jointly sufficient for defining all classical four-valued functions. The first is the well-known *conjunction*:

$$\nu \models A \wedge B \quad \text{iff} \quad \nu \models A \text{ and } \nu \models B$$
$$\nu \dashv A \wedge B \quad \text{iff} \quad \nu \dashv A \text{ or } \nu \dashv B.$$

Next, there are two negation connectives that can be seen as two alternative extensions of classical negation to the four-valued setting:

$$\nu \models \neg A \quad \text{iff} \quad \nu \not\models A \qquad \nu \dashv \neg A \quad \text{iff} \quad \nu \not\dashv A$$
$$\nu \models \sim A \quad \text{iff} \quad \nu \dashv A \qquad \nu \dashv \sim A \quad \text{iff} \quad \nu \models A.$$

We will call $\neg$ and $\sim$ a *local* and *global negation*, respectively. Each of them can be used together with the conjunction to define a *disjunction*:

$$A \vee B \equiv \sim(\sim A \wedge \sim B) \equiv \neg(\neg A \wedge \neg B).$$

Finally, the unary connective $\mathbf{A}$ can be seen as a kind of a modal operator, closely related to the modal operator L that will be used in the modal extension of our formalism.

$$\nu \models \mathbf{A}A \quad \text{iff} \quad \nu \not\dashv A \qquad \nu \dashv \mathbf{A}A \quad \text{iff} \quad \nu \dashv A.$$

From the classical point of view, the most natural subclass of the classical four-valued connectives is formed by connectives that behave as ordinary classical connectives with respect to each of the two contexts. Connectives of this kind will be called *locally classical*. The conjunction $\wedge$ and local negation $\neg$ form a functionally complete basis for all such connectives.

Having the four-valued connectives at our disposal, we can transform bisequents into more familiar inference rules, and even to ordinary logical formulas. Let $\neg u$ denote the set $\{\neg A \mid A \in u\}$, and similarly for $\sim u$, etc. Then any bisequent $a : b \Vdash c : d$ is equivalent to each of the following:

$$a, \sim b : \Vdash c, \sim d : \tag{1}$$
$$\Vdash \neg a, c : d, \neg b \tag{2}$$

Bisequents of the form (1) can be seen as ordinary sequents. In fact, this is a common trick used for representing many-valued logics in the form of a sequent calculus. If the language contains also conjunction (or, equivalently, disjunction), the set of premises can be replaced by their conjunction, while the set of conclusions - by their disjunction. Consequently, we can transform bisequents into Tarski-type rules $A \vdash B$. Actually, the resulting system will coincide with a (flat) theory of relevant entailment [Dunn, 1976].

An important alternative possibility arises from using the representation
(2). This time we can use the local connectives $\{\wedge, \neg\}$ in order to reduce
such bisequents to that of the form $\Vdash A : B$, where A and B are classical
propositions. The latter bisequents will correspond to production rules $B \Rightarrow A$ of production and causal inference relations, discussed later.

3.3 Nonmonotonic Semantics

Now we give a formalization of nonmonotonic reasoning in this logical framework. Nonmonotonic semantics of a biconsequence relation will be defined
as certain sets of its theories. Namely, such theories will be *explanatory
closed* in the sense that presence and absence of propositions in the main
context will be explained (i.e., derived) using the rules of the biconsequence
relation when the theory itself is taken as the assumption context.

Exact semantics

The notion of an exact theory of a biconsequence relation provides us with
a simplest and most general kind of nonmonotonic reasoning.

DEFINITION 9. A theory u of a biconsequence relation $\Vdash$ will be called
exact, if there is no other set $v \neq u$ such that (v, u) is a bitheory of $\Vdash$. The
set of exact theories will be called an *exact nonmonotonic semantics* of $\Vdash$.

Exact theories correspond to bitheories, for which the assumption context determines itself as a unique objective state compatible with it. Such
theories can be given the following syntactic description.

Lemma 3.1. A set u of propositions is an exact theory of a biconsequence
relation $\Vdash$ if and only if, for any proposition A,

$$A \in u \ \text{ iff } \ : \overline{u} \Vdash A : u \qquad \text{and} \qquad A \notin u \ \text{ iff } \ A : \overline{u} \Vdash \, : u.$$

The exact semantics is extendable to arbitrary bisequent theories by considering their associated biconsequence relations.

The above definition of nonmonotonic semantics leaves us much freedom in determining *nonmonotonic consequences* of a bisequent theory. The
most obvious skeptical choice consists in taking propositions that belong
to all exact theories. As a credulous alternative, however, we can consider
propositions that belong to at least one theory. An even more general understanding amounts to a view that *all* the information that can be discerned
from the nonmonotonic semantics of a bisequent theory or a biconsequence
relation can be seen as nonmonotonically implied by the latter.

Exact theories determine a truly nonmonotonic semantics, since adding
new bisequents to a bisequent theory may not only eliminate exact theories,
but also add new ones (or both). In other words, the set of exact theories
does not change monotonically with the growth of the set of bisequents.

It turns out that regular biconsequence relations constitute a maximal logic suitable for the exact nonmonotonic semantics. This can be shown by demonstrating that equivalence with respect to regular biconsequence relations coincides with strong equivalence from Definition 7 above.

Finally, if a bisequent theory contains only default bisequents $a{:}b \Vdash c{:}$, then any its exact theory will also be a minimal propositional theory.

Default semantics

A more familiar class of nonmonotonic models, extensions, correspond to extensions of default logic and stable models of logic programs.

DEFINITION 10. A set u is an *extension* of a biconsequence relation $\Vdash$, if (u, u) is a positively minimal bitheory of $\Vdash$. A *default nonmonotonic semantics* of a biconsequence relation is the set of its extensions.

u is an extension if it is a theory of a biconsequence relation such that there is no smaller set $v \subset u$ such that (v, u) is a bitheory. Hence any exact theory of a biconsequence relation is an extension, though not vice versa.

Let Pr denote the set of all propositions of the language. The following lemma provides a syntactic description of extensions.

Lemma 3.2. A set u is an extension of a biconsequence relation $\Vdash$ if and only if $u = \{A \mid {:}\overline{u} \Vdash A, \overline{u}{:}u\}$ and either $u \neq Pr$, or $\nVdash{:}Pr$.

By the above description, extensions are theories of a biconsequence relation that explain only why they have the propositions they have. In other words, for extensions we are relieved from the necessity of explaining why propositions do *not* belong to the intended theory. This agreement constitutes the essence of Reiter's Closed World Assumption. Another consequence of the above description is that extensions are determined by the autoepistemic bisequents of a biconsequence relation.

It turns out that the default nonmonotonic semantics is precisely an exact semantics under a stronger logic of consistent biconsequence relations. The effect of Consistency $A : A \Vdash$ amounts to an immediate refutation of any proposition that is *assumed* not to hold. That is why absence of propositions from extensions does not require explanation, only their presence. This makes the minimality condition in the definition of extensions a consequence of the logical Consistency postulate, instead of an independent 'rationality principle' of nonmonotonic reasoning.

Bisequent interpretation of logic programs

Representations of general logic programs and their semantics in the logical formalism of biconsequence relations establish the ways of providing (or, better, restoring) a logical basis of logic programming.

A *general logic program* is a set of program rules **not** $d, c \leftarrow a, \textbf{not}\, b$, where a, b, c, d are finite sets of propositional atoms. Such program rules involve disjunctions and default negations in heads and subsume practically all structural extensions of Prolog rules suggested in the literature.

A program rule **not** $d, c \leftarrow a, \textbf{not}\, b$ can be directly interpreted as a bisequent $a : b \Vdash c : d$. Let $bc(\Pi)$ denote the bisequent theory corresponding to a general program Π under this interpretation. The following result shows that it provides an exact correspondence between stable models of a general program and extensions of the associated bisequent theory.

THEOREM 11. *Stable models of a general logic program* Π *coincide with the extensions of* $bc(\Pi)$.

Moreover, the correspondence turns out to be bidirectional, since any bisequent in a four-valued language is logically reducible to a set of bisequents without connectives that can be viewed as program rules.

As to other semantics of logic programs, a general approach to defining such semantics in terms of biconsequence relations has been suggested in [Bochman, 1998a]. As has been shown in [Bochman, 1998b], the majority of existing semantics for logic programs are subsumed by this approach. This description has also shown that common kinds of reasoning about logic programs are reducible, in effect, to different kinds of logical reasoning with respect to a single nonmonotonic semantics.

3.4 Production and Causal Inference

We have given above a structural description of explanatory nonmonotonic reasoning. Now we introduce a primary *logical* system for such a reasoning. The systems of production and causal inference, described below, can be viewed as a generalization of classical logic obtained by dropping the Reflexivity postulate of classical inference. They originate in input/output logics of [Makinson and van der Torre, 2000]. Biconsequence relations turn out to constitute a structural counterpart of this logical formalism.

Production inference relations are based on conditionals of the form $A \Rightarrow B$ saying 'A produces, or explains, B'.

DEFINITION 12. A *(regular) production inference relation* is a binary relation $\Rightarrow$ on the set of classical propositions satisfying the following postulates:

(Strengthening) If $A \vDash B$ and $B \Rightarrow C$, then $A \Rightarrow C$;

(Weakening) If $A \Rightarrow B$ and $B \vDash C$, then $A \Rightarrow C$;

(And) If $A \Rightarrow B$ and $A \Rightarrow C$, then $A \Rightarrow B \wedge C$;

(Cut) If $A \Rightarrow B$ and $A \wedge B \Rightarrow C$, then $A \Rightarrow C$;

(Truth) $t \Rightarrow t$;

(Falsity) $f \Rightarrow f$.

From a logical point of view, the most significant 'omission' of the above set of postulates is the absence of reflexivity $A \Rightarrow A$. It is this feature that creates a possibility of nonmonotonic reasoning.

Production rules are extended to rules with sets of propositions in premises using the familiar compactness recipe: for a set u of propositions, we define

$$u \Rightarrow A \ \equiv \ \bigwedge a \Rightarrow A, \text{ for some finite } a \subseteq u.$$

Let $\mathcal{C}(u)$ denote the set $\{A \mid u \Rightarrow A\}$. The production operator $\mathcal{C}$ is monotonic and continuous, and it plays the same role as the usual derivability operator for consequence relations. A *theory* of a production relation is a deductively closed set u such that $\mathcal{C}(u) \subseteq u$. Such theories have much the same properties as ordinary theories of consequence relations.

A semantics for production relations can be given in terms of pairs of deductive theories that will be called, as before, bimodels.

DEFINITION 13. A *classical bimodel* is a pair of consistent deductively closed sets. A *classical binary semantics* is a set of classical bimodels. A classical binary semantics is *consistent*, if $u \subseteq v$, for any bimodel (u, v).

The validity of production rules is defined as follows.

DEFINITION 14. A production rule $A \Rightarrow B$ is *valid* in a classical binary semantics $\mathcal{B}$ if, for any bimodel (u, v) from $\mathcal{B}$, $A \in v$ only if $B \in u$.

Consistent classical binary semantics provides an adequate representation of regular production inference relations (see [Bochman, 2004a]).

General nonmonotonic semantics

Production inference relations determine also a natural nonmonotonic semantics, and provide thereby a logical basis for nonmonotonic reasoning. Namely, the fact that the production operator $\mathcal{C}$ is not reflexive creates an important distinction among theories of a production relation.

DEFINITION 15. A *nonmonotonic production semantics* of a production inference relation is the set of all its *exact theories*, namely sets u of propositions such that $u = \mathcal{C}(u)$.

An exact theory describes an informational state in which every proposition is *explained* by other propositions accepted in this state. Accordingly, restricting our universe of discourse to exact theories amounts to imposing a kind of an explanatory closure assumption on intended models.

By a *causal theory* we will mean an arbitrary set of production rules. We will stipulate that the nonmonotonic semantics of a causal theory coincides with the nonmonotonic semantics of the least production relation including it. The nonmonotonic semantics of causal theories is globally nonmonotonic, since adding new production rules may lead to a nonmonotonic change of the associated semantics, even though production rules themselves are monotonic, since they satisfy Strengthening (the Antecedent). Still, exact theories are fixed points of $\mathcal{C}$, and since the latter operator is monotonic and continuous, exact theories always exist.

Regular production inference provides a maximal logical framework adequate for reasoning with exact theories. Any postulate that is not valid for regular production relations can be 'falsified' by finding a suitable extension of two nonmonotonically equivalent causal theories that would determine different nonmonotonic semantics (cf. Definition 7).

Production inference and abduction

The nonmonotonic production semantics allows us to provide a formal representation of abductive reasoning. Though this kind of abductive reasoning does not cover all potential forms of abduction (see [Gabbay and Woods, 2005]), it captures the main applications of abduction in AI.

To begin with, *abducibles* can be identified with self-explanatory propositions of a production relation, that is, propositions satisfying $A \Rightarrow A$. Then it turns out that traditional abductive systems are representable via a special class of *abductive* production inference relations that satisfy

(Abduction) If $B \Rightarrow C$, then $B \Rightarrow A \Rightarrow C$, for some abducible A.

It has been shown in [Bochman, 2005] that abductive inference relations provide a generalization of abductive reasoning in causal theories [Konolige, 1992; Poole, 1994], as well as of abduction in logic programming. On the other hand, any production inference relation includes a greatest abductive subrelation, and in many regular situations (e.g., when the production relation is well-founded) the latter determines the same nonmonotonic semantics. Summing up, the general nonmonotonic semantics of a production relation is usually describable by some abductive system, and vice versa.

Causal Inference

A regular production inference relation is *causal* if it satisfies

(Or)   If $A \Rightarrow C$ and $B \Rightarrow C$, then $A \vee B \Rightarrow C$.

The rule Or sanctions reasoning by cases, and hence causal production relations can be seen as systems of reasoning about complete worlds. Moreover, production rules of a causal inference relation can already be inter-

preted as truly *causal rules*, since they provide a natural representation of
ordinary causal assertions.

Causal inference sanctions the following decomposition of causal rules:

Lemma 3.3. $A \Rightarrow B$ holds if and only if $A \wedge \neg B \Rightarrow \mathbf{f}$ and $A \wedge B \Rightarrow B$.

A rule of the form $A \Rightarrow \mathbf{f}$ is a *factual constraint*; it says that A should not
hold in any intended model. Such rules represent purely factual information.
Thus, $A \wedge \neg B \Rightarrow \mathbf{f}$ says, in effect, that the classical implication $A \rightarrow B$ should
hold in any causally consistent world.

Causal rules of the form $A \wedge B \Rightarrow B$ are *explanatory rules*. Though logically trivial, they play an important explanatory role in causal reasoning.
Namely, they say that, in any intended model in which A holds, we can
freely accept B, since it is self-explanatory in this context.

The above decomposition neatly separates two kinds of information conveyed by causal rules. One is a factual information that constraints the set
of admissible models, while the other is an explanatory information about
what propositions are caused (explainable) in such models. This decomposition allows us to give a concise description of the informational content
of causal theories. Namely, the *factual content* of a causal relation can be
identified with the set of its constraints, while its *explanatory content* can
be defined as the set of its explanatory rules. The interplay of the two
kinds of content determines, eventually, the behavior and properties of the
corresponding causal nonmonotonic semantics (see below).

The semantic characterization of causal production relations can be obtained by restricting the classical binary semantics to bimodels of the form
(α, β), where α, β are worlds. This setting can be generalized to a usual
relational possible worlds model (W, R, V), where W is a set of possible
worlds, R a binary accessibility relation on W, and V a valuation function.
The validity of causal rules can now be defined as follows:

DEFINITION 16. A rule $A \Rightarrow B$ is *valid* in a possible worlds model (W, R, V)
if, for any $\alpha, \beta \in W$ such that $\alpha R \beta$, if A holds in α, then B holds in β.

Causal inference relations are determined by possible worlds models in
which the relation R is *quasi-reflexive*, that is, $\alpha R \beta$ holds only if $\alpha R \alpha$.

Causal inference relations constitute a logical counterpart of biconsequence relations for the language of local classical connectives. For a biconsequence relation $\Vdash$ in such a language, we can define the following set
of production rules that will be called the *production subrelation* of $\Vdash$:

$$\Rightarrow_{\Vdash} = \{A \Rightarrow B \mid \Vdash B : A\}$$

Then the production subrelation of a regular biconsequence relation will
form a causal inference relation. Moreover, for any causal inference relation

$\Rightarrow$ there is a unique regular biconsequence relation in the language with the local connectives that has $\Rightarrow$ as its production subrelation. By this correspondence, a production rule $A \Rightarrow B$ can be seen as an assumption-based conditional saying that if A is assumed, then B should hold.

Causal nonmonotonic semantics

For the causal understanding of production rules, it is only natural to restrict the nonmonotonic production semantics to exact theories that are worlds. This semantics is called the *causal nonmonotonic semantics*, and it coincides with the semantics of causal theories suggested in [McCain and Turner, 1997] for reasoning about actions and change. The principle of explanation closure says in this case that any fact should have a cause, so it can be called the *principle of universal causation* (see [Turner, 1999]).

As can be shown, a world α is an exact world of a production inference relation if and only if, for any propositional atom p,

$$p \in \alpha \ \text{ iff } \ \alpha \Rightarrow p \qquad \text{and} \qquad \neg p \in \alpha \ \text{ iff } \ \alpha \Rightarrow \neg p.$$

Causal inference relations provide an adequate logical framework for the causal nonmonotonic semantics. Moreover, in the general correspondence between causal and biconsequence relations, the causal semantics corresponds to the exact nonmonotonic semantics of biconsequence relations.

By the above description, the exact worlds of a causal relation are determined ultimately by rules of the form $A \Rightarrow l$, where l is a literal. Such rules are called *determinate*. The causal nonmonotonic semantics of a determinate causal theory Δ coincides with the classical semantics of its *completion*, $comp(\Delta)$, defined as the set of the following classical formulas:

$$p \leftrightarrow \bigvee \{A \mid A \Rightarrow p \in \Delta\} \qquad \neg p \leftrightarrow \bigvee \{A \mid A \Rightarrow \neg p \in \Delta\},$$

for any atom p, plus the set $\{\neg A \mid A \Rightarrow \mathbf{f} \in \Delta\}$. The classical models of $comp(\Delta)$ precisely correspond to exact worlds of Δ. As an illustration, we will give a causal representation of the Reiter's simple solution to the frame problem (see [Reiter, 1991; Reiter, 2001]). The representation contains the main ingredients of causal reasoning in temporal domains.

EXAMPLE 17. (*Reiter's simple solution*). The temporal behavior of a propositional fluent F is described using two propositional atoms F_0 and F_1 saying, respectively, that F holds now and F holds in the next moment.

$$C^+ \Rightarrow F_1 \qquad C^- \Rightarrow \neg F_1$$
$$F_0 \wedge F_1 \Rightarrow F_1 \qquad \neg F_0 \wedge \neg F_1 \Rightarrow \neg F_1$$
$$F_0 \Rightarrow F_0 \qquad \neg F_0 \Rightarrow \neg F_0.$$

The first pair of causal rules describes the actions or natural factors that can cause F and, respectively, $\neg F$ (C^+ and C^- normally describe the present situation). Second, we have a pair of *inertia axioms*, purely explanatory rules stating that if F holds (does not hold) now, then it is self-explanatory that it will hold (resp., not hold) in the next moment. The last pair of *initial axioms* states that F_0 is an exogenous parameter.

The above causal theory is determinate, and its completion is as follows:

$$F_1 \leftrightarrow C^+ \vee (F_0 \wedge F_1) \qquad \neg F_1 \leftrightarrow C^- \vee (\neg F_0 \wedge \neg F_1).$$

These formulas are equivalent to the conjunction of $\neg(C^+ \wedge C^-)$ and

$$F_1 \leftrightarrow C^+ \vee (F_0 \wedge \neg C^-).$$

The above formulas provide an abstract description of Reiter's simple solution: the first formula corresponds to his *consistency condition*, while the last one - to the *successor state axiom* for F.

As for biconsequence relations, the default nonmonotonic semantics of causal theories can be obtained by imposing a causal postulate corresponding to Consistency postulate for biconsequence relations:

(Default Negation) $\neg p \Rightarrow \neg p$, for any propositional atom p.

Default Negation stipulates that negations of atomic propositions are self-explanatory, and hence it provides a simple causal expression for Reiter's Closed World Assumption. This kind of causal inference can also be used as a logical basis for logic programming. Namely, a program rule **not** $d, c \leftarrow a, \mathbf{not}\, b$ can be faithfully interpreted as the causal rule $d, \neg b \Rightarrow \bigwedge a \rightarrow \bigvee c$. This interpretation provides an adequate description of logic programs under the stable model semantics (see [Bochman, 2004b]).

3.5   Epistemic Explanatory Reasoning

Epistemic formalisms of default and modal nonmonotonic logics find their natural place in the framework of supraclassical biconsequence relations, defined below. An epistemic understanding of biconsequence relations amounts to treating the main and assumption contexts, respectively, as the contexts of knowledge and belief: propositions that hold in the main context can be viewed as known, while propositions of the assumption context form the associated set of beliefs. This understanding will receive an explicit expression in the modal extension of the formalism. Even in a non-modal setting, the epistemic reading implies that both contexts should correspond not to complete worlds, but to incomplete deductive theories.

DEFINITION 18. A biconsequence relation in a classical language will be called *supraclassical*, if it satisfies

Supraclassicality If $a \vDash A$, then $a : \Vdash A :$ and $: A \Vdash : a$.

Falsity $\mathbf{f} : \Vdash$ and $\Vdash : \mathbf{f}$.

In supraclassical biconsequence relations both contexts respect the classical entailment. In addition, sets of positive premises and negative conclusions can be replaced by their conjunctions, but positive conclusion sets and negative premise sets are not replaceable in this way by classical disjunctions. Also, the deduction theorem, contraposition, and disjunction in the antecedent are not valid, in general, for each of the two contexts.

A semantics of supraclassical biconsequence relations is obtained from the general binary semantics by requiring that bimodels are pairs of consistent deductively closed sets (see Definition 13 of the classical binary semantics). Structural rules for biconsequence relations can also be used in the supraclassical case. As before, Consistency will correspond to the requirement that $u \subseteq v$, for any bimodel (u, v). Similarly, regular biconsequence relations will be determined by quasi-reflexive binary semantics.

A supraclassical biconsequence relation will be called *saturated*, if it is consistent, regular, and satisfies the following postulate:

Saturation $\Vdash A \vee B, \neg A \vee B : B$.

For a deductively closed set u, let $u \bot$ denote the set of all maximal subtheories of u, plus u itself. Then a classical bimodel (u, v) will be called *saturated*, if $u \in v \bot$. A classical binary semantics $\mathcal{B}$ will be called *saturated* if it is regular, and all its bimodels are saturated. Such a semantics provides an adequate interpretation for saturated biconsequence relations.

Classical nonmonotonic semantics

The notions of an exact theory and extension can be directly extended to supraclassical consequence relations, with the only, though important, qualification that they form now deductively closed sets. Still, practically all the results about such semantics remain valid for the supraclassical case.

The default nonmonotonic semantics of supraclassical biconsequence relations forms a generalization of default logic. Supraclassical biconsequence relations that are consistent and regular constitute a maximal logic adequate for extensions. For such biconsequence relations, extensions are described as sets satisfying the following fixpoint equality:

$$u = \{A \mid : \overline{u} \Vdash A : u\}.$$

Thus, an extension is a set of formulas that are provable on the basis of taking itself as the set of assumptions. As before, classical extensions of a default bisequent theory will be minimal theories. Actually, default bisequent theories under this nonmonotonic semantics give an exact representation for the *disjunctive default logic* [Gelfond *et al.*, 1991]. For singular default rules $a{:}b \Vdash C{:}$, it reduces to the original default logic of [Reiter, 1980].

The nonmonotonic semantics defined below constitutes an exact non-modal counterpart of Moore's autoepistemic logic.

DEFINITION 19. A theory u of a supraclassical biconsequence relation $\Vdash$ is a *classical expansion* of $\Vdash$, if, for any $v \in u\bot$ such that $v \neq u$, the pair (v, u) is not a bitheory of $\Vdash$. The set of classical expansions determines the *autoepistemic semantics* of $\Vdash$.

Any extension of a supraclassical biconsequence relation will be a classical expansion, though not vice versa. In fact, classical expansions can be precisely characterized as extensions of saturated biconsequence relations.

The next result states important sufficient conditions for coincidence of classical expansions and extensions of a bisequent theory. A bisequent theory Δ will be called *positively simple*, if positive premises and positive conclusions of any bisequent from Δ are sets of classical literals.

THEOREM 20. *If a bisequent theory is autoepistemic or positively simple, then its classical expansions coincide with classical extensions.*

Bisequents $a{:}b \Vdash c{:}d$ such that a, b, c, d are sets of classical literals, are logical counterparts of program rules of extended logic programs with classical negation (see [Gelfond and Lifschitz, 1991; Lifschitz and Woo, 1992]). The semantics of such programs is determined by *answer sets* that coincide with extensions of respective bisequent theories. Moreover, such bisequent theories are positively simple, so by Theorem 20 extended logic programs obliterate the distinction between extensions and classical expansions. This is the logical basis for a possibility of representing extended logic programs also in autoepistemic logic (see [Lifschitz and Schwarz, 1993]).

3.6 Modal Nonmonotonic Logics

The modal approach to nonmonotonic reasoning was initiated in [McDermott and Doyle, 1980; McDermott, 1982]. These papers have led to autoepistemic logic [Moore, 1985; Konolige, 1988] and to general study of modal nonmonotonic logics in the works of Marek, Schwarz and Truszczyński.

A general representation of modal nonmonotonic reasoning can be given in the framework of modal biconsequence relations. The role of the modal operator L in this setting consists in reflecting assumptions and beliefs as propositions in the main context.

DEFINITION 21. A supraclassical biconsequence relation in a modal language will be called *modal* if it satisfies the following postulates:

Positive Reflection $A : \Vdash LA:$,

Negative Reflection $: LA \Vdash :A$,

Negative Introspection $: A \Vdash \neg LA:$.

Any theory of a modal biconsequence relation is a *modal stable set* in the sense of [Moore, 1985], and hence extensions and expansions of modal biconsequence relations will always be stable theories.

For a modal logic $\mathcal{M}$, a modal biconsequence relation $\Vdash$ will be called an $\mathcal{M}$-*biconsequence relation*, if $\Vdash A:$, for every modal axiom A of $\mathcal{M}$. A possible worlds semantics for K-biconsequence relations is obtainable in terms of Kripke models having a last cluster, namely models of the form $M = (W, R, F, V)$, where (W, R, V) is an ordinary Kripke model, while $F \subseteq W$ is a non-empty *last cluster* of the model (see [Segerberg, 1971]). We will call such models *final Kripke models*. The relevance of such semantics for modal nonmonotonic reasoning has been shown in [Schwarz, 1992a].

For a final Kripke model $M = (W, R, F, V)$, let $|M|$ denote the set of modal propositions that are valid in M, while $\|M\|$ the set of propositions valid in the S5-submodel of M generated by F. For a set $\mathcal{S}$ of final Kripke models, we define the binary semantics $\mathcal{B}_{\mathcal{S}} = \{(|M|, \|M\|) \mid M \in \mathcal{S}\}$, and say that a bisequent $a{:}b \Vdash c{:}d$ is *valid* in $\mathcal{S}$, if it is valid in $\mathcal{B}_{\mathcal{S}}$.

Biconsequence relations described in the next definition play a crucial role in a modal representation of extension-based nonmonotonic reasoning.

DEFINITION 22. A modal biconsequence relation is an *F-biconsequence relation*, if it is regular and satisfies

F $\Vdash A, LA{\rightarrow}B : B$.

F-biconsequence relations provide a concise representation for the modal logic S4F obtained from S4 by adding the axiom $(A \wedge MLB){\rightarrow}L(MA \vee B)$. A semantic characterization of S4F [Segerberg, 1971] is given in terms of final Kripke models (W, R, F, V), such that $\alpha R \beta$ iff either $\beta \in F$, or $\alpha \notin F$.

Any bisequent $a{:}b \Vdash c{:}d$ of an F-biconsequence relation is already reducible to a modal formula $\bigwedge(La \cup L\neg Lb) \rightarrow \bigvee(Lc \cup L\neg Ld)$. Consequently, any bisequent theory Δ in such a logic is reducible to an ordinary modal theory that we will denote by $\tilde{\Delta}$.

Modal nonmonotonic semantics

By varying the underlying modal logic, we obtain a whole range of *modal nonmonotonic semantics*.

DEFINITION 23. A set of propositions is an $\mathcal{M}$-*extension* ($\mathcal{M}$-*expansion*) of a bisequent theory Δ, if it is an extension (resp. expansion) of the least $\mathcal{M}$-biconsequence relation containing Δ.

If Δ is a plain modal theory, then $\mathcal{M}$-extensions of Δ coincide with $\mathcal{M}$-*expansions* in the sense of [Marek *et al.*, 1993][1]. Recall, however, that modal extensions and expansions are modal stable theories, and hence they are determined by their objective subsets. This suggests a possibility of reducing modal nonmonotonic reasoning to a nonmodal one, and vice versa.

For any set u of propositions, let u_o denote the set of all non-modal propositions in u. Similarly, if $\Vdash$ is a modal biconsequence relation, $_o\Vdash$ will denote its restriction to the non-modal sub-language. Then we have

THEOREM 24. *If $\Vdash$ is a modal biconsequence relation, and u a stable theory, then u is an extension of $\Vdash$ if and only if u_o is an extension of $_o\Vdash$.*

According to the above result, the objective biconsequence relation $_o\Vdash$ embodies all the information about the modal nonmonotonic semantics of $\Vdash$. In other words, the net effect of modal reasoning in biconsequence relations can be measured by the set of derived objective bisequents. Consequently, non-modal supraclassical biconsequence relations turn out to be sufficiently expressive to capture modal nonmonotonic reasoning.

In the other direction, in modal F-biconsequence relations any bisequent theory Δ is reducible to a usual modal theory $\tilde{\Delta}$. This allows us to use ordinary modal logical formalisms for representing non-modal nonmonotonic reasoning. Thus, the following result generalizes the corresponding result of [Truszczyński, 1991] about a modal embedding of default theories.

THEOREM 25. *If Δ is an objective bisequent theory, then classical extensions of Δ are precisely objective parts of S4F-extensions of $\tilde{\Delta}$.*

We end with considering modal expansions. Two kinds of expansions are important for a general description. The first is *stable expansions* of Moore's autoepistemic logic. They coincide with $\mathcal{M}$-expansions for any modal logic $\mathcal{M}$ in the range $5 \subseteq \mathcal{M} \subseteq$ KD45. The second kind of expansions is *reflexive expansions* of Schwarz' reflexive autoepistemic logic [Schwarz, 1992b]. They coincide with $\mathcal{M}$-expansions for any modal logic in the range $KT \subseteq \mathcal{M} \subseteq$ SW5. Both kinds of expansions can be computed either by transforming a bisequent theory into a modal theory and finding its modal extensions, or by reducing it to an objective bisequent theory and computing its classical expansions. Finally, 'normal' expansions in general can be viewed as a combination of these two kinds of expansions:

---

[1]This creates, of course, an unfortunate terminological discrepancy, which is compensated, however, by the conformity of our terminology with the rest of this study.

THEOREM 26. *A set of propositions is a K-expansion of a modal bisequent theory Δ iff it is both a stable and reflexive expansion of Δ.*

4 Conclusions

Despite clear success, twenty five years of nonmonotonic reasoning research have shown that we need a deep breath and long term objectives in order to make nonmonotonic reasoning a viable tool for the challenges posed by AI. There is still much to be done in order to meet the actual complexity of reasoning tasks required by the latter. In particular, the relation between the two principal paradigms of nonmonotonic reasoning has emerged as the main theoretical problem for a future development of the field.

By a presumably vague, but inspiring analogy, in nonmonotonic reasoning we have both a global relativity theory of preferential reasoning and a local quantum mechanics of explanatory reasoning. So, what we need is a unified theory of nonmonotonic reality. As in physics, however, this unified theory is not going to emerge as a straightforward juxtaposition of these components.

Preferential nonmonotonic reasoning has a clear logical semantics, and it provides a seemingly adequate interpretation for normality defaults. In particular, it provides a natural semantic support for the *specificity principle* of reasoning with default assumptions, according to which more specific defaults override more general ones. This principle has turned out to be difficult to express in explanatory formalisms. On the other hand, our previous discussion has shown that preferential nonmonotonic reasoning still does not have an established *nonmonotonic* semantics that would serve as a basis for the associated defeasible entailment from sets of default conditionals. Moreover, already suggested ways of constructing such a semantics indicate that it should involve important ingredients from the explanatory theory, such as argumentation and causation.

On the other side of the coin, explanatory nonmonotonic reasoning constitutes a well entrenched theory which resolves a lot of questions posed by AI. It provides natural and working representations for abductive and causal reasoning, logic programming and, most importantly, for reasoning about actions and change. In particular, it successfully resolves the frame and ramification problems of the latter (see [Shanahan, 1997]). Still, it falls short of resolving a related *qualification* problem, namely the problem of specifying (nonmonotonic) conditions for a successful performance of actions. The basic reason is that causal conditionals and rules of default logic are locally monotonic, since they freely admit strengthening of their premises. In contrast, our causal and action claims are usually *defeasible*, ceteris paribus claims that presuppose a lot of 'normality' assumptions; their representation naturally calls for a preferential analysis.

Taking into account the causal theory of explanatory reasoning, the difference between the two approaches to nonmonotonic reasoning boils down to the difference between two kinds of conditionals, preferential vs. causal ones. So the main problem can be reduced to the task of an intelligent 'fibring' of these two kinds of conditionals. Once it will be done, the unified theory of nonmonotonic reasoning should become a decent part of a general logical theory of practical rationality pioneered by Dov Gabbay.

Acknowledgement

I am grateful to David Makinson for instructive comments on this paper.

BIBLIOGRAPHY

[Adams, 1975] E. W. Adams. *The Logic of Conditionals*. Reidel, Dordrecht, 1975.

[Alchourrón et al., 1985] C. Alchourrón, P. Gärdenfors, and D. Makinson. On the logic of theory change: Partial meet contraction and revision functions. *Journal of Symbolic Logic*, 50:510–530, 1985.

[Andreka et al., 2002] H. Andreka, M. Ryan, and P.-Y. Schobbens. Operators and laws for combining preference relations. *Journal of Logic and Computation*, 12:13–53, 2002.

[Baral, 2003] C. Baral. *Knowledge Representation, Reasoning and Declarative Problem Solving*. Cambridge UP, 2003.

[Belnap, 1977] N. D. Belnap, Jr. A useful four-valued logic. In M. Dunn and G. Epstein, editors, *Modern Uses of Multiple-Valued Logic*, pages 8–41. D. Reidel, 1977.

[Benferhat et al., 1993] S. Benferhat, C. Cayrol, D. Dubois, J. Lang, and H. Prade. Inconsistency management and prioritized syntax-based entailment. In R. Bajcsy, editor, *Proceedings Int. Joint Conf. on Artificial Intelligence, IJCAI'93*, pages 640–645, Chambery, France, 1993. Morgan Kaufmann.

[Blamey, 1986] S. Blamey. Partial logic. In D. M. Gabbay and F. Guenthner, editors, *Handbook of Philosophical Logic, Vol. III*, pages 1–70. D. Reidel, 1986.

[Bochman, 1994] A. Bochman. On the relation between default and modal consequence relations. In *Proc Int. Conf. on Principles of Knowledge Representation and Reasoning*, pages 63–74, 1994.

[Bochman, 1998a] A. Bochman. A logical foundation for logic programming I: Biconsequence relations and nonmonotonic completion. *Journal of Logic Programming*, 35:151–170, 1998.

[Bochman, 1998b] A. Bochman. A logical foundation for logic programming II: Semantics of general logic programs. *Journal of Logic Programming*, 35:171–194, 1998.

[Bochman, 1998c] A. Bochman. On the relation between default and modal nonmonotonic reasoning. *Artificial Intelligence*, 101:1–34, 1998.

[Bochman, 2001] A. Bochman. *A Logical Theory of Nonomonotonic Inference and Belief Change*. Springer, 2001.

[Bochman, 2003] A. Bochman. Brave nonmonotonic inference and its kinds. *Annals of Mathematics and Artificial Intelligence*, 39:101–121, 2003.

[Bochman, 2004a] A. Bochman. A causal approach to nonmonotonic reasoning. *Artificial Intelligence*, 160:105–143, 2004.

[Bochman, 2004b] A. Bochman. A causal logic of logic programming. In D. Dubois, C. Welty, and M.-A. Williams, editors, *Proc. Ninth Conference on Principles of Knowledge Representation and Reasoning, KR'04*, pages 427–437, Whistler, 2004.

[Bochman, 2005] A. Bochman. *Explanatory Nonmonotonic Reasoning*. World Scientific, 2005.

[Clark, 1978] K. Clark. Negation as failure. In H. Gallaire and J. Minker, editors, *Logic and Data Bases*, pages 293–322. Plenum Press, 1978.

[Dix, 1991] J. Dix. Classifying semantics of logic programs. In A. Nerode, W. Marek, and V. S. Subrahmanian, editors, *Proceedings 1st International Workshop on Logic Programming and Nonmonotonic Reasoning*, pages 166–180, Cambridge, Mass., 1991. MIT Press.

[Dung and Son, 2001] P. M. Dung and T. C. Son. An argument-based approach to reasoning with specificity. *Artificial Intelligence*, 133:35–85, 2001.

[Dunn, 1976] J. M. Dunn. Intuitive semantics for first-degree entailment and coupled trees. *Philosophical Studies*, 29:149–168, 1976.

[Gabbay and Woods, 2005] D. M. Gabbay and J. Woods. *The Reach of Abduction: Insight and Trial*, volume 2 of *A Practical Logic of Cognitive Systems*. Elsevier, 2005.

[Gabbay, 1985] D. M. Gabbay. Theoretical foundations for non-monotonic reasoning in expert systems. In K. R. Apt, editor, *Logics and Models of Concurrent Systems*. Springer, 1985.

[Gabbay, 1998a] D. M. Gabbay. *Elementary Logic, A Procedural Pespective*. Prentice Hall, 1998.

[Gabbay, 1998b] D. M. Gabbay. *Fibring Logics*. Oxford University Press, 1998.

[Gärdenfors and Makinson, 1994] P. Gärdenfors and D. Makinson. Nonmonotonic inference based on expectations. *Artificial Intelligence*, 65:197–245, 1994.

[Geffner, 1992] H. Geffner. *Default Reasoning. Causal and Conditional Theories*. MIT Press, 1992.

[Gelfond and Lifschitz, 1988] M. Gelfond and V. Lifschitz. The stable model semantics for logic programming. In R. Kowalski and K. Bowen, editors, *Proc. 5th International Conf./Symp. on Logic Programming*, pages 1070–1080, Cambridge, MA, 1988. MIT Press.

[Gelfond and Lifschitz, 1991] M. Gelfond and V. Lifschitz. Classical negation and disjunctive databases. *New Generation Computing*, 9:365–385, 1991.

[Gelfond et al., 1991] M. Gelfond, V. Lifschitz, H. Przymusińska, and M. Truszczyński. Disjunctive defaults. In *Proc. Second Int. Conf. on Principles of Knowledge Representation and Reasoning, KR'91*, pages 230–237, Cambridge, Mass., 1991.

[Goldszmidt et al., 1993] M. Goldszmidt, P. Morris, and J. Pearl. A maximum entropy approach to nonmonotonic reasoning. *IEEE Transactions on Pattern Analysis and Machine Intelligence*, 15:220–232, 1993.

[Grosof, 1991] B. N. Grosof. Generalising prioritization. In J. Allen, R. Fikes, and E. Sandewall, editors, *Proc. Second International Conference on Principles of Knowledge Representation and Reasoning (KR'91)*, pages 289–300. Morgan Kaufmann, 1991.

[Horty, 1994] J. F. Horty. Some direct theories of nonmonotonic inheritance. In D. M. Gabbay, C. J. Hogger, and J. A. Robinson, editors, *Handbook of Logic in Artificial Intelligence and Logic Programming* **3**: *Nonmonotonic Reasoning and Uncertain Reasoning*. Oxford University Press, Oxford, 1994.

[Konolige, 1988] K. Konolige. On the relation between default and autoepistemic logic. *Artificial Intelligence*, 35:343–382, 1988.

[Konolige, 1992] K. Konolige. Abduction versus closure in causal theories. *Artificial Intelligence*, 53:255–272, 1992.

[Kratzer, 1981] A. Kratzer. Partition and revision: The semantics of counterfactuals. *Journal of Philosophical Logic*, 10:201–216, 1981.

[Kraus et al., 1990] S. Kraus, D. Lehmann, and M. Magidor. Nonmonotonic reasoning, preferential models and cumulative logics. *Artificial Intelligence*, 44:167–207, 1990.

[Lehmann and Magidor, 1992] D. Lehmann and M. Magidor. What does a conditional knowledge base entail? *Artificial Intelligence*, 55:1–60, 1992.

[Lehmann, 1989] D. Lehmann. What does a conditional knowledge base entail? In R. Brachman and H. J. Levesque, editors, *Proc. 2nd Int. Conf. on Principles of Knowledge Representation and Reasoning, KR'89*, pages 212–222. Morgan Kaufmann, 1989.

[Lehmann, 1995] D. Lehmann. Another perspective on default reasoning. *Annals of Mathematics and Artificial Intelligence*, 15:61–82, 1995.

[Lewis, 1981] D. Lewis. Ordering semantics and premise semantics for counterfactuals. *Journal of Philosophical Logic*, 10:217–234, 1981.

[Lifschitz and Schwarz, 1993] V. Lifschitz and G. Schwarz. Extended logic programs as autoepistemic theories. In L. M. Pereira and A. Nerode, editors, *Proc. Second Int. Workshop on Logic Programming and Nonmonotonic Reasoning*, pages 101–114. MIT Press, 1993.

[Lifschitz and Woo, 1992] V. Lifschitz and T. Woo. Answer sets in general nonmonotonic reasoning (preliminary report). In *Proc. Third Int. Conf. on Principles of Knowledge Representation and Reasoning, KR'92*, pages 603–614. Morgan Kauffman, 1992.

[Lifschitz et al., 2001] V. Lifschitz, D. Pearce, and A. Valverde. Strongly equivalent logic programs. *ACM Transactions on Computational Logic*, 2:526–541, 2001.

[Lifschitz, 1985] V. Lifschitz. Computing circumscription. In *Proc. 9th Int. Joint Conf. on Artificial Intelligence, IJCAI-85*, pages 121–127. Morgan Kaufmann, 1985.

[Makinson and van der Torre, 2000] D. Makinson and L. van der Torre. Input/Output logics. *Journal of Philosophical Logic*, 29:383–408, 2000.

[Makinson, 1989] D. Makinson. General theory of cumulative inference. In M. Reinfrank, editor, *Nonmonotonic Reasoning*, volume 346 of *LNAI*, pages 1–18. Springer, 1989.

[Makinson, 1994] D. Makinson. General patterns in nonmonotonic reasoning. In D. M. Gabbay and Others, editors, *Handbook of Logic in Artificial Intelligence and Logic Programming, Vol. 3, Nonmonotonic and Uncertain Reasoning*, volume 2, pages 35–110. Oxford University Press, Oxford, 1994.

[Makinson, 2005] D. Makinson. *Bridges from Classical to Nonmonotonic Logic*. King's College Publications, 2005.

[Marek and Truszczyński, 1993] W. Marek and M. Truszczyński. *Nonmonotonic Logic, Context-Dependent Reasoning*. Springer, 1993.

[Marek et al., 1993] V. W. Marek, G. F. Schwarz, and M. Truszchinski. Modal nonmonotonic logics: ranges, characterization, computation. *Journal of ACM*, 40:963–990, 1993.

[McCain and Turner, 1997] N. McCain and H. Turner. Causal theories of action and change. In *Proceedings AAAI-97*, pages 460–465, 1997.

[McCarthy, 1986] J. McCarthy. Applications of circumscription to formalizing common sense knowledge. *Artificial Intelligence*, 13:27–39, 1986.

[McDermott and Doyle, 1980] D. McDermott and J. Doyle. Nonmonotonic logic. *Artificial Intelligence*, 13:41–72, 1980.

[McDermott, 1982] D. McDermott. Nonmonotonic logic II: Nonmonotonic modal theories. *Journal of the ACM*, 29:33–57, 1982.

[Moore, 1985] R. C. Moore. Semantical considerations on non-monotonic logic. *Artificial Intelligence*, 25:75–94, 1985.

[Pearl, 1990] J. Pearl. System Z: A natural ordering of defaults with tractable applications to default reasoning. In *Proceedings of the Third Conference on Theoretical Aspects of Reasoning About Knowledge (TARK'90)*, pages 121–135, San Mateo, CA, 1990. Morgan Kaufmann.

[Poole, 1988] D. Poole. A logical framework for default reasoning. *Artificial Intelligence*, 36:27–47, 1988.

[Poole, 1989] D. Poole. Explanation and prediction: An architecture for default and abductive reasoning. *Computational Intelligence*, 5:97–110, 1989.

[Poole, 1994] D. Poole. Representing diagnosis knowledge. *Annals of Mathematics and Artificial Intelligence*, 11:33–50, 1994.

[Ramsey, 1978] F. P. Ramsey. *Foundations*. Routledge & Kegan Paul, London, 1978.

[Reiter, 1978] R. Reiter. On closed world data bases. In H. Gallaire and J. Minker, editors, *Logic and Data Bases*, pages 119–140. Plenum Press, 1978.

[Reiter, 1980] R. Reiter. A logic for default reasoning. *Artificial Intelligence*, 13:81–132, 1980.

[Reiter, 1991] R. Reiter. The frame problem in the situation calculus: A simple solution (sometimes) and a completeness result for goal regression. In V. Lifschitz, editor, *Artificial Intelligence and Mathematical Theory of Computation: Papers in Honor of Lohn McCarthy*, pages 318–420. Academic Press, 1991.

[Reiter, 2001] R. Reiter. *Knowledge in Action: Logical Foundations for Specifying and Implementing Dynamic Systems.* MIT Press, 2001.

[Schlechta, 2004] Karl Schlechta. *Coherent Systems*, volume 2 of *Studies in Logic and Practical Reasoning.* Elsevier, Amsterdam, 2004.

[Schwarz, 1992a] G. Schwarz. Minimal model semantics for nonmonotonic modal logics. In *Proceedings LICS-92*, pages 34–43, Santa Cruz, CA., 1992.

[Schwarz, 1992b] G. Schwarz. Reflexive autoepistemic logic. *Fundamenta Informaticae*, 17:157–173, 1992.

[Segerberg, 1971] K. Segerberg. *An Essay in Classical Modal Logic*, volume 13 of *Filosofiska Studier.* Uppsala University, 1971.

[Shanahan, 1997] M. P. Shanahan. *Solving the Frame Problem.* The MIT Press, 1997.

[Shoham, 1988] Y. Shoham. *Reasoning about Change.* Cambridge University Press, 1988.

[Tan and Pearl, 1995] S.-W. Tan and J. Pearl. Specificity and inheritance in default reasoning. In *Proceedings Int. Joint Conf. on Artificial Intelligence, IJCAI-95*, pages 1480–1486, 1995.

[Thomason, 2003] Richmond Thomason. Logic and artificial intelligence. In Edward N. Zalta, editor, *The Stanford Encyclopedia of Philosophy.* 2003.

[Truszczyński, 1991] M. Truszczyński. Modal interpretations of default logic. In J. Myopoulos and R. Reiter, editors, *Proceedings Int. Joint Conf. on Artificial Intelligence, IJCAI'91*, pages 393–398, San Mateo, Calif., 1991. Morgan Kaufmann.

[Turner, 1999] H. Turner. A logic of universal causation. *Artificial Intelligence*, 113:87–123, 1999.

[van Benthem, 1984] J. van Benthem. Foundations of conditional logic. *Journal of Philosophical Logic*, 13:303–349, 1984.

[van Emden and Kowalski, 1976] M. H. van Emden and R. A. Kowalski. The semantics of predicate logic as a programming language. *J. of ACM*, 23:733–742, 1976.

[Veltman, 1976] F. Veltman. Prejudices, presuppositions and the theory of conditionals. In J. Groenendijk and M. Stokhof, editors, *Amsterdam Papers on Formal Grammar*, volume 1, pages 248–281. Centrale Interfaculteit, Universiteit van Amsterdam, 1976.

[Wittgenstein, 1961] L. Wittgenstein. *Tractatus Logico-Philosophicus.* Routledge & Kegan Paul, London, 1961. English translation by D. F. Pears and B. F. McGuinness.

Compiled Labelled Deductive Systems for Access Control

KRYSIA BRODA AND ALESSANDRA RUSSO

1 Introduction

Labelled Deductive Systems (LDS) is a methodology initially proposed by Gabbay in [Gabbay, 1996] for both the theoretical study of logics and the development of logical systems suitable for the needs of specific applications. In the LDS approach, a basic unit of information is not just a formula but a *labelled formula*, where the label belongs to a given labelling algebra. Additional information, for example regarding the overall structure of the data or the underlying semantic properties of the logic, can be explicitly embodied, via the labels, in the object language of the system. Derivation rules act on both the labels and the formulae, according to certain fixed rules of propagation. The addition of this separate "object-level" dimension (i.e. labels and labelling algebra) enables the development of a *uniform* system for different logics within the same family, in that every deduction rule can be applied to each of these logics and their differences captured entirely by the labelling algebra. The general proposals in [Gabbay, 1996] acted initially as a manifesto or starting point for a wide programme of research. Detailed investigations were, for example, undertaken into the benefits of using the LDS methodology to reformulate intuitionistic modal logics [Sympson, 1993] and substructural logics [Broda *et al.*, 1999b; D'Agostino *et al.*, 1999; Gabbay and Olivetti, 2002]. Specialised frameworks based on LDS where also proposed [Artosi *et al.*, 2002; Blackburn, 2000; Broda and Gabbay, 2000; Russo, 1996]. Among these the *Compiled Labelled Deductive Systems* (CLDS) approach demonstrated how LDS techniques facilitates the reformulation and generalisation of a large class of modal logics and conditional logics [Broda *et al.*, 2002; Russo, 1996]. This chapter contributes to this programme of research by showing how the CLDS approach can be used to develop a *sound* and *complete* logical system for role-based access control, called AC_{CLDS} system, which enables the formalisation of and reasoning about access control in distributed systems.

Various logical formalisms for (role-based) access control have been proposed in the literature [Abadi *et al.*, 1993; Cholvy and Cuppens, 1997; Kamoda *et al.*, 2005; Massacci, 1997; Son and Lobo, 2001]. These all attempt to model and reason about the role-based hierarchical relationships between principals and between groups, where privileges can be inherited along the hierarchical structure. The approach in [Kamoda *et al.*, 2005] is based on a first-order logic representation of access control policies expressed in programming-style languages (such as Ponder [Damianou *et al.*, 2001] and Tower [Hitchens and Varadharajan, 2000]), the $\mathcal{PDL}$ approach in [Agrawal *et al.*, 2005; Son and Lobo, 2001] uses an *event-condition-action* formalisation and relative implementation in logic programming, and [Cholvy and Cuppens, 1997] deploys a deontic-based language, also translated into first-order logic for reasoning purposes. Abadiand Massacci, on the other hand, have proposed a modal logic formalisation of access control systems, referred in this chapter as the *logic for access control*. In particular, Abadi work in [Abadi *et al.*, 1993] provides a first axiomatisation of basic notions of access control systems, such as principals, privileges, authentication and delegation ([Sandhu, 1992]). This is based on multi-modal logic, where modalities are assumed to be abstractions of roles, and constraints over the different accessibility relations are used to define various delegation principles. The inference capability of this axiomatisation was shown by Massacci to be limited [Massacci, 1997]; for instance the so-called "hands-off axiom" $\neg(A \text{ says } \bot) \rightarrow (A \text{ controls } (P \implies A))$, used to express that a principal A hands-over its privileges to another principal P, is valid in Abadi's calculus, but not provable. This is partly due to the fact that universal formulae of the form $P \implies A$, although expressable in the languange, are not axiomatisable within a Hilbert System [van Benthem, 1986] and therefore not axiomatisable in Abadi's calculus. The labelled tableau system, proposed subsequently by Massacci in [Massacci, 1997], overcomes this limitation. This system enables explicit reasoning about formulas of the form $P \implies A$, so facilitating, in turn, dynamic inference of properties of the accessibility relations. For instance, the inference of $P \implies P|P$ forces *dynamically* the accessibility relation for the principal P to be transitive. However, as stated in [Massacci, 1997], Massacci's labelled tableaux system is only partially complete. This is because the tableaux rules do not allow these universal formula to be unfolded over compound principals.

This chapter aims to address the limitations of both Abadi's and Massacci's logic-based approaches. It builds upon Massacci's work to propose a CLDS system, called AC_{CLDS} system, where explicit reasoning about accessibility relations of both primitive and compound principals can be performed, so allowing the development of a sound and complete proof system for role-based access control in distributed system. In the CLDS approach a theory combines a logical theory written in an object *language* and a *labelling algebra*, written in a first-order *labelling language*. The latter is used to axiomatise semantic and/or proof theoretic properties that uniquely identify the underlying object logic. In the AC_{CLDS} system, the labelling algebra is defined as a binary first-order theory axiomatising the properties of atomic and compound principals, whereas the object language is the multi-modal language given in [Massacci, 1997]. The two languages are combined via the notion of a *declarative unit*, written as $\alpha : \lambda$, which expresses that the formula α is *true* at the label (i.e. point) λ. The inference rules include *basic* rules for the classical connectives, which appear in the formula part of the declarative units, rules, driven by the labelling algebra, for reasoning explicitly about the accessibility relations, as well as additional *specialised* rules for reasoning about modal connectives and their interaction with primitive and compound principals. The modularity of the CLDS approach is also reflected by the AC_{CLDS} system. The set of basic rules is in fact the same as the set of basic rules that any logical system would require when formalised within a CLDS framework [Broda *et al.*, 2004]. It is the set of specialised rules that provides the AC_{CLDS} system with the specific characteristic of a logical system for access control. It is this specialised set of rules that allows explicit unfolding of modal properties through compound principals so enabling the AC_{CLDS} system to be complete.

The chapter is organised as follows. Section 2 gives a brief overview of the language features and semantics of a CLDS approach. The general natural deduction style proof system for a CLDS system is described together with a general model–theoretic semantics, based on a translation method into classical logic and related notion of semantic entailment. Section 3 introduces basic concepts of access control for distributed systems and the features that a logic for this type of application should have. The specific AC_{CLDS} system is described in Section 4, and proofs of soundness and completeness are given in Section 5. The chapter ends in Section 6 with a general discussion on how to further improve the logic for access control so to capture more realistic behaviours of access control systems.

Notation Throughout the chapter predicate symbols, terms and meta-variable denoting principals begin with an upper-case letter, whereas other constants, variables and function symbols begin with a lower-case letter. Greek letters meta-variables are used to refer to terms different from the principals, and expressions in the system. Larger entities such as structures, sets, theories and languages are symbolised in calligraphic font, $\mathcal{A}, \mathcal{B}, \mathcal{C}$, etc.. The power set of a given set $\mathcal{A}$ will be denoted by $PW(\mathcal{A})$.

2 Overview of CLDS

This section introduces the CLDS approach within the context of modal logic, by giving a brief description of its basic language features, syntax and semantics[1]. A CLDS is a logical framework in which explicit reference can be made to specific possible worlds and to relationships between possible worlds, whilst retaining the conventional syntax of modal logic. In this approach, a logical theory, called a *configuration*, is a generalisation of standard modal theories, as it is defined as a structure of arbitrary modal theories, the proof-theoretical presentation is uniform across the different logics of the same family and retains the conciseness of standard modal logic proof systems. The semantics is defined through a translation into first-order logic, whereby the notions of model, satisfiability of a configuration and semantic entailment are given in terms of classical semantics.

2.1 Languages and Syntax

A CLDS language is, in general, defined as an ordered pair $\langle \mathcal{L}_P, \mathcal{L}_L \rangle$, where $\mathcal{L}_P$ is a given *object* language and $\mathcal{L}_L$ is a *labelling language* defined as a binary fragment of a first-order language. In the propositional case, the object language is composed of a countable set of propositional letters, $\{p, q, r, \ldots\}$ and boolean connectives $\{\neg, \wedge, \rightarrow\}$. The labelling language $\mathcal{L}_L$ includes, instead, a countable set of constant symbols $\{s_0, s_1, s_2, \ldots\}$, a countable set of variables $\{x, y, z, \ldots\}$, a set of binary predicate symbols $\{R_1, R_2, \ldots\}$, called *R-predicates*, the set of logical connectives $\{\neg, \wedge, \vee, \rightarrow, \leftrightarrow\}$ and the quantifiers $\forall$ and $\exists$.

For proof-theoretic purposes, the labelling language $\mathcal{L}_L$ is, in general, extended with additional sets of Skolem terms, which are used only in the definition of some labelled natural deduction rules of the system. The extended labelling language is then called a *semi-extended labelling language* and is defined as follows.

DEFINITION 1.1. Let $\mathcal{L}_P$ be a propositional language and $\{\alpha_1, \alpha_2, \ldots\}$ be the set of all wffs of $\mathcal{L}_P$. The *semi-extended labelling language* $Func(\mathcal{L}_P, \mathcal{L}_L)$ is defined as the language $\mathcal{L}_L$ extended with a set of *skolem* function symbols $\{sk^n_{\alpha_1}, sk^n_{\alpha_2}, \ldots\}$ where $n \geq 0$.

Ground terms of $Func(\mathcal{L}_P, \mathcal{L}_L)$ are called *labels*. The specific definition of the skolem symbols $sk^n_{\alpha_i}$ and their interpretation depend on the specific CLDS system. Similarly for the R-predicates. In the case of modal logics labels refer to possible worlds and R-predicates to the accessibility relation. For access control logics, R-predicates denote the accessibility relation for

---

[1]For a more general presentation of a CLDS, the reader is refererd to [Broda *et al.*, 2004].

the principals and they provide the direction to which privileges can be propagated amongst various roles.

Syntax. The CLDS language facilitates the formalisation of two types of information, (i) what holds at particular points, expressed by a syntactic entity called *declarative unit,* and (ii) which points are in relation with each other and which are not, expressed by the syntactic entity called an *R-literal.* A declarative unit is defined as a pair *"formula:label"*, where the label component is a ground term of the semi-extended labelling language $Func(\mathcal{L}_P, \mathcal{L}_L)$ and the formula is a wff of the language $\mathcal{L}_P$. An R-literal is any ground literal in the semi-extended labelling language involving a binary R-predicate, usually of the form $R_i(\lambda_1, \lambda_2)$ and $\neg R_i(\lambda_1, \lambda_2)$, where λ_1 and λ_2 are labels, expressing that λ_2 is or is not related to λ_1 by the accessibility relation R_i. For each R-literal Δ, the *conjugate* of Δ, written $\overline{\Delta}$, is the opposite in sign of Δ.

This combined aspect of the CLDS syntax yields a definition of a CLDS theory more general than the traditional definition of a logical theory. Informally, a CLDS theory, called a *configuration*, is composed of two sets, a set of R-literals and a set of declarative units. For example the pair of sets $\{R_Q(w_0, w_1))\}$ and $\{P \text{ says } r : w_0, P \implies Q : w_2, r : w_1\}$ is a consistent AC_{CLDS} configuration, which states that the principal P requests r (P says $r : w_0$) and this request is granted at the P accessible world w_1 ($r : w_1$) since w_1 is Q accessible ($R_Q(w_0, w_1)$)) and P inherits privileges from the principal Q ($P \implies Q : w_2$). The formal definition of a configuration is given below.

DEFINITION 1.2. Given a CLDS language, a configuration is a tuple $\langle \mathcal{D}, \mathcal{F} \rangle$ where $\mathcal{D}$, called a *diagram*, is a finite set of R-literals and $\mathcal{F}$ is a function from the set of ground terms of $Func(\mathcal{L}_P, \mathcal{L}_L)$ to the set $\text{PW}(\text{wff}(\mathcal{L}_P))$ of sets of wffs of $\mathcal{L}_P$.

To capture different classes of logics within the CLDS approach, an appropriate first-order theory $\mathcal{A}$, written in the language $\mathcal{L}_L$ and called a *labelling algebra*, needs to be defined. This is used by the proof system to infer properties about the diagram part of a configuration. Inference rules and the notion of a derivability relation are, in fact, defined between configurations. A set $\mathcal{R}$ of such inference rules, together with a CLDS language $\langle \mathcal{L}_P, \mathcal{L}_L \rangle$ and a labelling algebra $\mathcal{A}$, uniquely define a CLDS system (i.e. for any CLDS system $\mathcal{S}$, $\mathcal{S} = \langle \langle \mathcal{L}_P, \mathcal{L}_L \rangle, \mathcal{A}_S, \mathcal{R}_S \rangle$).

2.2 A "basic" natural deduction system

The "structural" aspect of a CLDS theory stimulated the idea of defining deductive processes that describe how configurations "evolve" by reasoning within and between the local theories associated with each point in a configuration or by reasoning about the diagram of a configuration. An inference rule of a CLDS is defined between configurations as follows.

DEFINITION 1.3. An inference rule $\mathcal{I}$ is a set of pairs of configurations, where each such pair is written as $\mathcal{C}/\mathcal{C}'$. If $\mathcal{C}/\mathcal{C}' \in \mathcal{I}$ then $\mathcal{C}$ is called an *antecedent* configuration of $\mathcal{I}$, and $\mathcal{C}'$ an *inferred* (or *consequence*) configuration of $\mathcal{I}$ with respect to $\mathcal{C}$.

All the rules except one have the effect of expanding the antecedent configuration. These rules can extend an antecedent configuration $\mathcal{C}$ with either a declarative unit, or with an R-literal, or with both. However, configurations equal or smaller than the antecedent one can also be inferred. This is facilitated by an inference rule called the $\mathcal{C}$-Reduction (C-R) rule. Only a graphical representation of the inference rules will be used throughout the chapter and the reader is referred to [Broda *et al.*, 2004; Russo, 1996] for a complete formal definition of a CLDS proof system. Tables 1.1 and 1.2 show, respectively, the classical rules for the $\rightarrow$, $\neg$ and $\wedge$ connectives[2] and the basic rules for R-literals, which are all common to any CLDS system. In this graphical representation, $\mathcal{C}\langle\alpha:\lambda\rangle$ (respectively

Table 1.1. Classical CLDS Rules.

$$\frac{\mathcal{C}\langle\alpha\rightarrow\beta:\lambda,\alpha:\lambda\rangle}{\mathcal{C}'\langle\beta:\lambda\rangle}\,(\rightarrow\mathcal{E})\qquad \frac{\begin{array}{c}\mathcal{C}\langle[\alpha:\lambda]\rangle\\ \vdots\\ \tilde{\mathcal{C}}\langle\beta:\lambda\rangle\end{array}}{\mathcal{C}'\langle\alpha\rightarrow\beta:\lambda\rangle}\,(\rightarrow\mathcal{I})$$

$$\frac{\mathcal{C}\langle\alpha\wedge\beta:\lambda\rangle}{\mathcal{C}'\langle\alpha:\lambda,\beta:\lambda\rangle}\,(\wedge\mathcal{E})\qquad \frac{\mathcal{C}\langle\alpha:\lambda,\beta:\lambda\rangle}{\mathcal{C}'\langle\alpha\wedge\beta:\lambda\rangle}\,(\wedge\mathcal{I})$$

$$\frac{\mathcal{C}\langle\neg\neg\alpha:\lambda\rangle}{\mathcal{C}'\langle\alpha:\lambda\rangle}\,(\neg\neg)\qquad \frac{\begin{array}{c}\mathcal{C}\langle[\alpha:\lambda]\rangle\\ \vdots\\ \tilde{\mathcal{C}}\langle\bot:\lambda'\rangle\end{array}}{\mathcal{C}'\langle\neg\alpha:\lambda\rangle}\,(\neg\mathcal{I})$$

$\mathcal{C}\langle\Delta\rangle$) denotes that $\mathcal{C}$ includes a declarative unit $\alpha:\lambda$ (respectively R–literal Δ). Declarative units and R-literals contained in square brackets are assumptions introduced within a derivation that are subsequently discharged.

---

[2]Introduction and elimination rules for $\vee$ can be derived using the equivalence $\alpha\vee\beta\equiv\neg(\neg\alpha\wedge\neg\beta)$.

$\mathcal{C}' \langle \pi \rangle$, where π is a declarative unit or an R-literal, represents that the inferred configuration $\mathcal{C}'$ is $\mathcal{C}$ extended with π. $\tilde{\mathcal{C}}$ are the configurations derived in subderivations after adding to the antecedent configuration $\mathcal{C}$ temporary assumptions.

In most CLDSs, the labels occurring in the classical rule are the same, except for the $(\neg \mathcal{I})$ rule. The notion of contradiction (or inconsistency) in the CLDS approach strictly depends on the type of logic. Modal logics, and therefore the logic for access control, include a classical notion of inconsistency, for which the symbol $\perp$ used in the $(\neg \mathcal{I})$ rule is a short-hand for any $\mathcal{L}_P$ wff of the form $\alpha \wedge \neg \alpha$. In CLDSs where the notion of classical inconsistency applies the particular labels λ and λ' in the $(\neg \mathcal{I})$ rule are not required to be the same – a contradiction in some part of a configuration introduces an inconsistency to the configuration as a whole.

The R-literals rules, given in Table 1.2, facilitate reasoning about the diagram of a configuration, using the particular labelling algebra $\mathcal{A}$ under consideration, and infering R-literals and declarative units which would not be inferred using only the logical connectives. For logics of the same family (*e.g.* different modal logics), the $(R\text{-}A)$ rule captures *entirely* the difference between one CLDS system and another, allowing all other inference rules to be equally applicable to any CLDS system. In the AC_{CLDS} system, the $(R\text{-}A)$ rule allows to reason about properties of principals explicitly within the derivation process. The rules $(\perp \mathcal{E})$ and $(R\mathcal{I})$ express additional forms of interactions between the R-literals and the declarative units. The $(\perp \mathcal{E})$ rule allows the inference of falsity (i.e. $\perp : \lambda$) whenever R-literals and its negations are present in a configuration. This is necessary because since no compound classical formulae with R-literals can be inferred in a configuration, inconsistency of this form would not otherwise be captured. The $(R\mathcal{I})$ rule enables instead the derivation of R-literals in the presence of a logical inconsistency.

DEFINITION 1.4. Given a CLDS system $\mathcal{S} = \langle \langle \mathcal{L}_P, \mathcal{L}_L \rangle, \mathcal{A}_S, \mathcal{R}_S \rangle$, a *proof* is a pair $\langle \mathcal{P}, m \rangle$, where $\mathcal{P}$ is a sequence of configurations $\{\mathcal{C}_0, \ldots, \mathcal{C}_n\}$, with $n > 0$, and m is a mapping from the set $\{0, \ldots, n-1\}$ to $\mathcal{R}_S$ such that for each i, $0 \leq i < n$, $\mathcal{C}_i / \mathcal{C}_{i+1} \in m(i)$.

DEFINITION 1.5. Given a CLDS system $\mathcal{S}$, and two configurations $\mathcal{C}$ and $\mathcal{C}'$, $\mathcal{C}'$ is *derivable* from $\mathcal{C}$ in $\mathcal{S}$, written $\mathcal{C} \vdash_s \mathcal{C}'$, if there exists a proof $\langle \{\mathcal{C}, \ldots, \mathcal{C}'\}, m \rangle$.

The above definition of derivability can be reformulated as a relation between a configuration and single declarative unit or R-literal. Given, for instance a CLDS $\mathcal{S}$, a configuration $\mathcal{C} = \langle \mathcal{D}, \mathcal{F} \rangle$ and a declarative unit or R-literal π, then $\mathcal{C} \vdash_s \pi$ if there exists a configuration $\mathcal{C}'$ such that $\mathcal{C} \vdash_s \mathcal{C}'$

Table 1.2. Rules for R-literals

$$\frac{\mathcal{C}\langle \Delta, \overline{\Delta} \rangle}{\mathcal{C}'\langle \alpha : \lambda \rangle} \quad (\bot \mathcal{E})$$

$$\frac{\mathcal{C}\langle [\; \overline{\Delta} \;] \rangle}{\vdots} \frac{\tilde{\mathcal{C}}\langle \bot : \lambda \rangle}{\mathcal{C}'\langle \Delta \rangle} \quad (R\mathcal{I})$$

$$\frac{\mathcal{C}}{\mathcal{C}'} \quad (C\text{-}R)$$
where $\mathcal{C}' \subseteq \mathcal{C}$

$$\frac{\mathcal{C}}{\mathcal{C}'\langle \Delta \rangle} \quad (R\text{-}A)$$
if $\mathcal{A} \cup \mathcal{D} \vdash_{FOL} \Delta$

and $\pi \in \mathcal{C}'$.

2.3 Semantics

A CLDS can be seen as a "semi-translated" approach to a given object logic, because of its feature of syntactically representing in the proof system semantic notions of the underlying logic. Its own model-theoretic semantics is therefore naturally defined in terms of a first-order semantics using a specific "compiled" translation method into first-order logic. This translation method is defined here and the notions of a CLDS model, satisfiability of a configuration and semantic entailment are given also in terms of classical semantics.

As mentioned before, a declarative unit $\alpha : \lambda$ represents that the formula α is verified (or holds) at the point λ, whose interpretation is strictly related to the type of underlying logic. In the CLDS approach these notions are expressed in terms of first-order statements of the form $[\alpha]^*(\lambda)$, where $[\alpha]^*$ is a predicate symbol. The *extended labelling language* $Mon(\mathcal{L}_P, \mathcal{L}_L)$ is an extension of the language $Func(\mathcal{L}_P, \mathcal{L}_L)$ given by adding a monadic predicate symbol $[\alpha]^*$ for each wff α of $\mathcal{L}_P$.

DEFINITION 1.6. Let $Func(\mathcal{L}_P, \mathcal{L}_L)$ be a semi-extended labelling language. Let $\alpha_1, \ldots, \alpha_n, \ldots$, be the ordered set of wffs of $\mathcal{L}_P$. The *extended labelling language* $Mon(\mathcal{L}_P, \mathcal{L}_L)$ is defined as the language $Func(\mathcal{L}_P, \mathcal{L}_L)$ extended with the set of unary predicate symbols $\{[\alpha_1]^*, \ldots, [\alpha_n]^*, \ldots\}$.

The relationships between the monadic predicates $[\alpha]^*$ in $Mon(\mathcal{L}_P, \mathcal{L}_L)$ are constrained by a set of first-order axiom schemas which capture the satisfiability conditions of each type of formula α. These schemas strictly depend on the underlying logic. For example, as shown in Section 4, the extended labelling algebra $\mathcal{A}^+$ of the AC_{CLDS} system includes the axiom schema $\forall x([A \implies B]^*(x) \rightarrow \forall y, z(R_B(y, z) \rightarrow R_A(y, z)))$ to capture the semantic meaning of the $\implies$ operator.

The translation method adopted by the CLDS approach associates syntactical expressions of the CLDS language with sentences of the language $Mon(\mathcal{L}_P, \mathcal{L}_L)$ and CLDS configurations with first-order theories in $Mon(\mathcal{L}_P, \mathcal{L}_L)$. Each declarative unit $\alpha : \lambda$ is associated with the sentence $[\alpha]^*(\lambda)$, and each R-literal is associated with itself. The first-order translation of a configuration is a first-order theory defined as follows.

DEFINITION 1.7. Let $\mathcal{C} = \langle \mathcal{D}, \mathcal{F} \rangle$ be a configuration. The *first-order translation* of $\mathcal{C}$, written $FOT(\mathcal{C})$, is a theory written in $Mon(\mathcal{L}_P, \mathcal{L}_L)$ and defined by the expression $FOT(\mathcal{C}) = \mathcal{D} \cup \mathcal{DU}$, where $\mathcal{DU} = \{[\alpha]^*(\lambda) \mid \alpha \in \mathcal{F}(\lambda), \lambda$ is a ground term of $Func(\mathcal{L}_P, \mathcal{L}_L)\}$.

Since labels in a configuration can only be ground terms of the language $Func(\mathcal{L}_P, \mathcal{L}_L)$, the first-order translation of a configuration is a set of *ground literals* of the language $Mon(\mathcal{L}_P, \mathcal{L}_L)$. The notions of model, satisfiability and semantic entailment are given in terms of classical semantics (where "$\mathcal{M} \Vdash_{FOL} \psi$" signifies that the classical formula ψ is true in the classical model $\mathcal{M}$, according to the standard classical logic definition).

DEFINITION 1.8. Given a CLDS system $\mathcal{S}$, the associated extended algebra $\mathcal{A}_{\mathcal{S}}^+$, a declarative unit $\alpha : \lambda$ and an R–literal Δ,

(1.1) $\mathcal{M}$ is a CLDS model of S $\Leftrightarrow_{\text{def}}$ $\mathcal{M}$ is a model of $\mathcal{A}_{\mathcal{S}}^+$

(1.2) $\mathcal{M} \Vdash_S \alpha : \lambda$ $\Leftrightarrow_{\text{def}}$ $\mathcal{M} \Vdash_{FOL} [\alpha]^*(\lambda)$

(1.3) $\mathcal{M} \Vdash_S \Delta$ $\Leftrightarrow_{\text{def}}$ $\mathcal{M} \Vdash_{FOL} \Delta$

In the above definition, the statement (1.1) defines the class of models of a CLDS system $\mathcal{S}$ in terms of models of the extended algebra $\mathcal{A}_{\mathcal{S}}^+$ associated with $\mathcal{S}$. Statements (1.2) and (1.3) define, respectively, the satisfiability of declarative units and R–literals in terms of classical satisfiability of their associated first–order translations. A CLDS model $\mathcal{M}$ satisfies a configuration $\mathcal{C}$, written $\mathcal{M} \Vdash_S \mathcal{C}$, if and only if for each $\pi \in \mathcal{C}$ (where π may be a declarative unit or an R–literal), $\mathcal{M} \Vdash_S \pi$. The notion of semantic entailment in a CLDS system is given here as a relation between configurations.

DEFINITION 1.9. Let $\mathcal{S} = \langle \langle \mathcal{L}_P, \mathcal{L}_L, \rangle, \mathcal{A}_S, \mathcal{R}_S \rangle$ be a CLDS and let $\mathcal{A}^+$ be the extended algebra of $\mathcal{S}$. Let $\mathcal{C} = \langle \mathcal{D}, \mathcal{F} \rangle$ and $\mathcal{C}' = \langle \mathcal{D}', \mathcal{F}' \rangle$ be two configurations of $\mathcal{S}$ and $FOT(\mathcal{C}) = \mathcal{D} \cup \mathcal{DU}$ and $FOT(\mathcal{C}') = \mathcal{D}' \cup \mathcal{DU}'$ be their respective first-order translations. The configuration $\mathcal{C}$ semantically entails C', written $C \models_S C'$, iff for each $\Delta \in \mathcal{D}'$, $\mathcal{A}^+ \cup FOT(C) \models_{FOL} \Delta$, and for each $[\alpha]^*(\lambda) \in \mathcal{DU}'$, $\mathcal{A}^+ \cup FOT(C) \models_{FOL} [\alpha]^*(\lambda)$.

Proving soundness. Given that the semantics is based on a first-order translation method, the proof of the soundness property of the $\vdash_S$ for a CLDS system $\mathcal{S}$ takes advantage of the soundness property of the first-order classical derivability relation $\vdash_{FOL}$. A diagrammatic representation of the soundness theorem of a CLDS system $\mathcal{S}$ is given in Figure 1.1. The

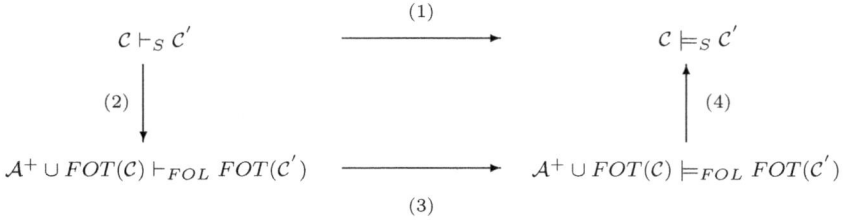

Figure 1.1. Proof of the soundness property of a CLDS system $\mathcal{S}$.

soundness statement, which corresponds to the arrow labelled with (1), is proved by the composition of three main steps, arrows (2), (3) and (4) respectively. The first step (arrow (2)) proves that the hypothesis, $C \vdash_S C'$, for a CLDS system $\mathcal{S}$, implies that $\mathcal{A}^+ \cup FOT(C) \vdash_{FOL} FOT(C')$. This trivially implies (by soundness of first-order logic) that $\mathcal{A}^+ \cup FOT(C) \models_{FOL} FOT(C')$, which gives the second step of the proof (arrow (3)). Arrow (4) is given by the definition of the semantic entailment between configurations given in Definition 1.9. Note that this methodology is generally applicable to any CLDS system. The first step is the only one that needs to be proved for each specific CLDS system.

Proving completeness. The completeness property of a CLDS system with respect to the semantics described in Section 2.3 can be proved using standard Henkin-style methodology [Hughes and Cresswell, 1996]. The theorem states that, given a CLDS system and two configurations C and C' such that $C' - C$ is finite, if C' is semantically entailed from C then C' is also derived from C, where $C' - C$ (formally defined in [Russo, 1996]) is the set of declarative units and R-literals in C' but not in C. The methodology adopted to prove the completeness of a CLDS system is diagrammatically represented in Figure 1.2 and can be informally described as follows.

The proof considers the contrapositive statement (arrow (1)), which states that, given a CLDS system and two configurations C and C' such that $C' - C$

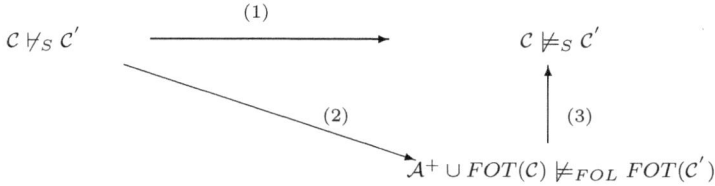

Figure 1.2. Proof of the completeness property of a CLDS system $\mathcal{S}$.

is a finite, if $\mathcal{C} \nvdash_s \mathcal{C}'$ then $\mathcal{C} \nvDash_s \mathcal{C}'$. This is proved by the composition of two main steps, arrows (2) and (3). Arrow (3) is already given by Definition 1.9, while arrow (2) represents the main part of the theorem. The proof of arrow (2) is based on the statement *if $\mathcal{C}$ is a consistent configuration then $\mathcal{C}$ is satisfiable*, known as the "Model Existence Lemma". It consists of the following reasoning steps.

- The hypothesis that $\mathcal{C}'$ is not derivable from $\mathcal{C}$, $\mathcal{C} \nvdash_s \mathcal{C}'$, implies that there exists a $\pi \in \mathcal{C}' - \mathcal{C}$ (where π is a declarative unit or an R-literal) such that $\mathcal{C} \nvdash_s \pi$.

- The above step implies that the configuration $\mathcal{C}$ extended with $\neg\pi$ (written $\mathcal{C} + [\neg\pi]$) is a consistent configuration.

- The above step implies then that the configuration $\mathcal{C} + [\neg\pi]$ is satisfiable. Therefore, there exists a semantic structure $\mathcal{M}$ of the CLDS system $\mathcal{S}$ which satisfies $\mathcal{C}$ and that also satisfies $\neg\pi$. It is then shown that $\mathcal{M}$ does not satisfy π. Thus, since $\pi \in \mathcal{C}'$, by definition of satisfiability of a configuration, $\mathcal{M}$ does not satisfy $\mathcal{C}'$. Hence $\mathcal{A}^+ \cup FOT(\mathcal{C}) \nvDash_{\text{FOL}} FOT(\mathcal{C}')$.

The above methods for proving soundness and completeness of a CLDS will be used in Section 5 to show the soundness and completeetness of the access control system AC_{CLDS}.

3 Overview of a logic for Access Control

Access control logics are used to formalize access rights in distributed systems. A collection of such authorization rights is called an *Access Control Policy* and individual rules of the collection are sometimes called *Policy Norms* [Cholvy and Cuppens, 1997], or more often referred to as policies [Kamoda *et al.*, 2005; Lupu and Sloman, 1999]. A large volume of work is present in the literature on the specification of access control policies, especially in the area of *Role Based Access Control* (RBAC) [Abadi *et al.*, 1993; Sandhu, 1992]. These include logic-based declarative descriptions like Abadi's Calculus for Access Control [Abadi *et al.*, 1993; Massacci, 1997] and the logic programming approaches proposed in [Agrawal *et al.*, 2005; Barker and Stuckey, 2003], as well as specialised programming language type descriptions, such as the PONDER [Damianou *et al.*, 2001] and Tower languages [Hitchens and Varadharajan, 2000]. Each method has its benefits and limitations. Logic-based specifications of access control policies, although sometime difficult to understand for the software engineer, are precise, generally succinct and amenable to formal analysis for inconsistency detection [Kamoda *et al.*, 2005]. On the other hand, the programming style approaches to RBAC can be quite verbose, but are more accessible to system managers and to those responsible for specifying such policies. The work described in this chapter falls into the first category of access control policy specification. It provides a new formalisation, called AC_{CLDS} system, of the logic for access control initially proposed by Abadi in [Abadi *et al.*, 1993] and subsequently enriched by Massacciwith a tableaux method [Massacci, 1997], which overcomes some of its current limitations. Before describing the AC_{CLDS} system in detail, an informal introduction of the main features and expressiveness of this logic is given below, together with an illustrative example.

In RBAC, specifications of policies make use of the notions of a *principal* (or *role*) and an associated set of *privileges* that grant holders of that role access to data or to a system. A principal may also represent a group; members of the group have then the privileges granted to the group. Formulas include atoms, representing satisfied requests, compound expressions stating the making of a request by a principal, relationships between principals and any Boolean combinations of these. For example, the information that a belongs to the manager-role A can be expressed as $a \Longrightarrow A$, the statements that a is requesting to read a file f and B is a sub-manager role of A can be, respectively, formalised as a says $read(f)$ and $A \Longrightarrow B$. Primitive operators can be defined to form compound principals which often imply combined privileges of different roles. For instance, the information that a role combines two different roles B and C can be formalised using a binary operator &. This combination of roles is often used to express particular rights of holders of a role B that can be kept private and not inherited. The Access Control Logic includes also the binary operator | used to state that a principal is a member of more than one different group or role and can make requests assuming any of these roles.

A simple example application drawn from a typical workplace organisation is given here to illustrate the formalisation of some of the basic features of a RBAC. This example consists of four different roles called $\{A, B, C, D\}$, where A could be seen as the role of charge-nurse, B and C as the role of staff nurse responsible for medications and patient welfare respectively, and D as the role of nurse. The privileges of a nurse are inherited by staff nurses, who also have some additional privileges, and the charge-nurse inherits all the privileges of staff nurses. This role dependency is expressed by the formulas $B \Longrightarrow D$, $C \Longrightarrow D$, $A \Longrightarrow C$ and $A \Longrightarrow B$, where $\Longrightarrow$ indicates the role hierarchy and is graphically represented in Figure 1.3(a). It is easy to show that $A \Longrightarrow B\&C$ follows from $A \Longrightarrow C$ and $A \Longrightarrow B$; that is, A has the union of the privileges of B and C.

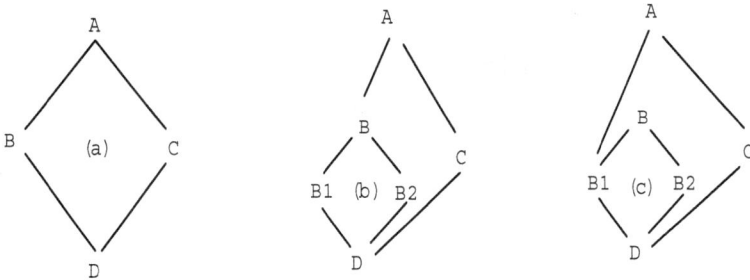

Figure 1.3. (a) Ward-nurse hierarchy, (b) privileges divided, (c) privileges restricted

In Abadi's logic for access control, access authorization is given by a formula of the form $(A \text{ says } r) \rightarrow r$, meaning that if A makes a request r then the request will be granted. This is short-handed as $A \text{ controls } r$. The role structure given in Figure 1.3(a), together with the assumption $B \text{ controls } r$ would allow the inference of $A \text{ controls } r$. Similarly if $C \text{ controls } r$. The above scenario could be elaborated by splitting B into two sub-roles $B1$ and $B2$, for example, to encompass the fact that members of staff-nurse of role B may access records of two different kinds. This new dependency, illustrated in Figure 1.3(b), can be expressed by the formulas $B \Longrightarrow B1\&B2$, $B1 \Longrightarrow D$ and $B2 \Longrightarrow D$ instead of $B \Longrightarrow D$. If now $B2 \text{ controls } r$ then it can be shown that $A \text{ controls } r$ also. In fact, since all members of $B2$ have authorization to access r and A inherits all the privileges of $B2$ through B, all members of A have authorization to access r as well. It may be, however, desirable for A to inherit only the privileges of role $B1$ (for example if the role $B2$ is for staff-nurse members who communicate only between themselves). In that case, the privilege inheritance, shown graphically in Figure 1.3(c), can be formalised by replacing $A \Longrightarrow B$ with $A \Longrightarrow B1$. Given this new statement, it is no longer possible to infer A controls r.

Another aspect of access control systems concerns *Separation of Duty* policies [Hitchens and Varadharajan, 2000]. Such a policy is useful, for example, when a role allows a member two types of access privileges for accessing the same data, but not both. For instance, members of the staff nurse roles might be allowed to issue a medication, but only if the request has been certified, which must be done by some other staff nurse. This kind of access control policy can be captured by the following formalization, where m is a medication: B controls $issue(m)$, B controls $certify(m)$, $b \Longrightarrow B$ and b says $issue(m) \rightarrow \neg b$ says $certify(m)$.

4 AC_{CLDS} – A CLDS for Access Control

In this section a CLDS for access control, called AC_{CLDS}, is formally defined following the approach described in Section 2.

Language of AC_{CLDS} . The language of the AC_{CLDS} system is a pair $\langle \mathcal{L}_{\mathcal{P}}, \mathcal{L}_L \rangle$, where $\mathcal{L}_{\mathcal{P}}$ and $\mathcal{L}_L$ share a common set of terms, called *principals*. These include a countable set of constants $\{P, Q, \ldots\}$, called *primitive principals*, and compound terms, called *compound principals*, constructed from the primitive principals using the operators $\{|, \&\}$. In the remainder of this chapter, meta-variables A, B, C will be used to denote principals. The labelling language $\mathcal{L}_L$ is a first-order language constructed from a countable set of constant symbols $\{w_0, w_1, \ldots\}$, representing possible worlds, a countable set of variables $\{x, y, z, \ldots\}$, the binary functors q_B^A for every principal A and B, and binary predicates R_A, for every principal A. The functor q_B^A is a skolem function used in the labelling algebra to express a property of the binary relation $R_{A|B}$. The second component of the AC_{CLDS} language, $\mathcal{L}_{\mathcal{P}}$, is a propositional multi-modal language consisting of a countable set of proposition letters $\{r, s, t, \ldots\}$, representing *access requests*, the constant symbol $\perp$, the classical connectives $\{\neg, \wedge, \rightarrow\}$, the binary operator $\Longrightarrow$ and the modality operators A says for every principal A. Well-formed formulae of $\mathcal{L}_{\mathcal{P}}$ are defined as follows.

DEFINITION 1.10. A well-formed formula of $\mathcal{L}_{\mathcal{P}}$ is defined inductively as:

- a propositional letter

- $A \Longrightarrow B$, for any principals A and B

- A says α, for any principal A and wff α

- boolean conbinations of wffs

The language $\mathcal{L}_L$ is extended into $Func(\mathcal{L}_P, \mathcal{L}_L)$ by adding for each wff α of $\mathcal{L}_P$ and principals A and B, the unary functor s_α^A and the constant symbols $sp_{A,B}^1$ and $sp_{A,B}^2$. The ground term $s_\alpha^A(\lambda)$ can be thought of as referring to any *arbitrary* world specifically associated with α and A, and used to express Kripke semantic notions of the form "for all possible worlds..." corresponding to a given formula A **says** α. The constant symbols $sp_{A,B}^1$ and $sp_{A,B}^2$ can also be thought of as *arbitrary* worlds and are used to capture the definition of the $\Longrightarrow$ operator within the proof rule.

The labelling algebra $\mathcal{A}_L$ is a set of first-order axioms expressing the properties of the binary relations R_A, for any principal A.

DEFINITION 1.11. The labelling algebra $\mathcal{A}_L$, written in $Func(\mathcal{L}_P, \mathcal{L}_L)$, is the first-order theory given by the following axioms:

$$\forall x, y(R_{A\&B}(x,y) \leftrightarrow (R_A(x,y) \vee R_B(x,y))) \qquad \text{(Union)}$$
$$\forall x, y(R_{A|B}(x,y) \rightarrow R_A(x, q_B^A(x,y)) \wedge R_B(q_B^A(x,y), y)) \qquad \text{(QuotingE)}$$
$$\forall x, y, z((R_A(x,z) \wedge R_B(z,y)) \rightarrow R_{A|B}(x,y)) \qquad \text{(QuotingI)}$$

An AC_{CLDS} system $\mathcal{L}$ is defined by the tuple $\mathcal{L} = \langle \langle \mathcal{L}_P, \mathcal{L}_L \rangle, \mathcal{A}_L, \mathcal{R}_L \rangle$ where the set $\mathcal{R}_L$ of inference rules is given in the following section.

Table 1.3. Additional Rules for the AC_{CLDS} system

$$
\frac{\begin{array}{c}\mathcal{C}\langle A \text{ says } \alpha : \lambda_1, \\ R_A(\lambda_1, \lambda_2)\rangle\end{array}}{\mathcal{C}'\langle \alpha : \lambda_2 \rangle} \ (\text{ says } \mathcal{E})
\qquad
\frac{\begin{array}{c}\mathcal{C}\langle [R_A(\lambda, s_\alpha^A(\lambda))]\rangle \\ \vdots \\ \tilde{\mathcal{C}}\langle \alpha : s_\alpha^A(\lambda)\rangle\end{array}}{\mathcal{C}'\langle A \text{ says } \alpha : \lambda \rangle} \ (\text{ says } \mathcal{I})
$$

$$
\frac{\mathcal{C}\langle B \Longrightarrow A : \lambda_1, R_A(\lambda_2, \lambda_3)\rangle}{\mathcal{C}'\langle R_B(\lambda_2, \lambda_3)\rangle} \ (\Longrightarrow\mathcal{E})
\qquad
\frac{\begin{array}{c}\mathcal{C}\langle [R_A(sp_{A,B}^1, sp_{A,B}^2)]\rangle \\ \vdots \\ \tilde{\mathcal{C}}\langle R_B(sp_{A,B}^1, sp_{A,B}^2)\rangle\end{array}}{\mathcal{C}'\langle B \Longrightarrow A : \lambda \rangle} \ (\Longrightarrow\mathcal{I})
$$

$$
\frac{\begin{array}{cc}& \mathcal{C}\langle R_{A\&B}(\lambda_1, \lambda_2)\rangle \\ \mathcal{C}\langle [R_A(\lambda_1, \lambda_2)]\rangle & \mathcal{C}\langle [R_B(\lambda_1, \lambda_2)]\rangle \\ \vdots & \vdots \\ \mathcal{C}'\langle \pi \rangle & \mathcal{C}'\langle \pi \rangle\end{array}}{\mathcal{C}'\langle \pi \rangle} \ (\&R\mathcal{E})
$$

4.1 Proof Rules for AC_{CLDS}

There are three types of reasoning step that can occur in an AC_{CLDS} system. Those of the first type are "classical" and occur within a particular local theory included in $\mathcal{C}$ and respect standard notions of inference for classical connectives. These are the rules given in Table 1.1. The second type of reasoning step concerns reasoning about the diagram part of a configuration using the properties of the labelling algebra. These are the rules described in Table 1.2 among which the (R-A) rule plays a prominent role. It allows, in particular, the inference of relationships about compound principals. For instance, given a configuration that includes the R-literals $R_A(\lambda_1, \lambda_2)$ and $R_B(\lambda_2, \lambda_3)$, the (R-A) would allow the inference of the R-literal $R_{A|B}(\lambda_1, \lambda_3)$ using the axiom (QuotingI) of the labelling algebra.

The third type of reasoning step is about the interaction between different theories in a configuration, and uses the rules given in Table 1.3. In particular, the rules (says $\mathcal{I}$) and (says $\mathcal{E}$) are similar to standard introduction and elimination rules for the modal $\Box$ operator. The term $s_\alpha^A(\lambda)$ used in the (says $\mathcal{I}$) rule refers to an arbitrary possible world uniquely associated with the formula α and accessible from λ via the accessibility relation R_A. The rules ($\Longrightarrow\mathcal{I}$) and ($\Longrightarrow\mathcal{E}$) allow dynamic inference of properties of the diagram part of a given configuration. For instance, the transitive property of an R_P relation can be implicitly introduced in a derivation from $P \Longrightarrow P|P:\lambda_1$ whenever the two R-literals $R_P(\lambda_1, \lambda_2)$ and $R_P(\lambda_2, \lambda_3)$ are given or inferred, as shown in the AC_{CLDS} derivation in Figure 1.4. The rule (&R$\mathcal{E}$) allow the inference process to implicitly reason about disjunctions of R-literals, as this type of formulae cannot be directly expressed in a configuration.

$$\frac{\mathcal{C}\langle P \Longrightarrow P|P:\lambda_1, R_P(\lambda_1, \lambda_2), R_P(\lambda_2, \lambda_3)\rangle}{\begin{array}{c}\mathcal{C}_1\langle R_{P|P}(\lambda_1, \lambda_3)\rangle\\\hline\mathcal{C}_2\langle R_P(\lambda_1, \lambda_3)\rangle\end{array}} \qquad \begin{array}{c}\text{Assumptions}\\(R\text{-A})\\(\Longrightarrow\mathcal{E})\end{array}$$

Figure 1.4. Implicit inference of R_P's transitivity from $P \Longrightarrow P|P:\lambda_1$

Finally, the rules given in Table 1.4 are "short-hand" derived rules, which allow compound principal terms involved in **says** formulas to be reduced to primitive principal terms. The first four of these rules mirror the tableau rules given in [Massacci, 1997]. The (*univ*) rule captures exactly the global property of the formula $A \Longrightarrow B$ and corresponds to Massacci's rule $[U_{gr}]$. The (*speaks*) rule is a generalisation of Massacci's rules $[L_{gr}]$ and $\langle R_{gr}\rangle$, in that neither principals A and B need to be restricted to be primitive whereas the ($\neg speaks$) rule corresponds to Massacci's $\langle U_{gr}\rangle$ rule but without requiring the introduction of new arbitrary requests since explicit reference to the accessibility relations can be made in the AC_{CLDS} system. Finally, the (*cons*) rule is a new rule that states the consistency of a principal A. This rule is, in particular, used for the derivation of the so-called " hands-off" axiom [Abadi *et al.*, 1993] (see Figure 1.5), which states that from

$\neg(Q \text{ says } \perp) : s_0$ can be derived $Q \text{ controls } P \Longrightarrow Q$ (i.e. $Q \text{ says } (P \Longrightarrow Q) \to P \Longrightarrow Q) : s_0$.

Table 1.4. A collection of useful derived rules in the AC_{CLDS} system

$$\frac{\mathcal{C}\langle(A|B) \text{ says } \alpha : \lambda\rangle}{\mathcal{C}'\langle A \text{ says } B \text{ says } \alpha) : \lambda\rangle} \quad (quote\mathcal{E}) \qquad \frac{\mathcal{C}\langle A \text{ says } B \text{ says } \alpha : \lambda\rangle}{\mathcal{C}'\langle(A|B) \text{ says } \alpha : \lambda\rangle} \quad (quote\mathcal{I})$$

$$\frac{\mathcal{C}\langle(A\&B) \text{ says } \alpha : \lambda\rangle}{\mathcal{C}'\langle A \text{ says } \alpha : \lambda, \atop B \text{ says } \alpha : \lambda\rangle} \quad (\&\mathcal{E}) \qquad \frac{\mathcal{C}\langle A \text{ says } \alpha : \lambda, \atop B \text{ says } \alpha : \lambda\rangle}{\mathcal{C}'\langle(A\&B) \text{ says } \alpha : \lambda\rangle} \quad (\&\mathcal{I})$$

$$\frac{\mathcal{C}\langle A \Longrightarrow B : \lambda_1, \atop A \text{ says } \alpha : \lambda_2\rangle}{\mathcal{C}'\langle B \text{ says } \alpha : \lambda_2\rangle} \quad (speaks) \qquad \frac{\mathcal{C}\langle\neg(A \Longrightarrow B) : \lambda\rangle}{\mathcal{C}'\langle R_B(sp^1_{A,B}, sp^2_{A,B}), \atop \neg R_A(sp^1_{A,B}, sp^2_{A,B})\rangle} \quad (\neg speaks)$$

$$\frac{\mathcal{C}\langle A \Longrightarrow B : \lambda_1\rangle}{\mathcal{C}'\langle A \Longrightarrow B : \lambda_2\rangle} \quad (univ) \qquad \frac{\mathcal{C}\langle\neg(A \text{ says } \perp) : \lambda\rangle}{\mathcal{C}'\langle R_A(\lambda, s^A_{\perp}(\lambda))\rangle} \quad (cons)$$

$$\frac{\mathcal{C}_0\langle\neg(Q \text{ says } \perp) : w_0\rangle}{} \quad (assumption)$$

$$\frac{\tilde{\mathcal{C}}_0\langle\neg(Q \text{ says } \perp) : w_0, [Q \text{ says } (P \Longrightarrow Q) : w_0]\rangle}{} \quad (assumption)$$

$$\frac{\tilde{\mathcal{C}}_1\langle Q \text{ says } (P \Longrightarrow Q) : w_0, R_Q(w_0, s^Q_{\perp}(w_0))\rangle}{} \quad (cons)$$

$$\frac{\tilde{\mathcal{C}}_2\langle P \Longrightarrow Q : s^Q_{\perp}(w_0)\rangle}{} \quad (\text{ says } \mathcal{E})$$

$$\frac{\tilde{\mathcal{C}}_3\langle P \Longrightarrow Q : w_0\rangle}{} \quad (univ)$$

$$\mathcal{C}'\langle Q \text{ says } (P \Longrightarrow Q) \to P \Longrightarrow Q : w_0\rangle \quad (\to\mathcal{I})$$

Figure 1.5. Derivation of $Q \text{ controls } P \Longrightarrow Q$

Two example derivations are given in Figures 1.5 and 1.6 showing, respectively, a proof of the "hands-off" axiom and a derivation of the (*cons*). These example derivations make use of the following convenient notation. Declarative units introduced in a proof as temporary assumptions, mainly by introduction rules, are written in square brackets, as they represent temporary assumptions that need to be discharged once the rule has been applied. In the

$$\frac{\mathcal{C}_0\langle\neg(A\ \textbf{says}\ \bot):\lambda\rangle}{\quad} \qquad\qquad \text{(Assumption)}$$

$$\frac{\tilde{\mathcal{C}}_0\langle[\neg R_A(\lambda, s_\bot^A(\lambda))]\rangle}{\quad} \qquad\qquad \text{(Assumption)}$$

$$\frac{\tilde{\mathcal{C}}_1\langle\neg R_A(\lambda, s_\bot^A(\lambda)), [R_A(\lambda, s_\bot^A(\lambda))]\rangle}{\quad} \qquad\qquad \text{(Assumption)}$$

$$\frac{\tilde{\mathcal{C}}_2\langle\bot : s_\bot^A(\lambda)\rangle}{\quad} \qquad\qquad (\bot\mathcal{E})$$

$$\frac{\tilde{\mathcal{C}}_3\langle\neg(A\ \textbf{says}\ \bot):\lambda, A\ \textbf{says}\ \bot:\lambda\rangle}{\quad} \qquad\qquad (\ \textbf{says}\ \mathcal{I})$$

$$\frac{\tilde{\mathcal{C}}_4\langle\bot:\lambda\rangle}{\quad} \qquad\qquad (\wedge\mathcal{I})$$

$$\mathcal{C}'\langle R_A(\lambda, s_\bot^A(\lambda))\rangle \qquad\qquad (R\mathcal{I})$$

Figure 1.6. Derivation of rule *cons*

example derivation in Figure 1.5, for instance, the given configuration $(\mathcal{C}_0)$ is extended to include the temporary assumption $Q\ \textbf{says}\ (P \Longrightarrow Q):w_0$. The configurations $\tilde{\mathcal{C}}_1$, $\tilde{\mathcal{C}}_2$ and $\tilde{\mathcal{C}}_3$ are then infered from $\tilde{\mathcal{C}}_0$ by applying (*cons*), (**says** $\mathcal{E}$) and (*univ*) respectively. In the last step of the derivation, the configuration C' is derived by the initial configuration C_0 using the $(\neg\mathcal{I})$ rule.

4.2 Semantics for AC_{CLDS}

The semantics of the AC_{CLDS} system is based on the model theoretic semantics defined in Section 2.3 for a general CLDS system. The *extended labelling algebra* $\mathcal{A}_{\mathrm{Ac}}^+$ of the AC_{CLDS} system is formally given below.

DEFINITION 1.12.

Given the extended labelling language $Mon(\mathcal{L}_P, \mathcal{L}_L)$ and the labelling algebra $\mathcal{A}_{\mathrm{Ac}}$ of the AC_{CLDS} system, the *extended algebra* $\mathcal{A}_{\mathrm{Ac}}^+$ is the theory in $Mon(\mathcal{L}_P, \mathcal{L}_L)$ given by the labelling algebra extended with the following axiom schemas (Ax1)-(Ax7). For any wffs α and β of $\mathcal{L}_P$ and principals A and B:

$$\forall x([\alpha \wedge \beta]^*(x) \leftrightarrow ([\alpha]^*(x) \wedge [\beta]^*(x))) \qquad\qquad \text{(Ax1)}$$

$$\forall x([\neg\alpha]^*(x) \leftrightarrow \neg[\alpha]^*(x)) \qquad\qquad \text{(Ax2)}$$

$$\forall x([\alpha \rightarrow \beta]^*(x) \leftrightarrow ([\alpha]^*(x) \rightarrow [\beta]^*(x))) \qquad\qquad \text{(Ax3)}$$

$$\forall x((R_A(x, s_\alpha^A(x)) \to [\alpha]^*(s_\alpha^A(x))) \to [A \text{ says } \alpha]^*(x)) \tag{Ax4}$$

$$\forall x([A \text{ says } \alpha]^*(x) \to (\forall y(R_A(x, y) \to [\alpha]^*(y)))) \tag{Ax5}$$

$$\forall x((R_B(sp_{A,B}^1, sp_{A,B}^2) \to R_A(sp_{A,B}^1, sp_{A,B}^2)) \to [A \Longrightarrow B]^*(x)) \tag{Ax6}$$

$$\forall x([A \Longrightarrow B]^*(x) \to (\forall y, z(R_B(y, z) \to R_A(y, z)))) \tag{Ax7}$$

The first three axiom schemas express the distributive properties of the logical connectives among the monadic predicates of $Mon(\mathcal{L}_P, \mathcal{L}_L)$. The axiom schemas (Ax4) and (Ax5) reflect the traditional Kripke semantic definition of satisfiability of modal $\Box$. The axiom schemas (Ax6) and (Ax7) together express the *specific* semantic meaning of the operator $\Longrightarrow$ for the logic for access control. These axioms facilitate the inference of dynamic properties of the accessibility relations (*i.e.* by simply using the semantic definition of the formula $P \Longrightarrow P|P$), as illustrated in the first-order proof derivation given in Figure 1.7. This derivation clearly shows this main feature of the AC_{CLDS} system.

$$\frac{\begin{array}{ll} [P \Longrightarrow P|P]^*(\lambda_1) & \text{(Assumption)} \\ \hline \forall x, y(R_{P|P}(x, y) \to R_P(x, y)) & \text{(Ax7)} \\ \hline \forall x, y, z(R_P(x, y) \land R_P(y, z) \to R_{P|P}(x, z)) & \text{(QuotingI)} \\ \hline \forall x, y, z(R_P(x, y) \land R_P(y, z) \to R_P(x, z)) & (\to \mathcal{I}) \end{array}}{}$$

Figure 1.7. First-order derivation of transitivity of R_P from $P \Longrightarrow P|P : \lambda_1$

The notions of satisfiability and semantic entailment of the AC_{CLDS} system, denoted with $\models_{\text{Ac}}$, are as specified in Definitions 1.8 and 1.9, but based on the extended algebra $\mathcal{A}_{\text{Ac}}^+$.

5 Soundness and Completeness Properties

The soundness and completeness proofs given in this section are based respectively on the two methodologies described in Section 2.3.

5.1 Soundness of AC_{CLDS}

As described in Section 2.3 it is sufficient to show Lemma 1.13, called "Soundness Lemma". The proof of this lemma is by induction on the size of the derivation of $\mathcal{C}'$ from $\mathcal{C}$, and makes use of the following property. Given any first-order theory T and formula ϕ:

(1.4) If $T \vdash \phi$ then $T \vdash T \cup \{\phi\}$.

Each derivation rule has an associated *size*. The rule $(C\text{-R})$ has size equal to 0. Rules such as (**says** $\mathcal{E}$) or $(\neg\neg)$, in which no assumptions are made, have size equal to 1. Rules in which a single assumption is made, such as $(\rightarrow\mathcal{I})$ or $(\Longrightarrow\mathcal{I})$, have size equal to 1 plus the size of the smallest possible derivation that could be made for the antecedent sub-proof. The rule $(\&\text{R}\mathcal{E})$ has size equal to 1 plus the maximum of the sizes of the smallest possible derivations that could be made for the two antecedent sub-proofs.

LEMMA 1.13. *Let $\mathcal{C}$ and $\mathcal{C}'$ be two configurations. If $\mathcal{C} \vdash_{AC} \mathcal{C}'$ then $\mathcal{A}_{Ac}^{+} \cup FOT(\mathcal{C}) \vdash FOT(\mathcal{C}')$,*

Proof. Let $\langle \mathcal{C}_0, \ldots, \mathcal{C}_n = \mathcal{C}' \rangle$ be a derivation of $\mathcal{C}'$ from $\mathcal{C}$ of size k, $k \geq 0$. The proof is by induction on the length of the derivation.

Base Case: The length of the derivation is $k = 0$. Then the derivation consists only of $(C\text{-R})$ steps. For each i, $0 < i \leq n$, $FOT(\mathcal{C}_i) \subseteq FOT(\mathcal{C}_{i-1})$ and hence, by the transitivity of $\subseteq$, $FOT(\mathcal{C}_n) \subseteq FOT(\mathcal{C}_0)$ and hence by first-order logic $FOT(\mathcal{C}_0) \vdash FOT(\mathcal{C}_n)$.

Inductive Hypothesis: Assume that if $\mathcal{C} \vdash_{AC} \mathcal{C}''$ has a derivation of size less than k then $\mathcal{A}_{Ac}^{+} \cup FOT(\mathcal{C}) \vdash FOT(\mathcal{C}'')$.

Inductive Step: The length of the derivation is $k > 0$. Then the proof is by cases on the last derivation step between $\mathcal{C}_{n-1}$ and $\mathcal{C}_n$. For each case it is shown that $FOT(\mathcal{C}_{n-1}) \vdash FOT(\mathcal{C}_n)$. Then by the induction hypothesis, since the size of the derivation $\mathcal{C}_0 \vdash_{AC} \mathcal{C}_{n-1}$ is $< k$, $FOT(\mathcal{C}_0) \vdash FOT(\mathcal{C}_{n-1})$ and hence, by the transitivity of $\vdash$, $FOT(\mathcal{C}_0) \vdash FOT(\mathcal{C}_n)$.

Case $(\wedge\mathcal{E})$: By assumption $[\alpha \wedge \beta]^*(\lambda) \in FOT(\mathcal{C}_{n-1})$. By the $(\forall\mathcal{E})$ and $(\wedge\mathcal{E})$ natural deduction rules of first-order logic $[\alpha]^*(\lambda)$ and $[\beta]^*(\lambda)$ are derivable from Axiom (Ax1). Therefore, by property (1.4), $FOT(\mathcal{C}_{n-1}) \vdash FOT(\mathcal{C}_n)$.

Case $(\Longrightarrow\mathcal{E})$: By assumption $[A \Longrightarrow B]^*(\lambda) \in FOT(\mathcal{C}_{n-1})$ and $R_B(\lambda', \lambda'') \in FOT(\mathcal{C}_{n-1})$. By first-order natural deduction $R_A(\lambda', \lambda'')$ is derivable using Axiom (Ax7). Therefore, $FOT(\mathcal{C}_{n-1}) \vdash FOT(\mathcal{C}_n)$.

The cases for the rules of $(\rightarrow\mathcal{E})$, $(\wedge\mathcal{I})$, $(\neg\neg)$, $(\perp\mathcal{E})$, (R-A) and (**says** $\mathcal{E}$) are similar to the above cases.

Case $(\rightarrow\mathcal{I})$: By assumption there is a sub-derivation of $\tilde{\mathcal{C}}\langle \beta : \lambda \rangle$ from $\mathcal{C}\langle[\alpha : \lambda]\rangle$ with size $< k$. Therefore, by the induction hypothesis there is a derivation of $FOT(\mathcal{C}) \cup \{[\beta]^*(\lambda)\}$ from $FOT(\mathcal{C}) \cup \{[\alpha]^*(\lambda)\}$ and hence by the $(\rightarrow\mathcal{I})$ natural deduction rule of first-order logic and property (1.4), there is a derivation of $FOT(\mathcal{C}) \cup \{[\alpha \rightarrow \beta]^*(\lambda)\}$.

*Case (**says** $\mathcal{I}$):* By assumption there is a sub-derivation of $\tilde{\mathcal{C}}\langle \alpha : s_\alpha^A(\lambda) \rangle$ from $\mathcal{C}\langle[R_A(\lambda, s_\alpha^A(\lambda))]\rangle$ with size $< k$. Therefore, by the induction hypothesis there is a derivation of $FOT(\mathcal{C}) \cup \{[\alpha]^*(s_\alpha^A(\lambda))\}$ from $FOT(\mathcal{C}) \cup \{R_A(\lambda, s_\alpha^A(\lambda))\}$ and hence by the $(\rightarrow\mathcal{I})$ natural deduction rule of first-order logic and property (1.4) there is a derivation of $FOT(\mathcal{C}) \cup \{[A \text{ **says** } \alpha]^*(\lambda)\}$.

Case (&R$\mathcal{E}$): By assumption there is a sub-derivation of $\tilde{\mathcal{C}}\langle\pi\rangle$ from each of $\mathcal{C}\langle[R_A(\lambda_1,\lambda_2)]\rangle$ and $\mathcal{C}\langle[R_B(\lambda_1,\lambda_2)]\rangle$ both with size $< k$. Therefore, by the induction hypothesis there are two derivations of $FOT(\mathcal{C})\cup\{FOT(\pi)\}$ from $FOT(\mathcal{C}) \cup \{R_A(\lambda_1,\lambda_2)\}$ and from $FOT(\mathcal{C}) \cup \{R_B(\lambda_1,\lambda_2)\}$. Hence by the $(\rightarrow\mathcal{I})$ and $(\vee\mathcal{E})$ natural deduction rules of first-order logic and property (1.4) there is a derivation of $FOT(\mathcal{C}) \cup \{FOT(\Psi)\}$.

The cases for the rules of $(\neg\mathcal{I})$, $(\Longrightarrow\mathcal{I})$ and $(R\mathcal{I})$ are similar to the above cases. ∎

Using the method described in Section 2.3, the Soundness Lemma allows to prove the soundness of AC_{CLDS}:

THEOREM 1.14. *Let $\mathcal{C}$ and $\mathcal{C}'$ be two configurations of the AC_{CLDS} system. If $\mathcal{C} \vdash_{AC} \mathcal{C}'$, then $\mathcal{C} \models_{AC} \mathcal{C}'$*

5.2   Completeness of AC_{CLDS}

The Completeness proof of the AC_{CLDS} system follows the steps of the completeness proof for a general CLDS outlined in Section 2.3. It makes use of a main lemma, called the *Model Existence Lemma* whose proof uses the results given in Propositions 1.15, 1.16 and 1.17. All the following proofs make use of an additional notation. Let $\mathcal{C} = \langle\mathcal{D},\mathcal{F}\rangle$ be a configuration and π be a declarative unit or a R-literal. If π is of the form $\alpha : \lambda$, $\mathcal{C} + [\alpha : \lambda]$ is the configuration $\langle\mathcal{D},\mathcal{F}'\rangle$, such that $\mathcal{F}'(\lambda) = \mathcal{F}(\lambda)\cup\{\alpha\}$ and for any λ different from λ, $\mathcal{F}'(\lambda') = \mathcal{F}(\lambda')$. If π is an R-literal Δ, then $\mathcal{C} + [\Delta]$ is the configuration $\langle\mathcal{D}',\mathcal{F}\rangle$ such that $\mathcal{D}' = \mathcal{D}\cup\{\Delta\}$.

PROPOSITION 1.15 (Consistency). *Let $\mathcal{C}$ be an AC_{CLDS} configuration, $\alpha : \lambda$ be an arbitrary declarative unit and Δ be an arbitrary R-literal. If $\mathcal{C} \nvdash_{AC} \alpha : \lambda$ then $\mathcal{C} + [\neg\alpha : \lambda] \nvdash_{AC} \perp : \lambda'$. Similarly, if $\mathcal{C} \nvdash_{AC} \Delta$ then $\mathcal{C} + [\overline{\Delta}] \nvdash_{AC} \perp:\lambda'$.*

Proof. Suppose by contradiction that $\mathcal{C} + [\neg\alpha : \lambda] \vdash_{AC} \perp : \lambda'$. Then by rule $(\neg\mathcal{I})$ $\mathcal{C} \vdash_{AC} \alpha:\lambda$, which contradicts the hypothesis. Similarly, assume that $\mathcal{C} + [\overline{\Delta}] \vdash_{AC} \perp:\lambda'$. Then by rule $(R\mathcal{I})$ $\mathcal{C} \vdash_{AC} \Delta$, which contradicts the hypothesis. ∎

A consistent configuration $\mathcal{C}$ can be extended into a *maximal consistent configuration* $\mathcal{M}_{\mathrm{ccc}}$ by enumerating every possible formula of AC_{CLDS} and, starting from $\mathcal{C}$ add each formula in the enumeration, depending whether it is consistent or not. It is easy to prove that by construction a $\mathcal{M}_{\mathrm{ccc}}$ is a consistent configuration (*i.e.* $\mathcal{M}_{\mathrm{ccc}} \nvdash_{AC} \perp:\lambda$ for any label λ).

The following two Propositions 1.16 and 1.17 prove that an $\mathcal{M}_{\mathrm{ccc}}$ is maximal and satisfies various semantic properties on the accessibility relation and logical operators.

PROPOSITION 1.16 (Maximality of $\mathcal{M}_{ccc}$). *Let $\mathcal{M}_{ccc}$ be a maximal consistent configuration. For any declarative unit $\alpha:\lambda$ and R-literal Δ:*

(1.5) *If $\alpha:\lambda \in \mathcal{M}_{ccc}$ then $\neg\alpha:\lambda \notin \mathcal{M}_{ccc}$,*
     and if $\Delta \in \mathcal{M}_{ccc}$ then $\overline{\Delta} \notin \mathcal{M}_{ccc}$.

(1.6) *Either $\alpha:\lambda \in \mathcal{M}_{ccc}$ or $\neg\alpha:\lambda \in \mathcal{M}_{ccc}$,*
     and either $\Delta \in \mathcal{M}_{ccc}$ or $\overline{\Delta} \in \mathcal{M}_{ccc}$.

Proof.
Property (1.5): Suppose by contradiction that $\neg\alpha : \lambda \in \mathcal{M}_{ccc}$. Then $\mathcal{M}_{ccc} \vdash_{AC} \bot : \lambda$ by $(\wedge\mathcal{I})$, which contradicts the fact that $\mathcal{M}_{ccc}$ is consistent. Similarly for any R-literal Δ and its conjugate.
Property (1.6): Suppose by contradiction that $\alpha:\lambda \notin \mathcal{M}_{ccc}$ and $\neg\alpha:\lambda \notin \mathcal{M}_{ccc}$. By construction, there exist configurations $\mathcal{C}' \subseteq \mathcal{M}_{ccc}$ and $\mathcal{C}'' \subseteq \mathcal{M}_{ccc}$ such that $\mathcal{C}' + [\alpha:\lambda] \vdash_{AC} \bot:\lambda$ and $\mathcal{C}'' + [\neg\alpha:\lambda] \vdash_{AC} \bot:\lambda$. By monotonicity, $\mathcal{M}_{ccc} + [\alpha : \lambda] \vdash_{AC} \bot : \lambda$ and $\mathcal{M}_{ccc} + [\neg\alpha : \lambda] \vdash_{AC} \bot : \lambda$. Therefore, by $(\neg\mathcal{I})$, $\mathcal{M}_{ccc} \vdash_{AC} \alpha : \lambda$ and $\mathcal{M}_{ccc} \vdash_{AC} \neg\alpha : \lambda$ and hence $\mathcal{M}_{ccc} \vdash_{AC} \bot : \lambda$, which contradicts the fact that $\mathcal{M}_{ccc}$ is consistent. Similarly for Δ and its conjugate. ∎

PROPOSITION 1.17 (Properties of $\mathcal{M}_{ccc}$). *Let $\mathcal{M}_{ccc}$ be a maximal consistent configuration. The following properties are satisfied, for any labels λ, λ', λ'', wff α and principals A and B:*

(1.7) $R_{A\&B}(\lambda,\lambda') \in \mathcal{M}_{ccc}$ *if, and only if,*
     $R_A(\lambda,\lambda') \in \mathcal{M}_{ccc}$ *or* $R_B(\lambda,\lambda') \in \mathcal{M}_{ccc}$

(1.8) *if* $R_A(\lambda,\lambda'') \in \mathcal{M}_{ccc}$ *and* $R_B(\lambda'',\lambda') \in \mathcal{M}_{ccc}$ *then* $R_{A|B}(\lambda,\lambda') \in \mathcal{M}_{ccc}$

(1.9) *if* $R_{A|B}(\lambda,\lambda') \in \mathcal{M}_{ccc}$ *then*
     $R_A(\lambda, q_B^A(\lambda,\lambda')) \in \mathcal{M}_{ccc}$ *and* $R_B(q_B^A(\lambda,\lambda'),\lambda') \in \mathcal{M}_{ccc}$

(1.10) *if A says $\alpha:\lambda \in \mathcal{M}_{ccc}$ and $R_A(\lambda,\lambda') \in \mathcal{M}_{ccc}$ then $\alpha:\lambda' \in \mathcal{M}_{ccc}$*

(1.11) *if* $\neg(A$ *says* $\alpha):\lambda \in \mathcal{M}_{ccc}$ *then*
     $R_A(\lambda, s_\alpha^A(\lambda)) \in \mathcal{M}_{ccc}$ *and* $\neg\alpha:s_\alpha^A(\lambda) \in \mathcal{M}_{ccc}$

(1.12) *if $A \Longrightarrow B:\lambda \in \mathcal{M}_{ccc}$ and $R_B(\lambda',\lambda'') \in \mathcal{M}_{ccc}$ then*
     $R_A((\lambda',\lambda'') \in \mathcal{M}_{ccc}$

(1.13) *if* $\neg(A \Longrightarrow B):\lambda \in \mathcal{M}_{ccc}$ *then*
     $R_B(sp_{A,B}^1, sp_{A,B}^2) \in \mathcal{M}_{ccc}$ *and* $\neg R_A(sp_{A,B}^1, sp_{A,B}^2) \in \mathcal{M}_{ccc}$

(1.14) $\alpha \wedge \beta : \lambda \in \mathcal{M}_{ccc}$ *if, and only if,*
$\alpha : \lambda \in \mathcal{M}_{ccc}$ *and* $\beta : \lambda \in \mathcal{M}_{ccc}$

(1.15) $\alpha \to \beta : \lambda \in \mathcal{M}_{ccc}$ *if, and only if,*
either $\neg \alpha : \lambda \in \mathcal{M}_{ccc}$ *or* $\beta : \lambda \in \mathcal{M}_{ccc}$

Proof.

Property (1.7): (Only if.) Suppose by contradiction that $R_{A\&B}(\lambda, \lambda') \in \mathcal{M}_{ccc}$, $R_A(\lambda, \lambda') \notin \mathcal{M}_{ccc}$ and $R_B(\lambda, \lambda') \notin \mathcal{M}_{ccc}$. By Property (1.6) the literals $\neg R_A(\lambda, \lambda') \in \mathcal{M}_{ccc}$ and $\neg R_B(\lambda, \lambda') \in \mathcal{M}_{ccc}$. By (R-A) rule $\mathcal{M}_{ccc} \vdash_{AC} \neg R_{A\&B}(\lambda, \lambda')$ which together with the assumption gives a contradiction. *(If.)* A similar argument can be applied in this case, using the assumptions either $R_A(\lambda, \lambda') \in \mathcal{M}_{ccc}$ or $R_B(\lambda, \lambda') \in \mathcal{M}_{ccc}$, and $R_{A\&B}(\lambda, \lambda') \notin \mathcal{M}_{ccc}$.

The proofs for Properties (1.8), (1.9), (1.10), (1.12) and (1.14) are similar to the proof for Property (1.7).

Property (1.11): Suppose by contradiction that $\neg(A \text{ says } \alpha) : \lambda \in \mathcal{M}_{ccc}$ and $R_A(\lambda, s_\alpha^A(\lambda)) \notin \mathcal{M}_{ccc}$. Therefore, by Property (1.6) $\neg R_A(\lambda, s_\alpha^A(\lambda)) \in \mathcal{M}_{ccc}$. Using the rules (says $\mathcal{I}$), ($\perp\mathcal{E}$) and ($\wedge\mathcal{I}$) it can be shown that $\mathcal{M}_{ccc} \vdash_{AC} \perp : \lambda$. On the other hand, suppose for contradiction that $\neg(A \text{ says } \alpha) : \lambda \in \mathcal{M}_{ccc}$ and $\neg\alpha : s_\alpha^A(\lambda) \notin \mathcal{M}_{ccc}$. Therefore, by Property (1.6) $\alpha : s_\alpha^A(\lambda) \in \mathcal{M}_{ccc}$. Using the rules (says $\mathcal{I}$) and ($\wedge\mathcal{I}$), it can be shown that $\mathcal{M}_{ccc} \vdash_{AC} \perp : \lambda$.

Property (1.13): Suppose by contradiction that $\neg(A \implies B) : \lambda \in \mathcal{M}_{ccc}$ and $R_B(sp_{A,B}^1, sp_{A,B}^2) \notin \mathcal{M}_{ccc}$. Therefore, by Property (1.6) the literal $\neg R_B(sp_{A,B}^1, sp_{A,B}^2) \in \mathcal{M}_{ccc}$. Using the derived rule ($\neg speaks$) it can be shown that $\mathcal{M}_{ccc} \vdash_{AC} R_B(sp_{A,B}^1, sp_{A,B}^2)$. The case for $\neg R_A(sp_{A,B}^1, sp_{A,B}^2)$ is similar.

Property (1.15): (Only If.) Suppose by contradiction that $\alpha \to \beta : \lambda \in \mathcal{M}_{ccc}$, $\neg\alpha : \lambda \notin \mathcal{M}_{ccc}$ and $\beta : \lambda \notin \mathcal{M}_{ccc}$. By Property (1.6) $\alpha : \lambda \in \mathcal{M}_{ccc}$ and $\neg\beta : \lambda \in \mathcal{M}_{ccc}$. By the ($\to\mathcal{E}$) and ($\wedge\mathcal{I}$) rules $\mathcal{M}_{ccc} \vdash_{AC} \perp : \lambda$. *(If.)* Suppose for contradiction that $\neg\alpha : \lambda \in \mathcal{M}_{ccc}$ and $(\alpha \to \beta) : \lambda \notin \mathcal{M}_{ccc}$. By Property 1.6 $neg(\alpha \to \beta) : \lambda \in \mathcal{M}_{ccc}$. Using the rules ($\to\mathcal{I}$), ($\neg\mathcal{I}$) and ($\neg\neg$) it can be shown that $\mathcal{M}_{ccc} \vdash_{AC} \perp : \lambda$. The case for $\beta : \lambda$ is similar. ∎

The following Model Existence Lemma shows how to construct a AC_{CLDS} model for a maximal consistent configuration $\mathcal{M}_{ccc}$ obtained from a initial consistent configuration $\mathcal{C}$.

LEMMA 1.18 (Model Existence Lemma). *Let $\mathcal{M}_{ccc}$ be an AC_{CLDS} maximal consistent configuration. There exists an AC_{CLDS} model M that satisfies $\mathcal{M}_{ccc}$.*

Proof. According to Definition 1.8 an AC_{CLDS} model is a first-order interpretation that satisfies the labelling algebra of AC_{CLDS} and the semantic axioms of AC_{CLDS}, and it satisfies the $\mathcal{M}_{\text{ccc}}$ if and only if (by definition) it satisfies the $FOT(\mathcal{M}_{\text{ccc}})$. Such an interpretation M can be constructed as follows.

- The domain of M is the set of ground labels in $Func(\mathcal{L}_P, \mathcal{L}_L)$.

- For each predicate $[\alpha]^*$ in the extended labelling language $Mon(\mathcal{L}_P, \mathcal{L}_L)$, the interpretation $M([\alpha]^*)$ is given by the set $\{\lambda | \alpha : \lambda \in \mathcal{M}_{\text{ccc}}\}$.

- For each R-literal R_A in $Mon(\mathcal{L}_P, \mathcal{L}_L)$, the interpretation $M(R_A)$ is given by the set $\{(\lambda, \lambda') | R_A(\lambda, \lambda') \in \mathcal{M}_{\text{ccc}}\}$.

It is shown below that the interpretation M satisfies the three axioms of the labelling algebra and the seven semantic axioms (Ax1) - (Ax7) and it is therefore an AC_{CLDS} model. It can then be easily shown that, by construction, M satisfies the $FOT(\mathcal{M}_{\text{ccc}})$.

Case (Union): Let λ and λ' be elements of $Func(\mathcal{L}_P, \mathcal{L}_L)$, the domain of M. (*If.*) Assume that $(\lambda, \lambda') \in M(R_A)$ or $(\lambda, \lambda') \in M(R_B)$. Then by construction either $R_A(\lambda, \lambda') \in \mathcal{M}_{\text{ccc}}$ or $R_B(\lambda, \lambda')) \in \mathcal{M}_{\text{ccc}}$. In the first case, by Property (1.7) of Lemma 1.17, $R_{A\&B}(\lambda, \lambda') \in \mathcal{M}_{\text{ccc}}$ and hence $(\lambda, \lambda') \in M(R_{A\&B})$ Similarly for the second case. (*Only if.*) Assume that $(\lambda, \lambda') \in M(R_{A\&B})$, then $R_{A\&B}(\lambda, \lambda') \in \mathcal{M}_{\text{ccc}}$. By Property (1.7) of Lemma 1.17 either $R_A(\lambda, \lambda') \in \mathcal{M}_{\text{ccc}}$ or $R_B(\lambda, \lambda')) \in \mathcal{M}_{\text{ccc}}$. Hence, by construction, $(\lambda, \lambda') \in M(R_A)$ or $(\lambda, \lambda') \in M(R_B)$.

The proof for cases (QuotingE) and (QuotingI) are similar to the case for (Union).

Case (Ax2): Let λ be an arbitrary element of $Func(\mathcal{L}_P, \mathcal{L}_L)$. (*Only if.*) Assume $\lambda \in M([\neg\alpha]^*)$, then $\neg\alpha : \lambda \in \mathcal{M}_{\text{ccc}}$ and by Property (1.6) of Lemma 1.16 $\alpha : \lambda \notin \mathcal{M}_{\text{ccc}}$ and $\lambda \notin M([\alpha]^*)$. (*If.*) The proof is similar.

The proofs for (Ax1) and (Ax3) are similar to the proof for (Ax2).

Case (Ax4): Let λ be an arbitrary element of $Func(\mathcal{L}_P, \mathcal{L}_L)$. Assume $\lambda \notin M([A \text{ says } \alpha]^*)$, then by construction A **says** $\alpha : \lambda \notin \mathcal{M}_{\text{ccc}}$ and by Property (1.11) of Lemma 1.17 $R_A(\lambda, s_\alpha^A(\lambda)) \in \mathcal{M}_{\text{ccc}}$ and $\alpha : s_\alpha^A(\lambda)) \notin \mathcal{M}_{\text{ccc}}$. Therefore $(\lambda, s_\alpha^A(\lambda)) \in M(R_A)$ and $s_\alpha^A(\lambda) \notin M([\alpha]^*)$, and hence M satisfies $\neg(R_A(\lambda, s_\alpha^A(\lambda)) \to [\alpha]^*(s_\alpha^A(\lambda)))$.

Case (Ax7): Let λ be an arbitrary ground term of $Func(\mathcal{L}_P, \mathcal{L}_L)$ and assume that $\lambda \in M([A \implies B]^*)$. Hence $A \implies B : \lambda \in \mathcal{M}_{\text{ccc}}$. Suppose for arbitrary λ' and λ'' that $(\lambda', \lambda'') \in M(R_B)$ and hence by construction $R_B(\lambda', \lambda'') \in \mathcal{M}_{\text{ccc}}$. By Property (1.12) of Lemma 1.17 $R_A(\lambda', \lambda'') \in \mathcal{M}_{\text{ccc}}$ and hence $(\lambda', \lambda'') \in M(R_A)$.

The proofs for cases (Ax5) and (Ax6) are similar to the proofs for Cases (Ax4) and (Ax7). ∎

Using the result of Lemma 1.18 and following the steps given in Section 2.3 the completeness of the AC_{CLDS} system with respect to its semantics is proved.

6 Discussion

This paper introduces the AC_{CLDS} system as a logic for access control and proves its soundness and completeness. This system is more flexible than the logic in [Massacci, 1997] in the following ways. It can allow reasoning about formulae of the form $P \Longrightarrow Q$ where neither P nor Q is restricted to the atomic case, and reasoning about implicit dynamic properties on the accessibility relations. These are facilitated by the use of a labelling algebra that allows explicit inferences about the accessibility relation of compound principals, which can only be done implicitly in Massacci's tableau system. As a consequence the AC_{CLDS} proof system is complete with respect to the semantics given in Section 2.3. This semantics captures the same semantic principles of Massacci's logic, for which his tableau system is only partially complete [Massacci, 1997]. A formal correspondence between Massacci's logic and the AC_{CLDS} system could be proved following the methodology described in [Broda *et al.*, 1999a; Broda *et al.*, 2004]. This would require to show that the derivability relation in Massacci's system is equivalent to the AC_{CLDS} derivability relation. Informally, this would require to prove the following two steps. If a closed tableau in his system exists for a formula $1 : \neg\phi$, then the declarative unit $\phi : w_0$, for any arbitrary world w_0, can be shown to be derivable from an empty configuration. And, if there is an open branch in the complete tableau for $1 : \neg\phi$, then it can be shown that there exists an AC_{CLDS} model where $\phi : w_0$ is false.

In both [Abadi *et al.*, 1993] and [Massacci, 1997] the definition of the `controls` operator causes peculiarities in the reasoning process, namely that "if a principal A is making a request r on which it is trusted, then every principal will be trusted on r". This arises from their definition of the formula A `controls` r as a shorthand for A `says` $r \to r$. In fact, assuming A `controls` r (read as A can be trusted on r) and A `says` r (read as A makes the request r), it can be proved that B `controls` r for any principal B as follows. From the definition of `controls` the two assumptions give r, which implies B `controls` r for any B. Clearly, this is a rather unsatisfactory result and should not be the case for such a plausible scenario when B has no relationship to A. The underlying problem is that in the definition of `controls` the consequence that r gets granted is unrelated to the principal making the request. This could be addressed by, for instance, redefining A `controls` r either as A `says` $r \to \Diamond(A)r$, or as A `says` $r(A) \to r(A)$. In the first case it would require the accessibility relation for each principal to satisfy the seriality property, whereas the second case is not so restrictive, but would require the parameter of the modality to be a ground term in the language, in a similar way as in Fitting's Term Modal Logics [Fitting *et al.*, 2001]. The current AC_{CLDS} system can be extended to include either of these solutions.

An additional peculiarity with the definition of `controls` is regarding the concept of denial of access. Assuming denial of access by a principal B on request r to be expressed by $\neg(B\ \text{controls}\ r)$, an inconsistency would immediately arise whenever a second unrelated principal A makes a request r on which it is trusted. Clearly this is an unexpected behaviour of an access control system. The above suggested solutions would also stop this problem from occurring and allow the concept of negative authorisation to be expressed using classical negation Other approaches, in which the concepts of control and denial are taken as independent primitives, don't seem to suffer from these problems [Barker and Stuckey, 2003; ?; Kamoda *et al.*, 2005; Lupu and Sloman, 1999].

An additional benefit of the AC_{CLDS} system is its first-order translation-based semantics, which provides already a "semi-compiled" formalisation of the multi-modal logic into first-order logic. This semi-compiled theory (*i.e.* the extended labelling algebra) can be used to develop a first-order automated theorem prover for the AC_{CLDS} system. The following example illustrates this process. Consider the derivation from $\neg(Q\ \text{says}\ \bot):w_0$ of $Q\ \text{says}\ (P \implies Q) \to P \implies Q$ given in Figure 1.5. The OTTER theorem prover [McCune, 2003] can be used to show the same derivation by means of resolution. Resolution is a refutation technique and requires the conclusion to be negated, which, after translation gives the first-order sentences $\neg[Q\ \text{says}\ \bot]^*(w_0)$, $[Q\ \text{says}\ (P \implies Q)]^*(w_0)$ and $\neg[P \implies Q]^*(w_0)$.

Using the extended labelling algebra the instantiations of the axiom schema needed for this proof are the following:

$$\forall x([P \implies Q]^*(x) \to \forall y, z(R_Q(y,z) \to R_P(y,z))) \qquad \text{(Ax 7)}$$
$$\forall x, y([Q\ \text{says}\ (P \implies Q)]^*(x) \to \forall y(R_Q(x,y) \to [P \implies Q]^*(y))) \qquad \text{(Ax 5)}$$
$$\forall x((R_Q(sp_{P,Q}^1, sp_{P,Q}^2) \to R_P(sp_{P,Q}^1, sp_{P,Q}^2)) \to [P \implies Q]^*(x)) \qquad \text{(Ax 6)}$$
$$\forall x((R_Q(x, s_\bot^Q(x)) \to [\bot]^*(s_\bot^Q(x))) \to [Q\ \text{says}\ \bot]^*(x)) \qquad \text{(Ax 4)}$$

In order to use OTTER, some further notational changes are needed; some terms and predicates are introduced as a short-hand, P and Q are replaced by p and q and variables x, y and z are replaced by X, Y and Z and are assumed to be universally quantified.

In particular, in the formulation shown in Figure 1.8, the term $s1(p,q)$ stands for $P \implies Q$ when it occurs as a sub-formula of $Q\ \text{says}\ (P \implies Q)$, while the predicate $s2(p,q,X)$ is shorthand for the atom $[P \implies Q]^*(X)$ and

1. $\neg says(q, b1, w0)$ data
2. $says(q, s1(p, q), w0)$ negation of goal(i)
3. $\neg s2(p, q, w0)$ negation of goal(ii)
4. $(s2(p, q, X) \wedge R(q, Y, Z)) \rightarrow R(p, Y, Z)$ instance of (Ax 7)
5. $(says(q, s1(p, q), X) \wedge R(q, X, Y)) \rightarrow s2(p, q, Y)$ instance of (Ax 5)
6. $(R(q, sp1, sp2) \rightarrow R(p, sp1, sp2)) \rightarrow s2(p, q, X)$ instance of (Ax 6)
7. $(R(q, X, s(X)) \rightarrow b2(s(X))) \rightarrow says(q, b1, X)$ instance of (Ax 4)

Figure 1.8. Compiled data required for an automatic derivation

```
list(sos).
1 [] -says(q,b1,w0).
2 [] says(q,s1(p,q),w0).
3 [] -s2(p,q,w0).
4 [] -s2(p,q,X)| -R(q,Y,Z)|R(p,Y,Z).
5 [] -says(q,s1(p,q),X)| -R(q,X,Y)| s2(p,q,Y).
6 [] s2(p,q,X)|R(q,sp1,sp2).
7 [] s2(p,q,X)| -R(p,sp1,sp2).
8 [] says(q,b1,X)|R(q,X,s(X)).
9 [] says(q,b1,X)|-b2(s(X)).
end_of_list.

1 [] -says(q,b,w0).
2 [] says(q,s1(p,q),w0).
3 [] -s2(p,q,w0).
4 [] -s2(p,q,X)| -R(q,Y,Z)|R(p,Y,Z).
5 [] -says(q,s1(p,q),X)| -R(q,X,Y)| s2(p,q,Y).
6 [] s2(p,q,X)|R(q,sp1,sp2).
7 [] s2(p,q,X)| -R(p,sp1,sp2).
8 [] says(q,b1,X)|R(q,X,s(X)).
11 [binary,6.1,3.1] R(q,sp1,sp2).
12 [binary,7.1,3.1] -R(p,sp1,sp2).
13 [binary,8.1,1.1] R(q,w0,s(w0)).
15 [binary,4.2,11.1,unit_del,12] -s2(p,q,A).
16 [binary,5.1,2.1,unit_del,15] -R(q,w0,A).
17 [binary,16.1,13.1] $F.
```

Figure 1.9. Proof using OTTER

the atom $says(q, s1(p,q), Z)$ stands for $[Q \text{ says } (P \implies Q)]^*(Z)$. Similarly, the constant $b1$ represents $\perp$ when it occurs as a sub-formula in $Q \text{ says } \perp$, the term $s(X)$ is a short-hand for the Skolem $s^q_\perp(X)$ and the atom $b2(s(X))$ stands for $[\perp]^*(s^Q_\perp(x))$. The constants $sp1$ and $sp2$ are short-hands for the Skolem terms in the instance of (Ax 6). Atoms such as $R(p, X, Y)$ stand for $R_P(x, y)$. More general instances of the axiom schema could have been used, such as $(s2(A, B, X) \wedge R(B, Y, Z)) \rightarrow R(A, Y, Z)$ for the instance of Axiom (Ax 7). After converting the data in Figure 1.8 into clausal form and submitting it to the OTTER theorem prover the refutation shown in Figure 1.9 was returned, which is equivalent to the natural deduction proofs in Figures 1.5 and 1.6.

BIBLIOGRAPHY

[Abadi et al., 1993] M. Abadi, M. Burrows, B. Lampson and G. Plotkin. A calculus for access control in distributed systems. *ACM Trans. on Prog. Lang. and Sys.* 15(4), pp 706-734, 1993.

[Agrawal et al., 2005] D. Agrawal, J. Giles, K-W Lee and J. Lobo. Policy Ratification, *Proceedings of IEEE Policy 2005*, IEEE Computer Society, 2005.

[Artosi et al., 2002] A. Artosi, G. Governatori and A. Rotolo. Labelled Tableaux for non-monotonic reasoning: Cumulative consequence relations, *Journal of Logic and Computation*, Oxford University Press, 12(6), pp 1027-1060, 2002.

[Barker and Stuckey, 2003] S. Barker and P.J. Stuckey. Flexible Access Control Policy Specification with Constraint Logic Programming. *ACM Trans. on Inf. and Sys. Security*, 6(4), pp 501-546, 2003.

[van Benthem, 1986] J. van Bentham. Correspondence Theory, in *Handbook of Philosophical Logic*, Vol II, Reidel, 1986.

[Blackburn, 2000] P. Blackburn. Internalising Labelled Deduction, *Journal of Logic and Computation*, 10(1), Oxford University Press, pp 137-168, 2000.

[Broda et al., 1999a] K. Broda, M. D'Agostino, and A. Russo. Transformation Methods in LDS in *Logic, Language and Reasoning: An Essay in Honour of Gabbay*, Ed. J.J. Ohlbach and U. Reyle, Kluwer, pp 335-376, 1999.

[Broda et al., 1999b] K. Broda, M. Finger and A. Russo. Labelled natural deduction for substructural logics, *Logic Journal of the IGPL*, 7(3), pp 283-318, 1999.

[Broda and Gabbay, 2000] K. Broda and D. Gabbay. Labelled Abduction - Compiled Labelled Abductive Systems, in *Labelled Deduction*, Eds. D. Basin et al, Kluwer Academic Publishers, 2000.

[Broda et al., 2000] K. Broda, A. Russo and D. Gabbay. A unified compilation style natural deduction system for modal, substructural and fuzzy logics. In *Discovering World with Fuzzy Logic: Perspectives and Approaches to Formalization of Human-consistent Logical Systems*, Springer, 2000.

[Broda et al., 2002] K. Broda, D. Gabbay, L. C. Lamb and A. Russo. Labelled natural deduction for conditional logics of normality, *Logic Journal of the IGPL*, 10(2), pp 123-163, 2002.

[Broda et al., 2004] K. Broda, D. M. Gabbay, L. C. Lamb and A. Russo. *Compiled Labelled Deductive Systems: A Uniform Presentation of Non-Classical Logics*, Research Studies Press Ltd, London, 2004.

[Cholvy and Cuppens, 1997] L. Cholvy and F. Cuppens. Analyzing Consistency of Security Policies, *Proc. of IEEE Symposium on Security and Privacy, SP-97*, pp 103-112, IEEE Computer Society, 1997.

[Clark, 1978] K. L. Clark, Negation as Failure, In *Logic and Databases*, Ed. H. Gallaire and J. Minker, Plenum Press, 1978.

[D'Agostino *et al.*, 1999] M. D'Agostino, D. Gabbay and K. Broda. Tableau Methods for Substructural Logics, in *Handbook of Tableau methods*, Eds. M. D'Agostino et al, Kluwer Academic Publishers, pp. 397-467, 1999.

[Damianou *et al.*, 2001] N. Damianou, N. Dulay, E. Lupu and M. Sloman. The Ponder Policy Specification Language, *Proc. of Policy 2001: Workshop on Policies for Distributed Systems and Networks*, LNCS 1995, pp18-39, 2001.

[Fitting *et al.*, 2001] M. Fitting, L. Thalmann and A. Voronkov. Term Modal Logics, *Studio Logica* 69(1), pp 133-169, 2001.

[Gabbay, 1996] D. Gabbay. *Labelled Deductive Systems: Volume I*, Oxford Logic Guides, Vol. 33, Oxford Clarendon Press, 1996.

[Gabbay and Olivetti, 2002] D. Gabbay and N. Olivetti. Goal oriented deduction, in D. Gabbay and F. Guenthner, Eds, *Handbook of Philosophical Logic*, Vol. 9, Kluwer Academic Publisher, 2nd Ed. pp 199-285, 2002.

[Hitchens and Varadharajan, 2000] M. Hitchens and V. Varadharajan. On the Design and Specification of Role Based Access Control Policies, *IEEE Process-Software*, Vol. 147(4), pp 117-129, 2000.

[Hughes and Cresswell, 1996] G. Hughes and M. Cresswell. *A new introduction to Modal Logics*, Methuen, London, 1996.

[Kamoda *et al.*, 2005] H. Kamoda, A. Hayakawa, M. Yamaoka, S. Matsuda, K. Broda, and M. Sloman. Policy Conflict Analysis Using Tableaux for On Demand VPN Framework, *Proc. of the Sixth IEEE International Symposium on a World of Wireless, Mobile and Multimedia Networks*, IEEE Computer Society, pp 565–569, 2005.

[Lupu and Sloman, 1999] E. Lupu and M. Sloman. Conflicts in Policy-Vased Distributed Systems Management. *IEEE Trans. on Software Eng.*, 25(6), pp 852-869, 1999.

[Massacci, 1997] F. Massacci, Tableaux Methods for Access Control in Distributed Systems, *TABLEAUX-97*, LNAI 1227, pp 246 -260, 1997.

[McCune, 2003] W. McCune. *Otter 3.3 reference Manual and Guide*, Argonne National Laboratory, Argonne, Illinois, 2003.

[Russo, 1996] A. Russo. *Modal Logics as Labelled Deductive Systems*, PhD Thesis, Dept. of Computing, Imperial College London, 1996.

[Sandhu, 1992] R.S.Sandhu. The Typed Access Matrix Model. *Proc. of IEEE Symposium on Security and Privacy, SP-92*, IEEE Computer Society, pp 122-136, 1992.

[Sympson, 1993] A. Sympson. *The Proof Theory and Semantics of Intuitionistic Modal Logics*, PhD Thesis, University of Edinburgh, 1993.

[Son and Lobo, 2001] T. C. Son and J. Lobo. Reasoning about Policies using Logic Programs, *Proc. of American Assoc. for Artificial Intelligence*, pp 210-216, 2001.

Quantum Logic of Semantic Space: An Exploratory Investigation of Context Effects in Practical Reasoning

PETER D. BRUZA AND RICHARD J. COLE

1 Introduction

The field of non-monotonic reasoning (NMR) has successfully provided an impressive symbolic account of human practical reasoning over the last two and half decades. There remains, however, a disappointment - the dearth of large-scale operational NMR systems on the ground. During Lora Morgenstern's keynote address at the International Joint Conference on Artificial Intelligence (IJCAI-97) with the title "Inheritance Comes of Age: Applying Non-monotonic Techniques to Problems in Industry" she warned researchers that NMR needs to go beyond the examination of toy examples and to tackle serious, large scale problems, or run the risk of NMR becoming a backwater at artificial intelligence conferences. That is getting on ten years ago. Since then NMR has largely crystallized and is well understood from a stratum of theoretical perspectives. Morgenstern's warning still lingers, in our opinion. Theoretical insight without corresponding reasoning systems on the ground belies NMR's promise of embodying human practical reasoning.

We feel that the symbolic characterization of practical reasoning is only part of the picture. Gärdenfors ([Gärdenfors, 2000], p127) argues that one must go under the symbolic level of cognition. In this vein, he states, "...information about an object may be of two kinds: *propositional* and *conceptual*. When the new information is propositional, one learns new *facts* about the object, for example, that x is a penguin. When the new information is conceptual, one *categorizes* the object in a new way, for example, x is *seen as* a penguin instead of as just a bird". Gärdenfors' mention of "conceptual" refers to the conceptual level of a three level model of cognition [Gärdenfors, 2000]. How information is represented varies greatly across the different levels. The sub-conceptual level is the lowest level within which information is carried by a connectionist representation. Within the uppermost level information is represented symbolically. It is the intermediate, *conceptual level*, or *conceptual space*, which is of particular relevance to this account. Here properties and concepts have a geometric representation in a dimensional space. For example, the property of "redness" is represented as a convex region in a tri-dimensional space determined by the dimensions hue, chromaticity and brightness. The point left dangling for the moment is that representation at the conceptual level is rich in associations, both explicit and implicit. We speculate that the dynamics of associations are primordial stimuli for practical inferences drawn at the symbolic level of cognition. For example, it seems that associations and analogies generated within conceptual space play an important role in hypothesis generation. Gärdenfors ([Gärdenfors, 2000], p48) alludes to this point when he states, "most of scientific theorizing takes place within the conceptual level." His conjecture is aligned with Gabbay and Woods' insights regarding the cognitive economic basis of abduction [Gabbay and Woods, 2005]. Put crudely, it is cheaper to "guess" than to pursue a deductive agenda in relation to a problem at hand. Gabbay and Woods' notion of cognitive economy rests on compensation strategies employed by a practical agent to alleviate the consequences of key cognitive resources such as *information, time*, and *computational capacity*. Practical reasoning is reasoning performed by practical agents, and is therefore subject to cognitive economy. In this connection, we put forward the following conjecture: It may well be that because such associations are formed below the symbolic level of cognition, significant cognitive economy results. This is not only interesting from a cognitive point of view, but also opens the door to providing a computationally tractable practical reasoning systems, for example, operational abduction to drive scientific discovery in biomedical literature [Bruza *et al.*, 2004; Bruza *et al.*, 2005]

The appeal of Gärdenfors' cognitive model is that it allows inference to be considered not only at the symbolic level, but also at the conceptual (geometric) level. Inference at the symbolic level is typically a linear, deductive process. Within a conceptual space, inference takes on a decidedly associational character because associations are often based on similarity (e.g., semantic or analogical similarity), and notions of similarity are naturally expressed within a dimensional space. For example, Gärdenfors' states that a more natural interpretation of "defaults" is to view them as "relations between concepts". This is a view which flows into the account which follows: the strength of associations between concepts change dynamically under the influence of context. This, in turn, influences the defaults haboured within the symbolic level of cognition.

It is important to note the paucity of representation at the symbolic level and reflect how symbolic reasoning systems are hamstrung as a result. In this connection, Gärdenfors ([Gärdenfors, 2000], p127) states, " ..information about categorization can be quite naturally transfered to propositional information: categorizing x as an emu, for example, can be expressed by the proposition "x is an emu". This transformation into the propositional form, however, tends to suppress the internal *structure* of concepts. Once one formalizes categorizations of objects by *predicates* in a first-order language, there is a strong tendency to view the predicates as primitive, atomic notions and to forget that there are rich relations among concepts that disappear when put into standard logical formalism."

The above contrast between the conceptual and symbolic levels raises the question as to what are the implications for providing an account of practical reasoning. Gärdenfors states that concepts generate "expectations that result in different forms of *non-monotonic reasoning*", which are summarized as follows:

Change from a general category to a subordinate

When shifting from a basic category, e.g., "bird" to a subordinate category, e.g., "penguin", certain default associations are given up (e.g., "Tweety flies"), and new default properties may arise (e.g., "Tweety lives in Antarctica").

Context effects

The context of a concept triggers different associations that "lead to non-monotonic inferences". For example, *Reagan* has default associations "Reagan is a president", "Reagan is a republican" etc., but *Reagan* seen in the context of *Iran* triggers associations of "Reagan" with "arms scandal", etc.

The effect of contrast classes

Properties can be relative, for example, "a tall Chihuahua is not a tall dog" ([Gärdenfors, 2000], p119). In the first contrast class "tall" is applied to Chihuahuas and the second instance it is applied to dogs in general. Contrast classes generate conceptual subspaces, for example, skin colours form a subspace of the space generated by colours in general. Embedding into a subspace produces non-monotonic effects. For example, from the fact that x is a white wine and also an object, one cannot conclude that x is a white object (as it is yellow).

Concept combination

Combining concepts results in non-monotonic effects. For example, *metaphors* ([Gärdenfors, 2000], p130) Knowing that something is a lion usually leads to inferences of the form that it is alive, that it has fur, and so forth. In the combination, *stone lion*, however, the only aspect of the object that is lion-like is its shape. One cannot conclude that a stone lion has the other usual properties of a lion, and thus we see the non-monotonicity of the combined concept.

An example of the non-monotonic effects of concept combination not involving metaphor is the following: *A guppy is not a typical pet, nor is guppy is a typical fish, but a guppy is a typical pet fish.*

In short, concept combination leads to conceptual change. These corre-spond to revisions of the concept and parallel belief revisions modelled at the symbolic level, the latter having received thorough examination in the artificial intelligence literature.

The preceding brief characterization of the dynamics of concepts and associated non-monotonic effects is intended to leave the impression that a lot of what is happening in relation with practical reasoning is taking place within a conceptual (geometric) space. In addition, this impression may provide a foothold towards realizing genuine operational systems. This would require at least three issues to be addressed. The first is that a com-putational variant of the conceptual level of cognition is required. Secondly, the non-monotonic effects surrounding concepts would need to be formal-ized and implemented. Thirdly, the connection between these effects and NMR at the symbolic level needs to be specified. This account will attempt to address the first two of these questions. Computational approximations of conceptual space will be furnished by semantic space models which are emerging from the fields of cognition and computational linguistics. Seman-tic space models not only provide a cognitively motivated basis to underpin human practical reasoning, but from a mathematical perspective, they are real-valued Hilbert spaces. This introduces the tantalizing and highly spec-ulative prospect of formalizing aspects of human practical reasoning via quantum mechanics. In this account will focus on a treatment of how to formalize context effects as well as keeping an eye on operational issues.

2 Semantic space: computational approximations of conceptual space

To illustrate how the gap between cognitive knowledge representation and actual computational representations, the Hyperspace Analogue to Lan-guage (HAL) model is employed [Lund and Burgess, 1996; Burgess *et al.*, 1998]. HAL produces representations of words in a high dimensional space that seem to correlate with the equivalent human representations. For ex-ample, "...simulations using HAL accounted for a variety of semantic and as-sociative word priming effects that can be found in the literature...and shed light on the nature of the word relations found in human word-association

```
def calculate_hal(documents, n)
  HAL = 2DArray.new()
  for d in documents {
    for i in 1 .. d.len {
      for j in max(1,i-n) .. i-1 {
        HAL[d.word(i),d.word(j)] += n+1-(i-j)
}}}
  return HAL
end
```

Figure 1. Algorithm to compute the HAL matrix for a collection of documents. It is assumed that the documents have been pruned of stop words and punctuation.

norm data"[Lund and Burgess, 1996]. Given an n-word vocabulary, a HAL space is an $n \times n$ matrix constructed by moving a window of length l over the corpus by one word increment ignoring punctuation, sentence and paragraph boundaries. All words within the window are considered as co-occurring with the last word in the window with a strength inversely proportional to the distance between the words. Each row i in the matrix represents accumulated weighted associations of word i with respect to other words which preceded i in a context window. Conversely, column i represents accumulated weighted associations with words that appeared after i in a window. For example, consider the text "President Reagan ignorant of the arms scandal", with $l = 5$, the resulting HAL matrix H would be:

	arms	ig	of	pres	reag	scand	the
arms	0	3	4	1	2	0	5
ig	0	0	0	4	5	0	0
of	0	5	0	3	4	0	0
pres	0	0	0	0	0	0	0
reag	0	0	0	5	0	0	0
scand	5	2	3	0	1	0	4
the	0	4	5	2	3	0	0

Table 1. A simple semantic space computed by HAL

If word precedence information is considered unimportant the matrix $S = H + H^T$ denotes a symmetric matrix in which $S[i, j]$ reflects the strength of association of word i seen in the context of word j, irrespective of whether

president (5259), administration (2859), trade (1451), house (1426), budget (1023), congress (991), bill (889), tax (795), veto (786), white (779),
japan (767), senate (726), iran (687), billion (666), dlrs (615), japanese
(597), officials (554), arms (547), tariffs (536) ...

Table 2. Example representation of the word "Reagan"

word i appeared before or after word j in the context window. The column
vector S_j represents the strengths of association between j and other words
seen in the context of the sliding window: the higher the weight of a word,
the more it has lexically co-occurred with j in the same context(s). For
example, table 2 illustrates the vector representation for "Reagan" taken
from a matrix S computed from a corpus of 21578 Reuters news feeds taken
from the year 1988.

HAL is an exemplar of a growing ensemble of computational models
emerging from cognitive science, which are generally referred to as *semantic spaces* [Lund and Burgess, 1996; Burgess *et al.*, 1998; Lowe, 2000; Lowe,
2001; Landauer and Dumais, 1997; Landauer *et al.*, 1998; Patel *et al.*, 1997;
Schütze, 1998; Levy and Bullinaria, 1999; Sahlgren, 2002]. Even though
there is ongoing debate about specific details of the respective models, they
all feature a remarkable level of compatibility with a variety of human information processing tasks such as word association. Semantic spaces provide
a geometric, rather than propositional, representation of knowledge. They
can be considered to be approximations of conceptual space proposed by
Gärdenfors [Gärdenfors, 2000].

Within a conceptual space, knowledge has a dimensional structure. For
example, the property colour can be represented in terms of three dimensions: hue, chromaticity, and brightness. Gärdenfors argues that a property
is represented as a convex region in a geometric space. In terms of the
example, the property "red" is a convex region within the tri-dimensional
space made up of hue, chromaticity and brightness. The property "blue"
would occupy a different region of this space. A domain is a set of integral dimensions in the sense that a value in one dimension(s) determines or
affects the value in another dimension(s). For example, the three dimensions defining the colour space are integral since the brightness of a colour
will affect both its saturation (chromaticity) and hue. Gärdenfors extends
the notion of properties into concepts, which are based on domains. The
concept "apple" may have domains taste, shape, colour, etc. Context is
modelled as a weighting function on the domains, for example, when eating
an apple, the taste domain will be prominent, but when playing with it,

the shape domain will be heavily weighted (i.e., it's roundness). One of the goals of this article is to provide both a formal and operational account of this weighting function.

Observe the distinction between representations at the symbolic and conceptual levels. At the symbolic level "apple" can be represented as the atomic proposition $apple(x)$, however, within a conceptual space (conceptual level), it has a representation involving multiple inter-related dimensions and domains. Colloquially speaking, the token "apple" (symbolic level) is the tip of an iceberg with a rich underlying representation at the conceptual level. Gärdenfors points out that the symbolic and conceptual representations of information are not in conflict with each other, but are to be seen as "different perspectives on how information is described".

Barwise and Seligman [Barwise and Seligman, 1997] also propose a geometric foundation to their account of inferential information content via the use of real-valued state spaces. In a state space, the colour "red" would be represented as a point in a tri-dimensional real-valued space. For example, brightness can be modelled as a real-value between white (0) and black (1). Integral dimensions are modelled by so called observation functions defining how the value(s) in dimension(s) determine the value in another dimension. Observe that this is a similar proposal, albeit more primitive, to that of Gärdenfors as the representations correspond to points rather than regions in the space.

Semantic space models are an approximation of Barwise and Seligman state spaces whereby the dimensions of the space correspond to words. A word j is a point in the space. This point represents the "state" in the context of the associated text collection from which the semantic space was computed. If the collection changes, the state of the word may also change. Semantic space models, however, do not make provision for integral dimensions. An important intuition for the following is the state of a word in semantic space is tied very much with its "meaning", and this meaning is context-sensitive. Further, context-sensitivity will be realized by state changes of a word.

In short, HAL, and more generally semantic spaces, are a promising, pragmatic means for knowledge representation based on text. They are computational approximations, albeit rather primitive, of Gärdenfors' conceptual space. Moreover, due to their cognitive track record, semantic spaces would seem to be a fitting foundation for considering realizing computational variants of human reasoning. Finally, a semantic space is a real-valued Hilbert space which opens the door to connections with quantum mechanics.

3   Context effects in Semantic Space

Human beings are adept at producing context-sensitive inferences. Shifts in context effect the inferences made, even to a dramatic degree. The well known "Tweety" problem exemplifies this. A rough account of this example in terms of Gärdenfors' model of cognition is as follows: When given "Tweety is a bird", a prototypical concept of bird is activated within conceptual space and default inferences at the symbolic level such as "Tweety flies" arise as a strong association is primed between "Tweety" and "flies" at the conceptual level. The prototypical "Tweety" would be a point in the centre of a convex region in conceptual space representing birds (see [Gärdenfors and Williams, 2001],[Gärdenfors, 2000], p139). Learning "Tweety is a penguin", shifts the representation of "Tweety" towards the edge of the region representing birds as penguins differ significantly to the prototypical bird. As a consequence the association with "flies" diminishes radically and new associations arise, e.g., with "Antarctica". Even though Gärdenfors characterizes this type of NMR as being driven by a change from a general category to a subordinate, we feel that the associations can be more generally considered as a product of a shift of context — in this case the context is being *refined* from broader to the more specific. Initially the object "Tweety" is placed in the context of the concept "bird". The context is then refined to "penguin" leading to a change in the associations being primed, and consequently a change in the inferences being drawn.

The "Reagan" example exhibits similar characteristics. The vector representation given in table 2 is almost the prototypical representation of "Reagan" in the context of the underlying corpus. This is because HAL accumulates the association weights as it goes along. In fact, the weights in table 2 need only be divided by the frequency of the term "Reagan" in the underlying corpus to produce the vector representing prototypical "Reagan". Highly weighted associations in the representation have the character of being default like - "Reagan was a president", "Reagan had an administration" etc. Such default associations reflect the run of the mill presidential Reagan dealing with trade, budgets, congress etc.

The above two examples exhibit very common, or "garden" variety of practical inference. In this section, we attempt to provide a formal account in terms of quantum mechanics (QM).

3.1   Bridging semantic space and QM

A semantic space is a vector space and these can be expressed in the notation of quantum mechanics (The following draws heavily from [van Rijsbergen, 2004]).

A semantic space S is a $m \times n$ matrix where the columns $\{1, \ldots, n\}$

correspond to a vocabulary V of n words. A typical method for deriving the vocabulary is to tokenize the associated corpus and remove non information bearing words such as "the", "a", etc. The letters u, v, w will be used to identify individual words.

The interpretation of the rows $\{1 \ldots m\}$ depends of the type of semantic space in question. For example, table 2 illustrates that HAL produces a square matrix in which the rows are also interpreted as representations of terms. In contrast, a row in the semantic space models produced by Latent Semantic Analysis [Landauer et al., 1998] corresponds to a text item, for example, a whole document, a paragraph, or even a fixed window of text, as above. The value $S[t, w] = x$ denotes the salience x of word w in text t. Information-theoretic approaches are sometimes use to compute salience. Alternatively, the frequency of word w in context t can be used.

For reasons of a more straightforward embedding of semantic space into QM, we will focus on square, symmetric semantic spaces ($m = n$). A word w is represented as a column vector in S:

$$
(1) \qquad |w\rangle = \begin{pmatrix} w_1 \\ \vdots \\ w_n \end{pmatrix}
$$

The notation on the LHS is called a *ket*, and originates from quantum physicist Paul Dirac. Conversely, a row vector $v = (v_1, \ldots, v_n)$ is denoted by the *bra* $\langle v|$.

Multiplying a ket by a scalar α is as would be expected:

$$
(2) \qquad \alpha|w\rangle = \begin{pmatrix} \alpha w_1 \\ \vdots \\ \alpha w_n \end{pmatrix}
$$

Addition of vectors $|u\rangle + |v\rangle$ is also as one would expect. In Dirac notation, the scalar product of two n-dimensional real[1] valued vectors u and v produces a real number:

$$
(3) \qquad \langle u|v\rangle = \sum_{i=1}^{n} u_i v_i
$$

The product $|u\rangle\langle u|$ produces a symmetric matrix. Vectors u and v are *orthogonal* iff $\langle u|v\rangle = 0$. Scalar product allows the length of a vector to

---

[1]QM is founded on complex vector spaces. We restrict our attention to finite vector spaces of real numbers.

be defined: $\|u\| = \sqrt{\langle u|u \rangle}$. A vector $|u\rangle$ can be normalized to unit length ($\|u\| = 1$) by dividing each of its components by the vector's length: $\frac{1}{\|u\|}|u\rangle$.

A Hilbert space is a complete[2] inner product space. In the formalization to be presented in ensuing sections, a semantic space S is an n-dimensional real-valued Hilbert space using Euclidean scalar product as the inner product.

A Hilbert spaces allows the state of a quantum system to be represented. It is important to note that a Hilbert space is an *abstract* state space meaning QM does not prescribe the the state space of specific systems such as electrons. This is the responsibility of a physical theory such as quantum electrodynamics. Accordingly, it is the responsibility of semantic space theory to offer the specifics: In a nutshell, a ket $|w\rangle$ describes the state of a word w. It is akin to a particle in QM. The state of a word changes due to context effects in a process somewhat akin to quantum collapse. This in turn bears on practical inferences drawn due to context effects of word seen together with other words as described above.

In QM, the state can represent a superposition of potentialities. By way of illustration consider the state σ of a quantum bit, or *qubit* as:

$$(4) \qquad |\sigma\rangle = \alpha|0\rangle + \beta|1\rangle$$

where $\alpha^2 + \beta^2 = 1$. The vectors $|0\rangle$ and $|1\rangle$ represent the potentialities, or *eigenstates* of "off" and "on". Eigenstates are sometimes referred to as *pure states*. They can be pictured as defining orthogonal axes in a 2-D plane:

$$(5) \qquad \alpha|0\rangle = \begin{pmatrix} 0 \\ 1 \end{pmatrix}$$

and

$$(6) \qquad \alpha|1\rangle = \begin{pmatrix} 1 \\ 0 \end{pmatrix}$$

The state σ is a linear combination of eigenstates. Hard though it is to conceptualize, the linear combination allows the state of the qubit to be a mixture of the potentialities of being "off" and "on" at the same time.

In summary, a quantum state encodes the probabilities of its measurable properties, or eigenstates. The probability of observing the qubit being off (i.e., $|0\rangle$ is α^2). Similarly, β^2 is the probability of observing it being "on".

The above detour into QM raises questions in relation to semantic space. What does it mean that a word is a superposition - a "mixture of potentialities"? What are the eigenstates of a word?

---

[2]The notion of a "complete" vector space should not be confused with "completeness" in logic. The definition of a completeness in a vector space is rather technical, the details of which are not relevant to this account.

3.2   Mixed and eigenstates of a word

Consider the following traces of text from the Reuters-21578 collection:

- *President Reagan was ignorant about much of the Iran arms scandal*

- *Reagan says U.S to offer missile treaty*

- *Reagan seeks more aid for Central America*

- *Kemp urges Reagan to oppose stock tax.*

Each of these is a window which HAL will process accumulating weighted word associations in relation to the word "Reagan", say. In other words, included in the HAL vector for "Reagan" are associations dealing with the Iran-contra scandal, missile treaty negotiations with the Soviets, stock tax etc. The point is the HAL vector for "Reagan" represents a mixture of potentialities.

Let us now generalize the situation somewhat. Consider once again the HAL matrix H computed from the text "President Reagan ignorant of the arms scandal". As mentioned before, $S = H + H^T$ is a symmetric matrix. Technically, S is a *Hermitian linear operator*. Consider a set of text windows of length l which are centred around a word w. Assume there are m such windows. Associated with each such text window $j, 1 \leq j \leq m$, is a semantic space S_j. It is assumed that the semantic space is n-dimensional, whereby the n dimensions correspond to a fixed vocabulary V as above. The semantic space around word w, denoted by S_w, can be calculated by the sum:

$$(7) \qquad S_w = \sum_{j=1}^{k} S_j$$

In other words, the semantic space around the word "Reagan" is a summation of n-dimensional Hermitian linear operators computed from text windows centred around "Reagan".

In turn, the semantic space of the associated corpus, termed the *global semantic space*, denoted $\mathcal{S}$ can be considered as a mixture of the semantic space of the words in the associated vocabulary V:

$$(8) \qquad \mathcal{S} = \sum_{w \in V} S_w$$

An important intuition drawn from QM is that a word meaning equates with a state. The state may be *mixed*, that is the state embodies different potentialities corresponding to different "senses" of the word Reagan. Here we use the word "sense" with some poetic licence, but we do so deliberately because the "Reagan" example is similar to the case of an ambiguous word. Consider the word "suit" in isolation. Is it an item of clothing or a legal procedure? We put forward the intuition that, both "Reagan" or "suit" are states involving mixtures of senses, which parallels the superposition of eigenstates in the qubit given above. More formally, let $|r\rangle$ be the vector representing the state of "Reagan" in a semantic space S, and $\{e_1, \ldots, e_k\}$ represent the eigenstates of S, then

$$(9) \qquad |r\rangle = \alpha_1 |e_1\rangle + \ldots + \alpha_n |e_k\rangle$$

where $\alpha_1^2 + \ldots + \alpha_n^2 = 1$. The preceding intuition connecting word "meanings" in semantic space to QM seems to have independently arisen. (See [Widdows, 2004; Widdows and Peters, 2003; Aerts and Gabora, 2005; Aerts and Czachor, 2004]). The eigenstates define the different senses of the word in question. In QM terms, these correspond to the eigenstates of "Reagan".

In QM, the interpretation of the eigenstates are clearly grounded, e.g., the "on", "off" states of a qubit, or the momentum eigenstate of a particle. The eigenstates of a word are more subtle. This subtlety is not due to subjective interpretations of word meanings. By using a semantic space constructed from a corpus of text, the "meanings" ultimately are derived from this corpus. It could be argued that such meanings are inter-subjective due to the track record of semantic space model in replicating human word association norms. The subtlety derives more from the range of potential eigenstates. We shall see as we go along, however, the state of a word is nevertheless amenable to a formal treatment.

Computing eigenstates of a word by Singular Value Decomposition

Singular value decomposition, a theorem from linear algebra, allows a matrix to be decomposed. In the following, many of the technical details of SVD will be skipped, and only the essential elements will be presented. See [Golub and Loan, 1993] for a comprehensive account. As S_w is a symmetric matrix, SVD decomposes it as follows:

$$(10) \qquad S_w = U D U^T$$

where U is a $n \times n$ unitary matrix, the columns of which are the orthonomal basis of S_w. This means, the columns of U are pairwise orthogonal. To remain consistent with our notation, the i-th column vector of U will be denoted by $|e_i\rangle$. Matrix D is a positive $n \times n$ diagonal matrix, the values of which are the eigen-values of S_w. The value $D[i, i]$ will be denoted d_i.

The spectral decomposition of SVD allows S_w to be reconstructed, where $k \leq n$:

$$S_w = \sum_{i=1}^{k} |e_i\rangle d_i \langle e_i| \tag{11}$$

$$= \sum_{i=1}^{k} d_i |e_i\rangle \langle e_i| \tag{12}$$

$$= d_1 |e_1\rangle\langle e_1| + \ldots + d_k |e_k\rangle\langle e_k| \tag{13}$$

This shows once again how a word w is a mixture of eigenstates $|e_i\rangle$. The eigenvalues are related to the probabilities of the eigenstates occurring after a quantum measurement. In the semantic space interpretation, the eigenstates $|e_i\rangle$ of S_w correspond to the senses of word w.

The spectral decomposition of S_w parallels the decomposition of a density state ρ [van Rijsbergen, 2004]:

$$(14) \qquad \rho = a_1 P_1 + \ldots + a_k P_k$$

where P_i is a projection operator and $a_1 + \ldots + a_k = 1$. Projection operators, in real valued state spaces, are idempotent, symmetric matrices. (Note that $P_i = |e_i\rangle\langle e_i|$ is a projection operator)

A *density state*, or *density operator*, or *density matrix* expresses the distribution of quantum states in an ensemble of particles. The intuition is that we will run with is that words are particles, and that context acts like a "measurement", e.g., on a particle. A density matrix is a Hermitian operator with all its eigenvalues between 0 and 1.

Aerts and Czachor [Aerts and Czachor, 2004] have shown how to render a semantic space computed by Latent Semantic Analysis into a density matrix. For the purposes of this article, however, equation 14 allows the notion of density matrix to be directly equated with semantic space when the weights in the space are normalized. This ensures the eigenvalues lie between 0 and 1.

3.3 An analysis of the eigenstates of "Reagan"

Table 3 contains the first four eigenstates, or eigenvectors, of S_{reagan}. The first eigenstate contains all positive values and can be seen as a kind of average of the space. The subsequent eigenstate has a single positive component and a collection of negative components. Eigenstates having both positive and negative components represent two contrasting aspects of the space. Individual word vectors may project onto either the positive or the negative portion of each eigenstate. If they were to project onto both then their

1: reagan (0.62), president (0.48), administration (0.22), house (0.17), trade (0.15), congress (0.11), budget (0.11), bill (0.10), veto (0.10), white (0.09), tax (0.09), japan (0.08), senate (0.08), billion (0.08), iran (0.07),
2: reagan (0.74), ...   bill (-0.04), congress (-0.05), trade (-0.07), house (-0.08), administration (-0.23), president (-0.55)
3: japan (0.25), trade (0.25), japanese (0.24), tariffs (0.21), administration (0.13), united (0.11), sanctions (0.11), exports (0.11) ...   tax (-0.11), senate (-0.13), veto (-0.14), budget (-0.19), white (-0.31), house (-0.38)
4: billion (0.44), dlrs (0.37), dlr (0.21), budget (0.18), veto (0.18), deficit (0.17), bill (0.14), highway (0.13), mln (0.10), ...   conference (-0.07), house (-0.08), baker (-0.09), scandal (-0.12), white (-0.14), arms (-0.24), iran (-0.25)

Table 3. First four eigenstates of S_{reagan}. Components are listed in order. Only the largest components (by magnitude) are for each eigenstate are shown.

projection into the subspace would be small, and since SVD maximises the average length of the projected vectors, the negative and positive parts of the eigenstates tend to be in opposition. The third eigenstate, for example, seems to indicate that reports about Reagan concerning exports, tariffs and Japan are in opposition to reports about the senate, vetos and the budget.

It is important to recognize that the eigenstates do not neatly partition the meanings of Reagan into distinct clusters but rather span a subspace describing the topics in which Reagan is involved. The space can be though of as being lumpy but continuous rather than being due to a small number of discrete and largely disjoint topics.

In short, the eigenstates computed by SVD do not seem to correspond to the intuitively expected eigenstates of the word "Reagan".

3.4 Summary

As this section has proceeded a sort of duality has emerged. Initially, the state of a word w was presented as a ket $|w\rangle$ in a n-dimensional semantic (Hilbert) space S. The ket $|w\rangle$ may represent a mixture of the senses of the word w. Later, a connection was made between the semantic space S_w constructed around a word w and a density matrix, a notion from QM. From a technical point of view, this is not a problem. A density matrix can represent both eigenstates and mixed states: If $|w\rangle$ represents an eigenstate, then $|w\rangle\langle w|$ represents the corresponding density matrix. As a consequence, the state of a word, whether pure or mixed, can be represented as density matrix.

However, this technical resolution, does not seem to fully resolve the perceived duality. For example Widdows [Widdows and Peters, 2003] has proposed a quantum of word meanings drawn from semantic space. The meanings are represented as kets with no recourse to density matrices. Aerts and Gabora [Aerts and Gabora, 2005], on the other hand employ kets for the pure states of a concept, and a density matrix for a mixed state of a concept. It would seem that more research is needed to resolve this duality.

Quantum Collapse and Context Effects in Semantic Space

A quantum system is usually not in an eigenstate of whatever observable (e.g., momentum) is intended to be measured. However, if the observable is measured, the state of the system will immediately become an eigenstate of that observable. This process is known as *quantum collapse*.

A parallel can be drawn with respect to words in semantic space. When a word is seen in context, the superposition (mixed) state of the word collapses onto one of its senses. The senses of a word are the observables. For example, when "Reagan" is seen in the context of "Iran", the mixture of potentialities of "Reagan" collapses onto the eigenstate representing the sense dealing with the Iran-Contra scandal. After collapse, weights of associations to words such as "Contra", "illegal", "arms", "scandal", "sale" will be high, whereas before collapse the weights of such associations may have been weak. The highly weighted associations may, for example, "bubble up" and give rise to defaults at the symbolic level of cognition. This intuition gives rise to the tantalizing possibility that context effects within the conceptual level of cognition may be formalized by quantum collapse. This change in weighting can be dramatic and thus produce non-monotonic effects in relation to the weights of associations. For the moment, the observables can be conceived of as the different senses of a word. Seeing a word in the context of other word(s) acts like a "measurement". This measurement collapses the word meaning into one of its potential senses.

The description above of the interaction between context and collapse is essentially the same as that of Aerts and Gabora [Aerts and Gabora, 2005]. They state: "A state [of a concept] that is not an eigenstate of the context is called a potentiality state with respect to this context. The effect of a context is to change a potentiality state of this context, and this change will be referred to as collapse".

The context effects that are being considered here are similar to Aerts and Gabora. For simplicity, the case of a word v seen in the context of word u will be considered, the prototype of the running example: "Reagan" in the context of "Iran".

3.5 Formalizing context effects by quantum collapse

Aerts and Gabora [Aerts and Gabora, 2005] state a measurement in QM is described by a Hermitian operator M. For the context word u, there is an associated operator M_u. It is assumed that the state of word v is represented by the $|v\rangle$ drawn from some density matrix ρ. The parallel with QM is the following — a particle (word) v is drawn from a quantum system, the state of which is ρ. It is subjected to a "measurement", which is a product of word v being seen in the context of word u. The state of v collapses as a result. This intuition is formalized as follows, where $|v_u\rangle$ denotes the state of word v after collapse:

$$(15) \qquad |v_u\rangle = \frac{M_u|v\rangle}{\sqrt{\langle v|M_u|v\rangle}}$$

The value $\sqrt{\langle v|M_u|v\rangle}$ is a normalizing factor. One way of inspecting non-montonic effects in relation to associations is is simply to compare $|v_u\rangle$ with $|v\rangle$. Recall the ket representation of a word is a vector whose components correspond to words. The value x of the component i represents the strength of association of $|v_u\rangle$ with the i-th word of vocabulary V. Examples will follow shortly.

The above equation is a more liberal interpretation of that proposed by Aerts and Gabora's equation 11 in [Aerts and Gabora, 2005]. Their equation requires $|v\rangle$ to be a pure state. Our more liberal proposal arises from the following intuition: Collapse due to context may not necessarily result in a pure state. For example, Reagan's involvement with Iran included the U.S. embassy hostage crisis as well as the Iran-contra scandal. Intuitively this phenomena corresponds to a *partial* collapse of "Reagan", whereby the resultant state is less mixed than originally. In other words, the context "Iran" has not fully led to a collapse of the "meaning" of "Reagan" onto an unambiguous sense. This phenomenon shows the embedding of semantic space into QM is not always straightforward.

3.6 Example: "Reagan" in the context of "Iran"

To illustrate the effect of equation 15, $|v\rangle$ is primed to be the state of the word "Reagan" extracted from the density matrix ρ_{Reagan} computed from the Reuters collection. This ket represents the prototypical presidential Reagan, and is illustrated in table 2. The measurement operator M_u is primed as the Hermitian operator S_w with w equal to the word "Iran". One interpretation of the resulting quantum collapse is that it promotes words occurring in the vicinity of "Iran" based on how similar their meaning in the context of "Iran" is to the meaning of the prototypical Reagan.

---

iran (59), reagan (27), arms (21), iraq (12), gulf (12), scandal (10), war (7), oil (7), iranian (7), sales (7), house (6), president (5), attack (5), contra (5), united (5), states (4), white (4), missiles (4), profits (4), action (3), military (3), officials (3), senate (3), new (3), tehran (3), shipping (3), news (3), offensive (3), sale (3), rebels (2), speech (2), secret (2), warned (2), iraqi (2), policy (2), fighting (2), commission (2), response (2), hussein (2), diversion (2), major (2), official (2), tower (2), ship (2), denied (2), foreign (2), deal (2), affair (2), administration (2), saddam (2), . . .

Table 4. "Reagan" in the context of "Iran".

Compare the above weighted associations with those of table 2. Observe how the above no longer represent the prototypical "Reagan", but where associations relevant to the Iran-contra scandal are apparent, e.g., "scandal", "arms", "sales", "contra". Therefore, the Iran-contra sense of Reagan is coming through. Also there are prominent associations to "Iraq" and "oil". These may be related to the sense of "Reagan" reflecting President Reagan's dealings with Iraq during the Iran-Iraq war. Therefore, table 4 seems to reflect two senses of "Reagan". In other words, the resultant state after collapse is mixed. The reason for this is that the context word "Iran" is also a mixture of senses.

Perhaps, for this reason, Aerts and Gabora [Aerts and Gabora, 2005] do not directly employ measurement operator M as a whole, but its spectral decomposition:

$$M = d_1|e_1\rangle\langle e_1| + \ldots + d_k|e_k\rangle\langle e_k|$$

where the projector $P_j = |e_j\rangle\langle e_j|$. They refer to projector P_j as a "piece of context". Take for example, $u = \{\text{Iran}\}$. This context is a mixture of senses involving oil trade, Iran-Iraq war, the US embassy siege etc. The intuition of the projector P_j is that it represents one of these senses, and this in turn is a "piece of context" which can be substituted in equation 15 instead of M_u.

Recourse to "pieces of context" does not satisfactorily remove an incongruence. Why is it that when presented with "Reagan" in the context of "Iran", most will readily assume the Iran-contra sense, which we argued earlier, is an eigenstate of "Reagan". This stands in contrast to the above mixed state of "Reagan" after collapse illustrated in Table 4. The progression is as follows. Initially the state of "Reagan" is mixed as reflected by the following ket $|r\rangle$ drawn from the Reagan density matrix ρ_r. Assume that the eigenstate of e_i corresponds to the Iran-contra sense of "Reagan":

$$|r\rangle = \alpha_1|e_1\rangle + \ldots + \alpha_i|e_i\rangle + \ldots + \alpha_k|e_k\rangle$$

After collapse due to context "Iran", the state of "Reagan" is still mixed but less mixed than before. The result computed above suggests two senses, denoted e_i and e_j:

$$|r\rangle = \beta_i|e_i\rangle + \beta_j|e_j\rangle$$

The eigenvalues β_i are related to probabilities. For the sake of argument, let us assume that $\beta_i > \beta_j$. We speculate that the reason that the eigenstate e_i is assumed by most, is because it is the more probable sense left after collapse. Bear in mind, these probabilities are furnished by the geometry of the space and not by a frequentist approach which dominates statistical language processing.

3.7   Another way to view collapse of word meanings

Let us assume that before any words are seen or uttered there is a global density state represented by ρ_S, where S signifies the global semantic space. This is akin to a quantum system with many particles, each particle corresponding to a word in the vocabulary V, which may number in the hundreds of thousands. Consider what happens when a word v is expressed in isolation of other words. This is transforming a situation without context into one where the context is simply given by the word v. We contend that this changes the density state from from ρ_S to ρ_v, which is a subspace of ρ_S. Generalizing from this intuition leads to the hypothesis that context can be represented as a projection of a density matrix ρ onto a subspace represented by another density matrix:

$$\rho_X = P_X\rho$$

P_X is a projection operator constructed from one or more context words represented by X. A word v collapses from $|v\rangle = \rho|e_i\rangle$ to $|v_X\rangle = \rho_X|e_i\rangle = P_X\rho|e_i\rangle = P_X|v\rangle$, where $|e_i\rangle$ selects the column from ρ that corresponds to word v. It is curious to note that the v column of ρ, namely $|v\rangle$, is invariant under the transform P_v, because the components of $|v\rangle$ are all drawn from contexts containing the word v, so restricting the context to those containing the word v, i.e. applying P_v, doesn't change $|v\rangle$.

The full technical details of P_X still need to be worked out in relation to semantic space, however the effect of one context word $X = \{u\}$ can nevertheless be illustrated as ρ_w can be constructed directly from the Reuters collection via equation 7.

Table 5 depicts the state of "Reagan" in ρ_{Iran} and table 6 depicts the state of "Iran" in ρ_{Reagan}. Both tables represent unnormalized kets with the strength of association to other words represented as values in brackets.

iran (827), president (538), arms (430), scandal (208), administration (118), sales (101), contra (97), speech (91), senate (83), house (80), profits (77), tower (77), contras (76), commission (73), deal (69), approved (60), conference (58), policy (58), aid (57), diversion (53), approval (52), fighting (51), poindexter (50), ronald (50), new (49), rating (48), decision (48), sale (47), funds (47), mistake (47), rebels (46), wrong (46), investigating (46), denied (45), knew (45), nation (44), news (42), secret (42), initiative (39), role (39), response (39), pct (38), defense (38), recollection (38), money (37), congress (37), televised (37), sell (35), adviser (35), gave (35), ...

Table 5. "Reagan" in the context of "Iran".

This ket shows an collapse of the prototypical "Reagan" onto a state where associations relevant to the Iran-contra sense are prominently weighted. The ket depicted in table 6, however, reflects "Iran" in the context of "Reagan". One can clearly discern by comparing both kets that context effects are not symmetric.

arms (1522), iraq (1494), gulf (1432), war (939), oil (864), reagan (827), scandal (639), missiles (620), iranian (594), president (540), attack (528), offensive (504), sales (463), new (424), shipping (399), united (396), military (395), states (379), house (370), iraqi (364), contra (355), action (327), silkworm (291), news (285), hormuz (280), launched (270), diplomats (268), warned (258), southern (248), sale (247), major (244), attacked (243), tehran (239), strait (239), officials (236), kuwait (233), fighting (232), profits (230), north (225), senate (216), forces (213), foreign (212), washington (203), shipments (197), soviet (197), strike (196), attacks (193), american (191), crude (188), mln (185), ...

Table 6. "Iran" in the context of "Reagan".

4   Summary and Outlook

This article began with speculation that important aspects of human practical reasoning are manifest within the conceptual level of cognition referred to as conceptual space. Within conceptual space, information is represented in a geometric space, and inference has a associational, rather than a deductive, linear character. Our investigation focused on providing a formal account of the non-monotonic effects on conceptual associations due to context. Our aim is to provide the foundations for operational practical reasoning systems. To this end, the conceptual space was approximated by a semantic space model which can be automatically derived from a corpus of text. Within semantic space, words, or concepts, are represented as vectors in a high dimensional space. Semantic space models have emerged from cognitive science and computational linguistics. They have an encouraging, and at times impressive, track record of cognitive compatibility with humans across a number of information processing tasks. Due to their cognitive credentials semantic space models would seem to be a fitting foundation for realizing computational variants of human practical reasoning. The particular focus was formalizing the non-monotonic dynamics of associations within semantic space due to context effects. Context is a notoriously slippery notion to pin down. Yet context effects seem to trigger many garden variety non-monotonic inferences.

It has recently been pointed out in a letter to the editor of a journal in physics and mathematics that semantic space models bear some interesting similarities with the framework of quantum mechanics (QM) [Aerts and Czachor, 2004]. We have explored the connection between the two in the light of human practical reasoning and our intention has been more to provoke thought than provide concrete answers. It was shown that there is a very close parallel between semantic space and the notion of a density operator in QM. In a nutshell, the non-monotonic dynamics of word associations due to context are formalized by means of the quantum collapse of the state of a word in semantic space onto a sense which is determined by context words. A product of the collapse is a change of state, or "meaning" of the word. As a consequence, word associations also change. QM is one of the few frameworks in which context is neatly integrated. Essentially, context is something akin to a quantum measurement which brings about collapse. We speculate these changes in word association are the primordial beginnings of non-monotonic inferences at the symbolic level of cognition.

The embedding of semantic space into QM is not perfect. A summary of the major problem areas is given as follows:

- In QM, eigenstates are orthogonal, whereas the senses of a word need not be.

- In QM, collapse results in an eigenstate, whereas the collapse of word meaning in semantic space may be partial.

Neither of these problems would seem to fatally undermine further research. Aerts , Broekaert and Gabaora [Aerts et al., 2005a] go so far to state "..generalizations of the mathematical formalisms of quantum mechanics are transferable to the modeling of the creative, contextual manner in which concepts are formed, evoked, and often merged together in cognition". The theory developed in this article is complemented by realistic illustrations in an operational setting. The non-monotonic effects witnessed in the illustrations allow for cautious optimism.

The title of this account includes the phrase "quantum logic". Where is the logic? The phrase "quantum logic" is promissory. It reflects our belief that quantum logic, or something akin to it, can be employed to provide an account of logics of "down below" meaning practical reasoning as it is transacted below the symbolic level of cognition. It is important to stress that the view of reasoning presented in this account does not rest on traditional conception of logic. Gabbay and Woods ([Gabbay and Woods, 2003], p63) speculate that a logic of "down below" could be "a logic of semantic processing without rules". We feel that collapse of word meanings in semantic space falls very much within the ambit of such speculation and actually reinforces it.

There are yet many stones that need be laid to provide an adequate bridge between semantic space and quantum logic. In this regard, Widdows [Widdows and Peters, 2003; Widdows, 2004; Widdows and Higgins, 2004] have provided an important contribution with his quantum logic of word meanings and initial explorations into the lattice structure of vector subspaces. Such lattices provide the meeting point for Gabbay and Engesser's pioneering investigation into the connection between non-monotonic logic and quantum logic [Engesser and Gabbay, 2002].

Finally, there is the bigger picture. QM is emerging out of physics and permeating into other areas, for example, information retrieval [van Rijsbergen, 2004], human language [Aerts et al., 2005b] and cognition [Aerts et al., 2005a]. This offers tantalizing possibilities and bizarre implications. (See Malin [Malin, 2003] for a wonderfully daring view of the philosophical implications of QM). In terms of semantic space, intriguing questions arise in relation to QM notions such as entanglement. For example, Aerts and Gabora [Aerts and Gabora, 2005] contend that the pet fish example mentioned in the introductions arises because the concepts "pet" and "fish" are entangled. If so, does entanglement manifest in semantic space and can it be exploited in an operational setting? Certainly we agree with Aerts and Czachor [Aerts and Czachor, 2004] that the emedding of semantic space models into QM is mostly unexplored. This article documents a tiny exploratory step.

Acknowledgments

The work reported in this paper has been funded in part by the Co-operative Research Centre for Enterprise Distributed Systems Technology (DSTC) through the Australian Federal Government's CRC Programme (Department of Education, Science, and Training).

In the year 2000, the first author attended a course at ESSLLI on abduction given by Dov Gabbay and John Woods. Since then he has been following their bold and inspiring trail into the "New Logic". The first author thanks Professor Shimon Malin and Dominic Widdows for their thought provoking discussion and deep insight.

BIBLIOGRAPHY

[Aerts and Czachor, 2004] D. Aerts and M. Czachor. Quantum Aspects of Semantic Analysis and Symbolic Artificial Intelligence. *Journal of Physics A-Mathematical and General*, 37:L123–L132, 2004. http://uk.arxiv.org/abs/quant-ph/0309022.

[Aerts and Gabora, 2005] D. Aerts and L. Gabora. A Theory of Concepts and Their Combinations II: A Hilbert Space Representation. *Kybernetes*, 34:176–205, 2005. http://uk.arxiv.org/abs/quant-ph/0402205.

[Aerts et al., 2005a] D. Aerts, J. Broekaert, and L. Gabora. A Case for Applying an Abstracted Quantum Formalism to Cognition. In M.H. Bickhard and R. Campbell, editors, *Mind in Interaction*. John Benjamins: Amsterdam, 2005.

[Aerts et al., 2005b] D. Aerts, M. Czachor, and B. D'Hooghe. Towards a quantum evolutionary scheme: violating Bell's inequalities in language . In N. Gontier, J. P. Van Bendegem, and D. Aerts, editors, *Evolutionary Epistemology, Language and Culture*. Springer, 2005.

[Barwise and Seligman, 1997] J. Barwise and J. Seligman. *Information flow: the logic of distributed systems*. Cambridge University Press, 1997.

[Bruza et al., 2004] P.D. Bruza, D.W. Song, and R.M. McArthur. Abduction in semantic space: Towards a logic of discovery. *Logic Journal of the IGPL*, 12(2):97–110, 2004.

[Bruza et al., 2005] P.D. Bruza, R.J. Cole, Z. Abdul Bari, and D. Song. Towards operational abduction from a cognitive perspective. In L. Magnani, editor, *Abduction and Creative Inferences in Science*, Studies in Logic and Practical Reasoning. Elsevier, 2005.

[Burgess et al., 1998] C. Burgess, K. Livesay, and K. Lund. Explorations in context space: words, sentences, discourse. *Discourse Processes*, 25(2&3):211–257, 1998.

[Engesser and Gabbay, 2002] K. Engesser and D.M. Gabbay. Quantum Logic, Hilbert Space, revision theory. *Artificial Intelligence*, 136(1):61–100, 2002.

[Gabbay and Woods, 2003] D. Gabbay and J. Woods. *Agenda Relevance: A Study in Formal Pragmatics*, volume 1 of *A Practical Logic of Cognitive Systems*. Elsevier, 2003.

[Gabbay and Woods, 2005] D. Gabbay and J. Woods. *The Reach of Abduction: Insight and Trial*, volume 2 of *A Practical Logic of Cognitive Systems*. Elsevier, 2005.

[Gärdenfors and Williams, 2001] P. Gärdenfors and M. Williams. Reasoning about Categories in Concepual Spaces. In *Proceedings of 17th International Joint Conference on Artificial Intelligence*, pages 385–392, 2001.

[Gärdenfors, 2000] P. Gärdenfors. *Conceptual Spaces: The Geometry of Thought*. MIT Press, 2000.

[Golub and Loan, 1993] G.H. Golub and C.F. van Loan. *Matrix Computations*. John Hopkins University Ptess, 1993.

[Landauer and Dumais, 1997] T.K. Landauer and S.T. Dumais. A solution to Plato's problem: The latent semantic analysis theory of acquisition, induction and representation of knowledge. *Psychological Review*, 104:211–240, 1997.

[Landauer et al., 1998] T.K. Landauer, P.W. Foltz, and D. Laham. An introduction to latent semantic analysis. *Discourse Processes*, 25(2&3):259–284, 1998.

[Levy and Bullinaria, 1999] J.P. Levy and J.A. Bullinaria. Learning lexical properties from word usage patterns: Which context words should be used? In R.F. French and J.P. Sounge, editors, *Connectionist Models of Learning, development and Evolution: Proceedings of the Sixth Neural Computation and psychology Workshop*, pages 273–282. Springer, 1999.

[Lowe, 2000] W. Lowe. What is the dimensionality of human semantic space? In *Proceedings of the 6th Neural Computation and Psychology workshop*, pages 303–311. Springer Verlag, 2000.

[Lowe, 2001] W. Lowe. Towards a theory of semantic space. In J. D. Moore and K. Stenning, editors, *Proceedings of the Twenty-Third Annual Conference of the Cognitive Science Society*, pages 576–581. Lawrence Erlbaum Associates, 2001.

[Lund and Burgess, 1996] K. Lund and C. Burgess. Producing high-dimensional semantic spaces from lexical co-occurrence. *Behavior Research Methods, Instruments & Computers*, 28(2):203–208, 1996.

[Malin, 2003] S. Malin. *Nature Loves to Hide: Quantum Physics and Reality, a Western Perspective*. Oxford University Press, 2003.

[Patel et al., 1997] M. Patel, J.A. Bullinaria, and J.P. Levy. Extracting semantic representations from large text corpora. In R.F. French and J.P. Sounge, editors, *Connectionist Models of Learning, Development and Evolution: Proceedings of the Fourth Neural Computation and Psychology Workshop*, pages 199–212. Springer, 1997.

[Sahlgren, 2002] M. Sahlgren. Towards a Flexible Model of Word Meaning. Paper presented at the AAAI Spring Symposium 2002, March 25-27, Stanford University, Palo Alto, California, USA, 2002.

[Schütze, 1998] H. Schütze. Automatic word sense discrimination. *Computational Linguistics*, 24(1):97–124, 1998.

[van Rijsbergen, 2004] C.J. van Rijsbergen. *The Geometry of Information Retrieval*. Cambridge University Press, 2004.

[Widdows and Higgins, 2004] D. Widdows and M. Higgins. Geometric ordering of Concepts, Logical Disjuntion, and Leaning by Induction. In *Compositional Connectionism in Cognitive Science*, 2004. AAAI Fall Symposium Series.

[Widdows and Peters, 2003] D. Widdows and S. Peters. Word Vectors and Quantum Logic: Experiments with negation and disjunction. In *Proceedings of Mathematics of Language 8*, pages 141–154, 2003.

[Widdows, 2004] D. Widdows. *Geometry and Meaning*. CSLI Publications, 2004.

Fibring Logics: Past, Present and Future

CARLOS CALEIRO, AMÍLCAR SERNADAS AND
CRISTINA SERNADAS

1 Introduction

Combination mechanisms are operations that take logics as arguments and
produce new logics as a result, including language, deductive calculi and
semantics. A thorough understanding of such mechanisms is, of course, interesting in itself and in purely theoretic grounds. Namely, one might be
tempted to look at predicate temporal logic as resulting from the combination of first-order logic and propositional temporal logic. But what is
more, combined logics can also have a deep practical significance, namely in
areas like knowledge representation in artificial intelligence, or the formal
specification and verification of algorithms and protocols within software
engineering and security. In complex application fields, the need for working with a combination of several different logics at the same time is the rule
rather than the exception. Take, for instance, a knowledge representation
problem. Depending on the universe of discourse, it may well be necessary
to encompass simultaneously temporal, spatial, deontic and probabilistic
aspects (e.g., for reasoning about mixed assertions like "with probability
greater than 0.99, sometime in the future smoking will be forbidden everywhere"). In general, one needs at least to be able to develop theories
with components in different logics or, even better, to work with theories
defined in the combination of those logics (where such mixed assertions
are allowed and meaningful). For all these reasons, the interest in combining logics has recently been growing, as reflected in the series [Baader and
Schulz, 1996; Gabbay and de Rijke, 2000; Kirchner and Ringeissen, 2000;
Armando, 2002], the special journal issues [de Rijke and Blackburn, 1996;
Gabbay and Pirri, 1997], and the dedicated workshop [Carnielli *et al.*,
2004a]. For an early overview of the issues raised by combining logics, both
from a practical and from a theoretical perspective, see also [Blackburn and
de Rijke, 1997].

Several mechanisms for combining logics have been introduced and studied. The most abstract mechanisms are the ones that work at the level of the language (no details about how formulas are constructed) and with abstract consequence systems both for derivation and semantics. A good example of this approach is synchronization of logics [Sernadas *et al.*, 1997a; Sernadas *et al.*, 1997b; Caleiro, 2000]. However, the most rewarding mechanisms require a more concrete description of the details of the logics in hand, in particular of their syntax, namely by setting-up with the notion of signature. Temporalization [Finger and Gabbay, 1992; Finger and Gabbay, 1996; Wolter and Zakharyaschev, 2000; Finger and Weiss, 2002] and parameterization [Caleiro *et al.*, 1999] already take some advantage of this added structure. Still, the full combination of signatures that one can get by putting together the symbols that one has in the signatures of all the logics being combined was only first considered in two mechanisms for combining modal logics, namely product [Marx, 1999; Gabbay and Shehtman, 1998; Gabbay and Shehtman, 2000; Gabbay and Shehtman, 2002] and, even more general, the fusion mechanism [Thomason, 1984; Kracht and Wolter, 1991; Kracht and Wolter, 1997; Wolter, 1998; Gabbay *et al.*, 2003]. In short, the fusion of two modal systems leads to a bimodal system including the two original modal operators and common propositional connectives. Several interesting properties of modal logic systems (like soundness, weak completeness, Craig interpolation and decidability) were shown to be preserved by fusion (see [Kracht and Wolter, 1991; Kracht, 1999]).

It is at this point that Dov M. Gabbay proposes fibring [Gabbay, 1996a; Gabbay, 1996b; Gabbay, 1999] as a generalization of such combination mechanisms. Indeed, fibring can be applied beyond the universe of modal systems and captures fusion as a special case. Syntactically, the language of the fibring is freely generated from the combined signature and includes formulas where the symbols of the logics being combined can be intertwined arbitrarily many times. In proof-theoretic terms, as recognized from the very beginning, fibring is relatively easy when the calculi are presented in the same way (homogeneous fibring): the inference rules of the fibring are the union of the inference rules of original calculi, provided that one uses schema variable (for making instantiations of schema variables in a schematic rule of one of the calculi with formulas from the other logic. Herein we shall concentrate just on Hilbert calculi but the fibring of semantic tableaux [Beckert and Gabbay, 1998], or general labelled deductive systems [Rasga *et al.*, 2002; Rasga, 2003] has also been carefully studied. Fibring of calculi presented in different ways (heterogeneous fibring) is a different matter. For instance, what should be the fibring of two calculi when one is presented in a Hilbert style and the other using sequents? An answer to this problem is given in

[Cruz-Filipe *et al.*, 2005] using the concept of abstract proof system and observing that all the conventional ways of presenting calculi are particular cases of abstract proof systems. From a semantic point of view, things are even more complex. The original operational idea of Gabbay was based on the assumption that both logics were endowed with a point-based semantics. In this way, a model of the fibring would also be point-based and each point would be closely related to a point in a model of each original logic. As a consequence the same model of the fibring would be related to several models in each of the original logics. As investigated in [Sernadas *et al.*, 1999], the models of the fibring would be obtained via a fixed-point construction corresponding to a universal construction (a colimit in a category with very complex morphisms). It is worthwhile to observe at this point that homogeneous fibring was only investigated in fusion of modal logics. In all other cases, some kind of common structure was defined to cope with logics with quite different semantics (of course, semantic entailment was preserved in the transformation). In [Zanardo *et al.*, 2001], it is shown that if the original point-based semantics is closed for unions then each model of the fibring will point to one model in each original logic. The interesting result is that every point-based semantics can be closed for unions without changing the entailment. With this result, a more abstract structure can be found, as we describe here, by abstracting away the point structure (like modal algebra can be seen as a more abstract structure than general Kripke frames).

Applying general mechanisms for combining logics will be significant only if general preservation results are available. Fibring is no exception. Indeed, fibring raises very nice and often very difficult preservation problems. For example, if it has been established that completeness is preserved by a combination mechanism • and it is known that logic system $\mathcal{L}$ corresponds to $\mathcal{L}' \bullet \mathcal{L}''$, then the completeness of $\mathcal{L}$ follows from the completeness of $\mathcal{L}'$ and $\mathcal{L}''$. This transference problem appears, of course, with respect to any relevant syntactic or semantic property of the logics at hand, possibly including, for instance the metatheorem of deduction, cut elimination, interpolation, decidability, complexity, or the finite model property. In some cases it is possible to identify sufficient conditions for the preservation of a certain property. These sufficient conditions typically involve specific properties of the original logics. No wonder that much effort has been dedicated to establishing preservation results, or finding preservation counterexamples, about different combination mechanisms.

In this paper, we will overview some of the main features and results of the theory of fibring. A more comprehensive survey about fibring can be found in [Caleiro *et al.*, 2005]. For the sake of simplicity we will adopt a simple but expressive basic universe of logic systems encompassing only propositional-based systems endowed with Hilbert calculi and ordered algebraic semantics. Nevertheless, this universe is rich enough to illustrate many interesting properties of fibring and to provide the basis for the combination of systems varying from intuitionistic to many-valued logics (and modal systems as special cases). Along the exposition we will also try to bridge the current status of the theory of fibring with the original ideas underlying Gabbay's initial formulation. Those interested in wider universes of logics (including first-order quantification, higher-order features, non-truth-functional semantics, etc.) where fibring can still be defined should look also at [Sernadas et al., 2000; Sernadas et al., 2002a; Coniglio et al., 2003; Caleiro et al., 2003a; Governatori et al., 2002].

As far as preservation results are concerned, we will concentrate our attention on soundness, completeness and interpolation, along with some necessary metatheoretical properties like the metatheorem of deduction. With respect to completeness, we will focus on finding sufficient conditions for the preservation of strong global completeness, in the lines of [Zanardo et al., 2001; Caleiro, 2000; Caleiro et al., 2001], where further completeness preservation results can also be found. We will also overview the basics of Craig interpolation and identify sufficient conditions for its preservation. The interested reader is referred to [Carnielli and Sernadas, 2004; Carnielli et al., 2004b] for further details. In order to make justice to the title, we will also briefly outline some of the current research being carried out, namely, in connection with the so-called *collapsing problem*. We will review the details of the problem, illustrate it, and explain in broad lines how it can be overcome [Sernadas et al., 2002b; Caleiro and Ramos, 2004; Caleiro and Ramos, 2005].

We proceed as follows. In Section 2 we will introduce our working universe of logic systems. The key ingredients of completeness and interpolation for these logic systems will be explored in Section 3. Section 4 will be devoted to defining and illustrating fibring in this context and to explain the connection between the fibred algebraic semantics being introduced and Gabbay's ideas for point-based fibred semantics using the fibring function. In Section 5 we will concentrate on preservation results for (strong global) completeness and some forms of interpolation. Section 6 will overview and illustrate the semantic collapsing problem and provide an outline of ongoing research aimed at its solution. We will conclude, in Section 7, with some final remarks, references to further work, and ideas for future research. Barring some examples and omitted proofs that can be found in the literature, the presentation will be self contained.

2   Logic systems

A logic system is composed by a signature, a Hilbert calculus and a semantic domain. Both the Hilbert calculus and the semantic domain generate consequence systems. Soundness and completeness have to do with the fact that it is desirable that both consequence systems are the same.

A *signature* C is an $\mathbb{N}$-indexed family of countable sets. The elements of each C_k are called *constructors* of arity k. The signature of classical logic is such that C_0 is a set of propositional constants, $C_1 = \{\sim\}$, $C_2 = \{\supset\}$ and $C_k = \emptyset$ for every $k > 2$. The intuitionistic signature is such that C_0 is a set of propositional constants, $C_1 = \{\neg\}$, $C_2 = \{\Rightarrow, \wedge, \vee\}$ and $C_k = \emptyset$ for every $k > 2$. The modal signature is such that C_0 is a set of propositional constants, $C_1 = \{\neg, \Box\}$, $C_2 = \{\Rightarrow\}$ and $C_k = \emptyset$ for every $k > 2$.

Let $T(C, \Xi)$ be the free algebra over C generated by Ξ. The *language* $L(C)$ is $T(C, \emptyset)$. We shall consider different signatures but we assume fixed once and for all a set Ξ of propositional variables. Fixed Ξ, the *schema language* $T(C, \Xi)$ is denoted by $sL(C)$. Given $\varphi \in sL(C)$ we will use var(φ) to denote the set of propositional variables occurring in φ. When var$(\varphi) = \{\xi_1, \dots, \xi_n\}$ and $\sigma : \Xi \to sL(C)$ is a substituition such that $\sigma(\xi_i) = \psi_i$ for $i = 1, \dots, n$, we may use $\varphi(\psi_1, \dots, \psi_n)$ to denote $\sigma(\varphi)$.

A (Hilbert) *calculus* is a set of rules for some signature C. A *rule* over C is a pair $r = \langle \Theta, \eta \rangle$ where $\Theta \cup \{\eta\} \subseteq sL(C)$. We shall work only with finitary rules, that is, we assume that the set Θ of premises is finite. We also distinguish between local rules and global rules.

We adopt an algebraic semantics. This means that the basic semantic structure is an algebra of truth values. An *ordered algebra* over C is a tuple $\mathbf{A} = \langle A, \leq, \top, \cdot_\mathbf{A} \rangle$ where $\langle A, \leq, \top \rangle$ is a topped partial order and $\langle A, \cdot_\mathbf{A} \rangle$ is an algebra over C. The relation $\leq$ allows the comparison between truth-values. We require the existence of a top to represent validity and also global semantic consequence.

A *logic system* is a tuple $\mathcal{L} = \langle C, \mathcal{A}, R_\ell, R_g \rangle$ where C is a signature, $\mathcal{A}$ is a class of ordered algebras over C (the models of the system) and both R_ℓ and R_g are sets of rules over C called local and global rules, respectively. We further assume that the set of local rules R_ℓ is included in the set R_g of global rules and that all the rules in $R_g \setminus R_\ell$ have a non-empty set of premises.

As an example, let us consider classical logic. The class of models includes every ordered algebra induced by a Boolean algebra. The local rules are the usual rules of a Hilbert calculus for the negation and implication fragment of classical logic. Finally, there are no extra global rules. Another example is the following intuitionistic system. The class of models includes every ordered algebra induced by a Heyting algebra (with $a \leq b$ iff

$a \wedge_{\mathbf{A}} b = a \sqcap b = a$). The local rules are the usual rules of a Hilbert calculus for intuitionistic propositional logic. Finally, there are no extra global rules. A detailed presentation of intuitionistic logic along these lines can be found in [Rybakov, 1997]. Consider also the example of the following modal system. The class of models includes every ordered algebra $\mathbf{B} = \langle \mathcal{B}, \subseteq, W, \cdot_{\mathbf{B}} \rangle$ induced by a general Kripke structure $\langle W, \mathcal{B}, \rho, V \rangle$ as follows:

- $\pi_{\mathbf{A}} = V(\pi)$ for each propositional constant π;

- $\neg_{\mathbf{A}}(b) = W \setminus b$;

- $\Rightarrow_{\mathbf{A}}(b_1, b_2) = (W \setminus b_1) \cup b_2$;

- $\Box_{\mathbf{A}}(b) = \{w \in W : \{w' : w\rho w'\} \subseteq b\}$.

The notion of general Kripke structure was proposed in [van Benthem, 1983] in order to obtain a completeness theorem for modal logic. A more direct approach would be to take as models the ordered algebras induced by modal algebras. The local rules include the classical propositional rules plus the normalization axiom

$$\langle \emptyset, (\Box(\xi_1 \Rightarrow \xi_2)) \Rightarrow ((\Box\xi_1) \Rightarrow (\Box\xi_2)) \rangle .$$

The unique extra global rule is the necessitation rule $\langle \{\xi_1\}, \Box\xi_1 \rangle$.

Many other interesting logics (even many-valued ones like Gödel's and Łukasiewicz's — see for instance [Gottwald, 2001]) are also logic systems in the sense given above.

Within the context of a logic system, the denotation $[\![\varphi]\!]_{\mathbf{A}}^{\alpha}$ of a schema formula φ on an ordered algebra $\mathbf{A}$ and for an assignment $\alpha : \Xi \to A$ is easily defined by induction on the structure of φ.

In any given logic system $\mathcal{L} = \langle C, \mathcal{A}, R_\ell, R_{\mathrm{g}} \rangle$ we are able to define the following four consequence operators:

- global entailment: $\Gamma \vDash_{\mathcal{L}}^{\mathrm{g}} \varphi$ iff, for every $\mathbf{A} \in \mathcal{A}$ and $\alpha : \Xi \to A$, if $\top \leq [\![\gamma]\!]_{\mathbf{A}}^{\alpha}$ for each $\gamma \in \Gamma$ then $\top \leq [\![\varphi]\!]_{\mathbf{A}}^{\alpha}$;

- local entailment: $\Gamma \vDash_{\mathcal{L}}^{\ell} \varphi$ iff, for every $\mathbf{A} \in \mathcal{A}$, $\alpha : \Xi \to A$ and $a \in A$, if $a \leq [\![\gamma]\!]_{\mathbf{A}}^{\alpha}$ for each $\gamma \in \Gamma$ then $a \leq [\![\varphi]\!]_{\mathbf{A}}^{\alpha}$;

- global derivation: $\Gamma \vdash_{\mathcal{L}}^{\mathrm{g}} \varphi$ iff φ can be derived from Γ using the rules in R_{g};

- local derivation: $\Gamma \vdash_{\mathcal{L}}^{\ell} \varphi$ iff φ can be derived from Γ and theorems (formulae globally derived from an empty set of assumptions) using only the rules in R_ℓ.

Observe that in the modal system described above we have

$$\{\varphi_1 \Rightarrow \varphi_2\} \vdash^{\mathrm{g}}_{\mathcal{L}} (\square\varphi_1) \Rightarrow (\square\varphi_2)$$

but

$$\{\varphi_1 \Rightarrow \varphi_2\} \nvdash^{\ell}_{\mathcal{L}} (\square\varphi_1) \Rightarrow (\square\varphi_2).$$

Semantically the same happens, that is:

$$\{\varphi_1 \Rightarrow \varphi_2\} \vDash^{\mathrm{g}}_{\mathcal{L}} (\square\varphi_1) \Rightarrow (\square\varphi_2)$$

but

$$\{\varphi_1 \Rightarrow \varphi_2\} \nvDash^{\ell}_{\mathcal{L}} (\square\varphi_1) \Rightarrow (\square\varphi_2).$$

The distinction between local and global reasoning appeared in the context of modal logic (local means carried out at a single world and global refers to reasoning about all worlds) but can be useful in other universes.

3 Metalogical properties

3.1 Soundness and completeness

Soundness and completeness are two recurrent themes in logic. In the present setting, a logic system $\mathcal{L}$ is said to be *strongly globally sound* when if $\Gamma \vdash^{\mathrm{g}}_{\mathcal{L}} \varphi$ then $\Gamma \vDash^{\mathrm{g}}_{\mathcal{L}} \varphi$. Of course, it is said to be *strongly globally complete* when if $\Gamma \vDash^{\mathrm{g}}_{\mathcal{L}} \varphi$ then $\Gamma \vdash^{\mathrm{g}}_{\mathcal{L}} \varphi$. When we only consider $\Gamma = \emptyset$ we get the corresponding weak notions. Mutatis mutandis, we define the local versions.

Proving that a certain calculus is sound for a given semantics is typically just a matter of showing that each of its rules is fulfilled by the interpretation structures. Completeness, on its turn, is well known to be a much harder problem. Still, it is possible to obtain some general completeness results. A logic system is said to be *full* when $\mathcal{A}$ is composed of all ordered algebras over C that fulfill the rules in both R_ℓ and R_{g}. Therefore, every full logic system is (weakly and strongly, locally and globally) sound. A logic system has *verum* if its language contains a theorem that denotes $\top$ in every model.

A logic system $\mathcal{L}$ is said to be *congruent* when for every Γ closed for global derivation, $c \in C_k$ and $\varphi_1, \ldots, \varphi_k, \psi_1, \ldots, \psi_k \in sL(C)$:

$$\frac{\begin{array}{c}\Gamma, \varphi_i \vdash^{\ell}_{\mathcal{L}} \psi_i \\ \Gamma, \psi_i \vdash^{\ell}_{\mathcal{L}} \varphi_i\end{array} \quad i = 1, \ldots, k}{\Gamma, c(\varphi_1, \ldots, \varphi_k) \vdash^{\ell}_{\mathcal{L}} c(\psi_1, \ldots, \psi_k)}$$

THEOREM 1. Every full and congruent logic system with verum is strongly globally complete.

The proof is carried out using a Lindenbaum-Tarski construction. A syntactic ordered algebra $\mathbf{A}_\Gamma$ can be built as follows from each Γ closed for $\vdash_{\mathcal{L}}^{g}$. First we define a congruence relation over $sL(C)$: $\varphi \cong_{\Gamma} \psi$ iff $\Gamma, \varphi \vdash_{\mathcal{L}}^{\ell} \psi$ and $\Gamma, \psi \vdash_{\mathcal{L}}^{\ell} \varphi$. Then, we choose A to be $sL(C)/\cong_{\Gamma}$. The partial order is defined as follows: $[\varphi]_{\Gamma} \leq [\psi]_{\Gamma}$ iff $\Gamma, \varphi \vdash_{\mathcal{L}}^{\ell} \psi$. The top $\top$ is the equivalence class of the verum. Finally, for each language constructor, $c_{\mathbf{A}_{\Gamma}}([\varphi_1]_{\Gamma}, \ldots, [\varphi_k]_{\Gamma}) = [c(\varphi_1, \ldots, \varphi_k)]_{\Gamma}$. Clearly, by construction, we infer that $[\![\varphi]\!]_{\mathbf{A}_{\Gamma}}^{\lambda\xi.[\xi]_{\Gamma}} = \top$ iff $\varphi \in \Gamma$ and that $\mathbf{A}_{\Gamma}$ fulfills the rules of the logic system.

Assume that $\Delta \not\vdash_{\mathcal{L}}^{g} \epsilon$. We have to show $\Delta \not\models_{\mathcal{L}}^{g} \epsilon$. It is sufficient to find an ordered algebra $\mathbf{A} \in \mathcal{A}$ such that $[\![\delta]\!]_{\mathbf{A}}^{\lambda\xi.[\xi]_{\Gamma}} = \top$ for each $\delta \in \Delta$ and $[\![\epsilon]\!]_{\mathbf{A}_{\Gamma}}^{\lambda\xi.[\xi]_{\Gamma}} \neq \top$. Consider $\Gamma = \Delta^{\vdash^{g}}$. Then, $\mathbf{A}_{\Gamma}$ globally satisfies each element of Δ (since $\Delta \subseteq \Gamma$) but $\mathbf{A}_{\Gamma}$ does not globally satisfy ϵ (since $\epsilon \notin \Gamma$). This concludes the proof of the completeness theorem.

Observe that any complete logic system can be made full without changing its entailments. And if verum is not present, it can be conservatively added to the language. Note also that through a mild strengthening of the requirements of the theorem we can ensure finitary strong local completeness (see for instance [Sernadas et al., 2002b]). A similar strong (local and global) completeness theorem is obtained in [Zanardo et al., 2001] without extra requirements for local reasoning but assuming a more complex semantics and using a Henkin construction. Other similar completeness results, namely with respect to generalized matrix semantics can be found in [Caleiro, 2000; Caleiro et al., 2001; Caleiro et al., 2003b]. If congruence fails, however, then there is nothing we can do within the scope of the basic theory of fibring outlined here. Still, in that case, we can resort to the theory of non-truth-functional fibring [Caleiro et al., 2003a] where general completeness preservation results also obtain, but using rather different techniques, involving algebraization.

3.2 Interpolation

In its various flavours, *interpolation* is a heritage of the classical results by W. Craig [Craig, 1957] for first-order logic from which several abstractions have emerged, either proof-theoretically (e.g. [Carbone, 1997]) or in (non-constructive) model-theoretical style (e.g. for modal and positive logics [Maksimova, 1997; Maksimova, 2002], but also for intuitionistic logic [Gabbay, 1977], or hybrid logic [Areces et al., 2001; Areces et al., 2003]). Note that general techniques for obtaining interpolation properties are not known. For instance, Craig interpolation fails unexpectedly for all the many-valued logics of Łukasiewicz and Gödel [Krzystek and Zachorowski, 1977; Baaz and Veith, 1999]. Developing constructive proofs of interpolation is still a harder problem.

Moreover, interpolation properties are known to be related with properties of model theory as exemplified by the correspondence between Craig interpolation and joint consistency properties for classical propositional logic. This correspondence is mediated in the classical case by finite algebraizability and by the familiar (global) metatheorem of deduction. In the general case of our deductive calculi, specially due to the peculiarities of local and global deduction, this correspondence opens difficult and challenging problems.

We recast here some forms of interpolation taking into account the distinction between local and global deduction, as well as some general results, following [Carnielli and Sernadas, 2004; Carnielli *et al.*, 2004b]. Whenever a notion applies both to local and global reasoning we will use d instead of ℓ or g. A logic system $\mathcal{L}$ has:

- the *d-Craig interpolation property* (d-CIP) whenever $\Gamma \vdash_{\mathcal{L}}^{d} \varphi$ and $\mathrm{var}(\Gamma) \cap \mathrm{var}(\varphi) \neq \emptyset$ implies that there exists $\Gamma' \subseteq T(C, \mathrm{var}(\Gamma) \cap \mathrm{var}(\varphi))$ such that $\Gamma \vdash_{\mathcal{L}}^{d} \Gamma'$ and $\Gamma' \vdash_{\mathcal{L}}^{d} \varphi$;

- the *d-extension interpolation property* (d-EIP) whenever $\Gamma, \Psi \vdash_{\mathcal{L}}^{d} \varphi$ implies that there exists $\Gamma' \subseteq T(C, \mathrm{var}(\Psi) \cup \mathrm{var}(\varphi))$ such that $\Gamma \vdash_{\mathcal{L}}^{d} \Gamma'$ and $\Gamma', \Psi \vdash_{\mathcal{L}}^{d} \varphi$;

- the *d-Maehara interpolation property* (d-MIP) whenever $\Gamma, \Psi \vdash_{\mathcal{L}}^{d} \varphi$ and $T(C, \mathrm{var}(\Gamma) \cap (\mathrm{var}(\Psi) \cup \mathrm{var}(\varphi))) \neq \emptyset$ implies that there exists $\Gamma' \subseteq T(C, \mathrm{var}(\Gamma) \cap (\mathrm{var}(\Psi) \cup \mathrm{var}(\varphi)))$ such that $\Gamma \vdash_{\mathcal{L}}^{d} \Gamma'$ and $\Gamma', \Psi \vdash_{\mathcal{L}}^{d} \varphi$.

As mentioned before, although Łukasiewicz logics do not have the CIP they do enjoy the EIP.

To deal with the peculiarities of local and global deduction we need the following notion. A logic system $\mathcal{L}$ is said to allow *careful-reasoning-by-cases* if whenever $\Gamma \vdash_{\mathcal{L}}^{g} \varphi$ then there exists Ψ such that $\Gamma \vdash_{\mathcal{L}}^{g} \Psi$, $\mathrm{var}(\Psi) \subseteq \mathrm{var}(\Gamma)$ and $\Psi \vdash_{\mathcal{L}}^{\ell} \varphi$. For instance, modal and first-order logics allow careful-reasoning-by-cases.

THEOREM 2. A logic system $\mathcal{L}$ allowing careful-reasoning-by-cases enjoys the g-CIP whenever it has the ℓ-CIP.

Following the spirit of [Font *et al.*, 2003], we say that a logic system $\mathcal{L}$ enjoys the *d-metatheorem of Modus Ponens* (d-MTMP) with respect to $\Delta \subseteq T(C, \{\xi_1, \xi_2\})$ if

$$\frac{\Gamma \vdash_{\mathcal{L}}^{d} \Delta(\varphi_1, \varphi_2)}{\Gamma, \varphi_1 \vdash_{\mathcal{L}}^{d} \varphi_2}.$$

Analogously, $\mathcal{L}$ enjoys the *d-metatheorem of deduction* (d-MTD) with respect to Δ if

$$\frac{\Gamma \vdash^g_{\mathcal{L}}, \varphi_1 \vdash^d_{\mathcal{L}} \varphi_2}{\Gamma \vdash^g_{\mathcal{L}} \vdash^d_{\mathcal{L}} \Delta(\varphi_1, \varphi_2)} \, .$$

In [Czelakowski and Pigozzi, 1999] it is shown that the g-MIP is equivalent to the conjunction of g-EIP and the g-CIP. The next result, together with the result in [Czelakowski and Pigozzi, 1999], shows that d-MTD and d-EIP are provable from each other.

THEOREM 3. A logic system enjoying d-MTD, d-MTMP and d-CIP has the d-MIP.

4 Fibring

Consider signatures C and C' such that $C'_k \subseteq C_k$ for each $k \in \mathbb{N}$. Given an ordered algebra $\mathbf{A}$ over C, we denote by $\mathbf{A}|_{C'}$ the reduct $\langle A, \leq, \top, \cdot_{\mathbf{A}}|_{C'}\rangle$ of $\mathbf{A}$ by the inclusion (where $\cdot_{\mathbf{A}}|_{C'}$ is the restriction of $\cdot_{\mathbf{A}}$ to C'). Clearly, $\mathbf{A}|_{C'}$ is an ordered algebra over C'.

Given logic systems $\mathcal{L}' = \langle C', \mathcal{A}', R_\ell', R_g' \rangle$ and $\mathcal{L}'' = \langle C'', \mathcal{A}'', R_\ell'', R_g'' \rangle$, their *fibring* $\mathcal{L}' \odot \mathcal{L}'' = \langle C, \mathcal{A}, R_\ell, R_g \rangle$ is defined as follows:

- $C_k = C'_k \cup C''_k$ for each $k \in \mathbb{N}$;

- $\mathcal{A}$ is the class containing every ordered algebra $\mathbf{A}$ over C such that $\mathbf{A}|_{C'} \in \mathcal{A}'$ and $\mathbf{A}|_{C''} \in \mathcal{A}''$;

- $R_\ell = R_\ell' \cup R_\ell''$; $R_g = R_g' \cup R_g''$.

This definition corresponds to the constrained version of fibring (as defined in [Sernadas *et al.*, 1999]) since any symbols common to both logic systems will be shared. Unconstrained fibring can be obtained by making sure that no symbols appear in the intersection of the two signatures. Fibring can be characterized using a universal construction in a suitable category of logic systems (as explored in [Sernadas *et al.*, 1999] where the categorical approach was important in fine tuning the semantics of fibring).

As a first example of fibring, consider the combination of two modal systems while sharing the propositional connectives. This constrained fibring is equivalent to the fusion of the two given modal systems. The result is a bimodal system. The combination of a modal system with a relevance system is similar from the point of view of fibring but beyond the scope of fusion. By sharing the propositional connectives we obtain a logic system with a modal box and a relevance implication. For details about relevance logic see for instance [Dunn, 1986]. Note that, even when no symbols are shared, fibring may impose unexpected interactions between the logical operations from the two given logics. For instance, consider the unconstrained fibring of classical propositional logic and intuitionistic propositional logic. Unexpectedly, in the resulting logic system the intuitionistic implication collapses into classical implication. In short, in the resulting logic system we have two copies of classical logic. This first example of collapsing was first identified in [Gabbay, 1996b; del Cerro and Herzig, 1996]. We will debate this so-called *collapsing problem* later on, in Section 6.

4.1   A bridge to the past

Conceptually, there is a substantial gap between the point based semantics of fibring, as suggested by the fusion of modal logics, and the algebraic semantics we propose here. Let us review this abstraction process. The relationship between the definition of fibring presented above and Gabbay's original idea in [Gabbay, 1996a] certainly deserves a careful analysis. Besides furnishing a historical account of the development of fibring as a methodology for combining logics, we hope that this digression will provide a better overall understanding of the problem of combining logics and the technical difficulties involved.

Since there are no differences with respect to the proof-theoretic dimension of fibring, we shall concentrate just on the semantic aspects. Hence, if $\mathcal{L} = \langle C, \mathcal{A}, R_\ell, R_g \rangle$ is a logic system then we shall dub $\mathcal{I} = \langle C, \mathcal{A} \rangle$ an interpretation system. To get into the level of abstraction considered in [Gabbay, 1996a], we need to restrict ourselves to interpretation systems based on Kripke-like structures (K-structures, for simplicity). A *K-structure* over C is a tuple $\mathbf{W} = \langle W, \cdot_{\mathbf{w}} \rangle$ where W is a non-empty set and $\langle 2^W, \cdot_{\mathbf{w}} \rangle$ is a C-algebra. A *K-interpretation system* is a pair $\mathcal{KI} = \langle C, \mathcal{W} \rangle$ where C is a signature and $\mathcal{W}$ is a class of K-structures over C.

In each K-structure, the set W of *worlds* induces the space of truth values 2^W, ordered by inclusion, whose top element is precisely W. In this way, we can recognize a K-interpretation structure $\mathbf{W}$ as a special way of presenting the ordered structure $\overline{\mathbf{W}} = \langle 2^W, \subseteq, W, \cdot_{\mathbf{w}} \rangle$. Of course, a K-interpretation system $\mathcal{KI} = \langle C, \mathcal{W} \rangle$ can also be seen as a special way of presenting the interpretation system $\overline{\mathcal{KI}} = \langle C, \overline{\mathcal{W}} \rangle$.

It is interesting to note that we can recover the usual Kripke-like notions of local and global reasoning for each $\mathcal{KI}$, if we adopt the corresponding general definitions for $\overline{\mathcal{KI}}$. Given $\mathbf{W} \in \mathcal{W}$, define the *local satisfaction* relation at $w \in W$ by $\mathbf{W}, w \Vdash^\ell \varphi$ if $w \in [\![\varphi]\!]_{\mathbf{W}}$. Analogously, given a global model $\mathbf{W} \in \mathcal{W}$ define the *global satisfaction* by $\mathbf{W} \Vdash^g \varphi$ if $\mathbf{W}, w \Vdash^\ell \varphi$ for every $w \in W$. Then, at $\overline{\mathcal{KI}}$ we have that:

- *globally*: $\Psi \vDash^g \varphi$ if and only if for every $\mathbf{W} \in \mathcal{W}$, if $\mathbf{W} \Vdash^g \psi$ for each $\psi \in \Psi$ then $\mathbf{W} \Vdash^g \varphi$;

- *locally*: $\Psi \models^\ell \varphi$ if and only if for every $\mathbf{W} \in \mathcal{W}$ and $w \in W$, if $\mathbf{W}, w \Vdash^\ell \psi$ for each $\psi \in \Psi$ then $\mathbf{W}, w \Vdash^\ell \varphi$.

Gabbay's original idea of fibred semantics [Gabbay, 1996a; Gabbay, 1996b; Gabbay, 1999] was based on the notion of *fibring function*, and assumed that both logics had a Kripke-like semantics. In this case, the fibring function F would provide, at any moment, a way to map models and worlds from one logic to the other, and back again. Suppose that φ' is a formula and c' a unary constructor of the first logic system, given by $\mathcal{KI}'$, and c'' a unary constructor of the second logic, given by $\mathcal{KI}''$. To evaluate $c'(c''(\varphi'))$ in the combined logic we should proceed as follows.

1. Take a model $\mathbf{W}' = \langle W', \cdot_{\mathbf{W}'} \rangle$ of the first logic.

2. Typically, the satisfaction of $c'(c''(\varphi'))$ at $\mathbf{W}'$ will depend on some condition involving the *unknown* satisfaction of $c''(\varphi')$ at $\mathbf{W}'$.

3. For each world $w' \in W'$, instead of $\mathbf{W}', w' \Vdash'_L c''(\varphi')$, apply the fibring function F to obtain $F(\mathbf{W}', w') = \langle \mathbf{W}'', w'' \rangle$ where $\mathbf{W}'' = \langle W'', \cdot_{\mathbf{W}''} \rangle$ is a model of the second logic and $w'' \in W''$, and use $\mathbf{W}'', w'' \Vdash''_L c''(\varphi')$.

4. Again, the satisfaction of $c''(\varphi')$ at $\mathbf{W}''$ will depend on some condition involving the *unknown* satisfaction of φ' at $\mathbf{W}''$.

5. For each world $u'' \in W''$, instead of $\mathbf{W}'', u'' \Vdash''_L \varphi'$, apply the fibring function F to obtain $F(\mathbf{W}'', u'') = \langle \mathbf{U}', u' \rangle$ where $\mathbf{U}' = \langle U', \cdot_{\mathbf{U}'} \rangle$ is a model of the first logic and $u' \in U'$, and use $\mathbf{U}', u' \Vdash'_L \varphi'$.

The idea behind this procedure is intuitive and appealing, but it is not obvious how to accommodate this operational view into a meaningful definition of fibred model. Things are even less clear if we further require fibring to be a universal construction between K-interpretation systems. In that case, how to characterize the resulting system $\mathcal{KI}' \odot \mathcal{KI}''$? And what is the relevant notion of K-interpretation systems morphism?

The first solution to these questions, proposed in [Sernadas *et al.*, 1999], considered that fibred models could be partitioned, simultaneously, into clouds of disjoint models from each of the logics. The category-theoretical approach was an essential ingredient in fine tuning the notion presented in [Sernadas *et al.*, 1999], although it involved a quite complicate notion of morphism. However, it can be much simplified if we just assume that the classes of K-structures of the logics being combined are already *closed for unions*. This simplification was first proposed in [Zanardo *et al.*, 2001], where it was also noted that closing a given K-interpretation system for unions simply does not change its entailment operators. When closing for unions, a K-interpretation structure $\mathbf{W}$ of the resulting system can be seen as being built from a cloud $\{\mathbf{W}_i : i \in I\}$ of pairwise disjoint structures of the original system. Rigorously, $\mathbf{W}$ is defined to be the unique K-structure with $W = \bigcup_{i \in I} W_i$, and $\nu_{\mathbf{W}}(c)(X_1, \ldots, X_k) \cap W_i = \nu_{\mathbf{W}_i}(c)(X_1 \cap W_i, \ldots, X_k \cap W_i)$ for each k-ary constructor in the signature and $X_1, \ldots, X_k \in 2^W$. To make the notion robust with respect to the particular "names" of the worlds in each structure, it is useful to work under the assumption that the K-interpretation systems being combined are *closed under isomorphisms*.

The *fibring of K-interpretation systems* $\mathcal{KI}' = \langle C', \mathcal{W}' \rangle$ and $\mathcal{KI}'' = \langle C'', \mathcal{W}'' \rangle$ *closed for unions* is the interpretation system $\mathcal{KI}' \odot \mathcal{KI}'' = \langle C' \cup C'', \mathcal{W} \rangle$ where $\mathcal{W}$ is the class of all K-interpretation structures $\mathbf{W}$ over $C' \cup C''$ that can be built from interpretations structures $\mathbf{W}' \in \mathcal{W}'$ and $\mathbf{W}'' \in \mathcal{W}''$ satisfying:

- $W = W' = W''$;

- if $c \in C'_k \cup C''_k$ and $X_1, \ldots, X_k \in 2^W$ then

$$c_{\mathbf{W}'}(X_1, \ldots, X_k) = c_{\mathbf{W}''}(X_1, \ldots, X_k),$$

by defining $\mathbf{W} = \langle W, \cdot_{\mathbf{W}} \rangle$ as follows:

- for each $c' \in C'_k$ and $X_1, \ldots, X_k \in 2^W$,

$$c'_{\mathbf{W}}(X_1, \ldots, X_k) = c'_{\mathbf{W}'}(X_1, \ldots, X_k);$$

- for each $c'' \in C''_k$ and $X_1, \ldots, X_k \in 2^W$,

$$c''_{\mathbf{W}}(X_1, \ldots, X_k) = c''_{\mathbf{W}''}(X_1, \ldots, X_k).$$

As before, the fibring is defined under the assumption that the common subsignature is shared. In fact, for every shared constructor $c \in C' \cap C''$, the definition above implies that the two clouds of models $\{\mathbf{W}'_i : i \in I\}$ and $\{\mathbf{W}''_j : j \in J\}$, corresponding to each pair of structures $\mathbf{W}'$ and $\mathbf{W}''$ being fibred, agree on their interpretation. Note that, according to Gabbay's operational description, we can recognize the fibring function F associated to the fibred model $\mathbf{W}$ as mapping each pair $\langle \mathbf{W}'_i, w \rangle$ such that $w \in W'_i$ to the pair $\langle \mathbf{W}''_j, w \rangle$ where j is the unique element of J such that $w \in W''_j$, and vice-versa.

To understand the definition of fibring of interpretation systems with algebraic semantics, what we need is an operation on interpretation systems that mimics the closure for unions of K-interpretation systems. Notably, given a set $\{\mathbf{W}_i : i \in I\}$ of K-structures, it is not difficult to conclude that $\bigcup_{i \in I} \mathbf{W}_i$ is isomorphic to $\prod_{i \in I} \overline{\mathbf{W}}_i$. Thus, if $\mathcal{KI}$ is closed for unions then it immediately follows that $\overline{\mathcal{KI}}$ is closed for products. Now, it is a small step to ckeck that $\overline{\mathcal{KI}' \odot \mathcal{KI}''}$ and $\overline{\mathcal{KI}'} \odot \overline{\mathcal{KI}''}$ coincide, which justifies our general definition of fibring in the wider setting of interpretation systems. For brevity we omit the (obvious) definition of product of interpretation structures. Still, note that closing a given interpretation system for products also does not change its entailment operators.

This Kripke-like semantic view is nevertheless restrictive. In general, there is no reason to suppose that interesting logics should be endowed with K-interpretation structures. Moreover, general completeness results even for modal logics are only possible if we consider general Kripke structures, or alternatively, modal algebras. Still, by now, it should be easy to bridge the intuitive gap to the broader algebraic setting initially proposed. To this end, the ideas in [Mateus et al., 2004] should be relevant.

5 Transference results

5.1 Soundness and completeness

We now turn our attention to transference results. We start by examining if soundness is preserved by fibring. Then we consider completeness. To this end we have to establish the preservation of other interesting properties, namely the metatheorem of deduction.

THEOREM 4. Soundness is preserved by fibring.

It is straightforward to prove that (strong and weak, global and local) soundness is unconditionally preserved by fibring in the basic universe of logic systems considered here. However, in larger universes things can be more complicated. For instance, when fibring logic systems with quantifiers and using rules with side provisos (like, provided that term θ is free for variable x in formula ξ), soundness is not always preserved [Sernadas et al., 2002a; Coniglio et al., 2003].

Also weak completeness is not always preserved by fibring, as shown in [Zanardo et al., 2001]. Herein we examine in detail if strong global completeness is preserved when fibring basic logic systems as defined above. Adapting the technique originally proposed in [Zanardo et al., 2001], we capitalize on the completeness theorem stated above about such logic systems. That is, when fibring two given logic systems that are full, congruent and with verum (and, therefore, strongly globally complete) we shall try to obtain the strong global completeness of the result by identifying the conditions under which fullness, congruence and verum are preserved by fibring.

LEMMA 5. Fullness is preserved by fibring.

LEMMA 6. The result of fibring has verum provided that at least one of the given logic systems has verum.

However, congruence is not always preserved by fibring. Consider the fibring of two logic systems $\mathcal{L}', \mathcal{L}''$ with the following signatures and rules:

$$C_0' = \{\pi_0, \pi_1, \pi_2\} \quad C_1' = \{c\} \quad C_k' = \emptyset \text{ for } k > 1$$

$$R_\ell{}' = \emptyset \quad R_{\mathrm{g}}{}' = \{\langle\{\xi\}, c(\xi)\rangle\}$$

$$C_0'' = \{\pi_0, \pi_1, \pi_2\} \quad C_k'' = \emptyset \text{ for } k > 0$$

$$R_\ell{}'' = R_{\mathrm{g}}{}'' = \{\langle\{\pi_0, \pi_1\}, \pi_2\rangle, \langle\{\pi_0, \pi_2\}, \pi_1\rangle\}$$

Clearly, both $\mathcal{L}'$ and $\mathcal{L}''$ are congruent, but their fibring $\mathcal{L} = \mathcal{L}' \odot \mathcal{L}''$ is not congruent. Indeed, consider $\Gamma = \{\pi_0\}^{\vdash^{\mathrm{g}}} = \{c^n(\pi_0) : n \geq 0\}$. So, from Γ, π_1 and π_2 are locally interderivable but, from Γ, $c(\pi_1)$ and $c(\pi_2)$ are not locally interderivable.

Fortunately, it is possible to establish a useful sufficient condition for the preservation of congruence by fibring. A logic system $\mathcal{L}$ is said to have *implication* if its signature contains a binary connective $\Rightarrow$ fulfilling the (local versions for $\Delta = \{\xi_1 \Rightarrow \xi_2\}$ of the) metatheorems of Modus Ponens and of deduction, that is

$$(\mathrm{MTMP}) \quad \frac{\Gamma \vdash_{\mathcal{L}}^{\ell} \delta_1 \Rightarrow \delta_2}{\Gamma, \delta_1 \vdash_{\mathcal{L}}^{\ell} \delta_2} \quad \text{and} \quad (\mathrm{MTD}) \quad \frac{\Gamma \vdash_{\mathcal{L}}^{\mathrm{g}}, \delta_1 \vdash_{\mathcal{L}}^{\ell} \delta_2}{\Gamma \vdash_{\mathcal{L}}^{\mathrm{g}} \vdash_{\mathcal{L}}^{\ell} \delta_1 \Rightarrow \delta_2}.$$

When fibring two logic systems with implication while sharing the implication symbol, it is straightforward to verify that the resulting logic system also has implication. Indeed:

THEOREM 7. The result of fibring has the MTMP for an implication provided that at least one of the given logic systems has the MTMP for that implication.

THEOREM 8. The result of fibring has the MTD for a shared implication provided that both given logic systems have the MTD for that implication.

The latter result is a direct corollary of the following fact:

LEMMA 9. The MTD for $\Rightarrow$ holds in a logic system iff:

- $\vdash^{\ell} (\xi \Rightarrow \xi)$;

- $\{\xi_1\}^{\vdash^{\mathrm{g}}} \vdash^{\ell} (\xi_2 \Rightarrow \xi_1)$; and

- $\{(\xi \Rightarrow \gamma_1), \ldots, (\xi \Rightarrow \gamma_k)\}^{\vdash^{\mathrm{g}}} \vdash^{\ell} (\xi \Rightarrow \gamma)$ for each local rule $\langle\{\gamma_1, \ldots, \gamma_k\}, \gamma\rangle$ and ξ that does not occur in the rule.

A logic system is said to have *equivalence* if it has implication and its signature contains a binary connective $\Leftrightarrow$ fulfilling the two metatheorems of biconditionality (relating implication with equivalence) and the metatheorem of substitution of equivalents.

THEOREM 10. A logic system with equivalence is congruent.

When fibring two logic systems with equivalence while sharing the implication symbol as well as the equivalence symbol we obtain a logic system with equivalence. Therefore:

THEOREM 11. The fibring while sharing implication and equivalence of full logic systems with equivalence and verum is strongly globally complete.

This preservation result is quite useful because many widely used logic systems do have equivalence in the sense above.

5.2 Interpolation

Preservation results for Craig interpolation in the context of fusion were obtained in [Kracht and Wolter, 1991]. We will overview some interpolation preservation results in the broader setting of fibring, again following [Carnielli and Sernadas, 2004; Carnielli *et al.*, 2004b].

Since the metatheorems of Modus Ponens and of deduction will play a central role, we first note that the preservation results already obtained for the case of implication extend smoothly to the general case.

THEOREM 12. The d-MTMP and d-MTD are preserved by fibring logic systems with respect to the same shared Δ.

Careful reasoning also transfers along fibring.

THEOREM 13. Careful-reasoning-by-cases is preserved by fibring.

A logic system $\mathcal{L}$ is said to have *conjunction* if its signature contains a binary connective $\wedge$ fulfilling $\{\varphi_1 \wedge \varphi_2\} \vdash^g_{\mathcal{L}} \varphi_1$, $\{\varphi_1 \wedge \varphi_2\} \vdash^g_{\mathcal{L}} \varphi_2$ and $\{\varphi_1, \varphi_2\} \vdash^g_{\mathcal{L}} \varphi_1 \wedge \varphi_2$.

THEOREM 14. The g-CIP is preserved by fibring two logic systems with conjunction, g-MTMP, g-MTMP and an axiom that can be instantiated with any finite number of variables.

For instance in modal logic, the axiom $(\xi_1 \Rightarrow (\xi_2 \Rightarrow \xi_1))$ can be instantiated with any finite number of variables: the instance $((\wedge^k_{j=1}\xi_j) \Rightarrow (\xi_2 \Rightarrow (\wedge^k_{j=1}\xi_j)))$ includes $\xi_1, \ldots, \xi_k$.

The preservation of ℓ-CIP requires more assumptions on the component logics, namely a refinement of the notion of careful-reasoning-by-cases. A logic system $\mathcal{L}$ allows *localized-careful-reasoning-by-cases* if whenever $\Gamma, \Psi \vdash^g_{\mathcal{L}} \varphi$ with Ψ finite and with a derivation where global rules are only applied to hypotheses in Ψ then there exists a finite $\Omega \in T(C, \mathrm{var}(\Psi))$ such that $\Omega \subseteq \Psi^{\vdash^g_{\mathcal{L}}}$ and $\Omega \vdash^\ell_{\mathcal{L}} \varphi$. Both modal and first-order logics have this property.

THEOREM 15. ℓ-CIP is preserved by the fibring logic systems with localized-careful-reasoning-by-cases, conjunction, ℓ-MTMP, ℓ-MTD and an axiom that can be instantiated with any finite number of variables.

THEOREM 16. d-MIP is preserved by fibring logic systems with localized-careful-reasoning-by-cases, conjunction, d-MTMP, d-MTD, and an axiom that can be instanced with any number of variables.

The importance of a general form of metatheorem of deduction is stressed, proving that Craig interpolation implies another form of interpolation proposed by S. Maehara [Maehara, 1960/1961], thus showing that the mediation of the metatheorem of deduction plays a central role.

6   The collapsing problem and beyond

Despite its many strong points, already illustrated in the previous sections, fibring suffers from an anomaly usually known as "the collapsing problem" [Gabbay, 1996b; del Cerro and Herzig, 1996]. Indeed, since the beginning, it could be noticed that fibring the semantics of classical with intuitionistic logic would collapse into just classical logic.

In this section we will review this concrete example and try to shed some light at the problem in the general case. The analysis will allow us to identify ways to avoid this collapses, and to introduce further related topics of the theory of fibring that need to be addressed in the near future.

Let $\mathcal{L}' = \langle C', \mathcal{A}', R_\ell{}', R_g{}' \rangle$ and $\mathcal{L}'' = \langle C'', \mathcal{A}'', R_\ell{}'', R_g{}'' \rangle$ be, respectively, the logic systems for classical and intuitionistic propositional logics as described in Section 2. In order to ensure that there are absolutely no shared constructors between the two systems we shall assume that their respective sets of propositional constants $\Pi' = C_0'$ and $\Pi'' = C_0''$ are disjoint, that is $\Pi' \cap \Pi'' = \emptyset$.

We are interested in investigating the fibred logic system $\mathcal{L}' \odot \mathcal{L}'' = \langle C, \mathcal{A}, R_\ell, R_g \rangle$. Note that the combined signature is such that $C_0 = \Pi' \cup \Pi''$, $C_1 = \{\sim, \neg\}$, $C_2 = \{\supset, \Rightarrow, \wedge, \vee\}$, and $C_k = \emptyset$ if $k > 2$. For a start, let us have a look at its class of interpretation structures $\mathcal{A}$. From the definition of fibring, as put forth in Section 4, it easily follows that an interpretation structure $\mathbf{A} = \langle A, \leq, \top, \cdot_{\mathbf{A}} \rangle$ over C is in $\mathcal{A}$ if and only if $\mathbf{A}' = \mathbf{A}|_{C'}$ is a Boolean algebra and $\mathbf{A}'' = \mathbf{A}|_{C''}$ is a Heyting algebra. Since the ordered structure $\langle A, \leq, \top \rangle$ is the same in both $\mathbf{A}'$ and $\mathbf{A}''$, it is obvious that the Heyting structure $\mathbf{A}''$ is also Boolean and thus $\neg_{\mathbf{A}} = \sim_{\mathbf{A}}$, $\Rightarrow_{\mathbf{A}} = \supset_{\mathbf{A}}$, and the interpretation of the intuitionistic connectives $\wedge$ and $\vee$ also coincides with the classical conjunction and disjunction operations that can be obtained from $\sim$ and $\supset$ by the usual abbreviations. Therefore, unexpectedly, $\mathcal{L}' \odot \mathcal{L}''$ does not enjoy classical and intuitionistic features simultaneously. Instead, it is just a classical system with two synonymous ways of writing negation and implication.

This was, however, just a semantic analysis. What happens at the deductive level? Certainly the same phenomenon was to be expected if only we could be sure that the completeness of $\mathcal{L}' \odot \mathcal{L}''$ would follow from the completeness of both $\mathcal{L}'$ and $\mathcal{L}''$. However, this is not obvious, specially because fullness (as introduced in Section 3 and used in Section 5) is not guaranteed in either case. If there would also be a collapse at the deductive level then one would have, in particular, $\Gamma \vdash_{\mathcal{L}}^{\ell} \varphi \supset \psi$ if and only if $\Gamma, \varphi \vdash_{\mathcal{L}}^{\ell} \psi$ if and only if $\Gamma \vdash_{\mathcal{L}}^{\ell} \varphi \Rightarrow \psi$. The same would happen globally, of course. But, the fact that both classical and intuitionistic implications per se enjoy the MTD does not imply that either MTD will hold in the fibring (note that this does not contradict Theorem 8 since the two implications are not shared). Indeed, showing that the MTD holds for any of the two implications in $\mathcal{L}$ does not seem possible.

Thus, we have an hint that perhaps the collapse does not happen at the deductive level. But how can we show this? If we focus just on the implications, one possibility would be to show that $\vdash_{\mathcal{L}}^{\ell} ((\varphi \Rightarrow \psi) \Rightarrow \psi) \Rightarrow \psi$ is not always the case. Note that we know that $\vdash_{\mathcal{L}}^{\ell} ((\varphi \supset \psi) \supset \psi) \supset \psi$ holds because the formula at hand is the well-known *Peirce axiom*. To show that $\vdash_{\mathcal{L}}^{\ell} ((\varphi \Rightarrow \psi) \Rightarrow \psi) \Rightarrow \psi$ may fail we could try, for instance, to use a semantic argument, but we have already seen that fibred semantics cannot help in this task. Still, it is perhaps a matter of looking beyond fibred models.

Let us consider an interpretation structure obtained by a natural extension of the usual Kripke semantics for intuitionistic logic. Given a partial-order Kripke structure $\langle W, \leq \rangle$, and letting B be the set of all filters (up-closed subsets) of W, any corresponding model $\langle W, \leq, V \rangle$ with $V : \Pi' \cup \Pi'' \to B$ induces an interpretation structure $\mathbf{B} = \langle B, \subseteq, \{W\}, \cdot_{\mathbf{B}}, \rangle$ with:

- $\pi'_{\mathbf{B}} = V(\pi')$ and $\pi''_{\mathbf{B}} = V(\pi'')$;

- $\sim_{\mathbf{B}} (b) = (W \setminus b)^c$;

- $\supset_{\mathbf{B}} (b_1, b_2) = ((W \setminus b_1) \cup b_2)^c$;

- $\neg_{\mathbf{B}}(b) = (W \setminus b)^i$;

- $\Rightarrow_{\mathbf{B}}(b_1, b_2) = ((W \setminus b_1) \cup b_2)^i$;

- $\wedge_{\mathbf{B}}(b_1, b_2) = b_1 \cap b_2$;

- $\vee_{\mathbf{B}}(b_1, b_2) = b_1 \cup b_2$,

where $X^c = \{w \in W : \text{there exists } x \in X \text{ such that } x \le w\}$ and $X^i = \{w \in W : \{w' : w \le w'\} \subseteq X\}$, given $X \subseteq W$.

In the particular case when $W = \{u, v\}$ and $u \le v$, only, we obtain the following 3-valued truth tables:

$\supset$	$\emptyset$	$\{v\}$	$\{u,v\}$
$\emptyset$	$\{u,v\}$	$\{u,v\}$	$\{u,v\}$
$\{v\}$	$\{u,v\}$	$\{u,v\}$	$\{u,v\}$
$\{u,v\}$	$\emptyset$	$\{v\}$	$\{u,v\}$

$\Rightarrow$	$\emptyset$	$\{v\}$	$\{u,v\}$
$\emptyset$	$\{u,v\}$	$\{u,v\}$	$\{u,v\}$
$\{v\}$	$\emptyset$	$\{u,v\}$	$\{u,v\}$
$\{u,v\}$	$\emptyset$	$\{v\}$	$\{u,v\}$

	$\sim$	$\neg$
$\emptyset$	$\{u,v\}$	$\{u,v\}$
$\{v\}$	$\{u,v\}$	$\emptyset$
$\{u,v\}$	$\emptyset$	$\emptyset$

$\wedge$	$\emptyset$	$\{v\}$	$\{u,v\}$
$\emptyset$	$\emptyset$	$\emptyset$	$\emptyset$
$\{v\}$	$\emptyset$	$\{v\}$	$\{v\}$
$\{u,v\}$	$\emptyset$	$\{v\}$	$\{u,v\}$

$\vee$	$\emptyset$	$\{v\}$	$\{u,v\}$
$\emptyset$	$\emptyset$	$\{v\}$	$\{u,v\}$
$\{v\}$	$\{v\}$	$\{v\}$	$\{u,v\}$
$\{u,v\}$	$\{u,v\}$	$\{u,v\}$	$\{u,v\}$

Despite the fact that this interpretation structure is a three-valued Heyting algebra and therefore not a Boolen algebra, it is a simple matter to check that all the axioms and rules of classical and intuitionistic logic are satisfied. Moreover, we can indeed confirm that $((\varphi \Rightarrow \psi) \Rightarrow \psi) \Rightarrow \psi$ does not hold in general. Take, for instance, $((\pi''_1 \Rightarrow \pi''_2) \Rightarrow \pi''_1) \Rightarrow \pi''_1$ with $\pi''_1, \pi''_2 \in \Pi''$ and set $V(\pi''_1) = \{v\}$ and $V(\pi''_2) = \emptyset$. It is straightforward to check that in the resulting structure $\mathbf{B}$, $[\![((\pi''_1 \Rightarrow \pi''_2) \Rightarrow \pi''_1) \Rightarrow \pi''_1]\!]_{\mathbf{B}} = \{v\} \ne \{u,v\}$. Similarly, we can show that the two negations do not collapse in this model. Indeed, $(\sim (\sim \varphi)) \supset \varphi$ is valid, but $(\neg(\neg \varphi)) \supset \varphi$ can be falsified. Namely, using the same structure as above, $[\![(\neg(\neg \pi''_1)) \supset \pi''_1]\!]_{\mathbf{B}} = \{v\}$.

This example makes clear that collapsing situations may be due to a limitation in the definition of fibred semantics. Of course, there will always be cases where collapses cannot be avoided, specially if there are shared constructors. However, unwanted collapsing situations can be dealt with if we just enlarge the definition of fibred model in order to encompass combined structures as the one above for classical and intuitionistic logic. What happens here is that whenever the valuation V is taken in such a way that $V(\Pi') \subseteq \{\emptyset, W\}$, it is possible to identify two structures $\mathbf{A}' \in \mathcal{A}'$ and $\mathbf{A}'' \in \mathcal{A}''$ and homomorphisms $h' : \mathbf{A}' \to \mathbf{B}'$ and $h'' : \mathbf{A}'' \to \mathbf{B}''$, of C'

and C''-algebras respectively, that preserve the ordered structure in a strict manner. It suffices to consider $\mathbf{A}'$ to be the 2-valued Boolean structure, $\mathbf{A}'' = \mathbf{B}''$, h' injecting top and bottom into W and $\emptyset$, and h'' the identity.

Such a possibility was first considered in [Sernadas *et al.*, 2002b], where *modulated fibring* was introduced and shown to avoid these collapses, by means of a very careful use of adjunctions between the ordered semantic structures. The work on *cryptofibring* [Caleiro and Ramos, 2004], now underway, proposes a structurally simpler alternative to solve the semantic collapse problem by adopting a generalization of fibred semantics using cryptomorphisms [Caleiro and Ramos, 2005]. Cryptomorphisms are precisely the strict homomorphisms that have appeared in the example above. It can be shown that the novel notion encompasses the original definition of fibred model, but admits many more models. Once again the categorial setting is essential to tune up the details of the construction.

The scope of these investigations is, nevertheless, somewhat wider. It directly addresses the question of *conservativeness*. As Gabbay puts it in [Gabbay, 1999], it would be desirable that $\mathcal{L}' \odot \mathcal{L}''$ could be shown to be the least logic system over the combined language that conservatively extends the original logic systems $\mathcal{L}'$ and $\mathcal{L}''$. It is clear that, in general, such a desideratum is unattainable. It may very well happen, namely due to shared constructors, that no conservative extension of the two logic systems exists. Still, we would like to guarantee that we can build it whenever it exists. The relationship between conservativeness and the collapsing problem should be clear: collapses are special cases of failure of conservativeness.

This line of work is not without other difficulties. Namely, if we consider combined structures that are strict homomorphic extensions of structures of the original logic, we may very well get into trouble as far as the associated proof-calculi are concerned. Namely, even if both $\mathcal{L}'$ and $\mathcal{L}''$ are sound, it may happen that we get some enlarged structures that violate the rules. To recover soundness it is essential to get rid of these "bad" structures. But then also conservativeness may be at stake.

7 Final remarks

In this guided tour, we defined fibring in a very simple (yet useful) context and established some prototypical transference results. In this respect note that, concerning conditions for the preservation of weak completeness, it is still an open problem if the ghost symbol technique (used in [Kracht and Wolter, 1991] for proving the preservation of weak completeness by fusion) can be generalized in order to be used for fibring. As already mentioned, fibring can and has been defined and analyzed in much more complex situations. Current research is directed at widening the universe where fibring can be defined and at establishing transference results for other interesting properties. It is clear that in order to encompass, for instance, substructural logics like linear logic, we need a significant revamp of the theory of fibring. A key idea is to replace formula entailment by sequent entailment, using sequents with arbitrary structure. In this direction, it seems worthwhile to look at a logic as a kind of generalized 2-category with formulae as objects, connectives as multimorphisms, and sequents as poly-2-cells which will also raise interesting new problems in the theory of multicategories and of polycategories.

Another topic worth pursuing is the preservation by fibring of algebraizability and related notions, in the sense of [Blok and Pigozzi, 1989; Font *et al.*, 2003], and keeping in mind the results in [Jánossy *et al.*, 1996]. Some preliminary results can be found in [Fernández and Coniglio, 2004]. This question has a deep relationship with the forthcoming development of cryptofibring [Caleiro and Ramos, 2004], and should also take into account the techniques used in [Caleiro *et al.*, 2003a].

Closer to the concerns of the target application area of software engineering, we should mention the effort to bring fibring to the realm of the general theory of logics as institutions [Goguen and Burstall, 1992], namely in connection to the combination of parchments [Mossakowski, 1996; Mossakowski *et al.*, 1997; Mossakowski *et al.*, 1998]. Work along these lines is reported in [Caleiro *et al.*, 2001; Caleiro *et al.*, 2003b] and, more recently, in [Caleiro and Ramos, 2005]. As for new transference results, a particularly interesting open problem is to extend to fibring the results about model checking for combined temporal logics in [Franceschet *et al.*, 2004]. Recent developments in the logic of security protocols, namely concerning the combination of "intruder theories" [Chevalier and Rusinowitch, 2005], suggest that fibring may also become relevant in this application area.

Despite all the work done so far, there still remain many unaddressed challenges to the discipline of combining logics beyond fibring, some of them raised by emerging application areas. The interest in probability logic has recently increased due to the growing importance of probability in cryptography, security, and quantum computation and information. Also in classical software and hardware systems probabilities play an important role, for instance in distributed co-ordination and routing and fault-tolerance problems (see [Rutten *et al.*, 2004]). For other motivations see also [Fox, 2003]. An essential issue in quantum logic is to accommodate the fourth postulate of quantum mechanics stating that when a physical quantity is measured using an observable on a system in a given state, the resulting outcomes are ruled by a probability space. For more details on a quantum logic encompassing the probability aspects of the fourth postulate see [Mateus and

Sernadas, 2004a; Mateus and Sernadas, 2004b], and [Mateus and Sernadas, 2005] for a complete axiomatization. The exogenous approach used to build this logic is quite interesting: the models of the quantum logic are superpositions of the models of the underlying logic and so quantum logic is a conservative extension of the underlying logic. Therefore, it seems tempting to define both probabilization and quantization as exogenous operations on an arbitrary base logic. For a detailed discussion of the the exogenous approach see [Mateus et al., 2005].

Acknowledgments

This work was partially supported by Fundação para a Ciência e a Tecnologia and EU FEDER through POCI, namely via the Center for Logic and Computation and the Project FEDER POCI/MAT/55796/2004 QuantLog. The authors wish to express their gratitude to all colleagues with whom they have had the privilege of working in the topic of combining logics, specially Alberto Zanardo and Walter Carnielli, as well as João Rasga and João Marcos for useful comments on an earlier version of this survey that was published in the Bulletin of CIM [Sernadas and Sernadas, 2003].

BIBLIOGRAPHY

[Areces et al., 2001] C. Areces, P. Blackburn, and M. Marx. Hybrid logics: characterization, interpolation and complexity. *Journal of Symbolic Logic*, 66(3):977–1010, 2001.

[Areces et al., 2003] C. Areces, P. Blackburn, and M. Marx. Repairing the interpolation theorem in quantified modal logic. *Annals of Pure and Applied Logic*, 124(1-3):287–299, 2003.

[Armando, 2002] A. Armando, editor. *Frontiers of Combining Systems*, volume 2309 of *Lecture Notes in Computer Science*, Berlin, 2002. Springer-Verlag. Lecture Notes in Artificial Intelligence.

[Baader and Schulz, 1996] F. Baader and K. U. Schulz, editors. *Frontiers of Combining Systems*, volume 3 of *Applied Logic Series*. Kluwer Academic Publishers, Dordrecht, 1996.

[Baaz and Veith, 1999] M. Baaz and H. Veith. Interpolation in fuzzy logic. *Archive for Mathematical Logic*, 38(7):461–489, 1999.

[Beckert and Gabbay, 1998] B. Beckert and D. M. Gabbay. Fibring semantic tableaux. In *Automated Reasoning with Analytic Tableaux and Related Methods*, volume 1397 of *Lecture Notes in Computer Science*, pages 77–92. Springer Verlag, 1998.

[Blackburn and de Rijke, 1997] P. Blackburn and M. de Rijke. Why combine logics? *Studia Logica*, 59(1):5–27, 1997.

[Blok and Pigozzi, 1989] W. Blok and D. Pigozzi. Algebraizable logics. *Memoirs of the American Mathematical Society*, 77(396), 1989.

[Caleiro and Ramos, 2004] C. Caleiro and J. Ramos. Cryptofibring. In Carnielli et al. [2004a], pages 87–92. Extended abstract.

[Caleiro and Ramos, 2005] C. Caleiro and J. Ramos. Cryptomorphisms at work. In J. Fiadeiro, P. Mosses, and F. Orejas, editors, *Recent Trends in Algebraic Development Techniques - Selected Papers*, volume 3423 of *Lecture Notes in Computer Science*, pages 45–60. Springer-Verlag, 2005.

[Caleiro et al., 1999] C. Caleiro, C. Sernadas, and A. Sernadas. Parameterisation of logics. In J. Fiadeiro, editor, *Recent Trends in Algebraic Development Techniques - Selected Papers*, volume 1589 of *Lecture Notes in Computer Science*, pages 48–62. Springer-Verlag, 1999.

[Caleiro et al., 2001] C. Caleiro, P. Mateus, J. Ramos, and A. Sernadas. Combining logics: Parchments revisited. In M. Cerioli and G. Reggio, editors, *Recent Trends in Algebraic Development Techniques - Selected Papers*, volume 2267 of *Lecture Notes in Computer Science*, pages 48–70. Springer-Verlag, 2001.

[Caleiro et al., 2003a] C. Caleiro, W. A. Carnielli, M. E. Coniglio, A. Sernadas, and C. Sernadas. Fibring non-truth-functional logics: Completeness preservation. *Journal of Logic, Language and Information*, 12(2):183–211, 2003.

[Caleiro et al., 2003b] C. Caleiro, P. Gouveia, and J. Ramos. Completeness results for fibred parchments: Beyond the propositional base. In M. Wirsing, D. Pattinson, and R. Hennicker, editors, *Recent Trends in Algebraic Development Techniques - Selected Papers*, volume 2755 of *Lecture Notes in Computer Science*, pages 185–200. Springer-Verlag, 2003.

[Caleiro et al., 2005] C. Caleiro, W. A. Carnielli, J. Rasga, and C. Sernadas. Fibring of logics as a universal construction. In D. M. Gabbay and F. Guenthner, editors, *Combination of logics*, volume 13 of *Handbook of Philosophical Logic*. Kluwer, 2005.

[Caleiro, 2000] C. Caleiro. *Combining Logics*. PhD thesis, IST, Universidade Técnica de Lisboa, 2000. Supervised by A. Sernadas.

[Carbone, 1997] A. Carbone. Interpolants, cut elimination and flow graphs for the propositional calculus. *Annals of Pure and Applied Logic*, 83(3):249–299, 1997.

[Carnielli and Sernadas, 2004] W. Carnielli and C. Sernadas. Preservation of interpolation features by fibring. Preprint, CLC, Department of Mathematics, Instituto Superior Técnico, 1049-001 Lisboa, Portugal, 2004. Submitted for publication.

[Carnielli et al., 2004a] W. A. Carnielli, F. M. Dionísio, and P. Mateus, editors. *Proceedings of CombLog'04, Workshop on Combination of Logics: Theory and Applications*, 1049-001 Lisboa, Portugal, 2004. Departamento de Matemática, Instituto Superior Técnico.

[Carnielli et al., 2004b] W. A. Carnielli, C. Sernadas, and A. Zanardo. Preservation of interpolation by fibring. In Carnielli et al. [2004a], pages 151–157. Extended abstract.

[Chevalier and Rusinowitch, 2005] Y. Chevalier and M. Rusinowitch. Combining intruder theories. In L. Caires, G. Italiano, L. Monteiro, C. Palamidessi, and M. Yung, editors, *Automata, Languages and Programming ICALP 2005*, volume 3580 of *Lecture Notes in Computer Science*, pages 639–651. Springer-Verlag, 2005.

[Coniglio et al., 2003] M. E. Coniglio, A. Sernadas, and C. Sernadas. Fibring logics with topos semantics. *Journal of Logic and Computation*, 13(4):595–624, 2003.

[Craig, 1957] W. Craig. Linear reasoning. A new form of the Herbrand-Gentzen theorem. *Journal of Symbolic Logic*, 22:250–268, 1957.

[Cruz-Filipe et al., 2005] L. Cruz-Filipe, A. Sernadas, and C. Sernadas. Heterogeneous fibring of deductive systems via abstract proof systems. Preprint, CLC, Department of Mathematics, Instituto Superior Técnico, 1049-001 Lisboa, Portugal, 2005. Submitted for publication.

[Czelakowski and Pigozzi, 1999] J. Czelakowski and D. Pigozzi. Amalgamation and interpolation in abstract algebraic logic. In *Models, Algebras, and Proofs*, volume 203 of *Lecture Notes in Pure and Applied Mathematics*, pages 187–265. Dekker, 1999.

[de Rijke and Blackburn, 1996] M. de Rijke and P. Blackburn, editors. *Special issue on combining logics*, volume 37(2) of *Notre Dame Journal of Formal Logic*. University of Notre Dame, 1996.

[del Cerro and Herzig, 1996] L. Fariñas del Cerro and A. Herzig. Combining classical and intuitionistic logic. In Baader and Schulz [1996], pages 93–102.

[Dunn, 1986] J. M. Dunn. Relevance logic and entailment. In D. M. Gabbay and F. Guenthner, editors, *Handbook of Philosophical Logic, Vol. III*, pages 117–224. D. Reidel Publishing Company, 1986.

[Fernández and Coniglio, 2004] V. L. Fernández and M. E. Coniglio. Fibring algebraizable consequence systems. In Carnielli et al. [2004a], pages 79–86. Extended abstract.

[Finger and Gabbay, 1992] M. Finger and D. M. Gabbay. Adding a temporal dimension
to a logic system. *Journal of Logic, Language and Information*, 1(3):203–233, 1992.

[Finger and Gabbay, 1996] M. Finger and D. M. Gabbay. Combining temporal logic
systems. *Notre Dame Journal of Formal Logic*, 37(2):204–232, 1996.

[Finger and Weiss, 2002] M. Finger and M. A. Weiss. The unrestricted combination of
temporal logic systems. *Logic Journal of the IGPL*, 10(2):165–189, 2002.

[Font et al., 2003] J. M. Font, R. Jansana, and D. Pigozzi. A survey of abstract algebraic
logic. *Studia Logica*, 74(1-2):13–97, 2003.

[Fox, 2003] J. Fox. Probability, logic and the cognitive foundations of rational belief.
Journal of Applied Logic, 1(3-4):197–224, 2003.

[Franceschet et al., 2004] M. Franceschet, A. Montanari, and M. de Rijke. Model checking for combined logics with an application to mobile systems. *Automated Software
Engineering*, 11:289–321, 2004.

[Gabbay and de Rijke, 2000] D. M. Gabbay and M. de Rijke, editors. *Frontiers of combining systems. 2*, volume 7 of *Studies in Logic and Computation*. Research Studies
Press Ltd., Baldock, 2000. Papers from the 2nd International Workshop (FroCoS'98)
held at the University of Amsterdam, Amsterdam, October 2-4, 1998.

[Gabbay and Pirri, 1997] D. M. Gabbay and F. Pirri, editors. *Special issue on combining
logics*, volume 59(1,2) of *Studia Logica*. Springer, 1997.

[Gabbay and Shehtman, 1998] D. M. Gabbay and V. Shehtman. Products of modal
logics. I. *Logic Journal of the IGPL*, 6(1):73–146, 1998.

[Gabbay and Shehtman, 2000] D. M. Gabbay and V. Shehtman. Products of modal logics. II. Relativised quantifiers in classical logic. *Logic Journal of the IGPL*, 8(2):165–
210, 2000.

[Gabbay and Shehtman, 2002] D. M. Gabbay and V. Shehtman. Products of modal
logics. III. Products of modal and temporal logics. *Studia Logica*, 72(2):157–183,
2002.

[Gabbay et al., 2003] D. M. Gabbay, A. Kurucz, F. Wolter, and M. Zakharyaschev.
Many-dimensional modal logics: Theory and applications. Studies in Logic and the
Foundations of Mathematics, 148. Elsevier, 2003.

[Gabbay, 1977] D. M. Gabbay. Craig interpolation theorem for intuitionistic logic and
extensions. III. *Journal of Symbolic Logic*, 42(2):269–271, 1977.

[Gabbay, 1996a] D. M. Gabbay. Fibred semantics and the weaving of logics: Part 1.
Journal of Symbolic Logic, 61(4):1057–1120, 1996.

[Gabbay, 1996b] D. M. Gabbay. An overview of fibred semantics and the combination
of logics. In Baader and Schulz [1996], pages 1–55.

[Gabbay, 1999] D. M. Gabbay. *Fibring logics*. Oxford University Press, 1999.

[Goguen and Burstall, 1992] J. Goguen and R. Burstall. Institutions: abstract model
theory for specification and programming. *Journal of the ACM*, 39(1):95–146, 1992.

[Gottwald, 2001] S. Gottwald. *A Treatise on Many-Valued Logics*. Research Studies
Press, 2001.

[Governatori et al., 2002] G. Governatori, V. Padmanabhan, and A. Sattar. On fibring
semantics for BDI logics. In S. Flesca and G. Ianni, editors, *Logics in Computer
Science - JELIA*, volume 2424 of *Lecture Notes in Artificial Intelligence*, pages 198–
210. Springer Verlag, 2002.

[Jánossy et al., 1996] A. Jánossy, Á. Kurucz, and Á. Eiben. Combining algebraizable
logics. *Notre Dame Journal of Formal Logic*, 37(2):366–380, 1996.

[Kirchner and Ringeissen, 2000] H. Kirchner and C. Ringeissen, editors. *Frontiers of
Combining Systems*, volume 1794 of *Lecture Notes in Computer Science*, Berlin, 2000.
Springer-Verlag. Lecture Notes in Artificial Intelligence.

[Kracht and Wolter, 1991] M. Kracht and F. Wolter. Properties of independently axiomatizable bimodal logics. *Journal of Symbolic Logic*, 56(4):1469–1485, 1991.

[Kracht and Wolter, 1997] M. Kracht and F. Wolter. Simulation and transfer results in modal logic – a survey. *Studia Logica*, 59(2):149–177, 1997.

[Kracht, 1999] M. Kracht. *Tools and Techniques in Modal Logic*, volume 142 of *Studies in Logic and the Foundations of Mathematics*. North-Holland Publishing Co., Amsterdam, 1999.

[Krzystek and Zachorowski, 1977] P. S. Krzystek and S. Zachorowski. Łukasiewicz logics have not the interpolation property. *Rep. Math. Logic*, (9):39–40, 1977.

[Maehara, 1960/1961] S. Maehara. On the interpolation theorem of Craig. *Sûgaku*, 12:235–237, 1960/1961.

[Maksimova, 1997] L. Maksimova. Interpolation in superintuitionistic predicate logics with equality. *Algebra i Logika*, 36(5):543–561, 600, 1997.

[Maksimova, 2002] L. Maksimova. Complexity of interpolation and related problems in positive calculi. *Journal of Symbolic Logic*, 67(1):397–408, 2002.

[Marx, 1999] M. Marx. Complexity of products of modal logics. *Journal of Logic and Computation*, 9(2):197–214, 1999.

[Mateus and Sernadas, 2004a] P. Mateus and A. Sernadas. Exogenous quantum logic. In Carnielli et al. [2004a], pages 141–149. Extended abstract.

[Mateus and Sernadas, 2004b] P. Mateus and A. Sernadas. Reasoning about quantum systems. In J. Alferes and J. Leite, editors, *Logics in Artificial Intelligence, Ninth European Conference, JELIA '04*, volume 3229 of *Lecture Notes in Artificial Intelligence*, pages 239–251. Springer-Verlag, 2004.

[Mateus and Sernadas, 2005] P. Mateus and A. Sernadas. Weakly complete axiomatization of exogenous quantum propositional logic. Preprint, CLC, Department of Mathematics, Instituto Superior Técnico, 1049-001 Lisboa, Portugal, 2005. ArXiv math.LO/0503453. Submitted for publication.

[Mateus et al., 2004] P. Mateus, A. Sernadas, C. Sernadas, and L. Viganò. Modal sequent calculi labelled with truth values: Completeness, duality and analyticity. *Logic Journal of the IGPL*, 12(3):227–274, 2004.

[Mateus et al., 2005] P. Mateus, A. Sernadas, and C. Sernadas. Exogenous semantics approach to enriching logics. In G. Sica, editor, *Essays on the Foundations of Mathematics and Logic*, volume 1 of *Advanced Studies in Mathematics and Logic*, pages 165–194. Polimetrica, 2005.

[Mossakowski et al., 1997] T. Mossakowski, A. Tarlecki, and W. Pawłowski. Combining and representing logical systems. In *Category Theory and Computer Science 97*, volume 1290 of *LNCS*, pages 177–196. Springer, 1997.

[Mossakowski et al., 1998] T. Mossakowski, A. Tarlecki, and W. Pawłowski. Combining and representing logical systems using model-theoretic parchments. In *Recent Trends in Algebraic Development Techniques - Selected Papers*, volume 1376 of *LNCS*, pages 349–364. Springer, 1998.

[Mossakowski, 1996] T. Mossakowski. Using limits of parchments to systematically construct institutions of partial algebras. In *Recent Trends in Data Type Specification*, volume 1130 of *LNCS*, pages 379–393. Springer, 1996.

[Rasga et al., 2002] J. Rasga, A. Sernadas, C. Sernadas, and L. Viganò. Fibring labelled deduction systems. *Journal of Logic and Computation*, 12(3):443–473, 2002.

[Rasga, 2003] J. Rasga. *Fibring Labelled First-order Based Logics*. PhD thesis, IST, Universidade Técnica de Lisboa, 2003. Supervised by C. Sernadas.

[Rutten et al., 2004] J. Rutten, M. Kwiatkowska, G. Norman, and D. Parker. *Mathematical Techniques for Analyzing Concurrent and Probabilistic Systems*. CRM Monograph Series, vol. 23, AMS, 2004.

[Rybakov, 1997] V. Rybakov. *Admissibility of logical inference rules*, volume 136 of *Studies in Logic and the Foundations of Mathematics*. North-Holland Publishing Co., Amsterdam, 1997.

[Sernadas and Sernadas, 2003] A. Sernadas and C. Sernadas. Combining logic systems: Why, how, what for? *CIM Bulletin*, 15:9–14, December 2003.

[Sernadas *et al.*, 1997a] A. Sernadas, C. Sernadas, and C. Caleiro. Synchronization of logics. *Studia Logica*, 59(2):217–247, 1997.

[Sernadas *et al.*, 1997b] A. Sernadas, C. Sernadas, and C. Caleiro. Synchronization of logics with mixed rules: Completeness preservation. In M. Johnson, editor, *Algebraic Methodology and Software Technology*, volume 1349 of *Lecture Notes in Computer Science*, pages 465–478. Springer-Verlag, 1997.

[Sernadas *et al.*, 1999] A. Sernadas, C. Sernadas, and C. Caleiro. Fibring of logics as a categorial construction. *Journal of Logic and Computation*, 9(2):149–179, 1999.

[Sernadas *et al.*, 2000] A. Sernadas, C. Sernadas, C. Caleiro, and T. Mossakowski. Categorial fibring of logics with terms and binding operators. In Gabbay and de Rijke [2000], pages 295–316.

[Sernadas *et al.*, 2002a] A. Sernadas, C. Sernadas, and A. Zanardo. Fibring modal first-order logics: Completeness preservation. *Logic Journal of the IGPL*, 10(4):413–451, 2002.

[Sernadas *et al.*, 2002b] C. Sernadas, J. Rasga, and W. A. Carnielli. Modulated fibring and the collapsing problem. *Journal of Symbolic Logic*, 67(4):1541–1569, 2002.

[Thomason, 1984] R. H. Thomason. Combinations of tense and modality. In *Handbook of philosophical logic, Vol. II*, volume 165 of *Synthese Library*, pages 135–165. Reidel, Dordrecht, 1984.

[van Benthem, 1983] J. van Benthem. *The logic of time*, volume 156 of *Synthese Library*. D. Reidel Publishing Co., Dordrecht, 1983.

[Wolter and Zakharyaschev, 2000] F. Wolter and M. Zakharyaschev. Temporalizing description logics. In Gabbay and de Rijke [2000], pages 379–401.

[Wolter, 1998] F. Wolter. Fusions of modal logics revisited. In *Advances in modal logic, Vol. 1*, volume 87 of *CSLI Lecture Notes*, pages 361–379. CSLI Publ., 1998.

[Zanardo *et al.*, 2001] A. Zanardo, A. Sernadas, and C. Sernadas. Fibring: Completeness preservation. *Journal of Symbolic Logic*, 66(1):414–439, 2001.

Splitting Logics

WALTER CARNIELLI AND MARCELO E. CONIGLIO

1 Splitting logics, splicing logics and their use

One of fundamental questions in the philosophy of logic, "Why there are so many logics instead of just one?" (or even, instead of none), is naturally counterposed by another: If there are indeed many logics, are they excluding alternatives, or are they compatible? Is it possible to combine them into coherent systems, with the purpose of using them in applications and of taking profit of this composionality capacity to better understand logics? And if we can compose, why not decompose logics?

One of the first, and one of the most general, approaches for the question of combining logics is the concept of fibring introduced by D. Gabbay in [Gabbay, 1996]. Fibring is able to combine logics creating new and expressive systems, in the direction of what we call *splicing logics*.

The other direction is called *splitting logics*. Though, as we shall argue, there is no essential distinction between splicing and splitting, there are important differences with respect to the aims one may have in mind. Splitting as a process for investigating logics has been under–appreciated, and we intend to stress here some results and some views that we believe to be of interest for the sake of splitting in the trade of combining logics. [1]

Possible–translations semantics were proposed in [Carnielli, 1990], and were designed to help solve the problem of assigning semantic interpretations to non–classical logics. The idea behind possible–translations semantics is to build an interpretation for a given logic by taking into account a specific set of translations from its formulas into a class of simpler logics, with known or acceptable semantics. For a certain time it was even called "non–deterministic semantics" (as in [Carnielli and D'Ottaviano, 1997]), due to the apparent ambiguity of having several translations from the same domain.

---

[1]The process tags "splicing" and "splitting" logics were introduced in [Carnielli and Coniglio, 1999]. As a noun, "splitting" is also used in the literature in a completely different sense, viz., to designate a "logic that splits a class", as e.g. in W.J. Blok, "On the degree of incompleteness of modal logics" (abstract). *Bulletin of the Section of Logic of the Polish Academy of Sciences*, 7(4):167-175, December 1978.

Such semantics comprise a flexible and widely applicable tool for endowing logics with recursive and palatable semantic interpretation: detailed examples will be given in Section 3, but it is worth mentioning that several paraconsistent logics (as fragments of classical logic) which are not characterizable by finite matrices can be characterized by suitable combinations of many–valued logics. The reader is invited to check details for the case of N. da Costa's hierarchy $\{\mathcal{C}_n\}_{n\in\mathbb{N}}$ in [Carnielli, 2000] and [Marcos, 1999].

Examples of possible–translations semantics go in the direction of splitting, illustrating how a complex logic can be analyzed into less complex factors.

We also analyze here the *nondeterministic semantics* (see Section 4) and the *direct union of matrices* and *plain fibring* (see Section 5).

The traditional notion of matrix semantics, due to J. Łukasiewicz and E. Post, is also briefly reviewed in Section 3. Matrix semantics generalize algebraic semantics, as used in algebraic logic. They constitute a method for assigning semantic meaning for logics, as well as a method for defining logical systems.

The fact that possible–translations semantics are a widely applicable tool is witnessed by our results below, which show that both nondeterministic semantics and matrix semantics are particular cases of possible–translations semantics. As the notion of matrix semantics proves to be adequate for any structural deductive system, so are possible–translations semantics.

Another application, better suited for many–valued logics, is the concept of *society semantics* (cf. [Carnielli and Lima-Marques, 1999] and [Fernández and Coniglio, 2003]) that we do not treat here.

Possible–translations semantics (and their particular cases) work not only as general tools for assigning semantics for logics, but also as a tool for splitting logics as well. In the same manner they work for the direct unions of matrices and the plain fibring, which are not reducible to possible–translations semantics.

The particular cases are not to be discounted by any means: on the contrary, they are significant, specially when regarded from the splitting standpoint, in the measure that they provide operative methods for computing factors.

Although there is no fundamental distinction between splitting and splicing logics, as much as there is no fundamental distinction between factoring a number into primes or multiplying primes to compose a number, there is difference of attitudes and expectations, which is reflected in the distinction between what we have and what we wish to obtain.

If we represent the result of a process of factoring logics as $\mathcal{L} = \mathcal{L}_1 \odot \mathcal{L}_2$, there are, by and large, two ways to read this equation:

- If we had started from known logics $\mathcal{L}_1$ and $\mathcal{L}_2$ and $\mathcal{L}$ is our *incognita*, then we have a typical case of splicing $\mathcal{L}_1$ and $\mathcal{L}_2$ to obtain $\mathcal{L}$. Example 47 exemplifies this where the underlying operation is direct union.

- On the other hand, if $\mathcal{L}$ is known, we then have a typical case of splitting $\mathcal{L}$ into (presumably simpler) factors $\mathcal{L}_1$ and $\mathcal{L}_2$. The factors may be new logics (or new fragments of known logics), in which case we have encountered novelty, or they may be known logics, in which case we have found new relations among $\mathcal{L}$, $\mathcal{L}_1$ and $\mathcal{L}_2$ (and this is as much splitting as it is splicing). An instance of the former case is found in Example 45, and of the latter in Example 25. In this sense, Example 41 is also an instance of factoring classical propositional logic into its fragments.

The whole enterprise of splitting and splicing logics has several predecessors, depending upon the particular guise we may have in mind: some ingredients of the possible–translations semantics, even if in incipient form, will also be recognized in some variants of Gabbay's fibring. Still earlier, traces of S. Jaśkowski's discussive logics (cf. [Jaśkowski, 1949]) are recognizable in the general idea of society semantics. Plain fibring of matrices, on their side, have as antecedent both the original (modal) fibring of Gabbay and a certain product of matrices introduced by J. Łukasiewicz in [Łukasiewicz, 1953] to study his four–valued modalities: he used truth–values $\{\langle 0,0 \rangle, \langle 0,1 \rangle, \langle 1,0 \rangle, \langle 1,1 \rangle\}$, in such a way that his algebra of truth–functions coincides with $B \times B$ (where B is the two–elements Boolean algebra) and thus his modal-free tautologies coincide with classical tautologies.

2 Basic concepts about signatures and logics

This section briefly describes the basic definitions, notation and facts concerning propositional signatures and logics that will be used throughout the paper.

DEFINITION 1 (i) A *set of propositional variables* is a countable set $\mathcal{V}$, which will keep fixed. The elements of $\mathcal{V}$ will be denoted by $p_1, p_2, \ldots$.
(ii) A (propositional) *signature* is a family $C = \{C^k\}_{k \in \mathbb{N}}$, where each C^k is a set of *connectives of arity* k. It will be assumed that $C^k \cap C^n = \emptyset = C^k \cap \mathcal{V}$ for every $k \neq n$. The *domain of the signature* C is the set $|C| = \bigcup_{k \in \mathbb{N}} C^k$. Given two signatures C_1 and C_2, we say that C_1 *is included in* C_2 (denoted by $C_1 \subseteq C_2$) if, for every $k \in \mathbb{N}$, $C_1^k \subseteq C_2^k$. The signature $C_1 \uplus C_2$ (the disjoint union of C_1 and C_2) is defined as expected, that is: $(C_1 \uplus C_2)^k = C_1^k \uplus C_2^k$ for every $k \in \mathbb{N}$, where $A \uplus B$ denotes the usual set–theoretic disjoint union

of the sets A and B.

(*iii*) A (propositional) *language with signature* C, denoted by $L(C)$, is the algebra of words freely generated by C over $\mathcal{V}$ such that C^k is the set of k–ary operations of $L(C)$. Elements of $L(C)$ are called C–*formulas* (or simply *formulas*).

(*iii*) Given a signature C and $n \in \mathbb{N}$, $L(C)[n]$ is the set of formulas φ such that the set of propositional variables occurring in φ is exactly $\{p_1, \ldots, p_n\}$. ∎

Observe that $L(C)[0]$ is the set of formulas without variables. It is worth noting that there may be signatures $C \neq C'$ such that $L(C) = L(C')$. For the sake of simplicity, a signature will be frequently identified with its domain. We now describe the category of signatures.

DEFINITION 2 Let C and C' be signatures. A *signature morphism* f from C to C', denoted $C \xrightarrow{f} C'$, is a mapping $f : |C| \to L(C')$ such that, if $c \in C^n$ then $f(c) \in L(C')[n]$. ∎

Given a signature morphism $C \xrightarrow{f} C'$, a mapping $\widehat{f} : L(C) \to L(C')$ can be defined as expected:

1. $\widehat{f}(p) = p$ if $p \in \mathcal{V}$;

2. $\widehat{f}(c) = f(c)$ if $c \in C^0$;

3. $\widehat{f}(c(\varphi_1, \ldots, \varphi_n)) = f(c)(\widehat{f}(\varphi_1), \ldots, \widehat{f}(\varphi_n))$ if $c \in C^n$ and $\varphi_1, \ldots, \varphi_n \in L(C)$.

Clearly the extension $\widehat{f}$ of f is unique. Moreover, if f, f' are signature morphisms such that $\widehat{f} = \widehat{f'}$ then $f = f'$. Additionally, the propositional variables occurring in φ and in $\widehat{f}(\varphi)$ are the same.

DEFINITION 3 Let $C \xrightarrow{f} C'$ and $C' \xrightarrow{g} C''$ be signature morphisms. The *composition* $g \cdot f$ of f and g is defined to be the signature morphism $C \xrightarrow{g \cdot f} C''$ given by the mapping $\widehat{g} \circ f : |C| \to L(C'')$. ∎

DEFINITION 4 The category **Sig** of (propositional) languages is defined as follows:

- Its objects are propositional signatures (see Definition 1);

- Its morphisms are signature morphisms (see Definition 2);

- The composition of morphisms is as in Definition 3;

- For every signature C, the identity morphism $C \overset{id_C}{\rightarrow} C$ is defined by $id_C(c) = c$ (for $c \in C^0$) and $id_C(c) = c(p_1, \ldots, p_n)$ (for $c \in C^n$, $n \geq 1$). ∎

The next result was proved in [Bueno-Soler *et al.*, 2005].

THEOREM 5 **Sig** is a category with arbitrary (small) products.

In the category **Sig** all the objects are restricted to sequences of sets, constraining us into considering just small diagrams in the theorem above.

We stipulate below a concept of *propositional logic* which is broad enough to encompass all logics that are usually found, though this does not of course include *all* possible propositional logics.

DEFINITION 6 Let C be a signature. A *consequence relation over the signature* C is a relation $\vdash \subseteq \wp(L(C)) \times L(C)$ satisfying the following properties (as usual, $(\Gamma, \alpha) \in \vdash$ will be denoted by $\Gamma \vdash \alpha$):

- If $\varphi \in \Gamma$ then $\Gamma \vdash \varphi$ (**Reflexivity**).

- If $\Gamma \vdash \varphi$ and $\Sigma \vdash \psi$, for every $\psi \in \Gamma$, then $\Sigma \vdash \varphi$ (**Transitivity**). ∎

Observe that, because of Reflexivity and Transitivity, any consequence relation $\vdash$ automatically satisfies the following:

- If $\Gamma \vdash \varphi$ and $\Gamma \subseteq \Sigma$ then $\Sigma \vdash \varphi$ (**Monotonicity**).

DEFINITION 7 A *(propositional) logic* is defined to be a pair $\mathcal{L} = \langle C, \vdash \rangle$ such that C is a signature and $\vdash$ is a consequence relation over C. A logic $\mathcal{L}$ is said to be *structural* if, additionally, it satisfies:

- For every substitution σ in C and every $\Gamma \cup \{\varphi\} \subseteq L(C)$: [2]
  If $\Gamma \vdash \varphi$ then $\hat{\sigma}(\Gamma) \vdash \hat{\sigma}(\varphi)$ (**Structurality**).

The logic $\mathcal{L}$ is said to be *finitary* if it also satisfies:

- For every $\Gamma \cup \{\varphi\} \subseteq L(C)$:
  If $\Gamma \vdash \varphi$ then $\Gamma' \vdash \varphi$ for some finite set $\Gamma' \subseteq \Gamma$ (**Finitariness**).

The logic $\mathcal{L}$ is said to be *standard* if it is structural and finitary. ∎

---

[2] Recall that a substitution in C is any function $\sigma : \mathcal{V} \rightarrow L(C)$. Since $L(C)$ is freely generated by C from $\mathcal{V}$, σ can be extended to a unique endomorphism $\hat{\sigma} : L(C) \rightarrow L(C)$.

DEFINITION 8

(i) Let $\mathcal{L} = \langle C, \vdash_{\mathcal{L}} \rangle$ be a logic, and let $C' \subseteq C$. The C'–fragment of $\mathcal{L}$ is the logic $\mathcal{L}|_{C'} := \langle C', \vdash_{\mathcal{L}|_{C'}} \rangle$ where $\vdash_{\mathcal{L}|_{C'}} = \vdash_{\mathcal{L}} \cap (\wp(L(C')) \times L(C'))$. This means that, for every $\Gamma \cup \{\varphi\} \subseteq L(C')$, $\Gamma \vdash_{\mathcal{L}|_{C}} \varphi$ iff $\Gamma \vdash_{\mathcal{L}} \varphi$.

(ii) The logic $\mathcal{L}' = \langle C', \vdash_{\mathcal{L}'} \rangle$ is a strong extension of $\mathcal{L} = \langle C, \vdash_{\mathcal{L}} \rangle$ if $C \subseteq C'$ and $\vdash_{\mathcal{L}} \subseteq \vdash_{\mathcal{L}'}$.

(iii) The logic $\mathcal{L}' = \langle C', \vdash_{\mathcal{L}'} \rangle$ is a weak extension of $\mathcal{L} = \langle C, \vdash_{\mathcal{L}} \rangle$ if $C \subseteq C'$ and $\vdash_{\mathcal{L}} \varphi$ implies that $\vdash_{\mathcal{L}'} \varphi$, for every $\varphi \in L(C)$.

(iv) The logic $\mathcal{L}' = \langle C', \vdash_{\mathcal{L}'} \rangle$ is a conservative extension of $\mathcal{L} = \langle C, \vdash_{\mathcal{L}} \rangle$ if $C \subseteq C'$ and $\mathcal{L} = \mathcal{L}'|_{C}$.

(v) The logic $\mathcal{L}' = \langle C', \vdash_{\mathcal{L}'} \rangle$ is a conservative weak extension of $\mathcal{L} = \langle C, \vdash_{\mathcal{L}} \rangle$ if $C \subseteq C'$ and $\vdash_{\mathcal{L}} \varphi$ iff $\vdash_{\mathcal{L}'} \varphi$, for every $\varphi \in L(C)$.   ∎

From the definitions above the next result is immediate.

THEOREM 9

(i) Each C–fragment of any (structural, finitary, standard) logic is also a (structural, finitary, standard) logic.

(ii) Every logic $\mathcal{L}$ is a conservative extension of any of its C–fragments.

Finally, the category of logics is specified.

DEFINITION 10 Let $\mathcal{L} = \langle C, \vdash_{\mathcal{L}} \rangle$ and $\mathcal{L}' = \langle C', \vdash_{\mathcal{L}'} \rangle$ be logics. A morphism between logics from $\mathcal{L}$ to $\mathcal{L}'$, denoted by $\mathcal{L} \xrightarrow{f} \mathcal{L}'$, is a **Sig**–morphism $C \xrightarrow{f} C'$ which satisfies, for every $\Gamma \cup \{\varphi\} \subseteq L(C)$:

$$\Gamma \vdash_{\mathcal{L}} \varphi \quad \text{implies} \quad \widehat{f}(\Gamma) \vdash_{\mathcal{L}'} \widehat{f}(\varphi).$$

∎

By defining composition of morphisms and identity morphisms, inheriting from what was done for the case of **Sig**, the category **Log** of (propositional) logics is defined. In this category, logics are presented by means of consequence relations. A fundamental property of **Log** is the following:

THEOREM 11 The category **Log** has arbitrary (small) products.

Proof. The argument can be easily adapted from that in [Bueno-Soler *et al.*, 2005] for the category of standard logics.   ∎

3   Possible–translations Semantics

In this section the method of possible–translations semantics ($PTSs$) is briefly summarized and reviewed, and some examples are addressed. A categorial characterization of the method is also offered.

The concept of $PTSs$ is based on the idea of defining a new global consequence relation by combining other, presumably simpler, consequence relations by means of translations. In this way, as commented in Section 1, $PTSs$ can be seen to work on two opposite directions: as a splitting procedure, and a splicing procedure.

The idea behind possible-translations semantics is to encompass two or more basic semantic models (of the same similarity type) in such a way as to define a new logic which depends upon the basic ones by means of a collection of translations. The basic models can be distinct copies of classical models, or distinct many-valued models, or even Kripke models (for intuitionistic or modal logics).

In [Carnielli and Coniglio, 1999] a somewhat more abstract account of possible–translations semantics was investigated, considering the basic models as organized through sheaf structures. As is well known, sheaves are used in mathematics as a tool for investigating the relationship between local and global phenomena, and seems to be an adequate framework to frame the idea of possible translations.

As mentioned in Section 1, instead of thinking of synthesizing some given logics through a combination process in order to obtain a new logic (as is done with fibring, for instance), a logic can be split into a family of other logics; this question can be examined in terms of categories, resulting in a universal construction. This section outlines a categorial characterization for this process, originally propounded in [Bueno et al., 2004].

DEFINITION 12 Let $\mathcal{L}_i = \langle C_i, \vdash_{\mathcal{L}_i} \rangle$ for $i = 1, 2$ be logics, and let $f : L(C_1) \to L(C_2)$ be a mapping.
(a) f is said to be a *translation* between $\mathcal{L}_1$ and $\mathcal{L}_2$ if it preserves deducibility, that is, for every $\Gamma \cup \{\varphi\} \subseteq L(C_1)$, $\Gamma \vdash_{\mathcal{L}_1} \varphi$ implies that $f(\Gamma) \vdash_{\mathcal{L}_2} f(\varphi)$.
(b) f is said to be a *conservative translation* between $\mathcal{L}_1$ and $\mathcal{L}_2$ if, for every $\Gamma \cup \{\varphi\} \subseteq L(C_1)$, $\Gamma \vdash_{\mathcal{L}_1} \varphi$ iff $f(\Gamma) \vdash_{\mathcal{L}_2} f(\varphi)$.
(c) A morphism $\mathcal{L}_1 \xrightarrow{f} \mathcal{L}_2$ in **Log** is said to be *conservative* if $\widehat{f} : L(C_1) \to L(C_2)$ is a conservative translation. ∎

Observe that each morphism f in **Log** induces a translation between logics $\widehat{f}$ in the sense of the definition above; we call it a *grammatical translation*, in the sense that n–ary connectives are mapped by f into n–ary formula schemas.

We begin by adapting the original definitions of [Carnielli, 2000] in order to make them suitable for categorial formalization.

DEFINITION 13 Let $\mathcal{L} = \langle C, \vdash_{\mathcal{L}} \rangle$ be a logic, and let $\{\mathcal{L}_i\}_{i \in I}$ be a family of logics such that $\mathcal{L}_i = \langle C_i, \vdash_{\mathcal{L}_i} \rangle$ for every $i \in I$. A *possible–translations frame for* $\mathcal{L}$ is a pair $P = \langle \{\mathcal{L}_i\}_{i \in I}, \{f_i\}_{i \in I} \rangle$ such that $f_i : L(C) \to L(C_i)$ is a translation between $\mathcal{L}$ and $\mathcal{L}_i$, for every $i \in I$. We say that $P = \langle \{\mathcal{L}_i\}_{i \in I}, \{f_i\}_{i \in I} \rangle$ is a *possible–translations semantics for* $\mathcal{L}$ (in short, a *PTS*) if, for every $\Gamma \cup \{\varphi\} \subseteq L(C)$,

$$\Gamma \vdash_{\mathcal{L}} \varphi \quad \text{iff} \quad f_i(\Gamma) \vdash_{\mathcal{L}_i} f_i(\varphi) \quad \text{for every } i \in I.$$

A frame $P = \langle \{\mathcal{L}_i\}_{i \in I}, \{f_i\}_{i \in I} \rangle$ is said to be *small* if the class I is a set, and is said to be *grammatical* if f_i is a morphism $\mathcal{L} \xrightarrow{f_i} \mathcal{L}_i$ in **Log**, for every $i \in I$. Analogously, a possible–translations semantics is said to be *small* (respectively, *grammatical*) if it is small (respectively, grammatical) regarded as a frame. ∎

REMARK 14 In order to obtain a categorial characterization of *PTS*s (see Theorem 15 below), possible–translations frames must here be restricted to small grammatical ones. ∎

As mentioned above, a *PTS* for a logic $\mathcal{L}$ can be seen as a way of splitting the logic $\mathcal{L}$ into the family $\{\mathcal{L}_i\}_{i \in I}$ of logics by means of the translations $\{f_i\}_{i \in I}$.

Using Theorem 11, a characterization of *PTS*s can be given in terms of products and conservative translations. The next result was originally stated in [Bueno *et al.*, 2004] for the category of standard logics.

THEOREM 15 Small grammatical possible–translations semantics for a logic $\mathcal{L}$ are the same as conservative morphisms $\mathcal{L} \xrightarrow{f} \mathcal{L}'$, where $\mathcal{L}'$ is a product in **Log** of some small family of logics.

Proof. Let $\mathcal{L} = \langle C, \vdash_{\mathcal{L}} \rangle$ be a logic and let P be a small grammatical *PTS* for $\mathcal{L}$. The idea is to define a conservative morphism $\mathcal{L} \xrightarrow{t(P)} \mathsf{L}(P)$ in **Log**, where $\mathsf{L}(P)$ is a product in **Log** of some family of logics, such that $t(P)$ encodes P. And, conversely, given a conservative morphism $\mathcal{L} \xrightarrow{f} \mathcal{L}'$ in **Log**, where $\mathcal{L}'$ is a product of logics, a small grammatical *PTS* for $\mathcal{L}$ encoding f, denoted $\mathsf{PTS}(f)$, can be defined, in such a manner that the assignments t and PTS are one inverse of the other.

Thus, assuming that $P = \langle \{\mathcal{L}_i\}_{i \in I}, \{f_i\}_{i \in I} \rangle$ is a small grammatical PTS for $\mathcal{L}$, consider the product $\langle \mathcal{L}^{\mathcal{F}}, \{\pi_i\}_{i \in I} \rangle$ in **Log** of the small family $\mathcal{F} = \{\mathcal{L}_i\}_{i \in I}$ (cf. Theorem 11). Since each f_i is a morphism in **Log** then, by the universal property of the product, there is a unique morphism $\mathcal{L} \xrightarrow{\mathsf{t}(P)} \mathcal{L}^{\mathcal{F}}$ in **Log** such that $f_i = \pi_i \cdot \mathsf{t}(P)$ for every $i \in I$. From this, it is not difficult to prove that

$$(*) \qquad \widehat{f_i} = \widehat{\pi_i} \circ \widehat{\mathsf{t}(P)} \,.$$

Using this, it can be proved that $\mathsf{t}(P)$ is a conservative morphism. Clearly, $\mathsf{t}(P)$ together with its codomain $\mathsf{L}(P) := \mathcal{L}^{\mathcal{F}}$ encodes all the information about P: every logic $\mathcal{L}_i$ is obtained as the codomain of π_i, and every morphism f_i is obtained as $f_i = \pi_i \cdot \mathsf{t}(P)$.

Conversely, let $\mathcal{L} \xrightarrow{f} \mathcal{L}'$ be a conservative morphism in **Log**, such that $\mathcal{L}'$ is a product in **Log** of a small family $\{\mathcal{L}_i\}_{i \in I}$ of logics, with canonical projections π_i for every $i \in I$. For every $i \in I$ consider the morphism $f_i = \pi_i \cdot f$ in **Log**, and define the small grammatical possible–translations frame $\mathsf{PTS}(f) = \langle \{\mathcal{L}_i\}_{i \in I}, \{f_i\}_{i \in I} \rangle$. Using $(*)$ again, it can be proven that $\mathsf{PTS}(f)$ is a (small and grammatical) PTS for $\mathcal{L}$. Moreover, all the information about f and $\mathcal{L}'$ can be recovered from $\mathsf{PTS}(f)$: in fact $f = \mathsf{t}(\mathsf{PTS}(f))$ and $\mathcal{L}'$ is the product of the family of logics of $\mathsf{PTS}(f)$. It is also clear that, if P is a small grammatical PTS for $\mathcal{L}$, then $\mathsf{PTS}(\mathsf{t}(P)) = P$. ∎

We now show that matrix semantics (referred to in Section 1) for propositional logics can be portrayed as a particular case of PTSs (see Theorem 22 below). In order to do this, we briefly recall some basic facts about matrix semantics.

DEFINITION 16 Given a signature C, a C–*matrix* is a pair $M = \langle \mathbf{A}, D \rangle$, where $\mathbf{A} = \langle A, C \rangle$ is an algebra over C, and $D \subseteq A$. The set D is usually referred to as the *set of designated values of* M. The M–*valuations of* $L(C)$ are the C–homomorphisms $v : L(C) \to A$. ∎

For simplicity, we sometimes write $M = \langle A, D \rangle$ instead of $M = \langle \mathbf{A}, D \rangle$ in concrete examples. Additionally, the interpretation of a connective c in M will be frequently written as c^M.

DEFINITION 17 Let C be a signature and let $\mathcal{K}$ be a class of C–matrices. The *matrix semantics for* $L(C)$ *induced by* $\mathcal{K}$ (denoted by $\vdash_{\mathcal{K}}$) is defined by: $\Gamma \vdash_{\mathcal{K}} \varphi$ iff, for every C–matrix $M = \langle \mathbf{A}, D \rangle$ belonging to $\mathcal{K}$ and every M–valuation v of $L(C)$, $v(\Gamma) \subseteq D$ implies that $v(\varphi) \in D$. ∎

A logic $\mathcal{L}$ is said to be a *matrix logic* if there exists a class $\mathcal{K}$ of $C_{\mathcal{L}}$–matrices such that $\vdash_{\mathcal{L}} = \vdash_{\mathcal{K}}$. In this case, we say that $\mathcal{K}$ is *adequate* for $\mathcal{L}$, and that $\mathcal{L}$ is *characterized by* $\mathcal{K}$. As shown in [Wójcicki, 1969], every structural logic is indeed a matrix logic (see Theorem 21 below). When $\mathcal{K} = \{M\}$ is a singleton then $\vdash_M$ will stand for $\vdash_{\{M\}}$.

DEFINITION 18 Let $\mathcal{L}$ be a logic and let M be a $C_{\mathcal{L}}$–matrix. If $\vdash_{\mathcal{L}} \subseteq \vdash_M$ we say that $\vdash_{\mathcal{L}}$ is *sound for* $\vdash_M$, or that M *is a matrix model for* $\mathcal{L}$. We define the class **MatMod**$(\mathcal{L})$ as being the class of all the matrix models for $\mathcal{L}$. ∎

Clearly, every matrix logic is a logic in the sense of Definition 7. Moreover, the following fundamental result due to J. Łoś and R. Suszko (see [Łoś and Suszko, 1958]) shows that a matrix logic is, in fact, structural:

THEOREM 19 Let $\mathcal{K}$ be a class of C–matrices. Then $\vdash_{\mathcal{K}}$ is a structural consequence relation and $\vdash_{\mathcal{K}} = inf\{\vdash_M : M \in \mathcal{K}\}$. [3]

Note that $\vdash_{\mathcal{K}}$ do not need to be finitary and, therefore, $\mathcal{L} = \langle C, \vdash_{\mathcal{K}} \rangle$ is not necessarily standard. The following sufficient condition for a matrix logic to be standard was obtained in [Wójcicki, 1973].

THEOREM 20 Every consequence relation induced by a finite class of finite matrices is finitary, and so defines a standard logic.

The next classical result is credited to A. Lindenbaum and R. Wójcicki (see [Wójcicki, 1969; Wójcicki, 1988]).

THEOREM 21 For every structural logic $\mathcal{L}$, the class **MatMod**$(\mathcal{L})$ is a complete matrix semantics for $\mathcal{L}$.

It is simple to see that, by just considering identity mappings as translations, the notion of matrix logics is nothing else than a special case of grammatical possible–translation semantics. Indeed:

THEOREM 22 Let $\mathcal{L} = \langle C, \vdash_{\mathcal{L}} \rangle$ be a matrix logic, and let $\mathcal{K}$ be a class of C–matrices adequate for $\mathcal{L}$. For every $M \in \mathcal{K}$ let $\mathcal{L}_M = \langle C, \vdash_M \rangle$ and let $\mathcal{L} \xrightarrow{f_M} \mathcal{L}_M$ be the morphism in **Log** induced by the identity morphism in the signature C. [4] Then the grammatical possible–translations frame

$$\mathsf{PTS}(\mathcal{K}) = \langle \{\mathcal{L}_M\}_{M \in \mathcal{K}}, \{f_M\}_{M \in \mathcal{K}} \rangle$$

---

[3] The infimum is taken with respect to the inclusion ordering $\subseteq$.
[4] Since $\vdash_{\mathcal{L}} \subseteq \vdash_M$ then f_M is, in fact, a morphism in **Log**.

is a grammatical possible–translations semantics for $\mathcal{L}$.

Proof. Immediate from Definition 13 and from the notion of adequate class of matrices.   ■

The last result can be recast as stating that a logic $\mathcal{L}$ characterized by a class of matrices $\mathcal{K}$ splits over the elements of $\mathcal{K}$. That is, every matrix in $\mathcal{K}$ acts as a legitimate factor of $\mathcal{L}$, and so an adequate matrix semantics works as a particular instance of the splitting method defined by possible–translations semantics. Note that, if $\mathcal{K}$ is a proper class (instead of a set), then $\mathsf{PTS}(\mathcal{K})$ is not small. [5]

As an illustrative example, we prove below that the set of theorems of the propositional intuitionistic logic Int can be characterized by a possible–translations semantics (with identity translations) splitting Int into Heyting algebras.

EXAMPLE 23 It is well–known that theoremhood in Int is characterized by the class of matrices

$$\mathcal{H} = \{\langle \mathbf{H}, \{\top\}\rangle \ : \ \langle \mathbf{H}, \{\top\}\rangle \text{ is a Heyting algebra with top element } \top\}$$

(see, for instance, [Rasiowa and Sikorski, 1968]). For every $H = \langle \mathbf{H}, \{\top\}\rangle$ in $\mathcal{H}$ let $\mathcal{L}_H := \langle C, \vdash_H\rangle$ and let f_H be the morphism in **Log** induced by the identity morphism in the signature of Int. Then the grammatical possible–translations frame

$$\mathsf{PTS}(\mathcal{H}) = \langle \{\mathcal{L}_H\}_{H\in\mathcal{H}}, \{f_H\}_{H\in\mathcal{H}}\rangle$$

characterizes theoremhood for propositional intuitionistic logic Int. That is, given a formula φ, to check whether $\vdash_{\mathsf{Int}} \varphi$ is equivalent to check whether $\vdash_H \varphi$ for every Heyting algebra H.   ■

REMARK 24 It is appropriate here to warn the reader that matrix logics do not see the fine distinction between local and global semantics (current in modal and first–order logics): they only convey the notion of global semantics, because they deal with non–ordered algebras. In the example above, the matrix semantics for Int just represents global entailments, which are not the usual ones in algebraic semantics. Thus, given a Heyting algebra H and a set of formulas $\Gamma \cup \{\varphi\}$, $\Gamma \vdash_H \varphi$ iff, for every homomorphism $v : L(C) \to \mathbf{H}$, $v(\Gamma) \subseteq \{\top\}$ implies $v(\varphi) = \top$. On the other hand, the usual

---

[5][Marcos, 2004] studies how possible-translations semantics characterize wider classes of propositional logics.

notion of entailment in an algebraic ordered structure is local, inasmuch as it requires that, for every v, $(\bigwedge_{\gamma \in \Gamma} v(\gamma)) \leq v(\varphi)$, where $\bigwedge$ denotes the infimum of a set. However, since Int is finitary and satisfies the Deduction meta–theorem, characterizing theoremhood is equivalent to characterizing the whole deducibility. ∎

This section concludes by showing some applications of possible–translations semantics to paraconsistent logics.

EXAMPLE 25 The *Logics of Formal Inconsistency*, **LFI**s, are paraconsistent logics that internalize the metalogic notions of consistency and inconsistency at the object–language level by means of unary connectives ∘ for *consistency* and • for *inconsistency* (cf. [Carnielli *et al.*, 2005]), appropriately constrained by specific axioms.

Some interesting **LFI**s are the logics **bC** and **Ci**, as well as its weaker versions **mCi** and **mbC**. In order to obtain PTSs for these systems, which are defined over the signature $C = \{\wedge, \vee, \Rightarrow, \neg, \circ\}$, consider the signature $C_1 = \{\wedge, \vee, \Rightarrow, \neg_1, \neg_2, \neg_3, \circ_1, \circ_2, \circ_3\}$ and the matrix M over C_1 defined by the truth–tables below, where T and t are the designated values.

$\wedge$	T	t	F
T	t	t	F
t	t	t	F
F	F	F	F

$\vee$	T	t	F
T	t	t	t
t	t	t	t
F	t	t	F

$\Rightarrow$	T	t	F
T	t	t	F
t	t	t	F
F	t	t	t

	$\neg_1$	$\neg_2$	$\neg_3$
T	F	F	F
t	F	t	t
F	T	t	T

	$\circ_1$	$\circ_2$	$\circ_3$
T	T	t	F
t	F	F	F
F	T	t	F

Now consider the clauses below for a mapping $f : L(C) \to L(C_1)$.

$(tr0)$ $f(p) = p$ for $p \in \mathcal{V}$;

$(tr1)$ $f(\varphi \# \psi) = (f(\varphi) \# f(\psi))$, for $\# \in \{\wedge, \vee, \Rightarrow\}$;

$(tr2)$ $f(\neg\varphi) \in \{\neg_1 f(\varphi), \neg_2 f(\varphi)\}$;

$(tr3)$ $f(\neg\varphi) \in \{\neg_1 f(\varphi), \neg_3 f(\varphi)\}$;

$(tr4)$ $f(\neg^{n+1}\circ\varphi) = \neg_1 f(\neg^n \circ\varphi)$, for $n \in \mathbb{N}$;

$(tr5)$ $f(\circ\varphi) \in \{\circ_2 f(\varphi), \circ_3 f(\varphi), \circ_2 f(\neg\varphi), \circ_3 f(\neg\varphi)\}$;

$(tr6)$ $f(\circ\varphi) \in \{\circ_1 f(\varphi), \circ_1 f(\neg\varphi)\}$;

$(tr7)$ if $f(\neg\varphi) = \neg_1 f(\varphi)$ then $f(\circ\varphi) = \circ_1 f(\neg\varphi)$.

In clause $(tr4)$ above, $\neg^n\varphi$ denotes n applications of $\neg$ over formula φ; in particular, $\neg^0\varphi = \varphi$. Now consider the following collections of mappings:
(a) Let $\{f_i^1\}_{i \in I_1}$ be the family of translations $f_i^1 : L(C) \to L(C_1)$ between **mbC** and $\langle C_1, \vdash_M \rangle$ satisfying clauses $(tr0)$, $(tr1)$, $(tr2)$ and $(tr5)$ above;
(b) Let $\{f_i^2\}_{i \in I_2}$ be the family of translations $f_i^2 : L(C) \to L(C_1)$ between **mCi** and $\langle C_1, \vdash_M \rangle$ satisfying clauses $(tr0)$, $(tr1)$, $(tr2)$, $(tr4)$ and $(tr6)$ above;
(c) Let $\{f_i^3\}_{i \in I_3}$ be the family of translations $f_i^3 : L(C) \to L(C_1)$ between **bC** and $\langle C_1, \vdash_M \rangle$ satisfying clauses $(tr0)$, $(tr1)$, $(tr3)$ and $(tr5)$ above;
(d) Let $\{f_i^4\}_{i \in I_4}$ be the family of translations $f_i^4 : L(C) \to L(C_1)$ between **Ci** and $\langle C_1, \vdash_M \rangle$ satisfying clauses $(tr0)$, $(tr1)$, $(tr3)$, $(tr6)$ and $(tr7)$ above.

Let $\mathcal{L}_i = \langle C_1, \vdash_M \rangle$ for $i \in \bigcup_{j=1}^4 I_j$. The next results are found in [Marcos, 2005] (see also [Carnielli *et al.*, 2005]):

(1) $\mathsf{PTS}_1 = \langle \{\mathcal{L}_i\}_{i \in I_1}, \{f_i^1\}_{i \in I_1} \rangle$ is a possible–translations semantics for the logic **mbC**.

(2) $\mathsf{PTS}_2 = \langle \{\mathcal{L}_i\}_{i \in I_2}, \{f_i^2\}_{i \in I_2} \rangle$ is a possible–translations semantics for the logic **mCi**.

(3) $\mathsf{PTS}_3 = \langle \{\mathcal{L}_i\}_{i \in I_3}, \{f_i^3\}_{i \in I_3} \rangle$ is a possible–translations semantics for the logic **bC**.

(4) $\mathsf{PTS}_4 = \langle \{\mathcal{L}_i\}_{i \in I_4}, \{f_i^4\}_{i \in I_4} \rangle$ is a possible–translations semantics for the logic **Ci**.

Examples (1)–(4) above illustrate the fact that the class of grammatical *PTSs*, even if quite wide, is not enough: for simple logics as the **LFIs** above mentioned no grammatical *PTSs* are known. There are in the literature other important examples of non–grammatical *PTSs*, as for instance the one for the hierarchy $\{C_n\}_{n\in\mathbb{N}}$ of da Costa's paraconsistent systems given in [Carnielli, 2000] (see also [Marcos, 1999]). The fact that in the mentioned examples no grammatical PTSs are known does not mean, of course, that they would be impossible to find. We conjecture, however, that in all those cases no grammatical PTS can be found. ∎

4　Nondeterministic semantics and a comparison

This section is devoted to reviewing the nondeterministic semantics introduced in [Avron and Lev, 2001] (see also [Avron and Lev, 2005]), proposing a generalization and comparing them with possible–translation semantics. The basic idea of nondeterministic semantics is to use matrices in which each entry consists of a set of truth–values instead of a single value.

DEFINITION 26 Let C be a signature. A *nondeterministic matrix* for C is a structure $\mathsf{M} = \langle \mathcal{T}, \mathcal{D}, \mathcal{O} \rangle$ such that $\mathcal{T}$ is a nonempty set (of *truth–values*), $\mathcal{D} \subseteq \mathcal{T}$ is a nonempty set (of *designated values*) and $\mathcal{O}$ is a mapping which assigns to every n–ary connective $c \in C^n$ a mapping $\mathcal{O}(c) : \mathcal{T}^n \to \wp(\mathcal{T}) \setminus \{\emptyset\}$. A *valuation* over M is a mapping $v : L(C) \to \mathcal{T}$ such that $v(c(\varphi_1, \ldots, \varphi_n)) \in \mathcal{O}(c)(v(\varphi_1), \ldots, v(\varphi_n))$ for every $c \in C^n$, $\varphi_i \in L(C)$ $(i = 1, \ldots, n)$ and $n \in \mathbb{N}$. The consequence relation $\models_\mathsf{M}$ induced by M is defined as follows: let $\Gamma \cup \{\varphi\} \subseteq L(C)$; then $\Gamma \models_\mathsf{M} \varphi$ if, for every valuation v over M, $v(\Gamma) \subseteq \mathcal{D}$ implies $v(\varphi) \in \mathcal{D}$. A logic $\mathcal{L}$ is *sound* (respectively, *complete*) for M if, for every $\Gamma \cup \{\varphi\} \subseteq L(C)$ it holds: $\Gamma \vdash_\mathcal{L} \varphi$ only if (respectively, if) $\Gamma \models_\mathsf{M} \varphi$. $\mathcal{L}$ is *adequate* for M if it is sound and complete for M. ∎

EXAMPLE 27 [Avron and Lev, 2005] Let C be the signature for the logic **Ci** (recall Example 25) and consider the following nondeterministic matrix $\mathsf{M}_{\mathbf{Ci}}$: $\mathcal{T} = \{\top, \bot, u\}$; $\mathcal{D} = \{\top, u\}$; and $\mathcal{O}$ is defined by the tables below.

$\mathcal{O}(\vee)$	$\bot$	u	$\top$
$\bot$	$\{\bot\}$	$\mathcal{D}$	$\mathcal{D}$
u	$\mathcal{D}$	$\mathcal{D}$	$\mathcal{D}$
$\top$	$\mathcal{D}$	$\mathcal{D}$	$\mathcal{D}$

$\mathcal{O}(\wedge)$	$\bot$	u	$\top$
$\bot$	$\{\bot\}$	$\{\bot\}$	$\{\bot\}$
u	$\{\bot\}$	$\mathcal{D}$	$\mathcal{D}$
$\top$	$\{\bot\}$	$\mathcal{D}$	$\mathcal{D}$

$\mathcal{O}(\Rightarrow)$	$\bot$	u	$\top$
$\bot$	$\mathcal{D}$	$\mathcal{D}$	$\mathcal{D}$
u	$\{\bot\}$	$\mathcal{D}$	$\mathcal{D}$
$\top$	$\{\bot\}$	$\mathcal{D}$	$\mathcal{D}$

	$\mathcal{O}(\neg)$	$\mathcal{O}(\circ)$
$\bot$	$\{\top\}$	$\{\top\}$
u	$\mathcal{D}$	$\{\bot\}$
$\top$	$\{\bot\}$	$\{\top\}$

Then the logic **Ci** is adequate for $\mathsf{M_{Ci}}$, as shown in [Avron and Lev, 2005]. ∎

In what follows we show that nondeterministic matrices are a particular case of possible–translations semantics. From now on, C will denote a fixed signature.

DEFINITION 28 Let $\mathsf{M} = \langle \mathcal{T}, \mathcal{D}, \mathcal{O} \rangle$ be a nondeterministic matrix for C.
(1) Let $c \in C^n$ be a n–ary connective. An *instance of c in* M is a mapping $\mathsf{i} : \mathcal{T}^n \to \mathcal{T}$ such that $\mathsf{i}(\vec{x}) \in \mathcal{O}(c)(\vec{x})$ for every $\vec{x} \in \mathcal{T}^n$. Let $\mathcal{I}_c^\mathsf{M}$ be the set of instances of c in M. Note that, if $c \in C^0$ then $\mathcal{I}_c^\mathsf{M} = \mathcal{O}(c)$.
(2) For each $c \in |C|$ and every $\mathsf{i} \in \mathcal{I}_c^\mathsf{M}$ let $\bar{\mathsf{i}}$ be a new symbol such that $\mathsf{i} \neq \mathsf{j}$ implies that $\bar{\mathsf{i}} \neq \bar{\mathsf{j}}$. The *signature derived from* M is the signature C_M such that $C_\mathsf{M}^n = \bigcup_{c \in C^n} \{ \bar{\mathsf{i}} \; : \; \mathsf{i} \in \mathcal{I}_c^\mathsf{M} \}$.
(3) The *matrix derived from* M is the C_M–matrix $M(\mathsf{M})$ with domain $\mathcal{T}$ and set of designated values $\mathcal{D}$ such that, for every $n \in \mathbb{N}$ and every $\bar{\mathsf{i}} \in C_\mathsf{M}^n$, $\bar{\mathsf{i}}^{M(\mathsf{M})} = \mathsf{i}$. That is, the interpretation of the n–ary connective $\bar{\mathsf{i}}$ in the algebra $M(\mathsf{M})$ is the mapping $\mathsf{i} : \mathcal{T}^n \to \mathcal{T}$. Let $\mathcal{L}_\mathsf{M} = \langle C_\mathsf{M}, \vdash_{M(\mathsf{M})} \rangle$ be the matrix logic associated to the matrix $M(\mathsf{M})$. ∎

EXAMPLE 29 Consider again the nondeterministic matrix $\mathsf{M_{Ci}}$ (see Example 27). The mappings $\vee_i : \mathcal{T}^2 \to \mathcal{T}$ $(i = 1, 2)$ defined by the tables below are two instances of disjunction $\vee$ in $\mathsf{M_{Ci}}$.

$\vee_1$	$\bot$	u	$\top$
$\bot$	$\bot$	t	u
u	u	$\top$	u
$\top$	$\top$	$\top$	$\top$

$\vee_2$	$\bot$	u	$\top$
$\bot$	$\bot$	u	u
u	$\top$	$\top$	$\top$
$\top$	u	u	u

On the other hand, there are two instances $\neg_1$ and $\neg_2$ of the negation $\neg$ and just one instance $\circ_1$ of the consistency operator $\circ$ in $\mathsf{M_{Ci}}$, displayed below.

	$\neg_1$	$\neg_2$	$\circ_1$
$\bot$	$\top$	$\top$	$\top$
u	$\top$	u	$\bot$
$\top$	$\bot$	$\bot$	$\top$

∎

Note that the signature $C^{\mathsf{M}_{\mathsf{Cl}}}$ contains 2^8 symbols for disjunction, 2^4 symbols for conjunction, 2^7 symbols for implication, two symbols for negation and one symbol for consistency. In general, if $c \in C^n$ and the set $\mathcal{T}$ has cardinal κ, let $\mathcal{T}^n = \{\vec{x}_\alpha \ : \ \alpha < \kappa^n\}$. Suppose that the set $\mathcal{O}(c)(\vec{x}_\alpha)$ has cardinal κ_α for every $\alpha < \kappa^n$. Then there are $\prod_{\alpha < \kappa^n} \kappa_\alpha$ instances of c in M.

Using the definitions above, we can obtain a possible–translations semantics PTS for $\mathcal{L} = \langle C, \models_{\mathsf{M}} \rangle$. The central idea is to substitute each nondeterministic operation $\mathcal{O}(c)$ by the set of operations $\mathcal{I}_c^{\mathsf{M}}$. This is done by means of a set of translations such that the formula $c(\varphi_1, \ldots, \varphi_n)$ is translated by f as $\bar{\mathsf{i}}(f(\varphi_1), \ldots, f(\varphi_n))$ for some $\mathsf{i} \in \mathcal{I}_c^{\mathsf{M}}$.

DEFINITION 30 Given a nondeterministic matrix M over C, let $\mathcal{F}_{\mathsf{M}}$ be the family $\{f_j\}_{j \in I}$ of all the mappings $f_j : L(C) \to L(C_{\mathsf{M}})$ such that:

   ($tr0$) $f_j(p) = p$ for $p \in \mathcal{V}$;

   ($tr1$) $f_j(c) \in \{\bar{\mathsf{i}} \ : \ \mathsf{i} \in \mathcal{I}_c^{\mathsf{M}}\}$, for $c \in C^0$;

   ($tr2$) $f_j(c(\varphi_1, \ldots, \varphi_n)) \in \{\bar{\mathsf{i}}(f_j(\varphi_1), \ldots, f_j(\varphi_n)) \ : \ \mathsf{i} \in \mathcal{I}_c^{\mathsf{M}}\}$,
             for $c \in C^n$, $n \geq 1$, $\varphi_1, \ldots, \varphi_n \in L(C)$.   ∎

LEMMA 31 For every valuation $w : L(C_{\mathsf{M}}) \to \mathcal{T}$ for $\mathcal{L}_{\mathsf{M}}$ and every mapping f_j in $\mathcal{F}_{\mathsf{M}}$ there exists a valuation v over M such that $v(\varphi) = w(f_j(\varphi))$ for every formula $\varphi \in L(C)$.

Proof. Given w and f_j, consider the mapping $v : L(C) \to \mathcal{T}$ such that $v(\varphi) = w(f_j(\varphi))$ for every formula $\varphi \in L(C)$. It is clear by definition of $\mathcal{F}_{\mathsf{M}}$ that v is a valuation over M.   ∎

COROLLARY 32 Let M be a nondeterministic matrix over C, let $\mathcal{L} = \langle C, \models_{\mathsf{M}} \rangle$ and let $\mathcal{F}_{\mathsf{M}} = \{f_j\}_{j \in I}$ as in Definition 30. Then every mapping f_j in $\mathcal{F}_{\mathsf{M}}$ is in fact a translation between $\mathcal{L}$ and $\mathcal{L}_{\mathsf{M}}$ (recall Definition 28(3)).

DEFINITION 33 Let M be a nondeterministic matrix over C, let $\mathcal{L} = \langle C, \models_{\mathsf{M}} \rangle$ and let $\mathcal{F}_{\mathsf{M}} = \{f_j\}_{j \in I}$ as in Definition 30. The *possible–translations frame for* M is defined as $\mathsf{PTS}(\mathsf{M}) = \langle \{\mathcal{L}_j\}_{j \in I}, \{f_j\}_{j \in I} \rangle$ such that $\mathcal{L}_j = \mathcal{L}_{\mathsf{M}}$ for every $j \in I$.   ∎

From Corollary 32 it follows that the just defined $\mathsf{PTS}(\mathsf{M})$ is indeed a possible–translations frame.

LEMMA 34 For every valuation v over M there exists a valuation w : $L(C_M) \to \mathcal{T}$ for $\mathcal{L}_M$ and a translation f in PTS(M) such that $v(\varphi) = w(f(\varphi))$ for every formula $\varphi \in L(C)$.

Proof. Given the valuation v, define $w(p) = v(p)$ for every $p \in \mathcal{V}$, and extend the mapping w to $L(C_M)$ homomorphically. The mapping f is recursively defined as follows: $f(p) = p$ if $p \in \mathcal{V}$ and $f(c) = \bar{\mathsf{i}}$ for some element i of $\mathcal{I}_c^M$, if $c \in C^0$. Note that $v(\varphi) = w(f(\varphi))$ for every formula $\varphi \in L(C)$ with complexity 1. Suppose that the mapping f was already defined for every formula φ with complexity $\leq n$ $(n \geq 1)$ such that $v(\varphi) = w(f(\varphi))$. Let $c \in C^k$ and $\varphi_1, \ldots, \varphi_k \in L(C)$ (for $k \geq 1$) such that every φ has complexity at most n. Let $\mathsf{i} \in \mathcal{I}_c^M$ such that $\mathsf{i}(v(\varphi_1), \ldots, v(\varphi_k)) = v(c(\varphi_1, \ldots, \varphi_k))$. It is clear that there exists such an instance because $v(c(\varphi_1, \ldots, \varphi_k)) \in \mathcal{O}(c)(v(\varphi_1), \ldots, v(\varphi_k))$. Then let us define $f(c(\varphi_1, \ldots, \varphi_k)) = \bar{\mathsf{i}}(f(\varphi_1), \ldots, f(\varphi_k))$. By induction hypothesis, $v(\varphi_i) = w(f(\varphi_i))$ for $i = 1, \ldots, k$. Then

$$
\begin{aligned}
v(c(\varphi_1, \ldots, \varphi_k)) &= \mathsf{i}(v(\varphi_1), \ldots, v(\varphi_k)) \\
&= \mathsf{i}(w(f(\varphi_1)), \ldots, w(f(\varphi_k))) \\
&= w(\bar{\mathsf{i}}(f(\varphi_1), \ldots, f(\varphi_k))) \\
&= w(f(c(\varphi_1, \ldots, \varphi_k))).
\end{aligned}
$$

■

THEOREM 35 Nondeterministic semantics can be simulated by possible–translations semantics.

Proof. From Lemmas 34 and 31, it is immediate that the possible–translations frame PTS(M) is a possible–translations semantics for M. That is, for every $\Gamma \cup \{\varphi\} \subseteq L(C)$ it holds: $\Gamma \models_M \varphi$ if and only if $\Gamma \models_{PTS(M)} \varphi$. ■

A weaker converse relation (the fact that every possible–translations semantics PTS for a structural logic can be simulated by a nondeterministic matrix) is easily seen to be valid, provided that the logic $\mathcal{L}$ represented by the possible–translations frame PTS is structural and that we expand the definition of a nondeterministic matrix, allowing for *classes* of nondeterministic matrices. Indeed, in formal terms, consider the following:

DEFINITION 36 Let C be a signature. A *multiple nondeterministic matrix* is a class $\mathcal{M}$ of nondeterministic matrices for C. The *multiple nondeterministic matrix semantics for* $L(C)$ *induced by* $\mathcal{M}$ (denoted by $\models_{\mathcal{M}}$) is defined by: $\Gamma \models_{\mathcal{M}} \varphi$ iff $\Gamma \models_M \varphi$ for every $M \in M$. A logic $\mathcal{L} = \langle C, \vdash_{\mathcal{L}} \rangle$ over C is *sound* (respectively, *complete*) for $\mathcal{M}$ if $\vdash_{\mathcal{L}} \subseteq \models_M$ (respectively, $\models_M \subseteq \vdash_{\mathcal{L}}$). $\mathcal{L}$ is *adequate* for $\mathcal{M}$ if it is sound and complete for $\mathcal{M}$. ■

THEOREM 37 Possible–translations semantics for structural logics can be simulated by multiple nondeterministic matrix semantics.

Proof. Let PTS be a possible–translations semantics for a structural logic $\mathcal{L}$. Then $\mathcal{L}$ must have a matrix semantics, by Theorem 21. Since matrix semantics are particular cases of multiple nondeterministic matrix semantics, the result follows. ∎

As a matter of fact, Theorem 35 can be extended to cover multiple nondeterministic matrix semantics. Indeed:

THEOREM 38 Multiple nondeterministic matrix semantics can be simulated by possible–translations semantics.

Proof. Let $\mathcal{M}$ be a multiple nondeterministic matrix for C. For each $\mathsf{M} \in \mathcal{M}$ let $\mathsf{PTS(M)} = \langle \{\mathcal{L}_j\}_{j \in I_\mathsf{M}}, \{f_j\}_{j \in I_\mathsf{M}} \rangle$ be the possible–translations frame for M. Assume, without loss of generality, that $I_\mathsf{M} \cap I_\mathsf{N} = \emptyset$ if $\mathsf{M} \neq \mathsf{N}$. Let $J = \bigcup_{\mathsf{M} \in \mathcal{M}} I_\mathsf{M}$. Then it is immediate that the possible–translations frame $\mathsf{PTS}(\mathcal{M}) = \langle \{\mathcal{L}_j\}_{j \in J}, \{f_j\}_{j \in J} \rangle$ is a possible–translations semantics for $\mathcal{M}$. ∎

As a quick assessment of what has been accomplished so far, Theorems 21, 38 and 37 prove that, for structural logics, matrix semantics, possible–translations semantics and multiple nondeterministic matrix semantics are essentially equivalent.

For the original notion of nondeterministic matrices (recall Definition 26), however, the converse of Theorem 35 seems not to be true. Indeed, possible–translations semantics, and even the grammatical ones, are apparently more expressive than nondeterministic matrices, as the following argument intends to endorse.

As we saw above, given a nondeterministic semantics, a valuation v and a formula $c(\varphi_1, \ldots, \varphi_n)$ then, for some appropriate connective c', the value $v(c(\varphi_1, \ldots, \varphi_n))$ is of the form $c'(v(\varphi_1), \ldots, v(\varphi_n))$. This means that, if we think about the values of $v(\varphi)$ as being of the form $w(\widehat{f}(\varphi))$ for some morphism f and some valuation w onto a suitable logic, there is clearly no restriction on such morphisms, contrary to what is expected from the definition of $PTSs$, where some restrictions should apply to the translations in order to define expressive semantics. More specifically, suppose for simplicity that $\mathsf{PTS} = \langle \{\mathcal{L}_i\}_{i \in I}, \{f_i\}_{i \in I} \rangle$ is a given grammatical possible–translations semantics for $\mathcal{L}$ such that $\mathcal{L}_i = \mathcal{L}_j$ for every $i, j \in I$. Suppose that we want to define a nondeterministic matrix M representing PTS. In order to prove this, it would be necessary to recover Lemmas 34 and 31.

Thus, assuming that there is a given matrix $\langle \mathbf{A}, \mathcal{D} \rangle$ defining every $\mathcal{L}_i$ such that $\mathbf{A}$ is an algebra over signature C_1 with domain $\mathcal{T}$ such that $\mathcal{D} \subseteq \mathcal{T}$ and $\mathcal{L} \xrightarrow{f_i} \mathcal{L}_i$ for every $i \in I$, it seems clear that M should be defined over the set $\mathcal{T}$ with set $\mathcal{D}$ of designated values. Given $c \in C^n$ and $c' \in C_1^n$ we say that they are *compatible with respect to* PTS, written $comp_{\mathsf{PTS}}(c, c')$, if $\widehat{f}(c(\varphi_1, \ldots, \varphi_n)) = c'(\widehat{f}(\varphi_1), \ldots, \widehat{f}(\varphi_n))$ for some morphism f in PTS and some formulas $\varphi_1, \ldots, \varphi_n \in L(C)$. Then, the mapping $\mathcal{O}(c)$ should be defined as follows:

$$\mathcal{O}(c)(\vec{x}) = \{z \in \mathcal{T} \ : \ z = c'(\vec{x}) \text{ for some } c' \text{ such that } comp_{\mathsf{PTS}}(c, c')\}$$

for every $c \in C^n$ and every $\vec{x} \in \mathcal{T}^n$.

Let $\mathsf{M} = \langle \mathcal{T}, \mathcal{D}, \mathcal{O} \rangle$. It is easy to see that Lemma 31 holds good, that is: if w is a valuation over matrix $\langle \mathbf{A}, \mathcal{D} \rangle$ and f is a morphism in PTS then there exists a valuation v over M such that $v(\varphi) = w(\widehat{f}(\varphi))$ for every formula $\varphi \in L(C)$. But, in order to recover Lemma 34, we found an unbridgeable barrier. In fact, given a valuation v over M, it is clear that the valuation w must be defined as $w(p) = v(p)$ for every $p \in \mathcal{V}$. Assuming that a morphism f was defined for $\varphi_1, \ldots, \varphi_n$ such that $w(\widehat{f}(\varphi_i)) = v(\varphi_i)$ for $i = 1, \ldots, n$, suppose that $v(c(\varphi_1, \ldots, \varphi_n)) = c'(v(\varphi_1), \ldots, v(\varphi_n))$. Then $\widehat{f}(c(\varphi_1, \ldots, \varphi_n))$ must be defined as $c'(\widehat{f}(\varphi_1), \ldots, \widehat{f}(\varphi_n))$. The problem is that there is no guarantee that this definition is possible in PTS! That is, nothing guarantees that f defined as above is a morphism in PTS.

The results and arguments above provide an answer to a question posed in [Avron, 2005] about the relationship between nondeterministic matrices and possible–translations semantics.

5   Direct unions of matrices

This section discusses two simple mechanisms for combining matrix logics introduced in [Coniglio and Fernández, 2005] and [Fernández, 2005] called *direct union of matrices* and *plain fibring*. The fundamental idea is that, given two matrix logics $\mathcal{L}_1$ and $\mathcal{L}_2$ such that $\mathcal{L}_i$ is characterized by a single matrix M_i with domain A_i and designated values D_i ($i = 1, 2$), it is possible to extend the original operators of the algebra M_i to the disjoint union $A_1 \uplus A_2$ by means of mappings $f_i : A_j \rightarrow A_i$ ($i \neq j$).

Such mappings f_i allow us to 'transport' the 'foreign' truth–values of the matrix M_j into the truth–values of M_i. This approach follows similar lines as those in the original formulation of fibring (cf. [Gabbay, 1996; Gabbay, 1999]) in which, in order to evaluate, in a Kripke frame F_1, a modal operator $\Box_1$ in a world w_2 belonging to a Kripke frame F_2 (used for evaluating another modal operator $\Box_2$), the world w_2 is 'transported' (by means of a mapping

f) into a world $w_1 = f(w_2)$ of F_1, and vice–versa. The following definitions and results are taken from [Coniglio and Fernández, 2005].

Since the matrix defining the matrix logic is relevant for the operations to be defined below, in the sequel we will write $\langle C, M \rangle$ instead of $\langle C, \vdash_M \rangle$. Let us begin with the simplest case. Given two matrix logics having the same domain and the same sets of designated values, their *direct union* is obtained just by putting together both matrices. Formally:

DEFINITION 39 Let $\mathcal{L}_i = \langle C_i, M_i \rangle$ (with $i = 1, 2$) be two matrix logics, where each $M_i = \langle \mathbf{A}_i, D \rangle$ is a C_i–matrix such that $A_1 = A_2$. [6] Let $A = A_1$. The *direct union of* $\mathcal{L}_1$ *and* $\mathcal{L}_2$ is the logic $\mathcal{L}_1 + \mathcal{L}_2 = \langle C_1 \uplus C_2, \vdash_{M_1 + M_2} \rangle$ where $C_1 \uplus C_2$ is the disjoint union of C_1 and C_2 (recall Definition 1) and $\vdash_{M_1 + M_2}$ is the consequence relation induced by the $C_1 \uplus C_2$–matrix $M_1 + M_2 = \langle \mathbf{A}, D \rangle$. The matrix $M_1 + M_2$ is defined as follows: if $c \in C_i^k$ and $a_1, \ldots, a_k \in A$, then $c^{M_1 + M_2}(a_1, \ldots, a_k) = c^{M_i}(a_1, \ldots, a_k)$ ($k \geq 0$; $i = 1, 2$). ∎

THEOREM 40 Let $\mathcal{L} = \langle C, \vdash \rangle$ be a logic characterized by a C–matrix M. Let $\mathcal{L}_1$ and $\mathcal{L}_2$ be two fragments of $\mathcal{L}$ over C_1 and C_2, respectively, such that $C_1 \uplus C_2 = C$. Then $\mathcal{L}_1 + \mathcal{L}_2 = \mathcal{L}$.

The result above shows that the direct union of logics can be seen as a method for splitting and splicing logics. In particular, a given logic $\mathcal{L}$ can be split into two simpler factors $\mathcal{L}_1$ and $\mathcal{L}_2$ whenever $\mathcal{L} = \mathcal{L}_1 + \mathcal{L}_2$.

EXAMPLE 41 Let $\mathcal{L}_i = \langle C_i, M_i \rangle$ (with $i = 1, 2$) such that C_1 just contains a symbol $\neg$ for negation and C_2 just contains a symbol $\vee$ for disjunction and a symbol $\Rightarrow$ for implication. Suppose that M_1 is the matrix for classical negation, and that M_2 is the matrix for classical disjunction and implication over $A = \{1, 0\}$ where $D = \{1\}$. Then $\mathcal{L}_1 + \mathcal{L}_2$ turns out to be the matrix presentation $\mathcal{L}$ of classical propositional logic over A and D and signature $\{\neg, \vee, \Rightarrow\}$. The logics $\mathcal{L}_1$ and $\mathcal{L}_2$ are two (simpler) factors of $\mathcal{L}$. By its turn, $\mathcal{L}_2$, can of course be split into two elementary logics $\mathcal{L}_2^1$ (the logic of classical disjunction) and $\mathcal{L}_2^2$ (the logic of classical implication), that is, $\mathcal{L}_2 = \mathcal{L}_2^1 + \mathcal{L}_2^2$. Therefore, $\mathcal{L}$ splits into $\mathcal{L}_1$, $\mathcal{L}_2^1$ and $\mathcal{L}_2^2$, and so $\mathcal{L} = \mathcal{L}_1 + \mathcal{L}_2^1 + \mathcal{L}_2^2$. ∎

The possibility of recovering a logic from its components, as in the example above, was already addressed in [Coniglio, 2005], where some limitations of the usual notion of translation were pointed. In that paper it was shown that the operation of fibring as a coproduct (in a category of logics based

---

[6]It is worth noting that this condition *does not mean* that the operations defined in M_1 and M_2 coincide.

on translations, see, e.g., [Sernadas *et al.*, 1999]) cannot recover, in general, a logic from its fragments; this can be done by means of a stronger notion of translation that preserves meta–properties.

A more attractive case is to combine two matrix logics defined over different domains. In this case, each matrix logic is extended to the disjoint union of the domains by means of a pair of mappings, and then the direct union of the extensions is computed. The set of matrices obtained in this way is the matrix semantics of the so-called *plain fibring* (see Definition 43 below).

DEFINITION 42 Let $\mathcal{L}_i = \langle C_i, M_i \rangle$ (with $i = 1, 2$) be two matrix logics, where each $M_i = \langle \mathbf{A}_i, D_i \rangle$ is a C_i–matrix. Let A_i be the domain of the algebras $\mathbf{A}_i$.

(i) A pair $(f_1, f_2) \in A_1^{A_2} \times A_2^{A_1}$ is said to be *admissible* if it satisfies: $f_i(x) \in D_i$ iff $x \in D_j$, for every $x \in A_j$ ($i \neq j$).

(ii) Given an admissible pair $\mathsf{a} = (f_1, f_2)$ then the *extension of M_i by* a is the C_i–matrix $M_i^{\mathsf{a}} = \langle \mathbf{A}, D_1 \uplus D_2 \rangle$ such that $A = A_1 \uplus A_2$ and, for every $c \in C_i^n$ and every $x_1, \ldots, x_n \in A$, $c^{M_i^{\mathsf{a}}}(x_1, \ldots, x_n) = c^{M_i}(\tilde{x}_1, \ldots, \tilde{x}_n)$ where, for every $k = 1, \ldots, n$:

- If $x_k \in A_i$, then $\tilde{x}_k = x_k$.
- If $x_k \in A_j$, then $\tilde{x}_k = f_i(x_k)$ (for $j \neq i$). ∎

It is not hard to prove that the matrix logic $\mathcal{L}_i^{\mathsf{a}} = \langle C_i, M_i^{\mathsf{a}} \rangle$ coincides with $\mathcal{L}_i$, provided that a is admissible: i.e., $\Gamma \vdash_{M_i} \varphi$ iff $\Gamma \vdash_{M_i^{\mathsf{a}}} \varphi$, for every $\Gamma \cup \{\varphi\} \subseteq L(C_i)$ and $i = 1, 2$. This means that, by extending each logic by means of an admissible pair, the logics remain unchanged.

DEFINITION 43 Let $\mathcal{L}_i = \langle C_i, M_i \rangle$ (with $i = 1, 2$) be two matrix logics as in Definition 42. The *plain fibring* of $\mathcal{L}_1$ and $\mathcal{L}_2$ is the pair $\mathcal{L}_1 \odot \mathcal{L}_2 = \langle C_1 \uplus C_2, \vdash_{M_1 \odot M_2} \rangle$ such that $M_1 \odot M_2$ is the set of $C_1 \uplus C_2$–matrices $M_1 \odot M_2 = \{M_1^{\mathsf{a}} + M_2^{\mathsf{a}} : \mathsf{a}$ is admissible$\}$. ∎

We say that $\mathcal{L}_1$ and $\mathcal{L}_2$ are *compatible* if there exist admissible pairs in $A_1^{A_2} \times A_2^{A_1}$. Clearly, $\mathcal{L}_1$ and $\mathcal{L}_2$ are compatible iff:

(*i*) $D_1 \neq \emptyset$ iff $D_2 \neq \emptyset$; and

(*ii*) $A_1 - D_1 \neq \emptyset$ iff $A_2 - D_2 \neq \emptyset$.

THEOREM 44 Let $\mathcal{L}_i = \langle C_i, M_i \rangle$ (with $i = 1, 2$) be two matrix logics as in Definition 42 such that $\mathcal{L}_1$ and $\mathcal{L}_2$ are compatible. Then $\mathcal{L}_1 \odot \mathcal{L}_2$ is a conservative extension of both $\mathcal{L}_1$ and $\mathcal{L}_2$.

EXAMPLE 45 The 3–valued paraconsistent matrix logic P^1 was introduced in [Sette, 1973] and widely studied afterwards. The signature C_{P^1} of P^1 is such that its domain is $|C_{P^1}| = \{\neg_{P^1}, \Rightarrow_{P^1}\}$, and its matrix $M_{P^1} = \langle \mathbf{A}_{P^1}, \{T, T_1\} \rangle$ is such that $A_{P^1} = \{T, T_1, F\}$. The corresponding operations are displayed in the tables below.

	T	T_1	F
$\neg_{P^1}$	F	T	T

$\Rightarrow_{P^1}$	T	T_1	F
T	T	T	F
T_1	T	T	F
F	T	T	T

It is immediate to see that $P^1 = \mathcal{L}_1 + \mathcal{L}_2$, where $\mathcal{L}_1$ is the logic for $\neg_{P^1}$ and $\mathcal{L}_2$ is the logic for $\Rightarrow_{P^1}$ defined by the corresponding truth–tables. On the other hand, by computing the reduced matrix for $\mathcal{L}_2$ (observing that T and T_1 are congruent) it is clear that $\mathcal{L}_2$ is, in fact, the logic $\mathcal{L}_2^2$ for classical implication over $\{1, 0\}$ with 1 as the only designated value (recall Example 41). This shows that P^1 is obtained by composition of (or, equivalently in this case, can be decomposed into) the simpler logics $\mathcal{L}_1$ (the 3–valued logic of the P^1–negation) and $\mathcal{L}_2^2$ (the 2–valued logic of the classical implication). ∎

EXAMPLE 46 The splitting for P^1 exhibited above can be directly obtained as follows: let $\mathcal{L}_1$ the 3–valued logic of the P^1–negation and let M_1 be its matrix (see Example 45). Let $\mathcal{L}_2^2$ be the matrix logic of classical implication given by the matrix M_2 below (recall Example 41).

$\Rightarrow$	1	0
1	1	0
0	1	1

Let $A = \{T, T_1, F, 1, 0\}$ and $D = \{T, T_1, 1\}$, and let $\mathbf{a} = (f_1, f_2)$ such that $f_1(1) = T$, $f_1(0) = F$, $f_2(T) = f_2(T_1) = 1$ and $f_2(F) = 0$. Then $\mathbf{a}$ is admissible and $M_1^{\mathbf{a}}$ and $M_2^{\mathbf{a}}$ are given by the tables below.

	T	T_1	1	F	0
$\neg$	F	T	F	T	T

$\Rightarrow$	T	T_1	1	F	0
T	1	1	1	0	0
T_1	1	1	1	0	0
1	1	1	1	0	0
F	1	1	1	1	1
0	1	1	1	1	1

Let $\mathcal{L}$ be the logic over $\{\neg, \Rightarrow\}$ characterized by the matrix $M_1^{\mathsf{a}} + M_2^{\mathsf{a}}$ given by the two tables above, with $\{T, T_1, 1\}$ as the set of designated values. Since T and 1 are congruent, and F and 0 are also congruent, the reduced matrix for $\mathcal{L}$ produces the 3–valued logic P^1. The details of this construction are left to the reader. ∎

EXAMPLE 47 With the same notation as above we can consider, given M_1 and M_2, another admissible pair $\mathsf{a}' = (g_1, f_2)$ such that $g_1(1) = T_1$ and $g_1(0) = F$. It is worth noting that a and a' are the unique admissible pairs. The matrix $M_1^{\mathsf{a}'}$ is displayed below (observe that $M_2^{\mathsf{a}'} = M_2^{\mathsf{a}}$).

$$
\begin{array}{c|c|c|c|c|c}
  & T & T_1 & 1 & F & 0 \\
\hline
\neg & F & T & T & T & T
\end{array}
$$

Therefore the logic $\mathcal{L}_1 \odot \mathcal{L}_2^2$ is characterized by the set of matrices

$$M_1 \odot M_2 = \{M_1^{\mathsf{a}} + M_2^{\mathsf{a}}, \ M_1^{\mathsf{a}'} + M_2^{\mathsf{a}'}\}.$$

As we saw in Example 46, the logic characterized by $M_1^{\mathsf{a}} + M_2^{\mathsf{a}}$ is P^1. On the other hand, the logic characterized by $M_1^{\mathsf{a}'} + M_2^{\mathsf{a}'}$ does not satisfy the formula $(p_1 \Rightarrow p_2) \Rightarrow \neg\neg(p_1 \Rightarrow p_2)$. In fact, taking any valuation v over the matrix $M_1^{\mathsf{a}'} + M_2^{\mathsf{a}'}$ such that $v(p_1) = v(p_2)$ then $v(p_1 \Rightarrow p_2) = 1$ and so $v(\neg\neg(p_1 \Rightarrow p_2)) = \neg\neg 1 = \neg T = F$. Consequently $v((p_1 \Rightarrow p_2) \Rightarrow \neg\neg(p_1 \Rightarrow p_2)) = (1 \Rightarrow F) = 0$, a non–designated value. This shows that $(p_1 \Rightarrow p_2) \Rightarrow \neg\neg(p_1 \Rightarrow p_2)$ is not a theorem of the logic $\mathcal{L}_1 \odot \mathcal{L}_2^2$. ∎

6   On what is left open

We have analyzed here three methods for providing semantics for logical systems from the point of view of combinations of logics: possible–translations semantics in Section 3, nondeterministic semantics in Section 4 and direct union of matrices and the related plain fibring in Section 5. Such methods make up relevant procedures for combining logics, an area of increasing interest due to the needs of formalization from variate branches of investigation, as linguistics, software engineering, logic programming and others in formal science. In philosophy and logic themselves we will not fail to find appealing possibilities in combining logics, and specially in splitting logics

From the methods investigated, possible–translations semantics have been shown to be conceptually broader, embodying matrix semantics and nondeterministic semantics. There is, however, no known general procedure for deciding whether a given logic can be characterized by possible–translations semantics, nor there seems to be: general semantic tools, like the well-known relational semantics for modal logics, are just this way.

This amply justifies more restrained methods as nondeterministic semantics, direct union of matrices and plain fibring. More so from the outlook of splitting logics, as factors can be effectively computed, according to the examples given.

Attention to combining logics has been delimited to the bottom–up perspective of splicing logics. The top–down perspective of splitting deserves equal attention, as the potentialities for its applications are really significant.

Decomposing a logic $\mathcal{L}$ into simpler components offers optional tools for attacking problems of complexity of algorithms (via the satisfiability problem), questions in proof–theory and in algebraization of logics. We may even be able to define and to characterize which are prime logics, viz,. the ones that cannot be further split (up to a given method). These logics may be really interesting, as stimulating are the posed problems; our outcomes here advance in this direction.

Acknowledgements

This research was supported by FAPESP (Brazil), Thematic Project ConsRel 2004/14107-2. The first author was also supported by a Research Grant level 1 from CNPq (Brazil). We thank Richard L. Epstein and João Marcos for comments and remarks.

BIBLIOGRAPHY

[Avron and Lev, 2001] A. Avron and I. Lev. Canonical propositional Gentzen-type systems. In *Proceedings of the 1st International Joint Conference on Automated Reasoning (IJCAR 2001)*, volume 2083 of *Lecture Notes in Artificial Intelligence*, pages 529–544. Springer Verlag, 2001.

[Avron and Lev, 2005] A. Avron and I. Lev. Non-deterministic multiple-valued structures. *Journal of Logic and Computation*, 15(3):241–261, 2005.

[Avron, 2005] A. Avron. Non-deterministic semantics for paraconsistent C-systems. In *Proceedings of ECSQARU (Euro. Conf. Symb. and Quant. Approaches to Reasoning on Uncertainty)*, 2005.

[Bueno et al., 2004] J. Bueno, M.E. Coniglio, and W.A. Carnielli. Finite algebraizability via possible-translations semantics. In W.A. Carnielli, F.M. Dionísio, and P. Mateus, editors, *Proceedings of CombLog'04 - Workshop on Combination of Logics: Theory and Applications*, pages 79–86, Lisboa, Portugal, 2004. Departamento de Matemática, Instituto Superior Técnico.

[Bueno-Soler et al., 2005] J. Bueno-Soler, M.E. Coniglio, and W.A. Carnielli. Possible-translations algebraizability. In J.-Y. Béziau and W.A. Carnielli, editors, *Paraconsistent Logic Without Frontiers*, Studies in Logic and Practical Reasoning. North-Holland, 2005. To appear.

[Carnielli and Coniglio, 1999] W.A. Carnielli and M.E. Coniglio. A categorial approach to the combination of logics. *Manuscrito*, 22(2):69–94, 1999.

[Carnielli and D'Ottaviano, 1997] W. Carnielli and I.M.L. D'Ottaviano. Translations between logical systems: a manifesto. *Logique et Analyse (N.S.)*, 40(157):67–81, 1997.

[Carnielli and Lima-Marques, 1999] W.A. Carnielli and M. Lima-Marques. Society semantics for multiple-valued logics. In W. A. Carnielli and I. M. L. D'Ottaviano, editors, *Advances in Contemporary Logic and Computer Science*, volume 235 of *Contemporary Mathematics Series*, pages 33–52. American Mathematical Society, 1999.

[Carnielli et al., 2005] W.A. Carnielli, M.E. Coniglio, and J. Marcos. Logics of Formal Inconsistency. In D. Gabbay and F. Guenthner, editors, *Handbook of Philosophical Logic*, volume 14. Kluwer Academic Publishers, 2005. In Print. Preliminary version available at *CLE e-Prints*, Vol. 5(1), 2005.
URL = http://www.cle.unicamp.br/e-prints/articles.html.

[Carnielli, 1990] W. Carnielli. Many–valued logics and plausible reasoning. In *Proceedings of the XX International Congress on Many–Valued Logics, University of Charlotte, USA*, pages 328–335. IEEE Computer Society, 1990.

[Carnielli, 2000] W.A. Carnielli. Possible-Translations Semantics for Paraconsistent Logics. In D. Batens, C. Mortensen, G. Priest, and J. P. Van Bendegem, editors, *Frontiers of Paraconsistent Logic: Proceedings of the I World Congress on Paraconsistency*, Logic and Computation Series, pages 149–163. Baldock: Research Studies Press, King's College Publications, 2000.

[Coniglio and Fernández, 2005] M.E. Coniglio and V.L. Fernández. Plain fibring and direct union of logics with matrix semantics. In B. Prasad, editor, *Proceedings of the 2nd Indian International Conference on Artificial Intelligence (IICAI 2005), Pune, India*. IICAI, 2005.

[Coniglio, 2005] M.E. Coniglio. The meta-fibring environment: Preservation of meta-properties by fibring. *CLE e-Prints*, 5(4), 2005.
URL = http://www.cle.unicamp.br/e-prints/vol_5,n_4,2005.htm.

[Fernández and Coniglio, 2003] V.L. Fernández and M.E. Coniglio. Combining valuations with society semantics. *Journal of Applied Non-Classical Logics*, 13(1):21–46, 2003.

[Fernández, 2005] V. L. Fernández. *Fibrilação de lógicas na Hierarquia de Leibniz (Fibring Logics in the Leibniz Hierarchy, in Portuguese)*. PhD thesis, IFCH – Universidade Estadual de Campinas, Brazil, 2005.

[Gabbay, 1996] D. Gabbay. Fibred semantics and the weaving of logics: Part 1. *The Journal of Symbolic Logic*, 61(4):1057–1120, 1996.

[Gabbay, 1999] D. Gabbay. *Fibring Logics*. Clarendon Press - Oxford, 1999.

[Jaśkowski, 1949] S. Jaśkowski. O koniunkcji dyskusyjnej w rachunku zdań dla systemów dedukcyjnych sprzecznych. *Studia Societatis Scientiarun Torunesis, Sectio A*, I(8):171–172, 1949. Translated as 'On the discussive conjunction in the propositional calculus for inconsistent deductive systems' in *Logic and Logic Philosophy*, 7:57–59, 1999, Proceedings of the Stanisław Jaśkowski's Memorial Symposium, held in Toruń, Poland, July 1998.

[Łoś and Suszko, 1958] J. Łoś and R. Suszko. Remarks on sentential logics. *Indagationes Mathematicae*, 20:177–183, 1958.

[Łukasiewicz, 1953] J. Łukasiewicz. A system of modal logic. *J. Computing systems*, 1:111–149, 1953.

[Marcos, 1999] J. Marcos. Semânticas de Traduções Possíveis (Possible–Translations Semantics, in Portuguese). Master's thesis, IFCH-UNICAMP, Campinas, Brazil, 1999.
URL = http://www.cle.unicamp.br/pub/thesis/J.Marcos/.

[Marcos, 2004] M. Marcos. Possible-translations semantics. In W.A. Carnielli, F.M. Dionísio, and P. Mateus, editors, *Proceedings of CombLog'04 - Workshop on Combination of Logics: Theory and Applications*, pages 119–128, Lisboa, Portugal, 2004. Departamento de Matemática, Instituto Superior Técnico.

[Marcos, 2005] J. Marcos. Possible–translations semantics for some weak classically–based paraconsistent logics, 2005. *Journal of Applied Non-Classical Logics* (to appear).

[Rasiowa and Sikorski, 1968] H. Rasiowa and R. Sikorski. *The Mathematics of Meta-mathematics*. Polish Scientific Publishers, Warszawa, 2nd edition, 1968.

[Sernadas *et al.*, 1999] A. Sernadas, C. Sernadas, and C. Caleiro. Fibring of logics as a categorical construction. *Journal of Logic and Computation*, 9 (2):149–179, 1999.

[Sette, 1973] A. M. Sette. On the propositional calculus P^1. *Mathematica Japonicae*, 18:173–180, 1973.

[Wójcicki, 1969] R. Wójcicki. Logical matrices strongly adequate for structural sentential calculi. *Bulletin de l'Académie Polonaise des Sciences, Série des Sciences Mathématiques, Astronomiques et Physiques*, 17:333–335, 1969.

[Wójcicki, 1973] R. Wójcicki. Matrix approach in methodology of sentential calculi. *Studia Logica*, 32:7–37, 1973.

[Wójcicki, 1988] R. Wójcicki. *Theory of Logical Calculi - Basic theory of consequence operations*. Number 199 in Synthese Library. Kluwer Academic Publishers, 1988.

Indistinguishability by Default

Ariel Cohen, Michael Kaminski and
Johann A. Makowsky

For Dov Gabbay on the occasion of his 60th birthday

All animals are equal,
But some animals are more equal than others.
George Orwell, Animal Farm ([Orwell, 1945])

1 Introduction

As far back as 1973, Dov Gabbay and Julius Moravcsik, [Gabbay and Moravcsik, 1973], raised the following question: "Is 'same' always used in the same sense? Do the truth conditions of statements of the form 'x is the same as y' remain constant throughout the variety of uses in which they may be embedded?" They go on to consider several relations, each of which, in some sense, can be interpreted as saying that x and y are the same. In this paper we concentrate on yet another "sameness" relation: *indistinguishability by default*.

The idea underlying this notion has been proposed, though not formalized, in [Charniak, 1988]:[1] two objects are considered indistinguishable unless *proved* different, in which case they are, of course, distinguishable. By the axioms of ordinary equality, proving that t_1 and t_2 are different typically involves proving that for some formula ϕ, $\phi(t_1)$ yet $\neg\phi(t_2)$. Charniak's approach was further explored by A. Cohen and J.A. Makowsky in [Cohen and Makowsky, 1993], but the paper remained incomplete.

A considerable body of work in logic programming is based on the conceptually opposite principle, the *unique name assumption* ([Reiter, 1980a]), stating that two objects are different unless they can be proved to be equal. This fact, and the lack of formalization, may help explain why [Charniak, 1988] has not been followed up by the logic programming community. It therefore behooves us to provide some motivation for the notion of indistinguishability be default.

---

[1] Charniak calls the relation "nonmonotonic equality".

Such a relation may be useful for several applications, of which [Charniak, 1988] gives two examples. One is the task of resolving anaphora in natural language texts. A plausible way to do this is to reason by default — unless there is some incongruity between the pronoun and a candidate noun phrase (e.g., the pronoun is singular but the noun phrase is plural), assume that the pronoun refers to the noun phrase. Another potential use occurs in systems which attempt to understand natural language stories — in particular, when they face the task of identifying a motivation for some action described in the story. This is done by matching specific actions to stored plans. For example, suppose a story mentions a man lighting a specific match, and the program knows about a plan for lighting cigarettes, which contains the step of lighting some flammable object. The match will then be identified as the flammable object used, unless there are some facts to the contrary.

A different application where indistinguishability by default may be useful is McCarthy's Domain Circumscription ([McCarthy, 1977]): the idea that the domain should be kept as small as possible, i.e., as small as is consistent with what is provable. McCarthy argues that "humans often use circumscription, and robots must too" (see [McCarthy, 1977]). A deeper justification is given in [Davis, 1980], where the author sees circumscription as "an idea inherent in Occam's razor: only those objects should be assumed to exist which are minimally required by the context."

Yet another motivation for indistinguishability by default is more theory specific. It involves *Rough Set Theory* ([Pawlak, 1982]). The basic idea underlying Rough Set Theory is not to treat all properties as equally important: some properties are defined to be "distinguished." Two elements are considered *indiscernible* if they agree on all their distinguished properties. Relative to a given set of distinguished properties, the indiscernibility relation is clearly an equivalence relation, so it induces equivalence classes. For example, if the set of distinguished properties is $\{P, Q\}$, there are four equivalence classes: elements which satisfy $P \wedge Q$, elements which satisfy $P \wedge \neg Q$, elements which satisfy $\neg P \wedge Q$, and elements which satisfy $\neg P \wedge \neg Q$. Rough Set Theory, which deals with equivalence classes rather than directly with the elements of the domain, turns out to be useful for a number of applications, including data mining and uncertain reasoning.

Suppose, now, that we do not have perfect information. For instance, we know that $P(t_1)$ and $Q(t_1)$, but we only know that $P(t_2)$; we do not know whether $Q(t_2)$ or $\neg Q(t_2)$. Intuitively, as far as we know, t_1 and t_2 are indiscernible. However, the resulting relation is no longer an equivalence relation, because it is not, in general, transitive ([Kryszkiewicz, 1999]).

This indiscernibility relation is, intuitively, indistinguishability by default: we decide that t_1 and t_2 are the same if this is consistent with all their known (distinguished) properties. As we will show below, indistinguishability by default has formal properties similar to uncertain indiscernibility: it is reflexive, symmetric, but not transitive. Hence, indistinguishability by default can be seen as a way to formalize the relation of rough indiscernibility in the context of imperfect information.

A very different motivation for indistinguishability by default is a philosophical one: the ancient Greek *paradox of the Sorites*. Consider a sequence of objects $t_1, t_2, \ldots, t_n$ such that every two adjacent objects are indistinguishable from one another, yet the first and last objects are quite distinct. For example, imagine a sequence of tiles such that the leftmost tile is red, the rightmost tile is pink, whereas every two adjacent tiles are so similar in color as to be indistinguishable from one another. The paradox is that, since t_1 is red but t_n is not, there appears to be some cut-off point such that tiles to its left are red, and tiles to its right are not. However, any such arbitrarily chosen point will discriminate between two adjacent tiles that are, by hypothesis, indistinguishable. Any solution to this paradox will have to incorporate an appropriate definition of indistinguishability: one that will do justice to our intuitions, yet will not lead to paradox. It is easy to see, for example, that the relevant notion of indistinguishability cannot be transitive, or we would get the paradoxical result that red tile t_1 is indistinguishable from pink tile t_n. The notion of indistinguishability by default has turned out to be very useful in formalizing an appropriate indiscriminability relation ([Cohen, 2005]).

In this paper we provide a formalization of the intuitive notion of indistinguishability by default. Earlier work ([Cohen and Makowsky, 1993]) proposes a formalization that is appropriate for closed terms, but has undesirable consequences for terms containing variables; the formalization proposed here captures the idea of indistinguishability by default in its full generality. The paper is organized as follows. In the next section we discuss first-order default logic, which will be used in our formalization. Indistinguishability by default itself is discussed in Section 3. Finally, the last section contains some concluding remarks.

2 Background: default logic

This section consists of two parts. The first part is a non-formal description of the first-order default logic, whereas the second part contains the formal definitions.

2.1 First-order default logic

Non-monotonic logics are intended to simulate the process of human reasoning by providing a formalism for deriving consistent conclusions from an incomplete description of the world.

Reiter's *default logic* ([Reiter, 1980b]) is one of the widely used non-monotonic formalisms and maybe the only non-monotonic formalism that has a clearly useful contribution to the wider field of computer science through logic programming and database theory. This logic deals with rules of inference called *defaults* which are expressions of the form

$$(1) \quad \delta(\boldsymbol{x}) = \frac{\alpha(\boldsymbol{x}) : M\beta_1(\boldsymbol{x}), \ldots, M\beta_m(\boldsymbol{x})}{\gamma(\boldsymbol{x})},$$

where $\alpha(\boldsymbol{x})$, $\beta_1(\boldsymbol{x}), \ldots, \beta_m(\boldsymbol{x})$, $m \geq 1$, and $\gamma(\boldsymbol{x})$ are formulas of first-order logic whose free variables are among $\boldsymbol{x} = x_1, \ldots, x_n$. A default is *closed* if none of $\alpha, \beta_1, \ldots, \beta_m$, and γ contains a free variable. Otherwise it is *open*. Roughly speaking, the intuitive meaning of a default is as follows. For every n-tuple of objects $\boldsymbol{t} = t_1, \ldots, t_n$, if $\alpha(\boldsymbol{t})$ is believed, and the $\beta_i(\boldsymbol{t})$s are consistent with one's beliefs, then one is permitted to deduce $\gamma(\boldsymbol{t})$ and add it to the "belief set." Thus, an open default can be thought of as a kind of "default scheme," where free variables $\boldsymbol{x}$ can be replaced by any of the theory's objects. Various examples of deduction by defaults can be found in [Reiter, 1980b].

Whereas closed defaults have been quite thoroughly investigated, not much is known about open ones. However, interesting cases of default reasoning usually deal with open defaults, because the intended use of defaults is to determine whether an object possesses a given property, rather than accepting or rejecting a "fixed statement".

It was pointed out in [Lifschitz, 1990] that when applying open defaults one must specify *all* the objects of the underlying theory. Also, it was argued in [Kaminski, 1995] that one must distinguish between objects defined explicitly (closed terms) and objects introduced implicitly (by existential formulas, say).

In this paper we use the semantical definition of extensions for open default theories proposed in [Lifschitz, 1990] and [Kaminski, 1995], where, in contrast to the syntactical definitions in [Poole, 1988] and [Reiter, 1980b], the default free variables are treated as object variables, rather than metavariables for the closed terms of the theory. The reason for choosing a semantical definition of extensions is that, on the one hand, it provides a complete description of the theory objects, while distinguishing between explicitly and implicitly defined ones, on the other.

Since the semantical treatment of open default theories allows one to describe all the elements of the domain under consideration, it has no syntactical counterpart within the ordinary first-order default logic, unless the domain is explicitly defined by the *domain closure assumption*, i.e., the axiom

$$(2) \quad \forall x \bigvee_{i=1}^{m} x = t_i,$$

where $t_1, \ldots, t_m$ are closed. Under the *domain closure assumption*, extensions can be described syntactically by extending the underlying language of default theory with an infinite set of new constant symbols and replacing each open default with the set of all its closed instances.

It was shown in [Kaminski *et al.*, 1998] that extensions for open default theories depend on the domain cardinality (cf. [Kaminski, 1997]) and that over countable[2] or finite domains, extensions for open default theories can be described syntactically in first-order logic extended with an infinitary Carnap rule of inference

$$(3) \quad \frac{\{\varphi(t)\}_{t \in \boldsymbol{T}_{\mathcal{L}}}}{\forall x \varphi(x)}.$$

Here and hereafter $\boldsymbol{T}_{\mathcal{L}}$ denotes the set of all closed terms of a language $\mathcal{L}$.[3]

2.2 Definitions

In this section we briefly recall the definitions of default theories and the Herbrand semantics of first-order logic. *In what follows we assume that language $\mathcal{L}$ of the underlying first-order logic is countable.*

2.3 Default theories

Reiter's *default logic* ([Reiter, 1980b]) deals with rules of inference called *defaults* which are expressions of the form (1).

A *default theory* is a pair (D, A), where D is a set of defaults and A is a set of first-order sentences (axioms). A default theory is *closed*, if all its defaults are closed. However, in general, default theories are referred to as *open*.[4]

2.4 Extensions for closed default theories

Here we present the syntactical and semantical definitions of extensions for closed default theories.

Recall that closed defaults are expressions of the form

$$\frac{\alpha : M\beta_1, \ldots, M\beta_m}{\gamma},$$

where $\alpha, \beta_1, \ldots, \beta_m$, $m \geq 1$, and γ are closed formulas.

---

[2]In this paper, "countable" means *infinite* countable.

[3]Obviously, (2) implies (3), and, if $\boldsymbol{T}_{\mathcal{L}}$ is finite, then the converse implication also holds.

[4]That is, a closed default theory is a particular case of an open one.

DEFINITION 1. ([Reiter, 1980b, Definition 1]) Let (D, A) be a closed default theory. For any set of sentences S let $\Gamma_{(D,A)}(S)$ be the smallest set of sentences B (beliefs) that satisfies the following three properties.

D1. $A \subseteq B$.

D2. $\boldsymbol{Th}(B) = B$, i.e., B is deductively closed.

D3. If $\dfrac{\alpha : M\beta_1, \ldots, M\beta_m}{\gamma} \in D$, $\alpha \in B$, and $\neg\beta_1, \ldots, \neg\beta_m \notin S$, then $\gamma \in B$.

A set of sentences E is an *extension* for (D, A) if $\Gamma_{(D,A)}(E) = E$, i.e., if E is a fixed point of the operator $\Gamma_{(D,A)}$.

Next, we present a semantical definition of extension for closed default theories. Here and hereafter, for any class of interpretations W, by $\boldsymbol{Th}_{\mathcal{L}}(W)$ we mean the set of all closed formulas over $\mathcal{L}$ satisfied by all elements of W.

DEFINITION 2. ([Guerreiro and Casanova, 1990]) Let (D, A) be a closed default theory over $\mathcal{L}$. For any class of interpretations W, let $\Sigma_{(D,A)}(W)$ be the largest class V of models of A that satisfies the following condition.
    If $\dfrac{\alpha : M\beta_1, \ldots, M\beta_m}{\gamma} \in D$, $\alpha \in \boldsymbol{Th}_{\mathcal{L}}(V)$, and $\neg\beta_1, \ldots, \neg\beta_m \notin \boldsymbol{Th}_{\mathcal{L}}(W)$, then $\gamma \in \boldsymbol{Th}_{\mathcal{L}}(V)$.

Theorem 3 below states that the definition of extensions as the theories of the fixed points of the operator Σ is equivalent to Reiter's original definition (Definition 1).

Theorem 3 below is from [Guerreiro and Casanova, 1990], see also [Marek and Truszczyński, 1993, Theorem 3.45, p. 65].

THEOREM 3. *A set of sentences E is an extension for a closed default theory (D, A) if and only if $E = \boldsymbol{Th}_{\mathcal{L}}(W)$ for some fixed point W of $\Sigma_{(D,A)}$.*

2.5   Herbrand semantics of first-order logic

In this section we define Herbrand semantics of first-order logic that is the basis of the semantical approach to open default theories.

Let b be a set that contains no symbols of the underlying language $\mathcal{L}$. We denote by $\mathcal{L}_b$ the language obtained from $\mathcal{L}$ by augmenting its set of constants with all elements of b.[5] The set of all closed terms of the language $\mathcal{L}_b$ is called the *Herbrand universe* of $\mathcal{L}_b$. A *Herbrand b-interpretation* is a set of ground (closed) atomic formulas of $\mathcal{L}_b$. Note that closed formulas over $\mathcal{L}_b$ are of the form $\varphi(t_1, \ldots, t_n)$, where $t_1, \ldots, t_n$ are closed terms of language $\mathcal{L}_b$ and $\varphi(x_1, \ldots, x_n)$ is a formula over $\mathcal{L}$ whose free variables are

---

[5]Note that if b is uncountable, then so is $\mathcal{L}_b$.

among $x_1, \ldots, x_n$. The set b is called the *base* of Herbrand b-interpretation.

Let w be a Herbrand b-interpretation and let φ be a closed formula over $\mathcal{L}_b$. We say that w *satisfies* φ, denoted $w \models \varphi$, if the following holds.

- If φ is an atomic formula, then $w \models \varphi$ if and only if $\varphi \in w$;

- $w \models \varphi \supset \psi$ if and only if $w \not\models \varphi$ or $w \models \psi$;

- $w \models \neg\varphi$ if and only if $w \not\models \varphi$; and

- $w \models \forall x\varphi(x)$ if and only if for each $t \in \boldsymbol{T}_{\mathcal{L}_b}$, $w \models \varphi(t)$.

For a Herbrand b-interpretation w we define the $\mathcal{L}$-*theory* ($\mathcal{L}_b$-*theory*) of w, denoted $\boldsymbol{Th}_{\mathcal{L}}(w)$ ($\boldsymbol{Th}_{\mathcal{L}_b}(w)$), as the set of all closed formulas of $\mathcal{L}$ ($\mathcal{L}_b$) satisfied by w. For a set of Herbrand b-interpretations W we define the $\mathcal{L}$-*theory* ($\mathcal{L}_b$-*theory*) of W, denoted $\boldsymbol{Th}_{\mathcal{L}}(W)$ ($\boldsymbol{Th}_{\mathcal{L}_b}(W)$), as the set of all closed formulas of $\mathcal{L}$ ($\mathcal{L}_b$) satisfied by all elements of W. That is, $\boldsymbol{Th}_{\mathcal{L}}(W) = \bigcap\limits_{w \in W} \boldsymbol{Th}_{\mathcal{L}}(w)$ ($\boldsymbol{Th}_{\mathcal{L}_b}(W) = \bigcap\limits_{w \in W} \boldsymbol{Th}_{\mathcal{L}_b}(w)$).

Let X be a set of closed formulas over $\mathcal{L}_b$. We say that the Herbrand b-interpretation w is a *Herbrand b-model* of X, denoted by $w \models X$, if $X \subseteq \boldsymbol{Th}_{\mathcal{L}_b}(w)$. The set of all Herbrand b-models of X will be denoted by $\boldsymbol{Mod}_b(X)$. Finally, we say that X *semantically entails* a $\mathcal{L}_b$-formula φ, denoted $X \models_b \varphi$, if each Herbrand b-model of X satisfies φ.

REMARK 4. Note that the assignment to the equality relation in interpretations is a binary relation that does not have to be identity in the domain of the interpretation, but satisfies the equality first-order axioms. (Interpretations where the assignment to the equality relation is identity in the domain of the interpretation are called *normal*, see [Mendelson, 1997, p. 78] for details.) That is, the equality relation is treated as an ordinary dyadic predicate that satisfies the equality axioms.

REMARK 5. It is well-known that for an infinite set of new constant symbols b, Herbrand b-interpretations are complete and sound for first-order logic. That is, for a set of formulas X over $\mathcal{L}$ and a formula φ over $\mathcal{L}$, $X \vdash \varphi$ if and only if φ is satisfied by all Herbrand b-interpretations which satisfy X. In particular, Herbrand b-interpretations with an infinite base naturally arise in the Henkin proof of the completeness theorem ([Mendelson, 1997, Lemma 2.16, p. 89]).

2.6   Extensions for open default theories

In this section, departing from Definition 2 and following [Lifschitz, 1990] and [Kaminski, 1995] we present a definition of extensions for *open* default theories. It is known from [Kaminski, 1995] (see also Remark 7 below) that for closed default theories this definition is equivalent to the original Reiter's definition (Definition 1).

We start with the intuition underlying the definition. There are two types of objects in the domain of a default theory. One type consists of the *fixed* built-in objects which belong to $\boldsymbol{T}_{\mathcal{L}}$ and must be present in any Herbrand interpretation, and the other type consist of implicitly defined *unknown* objects which may vary from one Herbrand interpretation to other, e.g., objects introduced by existentially quantified formulas. These objects generate other unknown objects by means of the function symbols of $\mathcal{L}$. Thus, it seems natural to assume that the theory domain is a Herbrand universe of the original language augmented with a set of new (unknown) objects, cf. [Lloyd, 1993, Chapter 1, §3].

The following definition of extensions for open default theories is a relativization of Definition 2 to Herbrand b-interpretations with an infinite set of new constant symbols b. The reason for passing to a semantical definition is that, in general, it is impossible to describe a Herbrand universe by means of the standard proof theory. The only exception is the cases when the theory domain is explicitly finite (see [Kaminski, 1999]), i.e., contains axiom (2).

DEFINITION 6. (Cf. [Kaminski, 1995, Definition 27]) Let b be a set of new constant symbols and let (D, A) be a default theory. For any set of Herbrand b-interpretations W let $\Delta^b_{(D,A)}(W)$ be the largest set V of Herbrand b-models of A that satisfies the following condition.

For any default $\dfrac{\alpha(\boldsymbol{x}) : M\beta_1(\boldsymbol{x}), \ldots, M\beta_m(\boldsymbol{x})}{\gamma(\boldsymbol{x})} \in D$ and any tuple $\boldsymbol{t}$ of elements of $\boldsymbol{T}_{\mathcal{L}_b}$ if $V \models \alpha(\boldsymbol{t})$ and $W \not\models \neg\beta_i(\boldsymbol{t})$, $i = 1, 2, \ldots, m$, then $V \models \gamma(\boldsymbol{t})$.

A set of $\mathcal{L}_b$ sentences E is called a *b-extension* for (D, A) if $E = \boldsymbol{Th}_{\mathcal{L}_b}(W)$ for some fixed point W of $\Delta^b_{(D,A)}$.

We will also refer to the set b as the *base* of E.

REMARK 7. It follows from the Löwenheim-Skolem theorem, see [Chang and Keisler, 1990], that, for a closed default theory (D, A) and an infinite base b, a set of sentences is a b-extension for (D, A) if and only if it is an "ordinary" Reiter's extension for (D, A).

From now on, unless we state otherwise, we deal with infinite bases, because the cardinality of a finite base b can be extracted from the b-extension, which is undesirable in the general case.

REMARK 8. Note that for two bases b and b' of different cardinality the sets of b- and b'-extensions for an open default theory do not necessarily coincide, see [Kaminski *et al.*, 1998, Example 7.1].

3   Indistinguishability by default

Here we formalize the notion of *indistinguishability by default*. The formal definition is given in Section 3.1 and its properties are listed in Section 3.2.

Below $\sim$ is a binary predicate symbol that does not belong to the underlying language $\mathcal{L}$. It is intended to denote "indistinguishability by default."

3.1   Relation $\sim$

A two element set of defaults $\left\{ \dfrac{: Mx = y}{x \sim y}, \dfrac{x \neq y :}{x \not\sim y} \right\}$ will be denoted $D_\sim$.

REMARK 9.   In the modal context, the intuitive meaning of $\sim$ is that $x \sim y$ if and only if "possibly $x = y$," cf. [Kaminski and Rey, 2000, Section 7].

REMARK 10.   Let E be an extension for default theory $(D_\sim, A)$. Then E "is complete for $\sim$" in the following sense. For each two closed terms t_1 and t_2, either $t_1 \sim t_2$ or $t_1 \not\sim t_2$ belongs to E.

Let us now demonstrate that, for a given Herbrand base, the definition of '$\sim$' "works", in the sense that two terms are indistinguishable by default iff they cannot be proved to be different.

THEOREM 11.   *Operator* $\Delta^b_{(D_\sim, A)}$ *has a unique fixed point*

$$(4) \qquad \begin{aligned} V_b = \boldsymbol{Mod}_b(A \quad &\cup \quad \{t_1 \sim t_2 : t_1, t_2 \in \boldsymbol{T}_{\mathcal{L}_b} \quad and \quad A \not\models_b t_1 \neq t_2\} \\ &\cup \quad \{t_1 \not\sim t_2 : t_1, t_2 \in \boldsymbol{T}_{\mathcal{L}_b} \quad and \quad A \models_b t_1 \neq t_2\}). \end{aligned}$$

For the proof of Theorem 11 we shall need the following notation. For a Herbrand b-interpretation w of $\mathcal{L} \cup \{\sim\}$, we denote by w^- the restriction of w to $\mathcal{L}_b$, and for a set W of Herbrand b-interpretation of $\mathcal{L} \cup \{\sim\}$, we define a set W^- of Herbrand b-interpretations of $\mathcal{L}$ by $W^- = \{w^- : w \in W\}$.

Proof of Theorem 11.   Let W be a set of Herbrand b-interpretations of $\mathcal{L} \cup \{\sim\}$ satisfying A. Then $W \not\models t_1 \neq t_2$ implies $A \not\models_b t_1 \neq t_2$, which, in turn, implies $V_b \models t_1 \sim t_2$. In addition, if $V_b \models_b t_1 \neq t_2$, then $A \models_b t_1 \neq t_2$, implying $V_b \models_b t_1 \not\sim t_2$. This together with Remark 10 and (4) (in Theorem 11) implies $\Delta^b_{(D_\sim, A)}(W) = V_b$, and the theorem follows.   ∎

COROLLARY 12.   *For each infinite set of new constant symbols b, default theory $(D_\sim, A)$ has a unique b-extension $E_b = \boldsymbol{Th}_{\mathcal{L}}(V_b)$.*

Theorem 11 and its corollary seemingly leave open the undesirable possibility that indistinguishability by default depends on the particular choice of the Herbrand base b. However, this is only an apparent possibility. Theorem 13 below shows that restrictions of b-extensions for $(D_\sim, A)$ to $\mathcal{L}$ do not depend on the cardinality of base b, cf. [Kaminski et al., 1998, Example 7.1].

THEOREM 13. *Let b_1 and b_2 be infinite sets of new constant symbols and let V_{b_1} and V_{b_2} be the fixed points of $\Delta^{b_1}_{(D_\sim, A)}$ and $\Delta^{b_2}_{(D_\sim, A)}$, respectively. Then $\boldsymbol{Th}_\mathcal{L}(V_{b_1}) = \boldsymbol{Th}_\mathcal{L}(V_{b_2})$.*

The proof of Theorem 13 is based on the following observation.

LEMMA 14. *Let b_1 and b_2 be infinite sets of new constant symbols such that $b_1 \subseteq b_2$ and let φ be a closed formula over $\mathcal{L}_{b_1}$. Then $A \models_{b_1} \varphi$ if and only if $A \models_{b_2} \varphi$.*

Proof. We start with the proof of the "only if" part of the lemma. Let $A \models_{b_1} \varphi$ and assume to the contrary that $A \not\models_{b_2} \varphi$. Let w_2 be a Herbrand b_2-model of A such that $w_2 \not\models \varphi$. Let b' be the set of new constant symbols which appear in φ and let w_1 be an elementary subinterpretation of w_2 of cardinality $\|b_1\|$ containing b'.[6] In particular, w_1 is a Herbrand b'_1-interpretation for some subset b'_1 of b_2 of cardinality $\|b_1\|$. Renaming the elements of $b'_1 \setminus b'$, if necessary, we may assume that $b'_1 = b_1$. Then, being an elementary subinterpretation of w_2, $w_1 \models A \cup \{\neg\varphi\}$, which contradicts $A \models_{b_1} \varphi$.

For the proof of the "if" part of the lemma, let $A \models_{b_2} \varphi$ and assume to the contrary that $A \not\models_{b_1} \varphi$. Let w_1 be a Herbrand b_1-model of A such that $w_1 \not\models \varphi$. Let b' be the set of new constant symbols which appear in φ and let $\iota : b_2 \to b_1$ be a surjective mapping that is an identity on b_1. Consider a Herbrand b_2-interpretation

$$w_2 = \{P(c_1, c_2, \ldots, c_n) : P(\iota(c_1), \iota(c_2), \ldots, \iota(c_n)) \in w_1\}.$$

A simple induction on the complexity of an $\mathcal{L}$ formula $\psi(x_1, x_2, \ldots, x_n)$ shows that for any $c_1, c_2, \ldots, c_n \in b_2$ $w_2 \models \psi(c_1, c_2, \ldots, c_n)$ if and only if $w_1 \models \psi(\iota(c_1), \iota(c_2), \ldots, \iota(c_n))$.[7] Thus, w_2 is a Herbrand b_2-model of A that does not satisfy φ, which contradicts $A \models_{b_2} \varphi$. ∎

COROLLARY 15. *Let b_1 and b_2 be infinite sets of new constant symbols such that $b_1 \subseteq b_2$. Let $w_2 \in V_{b_2}$ and let w_1 be the Herbrand b_1-interpretation constructed in the proof of the "only if" part of Lemma 14. Then $w_1 \in V_{b_1}$.[8]*

---

[6]Recall that $\mathcal{L}$ is countable.

[7]In particular, w_1 is an elementary subinterpretation of w_2.

[8]See (4) (in Theorem 11) for the definition of V_b.

Proof. Since w_1 is an elementary subinterpretation of w_2, $w_1 \models A$.

Let $t_1, t_2 \in T_{\mathcal{L}_{b_1}}$ and assume that $A \not\models_{b_1} t_1 \neq t_2$. Then, by Lemma 14, $A \not\models_{b_2} t_1 \neq t_2$ either, which together with $w_2 \in V_{b_2}$ implies $w_2 \models t_1 \sim t_2$. Since w_1 is an elementary subinterpretation of w_2, $w_1 \models t_1 \sim t_2$ as well.

Similarly, $A \models_{b_1} t_1 \neq t_2$ implies $w_1 \models t_1 \not\sim t_2$, and $w_1 \in V_{b_1}$ follows. ∎

COROLLARY 16. *Let b_1 and b_2 be infinite sets of new constant symbols such that $b_1 \subseteq b_2$. Let $w_1 \in V_{b_1}$ and let w_2 be the Herbrand b_1-interpretation constructed in the proof of the "if" part of Lemma 14. Then $w_2 \in V_{b_2}$.*

Proof. The proof is similar to that of Corollary 15.

Since w_1 is an elementary subinterpretation of w_2 and $w_1 \in w_2$, $w_2 \models A$.

Let $t_1, t_2 \in T_{\mathcal{L}_{b_2}}$ and assume that $A \not\models_{b_2} t_1 \neq t_2$. Then $A \not\models_{b_2} \iota(t_1) \neq \iota(t_2)$.[9] Consequently, $w_1 \models \iota(t_1) \sim \iota(t_2)$, implying $w_2 \models t_1 \sim t_2$.

Similarly, $A \models_{b_2} t_1 \neq t_2$ implies $w_2 \models t_1 \not\sim t_2$, and $w_2 \in V_{b_2}$ follows. ∎

Proof of Theorem 13. Of course, we may assume that $b_1 \subseteq b_2$.

Let φ be a closed formula over $\mathcal{L}$. Assume that $\varphi \notin \boldsymbol{Th}_{\mathcal{L}}(V_{b_2})$. Let w_2 be an element of V_{b_2} that does not satisfy φ and let w_1 be an elementary subinterpretation of w_2 constructed in the proof of the "only if" part of Lemma 14. Then $w_1 \not\models \varphi$ and, by Corollary 15, $w_1 \in V_{b_1}$. Thus, $\varphi \notin \boldsymbol{Th}_{\mathcal{L}}(V_{b_1})$ either.

Conversely, assume that $\varphi \notin \boldsymbol{Th}_{\mathcal{L}}(V_{b_1})$. Let w_1 be an element of V_{b_1} that does not satisfy φ and let w_2 be the Herbrand b_2-interpretation constructed in the proof of the "if" part of Lemma 14. Then $w_2 \not\models \varphi$ and, by Corollary 16, $w_2 \in V_{b_2}$. Thus, $\varphi \notin \boldsymbol{Th}_{\mathcal{L}}(V_{b_2})$, which completes the proof. ∎

By Theorems 11 and 13, all restrictions of b-extensions for $(D_\sim, A)$ to $\mathcal{L}$ coincide, i.e., $\sim$ is well defined. So, regardless of the Herbrand base, $(D_\sim, A)$ has a unique extension, which we shall denote E_A.

3.2   Properties of $\sim$

In this section we observe a number of properties of $\sim$ which follows easily from its definition. First we note that indistinguishability by default complies with the intuitive notion described in the introduction. That is, that two domain elements are indistinguishable if and only if there is no property that can be proved to be true of one yet false of the other. In particular, if t_1 and t_2 are closed terms, then $t_1 \sim t_2$ if and only if there exists no formula $\varphi(x)$ with only one free variable x such that $A \vdash \varphi(t_1)$, but $A \vdash \neg\varphi(t_2)$.

---

[9]As usual, for a term $t \in T_{\mathcal{L}_{b_2}}$, $\iota(t)$ is obtained from t by each $c \in b_2$ with $\iota(c)$. We leave the formal definition to the reader.

Below (in addition to Remark 10) we list some properties of $\sim$ which immediately follow from the definition of E_A.

- Indistinguishability by default is *non-monotonic*. For example, obviously, $\forall x \forall y x \sim y \in E_\emptyset$, However, $\forall x \forall y x \sim y \notin E_{c_1 \neq c_2}$, because of the instance $\dfrac{c_1 \neq c_2 :}{c_1 \not\sim c_2}$ of default $\dfrac{x \neq y :}{x \not\sim y}$, see Theorem 11.

- By definition, $\sim$ is reflexive and symmetric. However, it is not transitive. For example, both $c_1 \sim c_2$ and $c_2 \sim c_3$ belong to $E_{c_1 \neq c_3}$, whereas $c_1 \sim c_3$ does not.

- On the other hand $\sim$ possesses the following "semi-transitivity" property, which looks almost like a "duble negation" of the usual transitivity of equality.

  Let $t_1, t_2, t_3 \in \boldsymbol{T}_\mathcal{L}$. Then $A \not\vdash t_1 \neq t_2 \vee t_2 \neq t_3$ implies $t_1 \sim t_3 \in E_a$.[10]

  For the proof assume $A \not\vdash t_1 \neq t_2 \vee t_2 \neq t_3$. By Theorem 11, it suffices to show that $A \not\vdash t_1 \neq t_3$. Were $A \vdash t_1 \neq t_3$, by transitivity of $=$, we would have $A \vdash t_1 \neq t_2 \vee t_2 \neq t_3$, which contradicts our assumption.

Finally we show that extensions for $(D_\sim, A)$ have a "generic" element given by Proposition 17 below.

PROPOSITION 17. *Let V_b be the fixed point of $\Delta^b_{(D_\sim, A)}$. Then $V_b \models \exists x \forall y (x \sim y)$.*

Proof. Let $c \in b$ be a new constant symbol. Then for each $t \in \boldsymbol{T}_{\mathcal{L}_b}$ there is a Herbrand b-interpretation $w \in V_b$ such that $w \models c = t$. Thus, $V_b \models \forall x (c \sim x)$, and the proposition follows. ∎

4 Conclusion

In this paper we have formalized the idea of considering two objects indistinguishable unless there is some property that distinguishes between them. We have shown that our formalization has some desirable properties. The relation is reflexive, symmetric, and has the property of semi-transitivity. We believe, therefore, that our formalization does justice to the intuitive notion of indistinguishability by default.

BIBLIOGRAPHY

[Chang and Keisler, 1990] C.C. Chang and H.J. Keisler. *Model Theory*. North Holland Publishing Co., Amsterdam, 1990.

---

[10]Obviously, this property would hold, were $\sim$ transitive.

[Charniak, 1988] E. Charniak. Motivation analysis, abductive unification and nonmonotonic equality. *Artificial Intelligence*, 34:275–295, 1988.

[Cohen and Makowsky, 1993] A. Cohen and J. A. Makowsky. Two approaches to nonmonotonic equality. Technical Report CIS-9317, Technion—Israel Institute of Technology, 1993.

[Cohen, 2005] A. Cohen. Vagueness and indiscriminability by failure. Presented at *the 33rd Annual Meeting of the Society for Exact Philosophy*, 2005.

[Davis, 1980] M. Davis. The mathematics of non-monotonic reasoning. *Artificial Intelligence*, 13:73–80, 1980.

[Gabbay and Moravcsik, 1973] D. Gabbay and J. M. E. Moravcsik. Sameness and individuation. *Journal of Philosophy*, 70:16, 1973. Reprinted in F. J. Pelletier (ed.), *Mass Terms: Some Philosophical Problems*, Dordrecht, 1979. Reidel.

[Guerreiro and Casanova, 1990] R. Guerreiro and M. Casanova. An alternative semantics for default logic. Presented at *the 3rd International Workshop on Nonmonotonic Reasoning*, 1990.

[Kaminski and Rey, 2000] M. Kaminski and G. Rey. First–order non-monotonic modal logics. *Fundamenta Informaticae*, 42:303–333, 2000.

[Kaminski et al., 1998] M. Kaminski, J.A. Makowsky, and M. Tiomkin. Extensions for open default theories via the domain closure assumption. *Journal of Logic and Computation*, 8:169–187, 1998.

[Kaminski, 1995] M. Kaminski. A comparative study of open default theories. *Artificial Intelligence*, 77:285–319, 1995.

[Kaminski, 1997] M. Kaminski. A note on the stable model semantics for logic programs. *Artificial Intelligence*, 96:467–479, 1997.

[Kaminski, 1999] M. Kaminski. Open default theories over closed domains. *Logic Journal of the IGPL*, 7:577–589, 1999.

[Kryszkiewicz, 1999] M. Kryszkiewicz. Rules in incomplete information systems. *Information Sciences*, 113:271–292, 1999.

[Lifschitz, 1990] V. Lifschitz. On open defaults. In J.W. Lloyd, editor, *Computational Logic: Symposium Proceedings*, pages 80–95, Berlin, 1990. Springer–Verlag.

[Lloyd, 1993] J.W. Lloyd. *Foundation of logic programming, second extended edition*. Springer–Verlag, Berlin, 1993.

[Marek and Truszczyński, 1993] W. Marek and M. Truszczyński. *Nonmonotonic Logic*. Springer–Verlag, Berlin, 1993.

[McCarthy, 1977] J. McCarthy. Epistemological problems of artificial intelligence. In *Proceedings of the 5th International Joint Conference on Artificial Intelligence*, pages 1038–1044, Los Angeles, CA, 1977. Kaufmann.

[Mendelson, 1997] E. Mendelson. *Introduction to mathematical logic*. Chapman and Hall, London, 1997.

[Orwell, 1945] G. Orwell. *Animal Farm*. Penguin, London, 1945.

[Pawlak, 1982] Z. Pawlak. Rough set. *International Journal of Computer and Information Sciences*, 11:341–356, 1982.

[Poole, 1988] D. Poole. A logical framework for default reasoning. *Artificial Intelligence*, 36:27–47, 1988.

[Reiter, 1980a] R. Reiter. Equality and domain closure in first order databases. *Journal of the ACM*, 27:235–249, 1980.

[Reiter, 1980b] R. Reiter. A logic for default reasoning. *Artificial Intelligence*, 13:81–132, 1980.

Classical Natural Deduction[1]

MARCELLO D'AGOSTINO

1 Introduction

In the tradition which considers formal logic as an *organon* of thought a
central role has been played by the "method of analysis", which amounts to
what today, in computer science circles, is called a "bottom-up" or "goal-
oriented" procedure[1]. Though the method was largely used in the mathematical practice of the ancient Greeks, its fullest description can be found
in Pappus (3rd century A.D.), who writes:

> Now analysis is a method of taking that which is sought as
> though it were admitted and passing from it through its con-
> sequences in order, to something which is admitted as a result
> of synthesis; for in analysis we suppose that which is sought be
> already done, and we inquire what it is from which this comes
> about, and again what is the antecedent cause of the latter, and
> so on until, by retracing our steps, we light upon something al-
> ready known or ranking as a first principle; and such a method
> we call analysis, as being a solution backwards[2].

---

[1]This paper is based on some ideas I had the privilege to discuss thoroughly with my
mentor and friend Marco Mondadori. Although he left us unexpectedly on 4 April 1999,
his scientific and philosophical contributions are as alive as ever. I will never be able
to express adequately how much I am indebted to him. I am also very much indebted
to Dov Gabbay, to whom this paper is dedicated on his 60th birthday, for giving me a
job in the first place (back in 1990), for sharing his deep and thrilling views about logic
and for teaching me to "think big". Thanks are due also to Krysia Broda with whom I
had the opportunity to work on problems related to the contents of this paper. Some of
the examples in Sections 2 and 3 are taken from a joint paper with her and Mondadori
([Broda *et al.*,]). We assume the reader is familiar with the natural deduction calculus
(in its several variants) as well as with the basic notions of Gentzen's sequent calculi and
Smullyan's analytic tableaux. For the reader's convenience the rules of these systems
are listed in the Appendix. For the tableau method see [Smullyan, 1968], for natural
deduction and its variants the classical references are [Prawitz, 1965] and [Tennant, 1978].

[1]For a general treatment of goal-oriented logical algorithms see [Gabbay and Olivetti, 2000].

[2][Thomas, 1941] pp. 596–599.

This is the so-called *directional* sense of analysis. The idea of an "analytic method", however, is often associated with another sense of "analysis" which is related to the "purity" of the concepts employed to obtain a result and, in the framework of proof-theory, to the *subformula principle* of proofs. In Gentzen's words: "No concepts enter into the proof other than those contained in its final result, and their use was therefore essential to the achievement of that result"[Gentzen, 1935, p. 69] so that "the final result is, as it were, gradually built up from its constituent elements" [Gentzen, 1935, p.88]. Both meanings of analysis have been represented, in the last fifty years, by the notion of a *cut-free* proof in Gentzen-style sequent calculi.

Gentzen introduced the sequent calculi **LK** and **LJ** as well as the natural deduction calculi **NK** and **NJ** in his famous 1935 paper [Gentzen, 1935]. Apparently, his primary interest was in natural deduction, although he considered the sequent calculi as technically more convenient for meta-logical investigation[3]. In the first place he wanted to set up a formal system which "comes as close as possible to actual reasoning"[4]. In this context he introduced the natural deduction calculi where each operator is associated with suitable rules for *introducing* and *eliminating* it. He also suggested that such rules could be seen as *definitions* of the operators themselves. In fact, he argued that the introduction rules alone are sufficient for this purpose and that the elimination rules are "no more, in the final analysis, than consequences of these definitions"[Gentzen, 1935, p. 80]. However, he observed that this "harmony" is exhibited by the intuitionistic calculus but breaks down in the classical case.[5] Indeed, in natural deduction inferences are analysed essentially in a *constructive* way and classical logic is obtained by adding the law of excluded middle in a purely external manner. Gentzen recognized that the special position occupied by this law would have prevented him from proving the *Hauptsatz* (i.e. the subformula principle) in the case of classical logic. So he introduced the sequent calculi as a technical device in order to enunciate and prove the *Hauptsatz* in a convenient form[6] both for intuitionistic and classical logic. These calculi still have a strong deduction-theoretical flavour and Gentzen did not show any sign of considering the relationship between the classical calculus and the semantic notion of entailment which, at the time, was considered as highly suspicious.

Whereas in the natural deduction calculi there are, for each operator, an introduction and an elimination rule, in the sequent calculi there are only

---

[3][Gentzen, 1935], p. 69.

[4][Gentzen, 1935], p. 68.

[5]For a thorough discussion of this subtle meaning-theoretical issue the reader is referred to the writings of Michael Dummett and Dag Prawitz, in particular [Dummett, 1978] and [Prawitz, 1978].

[6][Gentzen, 1935], p. 69.

introduction rules and the eliminations take the form of introductions in the antecedent. Gentzen seemed to consider the difference between the two formulations as a purely technical aspect.

Gentzen's sequent rules have become a paradigm both in proof-theory and in its applications. This is not without reason. First, like the natural deduction calculi, they provide a precise analysis of the logical operators by specifying how each operator can be introduced in the antecedent or in the succedent of a sequent. Second, their form ensures the validity of the *Hauptsatz*: each proof can be transformed into one which is cut-free, and cut-free proofs enjoy the *subformula property*, that is every sequent in the proof tree contains only subformulas of the formulas in the sequent to be proved.

Therefore, cut-free proofs comply with the *first* meaning of logical analysis, in that they employ no concepts outside those contained in the statement of the theorem. But they comply also with the *second* meaning, in that they can be discovered by means of simple "backwards" procedures like that described by Pappus. Hence, Gentzen-style cut-free methods (which include their semantic-flavoured descendant, known as "the tableau method") were among the first used in automated deduction[7] and are still today very popular in computer science circles and seem to have overturned the once unchallenged supremacy of resolution and its refinements in the area of automated deduction and artificial intelligence.

As a result of Dag Prawitz's later work on normalization ([Prawitz, 1965]) natural deduction also established itself as a respectable "analytic" method in its own right. Not only do normal proofs obey the subformula principle, but they can also be discovered by means of "backwards" procedures similar to those employed for the sequent calculus and there is a close, though far from trivial, relationship between cut-free proofs in the sequent calculus and normal proofs in the natural deduction system. Moreover, natural deduction methods have been devised for a wide variety of logics and constitute the proof-theoretical basis of the taxonomy of logical systems investigated in Dov Gabbay's book on Labelled Deductive Systems.

As Gabbay remarked in that book, "logical systems which may be conceptually far apart (in their philosophical motivation and mathematical definitions), when it comes to automated techniques and proof-theoretical presentation turn out to be brother and sister"[Gabbay, 1996, Section 1.1]. On the other hand, the same logical system (intended as a set-theoretical definition of a consequence relation) usually admits of a wide variety of proof-theoretical representations which may reveal or hide its similarities

---

[7]See, for example, [Beth, 1958], [Wang, 1960], [Kanger, 1963]. See also [Davis, 1983] and [Maslov *et al.*, 1983].

with other logical systems. As an example of this approach, Gabbay observes that "it is very easy to move from the truth-table presentation of classical logic to a truth-table system for Lukasiewicz's n-valued logic. It is not so easy to move to an algorithmic system for intuitionistic logic. In comparison, for a Gentzen system presentation, exactly the opposite is true. Intuitionistic and classical logic are neighbours, while Lukasiewicz's logics seem completely different."[Gabbay, 1996, Section 1.1].

The consideration of these phenomena prompted Gabbay to put forward the view that

> [...] a logical system **L** is not just the traditional consequence relation $\vdash$, but a pair $(\vdash, \mathbf{S}_\vdash)$, where $\vdash$ is a mathematically defined consequence relation (i.e. a set of pairs (Δ, Q) such that $\Delta \vdash Q$) and $\mathbf{S}_\vdash$ is an algorithmic system for generating all those pairs. Thus, according to this definition, classical logic $\vdash$ perceived as a set of tautologies together with a Gentzen system $\mathbf{S}_\vdash$ is not the same as classical logic together with the two-valued truth-table decision procedure $\mathbf{T}_\vdash$ for it. In our conceptual framework, $(\vdash, \mathbf{S}_\vdash)$ is *not the same logic* as $(\vdash, \mathbf{T}_\vdash)$[8].

If one takes Gabbay's idea seriously, even civilized areas, such as the elementary proof-theory of classical logic, may still reveal unexplored corners. For instance let us consider classical logic intended as the pair $(\vdash, \mathbf{T}_\vdash)$ where $\vdash$ is the set of all (Δ, Q) such that Δ classically implies Q, and $\mathbf{T}_\vdash$ is the two-valued truth-table decision procedure; we can call this logical system *standard classical logic*. Let us also consider classical logic intended now as the pair $(\vdash, \mathbf{N}_\vdash)$ where $\vdash$ is the same consequence relation as before and $\mathbf{N}_\vdash$ is Gentzen's natural deduction system for it. We shall argue in the sequel (and we think Gabbay would agree) that these two "logical systems" (in his stricter sense) are conceptually, philosophically and algorithmically "far apart". Moreover, natural deduction is not only a specific proof system but a philosophical approach to the meaning of the logical operators. The main problem we address in this paper is the following: can we devise a proof system which can be legitimately called a "natural deduction system" — in that it is constructed in accordance with the basic philosophical principles of Gentzen's and Prawitz' natural deduction — and yet can be perceived as a close "neighbour" of standard classical logic, i.e. the familiar system $(\vdash, \mathbf{T}_\vdash)$?

The interest of this problem is not only philosophical. Natural deduction systems, as remarked above, *do* lend themselves to automated proof search

---

[8][Gabbay, 1996, Section 1.1].

procedure and are at least as "portable" (i.e. easily adaptable to various logics) as the sequent calculus or the tableau method — as shown in Gabbay's
book on Labelled Deductive Systems ([Gabbay, 1996]). Moreover, they are
built on a solid meaning-theoretical basis and are extensively used in teaching. One of the reasons that their use in computer science and teaching
applications is not as widespread as it could be lies perhaps in the fact that
their *classical* variants appear to be technically and conceptually contrived,
a difficulty which stretches to all the "neighbours" (in Gabbay's sense) of
standard classical logic, such as classical modal logics or many-valued logics.

In the sequel we shall argue that *the main drawback of natural deduction*,
as a proof-theoretical presentation of classical logic and of its neighbours,
does not lie in its being unsuitable to backwards or goal-oriented proof-
search, but in the fact that its rules do *not* capture the *classical meaning* of
the logical operators. This was already remarked in Prawitz [Prawitz, 1965],
where the author pointed out that natural deduction rules are nothing but a
reading of Heyting's explanations of the *constructive meaning* of the logical
operators ([Heyting, 1956]). In particular, the rules for the *conditional* are
based on an interpretation of this operator which has nothing to do with
the classical interpretation based on the truth-tables.

This mismatch between the natural deduction rules and the classical
meaning of the logical operators is responsible for the fact that rather simple classical tautologies become quite hard to prove within the natural deduction framework, in the sense that their simplest proofs involve "tricks"
which are far from being natural.

On the other hand, *the main drawback of the cut-free sequent calculus in
all its variants* — still as a proof-theoretical presentation of logics in the
neighbouroods of standard classical logic — is that its rules do *not* capture
an essential feature of the classical notions of truth and falsity, namely the
principle of bivalence, according to which every proposition is determinately
true or false, independent of our epistemic means for establishing its truth-
value.[9] The lack of a rule expressing bivalence has two main consequences.
First, it does not allow the use of "lemmas". As a result, the construction
of a cut-free sequent proof (or of a tableau-proof) is a semi-mechanical procedure where the only choice involved concerns the order in which the rules
are applied, so restricting significantly the user's (or the computer's) free-

---

[9]The principle of bivalence is nothing but a semantic reading of the *cut rule* of the
sequent calculus. In the Gentzenian tradition, analytic arguments (satisfying the subformula principle) are construed as cut-free proofs. As far as classical logic is concerned, this
elimination of cut is, therefore, the *elimination of bivalence*. Indeed, there are 3-valued
semantics that characterize the cut-free sequent calculus (and are therefore equivalent
to the standard two-valued semantics as far as theoremhood is concerned, but do not
validate the cut rule). On these points see [Girard, 1987a] and [D'Agostino, 1990].

dom in devising and representing proofs. A second, related, consequence is that, as emerges from over twenty years of efforts in the area of "automated theorem-proving", the lack of a proper rule expressing the classical principle of bivalence (or "cut", or "transitivity") is responsible for severe inefficiency even if we restrict ourselves to the domain of "analytic" proofs, i.e. proofs satisfying the subformula principle[10].

In the sequel we shall attempt to develop a theory of classical natural deduction which avoids the drawbacks of both standard natural deduction and the cut-free sequent calculus.[11] The underlying philosophical ideas were put forward by Marco Mondadori in the late 1980's, and some of their proof-theoretical consequences were investigated in a series of joint papers by him and the present author.[12] We shall restrict our attention to the propositional operators, since the treatment of quantifiers, though not particularly difficult, requires a good deal of technical details, especially in connection with the introduction rules[13] and will therefore be postponed to a subsequent paper.

Our theory of classical natural deduction makes a neat distinction between *operational rules*, governing the use of logical operators, and *structural rules* dealing with the metaphysical assumptions governing the (classical) notions of truth and falsity, namely the *principle of bivalence* and the *principle of non-contradiction*. The operational rules are, like in standard natural deduction, introduction and elimination rules for the logical operators. Although all the rules have a distinctively "classical" flavour, the elimination rules are shown to be in "harmony" with the introduction rules, in the sense that they can be justified by means of an inversion principle which is very close to Prawitz' one,[14] except that it concerns classical style inference rules, i.e. rules whose premises and conclusions are stated in terms of the truth and falsity of sentences (where truth and falsity are interpreted in a classical, non-epistemic, way), rather than in terms of proofs. These rules can be

---

[10]On this point and on the advantages on "analytic cut" systems, see [D'Agostino and Mondadori, 1994], [D'Agostino, 1990], [D'Agostino, 1992], [D'Agostino and Mondadori, 1993] and [D'Agostino, 1999].

[11]Another, very different, approach to classical natural deduction which exploits the idea of multiple-conclusion sequents typical of the classical sequent calculus, see [Cellucci, 1988].

[12]See the references in footnote 24.

[13]For a discussion of introduction rules for quantifiers see [Mondadori, 1996].

[14]For a different attempt to develop a concept of "classical harmony" the reader is referred to [Weir, 1986]. We agree with Alan Weir's view that "it would be blind dogmatism to suppose that the classical conception of truth as independent of proof stands in no need of justification. The rebuttal of skeptical attempts to show that the classical notions are incompatible with plausible assumptions in the theory of meaning is a challenge which the classicist ought to try to meet."(p. 461).

seen as assigning a classical meaning to the logical operators, but they are not sufficient to yield the full deductive power of classical logic. The latter can be obtained only with the addition of the metaphysical assumptions expressed by the structural rules. So, the direct inferential role played by the logical operators, via their defined meaning, is clearly separated by the indirect inferential role played by the metaphysical assumptions.

Even with the addition of the structural rules, it is possible to show that the resulting proof system satisfies, like standard natural deduction, the (weak) subformula property. However, since the harmony constraint is satisfied only by the operational rules, a stricter notion of "analytic" proof is obtained with reference to proofs obtained by means of these rules only, without essential applications of the structural rules. We shall argue that the typical idea of an analytic argument as one in which the conclusion is "contained" in the premises applies to proofs which are analytic in this stricter sense. As a sign of the soundness of this approach, it turns out that the recognition of the conclusion's being "contained" in the premises, that is the discovery of a strictly analytic proof of the conclusions from the premises, is a *tractable* problem, as one would expect for any reasonable notion of "containment".

In sections 2 and 3 we shall point out the shortcomings of natural deduction and of the cut-free sequent calculus as formal representations of the notion of classical analytic proof.[15] Next, we shall present our own approach based on classical "intelim" (*intro*duction and *elim*ination) rules, first (section 4) in a format which resembles that of standard natural deduction and then (section 5) in a more convenient format which allows for a more concise representation of proofs. In section 6 we shall address typical normalization issues and show the subformula property for intelim proofs. Finally, in section 7, we shall investigate *strictly analytic refutations* — refutations which do not require applications of the principle of bivalence. We shall then propose a graph-based decision procedure for them and show that it runs in quadratic time. Finally, in section 7 we shall point at some future research which may be carried out in connection with this work.

2 Why Gentzen's natural deduction is not a natural formalization of classical logic

The difficulties of natural deduction with classical tautologies depend on the fact that a sentence and its negation are not treated symmetrically, whereas this is exactly what one would expect from the classical meaning of the logical operators.

---

[15]The example in Sections 2 and 3 are taken from [Broda *et al.*,].

Let us consider one of *de Morgan's laws*

$$\neg(P \wedge Q) \to (\neg P \vee \neg Q)$$

and try to prove it in the classical natural deduction system with the *classical reductio ad absurdum* rule. It is not difficult to check that the simplest proof one can obtain is the following:[16]

$$
\cfrac{
\neg(P \wedge Q)^1 \qquad
\cfrac{
\cfrac{
\cfrac{\neg(\neg P \vee \neg Q)^2 \quad \cfrac{\cfrac{\neg P^3}{\neg P \vee Q}\ 3}{}}{\mathbf{F}} \\[2pt]
\cfrac{}{P}\ 2
\qquad
\cfrac{\neg(\neg P \vee \neg Q)^2 \quad \cfrac{\cfrac{\neg Q^4}{\neg P \vee \neg Q}\ 4}{}}{\mathbf{F}} \\[2pt]
\cfrac{}{Q}\ 2
}{P \wedge Q}\ 2
}{\ }\ 2,1
}{
\cfrac{
\cfrac{\mathbf{F}}{\neg P \vee \neg Q}\ 1
}{\neg(P \wedge Q) \to (\neg P \vee \neg Q)}\ \emptyset
}
$$

where the numerals on the right of the inference lines mean that the conclusion depends on the assumptions marked with the same numerals (see the appendix for a listing of the natural deduction rules).

Notice that we obtain a rather unnatural deduction whatever variant of "natural" deduction we may decide to use, although probably the least bad of all is obtained by means of the variant which *explicitly* incorporates the principle of bivalence in the form of the "classical dilemma" rule:[17]

---

[16] We adopt a standard propositional language with P, Q, R, etc., possibly with subscripts, as propositional variables, augmented with a constant $\mathbf{F}$ standing for "the absurd".

[17] See [Tennant, 1978].

$$\cfrac{\neg(P \wedge Q)^1 \quad \cfrac{\cfrac{P^2 \quad Q^3}{P \wedge Q} \, 2,3}{\mathbf{F}} \, 1,2,3}{\cfrac{\cfrac{\mathbf{F}}{\neg P} \, 1,3}{\cfrac{\neg P \vee \neg Q}{\neg P \vee \neg Q} \, 1,3 \qquad \cfrac{\cfrac{\neg Q^3}{\neg P \vee \neg Q} \, 3}{1}}{\cfrac{\neg P \vee \neg Q}{\neg(P \wedge Q) \to (\neg P \vee \neg Q)} \, \emptyset}}$$

Another example of a tautology whose natural deduction proofs are extremely contrived is Peirce's law: $((P \to Q) \to P) \to P$. The reader can verify that any attempt to prove this classical tautology within the natural deduction framework leads to logical atrocities. Here is a typical proof:

$$\cfrac{\cfrac{(P \to Q) \to P^1 \quad \cfrac{\cfrac{\cfrac{\neg P^2 \quad P^3}{\mathbf{F}} \, 2,3}{Q} \, 2,3}{P \to Q} \, 2}{P} \, 1,2 \qquad \neg P^2}{\cfrac{\cfrac{\mathbf{F}}{P} \, 1}{((P \to Q) \to P) \to P} \, \emptyset} \, 1,2$$

It should be emphasized that the main problem here does not lie in the complexity of proofs (i.e. their length with respect to the length of the tautology to be proved): in fact, if we look at the class of tautologies which generalizes the de Morgan law proved above, namely $T_i = \neg(P_1 \wedge \ldots \wedge P_i) \to \neg P_1 \vee \ldots \vee \neg P_i$, it is easy to verify that the length of their shortest proofs is linear in the length of the tautology under consideration, both in the system with the classical dilemma and in the system with classical *reductio*. It is rather their *contrived character* which is disturbing, and this depends on the fact that the natural deduction rules do not capture the *classical* meaning of the logical operators. Since these rules are closely related to the *constructive* meaning of the operators, we would not expect such logical atrocities when we use them to prove *intuitionistic* tautologies. (Notice that both the de Morgan law we have considered above and Peirce's law are *not* intuitionistically valid.) Consider, for instance, the other (intuitionistically valid) de Morgan law:

$$\neg P \vee \neg Q \to \neg(P \wedge Q).$$

One can easily obtain a "really natural" intuitionistic proof of this tautology as follows:

$$
\cfrac{\neg P \vee \neg Q^1 \qquad \cfrac{\neg P^3 \quad \cfrac{\cfrac{P \wedge Q^2}{P}2}{\mathbf{F}}\;2,3}{\neg(P \wedge Q)}3 \qquad \cfrac{\neg Q^5 \quad \cfrac{\cfrac{P \wedge Q^4}{Q}4}{\mathbf{F}}\;4,5}{\neg(P \wedge Q)}5}{\cfrac{\neg(P \wedge Q)}{\neg P \vee \neg Q \to \neg(P \wedge Q)}\emptyset}1
$$

These examples illustrate our claim that *"natural" deduction is really natural only for intuitionistic tautologies* (and not necessarily for all of them).

3   Why cut-free systems are not natural formalizations of classical logic

Contrary to natural deduction, the systems based on the classical (cut-free) sequent calculus treat a sentence and its negation symmetrically. For instance, the analytic tableau method allows for a very simple and natural proof of the non-intuitionistic de Morgan law (see Figure 1, where × indicates, as usual, a closed branch). We stress once again that it is not the length of proofs which is at issue here. Although the tableau proofs for the class of tautologies which generalizes this de Morgan law are, strictly speaking, shorter than the corresponding natural deduction proofs, the difference in length is negligible, since in both cases one obtains proofs whose length is linear in the length of the given tautology. What is at issue is that the tableau proofs are more natural than those based on the "natural" deduction rules, as a result of the symmetrical treatment of sentences with respect to their negations. The symmetry of the tableau rules, however, is by no means sufficient to guarantee, in general, that such rules always lead to proofs which are natural from the classical point of view. Consider the class of tautologies T_i^* defined as the disjunction of all possible conjunctions which can be obtained from i propositional variables or their negations. For instance $T_2^* = (P_1 \wedge P_2) \vee (P_1 \wedge \neg P_2) \vee (\neg P_1 \wedge P_2) \vee (\neg P_1 \wedge \neg P_2)$. Each T_i^* contains i distinct propositional variables, but 2^i conjunctions. In [D'Agostino, 1992] it is shown that this class of tautologies is sufficient to separate the tableau method (and the cut-free sequent calculus in tree form) from the truth-table method. In fact, it is not difficult to prove that the length of

$$\neg(\neg(P \wedge Q) \to (\neg P \vee \neg Q))$$
$$|$$
$$\neg(P \wedge Q)$$
$$|$$
$$\neg(\neg P \vee \neg Q)$$
$$|$$
$$\neg\neg P$$
$$|$$
$$\neg\neg Q$$
$$|$$
$$P$$
$$|$$
$$Q$$

$$\diagup \quad \diagdown$$

$$\neg P \qquad \neg Q$$

$$\times \qquad \times$$

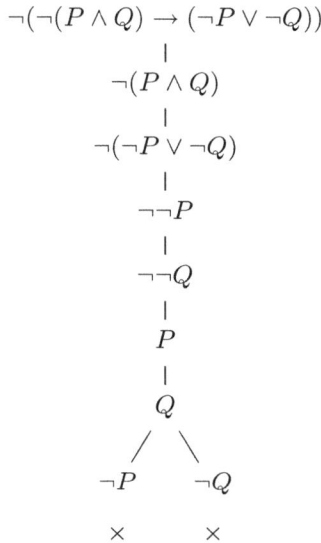

Figure 1. A tableau proof of the non-intuitionistic de Morgan Law.

the shortest tableau proofs for the tautologies T_i^* is not bounded above by any polynomial in the length of the tautologies. This implies that analytic tableaux (and the cut-free sequent calculus) cannot p-simulate the truth-table method, since the truth-tables for T_i^* have polynomial size. It also implies that analytic tableaux cannot p-simulate the natural deduction calculus (in any of its variants) since such tautologies do have polynomial size proofs in the latter and such proofs are not at all "difficult" to find.[18]

This embarrassing situation stems from the lack, both in the tableau method and in the cut-free sequent calculus, of a rule expressing bivalence. None of its rules forces a bivalent interpretation of the notion of truth: all the rules are compatible with many-valued interpretations. The fact that all classical tautologies can be proved is the result of the equivalence between such many-valued interpretations and the standard two-valued one. The non-bivalent character of the tableau rules allows for the generation of branches which do not represent *mutually incompatible* valuations and, therefore, give rise to a great deal of undesired redundancy in the enumeration of possible cases. (For a more detailed discussion of

---

[18] For an excellent survey of the relative complexity of propositional proof systems, the reader is referred to [Urquhart, 1995].

this point the reader is referred to [D'Agostino, 1990; D'Agostino, 1992; D'Agostino and Mondadori, 1994; D'Agostino, 1999].)

4 Classical natural deduction

In order to devise a truly classical variant of natural deduction, we must introduce a set of introduction and elimination rules which correspond closely to the *classical*, rather than to the intuitionistic, reading of the logical operators. From this point of view, as remarked above, the main problematic aspect of natural deduction calculi in the Gentzen-Prawitz tradition is the non-symmetrical treatment of sentences with respect to their negations. The typical involutive character of classical negation is captured in a rather roundabout way, via an *ad hoc* axiom such as the law of excluded middle or a similarly *ad hoc* rule for classical negation. Other problematic aspects are the "discharge rules" for introducing the conditional operator "→" and for eliminating the disjunction operator "∨", which are both based on the standard *intuitionistic* explanation of these operators. The remaining rules, on the other hand, can apparently be accommodated within the classical truth-functional view.

The problem about negation lies in the fact that, according to the intuitionistic interpretation, $\neg P$ is defined as having the same meaning as $P \rightarrow \mathbf{F}$, where $\mathbf{F}$ is "the absurd". The negation of P can be legitimately asserted only if one is able to provide a refutation of P. On the other hand, classical negation cannot be naturally defined in such a way. Classically speaking, the negation of a proposition P is true if and only if P is false. Thus, the meaning of $\neg P$ is intertwined with the "ontological" notion of falsity, governed by the principle of bivalence (as well as by the principle of non-contradiction), and is not amenable to epistemic reductions.

In intuitionistic natural deduction introduction and elimination rules specify, respectively, the conditions for validly *asserting* a sentence P containing an operator ♯ as its main operator and the consequences of such an assertion. From a classical perspective there should be symmetrical introduction and elimination rules specifying, respectively, the conditions under which P *is true or false* and the consequences of its being true or false. Hence, besides specifying introduction rules determining the conditions under which the truth of P can be inferred, we also need to specify suitable introduction rules determining the conditions under which its *falsity* — i.e. the truth of its negation $\neg P$ — can be inferred. Similarly, we have to specify suitable elimination rules determining the consequences of the falsity, as well as of the truth, of such a sentence P.

Following this approach, the obvious introduction rules for *classical* conjunction and disjunction, which can immediately be read off the classical two-valued truth-tables, are the following:

$$\frac{P \quad Q}{P \wedge Q} \quad \frac{\neg P}{\neg(P \wedge Q)} \quad \frac{\neg Q}{\neg(P \wedge Q)} \quad \frac{P}{P \vee Q} \quad \frac{Q}{P \vee Q} \quad \frac{\neg P \quad \neg Q}{\neg(P \vee Q)}$$

As for negation, since from the classical viewpoint the negation of P means that P is false and, as argued above, classical falsity is a primitive (ontological) notion, there cannot be a rule specifying the conditions under which a proposition $\neg P$ is true apart from the trivial one of the form $\neg P/\neg P$.[19] On the other hand, there is a rule specifying the conditions under which $\neg P$ is *false*, that is $\neg\neg P$ is true:

$$\frac{P}{\neg\neg P}$$

From a natural deduction standpoint such introduction rules should be taken as *definitions* of the logical operators. We claim that they provide a definition of the *classical* logical operators — the operators of what we have called *standard* classical logic — in that they are the only truth-conditional introduction rules which can be read off the two-valued truth tables. Classical elimination rules have therefore to be justified in terms of these definitions. Such rules, whichever they may be, should satisfy Prawitz' *inversion principle*: essentially, by an application of an elimination rule one only recovers a sentence which would have been established if the premiss(es) of the application had been inferred by an application of an introduction rule ([Prawitz, 1965]).

Let us consider the case of disjunction. First we justify the elimination rule for negated disjunctions. Since the conditions for inferring the falsity of a disjunction $P \vee Q$ are that *both* P and Q are false, the elimination rules for negated disjunctions are the following:

$$\frac{\neg(P \vee Q)}{\neg P} \quad \frac{\neg(P \vee Q)}{\neg Q}$$

Moreover, by the same reasoning, the elimination rules for true conjunctions are determined as follows:

---

[19] If we used *signed formulas* such a rule would no longer be redundant and would have the form: $FP/T\neg P$. The use of signed formulas, though unusual in the context of natural deduction calculi — which is the reason why we have not adopted it in this paper — may have some advantages especially for adapting them to some non-classical logics. On this point see also the final section on "future work".

$$\frac{P \land Q}{P} \quad \frac{P \land Q}{Q}$$

Determining the elimination rules for true disjunctions and false conjunctions requires a slightly more complex argument. In order to infer $P \lor Q$ by means of an introduction rule we need either P or Q as premiss. So, what we can infer from $P \lor Q$ is that *either* P is true *or* Q is true, but we have no means for determining which. In some analytic methods developed within the tradition of the cut-free sequent calculus — such as Smullyan's "analytic tableaux" — the "elimination rule" for $\lor$ reflects this non-determinism. For instance, in analytic tableaux, $\lor$-elimination is the following:

$$P \lor Q$$
$$/ \quad \backslash$$
$$P \qquad Q$$

However, such a non-deterministic elimination rule is not satisfactory from a natural deduction perspective. One could argue that disjunction here is hardly "eliminated" at all. The rule is not an *inference* rule in the standard sense: one cannot describe this rule as establishing that from the truth of $P \lor Q$ we can either infer the truth of P or infer the truth of Q, since this interpretation would be plainly incorrect. The only way of describing this rule as an inference rule would be to say something like: "from $P \lor Q$ we can infer $P \lor Q$" which is a sheer triviality.[20]

The solution of this problem in the framework of standard natural deduction is the well-known "constructive dilemma" elimination rule for disjunction. This rule, like the introduction rule for conditionals and negations, involves the discharge of assumptions. Strictly speaking, such "discharge rules" are not *inference* rules either, since they do not allow to infer a certain sentence as conclusion from the assumption of certain other sentences as premises, but involve further manipulation of the assumption formulas. They are better described as *proof rules* asserting that certain proofs can be combined to yield another proof. In other words, the premises of such proof rules are *sequents*, rather than formulas. Since the truth of a sentence, in intuitionistic terms, is tantamount to its provability, it is quite appropriate to construe logical deduction in terms of such proof rules: understanding the meaning of a sentence means understanding how it can be proved, and since logical deduction is a type of inference which is justified on the sole

---

[20]For a similar remark, referred to the multiple-conclusion sequent calculus, see [Dummett, 1991], p. 195.

basis of the meaning of the logical operators, it is perfectly natural to see it as a combinatorics of proofs. Hence, the very distinction we have just made between inference rules and proof rules tends to fade, especially considering that in any standard consequence relation (closed under transitivity) every inference rule can be reinterpreted as a proof rule[21] (but not vice-versa).

From the classical viewpoint, however, the distinction makes perfectly good sense. Understanding the meaning of a sentence means understanding how the world should be for the sentence to be true and how it should be for the sentence to be false. Therefore, the premises and the conclusions of logical rules are naturally construed in terms of the truth or falsity of sentences rather than in terms of proofs. As a result, the most direct logical rules are *inference rules* rather than proof rules.

We therefore claim that a *classical inference rule* must be a rule which allows us to infer, as conclusion, the truth or the falsity of a sentence from premises asserting the truth or the falsity of other sentences. *Classical introduction rules* for a logical operator $\sharp$ will be rules expressing the conditions for the truth or falsity of a sentence P containing $\sharp$ as main operator and whose premises assert the truth or the falsity of immediate subformulas of P. *Classical elimination rules* for $\sharp$ will be rules expressing the consequences of the truth or falsity of a sentence P containing $\sharp$ as main operator — possibly with help of auxiliary premises asserting the truth or falsity of some immediate subformula of P — and whose conclusion asserts the truth or falsity of another immediate subformula of P. According to this definition, constructive dilemma is not a classical inference rule.

This does not imply, of course, that constructive dilemma, like other similar discharge rules, is not classically sound. It only implies that if the meaning of the logical operators is to be explained truth-conditionally, in terms of truth and falsity of sentences rather than in terms of proofs (from given assumptions), constructive dilemma is not an argument that stems *directly* from the meaning of disjunction, but one that can be justified only indirectly by (meta-level) reasoning on the concept of classical proof.[22]

However, constructive dilemma is not the *only* elimination rule for true disjunctions that can be justified in terms of the corresponding introduction rules in accordance with the inversion principle. A better solution, in classical terms, arises from the following argument. In order to infer $P \vee Q$ by means of an introduction rule we need either P or Q as premiss. So, if we have valid grounds (in truth-conditional terms) for asserting the truth of

---

[21] For instance, the validity of the "horizontal" inference rule $P, Q \vdash P \wedge Q$ implies the validity of the "vertical" proof rule which from sequents $\Gamma \vdash P$ and $\Delta \vdash Q$ as premises leads to the sequent $\Gamma, \Delta \vdash P \wedge Q$ as conclusion.

[22] Such an indirect "natural deduction" system for full classical logic is presented in [Smullyan, 1965].

$P \vee Q$, in accordance with the meaning of $\vee$ fixed by the appropriate introduction rules, we must also have valid grounds for asserting the truth of P or for asserting the truth of Q. Suppose now that we have valid grounds for asserting the truth of $\neg P$ (that is the falsity of P). Then, by consistency, the only justification we may have for asserting $P \vee Q$ can only be the truth of Q. This analysis leads to a pair of elimination rules for true disjunctions that correspond to the traditional disjunctive syllogism argument:

$$\frac{P \vee Q \quad \neg P}{Q} \quad \frac{P \vee Q \quad \neg Q}{P}$$

Similar considerations lead to the dual rules for eliminating false conjunctions:

$$\frac{\neg(P \wedge Q) \quad P}{\neg Q} \quad \frac{\neg(P \wedge Q) \quad Q}{\neg P}$$

and to the double negation rule for eliminating false negations:

$$\frac{\neg\neg P}{P}$$

Let us now turn our attention to the conditional operator. Since the boolean conditional $P \rightarrow Q$ can be defined in terms of the other boolean operators, for instance as $\neg(P \wedge \neg Q)$, the introduction and elimination rules for conditionals are determined as follows:

$$\frac{Q}{P \rightarrow Q} \qquad \frac{\neg P}{P \rightarrow Q} \qquad \frac{P \quad \neg Q}{\neg(P \rightarrow Q)}$$

$$\frac{P \rightarrow Q \quad P}{Q} \qquad \frac{P \rightarrow Q \quad \neg Q}{\neg P}$$

$$\frac{\neg(P \rightarrow Q)}{P} \qquad \frac{\neg(P \rightarrow Q)}{\neg Q}$$

(Notice the presence of *modus tollens* as a second elimination rule, besides *modus ponens*, for true conditionals.) Interestingly enough, the elimination rules we have discussed in this section were already discovered by Chrysippus who claimed them to be the fundamental rules of reasoning ("anapodeiktoi"), except that disjunction was interpreted by him in an exclusive sense.

Chrysippus also maintained that his "anapodeiktoi" formed a complete set of inference rules.[23]

We have so specified a set of introduction and elimination rules such that:

1. the introduction rules can be taken as definitions of the classical meaning of the logical operators, specifying — in accordance with the standard two-valued semantics — the conditions for inferring the truth or the falsity of a sentence in terms of the truth or falsity of their immediate subformulas;

2. the elimination rules are justified in terms of this meaning by means of arguments based on a form of the inversion principle;

3. the rules for conjunction and disjunction are dual of each other;

4. sentences and their negations are treated in a symmetric way;

Our approach can be seen as falling half-way between the standard truth-conditional account of classical semantics and a proof-theoretical account, based on the Wittgensteinian meaning-as-use view, which is more typical of intuitionistic logic and its "neighbours".[24] We shall call our inference rules *intelim rules* for the classical operators. When dealing with the two-premiss elimination rules — such as the rules for eliminating true disjunctions, false conjunctions and true conditionals — we call the more complex premiss *major premiss*, while the auxiliary premiss is called *minor premiss*.

However, this is not the end of the story. Despite our claim that the introduction rules can be taken as satisfactory definitions of the classical meaning of the logical operators, and that the elimination rules are in "harmony" with them, as prescribed by the inversion principle, the intelim rules are *not* a complete set of rules for classical propositional logic. The reason, according to our analysis, is that *classical semantics is not completely characterized by the meaning of the logical operators* as it can be defined via the inferential approach. In other words: if we accept (i) the natural deduction stand that the meaning of the logical operators is fixed by specifying the introduction rules for them, (ii) that the introduction rules specified

---

[23]"The indemonstrables are those of which the Stoics say that they need no proof to be maintained. [...] They envisage many indemonstrable but especially five, from which it seems all others can be deduced". See [Blanché, 1970], pp.115–119 and [Bochensky, 1961], p.126.

[24]Variations on this proof-theoretical theme concerning classical propositional logic, the subformula property and the inversion principle are discussed also in [Mondadori, 1988a], [Mondadori, 1988b], [Mondadori, 1988d], [Mondadori, 1988c], [Mondadori, 1989], [D'Agostino and Mondadori, 1991],[D'Agostino and Mondadori, 1992], [D'Agostino and Mondadori, 1993], [Broda *et al.*,].

above are the "natural" introduction rules for the *classical* operators, and
(iii) that the elimination rules specified above are the only *inference* rules
(in the strict sense) that can be justified on the basis of the introduction
rules by means of the inversion principle, we are bound to conclude that
the inferential approach to the meaning of the classical operators *does not
characterize classical semantics completely.*

In fact, this conclusion is hardly surprising. The fundamental contrast
between classical and intuitionistic logic lies in the underlying notion of
truth. While intuitionistic truth is "internal" and can therefore be completely characterized by reference to the meaning of the logical words, as
defined by their use in logical inference, classical truth is "external" and requires a reference to a reality which exists independently of human reasoning
and perception. So there is no reason to expect the inferential approach to
the meaning of the logical operators to provide an exhaustive account of
classical deduction. What is missing is the inferential role played by an
extra metaphysical assumption — which dates back to Aristotle — namely
the assumption of an external reality which makes sentences determinately
true or false quite independently of our means of determining their truth
value. This assumption is expressed by the classical *principle of bivalence*
and we stress that it should not be regarded as an assumption about the
meaning of the logical operators — at least if we accept the philosophical
stand that this meaning is defined by appropriate introduction rules — but
as an assumption about the notions of truth and falsity and their ontological bearing. Bivalence is therefore a metaphysical assumption which plays
an *indirect* inferential role in justifying inferences, but cannot be justified
by reference to the meaning of the logical operators, even if this meaning
is construed in a classical way (i.e. truth-conditionally), unless we give up
the basic tenet of natural deduction and of proof-theoretic semantics in general, namely that the introduction rules should be taken as definitions of
the corresponding operators and the elimination rules should be in harmony
with them. Bivalence, therefore, is not really a logical rule but a *structural*
assumption about the relationship between language and world. So, our
account neatly separates the inferential role played by the logical operators,
whose meaning is defined by their use in inference, and the role played by
the bivalence principle.

Bivalence is not the only "metaphysical" assumption of classical logic.
Another Aristotelian assumption about the notions of truth and falsity is
expressed by the *principle of non-contradiction*, which says that the world
is such that a given sentence cannot be true and false at the same time. Observe that non-contradiction is also an intuitionistic principle which stems
immediately from the meaning of the negation operator. In intuitionistic

terms, however, this principle refers to the internal consistency of our mental representations. In classical terms, the negation of P means that P is false and the falsity of a sentence, like its truth, is a relation between the sentence itself and the postulated external world. Hence, in classical logic the principle of non-contradiction is, like the principle of bivalence, a *structural* assumption about the relationship between language and world.

Let us say that a set Γ of assumptions is *inconsistent* if for some sentence P, both P and $\neg P$ are deducible from Γ. Formally, we can introduce a rule like the following:

$$\frac{P \quad \neg P}{\times}$$

meaning that the tree above $\times$ is a proof that the assumptions on which it depends are mutually inconsistent.

An appropriate rule expressing the classical principle of bivalence is then the following *proof rule*:

$$\frac{\begin{array}{cc} [P] & [\neg P] \\ \vdots & \vdots \\ Q/\times & Q/\times \end{array}}{Q/\times}$$

where the notation "$Q/\times$" means that the above tree is either a proof of Q from the undischarged assumptions or a proof that the undischarged assumptions are *inconsistent*. Moreover, the symbol below the line may be equal to "$\times$" only if *both* the symbols immediately above it are equal to "$\times$", that is the whole tree is a a refutation of its undischarged assumption formulas only if both the subtrees above the line are refutations of their undischarged assumption formulas.

The special format of this rule — a proof rule rather than an inference rule — shows that the inferential role it plays is on a different level from that of the other rules. Arguments essentially involving this rule are *indirect* and are not completely justified by the meaning of the logical operators.[25] On the other hand, if the rules of bivalence and non-contradiction are added to the system consisting of the introduction rules, then the elimination rules are not merely "justified" in terms of the introduction rules, but become *derived rules*, as the reader can easily verify.[26]

---

[25] Unlike bivalence, constructive dilemma is an indirect argument that can be ultimately justified in terms of the meaning of the classical logical operators, though only by means of meta-level reasoning.

[26] Similarly the introduction rules are derived rules in the system consisting of the elimination rules plus the two structural rules. On this point see [D'Agostino, 1990] and [D'Agostino and Mondadori, 1991].

Augmenting the intelim rules with the above rules of *bivalence* and *non-contradiction* we obtain a complete set of rules for classical propositional logic. The completeness proof is entirely routine and will therefore be omitted. The resulting proof system is our candidate for a *classical* theory of natural deduction as opposed to the intuitionistic theory represented by the Gentzen-Prawitz calculi.

5   A more convenient format for classical natural deduction: intelim sequences and trees

The standard tree format of natural deduction proofs, with the conclusion as root and the assumptions as leaves, is certainly perspicuous since it allows us to visualize immediately the inner structure of the proof — the premises and the conclusion of each rule application, the assumptions on which each formula depends, etc.

However, this format involves a good deal of redundancy in the representation of proofs. Whenever a formula, which can be inferred from the assumptions, is used more than once as premiss of further inferences, its proof-tree has to be replicated and this leads to an unnecessary growth of the size of the whole proof. Moreover, the format of the rule of bivalence — the only one involving discharge of assumptions — is not particularly convenient for the transformation of proofs and for the implementation of efficient proof-search algorithms. In this section we propose a different format which, although slightly less perspicuous, appears to be better suited to algorithmic treatment.

In this new format, the tree grows upside-down, the premiss of a rule application do not occur in adjacent branches as immediate predecessors of the conclusion, but on the same branch as the conclusion. So, the application of the intelim rules is *sequential*.

DEFINITION 1. An *intelim sequence* is a sequence of formulas such that each formula is either an assumption or is the conclusion of the application of an intelim rule to preceding formulas. An intelim sequence *based on a set* Γ of assumptions is an intelim sequence such that all its assumption formulas are in Γ.

The *rule of bivalence* (RB for short) splits an intelim sequence into two branches that are developed in parallel:

$$\swarrow \quad \searrow$$
$$P \qquad \neg P$$

When an intelim tree is expanded in this way we say that RB has been applied to the formula P and will also say that P is *the RB-formula* of this application of RB.

DEFINITION 2. An *intelim tree* is a tree of occurrences of formulas such that each occurrence is either an assumption, or is obtained from the application of an intelim rule to formulas occurring above it in the same branch, or results from an application of RB. An intelim tree *based on a set* Γ of assumptions is a intelim tree such that all its assumption formulas are in Γ.

Observe that each branch of an intelim tree is an intelim sequence.

DEFINITION 3. A branch of an intelim tree is *closed* if it contains both P and $\neg P$ for some sentence P, otherwise it is *open*.

DEFINITION 4. A *classical intelim proof* of P from the assumptions Γ is an intelim tree based on Γ such that P occurs in every *open* branch.

The intelim rules are displayed in Table 1, while the structural rules are displayed in Table 2.

6   Normalization issues

In the sequel we shall use the lower case letters p, q, r, etc., possibly with subscripts, to denote arbitrary *atomic* formulas. By the *complexity* of a formula we shall mean the number of occurrences of logical operators in it (so, for instance, the complexity of atomic formulas is 0 and the complexity of $(p \vee q) \vee (r \wedge s)$ is 3).

DEFINITION 5. A branch is *atomically closed* if it contains both p and $\neg p$ for some atomic formula p.

PROPOSITION 6. *Every closed branch can be expanded into an atomically closed branch by means of applications of intelim rules only.*

Proof. If a branch is closed, it contains both P and $\neg P$ for some sentence P. The proof is an easy induction on the complexity of P and requires the discussion of several cases depending on the logical form of P. We discuss only the case $P = Q \vee R$, the other cases being similar.

If both $Q \vee R$ and $\neg(Q \vee R)$ occur in a branch, then we can apply the elimination rules to $\neg(Q \vee R)$ and append both $\neg Q$ and $\neg R$ to the end of branch. Then the extended branch will contain both $Q \vee R$ and $\neg Q$, so we can apply the appropriate elimination rule to these formulas and append R to the end of the branch. The resulting branch will contain both R and $\neg R$ and the complexity of R is strictly less than the complexity of P.   ∎

Notice that when a closed branch containing P and $\neg P$ is expanded into an atomically closed one following the procedure described in the proof of the

INTRODUCTION RULES

$$\frac{\begin{array}{c}P\\Q\end{array}}{P \wedge Q} \qquad \frac{\neg P}{\neg(P \wedge Q)} \qquad \frac{\neg P}{\neg(P \wedge Q)}$$

$$\frac{\begin{array}{c}\neg P\\\neg Q\end{array}}{\neg(P \vee Q)} \qquad \frac{P}{P \vee Q} \qquad \frac{P}{P \vee Q}$$

$$\frac{\begin{array}{c}P\\\neg Q\end{array}}{\neg(P \to Q)} \qquad \frac{\neg P}{P \to Q} \qquad \frac{Q}{P \to Q}$$

$$\frac{P}{\neg\neg P}$$

ELIMINATION RULES

$$\frac{\begin{array}{c}P \vee Q\\\neg P\end{array}}{Q} \qquad \frac{\begin{array}{c}P \vee Q\\\neg Q\end{array}}{P} \qquad \frac{\neg P \vee Q}{\neg P} \qquad \frac{\neg P \vee Q}{\neg Q}$$

$$\frac{\begin{array}{c}\neg(P \wedge Q)\\P\end{array}}{\neg Q} \qquad \frac{\begin{array}{c}\neg(P \wedge Q)\\Q\end{array}}{\neg P} \qquad \frac{P \wedge Q}{P} \qquad \frac{P \wedge Q}{Q}$$

$$\frac{\begin{array}{c}P \to Q\\P\end{array}}{Q} \qquad \frac{\begin{array}{c}P \to Q\\\neg Q\end{array}}{\neg P} \qquad \frac{\neg(P \to Q)}{P} \qquad \frac{\neg(P \to Q)}{\neg Q}$$

$$\frac{\neg\neg P}{P}$$

Table 1. The intelim rules.

$$\frac{}{P \mid \neg P} \text{ RB} \qquad \frac{\begin{array}{c}P\\\neg P\end{array}}{\times} \text{ RNC}$$

Table 2. The structural rules.

above proposition, its length grows at most linearly in the complexity of P.

PROPOSITION 7. *Every proof π of P from the set Γ of assumptions can be transformed into a proof π' of P from Γ such that RB is applied only to atomic formulas.*

Proof. We use the notation

$$T$$
$$\mathbf{n}$$

to denote either the empty intelim tree or a non-empty intelim tree such that $\mathbf{n}$ is one of its terminal nodes.

The proof is by lexicographic induction on $(\gamma(\pi), \lambda(\pi))$ where $\gamma(\pi)$ is the maximum complexity of an RB-formula in π and $\lambda(\pi)$ is the number of occurrences of RB-formulas of complexity $\gamma(\pi)$.

Let $\gamma(\pi) = k > 0$ and let Q be an RB-formula of complexity k. There are several cases depending on the logical form of Q. We discuss only the case $Q = R \vee S$, the other cases being similar.

If $Q = R \vee S$, then π has the following form:

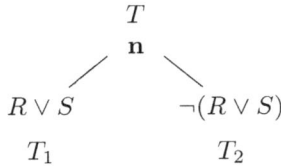

where T_1 and T_2 are intelim trees such that each of their open branches contains P.

Let π' be the following intelim tree:

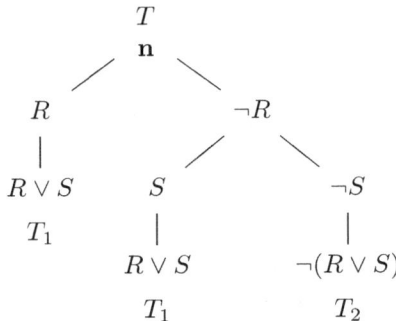

Clearly π' is still a classical intelim proof of P from Γ and either $\gamma(\pi') < \gamma(\pi)$ or $\gamma(\pi') = \gamma(\pi)$ and $\lambda(\pi') < \lambda(\pi)$. ∎

DEFINITION 8. An occurrence of a formula P is *idle* in a proof π if it is not used in π as premiss of some application of an intelim rule or of RNC.

DEFINITION 9. Given an intelim tree T, a *path* in T is a finite sequence of nodes such that the first node is the root of T and each of the subsequent nodes is an immediate successor of the previous one. A path is *closed* if it contains both P and $\neg P$ for some formula P.

Notice that, according to the above definition, every branch is a maximal path.

DEFINITION 10. A proof π of P from Γ is *pruned* if it satisfies the following conditions:

1. it contains no idle occurrences of formulas;

2. no branch ϕ of π contains more than one occurrence of the same formula;

3. no branch ϕ of π contains a path ending with P which is properly included in ϕ;n

4. no branch ϕ of π contains a closed path which is properly included in ϕ.

We omit the (routine, but long-winded) proof of the following proposition.

PROPOSITION 11. *Every proof π of P from Γ can be transformed into a pruned proof π' of P from Γ.*

DEFINITION 12. An occurrence of a formula P is a *detour* in a proof π if it is at the same time the conclusion of an application of an introduction rule and either the major premiss of an application of an elimination rule or one of the premises of an application of RNC.

Then, we are able to show that:

PROPOSITION 13. *If π is a pruned proof of P from Γ and every closed branch of π is atomically closed, then π contains no detours.*

Proof. Suppose a branch ϕ contains a detour, say an occurrence of $P \vee Q$. Since every closed branch in π is atomically closed, then $P \vee Q$ cannot be a premiss of RNC. So it must be at the same time the conclusion of an introduction and the major premiss of an elimination. Then either (i) ϕ contains two occurrences of P (one as premiss of the introduction of $P \vee Q$ and the other as conclusion of its elimination), or (ii) ϕ contains two occurrences of Q (again one as premiss of its introduction and the other as conclusion of its elimination), or (iii) ϕ contains a closed path which is properly included in it (P as premiss of its introduction and $\neg P$ as minor premiss of its elimination; or Q as premiss of its introduction and $\neg Q$ as minor premiss of its elimination). Then, π is not pruned. ∎

DEFINITION 14. A formula Q is a *weak subformula* of a formula P if Q is a subformula of P or the negation of a subformula of P.

DEFINITION 15. Given a proof π of P from Γ, we say that a formula Q is *spurious* in π if Q is not a weak subformula either of P or of some formula in Γ. A proof π of P from Γ has the *weak subformula property* (WSFP for short) if it contains no spurious formulas.

DEFINITION 16. A proof π of P from Γ is *canonical* if (i) every closed branch of π is atomically closed, (ii) RB is applied only to atomic formulas in π , (iii) π is pruned and (iv) π has the WSFP.

LEMMA 17. *Let π be a proof of P from Γ such that (i) π contains no idle occurrences of formulas and (ii) every closed branch of π is atomically closed. If Q is a spurious formula of maximal complexity in π, then Q may only occur in π either as an RB-formula or as a detour.*

Proof. The proof is left to the reader. ∎

PROPOSITION 18. *Every proof π of P from Γ can be transformed into a canonical proof π' of P from Γ.*

Proof. Given an arbitrary proof of P from Γ we first expand its closed branches into atomically closed ones (Proposition 6); next we transform it into a proof such that RB is applied only to atomic formulas (Proposition 7); finally, we transform it into a pruned proof (Proposition 11). Thus, we can assume without loss of generality that (i) every closed branch of π is atomically closed, (ii) RB is applied in π only to atomic formulas, and (iii) π is pruned. We then have to show that π can be turned into a proof of P from Γ with the WSFP.

Since π is pruned and every closed branch of π is atomically closed, then π contains no detours (Proposition 13). So, by Lemma 17, the only occurrences of spurious formulas in π may only be (atomic) RB-formulas. We show that all such disturbing occurrences of RB-formulas can be eliminated one by one.

Let $\sigma(\pi)$ be the number of occurrences of spurious RB-formulas in π. Suppose $\sigma(\pi) = k > 0$. Then π has the following form:

$$T$$
$$\mathbf{n}$$

$$\swarrow \qquad \searrow$$

$$q \qquad\qquad \neg q$$
$$T_1 \qquad\qquad T_2$$

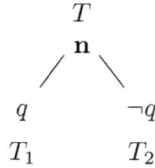

where (i) q is atomic and spurious, (ii) T_1 and T_2 are intelim trees such that each of their open branches contains P. Since π is pruned, q is used in T_1. Given that q is spurious, it follows that q cannot be used in T_1 as premiss of an introduction rule, because in such a case the conclusion would be a spurious formula, against the conclusion we have just reached that the only spurious formulas in π are atomic RB-formulas. For the same reason, q cannot be used as minor premiss of an elimination rule. Moreover, it is obvious that q cannot be used as major premiss of an elimination rule. Hence, q can be used in a closed branch of T_1 only as premiss of an application of RNC with $\neg q$ as the other premiss. This means that one of the terminal nodes of T_1, say $\mathbf{m}$, contains an occurrence of $\neg q$. Let π' be the following intelim tree:

$$T$$
$$\mathbf{n}$$
$$T_1$$
$$\mathbf{m}$$
$$T_2 \ .$$

It is easy to check that π' is still a classical intelim proof of P from Γ. Moreover, $\sigma(\pi') < \sigma(\pi)$.   ∎

The following proposition states the property we anticipated in section 4, namely that if we allow for RB and RNC, the elimination rules become derived rules.

PROPOSITION 19. *If there is a proof of P from Γ than there is a canonical proof π' of P from Γ which contains no application of the elimination rules.*

The above proposition implies that the system consisting of the introduction rules only plus RB and RNC is complete for classical logic and enjoys the WSFP.[27]

The following proposition states a remarkable property of canonical proofs showing that purely introductive canonical proofs of classical tautologies are nothing but a more concise form of the familiar truth-table procedure.

---

[27]This is the **KI** system, proposed in [Mondadori, 1988a] and further investigated in [Mondadori, 1996].

PROPOSITION 20. *If P is a tautology, every canonical proof of P from the empty set of assumptions is such that (i) π contains no applications of the elimination rules and (ii) all branches of π are open.*

EXAMPLE 21. The set of formulas

$$\{p \vee q, r \vee s, p \rightarrow t \wedge u, r \rightarrow v \wedge u, \neg u\}$$

classically implies $q \wedge s$. Figure 2(a) shows an intelim sequence for this inference.

EXAMPLE 22. The set

$$\{p \vee q, p \vee \neg q, r \vee s, r \vee \neg s\}$$

classically implies $p \wedge r$. Figure 2(b) shows a classical intelim proof of this inference (for the sake of readability we have omitted the justification of the inference steps).

In most practical applications, canonical proofs may be too restrictive. Indeed, turning a proof into one satisfying the requirement that RB is applied only to atomic formulas may lead to exponential growth in the size of the proof tree, depending on the structure of the conclusion and of the assumption formulas. The weaker requirement that RB-applications are *analytic*, i.e. that their RB formulas are not spurious, may be sufficient in most interesting cases and may lead to essentially shorter proofs.

DEFINITION 23. We say that an application of RB in a proof π of P from Γ is *analytic* if its RB-formula is not spurious in π.

It is not difficult to adapt the proofs of Propositions 7 and 18 to yield a proof of the following:

LEMMA 24. *Every proof π of P from the set Γ of assumptions can be transformed into a proof π′ of P from Γ such that all the applications of RB in π′ are analytic.*

Then, one can easily show that:

PROPOSITION 25. *Let π be a proof of P from Γ such that (i) every closed branch of π is atomically closed, (ii) every application of RB in π is analytic, and (iii) π is pruned. Then π has the WSFP.*

Since the method of proof *ex-absurdo* is the hallmark of classical logic, a systematic use of such proofs, rather than direct ones, is perfectly in tune with the philosophy of *classical* natural deduction. Using intelim trees as a *refutation* method also prompts for comparison with other familiar refutation methods such as Smullyan's analytic tableaux. We have already

Marcello D'Agostino

1.	$p \lor q$	
2.	$r \lor s$	
3.	$p \to t \land u$	
4.	$r \to v \land u$	
5.	$\neg u$	
6.	$\neg(t \land u)$	I-rule for $\neg\land$ to 5
7.	$\neg(v \land u)$	I-rule for $\neg\land$ to 5
8.	$\neg r$	E-rule for $\to$ to 4 and 7
9.	$\neg p$	E-rule for $\to$ to 3 and 6
10.	q	E-rule for $\lor$ to 1 and 9
11.	s	E-rule for $\lor$ to 2 and 8
12.	$q \land s$	I-rule for $\land$ to 10 and 11

(a)

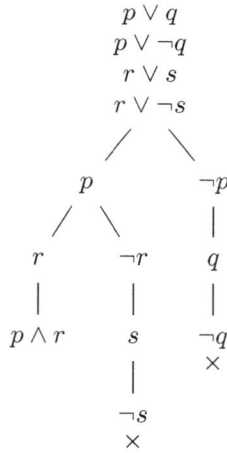

(b)

Figure 2. Examples of intelim refutations.

discussed the shortcomings of analytic tableaux in Section 3 and have developed our theory of classical natural deduction so as to avoid them.

It is immediate to check that the completeness of canonical restrictions of the intelim tree method is preserved when the latter is used as a refutation method:

PROPOSITION 26. *If there is a refutation π of Γ (i.e. an intelim tree based on Γ such that all its branches are closed), then there is a canonical refutation π' of Γ.*

So, interpreted as a refutation method, intelim trees can be fruitfully employed as a propositional basis for automated proof techniques. The next section will focus on this aspect and show how this method may be well-suited to the development of efficient and perspicuous refutation algorithms.

Looking at intelim trees from this perspective, we soon discover that we can do without the introduction rules:

PROPOSITION 27. *If there is an intelim proof of P from Γ than there is an intelim refutation π' of $\Gamma, \neg P$ such that (i) all applications of RB in π' are analytic and, (ii) π' contains no application of the introduction rules.*

The above proposition implies that the refutation system consisting of the elimination rules only plus the rule of bivalence (restricted to analytic applications) is a complete system for classical logic. Clearly such a system has the WSFP.[28] Observe that if we have a purely eliminative refutation of $\Gamma, \neg P$ with the WSFP, in order to obtain a purely eliminative proof of P from Γ with the WSFP we only need to apply RB once to the proposition P.

7   Feasible analytic proofs and refutations

Although being incomplete for classical propositional logic, the proof system consisting only of the intelim rules, without RB, is still quite powerful. In accordance with our truth-conditional reformulation of the theory of meaning underlying the natural deduction approach, the consequence relation associated with this system may be construed as the fragment of classical logic which can be justified only by means of the classical meaning of the logical operators as determined by the truth-conditions specified in the introduction rules. The resulting notion of inference, despite being based on "external" concepts of truth and falsity, is not committed to the principle

---

[28]This is the (analytic restriction) of the system **KE** first proposed in [Mondadori, 1988b] and whose proof-theoretical and computational advantages — especially with respect to standard cut-free refutation systems, such as Smullyan's Tableaux — are discussed extensively in [D'Agostino, 1990], [D'Agostino, 1992] and [D'Agostino and Mondadori, 1994]. See also [D'Agostino, 1999] and [Hähnle, 2001].

of bivalence, namely to what appears to be the most recalcitrant aspect of classical semantics from the point of view of the meaning-as-use approach.

Recall that, for every set of formulas Γ, we call *intelim sequence* based on Γ a (possibly infinite) sequence of formulas each of which is either a formula in Γ or is obtained by preceding formulas by an application of an intelim rule. An intelim sequence is *closed* if it contains both P and $\neg P$ for some formula P.

DEFINITION 28. A proof π of P from Γ is a *strict intelim proof* if it contains no applications of RB, i.e. if π is an intelim sequence. A formula P is *strictly intelim provable* from a set Γ of formulas if there is a strict intelim proof of P from Γ. A set Γ of formulas is *strictly intelim refutable* if there is a closed intelim sequence based on Γ.

Strict intelim proofs[29] and refutations are our candidates for the the set of classical logical arguments whose validity can be determined on the sole basis of the (classical) meaning of the logical operators. They are "analytic" in the particularly strict sense of being justified only by virtue of the way in which the language is immediately used in inference, with no appeal to metaphysical assumptions about the "world" or to forms of indirect reasoning. From this point of view, one could say that the inference process they represent conveys no information at all and therefore complies with the typical requirement of "analytic" arguments, namely that "the conclusion is contained in the premises". Indeed, one cannot get anything out of a set of formulas, by using elimination rules, that is not already in the set, in the sense of being obtainable from what we grasp as the meaning of the formulas involved by means of their definitions in terms of the introduction rules. As we shall show in the remaining of this section, this stricter sense of "analytic inference", unlike the general one,[30] is *tractable*: whether a formula P is strictly intelim provable from a finite set Γ of formulas or whether Γ is inconsistent are both questions that can be decided in polynomial time.

The discussion in the previous section can be easily adapted to show that if there is a strict intelim proof of P from Γ then there is a strict intelim proof π' of P from Γ such that π' has WSFP. In the remaining of this section we shall concentrate on strict analytic refutations rather than proofs, in order to emphasize their possible use for developments in the area of automated reasoning.

---

[29] Strict intelim provability (from Γ) is a standard Tarskian consequence relation. It is easy to check that it is closed under identity ($P \vdash P$), monotonicity ($\Gamma \vdash P$ implies that $\Gamma, \Delta \vdash P$) and cut ($\Gamma \vdash P$ and $\Delta, P \vdash Q$ imply $\Gamma, \Delta \vdash Q$). Notice that in this consequence relation there are no tautologies: there is no formula which is provable from the empty set of assumptions.

[30] At least according to the widespread conjecture that $P \neq NP$.

We shall say that an intelim sequence based on Γ is *E-saturated* if it contains all the formulas in Γ and is closed under the application of elimination rules. More precisely,

DEFINITION 29. An intelim sequence ϕ based on Γ is *E-saturated* whenever the following conditions hold true:

1. for every $P \in \Gamma$, P belongs to ϕ;

2. if the premiss (or the premisses) of an elimination rule belong(s) to ϕ, then its conclusion also belongs to ϕ.

Observe that E-saturated intelim sequences based on a finite set Γ of formulas are finite. Moreover, their length is at most linear in the length of Γ (the sum of the lengths of the formulas in Γ). Moreover, E-saturation obviously preserves the WSFP.

We shall also say that an intelim sequence based on Γ is *I-saturated* when it contains all the formulas in Γ and is closed under the application of introduction rules. More precisely:

DEFINITION 30. An intelim sequence ϕ based on Γ is *I-saturated* whenever the following conditions hold true:

1. for every $P \in \Gamma$, P belongs to ϕ;

2. if the premiss(es) of an application of an introduction rule belong to ϕ, then its conclusion also belongs to ϕ.

I-saturated intelim sequence, unlike E-saturated ones, are necessarily infinite, even if based on a finite set Γ, since there is no bound on the application of introduction rules. So, I-saturation clearly does not preserve the WSFP.

We shall say that an intelim sequence based on Γ is *saturated* if it is both E-saturated and I-saturated. Therefore, saturated intelim sequences are necessarily infinite, even if based on a finite set Γ and do not have the WSFP. Clearly, for every Γ, there is exactly one saturated intelim sequence based on Γ and:

PROPOSITION 31. *If the saturated intelim sequence based on Γ is open, then every intelim sequence based on Γ is also open.*

However, Proposition 26 ensures that if there is a closed intelim sequence for Γ, then there is also a closed intelim sequence for Γ with the WSFP.

We can therefore assume, without loss of completeness, that the application of introduction rules is restricted to the cases in which the conclusion is not a spurious formula.

DEFINITION 32. We say that an intelim sequence based on Γ is *analytically I-saturated* if it is closed under all applications of the introduction rules such that their conclusion is a weak subformula of some formula in Γ.

DEFINITION 33. We say that an intelim sequence based on Γ is *analytically saturated* if it is E-saturated and analytically I-saturated.

Then the propositions in the previous section immediately imply that:

PROPOSITION 34. *If there is a closed intelim sequence based on Γ, then the analytically saturated intelim sequence based on Γ is also closed.*

It follows that if the analytically saturated intelim sequence based on Γ is open, then there is no closed intelim sequence based on Γ.

DEFINITION 35. We say that P is an immediate subformula of Q if P is a proper subformula of Q (i.e. other than Q itself) and P is not a proper subformula of any proper subformula of Q.

DEFINITION 36 (labelled subformula graph). Let Γ be a set of formulas. The *subformula graph for Γ*, denoted by $S(\Gamma)$ is a directed graph whose vertices are all the subformulas of formulas in Γ and whose arcs are all the pairs

$$\{\langle P, Q \rangle \mid P \text{ is an immediate subformula of } Q\}.$$

A *labelled subformula graph for Γ* is a pair $\langle S(\Gamma), l \rangle$, where $S(\Gamma)$ is the subformula graph for Γ and l is its labelling function, i.e. a partial function mapping vertices into $\{0, 1\}$. A formula is *signed* if $l(P)$ is defined; if $l(P) = 1$ we say that P is *signed as true*, if $l(P) = 0$ we say that P is *signed as false*. For every signed formula P, we define its *unsigned* version P^* as follows: $P^* = P$ if $l(P) = 1$ and $P^* = \neg P$ if $l(P) = 0$.

DEFINITION 37 (intelim graph). An *intelim graph G for Γ* is a labelled subformula graph $\langle S(\Gamma), l \rangle$ for Γ satisfying the following conditions:

1. for every $Q \in \Gamma$, $l(Q) = 1$;

2. for every $Q \in S(\Gamma)$ such that $Q \notin \Gamma$, $l(Q)$ is defined *only if* either (i) there is a formula $P \in S(\Gamma)$ such that $l(P)$ is defined and Q^* is the conclusion of a one-premiss intelim rule applied to P^* or (ii) there are formulas $P_1, P_2 \in S(\Gamma)$ such that $l(P_1)$ and $l(P_2)$ are both defined and Q^* is the conclusion of a two-premiss intelim rule applied to P_1^* and P_2^*.

An intelim graph G for Γ is *completed* if it satisfies the two conditions above with the expression "only if" in the second condition being replaced by "if and only if".

EXAMPLE 38. The set of formulas

$$\{p \vee q, r \vee \neg q, p \rightarrow s \wedge t, r \rightarrow u \wedge t, \neg t\}$$

is classically inconsistent and strictly intelim refutable. The initial intelim graph for this set is shown in Figure 3.

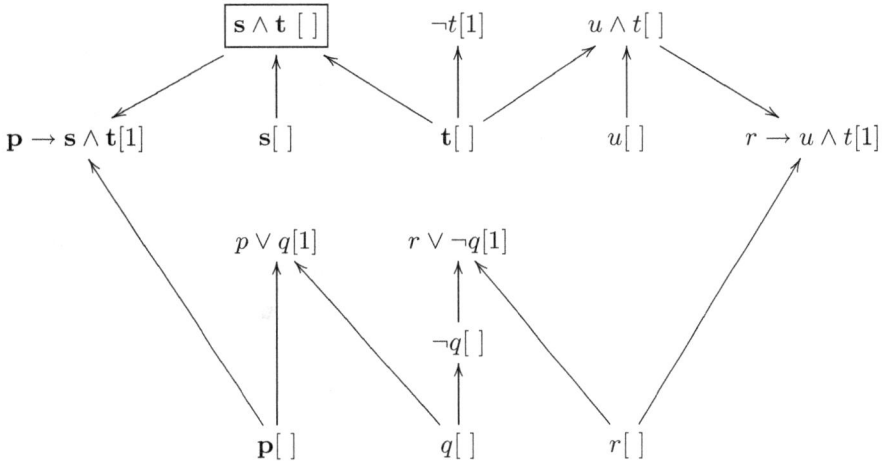

Figure 3. Initial intelim graph for the inconsistent set of formulas in Example 38. The values of the labelling function, when defined, are shown between square brackets. The boldface formulas are the *neighbours* of the boxed formula $s \wedge t$.

An *intelim* graph for Γ corresponds to an open *intelim sequence* based on Γ and a completed intelim graph for Γ to an open intelim sequence based on Γ which is analytically saturated. In both cases, the intelim sequence can be obtained from the intelim graph simply by means of a suitable sequential arrangement of the unsigned versions of the formulas that are signed in G. So, there is a closed intelim sequence based on Γ if and only if there exists no completed intelim graph for Γ.

Given the subformula graph $S(\Gamma)$ of a set of formulas Γ and a formula P in it, the *neighbouroods of* P, abbreviated as $N(P)$, is defined as the smallest subgraph of G containing, besides P itself, (i) the immediate subformulas of P, (ii) the immediate superformulas of P, and (iii) the immediate subformulas of the immediate superformulas of P. Each neighbouroods of a formula in an intelim graph is itself an intelim graph (although maybe not *for* the same set of initial formulas) and the overall graph is a completed intelim graph if and only if all the neighbouroods of the formulas in it are completed intelim graphs. So, an attempt to construct a completed intelim graph for the set of initial formulas Γ consists in trying to "complete", one by one, all the partial intelim graphs corresponding to the neighbouroods of each signed formula, until we stop either because the completion is impossible or because the graph has become completed.

In order to complete a partial intelim graph we start from a signed formula P, i.e. one for which the labelling function l is defined, and try to complete its neighbouroods, possibly generating new signed formulas (that is expanding the domain of definition of l). Next we visit another signed formula and repeat the process. If the attempt to complete the neighbouroods $N(Q)$ of some signed formula Q leads to conflicting assignments of values for the labelling function l, we say that the attempt to build a completed intelim graph has *failed* and stop. Otherwise, we proceed until the neighbouroods of all signed formulas have been visited (including the ones which have been signed on the way). We call the procedure we have just informally sketched the *completion procedure* for intelim graphs.

Let us denote by Γ_0 the class of all finite sets of formulas which are strictly intelim refutable. Then, on the basis of the completion procedure, it is not difficult to show the following:

THEOREM 39. *The complexity of the decision problem for* Γ_0 *is* $\mathcal{O}(n^2)$.

Proof.[Sketched] Given any finite set of formulas Γ and an *intelim* graph for Γ, the completion procedure, informally sketched above, always terminates either with a completed graph or with a graph such that for some formula P, its neighbouroods $N(P)$ cannot be completed. Moreover, the procedure runs in time $\mathcal{O}(n^2)$ where n is the size of the graph.

It is not difficult to see that the procedure always terminates, since the subformula graph is finite and the procedure stops when all the signed formulas of the graph have been visited once. Moreover, if it does not stop before, because the neighbouroods of some signed formula cannot be completed, the output is completed intelim graph. The simplest way to prove this is by contradiction. Suppose the procedure has terminated (all the signed formulas have been visited) and yet its output graph G is not completed. This means that there must be some formula Q in G such that (i) $l(Q)$ is not defined and (ii) Q is the conclusion of some intelim rule applied to a (pair of) premiss(es), (each of) which is the unsigned version of some signed formula in G. There are, as usual, several cases and, as usual, we shall discuss only of one of them, which we take as being representative of all, leaving the others to the reader.

Suppose both $P \lor Q$ and P are in G. Moreover, $l(P \lor Q) = 1$ and $l(P) = 0$, but $l(Q)$ is undefined. Now since P is signed in G, it must have been already visited once, say at the k-th step of the completion procedure. So, at the end of the k-th step $l(P \lor Q)$ must have been still undefined, otherwise, since both $P \lor Q$ and Q are in the neighbouroods of P, $l(Q)$ should have been defined and equal to 1 as a result of the completion step. Moreover, since $P \lor Q$ is also signed in G, it must have already been visited once, say at the n-th step of the procedure $(n > k)$. So, since P and Q are both in the neighbouroods of $P \lor Q$, at the end of the n-step, $l(P)$ must have been still undefined, which is impossible since $l(P)$ was assumed to be defined at step k.

The complexity upper bound is a very crude estimate whose justification can be briefly given as follows. Completing the neighbouroods $N(P)$ of a formula P requires (i) visiting each formula Q in $N(P)$ such that $Q \neq P$ and $l(Q)$ is not defined; (ii) try to assign a value to it according to the *intelim* rules. This, in turn, requires checking at most two "neighbours". More precisely:

1. if Q is an immediate subformula of P, only P and the other immediate subformula of P have to be checked;

2. if Q is an immediate superformula of P, only P and the other immediate subformula of Q have to be checked;

3. if Q is an immediate subformula of an immediate superformula of P, only P itself and the common immediate superformula of Q and P have to be checked.

Hence, completing the neighbouroods of P requires $\mathcal{O}(n)$ steps, where n is the size of the graph measured by the number of its vertices. Since completing the whole graph requires completing the neighbouroods of each signed formula in it, the whole completion procedure terminates in $\mathcal{O}(n^2)$ steps. Moreover, the size of the graph is bounded above by the total size of the input Γ (the sum of the number of symbols occurring in each formula). So, the procedure is still $\mathcal{O}(n^2)$ if n is taken to be the input size. Since building the initial subformula graph and initializing the labelling function l also takes $\mathcal{O}(n^2)$ steps, where n is again the input size, it follows that, for any finite set of formulas Γ, whether or not $\Gamma \in \mathbf{\Gamma}_0$ can be decided in $\mathcal{O}(n^2)$ steps. ∎

8 Future work

This work could be expanded in several directions, e.g.:

- clarifying the relationship between analytic intelim proofs and normal proofs in Prawitz' style natural deduction;

- comparing intelim trees and graphs, intended as a refutation method, with other known refutation methods (e.g. connection tableaux, see [Hähnle, 2001], Bibel's connection method [Bibel, 1982; Wallen, 1990], and Maslov's inverse method [Maslov *et al.*, 1983]);

- adapting intelim trees and graphs to provide an alternative natural deduction presentation of non-classical systems in the "neighbouroods" of classical logic, in particular those with an involutive type of negation, such as Girard's linear logic;[31]

- adapting intelim trees and graphs to the neighbours of intuitionistic logic, for instance by making appropriate use of *labelled signed formulas*, as in [D'Agostino and Gabbay, 1994] and [D'Agostino *et al.*, 1999];

- adding suitable rules for quantifiers and modal operators.

Moreover, we are currently investigating "external" semantics (as opposed to the "internal" proof-theoretic semantics discussed in this paper) for the sets of formulas in Γ_0, and working out the proof-theoretical and semantical properties of intelim trees with bounded applications of RB (on this point the reader is referred to the forthcoming [D'Agostino and Cilluffo, 2005]).

---

[31] See [Girard, 1987b] and [Avron, 1988] where a proof-theoretical presentation of (classical) linear logic is given in the framework of Prawitz-style natural deduction.

Appendix

NATURAL DEDUCTION RULES:

$$\frac{P \wedge Q}{P} \qquad \frac{P \wedge Q}{Q} \qquad \frac{P \quad Q}{P \wedge Q}$$

$$\frac{P}{P \vee Q} \qquad \frac{Q}{P \vee Q} \qquad P \vee Q \quad \begin{array}{c} \Psi_1, [P] \\ \vdots \\ R \end{array} \quad \begin{array}{c} \Psi_2, [Q] \\ \vdots \\ R \end{array}$$

$$\frac{}{R}$$

$$\begin{array}{c} \Psi, [P] \\ \vdots \\ \dfrac{Q}{P \to Q} \end{array} \qquad \frac{P \to Q \quad P}{Q}$$

$$\begin{array}{c} \Psi, [P] \\ \vdots \\ \dfrac{\mathbf{F}}{\neg P} \end{array} \qquad \frac{P \quad \neg P}{\mathbf{F}} \qquad \frac{\mathbf{F}}{Q}$$

The notation $[P]$ means that the conclusion of the rule does not depend on the assumption P. The rules listed above are the intuitionistic rules. Classical logic is obtained by adding one of the following three rules:

$$\frac{}{P \vee \neg P} \qquad \frac{\neg\neg P}{P} \qquad \begin{array}{c} [\neg P] \\ \vdots \\ \dfrac{\mathbf{F}}{P} \end{array}$$

The last of these rules is the *classical reductio ad absurdum* referred to in the text. Another variant of the classical calculus is described in Tennant 1978 where the author uses the following classical *rule of dilemma*:

$$\begin{array}{cc} [P] & [\neg P] \\ \vdots & \vdots \\ Q & Q \end{array}$$
$$\frac{}{Q}$$

TABLEAU RULES

$$\frac{P \wedge Q}{\begin{array}{c} P \\ Q \end{array}} \qquad \frac{\neg(P \wedge Q)}{\neg P \mid \neg Q} \qquad \frac{P \vee Q}{P \mid Q} \qquad \frac{\neg(P \vee Q)}{\begin{array}{c} \neg P \\ \neg Q \end{array}}$$

$$\frac{P \rightarrow Q}{\neg P \mid Q} \qquad \frac{\neg(P \rightarrow Q)}{\begin{array}{c} P \\ \neg Q \end{array}} \qquad \frac{\neg \neg P}{P}$$

BIBLIOGRAPHY

[Avron, 1988] A. Avron. The semantics and proof theory of linear logic. *Theoretical Computer Science*, 57:161–184, 1988.

[Beth, 1958] E.W. Beth. On machines which prove theorems. *Simon Stevin Wissen Natur-kundig Tijdschrift*, 32:49–60, 1958. Reprinted in [Siekmann and Wrightson, 1983], vol. 1, pages 79–90.

[Bibel, 1982] W. Bibel. *Automated Theorem Proving*. Vieweg, Braunschweig, 1982.

[Blanché, 1970] R. Blanché. *La logique et son histoire*. Armand Colin, Paris, 1970.

[Bochensky, 1961] I. M. Bochensky. *A History of Formal Logic*. University of Notre Dame, Notre Dame (Indiana), 1961.

[Broda et al.,] K. Broda, M. D'Agostino, and M. Mondadori. I sistemi logici di Karl Popper e l'insegnamento della logica. Technical report.

[Cellucci, 1988] C. Cellucci. Efficient natural deduction. In *Temi e prospettive della logica e della filosofia della scienza contemporanee*, volume I, pages 29–57, Bologna, 1988. CLUEB.

[D'Agostino and Cilluffo, 2005] M. D'Agostino and A. Cilluffo. Tractable analytic methods and non-deterministic semantics: a case study. Forthcoming. Technical Report available in September 2005, 2005.

[D'Agostino and Gabbay, 1994] M. D'Agostino and D.M. Gabbay. A generalization of analytic deduction via labelled deductive systems.Part I: Basic substructural logics. *Journal of Automated Reasoning*, 13:243–281, 1994.

[D'Agostino and Mondadori, 1991] M. D'Agostino and M. Mondadori. Deduzione analitica classica, significato degli operatori logici e complessitá. In Arturo Carsetti, Marco Mondadori, and Giorgio Sandri, editors, *Semantica, complessità e linguaggio naturale*, pages 21–57. CLUEB, Bologna, 1991.

[D'Agostino and Mondadori, 1992] M. D'Agostino and M. Mondadori. Carnap and logical truth. Annali dell'Università di Ferrara; Sez. III; Discussion paper 25, Università di Ferrara, 1992.

[D'Agostino and Mondadori, 1993] M. D'Agostino and M. Mondadori. Il dilemma dell'ATP(the ATP dilemma). *Discipline Filosofiche*, 3:179–224, 1993.

[D'Agostino and Mondadori, 1994] Marcello D'Agostino and Marco Mondadori. The taming of the cut. *Journal of Logic and Computation*, 4:285–319, 1994.

[D'Agostino et al., 1999] M. D'Agostino, D.M. Gabbay, and K. Broda. Tableau methods for substructural logics. In *[?]*, pages 397–467. Kluwer, Dordrecht, 1999.

[D'Agostino, 1990] Marcello D'Agostino. Investigations into the complexity of some propositional calculi. PRG Technical Monographs 88, Oxford University Computing Laboratory, 1990.

[D'Agostino, 1992] M. D'Agostino. Are tableaux an improvement on truth tables? Cut-free proofs and bivalence. *Journal of Logic, Language and Information*, 1:235–252, 1992.

[D'Agostino, 1999] M. D'Agostino. Tableau methods for classical propositional logic. In [?], pages 45–123. Kluwer, Dordrecht, 1999.

[Davis, 1983] M. Davis. The prehistory and early history of automated deduction. In [Siekmann and Wrightson, 1983], pages 1–28. 1983.

[Dummett, 1978] M. Dummett. Truth and other Enigmas. Duckworth, London, 1978.

[Dummett, 1991] M. Dummett. The logical basis of metaphysics. Duckworth, 1991.

[Gabbay and Olivetti, 2000] D.M. Gabbay and N. Olivetti. Goal-Directed Algorithmic Proof Theory. Applied Logic Series. Kluwer Academic Publisher, 2000.

[Gabbay, 1996] D.M. Gabbay. Labelled Deductive Systems, Volume 1 - Foundations. Oxford University Press, 1996.

[Gentzen, 1935] G. Gentzen. Unstersuchungen über das logische Schliessen. Math. Zeitschrift, 39:176–210, 1935. English translation in [?].

[Girard, 1987a] I. Girard. Proof Theory and Logical Complexity. Bibliopolis, Napoli, 1987.

[Girard, 1987b] J.-Y. Girard. Linear logic. Theoretical Computer Science, 50:1–102, 1987.

[Hähnle, 2001] R. Hähnle. Tableaux and related methods. In J. A. Robinson and A. Voronkov, eds., Handbook of Automated Reasoning, MIT Press, Cambridge, MA, 2001.

[Heyting, 1956] A. Heyting. Intuitionism. North-Holland, Amsterdam, 1956.

[Kanger, 1963] S. Kanger. A simplified proof method for elementary logic. In Computer Programming and Formal Systems, pages 87–94. North-Holland, Amsterdam, 1963. Reprinted in [Siekmann and Wrightson, 1983], pp. .

[Maslov et al., 1983] S. Yu. Maslov, G. E. Mints, and V.P Orevkov. Mechanical proof-search and the theory of logical deduction in the USSR. In [Siekmann and Wrightson, 1983], pages 29–38. 1983.

[Mondadori, 1988a] M. Mondadori. Classical analytical deduction. Annali dell'Università di Ferrara; Sez. III; Discussion paper 1, Università di Ferrara, 1988.

[Mondadori, 1988b] M. Mondadori. Classical analytical deduction, part II. Annali dell'Università di Ferrara; Sez. III; Discussion paper 5, Università di Ferrara, 1988.

[Mondadori, 1988c] M. Mondadori. On the notion of a classical proof. In Temi e prospettive della logica e della filosofia della scienza contemporanee, volume I, pages 211–224, Bologna, 1988. CLUEB.

[Mondadori, 1988d] M. Mondadori. Sulla nozione di dimostrazione classica. Annali dell'Università di Ferrara; Sez. III; Discussion paper 3, Università di Ferrara, 1988.

[Mondadori, 1989] M. Mondadori. An improvement of Jeffrey's deductive trees. Annali dell'Università di Ferrara; Sez. III; Discussion paper 7, Università di Ferrara, 1989.

[Mondadori, 1996] M. Mondadori. Efficient inverse tableaux. Journal of the IGPL, 3:939–953, 1996.

[Prawitz, 1965] D. Prawitz. Natural Deduction. A Proof-Theoretical Study. Almqvist & Wilksell, Uppsala, 1965.

[Prawitz, 1978] D. Prawitz. Proofs and the meaning and completeness of the logical constants. In J. Hintikka, I. Niinduoto, and E. Saarinen, editors, Essays on Mathematical and Philosphical Logic, pages 25–40. Reidel, Dordrecht, 1978.

[Siekmann and Wrightson, 1983] J. Siekmann and G. Wrightson. Automationof Reasoning. Springer-Verlag, New York, 1983.

[Smullyan, 1965] R.M. Smullyan. Analytic natural deduction. The Journal of Symbolic Logic, 30:549–559, 1965.

[Smullyan, 1968] R. Smullyan. First-Order Logic. Springer, Berlin, 1968.

[Tennant, 1978] N. Tennant. Natural Logic. Edimburgh University Press, Edinburgh, 1978.

[Thomas, 1941] I. Thomas, editor. Greek Mathematics, volume 2. William Heinemann and Harvard University Press, London and Cambridge, Mass., 1941.

[Urquhart, 1995] A. Urquhart. The complexity of propositional proofs. The Bulletin of Symbolic Logic, 1(IV):425–467, 1995.

[Wallen, 1990] L.A. Wallen. *Automated Deduction in Non-Classical Logics*. The MIT Press, Cambridge, Mass., 1990.

[Wang, 1960] Hao Wang. Towards mechanical mathematics. *IBM Journal for Research Development*, 4:2–22, 1960. Reprinted in [Siekmann and Wrightson, 1983], pp. 244-264.

[Weir, 1986] A. Weir. Classical harmony. *Notre Dame Journal of Formal Logic*, 27(4):459–482, 1986.

Neural-Symbolic Systems and the Case for Non-Classical Reasoning

Artur S. d'Avila Garcez and Luis C. Lamb

Dedicated to Dov Gabbay on his 60th birthday.

1 Introduction

The importance of the efforts to bridge the gap between the symbolic and connectionist paradigms of artificial intelligence has been widely recognised [Ajjanagadde, 1997; Cloete and Zurada, 2000; Shastri, 1999; Sun and Alexandre, 1997]. The merging of theory (known as background knowledge in machine learning systems) and data learning (learning from examples) into neural networks has been shown to provide a learning system that is more effective than purely symbolic or purely connectionist systems, especially when data are noisy [Towell and Shavlik, 1994].

Such developments have contributed to the growing interest in developing *Neural-Symbolic Systems*, i.e. hybrid systems based on neural networks that are capable of learning from examples and background knowledge [d'Avila Garcez *et al.*, 2002a]. Typically, translation algorithms from a symbolic to a connectionist representation and vice-versa are employed to provide either (*i*) a neural implementation of a logic, (*ii*) a logical characterisation of a neural system, or (*iii*) a hybrid learning system that brings together features from connectionism and symbolic artificial intelligence.

Until recently, neural-symbolic systems were not able to represent, compute, and learn languages other than propositional and some fragments of first-order logic [Browne and Sun, 2001; Cloete and Zurada, 2000]. However, in [d'Avila Garcez *et al.*, 2002b; d'Avila Garcez *et al.*, 2003a; d'Avila Garcez *et al.*, 2003b; d'Avila Garcez and Lamb, 2004; d'Avila Garcez and Gabbay, 2004; d'Avila Garcez *et al.*, 2004a; d'Avila Garcez *et al.*, 2006], a new approach to knowledge representation and reasoning in neural-symbolic systems based on neural network ensembles has been proposed. This new approach defined a class of connectionist non-classical logics, including connectionist modal, intuitionistic, temporal and epistemic logics [d'Avila Garcez *et al.*, 2003b; d'Avila Garcez and Lamb, 2004; d'Avila Garcez *et al.*, 2004a]. It shows that a variety of non-classical logics can be effectively represented in artificial neural networks. To the best of our knowledge, this has been the first approach to combine such non-classical logics with neural networks. Recently, it was also shown that value-based argumentation frameworks can be represented as neural networks, offering an integrated approach for learning and reasoning of arguments, including the computation of circular and accrual argumentation [d'Avila Garcez *et al.*, 2004c; d'Avila Garcez *et al.*, 2005].

As argued in [Browne and Sun, 2001], if connectionism is an alternative paradigm for artificial intelligence, neural networks must be able to compute symbolic reasoning in an efficient and effective way. Moreover, in hybrid learning systems usually the connectionist component is fault-tolerant, whilst the symbolic component may be "brittle and rigid". By integrating connectionist systems and non-classical logics we tackle this problem and offer a principled way to effectively compute, represent, and learn various non-classical logics within neural networks.

A historical criticism of neural networks has been raised by McCarthy already back in 1988 [McCarthy, 1988]. McCarthy referred to neural networks as having a "propositional fixation", in the sense that they were not able to represent first-order logic. This per se has remained a challenge for a decade, but several approaches have now dealt with first-order reasoning in neural networks, see e.g. [Browne and Sun, 2001].

Perhaps in an attempt to address McCarthy's criticism, many researchers in the area have focused attention only on first-order logic. This *first-order fixation* has suppressed developments in other important fronts, mainly in non-classical, practical reasoning, which we are convinced should be the real focus of neural-symbolic integration due to the practical nature of neural networks research. We have shown recently that non-classical reasoning can be used in a number of applications in neural-symbolic systems [d'Avila Garcez *et al.*, 2003b; d'Avila Garcez and Lamb, 2004; d'Avila Garcez *et al.*, 2004a; d'Avila Garcez *et al.*, 2004c; d'Avila Garcez *et al.*, 2005]. This has been possible through the integration of non-classical logics and neural networks. Non-classical logics have been shown to be adequate in expressing several reasoning features, allowing for the representation of temporal, epistemic and probabilistic abstractions in computer science and artificial intelligence, as shown e.g. in [Fagin *et al.*, 1995; Gabbay *et al.*, 2003; Halpern, 2003]. Other applications of non-classical logics include the characterisation of timing analysis in combinatorial circuits [Mendler, 2000] and in spatial reasoning [Bennett, 1994], with possible use in geographical information systems. For instance, Bennett's propositional intuitionistic approach provided for tractable yet expressive reasoning about topological and spatial relations.

We believe that for neural computation to achieve its promise, connectionist models must be able to cater for non-classical reasoning. Gabbay has been a pioneer and research leader in non-classical reasoning for the past 30 years. He has made several outstanding contributions to modal, temporal, intuitionistic, non-monotonic logic and their combinations, either using Labelled Deductive Systems or Fibring Logics [Gabbay, 1996; Gabbay, 1999]. We believe that the neural-symbolic community cannot ignore the achievements and impact that non-classical logics have had in computer science. Temporal logic, for instance, has had large an impact in both academia and industry [Gabbay *et al.*, 1994; Pnueli, 1977].

Modal logics, in turn, have become a *lingua franca* for the specification and analysis of knowledge and communication in Multi-Agent and Distributed Systems [Fagin *et al.*, 1995]. Non-monotonic reasoning has dominated research on AI in the eighties and nineties, and intuitionistic logic is considered by many as an adequate logical foundation in several core areas of theoretical computer science, including type theory and functional programming [van Dalen, 2002]. Notwithstanding all this evidence, very little attention has been given to non-classical reasoning and their integration with neural networks.

In this paper, we show that the representational power of artificial neural networks can go beyond the classical propositional level. We do so by integrating uncertainty reasoning with temporal and epistemic logics (which have also found a large number of applications, notably in game theory and in models of knowledge and interaction in multi-agent systems [Fagin *et al.*, 1995; Gabbay *et al.*, 1994; Pnueli, 1977]). We start by recalling how one can represent modal logics in neural networks. It is well-known that Modal Logics correspond, in terms of expressive power, to the two-variable fragment of first-order logic [Gabbay, 1981; van Benthem, 1984; Vardi, 1997]. Further, as the two-variable fragment of predicate logic is decidable, this explains why modal logics are so "robustly decidable" and amenable to applications. Both artificial intelligence and computer science have made extensive use of decidable modal logics, including the analysis and model checking of distributed and multi-agent systems, program verification and specification, and hardware model checking.

By extending the *Connectionist Modal Logic* (CML) framework, we allow reasoning and learning about probabilities dealing with different reasoning dimensions of an idealised agent [Friedman and Koller, 2003; Pearl, 2000]. Learning is achieved by training each individual network in the ensemble, which in turn corresponds to the current knowledge of an agent within a possible world. Such a form of learning aims to attend to the need for learning mechanisms in multi-agent systems in which modal logics are an essential feature to represent several kinds of knowledge dimensions an agent is typically endowed with.

In the long run, our work seeks to achieve a characterisation of a rich semantics for cognitive computation. This has been recently identified as a major challenge for computer science [Valiant, 2003]. We are proposing a methodology for the representation of several non-classical logics in artificial neural networks catering for integrated reasoning, knowledge representation, and learning.

According to Valiant, the two most fundamental phenomena of intelligent cognitive behaviour are the ability to learn from experience and the ability to reason from what has been learned. Aiming to attend to such requirements, this paper provides a robust computational model for reasoning about time and uncertainty. Knowledge is represented by a symbolic language, while deduction and learning are carried out by a connectionist engine.

In summary, we argue that non-classical reasoning is fundamental to neural-symbolic systems research. The combination of knowledge, time and probability in a connectionist system provides support for integrated knowledge representation and learning in a distributed environment, so far lacking in the literature. If one assumes that neural networks can represent rich models of human reasoning, it is undeniable that non-classical logics should be at the core of this enterprise.

The remainder of the paper is organised as follows. Section 2 presents the basics of *CML* and its extension to deal with time and knowledge. Section 3 shows how we can model uncertainty using *CML*, presents a new translation algorithm from probabilistic knowledge to neural networks, and describes the computation of reasoning under uncertainty using well-known problems on probabilistic reasoning. Section 4 concludes the paper and discusses directions for future work.

2 Preliminaries

In this section, we introduce neural networks and Connectionist Modal Logic (CML) concepts that shall be used in this paper.

2.1 Neural-Symbolic Systems

An artificial neural network is a directed graph. A node (or neuron) in this graph is characterised, at time t, by its input vector $I_i(t)$, its input potential $U_i(t)$, its activation state $A_i(t)$, and its output $O_i(t)$. The nodes of the network are interconnected via a set of directed and weighted connections. If there is a connection from node i to node j, then $W_{ji} \in \mathbb{R}$ denotes the weight of this connection (see Figure 1). The input potential of neuron i at time t $(U_i(t))$ is obtained by computing a weighted sum for neuron i

such that $U_i(t) = \sum_j W_{ij} I_i(t)$. The activation state of a neuron i at time t $(A_i(t))$ is a bounded real or integer number; $A_i(t)$ is given by the neuron's *activation rule* h_i, which is a function of the neuron's input potential, i.e. $A_i(t) = h_i(U_i(t))$. Typically, h_i is either a linear, a non-linear or a sigmoid activation function, e.g. $tanh(x)$. In addition, θ_i is known as the threshold of neuron i (think of $-\theta_i$ as an extra weight with input always fixed at 1; $-\theta_i$ is known as the bias of neuron i). We say that neuron i is *activated* at time t if $A_i(t) > \theta_i$. Finally, the neuron's output value $O_i(t)$ is given by $f_i(A_i(t))$; usually, f_i is the identity function.

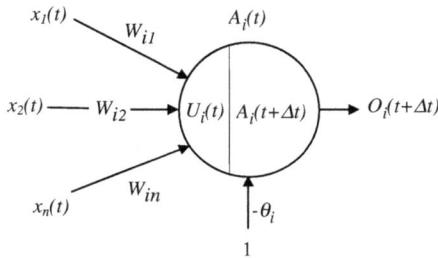

Figure 1. A representation of a neuron.

The nodes of a neural network can be organised in layers. A n-layer feedforward network is an acyclic graph containing one input layer, $n-2$ hidden layers, and one output layer. It computes a function $\varphi : \mathbb{R}^r \to \mathbb{R}^s$, where r and s denote the number of nodes occurring in the input and output layers, respectively. Most neural models also have a *learning rule*, responsible for changing the weights of the network so that it learns to approximate φ given a number of *training examples*, e.g. input vectors and their respective target output vectors.

The *Connectionist Inductive Learning and Logic Programming (C-ILP) System* [d'Avila Garcez *et al.*, 2002a] is a computational model based on neural networks that integrates inductive learning from examples and background knowledge with deductive learning using logic programming. In *C-ILP*, a translation algorithm maps a logic program $\mathcal{P}$ into a single hidden layer neural network $\mathcal{N}$ such that $\mathcal{N}$ computes the fixed-point operator $T_{\mathcal{P}}$ of $\mathcal{P}$. This provides a massively parallel model for computing the stable model semantics of $\mathcal{P}$. In addition, $\mathcal{N}$ can be trained with examples using a neural learning algorithm, having $\mathcal{P}$ as background knowledge. The knowledge acquired by training can then be extracted, closing the learning cycle [d'Avila Garcez *et al.*, 2002a].

Let us exemplify how the *C-ILP* translation algorithm works. Each rule (r_l) of $\mathcal{P}$ is mapped from the input layer to the output layer of $\mathcal{N}$ through one neuron (N_l) in the single hidden layer of $\mathcal{N}$. Intuitively, the translation produces an and-or network as follows. The algorithm from $\mathcal{P}$ to $\mathcal{N}$ implements the following conditions: (c_1) the input potential of a hidden neuron (N_l) can only exceed N_l's threshold (θ_l), activating N_l, when all the positive antecedents of r_l are assigned truth-value *true* (i.e. such input neurons are active), while all the negative antecedents of r_l are assigned *false* (i.e. such input neurons are not active); and (c_2) the input potential of an output neuron (A) can only exceed A's threshold (θ_A), activating A, when at least one hidden neuron N_l that is connected to A is activated.

EXAMPLE 1. (*C-ILP*) Consider the logic program $\mathcal{P} = \{r_1 : B \wedge C \wedge \sim D \rightarrow A; r_2 : E \wedge F \rightarrow A; r_3 : B\}$ where $\sim$ stands for *default negation*. The translation algorithm derives the network $\mathcal{N}$ of Figure 2, setting weights (W) and thresholds (θ) in such a way that conditions (c_1) and (c_2) above are satisfied. Note that, if $\mathcal{N}$ ought to be fully-connected, any other link (not shown in Figure 2) should receive weight zero initially. Each input and output neuron of $\mathcal{N}$ is associated with an atom of $\mathcal{P}$. As a result, each input and output vector of $\mathcal{N}$ can be associated with an interpretation for $\mathcal{P}$, so that an atom (e.g. A) is true iff its corresponding neuron (neuron A) is activated. Note also that each hidden neuron N_l corresponds to a rule r_l of $\mathcal{P}$. By construction the input potential of neuron N_3 will always be zero, regardless of the input vector. The algorithm then sets $\theta_3 < 0$ so that N_3, and then B, is always activated (true).

In order to compute a stable model, output neuron B should feed input neuron B such that $\mathcal{N}$ is used to iterate the fixed-point operator $T_{\mathcal{P}}$ of $\mathcal{P}$ [d'Avila Garcez et al., 2002a; Holldobler and Kalinke, 1994]. This is done by transforming $\mathcal{N}$ into a recurrent network, by connecting the output to the input layer of $\mathcal{N}$ using fixed weights with value 1. For instance, in Figure 2, output neuron B needs to be connected to input neuron B so that the network iterates $T_{\mathcal{P}}$. In the case of $\mathcal{P}$ above, the network converges to the following stable state: $A = false, B = true, C = false, D = false, E = false$, and $F = false$, which represents the unique fixed-point of $\mathcal{P}$.

2.2   The Language and Semantics of CML

In *CML*, the language of Modal Logic Programming is extended to allow modalities such as necessity ($\Box$) and possibility ($\Diamond$) to occur not only in the body, but also in the head of clauses. A modal translation algorithm then sets up an ensemble of *C-ILP* neural networks [d'Avila Garcez et al., 2002a], each network representing a possible world. Each network in the ensemble can be trained by examples just like *C-ILP* networks. The different net-

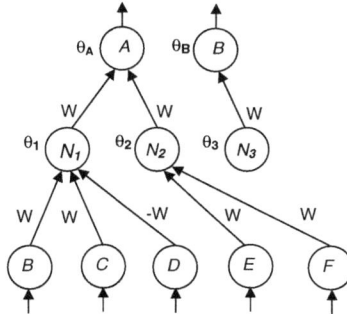

Figure 2. A neural network for program $\mathcal{P}$.

works in the ensemble are interconnected by following the semantics of the modalities given below. The ensemble computes a fixed-point semantics of modal theories, thus working as a massively parallel system for modal logic [d'Avila Garcez *et al.*, 2004a].

A main feature of modal logics is the use of Kripke's *possible world semantics*: a proposition is necessary in a world if it is true in all worlds which are possible in relation to that world, whereas it is possible in a world if it is true in at least one world which is possible in relation to that same world. This is expressed in the semantics formalisation by a (binary) relation between possible worlds. In modal logic programming, a *modal atom* is of the form MA, where $M \in \{\Box, \Diamond\}$ and A is an atom. A *modal literal* is of the form ML, where L is a literal. A *modal program* is a finite set of clauses of the form $\alpha_1, ..., \alpha_n \to \alpha_{n+1}$ where α_i, $1 \leq i \leq n$, is either an atom or a modal atom. We define *extended modal programs* as modal programs extended to allow modalities $\Box$ and $\Diamond$ in the head of clauses as well, and negation by finite failure $\sim$ in the body of clauses. In addition, each clause is labelled by the possible world in which it holds, similarly to Gabbay's Labelled Deductive Systems [Broda *et al.*, 2004]. Thus, an *extended modal program* is a finite set of clauses C of the form $\omega_i : \beta_1, ..., \beta_n \to \beta_{n+1}$, where ω_i is a label representing a world in which the associated clause holds, β_i, $1 \leq i \leq n$, is either a literal or a modal literal, β_{n+1} is either an atom or a modal atom, and a finite set of relations $\mathcal{R}(\omega_i, \omega_j)$ between worlds ω_i and ω_j in C.

A (Kripke) model M is a tuple $M = (\mathcal{W}, \mathcal{R}, \pi)$, where (i) $\mathcal{W}$ is a set of possible worlds; (ii) $\mathcal{R}$ is a binary accessibility relation over worlds; and (iii) π is a mapping associating worlds to formulas. We write $(M, \omega) \models \varphi$ if φ is true at ω in M. Formally:

$(M, \omega) \models p$ iff $\omega \in \pi(p)$ for a propositional letter p;

$(M, \omega) \models \neg\varphi$ iff $(M, \omega) \not\models \varphi$;

$(M, \omega) \models \varphi \wedge \psi$ iff $(M, \omega) \models \varphi$ and $(M, \omega) \models \psi$;

$(M, \omega) \models \Box\varphi$ iff $\forall\omega' \in \mathcal{W}$, if $\mathcal{R}(\omega, \omega')$ then $(M, \omega') \models \varphi$;

$(M, \omega) \models \Diamond\varphi$ iff $\exists\omega'$ such that $\mathcal{R}(\omega, \omega')$ and $(M, \omega') \models \varphi$.

When computing the semantics of a modal program, we have to consider both the fixed-point at a particular world, and the fixed-point of the program as a whole. When computing the fixed-point at each world, we have to consider the consequences derived locally and the consequences derived from the interaction between worlds. Locally, fixed-points are computed as before, by simply renaming each modal literal ML_i by a new literal L_j not in the language, and computing stable models. When considering interacting worlds, there are two cases to address, according to the $\Box$ and $\Diamond$ modalities and the accessibility relation $\mathcal{R}$, which might render additional consequences in each world.

Briefly, whenever $\Box A$ is true in a world (i.e. a neuron labelled $\Box A$ is activated in the corresponding neural network), A must be true in every world related to that (i.e. connections in the ensemble of networks must be established so that the firing of neuron $\Box A$ activates all neurons A in the related networks). Whenever $\Diamond A$ is true in a world (neuron $\Diamond A$ is activated), A must be true in one world related to that (connections must be established so that the firing of $\Diamond A$ activates A in one related world). The choice of the world in which to have A activated is arbitrary, reflecting the semantics of the $\Diamond$ modality. The following example illustrates this.

EXAMPLE 2. Let $\mathcal{P} = \{\omega_1 : r \rightarrow \Box q; \ \omega_1 : \Diamond s \rightarrow r; \ \omega_2 : s; \ \omega_3 : q \rightarrow \Diamond p; \ \mathcal{R}(\omega_1, \omega_2); \ \mathcal{R}(\omega_1, \omega_3)\}$. We start by creating three C-ILP neural networks to represent the worlds ω_1, ω_2, and ω_3 (see Figure 3). Then, we interconnect the networks according to the meaning of $\Box$ and $\Diamond$. Hidden neurons labelled $\{M, \vee, \wedge\}$ are created to do this (the remaining neurons are all created by C-ILP). For example, whenever neuron $\Box q$ is activated in ω_1, neuron q should be activated in both ω_2 and ω_3; whenever neuron $\Diamond s$ is activated in ω_1, neuron s should be activated in ω_2. This is implemented by using the hidden neurons labelled as M in the network. Dually, whenever q is activated in both ω_2 and ω_3, $\Box q$ should be activated in ω_1; whenever s is activated in ω_2, $\Diamond s$ should be activated in ω_1. This is implemented by using neurons labelled as $\wedge$ and $\vee$, respectively. Notice that, if we had s in ω_3

only, we would also derive $\Diamond s$ in ω_1, whereas we would need to have q in both ω_2 and ω_3 to derive $\Box q$ in ω_1 (hence the use of symbols $\vee$ and $\wedge$ in Figure 3.

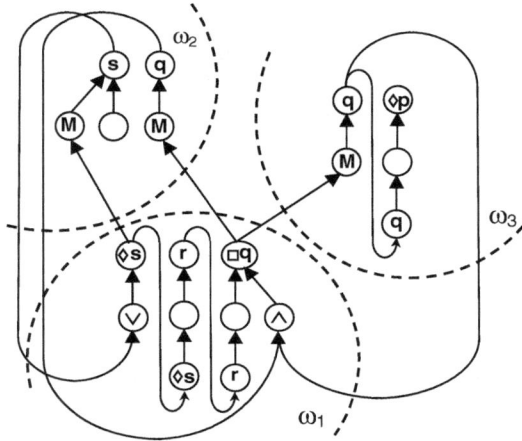

Figure 3. The ensemble $\{\mathcal{N}_1, \mathcal{N}_2, \mathcal{N}_3\}$ for program $\mathcal{P}$.

2.3   Reasoning about Time and Knowledge

In order to reason about the truth of sentences in time, and represent knowledge evolution through time, we need to add temporal operators to the language of *CML*, as described below. We consider a temporal logic of knowledge (see [Fagin *et al.*, 1995] for complete axiomatisations of such logics). The language of *CML* is extended with a set of agents $\mathcal{A} \subseteq \mathbb{N}$, a set of unary connectives: K_i, $i \in \mathcal{A}$, where $K_i p$ reads "agent i knows p", and the temporal operator $\bigcirc$ (next time). A temporal translation algorithm then is responsible for converting temporal rules of the form $t : K_{[\mathcal{A}]} L_1, ..., K_{[\mathcal{A}]} L_k \rightarrow \bigcirc K_{[\mathcal{A}]} L_{k+1}$, into neural network ensembles, where $[\mathcal{A}]$ denotes an element selected from $\mathcal{A}$ for each literal L_j, t is a timepoint in which the rule holds, $1 \leq j \leq k+1$, $1 \leq t \leq n$; $k, n \in \mathbb{N}$ [d'Avila Garcez and Lamb, 2004].

To each timepoint we associate the set of formulas holding at that point, and extend the definition of a model M as follows: $M = (T, \mathcal{R}_1, ..., \mathcal{R}_n, \pi)$, where (i) T is a set of (linearly) ordered points; (ii) $\mathcal{R}_i$ ($i \in A$) is an agent accessibility relation over points; (iii) π is a mapping associating time points to formulas. We write $(M, t) \models \alpha$ if α is true at point t in M. Formally:

$(M, t) \models \bigcirc \alpha$ iff $(M, t+1) \models \alpha$;

$(M, t) \models K_i \alpha$ iff $\forall u \in T$, if $R_i(t, u)$ then $(M, u) \models \alpha$.

It is worth noting that, whenever a rule's consequent is preceded by $\bigcirc$, a forward connection from t to $t + 1$ and a feedback connection from $t + 1$ to t need to be added to the ensemble. For example, if $t : a \rightarrow \bigcirc b$ is a rule in $\mathcal{P}$ then not only must the activation of neuron a at t activate neuron b at $t+1$, but the activation of neuron b at $t+1$ must also activate neuron $\bigcirc b$ at t.

EXAMPLE 3. One of the typical axioms of temporal logics of knowledge is $K_i \bigcirc \alpha \rightarrow \bigcirc K_i \alpha$ [Fagin *et al.*, 1995], which means that an agent does not forget tomorrow what he knew today. This can be represented in an ensemble of *C-ILP* networks by connecting output neuron $K \bigcirc \alpha$ of agent i at time t to a hidden neuron that connects to output neuron $K\alpha$ of agent i at time $t + 1$. In Figure 4, the black circle denotes a neuron that is always activated (*true*), and the activation value of output neuron $K \bigcirc \alpha$ at time t propagates to output neuron $K\alpha$ at time $t + 1$ via the above mentioned hidden neuron. Weights must be such that $K\alpha$ at $t + 1$ is also activated (*true*).

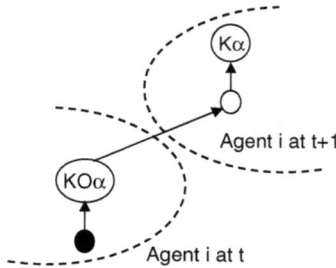

Figure 4. An example of temporal reasoning.

3 Representing Uncertainty

To represent uncertainty, we add a probabilistic dimension to the Connectionist Modal Logic framework presented in the previous section. We follow Halpern's work in order to do so [Halpern, 2003], and associate probabilities to possible worlds. In this way, the different probabilities that agents envisage are not captured by different probability distributions, but by the set of worlds that they consider possible. As a result, we need a (temporal) translation algorithm [d'Avila Garcez and Lamb, 2004] that caters for an agent that envisages multiple possible worlds at the same time point. Let us first consider the *two coin problem* as described in [Halpern, 2003] as an example of how to represent uncertainty in neural-symbolic systems. In what follows, we say that $p_i(\alpha) = x$ if the probability of formula α, according to agent i, is equal to x, where x is a real number in $[0, 1]$. We then allow neurons to be labelled as $\alpha = x$, indicating that x is the probability of α when the neuron is activated. There are two agents involved in the problem: Agent 1 holds two coins, one of them is fair (coin 1) and the other is biased (coin 2). Agent 1 is able to identify which coin is fair and which coin is biased. Agent 2, however, only knows that one coin is fair and the other is twice as likely to land heads as tails, but cannot tell them apart. Thus, the probability p_{c_1} of coin 1 landing heads is $p_{c_1}(head) = 1/2$, and the probability of coin 1 landing tails is $p_{c_1}(tail) = 1/2$. The probability p_{c_2} of coin 2 landing heads is $p_{c_2}(heads) = 2/3$, and the probability of coin 2 landing tails is $p_{c_2}(tails) = 1/3$.

Agent 1 chooses one of the coins to toss. Agent 2 does not know which coin Agent 1 has chosen, neither the probability that the fair coin has been chosen. What is the probability, according to Agent 2 that the result of the coin toss will be heads? What is the probability according to Agent 1? Following [Halpern, 2003], the difference between Agent 1 and Agent 2 is not captured by the probability distribution that they use, but by the set of worlds that they consider possible. We argue, as a result, that the Connectionist Modal Logic framework extended to reason about knowledge and time is suitable for representing the kinds of problems discussed in [Halpern, 2003]. Let us discuss the background knowledge of the agents involved in the problem. Below, we use **f** for *fair*, **b** for *biased*, **h** for *heads*, and **t** for *tails*. Moreover, as one would expect, knowledge persists through time.

Suppose that at time t_0, no coin has been chosen yet. At time t_1, Agent 1 has chosen a coin to toss, and at time t_2, Agent 1 has tossed the coin. At t_1, Agent 1 must know if she has selected a biased coin ($t_1 : K_1\mathbf{b}$) or a fair coin ($t_1 : K_1\mathbf{f}$). We say that Agent 1 is in one of the states B or F depending on the coin that she has selected (biased or fair, respectively). We represent this as a neural-symbolic system by having ω_{1B} as a possible world in which Agent 1 has selected the biased coin at time t_1. Similarly, we have ω_{1F} as a possible world in which Agent 1 has selected the fair coin at time t_1. Obviously, the associated probabilities $p(\mathbf{h}) = 2/3$ and $p(\mathbf{t}) = 1/3$ must hold in ω_{1B}, while $p(\mathbf{h}) = 1/2$ and $p(\mathbf{t}) = 1/2$ must hold in ω_{1F}. In contrast, Agent 2 is uncertain about the situation. All Agent 2 knows at t_1 is that $p(\mathbf{h}) = 1/2$ or $p(\mathbf{h}) = 2/3$, and that $p(\mathbf{t}) = 1/2$ or $p(\mathbf{t}) = 1/3$. We represent this by using two worlds: ω_{2B} as a possible world in which Agent 2 considers that $p(\mathbf{h}) = 2/3$ and $p(\mathbf{t}) = 1/3$, and ω_{2F} as a possible world in which Agent 2 considers that $p(\mathbf{h}) = 1/2$ and $p(\mathbf{t}) = 1/2$.

Notice how what is known by the agents should be implemented in the different neural networks: at t_1, either $p(\mathbf{h}) = 2/3$ or $p(\mathbf{h}) = 1/2$, but not both, will be activated for Agent 1 (according to the inputs to the networks representing ω_{1B} and ω_{1F}); at the same time, both of $p(\mathbf{h}) = 2/3$ and $p(\mathbf{h}) = 1/2$ should be activated for Agent 2 in worlds ω_{2B} and ω_{2F}, respectively, indicating the fact that Agent 2 is uncertain about the situation.

At time t_2, Agent 1 knows the result of the coin toss, and she assigns probability 1 to it. Agent 2 also knows the result of the toss (we assume that he can see it), but "never learns what happened" [Halpern, 2003]. As a result, at time t_2, there are two possible worlds: ω_H in which $p(\mathbf{h}) = 1$ and $p(\mathbf{t}) = 0$, and ω_T in which $p(\mathbf{h}) = 0$ and $p(\mathbf{t}) = 1$.

Figure 5 represents the reasoning process at t_1 and t_2, where $\mathbf{b}$ indicates that the biased coin has been selected, $\mathbf{f}$ indicates that the fair coin has been selected, $\mathbf{ct}$ indicates that the selected coin has been tossed, $\mathbf{h}$ stands for heads and $\mathbf{t}$ for tails. As discussed above, in ω_H, $\mathbf{h}$ is activated, and if $\mathbf{ct}$ is also activated, neurons $\mathbf{h} = 1$ and $\mathbf{t} = 0$ will be activated (using rules $\mathbf{ct} \wedge \mathbf{h} \rightarrow (p(\mathbf{h}) = 1)$ and $\mathbf{ct} \wedge \mathbf{h} \rightarrow (p(\mathbf{t}) = 0)$). Similarly, in ω_T, $\mathbf{t}$ is activated, and if $\mathbf{ct}$ is also activated, neurons $\mathbf{h} = 0$ and $\mathbf{t} = 1$ will be activated (using rules $\mathbf{ct} \wedge \mathbf{t} \rightarrow (p(\mathbf{h}) = 0)$ and $\mathbf{ct} \wedge \mathbf{t} \rightarrow (p(\mathbf{t}) = 1)$). In Figure 5, the connection between worlds can be established by the use of the *tomorrow* operator $\bigcirc$ as detailed in what follows.

The above example illustrates the need to consider a pair (ω, t), where ω is one of the worlds envisaged by an agent at time t. This allows us to reason about situations like the one described in the two-coin problem, where each agent i can reason about different possible worlds ω at a time point t, represented by $t : K_i^\omega \alpha$, where α is qualitative information about the probability of a literal. More precisely, in what follows, we consider *knowledge and probability rules* of the form $t : \bigcirc K_{[\mathcal{A}]}^{[\mathcal{W}]} L_1, ..., \bigcirc K_{[\mathcal{A}]}^{[\mathcal{W}]} L_k \rightarrow \bigcirc K_{[\mathcal{A}]}^{[\mathcal{W}]} L_{k+1}$, where $[\mathcal{W}]$ denotes an element selected from the set of possible worlds $\mathcal{W}$, for each literal L_j, $(1 \leq j \leq k + 1)$; $(1 \leq t \leq n)$; $k, n \in \mathbb{N}$. In addition, L_j can be either a literal, or a probabilistic statement of the form $p(\alpha) = x$, as defined above. Notice also that $\bigcirc$ is not required to precede every single literal.

Normally, we will have rules such as $t : K_{[\mathcal{A}]}^\omega L_1, ..., K_{[\mathcal{A}]}^\omega L_k \rightarrow \bigcirc K_{[\mathcal{A}]}(p(\alpha) = x)$. This means that, although a single rule might refer to different worlds, normally (as in the examples given in this paper) a single ω is selected from $[\mathcal{W}]$ for each rule. Recall that $t : \bigcirc K_i \alpha$ denotes that

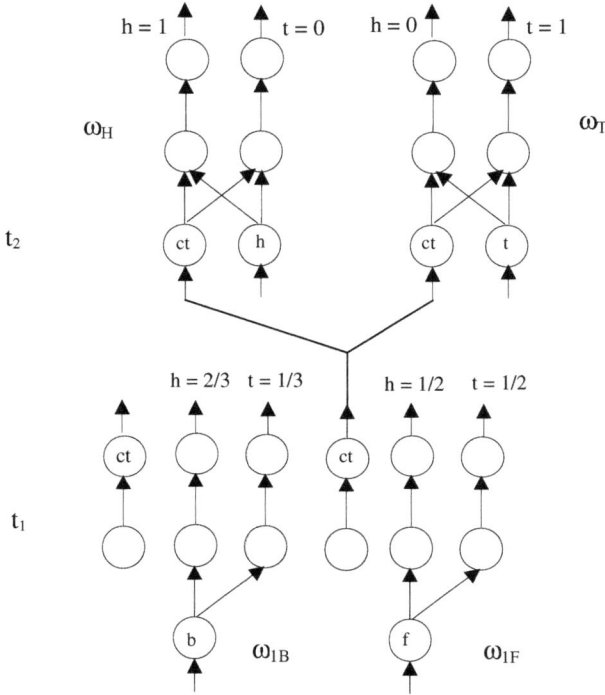

Figure 5. A representation of the two-coin problem.

α holds at every accessible possible world from the point of view of agent i at time $t+1$. Finally, notice that probabilistic statements are not generally preceded by negation.

Returning to the two-coin problem, the following would be the rules for the agents involved in the situation where ω_{iB} (respectively, ω_{iF}) denotes the world in which agent i considers that the coin is biased (respectively, fair), $i \in \{1, 2\}$.

$t_1 : K_i^{\omega_{iB}} \mathbf{b} \rightarrow K_i^{\omega_{iB}} (p(\mathbf{h}) = 2/3)$;

$t_1 : K_i^{\omega_{iB}} \mathbf{b} \rightarrow K_i^{\omega_{iB}} (p(\mathbf{t}) = 1/3)$;

$t_1 : K_i^{\omega_{iF}} \mathbf{f} \rightarrow K_i^{\omega_{iF}} (p(\mathbf{h}) = 1/2)$;

$t_1 : K_i^{\omega_{iF}} \mathbf{f} \rightarrow K_i^{\omega_{iF}} (p(\mathbf{t}) = 1/2)$;

$t_1 : K_1^{\omega_{1B}} \mathbf{ct}$;

$t_1 : K_1^{\omega_{1F}} \mathbf{ct}.$

In addition, once agent i knows that a coin has been tossed, the following rules would model this situation.

$t_2 : K_i^{\omega_H} \mathbf{ct} \wedge K_i^{\omega_H} \mathbf{h} \rightarrow (p(\mathbf{h}) = 1);$

$t_2 : K_i^{\omega_H} \mathbf{ct} \wedge K_i^{\omega_H} \mathbf{h} \rightarrow (p(\mathbf{t}) = 0);$

$t_2 : K_i^{\omega_T} \mathbf{ct} \wedge K_i^{\omega_T} \mathbf{h} \rightarrow (p(\mathbf{h}) = 0);$

$t_2 : K_i^{\omega_T} \mathbf{ct} \wedge K_i^{\omega_T} \mathbf{h} \rightarrow (p(\mathbf{t}) = 1).$

Finally, to interconnect the networks, we have the following rule expressing the fact that when $\mathbf{ct}$ holds in ω_{1F} then $\mathbf{ct}$ holds in both ω_H and ω_T:

$t_1 : K_i^{\omega_{1F}} \mathbf{ct} \rightarrow \bigcirc K_i \ \mathbf{ct}, \ \mathcal{R}(\omega_{1F}, \omega_H), \mathcal{R}(\omega_{1F}, \omega_T).$

We now present the algorithm to translate knowledge and probability rules into neural network ensembles. In the algorithm, the equations to calculate the values of weights W and thresholds were obtained from the *C-ILP* translation algorithm presented in [d'Avila Garcez *et al.*, 2002a]. The main difference between the two algorithms is that here probabilities can be associated with neurons and that probabilities can be combined with time and knowledge operators, along the lines of [Halpern, 2003]. The algorithm distinguishes the case in which the tomorrow operator occurs in the head of a clause (step 3) and the case in which it does not (step 4).

Notation: Let $\mathcal{P}$ contain q knowledge and probability rules. We number each L_j from 1 to $k + 1$ such that, when a neural network $\mathcal{N}$ is created, input and output neurons are created to represent each L_j. We also number worlds $\omega_\iota \in \mathcal{W}$ from 1 to u, $u \in \mathbb{N}$. We use $A_{min} \in (0, 1)$ to denote the minimum activation for a neuron to be considered active (i.e. its associated literal is considered *true*), a bipolar semi-linear activation function $h(x) = 2/(1 + e^{-\beta x}) - 1$, and inputs in $\{-1, 1\}$. Let k_l denote the number of literals in the body of rule r_l; μ_l, the number of rules in $\mathcal{P}$ with the same literal as consequent, for each rule r_l; $MAX_{r_l}(k_l, \mu_l)$, the greater element between k_l and μ_l for rule r_l; and $MAX_{\mathcal{P}}(k_1, ..., k_q, \mu_1, ..., \mu_q)$, the greatest element among all k_l's and μ_l's of $\mathcal{P}$. We also use $\overrightarrow{k}$ as shorthand for $(k_1, ..., k_q)$, and $\overrightarrow{\mu}$ as shorthand for $(\mu_1, ..., \mu_q)$.[1]

Knowledge and Probability Translation Algorithm

---

[1] For example, for $\mathcal{P} = \{r_1 : b \wedge c \wedge \neg d \rightarrow a; r_2 : e \wedge f \rightarrow a; r_3 : b\}$, $k_1 = 3$, $k_2 = 2$, $k_3 = 0$, $\mu_1 = 2$, $\mu_2 = 2$, $\mu_3 = 1$, $MAX_{r_1}(k_1, \mu_1) = 3$, $MAX_{r_2}(k_2, \mu_2) = 2$, $MAX_{r_3}(k_3, \mu_3) = 1$ and $MAX_{\mathcal{P}}(\overrightarrow{k}, \overrightarrow{\mu}) = 3$.

1. For each world $\omega_\iota \in \mathcal{W}$, for each time point t, do: Create a C-ILP Neural Network $\mathcal{N}_{\omega_\iota,t}$.

2. Calculate the weight W such that:

$$W \geq \frac{2}{\beta} \cdot \frac{\ln(1 + A_{\min}) - \ln(1 - A_{\min})}{MAX_{\mathcal{P}}(\overrightarrow{k}, \overrightarrow{\mu}) \cdot (A_{\min} - 1) + A_{\min} + 1};$$

3. For each rule r_l in $\mathcal{P}$ of the form

$$t: \bigcirc K_{[\mathcal{A}]}^{\omega_1} L_1, ..., \bigcirc K_{[\mathcal{A}]}^{\omega_{u-1}} L_k \rightarrow \bigcirc K_{[\mathcal{A}]}^{\omega_u} L_{k+1}:$$

(a) Add a hidden neuron L° to $\mathcal{N}_{\omega_u,t+1}$ and set $h(x)$ as the activation function of L°;

(b) Connect each neuron $\bigcirc K_{[\mathcal{A}]}^{\omega_\iota} L_i$, $(1 \leq i \leq k)$, $(1 \leq \iota \leq u-1)$ in $\mathcal{N}_{\omega_\iota,t}$ to L°. If L_i is a positive literal then set the connection weight to W; otherwise, set the connection weight to $-W$. Set the threshold of L° to $((1 + A_{\min}) \cdot (k_l - 1)/2) \cdot W$;

(c) Connect L° to $K_{[\mathcal{A}]}^{\omega_u} L_{k+1}$ in $\mathcal{N}_{\omega_u,t+1}$ and set the connection weight to W. Set the threshold of $K_{[\mathcal{A}]}^{\omega_u} L_{k+1}$ to $((1 + A_{\min}) \cdot (1 - \mu_l)/2) \cdot W$;

(d) Add a hidden neuron $L^\bullet$ to $\mathcal{N}_{\omega_\iota,t}$ and set h(x) as the activation function of $L^\bullet$;

(e) Connect neuron $K_{[\mathcal{A}]}^{\omega_u} L_{k+1}$ in $\mathcal{N}_{\omega_u,t+1}$ to $L^\bullet$ and set the connection weight to W; Set the threshold of $L^\bullet$ to zero;

(f) Connect $L^\bullet$ to $\bigcirc K_{[\mathcal{A}]}^{\omega_\iota} L_i$ in $\mathcal{N}_{\omega_\iota,t}$ and set the connection weight to W. Set the threshold of $K_{[\mathcal{A}]}^{\omega_\iota} L_i$ to $((1 + A_{\min}) \cdot (1 - \mu_l)/2) \cdot W$;

4. For each rule in $\mathcal{P}$ of the form

$$t: \bigcirc K_{[\mathcal{A}]}^{\omega_1} L_1, ..., \bigcirc K_{[\mathcal{A}]}^{\omega_{u-1}} L_k \rightarrow K_{[\mathcal{A}]}^{\omega_u} L_{k+1}:$$

(a) Add a hidden neuron L° to $\mathcal{N}_{\omega_u,t}$ and set h(x) as the activation function of L°;

(b) Connect each neuron $\bigcirc K_{[\mathcal{A}]}^{\omega_\iota} L_i$, $(1 \leq i \leq k)$, $(1 \leq \iota \leq u-1)$ in $\mathcal{N}_{\omega_\iota,t}$ to L°. If L_i is a positive literal then set the connection weight to W; otherwise, set the connection weight to $-W$. Set the threshold of L° to $((1 + A_{\min}) \cdot (k_l - 1)/2) \cdot W$;

(c) Connect L° to $K_{[\mathcal{A}]}^{\omega_u} L_{k+1}$ in $\mathcal{N}_{\omega_u,t}$ and set the connection weight to W. Set the threshold of $K_{[\mathcal{A}]}^{\omega_u} L_{k+1}$ to $((1 + A_{\min}) \cdot (1 - \mu_l)/2) \cdot W$;

THEOREM 4. (Correctness of Knowledge and Probability Algorithm) *For each set of knowledge and probability rules* $\mathcal{P}$, *there exists a neural network ensemble* $\mathcal{N}$ *such that* $\mathcal{N}$ *computes* $\mathcal{P}$.

Proof.*(sketch) This follows from the proof of an analogous theorem for single C-ILP networks given in [d'Avila Garcez et al., 2002a], and the proof that CML computes a fixed-point semantics for modal programs [d'Avila Garcez et al., 2004a], extended to cater for pairs* (ω, t) *instead of* ω *alone.* ∎

We now consider the Monty Hall puzzle. *Suppose you are in a game show and given a choice of three doors. Behind one is a car; behind the others are goats. You pick door 1. Before opening door 1, Monty Hall, the host (who knows what is behind each door), opens door 2, which has a goat. He then asks if you still want to take what is behind door 1, or what is behind door 3 instead. Should you switch?* [Halpern, 2003]

At time t_0, two goats and a car are placed each behind one door. At time t_1, a door is randomly selected by you, and another door, always having a goat behind it, is opened by Monty Hall. At time t_2, you have the choice of whether or not to change your selected door, and depending on your choice, you will have different probabilities of winning the car, as outlined below (for details, see [Halpern, 2003]).

Your chances of picking the right door are $1/3$, and your chances of picking the wrong door are $2/3$. When Monty Hall opens door 2, it becomes known that the probability of the car being behind door 2 is zero ($t_2 : p(door_2) = 0$). The probability of the car not being behind door 1 remains $2/3$ and, as a result, the probability of the car being hidden behind door 3 is then $2/3$ ($t_2 : p(door_3) = 2/3$). Therefore, once you learn that $p(door_2) = 0$, you ought to change from door 1 to door 3. This is summarised below.

$t_k : p(door_1) = 1/3, k \in \{0, 1, 2\}$;

$t_j : p(door_2) = 1/3, j \in \{0, 1\}$;

$t_j : p(door_3) = 1/3, j \in \{0, 1\}$;

$t_2 : K(goat_2) \rightarrow p(door_2) = 0$;

$t_2 : K(goat_2) \rightarrow p(door_3) = 2/3$.

We model the puzzle as follows. At t_2, there are two possible worlds, one in which your policy is to change (ω_c), and another one in which your policy is to stick to your original option (ω_s). In ω_c, your chances of getting the car are $2/3$, and in ω_s your chances are $1/3$. Figure 6 below implements the rules for the Monty Hall puzzle in a neural-symbolic system where g_2 denotes the fact that there is a goat behind door 2, D_i denotes the probability that the car is behind door i, win indicates your chances of winning the car, and c denotes that you have chosen to change from door 1 to door 3. So, if you choose to change, your chances of winning are $2/3$, and if you do not, your chances are $1/3$. This fact can be represented as follows:

$t_1 : g_2;$

$t_1 : g_2 \rightarrow \bigcirc g_2;$

$t_2 : K_i^{\omega_c} \mathbf{c} \rightarrow K_i^{\omega_c}(win = 2/3);$

$t_2 : K_i^{\omega_s} \neg \mathbf{c} \rightarrow K_i^{\omega_s}(win = 1/3).$

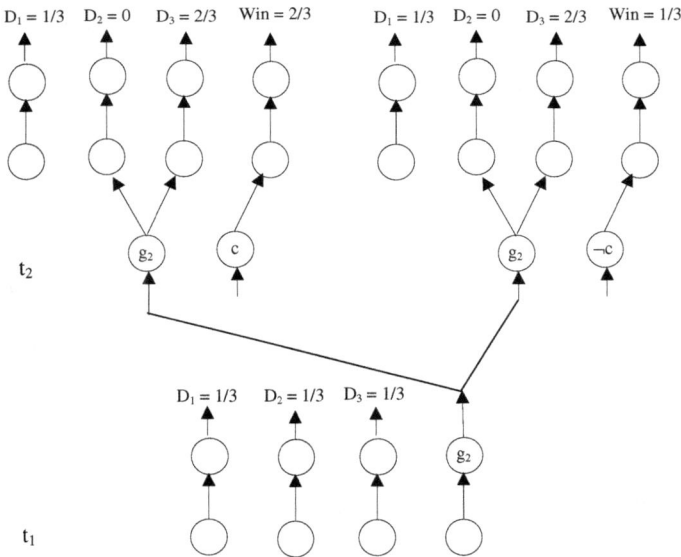

Figure 6. A representation of the Monty Hall puzzle.

4   Conclusion

The knowledge representation formalism presented here provides neural-symbolic learning systems with the ability to reason about time, knowledge and uncertainty. The need for rich, logic-based knowledge representation formalisms to be incorporated into learning systems has been argued since Valiant's seminal paper [Valiant, 1984]. The approach proposed here aims to attend to such a need, yet complying with important principles of connectionism such as massive parallelism and distributed learning. An important feature of our approach is that a temporal dimension can be combined with an epistemic dimension, at the same time allowing for probabilistic reasoning. We have illustrated this by showing that the formalism is applicable to well-known problems of reasoning about uncertainty.

Although the approach proposed here allows for reasoning with probabilities, it does not cater for computing probabilities; our next step is to process qualitative probabilistic statements, e.g. $(i > 0.8) \rightarrow x$, and to apply the approach to large case studies. Further, the integration of other logical systems which are relevant in artificial intelligence such as conditional [Broda et al., 2002] and BDI logics [Rao and Georgeff, 1998] would offer interesting foundational and application-oriented developments in neural-symbolic integration. The use of neural-symbolic systems also facilitates knowledge evolution and revision through learning. It would be interesting to apply the formalism to knowledge evolution in the context of distributed, multi-agent systems.

5   Afterword

Dov has played an important role in ideas that led to this paper, and in making a strong case for non-classical logics and their applications in computer science, artificial intelligence, law, and linguistics over the past thirty years. He has shown that many non-classical logics have numerous applications, and that they should be regarded as first-class languages in computer science, philosophy, artificial intelligence and linguistics. We believe that non-classical logics have also an important role in cognitive and neural computation, since those aim at explaining human cognition, which is undeniably permeated by several forms of non-classical reasoning.

This paper is dedicated to Dov on his 60th birthday. Dov was our PhD supervisor and we have learned a great deal from him, from his unsurpassable knowledge of logics, from his good-spirited approach to student supervision, teaching, mentoring and above all, from his friendship. He is also concerned about all aspects of his students' well-being, including their health and physical fitness. As many of you know, Dov owns thousands of books, and when he decided to repaint his office back in 1997, we helped him carry all of his books seven floors down from his office at Imperial College's Huxley building to a basement storage room. Of course, this also helped us find a couple of books from the top shelf which proved useful for our PhDs. We are very pleased to contribute with one of these 60 papers in honour of Dov's 60th birthday. Happy birthday, Dov.

Acknowledgements

Luís Lamb is partly supported by CAPES, CNPq and FAPERGS.

BIBLIOGRAPHY

[Ajjanagadde, 1997] V. Ajjanagadde. *Rule-Based Reasoning in Connectionist Networks.* PhD thesis, University of Minnesota, 1997.

[Bennett, 1994] B. Bennett. Spatial reasoning with propositional logics. In *4th International Conference on Principles of Knowledge Representation and Reasoning (KR'94)*, pages 51–62, 1994.

[Broda et al., 2002] K. Broda, D. Gabbay, L. C. Lamb, and A. Russo. Labelled natural deduction for conditional logics of normality. *Logic Journal of the IGPL*, 10(2):123–163, 2002.

[Broda et al., 2004] K. Broda, D. M. Gabbay, L. C. Lamb, and A. Russo. *Compiled Labelled Deductive Systems: A Uniform Presentation of Non-Classical Logics.* Research Studies Press, 2004.

[Browne and Sun, 2001] A. Browne and R. Sun. Connectionist inference models. *Neural Networks*, 14:1331–1355, 2001.

[Cloete and Zurada, 2000] I. Cloete and J. M. Zurada, editors. *Knowledge-Based Neurocomputing.* The MIT Press, 2000.

[d'Avila Garcez and Gabbay, 2004] A. S. d'Avila Garcez and D. Gabbay. Fibring neural networks. In *Proceedings of 19th National Conference on Artificial Intelligence (AAAI 04)*, San Jose, CA, 2004.

[d'Avila Garcez and Lamb, 2004] A. S. d'Avila Garcez and L. C. Lamb. Reasoning about time and knowledge in neural-symbolic learning systems. In S. Thrun, L. Saul, and B. Schoelkopf, editors, *Advances in Neural Information Processing Systems 16*, Proc. NIPS 2003, pages 921–928. MIT Press, 2004.

[d'Avila Garcez et al., 2002a] A. S. d'Avila Garcez, K. Broda, and D. M. Gabbay. *Neural-Symbolic Learning Systems: Foundations and Applications.* Springer-Verlag, 2002.

[d'Avila Garcez et al., 2002b] A. S. d'Avila Garcez, L. C. Lamb, and D. M. Gabbay. A connectionist inductive learning system for modal logic programming. In *Proc. ICONIP'02*, pages 1992–1997. IEEE Press, 2002.

[d'Avila Garcez et al., 2003a] A. S. d'Avila Garcez, L. C. Lamb, K. Broda, and D. M. Gabbay. Distributed knowledge representation in neural-symbolic learning systems: A case study. In *Proceedings of AAAI International FLAIRS Conference*, Florida, USA, 2003.

[d'Avila Garcez et al., 2003b] A. S. d'Avila Garcez, L. C. Lamb, and D. M. Gabbay. Neural-symbolic intuitionistic reasoning. In A. Abraham, M. Köppen, and K. Franke, editors, *Design and Application of Hybrid Intelligent Systems, HIS'03*, Frontiers in Artificial Intelligence and Applications Vol. 104, pages 399–408. IOS Press, 2003.

[d'Avila Garcez et al., 2004a] A. S. d'Avila Garcez, L. C. Lamb, K. Broda, and D. M. Gabbay. Applying connectionist modal logics to distributed knowledge representation problems. *Intl. Jnl. on Artificial Intelligence Tools*, 13(1):115–139, 2004.

[d'Avila Garcez et al., 2004c] Artur S. d'Avila Garcez, Dov M. Gabbay, and Luís C. Lamb. Argumentation neural networks. In *11th International Conference on Neural Information Processing, ICONIP'2004*, volume 3316 of *Lecture Notes in Computer Science*, pages 606–612. Springer, 2004.

[d'Avila Garcez et al., 2005] Artur S. d'Avila Garcez, Dov M. Gabbay, and Luís C. Lamb. Value-based argumentation frameworks as neural-symbolic learning systems. *Journal of Logic and Computation*, 2005. To appear.

[d'Avila Garcez et al., 2006] A. S. d'Avila Garcez, L. C. Lamb, and D. M. Gabbay. A connectionist model for constructive modal reasoning. In *Advances in Neural Information Processing Systems 18*, Proc. NIPS 2005. MIT Press, 2006. To appear.

[Fagin et al., 1995] R. Fagin, J. Halpern, Y. Moses, and M. Vardi. *Reasoning about Knowledge*. MIT Press, 1995.

[Friedman and Koller, 2003] N. Friedman and D. Koller. Being bayesian about network structure: A bayesian approach to structure discovery in bayesian networks. *Machine Learning*, 50:95–126, 2003.

[Gabbay, 1981] D. Gabbay. Expressive functional completeness in tense logic. In U. Mönnich, editor, *Aspects of Philosophical Logic*, pages 91–117. Reidel, 1981.

[Gabbay et al., 1994] D. M. Gabbay, I. Hodkinson, and M. Reynolds. *Temporal logic: mathematical foundations and computational aspects*. Oxford University Press, 1994.

[Gabbay et al., 2003] D. Gabbay, A. Kurucz, F.Wolter, and M. Zakharyaschev. *Many-dimensional modal logics: theory and applications*. Elsevier Science, 2003.

[Gabbay, 1996] D. M. Gabbay. *Labelled Deductive Systems*, volume 1. Clarendom Press, Oxford, 1996.

[Gabbay, 1999] D. M. Gabbay. *Fibring Logics*. Oxford University Press, Oxford, 1999. Oxford Logic Guides, Vol. 38.

[Halpern, 2003] J. Y. Halpern. *Reasoning about Uncertainty*. MIT Press, 2003.

[Holldobler and Kalinke, 1994] S. Holldobler and Y. Kalinke. Toward a new massively parallel computational model for logic programming. In *Proceedings of the Workshop on Combining Symbolic and Connectionist Processing, ECAI 94*, pages 68–77, 1994.

[McCarthy, 1988] J. McCarthy. Epistemological challenges for connectionism. *Behavioral and Brain Sciences*, 11(1):44, 1988.

[Mendler, 2000] M. Mendler. Characterising combinational timing analyses in intuitionistic modal logic. *Logic Journal of the IGPL*, 8(6):821–852, 2000.

[Pearl, 2000] J. Pearl. *Causality*. Cambridge University Press, New York, 2000.

[Pnueli, 1977] A. Pnueli. The temporal logic of programs. In *Proceedings of 18th IEEE Annual Symposium on Foundations of Computer Science*, pages 46–57, 1977.

[Rao and Georgeff, 1998] A. S. Rao and M. P. Georgeff. Decision procedures for BDI logics. *Journal of Logic and Computation*, 8(3):293–343, 1998.

[Shastri, 1999] L. Shastri. Advances in SHRUTI: a neurally motivated model of relational knowledge representation and rapid inference using temporal synchrony. *Applied Intelligence*, 11:79–108, 1999.

[Sun and Alexandre, 1997] R. Sun and F. Alexandre. *Connectionist Symbolic Integration*. Lawrence Erlbaum Associates, 1997.

[Towell and Shavlik, 1994] G. G. Towell and J. W. Shavlik. Knowledge-based artificial neural networks. *Artificial Intelligence*, 70(1):119–165, 1994.

[Valiant, 1984] L. G. Valiant. A theory of the learnable. *Communications of the ACM*, 27(11), 1984.

[Valiant, 2003] L. G. Valiant. Three problems in computer science. *Journal of the ACM*, 50(1):96–99, 2003.

[van Benthem, 1984] Johan van Benthem. Correspondence theory. In Dov M. Gabbay and Franz Guenthner, editors, *Handbook of Philosophical Logic*, chapter II.4, pages 167–247. D. Reidel Publishing Company, Dordrecht, 1984.

[van Dalen, 2002] D. van Dalen. Intuitionistic logic. In D. M. Gabbay and F. Guenthner, editors, *Handbook of Philosophical Logic*, volume 5. Kluwer, 2nd edition, 2002.

[Vardi, 1997] M. Y. Vardi. Why is modal logic so robustly decidable? In N. Immerman and P. Kolaitis, editors, *Descriptive Complexity and Finite Models*, volume 31 of *Discrete Mathematics and Theoretical Computer Science*, pages 149–184. DIMACS, 1997.

How Many First-order Variables are Needed on Finite Ordered Structures?

Anuj Dawar

1 Introduction

Let L^k denote the fragment of first-order logic in which each formula contains at most k variables, free or bound. We are interested in interpreting this language on finite ordered structures. That is, the interpretation is in finite structures $\mathbb{A}$ with a distinguished binary relation symbol $<$ which is to be interpreted as a linear order of the universe. The question that concerns us is whether, for every k, L^{k+1} is strictly more expressive than L^k. That is, is the hierarchy of expressive power formed by the number of first-order variables strict? The question is easy enough to state but turns out to be quite difficult to answer. Indeed, it is not even known if the hierarchy formed is *infinite*. Unlikely as it may seem, to the best of our current knowledge, it is conceivable that there is a k (depending on the signature) such that L^k is as expressive as all of first-order logic in this case.

This paper offers an exposition of the problem and its background illustrating many of the key issues in the subject of finite model theory. Finite variable logics have played an important role in the subject, and the question we consider is central to understanding the power of limited numbers of variables. Ordered structures have also been of crucial importance in finite model theory and it has been particularly difficult to obtain any inexpressiveness results on ordered structures as is again highlighted by this simple example. Finally, the problem also illustrates some of the important connections of finite model theory with other fields of investigation: temporal logic on the one hand and computational complexity on the other.

In addition to establishing the background, the paper also presents a few previously unpublished observations and constructions which explore the precise boundary between cases where the hierarchy of expressive power formed by the number of variables (which we call the L^k hierarchy for short) collapses and those where it is not known to. It has long been known that three first-order variables suffice to express all properties on *temporal* structures, i.e. ordered structures with unary predicates. As we shall see, the question of whether more than three variables are needed when the signature contains one binary relation in addition to the order is equivalent to the general question. In the case when infinite structures are permitted, it is known that the L^k hierarchy is strict and this can be shown on structures where the only relation in addition to the order is an equivalence relation. This makes equivalence relations of particular interest. We show that on finite, ordered equivalence relations, three variables suffice to express all *order-invariant* first-order properties. However, dropping the order-invariance requirement and allowing unary relations in addition to the one equivalence relation leads us to a version that is equivalent to the general question. We also look at a version of the hierarchy question obtained by restricting quantifier alternation. We show that, for each k, there is an *existential* formula in L^{k+1} that is not equivalent to any *existential* formula with only k variables (even an infinitary one) on finite ordered graphs. Indeed, it is not equivalent to any existential formula with k variables, even if infinitary connectives are allowed.

1.1   The Main Question

We only consider first-order relational languages. That is, our vocabularies do not, in general, contain function or constant symbols. Let σ be a relational vocabulary that does not contain the symbol $<$. So, σ is a finite collection of relation symbols R each with an associated arity r. By a (finite) structure over σ, or simply a (finite) σ-structure $\mathbb{A}$, we mean a (finite) set A (the universe of the structure) together with an interpretation $R^{\mathbb{A}} \subseteq A^r$ of each R in σ by a relation over A of the appropriate arity. We write $\mathcal{S}_\sigma$ for the collection of finite σ-structures. We also write $\mathcal{O}_\sigma$ for the collection of finite *ordered* σ-structures. These are structures in the vocabulary $\sigma \cup \{<\}$ which interpret $<$, a binary relation symbol, as a linear order of the universe.

We write FO_σ for the collection of first-order formulas (formed in the usual way) over the vocabulary σ. Where it is clear from context, we will drop the subscript σ. We write L^k for the fragment of FO consisting of formulas with no more than k variables (free or bound). Without loss of generality, we may assume that these formulas use only the variables $x_1, \ldots, x_k$ though we will still use x, y, z, etc. as variables to range over this set. We are mainly concerned with issues of the expressive power of the language rather than its syntactic manipulation so the precise choice of the set of variables is not of concern. In order to have the logic be closed under the renaming of bound variables one may prefer to think of L^k as those formulas φ of FO such that no subformula of φ contains more than k free variables.

For any class $\mathcal{C}$ of σ-structures, we say that two formulas φ and ψ are equivalent *over* $\mathcal{C}$ if, for any structure $\mathbb{A} \in \mathcal{C}$, $\mathbb{A} \models \varphi$ if, and only if, $\mathbb{A} \models \psi$. We can now state the main conjecture we are concerned with as follows:

CONJECTURE 1. For any vocabulary σ which contains at least one relation of arity two or more and for each k, there is a sentence φ of $\text{FO}_{\sigma \cup \{<\}}$ for which there is no sentence ψ of L^k such that φ is equivalent to ψ over $\mathcal{O}_\sigma$.

Clearly, $\mathcal{O}_\sigma$ is a subclass of $\mathcal{S}_{\sigma \cup \{<\}}$ and there are pairs of formulas that may be equivalent over $\mathcal{O}_\sigma$ without being equivalent over $\mathcal{S}_{\sigma \cup \{<\}}$. This may be seen as the crux of the conjecture: to what extent can order be used to reduce the number of variables required to express properties of finite structures?

Conjecture 1 asserts that the hierarchy formed by the number of variables is infinite. There is also a stronger version of this conjecture which would require φ to be in L^{k+1} rather than just in FO. We sometimes use the terms 'the L^k hierarchy conjecture' to refer to Conjecture 1 and 'the strict L^k hierarchy conjecture' to refer to the stronger form. And, more generally, we say that the L^k hierarchy is strict (or infinite) over a class of structures $\mathcal{C}$, if for each k there is a formula of L^{k+1} (resp. FO) that is not equivalent *over* $\mathcal{C}$, to any formula of L^k. We also say the L^k hierarchy *collapses* on $\mathcal{C}$ if it is not infinite over $\mathcal{C}$.

The above question is formulated as a conjecture because intuition strongly suggests that the answer is positive. As we shall see, the failure of the conjecture would have many unexpected consequences. However, there is little in the way of concrete results supporting it. Indeed, it is not known if there is any vocabulary σ for which the conjecture is true. Thus, a weaker form of the conjecture would be the following.

CONJECTURE 2. For some vocabulary σ containing at least one relation of arity two or more and for each k, there is a sentence φ of $\text{FO}_{\sigma \cup \{<\}}$ for which there is no sentence ψ of L^k such that φ is equivalent to ψ over $\mathcal{O}_\sigma$.

This is the version of the conjecture we will primarily work with. The simplest case of the conjecture is when σ consists of a single binary relation E. In this case, it is an open question whether there is any formula of FO_σ that is not equivalent to one of L^3.

1.2 Background

A natural question that may arise in the reader's mind is why one would care about the number of variables a formula uses. Why, further, should one restrict consideration to ordered structures, and finite ones? We look at some of the reasons why these questions are of interest.

Temporal Logic Propositional temporal logics are usually interpreted over flows of time. These are (partially) ordered structures $(T, <)$ which interpret an infinite set of propositional symbols. That is, we associate with each element of T the set of propositional symbols that it interprets as true. Using our notation, flows of time are structures over the vocabulary $\sigma \cup \{<\}$ where σ is a collection of unary predicate symbols corresponding to propositional symbols. A temporal logic is formed from the set of propositional symbols and a set of *temporal connectives*. In general, for most temporal logics considered, there is a translation into $\text{FO}_{\sigma \cup \{<\}}$. That is, for each temporal formula φ, there is a formula φ^* of $\text{FO}_{\sigma \cup \{<\}}$ which is satisfied in a flow of time if, and only if, φ is. Moreover, as Gabbay [1981] observed, if $\mathcal{T}$ is a temporal logic formed from a finite set of temporal connectives, each of which is itself first-order definable, then there is a k such that every formula of $\mathcal{T}$ is equivalent to one of L^k. In other words, standard temporal logics can be translated into first-order logic using only a bounded number of variables. Gabbay introduces the notion of *functional completeness* of temporal logics. A logic $\mathcal{T}$ is functionally complete over a given class of flows of time if every first-order formula (with one free variable) is equivalent over the class to a formula of $\mathcal{T}$. One consequence is that if a flow of time T admits a functionally-complete finite set of temporal connectives then it has what Gabbay termed finite H-dimension. In particular, this implies that there is a k such that every first-order sentence is equivalent to one of L^k. Conversely, Gabbay also shows that any flow of time with finite H-dimension admits a finite set of temporal connectives that is functionally complete. The notion of H-dimension is intimately tied to the bounded variable fragments L^k of first-order logic (see [Hodkinson and Simon, 1997] for a detailed discussion of the relationship).

Kamp [1968] showed that the temporal connectives *until* U and *since* S are functionally complete for Dedekind-complete linearly ordered flows of time but not for a rationally ordered flow of time. Stavi showed that U and S along with two new connectives U' and S' are complete over all linearly ordered flows of time (see [Gabbay *et al.*, 1994] for a proof). As it can be easily checked that the temporal logic with U, S, U' and S' translates into L^3, it follows that any first-order sentence is equivalent over linear flows of time to one of L^3. In other words, Conjecture 1 is false if we drop the requirement that σ contain at least one relation of arity two or more. If σ has only unary symbols, three variables suffice to express all first-order properties of ordered σ-structures. In Section 2 we see an alternative proof of this collapse, due to Poizat [1982].

Descriptive Complexity In his work on descriptive complexity Immerman (see [1999]) has advocated the use of first-order logic for the study of computational complexity. In this view, the logic functions essentially as a model of computation, in which measures such as the number of variables and quantifier alternation become the resources through which we measure complexity. We consider decision problems on σ-structures for some signature σ. In order to simulate computations on structures it is also useful to assume that the structures are additionally equipped with an order $<$ on the universe. It is also common to assume that the first-order language has access not only to the relation $<$ but to arithmetic relations $+$ and $\times$ that it induces. Thus, for instance, the ternary relation $+(x, y, z)$ holds of the ith, jth and kth elements of the linear order $<$ if, and only if, $i + j = k$.

Now, given a decision problem D (i.e. an isomorphism-closed class) of finite σ-structures and a positive integer n, it is possible to write down a first-order formula φ_n which is true of a σ-structure $\mathbb{A}$ with n or fewer elements if, and only if, $\mathbb{A}$ is in D. Indeed, if we allow ourselves the use of the order $<$ in φ_n, we can write the formula so that the number of variables used is bounded, independently of n, and this will be significant in our later discussions. For now, we note that this implies that any decision problem D (whether decidable or not) can be characterised by a family $\mathcal{F} = (\varphi_n \mid n \in \omega)$ of first-order formulas. Clearly, the decidability and complexity of D are related to the regularity of $\mathcal{F}$. By results of Barrington, Immerman and Straubing [1990], it is known that the decision problems obtained when $\mathcal{F}$ consists of a single formula (i.e. all φ_n are the same) are exactly those in uniform AC_0. That is, they are the properties decidable by a uniform family of circuits of polynomial size and constant depth. Immerman considers families $\mathcal{F}$ which are obtained by iterating a single formula using a bounded number of variables and thus obtains a tight connection between the number of variables used and the space complexity of D [Immerman, 1991]. Families with a bounded number of variables and a polynomial bound on their size correspond to polynomial-time computation [Immerman, 1982].

Thus, we obtain a tight connection between the number of variables and more conventional measures of computational complexity. Moreover, the presence of an ordering on the structures is crucial to the connection. As mentioned above, the family $\mathcal{F}$ characterising a problem D can always be chosen so that the number of variables is bounded (by some constant that depends only on σ). However, this may be at the cost of the regularity of $\mathcal{F}$ otherwise. For instance, it is a consequence of Immerman's results that there are families $\mathcal{F}$ in which the φ_n are obtained by iterating a single formula with $k + 1$ variables that cannot be obtained by iterating a single formula with k variables. What if $\mathcal{F}$ is defined by a single formula, i.e. D is in uniform AC_0? Is there a bound, depending only on σ on the number of variables needed? If Conjecture 1 fails, then the answer is yes. We suspect that the answer is no, that is, there is a real trade-off between number of variables and quantifier depth beyond the constant, but showing this requires us to prove Conjecture 1.

Fixed-Point Logics Research in descriptive complexity has also led to an interest in the study of fixed-point logics. These are logics that extend first-order logic by means of operators for forming fixed points of relational operators. The most extensively studied such operator is the least-fixed-point operator. We write LFP for the extension of first-order logic by means of such an operator (a more formal definition is given in Section 4 below). It was proved by Immerman [1986] and Vardi [1982] that a property of finite, ordered, relational structures is definable in LFP if, and only if, it is decidable in polynomial time. Similarly, PFP, an extension of first-order logic by means of a so-called partial fixed-point operator, characterises exactly the properties of finite, ordered, structures that are decidable in polynomial space [Vardi, 1982; Abiteboul and Vianu, 1991]. The presence of a linear order in each structure is crucial to these results. In the absence of such an order, the logics are strictly weaker than the respective complexity classes. Nevertheless, as Abiteboul and Vianu [1995] showed, the relationship between LFP and PFP mirrors exactly the relationship between the complexity classes PTIME and PSPACE even without assuming that there is an order present. That is to say, the logics are equivalent on finite structures in general if, and only if, the complexity classes coincide. A proof of this result (see [Dawar *et al.*, 1995]) relies on an analysis of these logics in terms of bounded-variable types.

To make things more precise, let $L_{\infty\omega}$ denote the closure of first-order logic with infinitary conjunctions and disjunctions, which take arbitrary (rather than just finite) sets of formulas. $L_{\infty\omega}^k$ is the fragment of $L_{\infty\omega}$ which uses only k distinct variables and $L_{\infty\omega}^\omega = \bigcup_{k\in\omega} L_{\infty\omega}^k$, i.e. $L_{\infty\omega}^\omega$ is the collection of formulas of $L_{\infty\omega}$ in which each formula has only finitely many variables. Kolaitis and Vardi [1992b] showed that, when restricted to finite structures, any formula φ of LFP is equivalent to a formula of $L_{\infty\omega}^\omega$. Indeed, there is a k (which may depend on φ) such that φ is equivalent to a single infinite disjunction of formulas of L^k. This led to an investigation of McColm's conjectures (see [Kolaitis and Vardi, 1992a]) which essentially stated the equivalence of three conditions for any class $\mathcal{C}$ of finite structures: (1) that LFP and FO have equivalent expressive power on $\mathcal{C}$; (2) that $L_{\infty\omega}^\omega$ and FO have equivalent expressive power on $\mathcal{C}$; and (3) that every first-order definable induction closes in a bounded number of stages on $\mathcal{C}$. While it was shown [Kolaitis and Vardi, 1992a] that (2) and (3) are equivalent, Gurevich et al. [1994] constructed a class of structures on which (1) holds but (3) fails. Nevertheless, Kolaitis and Vardi [1992a] had put a forth a weaker version of

the conjecture, namely that (1) and (3) are equivalent on any *ordered* class
$\mathcal{C}$. That is, any class $\mathcal{C}$ that is contained in $\mathcal{O}_\sigma$ for some σ. This is known as
the *ordered conjecture* and it was shown in [Dawar *et al.*, 1996] that either
a positive or negative resolution to this conjecture would have significant
consequences in complexity theory. It also follows from results in [Dawar *et
al.*, 1996] that the failure of Conjecture 2 would imply a positive resolution
to the ordered conjecture. We return to this topic in Section 4.

2 Variables, Quantifiers and Games

We write $\mathbb{A} \cong \mathbb{B}$ to indicate that two structures $\mathbb{A}$ and $\mathbb{B}$ are isomorphic.
$\mathbb{A} \equiv \mathbb{B}$ denotes that $\mathbb{A}$ and $\mathbb{B}$ are elementarily equivalent, i.e. they are not
distinguished by any sentence of FO. We also write $\mathbb{A} \equiv^k \mathbb{B}$ to denote that
$\mathbb{A}$ and $\mathbb{B}$ agree on all sentences of L^k and $\mathbb{A} \equiv_p \mathbb{B}$ to denote that they agree
on all sentences of FO with quantifier rank at most p. The quantifier rank of
a formula is simply the maximal depth of nesting of quantifiers (irrespective
of alternations) occurring within it. We combine these last two notations in
$\mathbb{A} \equiv_p^k \mathbb{B}$ to denote that $\mathbb{A}$ and $\mathbb{B}$ are not distinguished by any sentence of L^k
with quantifier rank at most p. More generally, if $\mathbf{s}$ is a tuple of elements
from $\mathbb{A}$ and $\mathbf{t}$ a tuple from $\mathbb{B}$, we write $(\mathbb{A}, \mathbf{s}) \equiv^k (\mathbb{B}, \mathbf{t})$ to denote that, for
any formula φ of L^k, $\mathbb{A} \models \varphi[\mathbf{s}]$ if, and only if, $\mathbb{B} \models \varphi[\mathbf{t}]$ and similarly for
the other equivalences. We now briefly examine the relation between these
equivalences and definability.

It is an easy observation that when the structures involved are finite,
elementary equivalence coincides with isomorphism. Suppose σ is a *finite*
relational vocabulary and $\mathbb{A}$ a σ-structure with n elements. We can construct a sentence

$$\delta_\mathbb{A} = \exists x_1 \ldots \exists x_n (\psi \wedge \forall y \bigvee_{1 \leq i \leq n} y = x_i)$$

where, $\psi(x_1, \ldots, x_n)$ is the conjunction of all atomic and negated atomic
formulas that hold in $\mathbb{A}$. Now, for any structure $\mathbb{B}$, $\mathbb{B} \models \delta_\mathbb{A}$ if, and only if,
$\mathbb{A} \cong \mathbb{B}$. This implies that any isomorphism-closed class of finite σ- structures
$\mathcal{C}$ is defined by a formula $\varphi_\mathcal{C}$ of $L_{\infty\omega}$. Indeed a single infinitary disjunction
of first-order formulas, or a single infinitary conjunction, will suffice. Note,
however, that the number of variables required for $\varphi_\mathcal{C}$ is potentially infinite.
This cannot be avoided as there are structures for every k (for instance a
pure set with $k + 1$ elements) that cannot be characterised by a formula of
L^k (or indeed $L_{\infty\omega}^k$)

However, in the presence of order, we can do better. For each σ, there is
a k such that any structure $\mathbb{A}$ in $\mathcal{O}_\sigma$ can be characterised up to isomorphism
by a formula of L^k [Dawar *et al.*, 1996]. To see this, we first note that for

every element a of $\mathbb{A}$, there is a formula $\beta_a(x)$ of L^2 such that a is the unique element of $\mathbb{A}$ satisfying $\mathbb{A} \models \beta_a[a]$. This is established by inductively defining the following class of formulas.

$$
\begin{aligned}
\alpha_0(x) &\equiv \neg(x = x) \\
\alpha_{n+1}(x) &\equiv \forall y((y < x) \to (\alpha_n(y))))
\end{aligned}
$$

where $\alpha_n(y)$ is the formula obtained from $\alpha_n(x)$ by *interchanging* all occurrences, free or bound, of x and y. It is clear that $\mathbb{A} \models \alpha_n[a]$ if and only if there are at most n elements less than or equal to a in the linear order $<^{\mathbb{A}}$. Thus, the formula $\beta_n \equiv \neg\alpha_{n-1} \wedge \alpha_n$ identifies the nth element of the order uniquely.

Using these formulas, it is clear that any m-tuple can be uniquely identified by a formula of L^m, and we can therefore construct a sentence $\psi_{\mathbb{A}}$ of L^m that determines the structure $\mathbb{A}$ up to isomorphism among structures in $\mathcal{O}_\sigma$, where m is at least 3 and at least as great as the maximum arity of any relation in σ. This means that any isomorphism-closed subclass of $\mathcal{O}_\sigma$ is definable by a sentence of $L^m_{\infty\omega}$ for some large enough m, depending only on σ.

It was shown by Kolaitis and Vardi [1992b] that a class of finite structures is definable in $L^k_{\infty\omega}$ if, and only if, it is closed under the relation $\equiv^k$. It is also not difficult to see that, provided the signature σ is finite, a class of structures is definable in FO if, and only if, it is closed under $\equiv_p$ for some p. Indeed, p is the quantifier rank of the defining sentence. This is because there are, up to logical equivalence, only finitely many sentences of quantifier rank p and the equivalence relations $\equiv_p$ therefore have finite index (see [Ebbinghaus and Flum, 1999, Sec. 2.2] for details). In particular, this implies that any infinitary sentence of finite quantifier rank is equivalent to a finitary sentence. Similar arguments show that a class of σ-structures is definable in L^k if, and only if, it is closed under $\equiv^k_p$ for some p. It is worth remarking here that the restriction to finite σ is important. For instance, if $\sigma = (P_i \mid i \in \omega)$ is an infinite collection of unary relation symbols and $S \subseteq \omega$ is an infinite and co-infinite set, then the sentence $\exists x(\bigwedge_{i \in S} P_i(x))$ is of quantifier rank 1 but not equivalent to any finitary sentence.

A natural question to ask might be, if a class of structures is closed under $\equiv^k$ and $\equiv_p$ for some k and some p, do there exist k' and p' such that it is also closed under $\equiv^{k'}_{p'}$? This question is discussed in some detail in [Dawar *et al.*, 1996]. Here we simply observe that the answer is yes on *ordered* structures if, and only if, Conjecture 2 fails. Thus, we have reformulated the main question as a question about suitable equivalence relations. We next look at characterisations of these relations by means of games.

Games The relation $\equiv_p^k$ has an elegant characterisation in terms of Ehrenfeucht-Fraïssé style two player games, essentially given by Barwise [1977] though versions were also independently presented by Immerman [1982] and Poizat [1982]. The game board consists of two structures $\mathbb{A}$ and $\mathbb{B}$ and a supply of k pairs of pebbles $(a_i, b_i), 1 \leq i \leq k$. The pebbles $a_1, \ldots, a_l$ are initially placed on the elements of an l-tuple $\mathbf{s}$ of elements in $\mathbb{A}$, and the pebbles $b_1, \ldots, b_l$ on a tuple $\mathbf{t}$ in $\mathbb{B}$. There are two players, Spoiler and Duplicator. At each move of the game, Spoiler picks up a pebble (either an unused pebble or one that is already on the board) and places it on an element of the corresponding structure. For instance he[1] might take pebble b_i and place it on an element of $\mathbb{B}$. Duplicator must respond by placing the other pebble of the pair in the other structure. In the above example, she must place a_i on an element of $\mathbb{A}$. If at the end of the move the partial map $f : \mathbb{A} \rightarrow \mathbb{B}$ given by $a_i \mapsto b_i$ is not a partial isomorphism, then Spoiler has won the game, otherwise it can continue for another move.

The result that links this game with the equivalence relation $\equiv_p^k$ is the following:

> If Duplicator has a strategy for playing the k-pebble game for p moves starting with the position $(\mathbb{A}, \mathbf{s}), (\mathbb{B}, \mathbf{t})$, then $(\mathbb{A}, \mathbf{s}) \equiv_p^k (\mathbb{B}, \mathbf{t})$.

If Duplicator has a strategy for playing indefinitely without losing, then $(\mathbb{A}, \mathbf{s}) \equiv^k (\mathbb{B}, \mathbf{t})$.

Finite Coloured Orders As an application of the game, we reproduce the argument given by Poizat [1982] showing that on coloured linear orders every sentence of FO is equivalent to one of L^3.

Let σ be a *finite* vocabulary consisting solely of unary relations and let $\mathbb{A}$ and $\mathbb{B}$ be structures in $\mathcal{O}_\sigma$. We refer to such structures as *coloured orders*. The key combinatorial argument in establishing the collapse of the L^k hierarchy on such structures is the following lemma.

LEMMA 3. *Let* $\mathbf{s} = (a_1, \ldots, a_m)$ *be a tuple of elements from* $\mathbb{A}$ *and* $\mathbf{t} = (b_1, \ldots, b_m)$ *a tuple of elements from* $\mathbb{B}$ *with* $a_i <^{\mathbb{A}} a_{i+1}$ *and* $b_i <^{\mathbb{B}} b_{i+1}$ *for all* i. *If, for each* $i < m$, $(\mathbb{A}, a_i, a_{i+1}) \equiv_p^3 (\mathbb{B}, b_i, b_{i+1})$ *then* $(\mathbb{A}, \mathbf{s}) \equiv_p (\mathbb{B}, \mathbf{t})$.

Proof. The proof is by induction on p. For the base case, $p = 0$, we only need to establish that the map $a_i \mapsto b_i$ is a partial isomorphism preserving $<$ and all relations in σ. The order is preserved by the hypothesis that the tuples are in increasing order and the unary relations are preserved since $(\mathbb{A}, a_i, a_{i+1}) \equiv_0^3 (\mathbb{B}, b_i, b_{i+1})$ for all i.

---

[1]By convention, Spoiler is male and Duplicator female.

Assume now that the claim is true for p and that $(\mathbb{A}, a_i, a_{i+1}) \equiv_{p+1}^3$ $(\mathbb{B}, b_i, b_{i+1})$ for all $i < m$. We wish to use this to show that Duplicator can win the $p + 1$-move game (with an unlimited number of pebbles) played from the starting position $(\mathbb{A}, \mathbf{s}), (\mathbb{B}, \mathbf{t})$. Suppose, without loss of generality, that Spoiler plays his first move in the structure $\mathbb{A}$, choosing element a. The case when $a = a_i$ for some i is trivial as Duplicator plays on b_i, so we assume that a is distinct from all elements of $\mathbf{s}$. If there is an i such that $a_i < a < a_{i+1}$, then by hypothesis Duplicator has a strategy to win the *3-pebble*, $p + 1$-move game played from $(\mathbb{A}, a_i, a_{i+1}), (\mathbb{B}, b_i, b_{i+1})$. Let b be the response dictated by this strategy to Spoiler moving on a. Then, $(\mathbb{A}, a_i, a, a_{i+1}) \equiv_p^3 (\mathbb{B}, b_i, b, b_{i+1})$ and so, in particular, $(\mathbb{A}, a_i, a) \equiv_p^3 (\mathbb{B}, b_i, b)$ and $(\mathbb{A}, a, a_{i+1}) \equiv_p^3 (\mathbb{B}, b, b_{i+1})$. Therefore the tuples $a_1, \ldots, a_i, a, a_{i+1}, a_m$ and $b_1, \ldots, b_i, b, b_{i+1}, b_m$ satisfy the hypothesis of the lemma and by the inductive hypothesis, we have that $(\mathbb{A}, a_1, \ldots, a_i, a, a_{i+1}, a_m) \equiv_p (\mathbb{B}, b_1, \ldots, b_i, b, b_{i+1}, b_m)$, as required. This leaves the cases when either $a < a_1$ or $a_m < a$. In these cases, by a similar argument, we are guaranteed a winning response using the winning strategies in the 3-pebble games for $(\mathbb{A}, a_1, a_2), (\mathbb{B}, b_1, b_2)$ and $(\mathbb{A}, a_{m-1}, a_m)$, $(\mathbb{B}, b_{m-1}, b_m)$. In all cases, we have established that $(\mathbb{A}, \mathbf{s}) \equiv_{p+1} (\mathbb{B}, \mathbf{t})$. ∎

The result for coloured orders now follows.

THEOREM 4. *If σ is a finite vocabulary with only unary predicates then every sentence of FO_σ is equivalent, over $\mathcal{O}_\sigma$ to one of L^3.*

Proof. Let φ be a sentence of FO and p be its quantifier rank. Thus, the class of structures $\mathcal{C} \subseteq \mathcal{O}_\sigma$ defined by φ is closed under $\equiv_p$. By Lemma 3, if $\mathbb{A}, \mathbb{B} \in \mathcal{O}_\sigma$ are equivalent under $\equiv_p^3$, they are equivalent under $\equiv_p$. Thus, $\mathcal{C}$ is also closed under $\equiv_p^3$ and therefore, since σ is finite, definable by a sentence of L^3 of quantifier rank p. ∎

It is worth remarking that the proofs of Lemma 3 and Theorem 4 do not rely on the structures being finite and work equally well with infinite structures. The restriction to finite vocabularies is required for the construction of the sentence of L^3 in the proof of Theorem 4. Of course, any sentence of FO can only refer to finitely many symbols, so a simple re-phrasing would establish it for arbitrary vocabularies.

3   Binary Relations

Theorem 4 establishes that the hierarchy formed by the number of variables on ordered finite structures collapses to L^3 when all relations other than the order are unary. The next step up in considering Conjecture 2 would be to consider the case when σ consists of a single, binary relation symbol. This case turns out to be equivalent to the general form of Conjecture 2. That is, we can make the following observation.

OBSERVATION 5. If the L^k hierarchy collapses on $\mathcal{O}_\sigma$ in the case where σ consists of a single binary relation, it collapses on $\mathcal{O}_\tau$ for all vocabularies τ.

The reason for this is that structures in an arbitrary fixed vocabulary can be encoded as graphs (i.e. structures with one binary relation E). Such an encoding is given, for instance, by a bi-interpretation in [Hodges, 1993, Sec. 5.5]. It is an easy adaptation of that construction to allow a bi-interpretation for ordered structures. This provides, for each σ, a translation Γ that maps a σ-structure $\mathbb{A}$ to a graph $\Gamma\mathbb{A}$ along with translations of formulas both ways. That is, there is for each σ-formula φ a formula $\gamma^\rightarrow(\varphi)$ in the language of graphs such that $\Gamma\mathbb{A} \models \gamma^\rightarrow(\varphi)$ if, and only if, $\mathbb{A} \models \varphi$. Conversely, there is for each formula ψ in the language of graphs a σ-formula $\gamma^\leftarrow(\psi)$ such that $\mathbb{A} \models \gamma^\leftarrow(\psi)$ if, and only if, $\Gamma\mathbb{A} \models \psi$. The translations $\gamma^\leftarrow$ and $\gamma^\rightarrow$ are themselves obtained as simple substitutions, i.e. $\gamma^\rightarrow(\varphi)$ is obtained from φ by replacing each occurrence of a relation symbol R from σ by an (E)-formula defining R. One consequence is that any increase in the number of variables in going from φ to $\gamma^\rightarrow(\varphi)$ (or from ψ to $\gamma^\leftarrow(\psi)$) is bounded by a multiplicative constant. Now, suppose there were a k such that every FO formula over ordered graphs was equivalent to one of L^k. Then, for an arbitrary σ-formula φ, $\gamma^\rightarrow(\varphi)$ would be equivalent, over ordered graphs, to a formula θ of L^k. Hence $\gamma^\leftarrow(\theta)$ would be a formula, equivalent to φ with at most ck variables for some constant c, establishing that any formula over σ-structures is equivalent to one of L^{ck}.

Infinite Structures While in the case of vocabularies with unary relations, Theorem 4 settles the question for both finite and infinite structures, in the presence of binary relations, the distinction between finite and infinite becomes important. Indeed, it is known that on infinite ordered structures with one binary relation, the L^k hierarchy is strict. This is established by an argument due to Venema [1990].

THEOREM 6. *There is a sentence φ of L^{k+1} in a vocabulary with one binary relation that is not equivalent, on infinite ordered structures to any sentence of $L^k_{\infty\omega}$.*

Proof. Let Q_k be a structure whose universe is the set of rational numbers. It interprets $<$ as the usual order on the rationals and interprets E as an equivalence relation with k equivalence classes, all of which are dense in the order. That is, between any two elements we can find representatives of all k equivalence classes. It is not difficult to see that $Q_k \equiv^k_{\infty\omega} Q_{k+1}$. That is to say, Duplicator can play an infinitely long k-pebble game on this pair of structures. Indeed, in response to any move by Spoiler, she only needs to find, in the suitable interval, a representative of a suitable equivalence class, i.e. one that bears the same relation of equivalence to the $k - 1$ currently pebbled elements as the element selected in Spoiler's move. With at least k equivalence classes on each side, the existence of such a suitable element is guaranteed.

Thus, if φ is the sentence that asserts the existence of $k + 1$ pairwise inequivalent elements, $Q_{k+1} \models \varphi$, $Q_k \not\models \varphi$ and we can conclude that φ is not equivalent to any sentence of $L^k_{\infty\omega}$ ∎

Theorem 6 clearly does not hold on finite structures for, as previously observed (see [Dawar *et al.*, 1996]) *any* property of ordered finite graphs is definable in $L^3_{\infty\omega}$. Thus, if a version of this theorem is to settle Conjecture 2, we must replace $L^k_{\infty\omega}$ by L^k. This is to say that the game argument must also take quantifier rank into account. For any candidate sentence φ of L^{k+1} it will be the case that the finite ordered graphs satisfying φ are closed under $\equiv^k$ but we would like to show that for any p they are not closed under $\equiv^k_p$. However, as we see next, for the particular sentences used in the proof of Theorem 6, this is not possible.

Finite Equivalence Relations Let $\mathcal{E} \subseteq \mathcal{O}_{(E)}$ be the class of finite ordered equivalence relations. It is easily seen that the usual axioms for an equivalence relation give a sentence θ of L^3 that defines this class. We now define a sentence η_k using just two variables x and y that, on $\mathcal{E}$, defines the relations with at least k distinct equivalence classes. We first define the formula $\mu(x)$ as:

$$\mu(x) \equiv \forall y(E(x, y) \rightarrow x \leq y)$$

and write $\mu(y)$ to denote the formula obtained by interchanging in $\mu(x)$ all occurrences of x and y. Clearly, for any $\mathbb{A} \in \mathcal{E}$, $\mathbb{A} \models \mu[a]$ if, and only if, a is $<$-minimal in its equivalence class. We now define the following formulas δ_k inductively.

$$\begin{aligned} \delta_1(x) &\equiv \mu(x) \\ \delta_{k+1}(x) &\equiv \mu(x) \wedge \exists y\, (x < y \wedge \delta_k(y)), \end{aligned}$$

where, once again, $\delta_k(y)$ denotes the formula obtained from $\delta_k(x)$ by interchanging all occurrences of x and y. The formula $\delta_k(x)$ asserts that x is minimal in its equivalence class and there are at least $k - 1$ elements greater than x in the order $<$, each of which is minimal in its equivalence class. The sentence $\eta_k \equiv \exists x\, \delta_k$ therefore defines exactly those structures in $\mathcal{E}$ which have at least k equivalence classes. $\eta_k \wedge \theta$ is then a sentence of L^3 defining equivalence relations of index at least k.

The above construction shows that the formulas used in the proof of Theorem 6 cannot be used to establish the L^k hierarchy on finite ordered structures. Indeed, the above construction can be generalised to show that any order-invariant property of finite ordered equivalence relations is definable in L^3.

We say that a formula φ of $\mathrm{FO}_{(E,<)}$ is order invariant on finite equivalence relations if, for any finite equivalence relation $\mathbb{A}$ and any two orderings $<_1$ and $<_2$ of the universe of $\mathbb{A}$, $(\mathbb{A}, <_1) \models \varphi$ if, and only if, $(\mathbb{A}, <_2) \models \varphi$. For instance, the sentence η_k constructed above is order invariant as it asserts a property of the equivalence relation and not of the order.

THEOREM 7. *Any sentence φ of $\mathrm{FO}_{(E,<)}$ that is order invariant on the class of finite equivalence relations is equivalent, on $\mathcal{E}$, to a sentence of L^2.*

Proof. Let the quantifier rank of φ be p. Our aim is to show that the models of φ in $\mathcal{E}$ are closed under $\equiv_q^2$ for some suitable value of q.

We associate with each equivalence relation $\mathbb{A}$ a (2^p+1)-tuple of integers which we call its *p-footprint*. To be precise, the p-footprint of $\mathbb{A}$ is the tuple $(n_1, \ldots, n_{2^p+1})$ where:

- for $1 \leq i \leq 2^p$, $n_i = \min(2^p, N_i)$ where N_i is the number of equivalence classes of $\mathbb{A}$ with exactly i elements.

- $n_{2^p+1} = \min(2^p, N)$ where N is the number of equivalence classes of $\mathbb{A}$ with more than 2^p elements.

We say an ordered equivalence relation $(\mathbb{A}, <)$ is in *normal order* if:

- the equivalence classes occupy contiguous blocks along the order, i.e. $(\mathbb{A}, <)$ satisfies the axiom $\forall x, y\, (E(x,y) \wedge x < y) \Rightarrow (\forall z\, x < z < y \Rightarrow E(x,z))$; and

- the equivalence classes occur in non-decreasing order of size, i.e. if $x < y$ then the equivalence class of y contains at least as many elements as that of x.

We now observe that if $\mathbb{A}$ and $\mathbb{B}$ are two equivalence relations with the same p-footprint, they cannot be distinguished by any order-invariant sentence φ of quantifier rank p or less. As φ is order-invariant, it suffices to show that if $(\mathbb{A}, <)$ and $(\mathbb{B}, <')$ are normal-order orderings of $\mathbb{A}$ and $\mathbb{B}$, then $(\mathbb{A}, <) \equiv_p (\mathbb{B}, <')$. This can be established by a game argument. Essentially, Duplicator's strategy in the p-move game on this pair of structures is to ensure that if, after r rounds have been played, pebbles have been placed on the elements $a_1 < \cdots < a_r$ and $b_1 <' \cdots <' b_r$ then the following conditions hold:

- for each i, the number of elements in the equivalence class of a_i is the same as that in the equivalence class of b_i or both numbers are greater than 2^p;

- for each i, the number of elements between a_i and a_{i+1} is either equal to the number of elements between b_i and b_{i+1} or both numbers are greater than 2^{p-r};

- the number of elements in the equivalence class of a_i that precede it in the order is equal to the corresponding number for b_i or both numbers are greater than 2^{p-r}; and

- the number of elements in the equivalence class of a_i that follow it in the order is equal to the corresponding number for b_i or both numbers are greater than 2^{p-r}.

The definition of p-footprint essentially guarantees that these conditions can be maintained. The proof is an adaptation of the proof that shows Duplicator can win a p-move game on a pair of linear orders of length greater than 2^p (see [Ebbinghaus and Flum, 1999, Example 2.3.6]) and we omit the details.

To complete the proof we show that for each p-footprint t, there is a sentence φ_t of L^2 that defines exactly the equivalence relations that have p-footprint t. These formulas cannot rely on a normal order, they need to work for any possible ordering of the equivalence relation. Observe that, since a p-footprint is a $(2^p + 1)$-tuple of integers each between 0 and 2^p, there are only finitely many p-footprints for each p. To construct φ_t we first define formulas $\alpha_i(x)$ which assert that the equivalence class of x contains at least i elements that are greater than or equal to x in the order. They can be defined inductively as follows.

$$
\begin{aligned}
\alpha_1(x) &\equiv x = x \\
\alpha_{i+1}(x) &\equiv \exists y\,(x < y \wedge E(x,y) \wedge \alpha_i(y)).
\end{aligned}
$$

Let $\beta_i(x) \equiv \mu(x) \wedge \alpha_i(x)$ be the formula that asserts that x is minimal in its equivalence class and that equivalence class contains at least i elements. We can now use this (in a manner by now familiar) to define formulas that state that there exist at least n elements satisfying $\beta(x)$. We first define formulas $\xi_{i,n}(x)$ inductively as follows:

$$
\begin{aligned}
\xi_{i,1}(x) &\equiv \beta_i(x) \\
\xi_{i,n+1}(x) &\equiv \beta_i(x) \wedge \exists y\,(x < y \wedge \xi_{i,n}(y))
\end{aligned}
$$

It is now easily seen that the sentence $\chi_{i,n} \equiv \exists x\, \xi_{i,n}(x)$ is true in $\mathbb{A}$ if, and only if, $\mathbb{A}$ contains at least n equivalence classes with i or more elements. It is then easy to see that a finite Boolean combination of these sentences will yield the desired φ_t. Finally, our original sentence φ, since it is characterised by a finite set of p-footprints, is equivalent to a finite disjunction of the sentences φ_t. ∎

One reason why equivalence relations are significant is because they appear to be a rather weak excursion from the case of unary relations into the realm of the binary. An equivalence relation is really the coding in a single binary relation of a collection of sets, i.e. a unary vocabulary. As the number of equivalence classes is unbounded, this really codes an infinite unary vocabulary but, as Theorem 7 demonstrates, the methods of Theorem 4 do partially extend to this case. As long as our sentence is order-invariant, it can only talk about the sizes of the equivalence classes and then only up to some fixed finite limit (depending on the sentence) and, in the presence of order, all of this can be done with two variables. The proof of Theorem 7 is easily adapted to the case where, in addition to the one binary relation interpreted as an equivalence relation, σ contains any number of unary relations.

THEOREM 8. *Let σ be a vocabulary with one binary relation E in addition to any number of unary relations. Let $\mathcal{E} \subseteq \mathcal{O}_\sigma$ be the class of structures that interpret E as an equivalence relation. Then, any order-invariant FO_σ sentence φ is equivalent, over $\mathcal{E}$, to a sentence of L^2.*

We shall next show that, if we could drop the restriction to order-invariant sentences in Theorem 8, this would imply the failure of Conjecture 2.

Coloured Grids For the rest of this section, let σ be the vocabulary consisting of a binary relation E and two unary relations D and G. Also, let τ be the vocabulary consisting of a single binary relation *edge*. We write $\mathcal{E}$ for the class of structures in $\mathcal{O}_\sigma$ which interpret E as an equivalence relation. Our aim is to show that the question of whether the L^k hierarchy collapses on $\mathcal{E}$ is equivalent to asking whether the L^k hierarchy collapses on $\mathcal{O}_\tau$. We already know (see Observation 5) that the latter is equivalent to settling Conjecture 2.

LEMMA 9. *If every sentence of FO is equivalent to one of L^k on $\mathcal{E}$, then every sentence of FO is equivalent to one of L^{2k} on $\mathcal{O}_\tau$.*

Proof. The proof proceeds, like Observation 5, by establishing a bi-interpretation from $\mathcal{O}_\tau$ into $\mathcal{E}$. That is, we construct a map $\Gamma : \mathcal{O}_\tau \to \mathcal{E}$ such that for each $\mathbb{A} \in \mathcal{O}_\tau$, $\Gamma\mathbb{A}$ is first-order interpretable in $\mathbb{A}$ and $\mathbb{A}$ is first-order interpretable in $\Gamma\mathbb{A}$.

To be precise, for $\mathbb{A} = (A, edge, <^{\mathbb{A}})$ we define $\Gamma\mathbb{A} = (A^2, E, D, G, <^{\Gamma\mathbb{A}})$ to be the structure defined on the universe A^2 with the following relations:

- $E((a_1, a_2), (b_1, b_2))$ if, and only if, $a_2 = b_2$;

- $D(a_1, a_2)$ if, and only if, $a_1 = a_2$;

- $G(a_1, a_2)$ if, and only if, $edge(a_1, a_2)$; and

- $(a_1, a_2) <^{\Gamma \mathbb{A}} (b_1, b_2)$ if, and only if, $a_1 <^{\mathbb{A}} a_2$ or $a_1 = a_2$ and $b_1 <^{\mathbb{A}} b_2$.

In other words, $\Gamma \mathbb{A}$ interprets $<$ as the lexicographical order induced on A^2 by $<^{\mathbb{A}}$; E as the equivalence relation of equality in the second component; D as the diagonal through the square A^2 and G encodes the edge relation of the original graph $\mathbb{A}$. It is clear that $\Gamma \mathbb{A}$ is interpreted in $\mathbb{A}$ by first-order formulas. Indeed, the above definition of Γ explicitly gives *quantifier-free* formulas defining the relations in $\Gamma \mathbb{A}$. Thus, any $\sigma \cup \{<\}$-formula φ can be translated into a $\tau \cup \{<\}$-formula $\gamma^{\leftarrow}(\varphi)$ by substituting the above definitions for the relation symbols and replacing quantifiers $\exists x$ by pairs of quantifiers $\exists x_1, x_2$. The result is that $\Gamma \mathbb{A} \models \gamma^{\leftarrow}(\varphi)$ if, and only if, $\mathbb{A} \models \varphi$ and it is easily seen that if φ uses k distinct variables then $\gamma^{\leftarrow}(\varphi)$ requires at most $2k$ variables.

We now show how $\mathbb{A}$ can be interpreted inside $\Gamma \mathbb{A}$. As a first step, we show how the relation on A^2 of equality in the first component is definable from the relations E, D and $<^{\Gamma \mathbb{A}}$. Note that $\Gamma \mathbb{A}$ satisfies the axiom

$$\forall x \exists! y \, (E(x, y) \wedge D(y)).$$

That is, every element has a unique companion on the diagonal with which it shares the second component. We write $\delta(x)$ for the companion of x with the understanding that its use in formulas can be replaced with a proper first-order definition. We now define the following formulas.

$$
\begin{aligned}
U(x) &\equiv x < \delta(x) \\
L(x) &\equiv \delta(x) < x \\
Ze(x, y) &\equiv x < y \wedge \forall z \, (x < z \wedge z < y) \Rightarrow \neg D(z) \\
Si(x, y) &\equiv x < y \wedge \exists! z \, (x < z \wedge z < y \wedge D(z)).
\end{aligned}
$$

In other words, $U(x)$ defines in $\Gamma \mathbb{A}$ the region of A^2 above the diagonal, i.e. those elements (a_1, a_2) with $a_1 <^{\mathbb{A}} a_2$ and $L(x)$ defines the region below the diagonal. U, L and D form a partition of A^2. $Ze(x, y)$ is a binary predicate asserting that no elements of the diagonal occur between x and y in the order $<^{\Gamma \mathbb{A}}$ and $Si(x, y)$ asserts that there is a unique diagonal element between them. Note that if two elements x and y in A^2 share the same first component then there is at most one diagonal element that can occur between them in the order $<$. The following formula $\eta(x, y)$ now defines all cases where $x < y$ and x and y share the same first component.

$$\eta(x,y) \quad \equiv \quad [(U(x) \lor D(x)) \land U(y) \land Ze(x,y)] \lor$$
$$[L(x) \land (D(y) \lor L(y)) \land Ze(x,y)] \lor$$
$$[U(x) \land L(y) \land Si(x,y)] \lor$$
$$[L(x) \land U(y) \land Si(x,y)].$$

Thus, the formula $F(x,y) \equiv \eta(x,y) \lor \eta(y,x) \lor x = y$ defines the relation of equality in the first component.

We can now define (a structure isomorphic to) $\mathbb{A}$ inside $\Gamma\mathbb{A}$ by taking the universe to be the diagonal $\{a \mid D(a)\}$, the order to be the order induced on this by $<^{\Gamma\mathbb{A}}$ and defining the edge relation by the following formula

$$\epsilon(x,y) \quad \equiv \quad \exists z \, E(x,z) \land F(y,z) \land G(z).$$

Thus, we can translate any $(edge, <)$-formula φ into a formula $\gamma^{\rightarrow}(\varphi)$ by relativising all quantifiers to D and replacing all occurrences of $edge$ by the formula ϵ. The result is that $\mathbb{A} \models \varphi$ if, and only if, $\Gamma\mathbb{A} \models \gamma^{\rightarrow}(\varphi)$.

To complete the proof note that if the hypothesis of the theorem is true, then for an arbitrary formula φ, $\gamma^{\rightarrow}(\varphi)$ is equivalent over $\mathcal{E}$ (and, in particular, over $\{\Gamma\mathbb{A} \mid \mathbb{A} \in \mathcal{O}_{\tau}\}$) to a formula θ of L^k. Thus, $\gamma^{\leftarrow}(\theta)$ is a formula of L^{2k} that is equivalent to φ over $\mathcal{O}_{\tau}$. ∎

A similar, and simpler, argument is used to establish a converse statement.

LEMMA 10. *If every sentence of* FO *is equivalent to one of* L^k *on* $\mathcal{O}_{\tau}$, *then every sentence of* FO *is equivalent to one of* L^{k+2} *on* $\mathcal{E}$.

Proof. The idea once again is to establish a bi-interpretation, i.e. a map $\Delta : \mathcal{E} \to \mathcal{O}_{\tau}$ which is definable by first-order formulas, as is its inverse. If $\mathbb{A} = (V, E, D, G, <)$, we define $\Delta\mathbb{A}$ to be the ordered graph whose universe is $V \uplus \{0,1\}$, the extension of V with two new elements with the order $<$ extended so that $0 < 1 < v$ for all $v \in V$ and the relation $edge$ to include the pairs

- $(0,v)$ if $v \in V$ and $D(v)$;

- $(1,v)$ if $v \in V$ and $G(v)$; and

- (u,v) if $u,v \in V$ and $E(u,v)$.

It is then an easy exercise to see that the map Δ admits translations of first-order formulas between $\mathbb{A}$ and $\Delta\mathbb{A}$. In particular, in translating an $(edge, <)$ formula we replace occurrences of $edge(x,y)$ by the formula $(x =$

$0 \wedge D(y)) \vee (x = 1 \wedge G(y)) \vee E(x, y)$, where "$x = 0$" and "$x = 1$" are abbreviations for the formulas that state x is minimal in the order and the successor of the element minimal in the order respectively. It can be verified that this translation can be carried out so as to require at most two additional variables.   ∎

Lemmas 9 and 10 immediately yield the following theorem.

THEOREM 11. *There is a k such that every sentence of* FO *is equivalent to one of L^k on $\mathcal{O}_\tau$ if, and only if, there is a k such that every sentence of* FO *is equivalent to one of L^k on $\mathcal{E}$.*

Theorem 11 should be contrasted with Theorem 8. While the latter shows that the L^k hierarchy collapses on the class $\mathcal{E}$ if only order-invariant sentences are concerned, the former shows that such a collapse for non-order-invariant sentences would refute Conjecture 2.

We conclude this section with the observation that the equivalence relations used in the proof of Lemma 9 are of a special kind. Call a structure $\mathbb{A}$ in the vocabulary with one binary relation E a *grid-equivalence* if its universe is A^2 for some set A and E is interpreted as equality on the second component of the pairs. Let $\mathcal{GE}$ be the class of ordered grid-equivalences where the order is a lexicographic order of pairs. As any such structure is characterised up to isomorphism by the cardinality of A, we write G_n to denote a canonical such structure with $A = \{1, \ldots, n\}$.

OBSERVATION 12. *Every first-order sentence is equivalent, on $\mathcal{GE}$ to one of L^2.*

Proof. We aim to show that if $m, n > 2^p$ $G_m \equiv_p G_n$. That is, Duplicator can win the p-move Ehrenfeucht game played on this pair of structures. Duplicator's strategy is essentially an adaptation of the strategy used in the game on two linear orders (see the proof of Theorem 7). Recall that there Duplicator's strategy was to ensure that after r moves, the corresponding intervals formed by pebbled points in the two structures were either of equal length or both larger than 2^{p-r}. In the case of grid-equivalences the strategy is to maintain this condition simultaneously on both projections of the grid. Thus, suppose that after r moves have been played, pebbles are on $a_1, \ldots, a_r \in G_m$ and $b_1, \ldots, b_r \in G_n$. We write a_i^1 and a_i^2 for the two elements of $\{1, \ldots, m\}$ that make up the pair a_i and similarly b_i^1 and b_i^2 for the two components of b_i. Duplicator's strategy in the game is to ensure that after r moves have been played, the following condition holds: for each $i, j \in 1, \ldots, r$ and each $k \in \{1, 2\}$ either $b_j^k - b_i^k = a_j^k - a_i^k$, or $b_j^k - b_i^k > 2^{p-r}$ and $a_j^k - a_i^k > 2^{p-r}$. An easy inductive argument shows that this can be done.   ∎

While Theorem 8 shows that in the proof of Theorem 11 we cannot remove the use of non-order-invariant sentences, Observation 12 shows that we cannot remove the use of unary relations from that proof.

4   Fixed-Point Logics

In this section, we look at the relationship between the L^k hierarchy conjecture and certain questions on fixed-point logics. These logics, much studied in finite model theory, are extensions of first-order logic by means of operators for recursive or iterative definition of relations. In analysing their expressive power, finite variable logics have played an important role. Here we examine specifically the connection between the L^k hierarchy conjecture and what is known as Kolaitis and Vardi's ordered conjecture and some of the complexity-theoretic implications of it. We begin with a brief introduction to fixed-point logics. For more background, the reader is referred to [Ebbinghaus and Flum, 1999] or [Dawar and Gurevich, 2002].

Let $\varphi(R, \mathbf{x})$ be a formula, where R is a relation symbol, and $\mathbf{x}$ is a tuple of first order variables whose length r is the same as the arity of R. If $\mathbb{A}$ is a structure, with universe A, interpreting all symbols in φ other than those displayed, we think of φ as defining a map Φ from $\mathrm{Pow}(A^r)$—the set of all r-ary relations on A—to $\mathrm{Pow}(A^r)$ given by $\Phi(P) = \{\mathbf{a} \mid (\mathbb{A}, P, \mathbf{a}) \models \varphi\}$.

This view of a formula defining an operator on the space of relations gives a natural formalisation of inductive definitions. For instance, if Φ is a monotone map, we can speak of the least relation R such that $R(\mathbf{x}) \leftrightarrow \varphi(R, \mathbf{x})$. This relation is the *least fixed point* of the operator defined by φ. We can now define the logic LFP, obtained by closing first-order logic simultaneously under all the formula forming operations of first-order logic along with the rule:

if R is a k-ary relation variable, $\mathbf{x}$ is a k-tuple of first order variables, $\mathbf{t}$ is a k-tuple of terms and φ is a formula in which R occurs only positively, then

$$[\mathbf{lfp}[R, \mathbf{x}]\varphi](\mathbf{t})$$

is a formula, in which all occurrences of R are bound, and all occurrences of the variables in $\mathbf{x}$ except those occurring in $\mathbf{t}$ are bound.

The intended semantics of this formula formation rule is that, for any structure $\mathbb{A}$, $\mathbb{A} \models [\mathbf{lfp}[R, \mathbf{x}]\varphi](\mathbf{t})$ if, and only if, $\mathbf{t}^{\mathbb{A}}$—the tuple of elements of $\mathbb{A}$ defined by the terms $\mathbf{t}$—is in the least fixed point of the monotone operator defined by $\varphi(R, \mathbf{x})$ on A^k.

The least fixed point of a monotone operator on a structure $\mathbb{A}$ can be obtained by iterating the operator. That is, we define a sequence of stages R^n where $R^0 = \emptyset$ and $R^{n+1} = \Phi(R^n)$, the relation defined by the formula φ when R is interpreted by R^n. If $\mathbb{A}$ is infinite, the iteration may have to be carried into the transfinite, but this is a case that need not concern us here. The sequence of relations R^n is non-decreasing and there is, therefore, a stage N at which $R^N = R^{N+1}$. The relation defined at this point is the least fixed point and we call the least N for which this is true the *closure ordinal* of φ on $\mathbb{A}$.

If the formula φ is not positive in the relation R, it does not define a monotone operator. It is, nonetheless, possible to iterate the operator it defines, starting with the empty relation. One may do this in an inflationary or a non-inflationary fashion. Adding operators for the fixed-points so formed yields the logic IFP (of inflationary fixed points) and PFP (of partial fixed points) respectively. We omit the precise definitions and refer the reader to [Ebbinghaus and Flum, 1999] or [Dawar and Gurevich, 2002] for details. In terms of expressive power, it is known that IFP and LFP are equivalent [Gurevich and Shelah, 1986; Kreutzer, 2002] and that LFP and PFP are equivalent if, and only if, the complexity classes PTIME and PSPACE coincide [Abiteboul and Vianu, 1995].

As noted in Section 1.2, Kolaitis and Vardi [1992b] showed that any formula of LFP (and, indeed, PFP) is equivalent, over finite structures, to a formula of $L^{\omega}_{\infty\omega}$. In particular, for any formula $\varphi(R, \mathbf{x})$ in L^k, there is a formula φ^n of L^{2k} that defines the relation R^n in all structures. Thus, $[\mathbf{lfp}[R, \mathbf{x}]\varphi](\mathbf{t})$ is equivalent to $\bigvee_{n \in \omega} \varphi^n(\mathbf{t})$. Of course, as we have earlier observed, for every vocabulary σ, there is a k such that every isomorphism-closed property of ordered σ-structures is expressible in $L^k_{\infty\omega}$. Thus, in the presence of order, the observation that the fixed-point logics translate into $L^{\omega}_{\infty\omega}$ is not very informative. Still, the form of the formula $\bigvee_{n \in \omega} \varphi^n(\mathbf{t})$ suggests an infinitary formula of a particularly uniform kind, as the formulas φ^n are generated from a single pattern. It is true that the number of variables in this formula depends on the formula φ and is not determined solely by the vocabulary σ. We may use the order to reduce the number of variables we use but this may result in a loss of uniformity. It is this trade-off between the number of variables and the uniformity of the first-order formula that we wish now to explore (for a related discussion of issues of uniformity in connection with descriptive complexity see [Barrington *et al.*, 1990]).

McColm [1990a; 1990b] considered the relative expressive power of LFP and FO in relation to a property he called *proficiency*. A class of structures $\mathcal{C}$ is said to be proficient if there is some FO formula $\varphi(R, \mathbf{x})$ such that the closure ordinals of φ on structures in $\mathcal{C}$ are unbounded. As Mc-Colm observed, the existence of unbounded inductions allows one to define arithmetic relations and allow certain coding and diagonalisation constructions. This led to the conjecture that on any proficient class of structures LFP is more expressive than FO. Though this was refuted in [Gurevich *et al.*, 1994], Kolaitis and Vardi [1992a] put forward a weaker form of the conjecture.

CONJECTURE 13 (Kolaitis-Vardi). On any class of ordered structures containing arbitrarily large finite structures, there is a formula of LFP that is not equivalent to any formula of FO.

The reason this is a weak form of McColm's conjecture is that it is trivial to define, on a linearly ordered structure, a formula whose closure ordinal is equal to the size of the structure. Thus, a class of structures satisfying the hypothesis of Conjecture 13 is necessarily proficient. The ordered conjecture of Kolaitis and Vardi remains open and it was shown in [Dawar *et al.*, 1996] that its resolution either way would solve significant complexity theoretic questions. In particular, a proof of the conjecture would show that the complexity classes LIN-H and ETIME are different. Here, LIN-H is the linear time hierarchy and ETIME is exponential time, i.e. the class of problems solvable in time $2^{O(n)}$. On the other hand, a refutation of Conjecture 13 would yield a separation of PTIME from PSPACE.

The relevance of this to the present discussion is that a refutation of Conjecture 2 would yield a proof of Conjecture 13. This is because McColm [1990a] showed, by a diagonalisation argument, that on any proficient class of structures, for every k, there is a sentence of LFP that is not expressible in LFP^k, the fragment of LFP that uses only k first-order variables. Thus, in particular, if $\mathcal{C} \subseteq \mathcal{O}_\sigma$ is a class of ordered σ-structures containing structures of arbitrarily large finite size, the number of first-order variables required to express all LFP formulas is unbounded. If Conjecture 2 were refuted, it would follow that there is a k, depending only on σ such that every formula of FO is equivalent on $\mathcal{C}$ to a formula of L^k. Hence any LFP formula that witnesses the fact that LFP is not equivalent to LFP^k would not be equivalent to any formula of FO.

One way of understanding McColm's parametrisation result, which shows that the number of variables hierarchy formed inside LFP is infinite on any infinite class of ordered structures is that it shows a trade-off between uniformity and the number of variables in infinitary formulas. While there is a k, depending only on σ such that every sentence of LFP is equivalent on $\mathcal{O}_\sigma$ to one of $L^k_{\infty\omega}$, it is not the case that it can be expressed simultaneously in LFP and using only k variables. The uniformity of the infinitary formula obtained as a translation of an LFP sentence requires an unbounded number of variables. Conjecture 2 asks for a similar trade-off in the case of first-order logic. However, this comparison also raises the possibility of looking for weaker trade-offs. Can we force the number of variables to be bounded by looking at weaker notions of uniformity? For instance, is it possible that every sentence of FO is equivalent on $\mathcal{O}_\sigma$ to one of LFP^k? This would not refute Conjecture 2 but would be sufficient to establish Conjecture 13.

It is worth noting here that we can show that there is a k, depending only on σ, such that any sentence of FO is equivalent, on $\mathcal{O}_\sigma$ to one of PFP^k, the fragment of PFP that uses only k variables. This follows from the fact that the satisfaction problem: 'given a finite structure $\mathbb{A}$ and first-order sentence φ, is it the case that $\mathbb{A} \models \varphi$?' is itself decidable in PSPACE. Since every property computable in PSPACE on ordered structures is definable in PFP, there is a coding of sentences of FO as natural numbers and a PFP formula $\mu(x)$ such that for $\mathbb{A} \in \mathcal{O}_\sigma$, $\mathbb{A} \models \mu[n]$ where n denotes the nth element of the linear order in $\mathbb{A}$ if, and only if, $n = \overline{\varphi}$ and $\mathbb{A} \models \varphi$, where $\overline{\varphi}$ denotes the number coding the formula φ. The formula μ has some fixed number c of variables and replacing all occurrences of n by its L^2 definition gives us a formula of PFP^{c+2} that is equivalent to φ on all sufficiently large structures (i.e. those with n or more elements). A disjunction of this with the finitely many sentences $\psi_{\mathbb{A}}$ (see Section 2) describing the smaller structures satisfying φ yields the desired sentence of PFP (see [Dawar and Hella, 1995] for a similar argument). This observation essentially establishes an upper bound on the amount of uniformity that needs to be sacrificed in order to turn all FO sentences on $\mathcal{O}_\sigma$ into infinitary ones with a bounded number of variables. We can do it in PFP. If we could do it in LFP, we would establish Conjecture 13. If we could do it in FO itself, we would refute Conjecture 2.

5   Clique and Circuit Complexity

We stated in Section 1.1 that Conjecture 1 was formulated in its positive form because intuition strongly suggests it is true. Nonetheless, in Sections 2 and 3 we concentrated on cases where the L^k-hierarchy collapses. This was intended to explore the boundary between the cases where a collapse occurs and those where we expect the hierarchy to be strict. However, another reason for concentrating on the cases where the hierarchy collapses is that we have no results which actually exhibit an interesting L^k hierarchy. Such a result would prove Conjecture 2. In this section we pose the question of what a proof of the main conjecture might look like. We will examine in particular the case of ordered graphs, i.e. the case when the vocabulary σ consists of a single binary relation E.

In order to prove that the L^k hierarchy is strict on $\mathcal{O}_\sigma$ the class of ordered graphs, we need two ingredients. We need a family of candidate formulas $(\varphi_k \mid k \in \omega)$ such that φ_k requires k variables to express and we need a technique for proving that φ_{k+1} is not expressible in L^k (or, for the weaker version, that for each k there is a φ_l that is not expressible in L^k).

For the first part—the family of formulas—there is a natural candidate that suggests itself. Let φ_k be the first-order sentence that asserts the existence of a k-clique. This sentence can clearly be written using k variables. Indeed, it only requires k *existential* quantifiers in prenex normal form. Moreover, not only is the sentence order-invariant, it does not require the use of the order at all. It is also straightforward to show, using Ehrenfeucht-Fraïssé games that, in the absence of order, φ_k is not expressible by any sentence (even an infinitary one) with fewer than k variables. The question then is whether we can show that the presence of an order does not help us in reducing the number of variables required to express the property. Or, at least, that as k increases, an unbounded number of variables is required.

This leads us to the question of the second ingredient, a technique for proving that φ_{k+1} cannot be expressed in L^k even when using an order. The first technique that suggests itself is the Ehrenfeucht-Fraïssé game which succeeds so well in the absence of order. Once we are dealing with ordered structures, we can no longer show that φ_{k+1} is not expressible in $L^k_{\infty\omega}$ so we must exploit the finite quantifier rank of formulas of L^k. That is to say, we would have to demonstrate, for each p, a pair of ordered graphs $\mathbb{A}_p$ and $\mathbb{B}_p$, one of which contains a $k + 1$-clique and the other does not, such that $\mathbb{A}_p \equiv^k_p \mathbb{B}_p$. Unfortunately, constructing such Ehrenfeucht-Fraïssé games on ordered structures is notoriously difficult and so far such a construction has eluded us. Using the maxim that it is sometimes easier to aim to prove a more general result than the one you are interested in, we will consider a more powerful method, that of circuit complexity.

Methods based on circuit complexity have proved extremely useful in establishing inexpressibility results for first-order logic on ordered structures. A paradigmatic example is the result due to Furst, Saxe and Sipser [1984] that shows that the parity of a unary relation is not decidable by any family of bounded-depth, polynomial-size circuits. This implies that it is not definable in first-order logic on ordered structures, even in the presence of arbitrary numerical predicates. We now make these notions a little more precise.

As mentioned in Section 1.2, Barrington et al. [1990] showed that any first-order formula φ on ordered graphs can be turned into a family of circuits. Such a family is a sequence $(C_n \mid n \in \omega)$ of circuits, made of (unbounded fan-in) **and** and **or** gates and single input **not** gates. C_n takes at its inputs a binary string representing a structure $\mathbb{A}$ with universe $\{0, \ldots, n-1\}$ and produces a single bit output which is 1 if $\mathbb{A} \models \varphi$ and 0 if $\mathbb{A} \not\models \varphi$. Moreover, there is a constant d and a polynomial p depending only on φ such that the depth of C_n is bounded by d and the size of C_n is bounded by $p(n)$.

In addition to the size and depth of a family of circuits, there is one other crucial resource to consider for such a family and that is its *uniformity*. This refers to the complexity of the function that maps n to C_n. The family of circuits generated by a first-order formula is highly uniform. However, we can take an arbitrary family of constant depth and polynomial size and produce an equivalent first-order sentence if we allow ourselves arbitrary numerical predicates. A numerical predicate (of arity a) is a map from n to an a-ary relation on $\{0, \ldots, n-1\}$. Thus, the family of circuits is translated into a formula with additional predicate symbols for the numerical predicates. A cogent presentation of the relationship between the uniformity of circuits and the presence of numerical predicates can be found in [Denenberg *et al.*, 1986].

Circuit complexity, in particular the method of restrictions, has provided some of the most powerful techniques for establishing complexity lower bounds. The question we wish to consider is whether such a technique could be adapted to the problem at hand, of showing that the number of variables required to express the property of a graph having a clique is unbounded. In order to do this, we need to consider what measure on a circuit corresponds to the number of variables in a first-order formula.

A circuit of depth d can be thought of as being formed of d layers of gates with wires going from lower layers to higher ones. Define the width of a layer to be the number of gates in it and the *width* of the circuit to be the maximum width of any of its layers. It is then clear that the size s of a circuit (i.e. the total number of gates) is bounded by $d \cdot w$ where w is the width. The important point is that a careful analysis of the translation of a formula to a family of circuits (such as in [Barrington *et al.*, 1990]) shows that a formula with k variables and quantifier rank d can be translated into a family of circuits of depth d and width n^k. To understand why this is the case, consider a subformula ψ of φ with l free variables $x_1, \ldots, x_l$. Just as a circuit corresponding to the sentence φ has one output bit for the truth value of the formula φ on the structure on its input so, inductively, we construct a circuit C_ψ which has n^l output bits. These give the truth value of ψ for each of the n^l possible interpretations of its free variables in the universe $\{0, \ldots, n-1\}$. Thus, if the sentence φ has no more than k variables to begin with, none of its subformulas has more than k free variables and we can therefore limit the number of wires leaving any layer of the circuit to n^k. This gives us the required bound on the width.

It follows that in order to prove that φ_{k+1} cannot be expressed in L^k it would suffice to show that the problem of detecting a $k + 1$-clique in an ordered graph cannot be solved by any family of circuits of constant depth and size $O(n^k)$. The clique problem has been much studied in the context of circuit complexity and a trade-off between the depth and size of circuits needed to solve it has been established. In particular, Lynch [1986] has established that there is a constant c such that any family of circuits of depth d that solves the k-clique problem must have size $\Omega(n^{ck/d})$. This shows a trade-off between depth and size but still leaves open the possibility that, given sufficient (but still constant) depth, a family of circuits could be found of size $O(n^k)$ for solving the $k + 1$-clique problem.

Showing that no constant-depth family of circuits of size $O(n^k)$ could detect cliques of size $k + 1$ would require a major breakthrough in circuit complexity. Moreover, such a result would establish something much stronger than we seek. Not only would it establish Conjecture 2, it would show that the problem of finding a $k + 1$-clique could not be expressed in L^k even if we allow, in addition to the order, arbitrary arithmetic and numerical predicates. In the absence of such a breakthrough, might it be possible at least to use the methods of circuit lower bounds to establish the result in the restricted case where numerical predicates are not permitted? Barrington et al. [1990] are primarily concerned with uniform circuit, using a uniformity condition that is equivalent to having first-order logic with additional predicates for *order, addition* and *multiplication*. As Conjecture 2 only requires order, it can be seen as an extremely uniform version of a plausible complexity-theoretic conjecture, namely that there is no $O(n^k)$ size, constant-depth family of circuits for $k + 1$-clique. The question then is, can we exploit the uniformity to give a proof of Conjecture 2 that does not involve solving the harder question? While circuit complexity has produced powerful methods for proving lower bounds, these methods do not typically rely on any assumptions of uniformity. It is not therefore clear what techniques could exploit the uniformity requirement embodied in Conjecture 2. Of course, one could think of dropping the order altogether as the ultimate uniformity requirement and, in this case, we do have a technique that works—that of Ehrenfeucht-Fraïssé games. Whether that can be pushed to work in the case when order is present remains open.

6   Existential Fragments

We noted that the formula expressing the existence of a $k + 1$-clique in an ordered graph is one obvious candidate for showing a separation between L^{k+1} and L^k. This is a particularly simple property in that it is expressible in L^{k+1} using a purely existential formula (i.e. one that involves no universal quantification). While we are not able to show that there is no equivalent formula of L^k, we can show that no *existential* formula of L^k can express the existence of a $k + 1$-clique. Thus, if there is a formula of L^k that expresses this property, it must make essential use of quantifier alternation. Indeed, we show that there isn't even an infinitary existential formula using only k variables that can express the existence of a $k + 1$-clique. We write $\exists L^k$ for the fragment of L^k which involves no universal quantifiers and in which no existential quantifier appears within the scope of a negation. $\exists L^k_{\infty\omega}$ is the similarly defined fragment of $L^k_{\infty\omega}$. The logic $\exists L^k_{\infty\omega}$ of infinitary existential formulas with k variables has been studied (see [Kolaitis and Vardi, 1995; Kolaitis and Vardi, 2000]) as a tool for analysing the database language Datalog and in the context of constraint satisfaction problems.

For a pair of structures $\mathbb{A}$ and $\mathbb{B}$ we write $\mathbb{A} \Rightarrow^k \mathbb{B}$ to denote that every sentence of $\exists L^k$ that is true in $\mathbb{A}$ is also true in $\mathbb{B}$. Note that, as the logic $\exists L^k$ is not closed under negation, this relation is not symmetric. This relation is characterised by a non-symmetric version of the k-pebble game, namely one in which Spoiler is confined to placing pebbles on elements of $\mathbb{A}$ and Duplicator must respond with elements of $\mathbb{B}$ (see [Kolaitis and Vardi, 1995] for details). If Duplicator can play this game forever without losing, then $\mathbb{A} \Rightarrow^k \mathbb{B}$ and, indeed, any sentence of $\exists L^k_{\infty\omega}$ that is true in $\mathbb{A}$ is true in $\mathbb{B}$.

We now aim to define a single pair of ordered graphs $\mathbb{A}$ and $\mathbb{B}$ such that $\mathbb{A}$ contains a $k + 1$ clique, $\mathbb{B}$ does not and Duplicator wins the infinite k-pebble existential game in which Spoiler only plays in $\mathbb{A}$.

Let $\mathbb{A}$ be the structure with universe $A = \{1, \ldots, k + 1\}$ with $<$ interpreted as the natural order and $E(i, j)$ for all $i, j \in A$. $\mathbb{B}$ is a structure on the universe $B = \{1, \ldots, k + 1\} \times \{1, \ldots, k\}$ with $<$ being the natural lexicographic order on pairs and $E((i, j), (i', j'))$ if, and only if, $i \neq i'$ and $j \neq j'$. It is easy to see that $\mathbb{B}$ does not contain a $k + 1$-clique as any set of elements forming a clique must have pairwise disjoint second components, but there are only k distinct possibilities for the second component.

Now, it remains to show that Duplicator can play an infinitely long k-pebble game. Her strategy is to respond to any move by Spoiler on an element $i \in A$ not currently pebbled by playing on $(i, j) \in B$ where j is such that there is no pebble currently on any element (i', j). This is possible as there are k possible values of j and only $k - 1$ pebbles currently on the board. To see that this strategy maintains a partial isomorphism, note that as the order on B is lexicographic, $(i, j) < (i', j')$ if $i < i'$, so the order is maintained. Furthermore the relation E holds between all pairs of elements in A and by the choice of Duplicator's move it holds on all pairs of pebbled elements in B.

This argument immediately yields the following result.

THEOREM 14. *For each k, there is a formula of $\exists L^{k+1}$ that is not equivalent to any formula of $\exists L^k_{\infty\omega}$ on ordered graphs.*

7 Other Connections

Throughout this paper, in considering the ramifications of Conjecture 2, we have established connections with a number of different areas of investigation ranging from temporal logics through fixed-point logics to complexity theory. The apparently simple question posed turns out to be closely linked to a number of other open problems. In the present section we present a few more questions from the areas of logic and computational complexity which have links to the main conjecture considered in the paper but which we have only space to consider briefly. The links mainly occur in the form of implausible (but not known to be false) consequences of the failure of Conjecture 2. Since it is much harder to find such dramatic consequences for a confirmation of the conjecture, this reinforces our belief in the conjecture.

Circuit Complexity We discussed in Section 5 the possibility that methods from circuit complexity might resolve the main conjecture. We noted that it would require a breakthrough to show that there is no constant-depth circuit of size $O(n^k)$ for detecting a $k + 1$-clique (or for that matter, an l clique for some $l > k$). There is, of course, a flip side to that observation. If Conjecture 2 were to fail, this would imply that there is, indeed, a fixed k such that for any l, the l-clique problem is solvable by a family of constant depth circuits of size $O(n^k)$. This would be more than a breakthrough, it would be a stunning result. Moreover, this would hold not just for the problem of detecting an l-clique but for any first-order definable property of ordered graphs. This seems a most unlikely scenario.

Evaluating First-Order Logic Closely related to the above observation on the size of circuits corresponding to a first-order sentence is the fact that any formula of L^k can be turned into an algorithm for deciding the property it defines that runs in time $O(n^k)$ (see [Dawar *et al.*, 1996] for an argument). Thus, the failure of Conjecture 2 would imply that there is a k such that any property of ordered graphs definable in first-order logic is, in fact, decidable in time $O(n^k)$. This would settle (in a most unexpected fashion) an open question posed by Stolboushkin and Taitslin [1995].

Fixed-Parameter Tractability Of course, if it were the case that there is a k such that every first-order definable property is decidable in time $O(n^k)$, this would also be true of the existence of an l-clique for arbitrary l. This does not imply that there is a polynomial time algorithm for solving the CLIQUE problem as the L^k formulas (and hence the constants involved in the running time of the $O(n^k)$ algorithms) for the l-clique problem would depend on l. But, we can make two observations. Suppose Conjecture 2 fails and there is a k such that every first-order sentence is equivalent on ordered graphs to one of L^k. Suppose furthermore that this collapse of the L^k hierarchy is *effective*. That is, there is a computable function that maps any sentence of FO to an equivalent sentence of L^k. This would imply that there is an algorithm that, given a graph G and a parameter l, determines whether or not G contains an l-clique and runs in time $f(l)n^k$ where f is some computable function and k is independent of n and l. That is to say, that CLIQUE would be fixed-parameter tractable. This is considered very unlikely as CLIQUE is $W[1]$-complete (see [Downey and Fellows, 1999] for more on fixed-parameter tractability and $W[1]$-completeness). The second observation is that, if the collapse of the L^k-hierarchy were not only effective but *feasible*, i.e. there was a polynomial-time computable function that mapped any sentence of FO to an equivalent L^k sentence, then we would have a polynomial time algorithm for solving CLIQUE and this implies that P = NP.

Second-order Hierarchy A well-known result of Fagin [1974] shows that a property of finite relational structures is decidable in the complexity class NP if, and only if, it is expressible in existential second-order logic. Since it is also known that monadic second-order quantifiers alone do not suffice to express all properties in NP, Fagin [1990] also asks whether one can show that the hierarchy formed by the arities of second-order quantifiers is strict. Indeed, it is an open question whether more than one binary second-order quantifier is ever needed to express any NP property. We can show that if Conjecture 2 were to fail then one (or indeed any fixed finite number) of binary quantifiers (or again, quantifiers of any fixed finite arity) would not suffice to express all NP properties of ordered graphs. The reason is that, if we fix a finite number and a maximum arity of second-order variables, then the first-order matrix of any sentence of existential second-order logic can be seen as being in a fixed finite vocabulary σ. If Conjecture 2 were to fail, there would be a k such that each such matrix could be replaced by a sentence of L^k. Just as any sentence of L^k can be evaluated in time $O(n^k)$ on structures of size n, one could show that, with the number and arity of second-order quantifiers bounded and the first-order matrix limited to L^k, any existential second-order formula can be evaluated by a nondeterministic algorithm running in time $O(n^l)$ for some fixed l. This would, of course, contradict the nondeterministic time hierarchy theorem (see [Papadimitriou, 1994]).

8 Conclusion

The question of whether one can establish a hierarchy on the number of first-order variables needed to express properties of finite ordered graphs has been known for many years within a community of researchers working in finite model theory. Little has been written about this question, largely because there are not many concrete partial results indicating progress towards solving it. The present paper aims to bring the question to a wider audience by demonstrating the deep links this apparently simple problem has with a number of areas of logic and complexity. At the same time, the problem illustrates the fundamental difficulties in proving inexpressibility results for first-order logic on finite ordered structures. We have two basic techniques with which we can establish such results. On the one hand there are Ehrenfeucht-Fraïssé games, which establish fine-grained results on unordered structures but which we do not know how to wield in the presence of order. On the other hand, we have methods from circuit complexity which work not only on ordered structures but on structures where the order is enriched with arbitrary numerical predicates. However, in the latter case the results they yield are not sufficiently fine-grained for our purposes. In the intermediate ground, of ordered structures with a high degree of uniformity, it seems that entirely new methods would be required. It is to be hoped that the development of such methods would yield new insights into the relation between logic and computational complexity. The problem of establishing the L^k hierarchy on ordered graphs appears to be an important testing ground for new techniques.

BIBLIOGRAPHY

[Abiteboul and Vianu, 1991] S. Abiteboul and V. Vianu. Datalog extensions for database queries and updates. *Journal of Computer and System Sciences*, 43:62–124, 1991.

[Abiteboul and Vianu, 1995] S. Abiteboul and V. Vianu. Computing with first-order logic. *Journal of Computer and System Sciences*, 50(2):309–335, 1995.

[Barrington et al., 1990] D.M. Barrington, N. Immerman, and H. Straubing. On uniformity within NC_1. *Journal of Computer and System Sciences*, 41:274–306, 1990.

[Barwise, 1977] J. Barwise. On Moschovakis closure ordinals. *Journal of Symbolic Logic*, 42:292–296, 1977.

[Dawar and Gurevich, 2002] A. Dawar and Y. Gurevich. Fixed-point logics. *Bulletin of Symbolic Logic*, 8:65–88, 2002.

[Dawar and Hella, 1995] A. Dawar and L. Hella. The expressive power of finitely many generalized quantifiers. *Information and Computation*, 123(2):172–184, 1995.

[Dawar et al., 1995] A. Dawar, S. Lindell, and S. Weinstein. Infinitary logic and inductive definability over finite structures. *Information and Computation*, 119(2):160–175, 1995.

[Dawar et al., 1996] A. Dawar, S. Lindell, and S. Weinstein. First order logic, fixed point logic and linear order. In H. Kleine-Büning, editor, *Computer Science Logic '95*, volume 1092 of *LNCS*, pages 161–177. Springer-Verlag, 1996.

[Denenberg et al., 1986] L. Denenberg, Y. Gurevich, and S. Shelah. Definability by constant-depth polynomial-size circuits. *Information and Control*, 70:216–240, 1986.

[Downey and Fellows, 1999] R.G. Downey and M.R. Fellows. *Parametrized Complexity*. Springer-Verlag, 1999.

[Ebbinghaus and Flum, 1999] H-D. Ebbinghaus and J. Flum. *Finite Model Theory*. Springer, 2 edition, 1999.

[Fagin, 1974] R. Fagin. Generalized first-order spectra and polynomial-time recognizable sets. In R. M. Karp, editor, *Complexity of Computation, SIAM-AMS Proceedings, Vol 7*, pages 43–73, 1974.

[Fagin, 1990] R. Fagin. Finite model theory – a personal perspective. In S. Abiteboul and P. C. Kanellakis, editors, *Third International Conference on Database Theory (LNCS 470)*, pages 3–24. Springer-Verlag, 1990.

[Furst et al., 1984] M. Furst, J. Saxe, and M. Sipser. Parity, circuits, and the polynomial time hierarchy. *Math Systems Theory*, 17:13–27, 1984.

[Gabbay et al., 1994] D. Gabbay, I. Hodkinson, and M. Reynolds. *Temporal Logic, vol. 1*. Oxford University Press, 1994.

[Gabbay, 1981] D. Gabbay. Expressive functional completeness in tense logic. In U. Mönnich, editor, *Aspects of Philosophical Logic*, pages 91–117. Reidel, 1981.

[Gurevich and Shelah, 1986] Y. Gurevich and S. Shelah. Fixed-point extensions of first-order logic. *Annals of Pure and Applied Logic*, 32:265–280, 1986.

[Gurevich et al., 1994] Y. Gurevich, N. Immerman, and S. Shelah. McColm's conjecture. In *Proc. 9th IEEE Symp. on Logic in Computer Science*, 1994.

[Hodges, 1993] W. Hodges. *Model Theory*. Cambridge University Press, 1993.

[Hodkinson and Simon, 1997] I. Hodkinson and A. Simon. The k-variable property is stronger than h-dimension k. *J. Philosophical Logic*, 26:81–101, 1997.

[Immerman, 1982] N. Immerman. Upper and lower bounds for first-order expressibility. *Journal of Computer and System Sciences*, 25:76–98, 1982.

[Immerman, 1986] N. Immerman. Relational queries computable in polynomial time. *Information and Control*, 68:86–104, 1986.

[Immerman, 1991] N. Immerman. DSPACE$[n^k]$=VAR$[k+1]$. In *Proc. 6th IEEE Symp. on Structure in Complexity Theory*, pages 334–340, 1991.

[Immerman, 1999] N. Immerman. *Descriptive Complexity*. Springer, 1999.

[Kamp, 1968] H. Kamp. *Tense Logic and the Theory of Linear Order*. PhD thesis, University of California, Los Angeles, 1968.

[Kolaitis and Vardi, 1992a] Ph. G. Kolaitis and M. Y. Vardi. Fixpoint logic vs. infinitary logic in finite-model theory. In *Proc. 7th IEEE Symp. on Logic in Computer Science*, pages 46–57, 1992.

[Kolaitis and Vardi, 1992b] Ph. G. Kolaitis and M. Y. Vardi. Infinitary logics and 0-1 laws. *Information and Computation*, 98(2):258–294, 1992.

[Kolaitis and Vardi, 1995] Ph. G. Kolaitis and M. Y Vardi. On the expressive power of Datalog: Tools and a case study. *Journal of Computer and System Sciences*, 51:110–134, 1995.

[Kolaitis and Vardi, 2000] Ph. G. Kolaitis and M. Y Vardi. A game-theoretic approach to constraint satisfaction. In *Proc. 17th National Conference on Artificial Intelligence*, pages 175–181, 2000.

[Kreutzer, 2002] S. Kreutzer. Expressive equivalence of least and inflationary fixed-point logic. In *Proc. of the 17th IEEE Symp. on Logic in Computer Science (LICS).*, pages 403–410, 2002.

[Lynch, 1986] J.F. Lynch. A depth-size tradeoff for boolean circuits with unbounded fan-in. In *Proc. of the Conf. on Structure in Complexity Theory*, pages 234–248, 1986.

[McColm, 1990a] G. L. McColm. Parametrization over inductive relations of a bounded number of variables. *Annals of Pure and Applied Logic*, 48:103–134, 1990.

[McColm, 1990b] G. L. McColm. When is arithmetic possible? *Annals of Pure and Applied Logic*, 50:29–51, 1990.

[Papadimitriou, 1994] Ch. H. Papadimitriou. *Computational Complexity*. Addison-Wesley, 1994.

[Poizat, 1982] B. Poizat. Deux ou trois choses que je sais de L_n. *Journal of Symbolic Logic*, 47(3):641–658, 1982.

[Stolbouskin and Taitslin, 1995] A. Stolbouskin and M. Taitslin. Is first order contained in an initial segment of PTIME? In *Computer Science Logic 94*, volume 933 of *LNCS*. Springer-Verlag, 1995.

[Vardi, 1982] M. Y. Vardi. The complexity of relational query languages. In *Proc. of the 14th ACM Symp. on the Theory of Computing*, pages 137–146, 1982.

[Venema, 1990] Y. Venema. Expressiveness and completeness of an interval tense logic. *Notre Dame Journal of Formal Logic*, 31:529–547, 1990.

Planning in Answer Set Programming using Ordered Task Decomposition

JÜRGEN DIX, UGUR KUTER AND DANA NAU

1 Introduction

In the past few years, the availability of fast nonmonotonic systems based on logic programming (LP) made it possible to attack problems from other, non-LP areas, by translating these problems into logic programs and running a fast prover on them. One of the first such systems was *Smodels* [Niemelä and Simons, 1996] and one of the early applications [Dimopoulos *et al.*, 1997] was to transform planning problems in a suitable way and to run *Smodels* on them (see also [Dix *et al.*, 2001]).

Since then, additional systems with different properties for dealing with logic programs have become available: *DLV* [Eiter *et al.*, 1998], *XSB* [Chen and Warren, 1996; Rao *et al.*, 1997], to cite the most well-known. In addition, the paradigm of Answer Set Programming (ASP) has emerged (put forth in [Niemelä, 1999; Marek and Truszczyński, 1999], see also [Apt *et al.*, 1999]). It is based on two key ideas: (1) to solve problems by computing models for logic programs rather than by evaluating queries against logic programs (as used to be done in conventional logic programming), (2) to tackle the problems located on the second level of the polynomial hierarchy, which seem well suited for the machinery of answer sets. In particular, many planning problems fit in this picture.

In this chapter, we investigate how to formulate and solve *Hierarchical Task-Network* (HTN) planning problems using nonmonotonic logic programs under the ASP semantics. HTN planning is an AI-planning paradigm in which the goals of the planner are defined in terms of activities (tasks) and the planning process is performed by using the techniques of task decomposition.

HTN planning was first proposed more than 25 years ago [Sacerdoti, 1990; Tate, 1977]. Historically, most of the HTN planning research has focused on specific application domains. Examples include production-line scheduling [Wilkins, 1988], crisis management and logistics [Currie and Tate, 1991; Tate *et al.*, 1994; Biundo and Schattenberg, 2001], planning and scheduling for spacecraft [Aarup *et al.*, 1994; Estlin *et al.*, 1997], equipment configuration [Agosta, 1995], manufacturability analysis [Hebbar *et al.*, 1996; Smith *et al.*, 1997], evacuation planning [Muñoz-Avila *et al.*, 2001], and the game of bridge [Smith *et al.*, 1998a; Smith *et al.*, 1998b]. However, there are several domain-independent HTN planning systems, such as *Nonlin* [Tate, 1977], *Sipe-2* [Wilkins, 1990], *O-Plan* [Currie and Tate, 1991; Tate *et al.*, 1994], *UMCP* [Erol *et al.*, 1994], *SHOP* [Nau *et al.*, 1999], *ASHOP* [Dix *et al.*, 2003; Dix *et al.*, 2002], and *SHOP2* [Nau *et al.*, 2001].

In this work, we focus on the *SHOP* planning system, which is a domain-independent HTN planning system that is built around a concept called *ordered task decomposition*. In particular:

- We describe a systematic translation method $\mathfrak{Trans}(\cdot)$ which transforms HTN planning problems as formalised in *SHOP* into logic programs with negation. Our basic goal is that an appropriate semantics of the logic program captures the solutions (plans) of the planning problem.

- We establish soundness and completeness results for our method: answer sets of the transformation are in one-to-one correspondence with solutions of the original planning problem.

- We propose to use established benchmarks for planning problems as benchmarks for testing ASP systems, by transforming the former using our translation into logic programs.

- Although we describe our transformation using the syntax of the *Smodels* software, our translation does not depend on the system used. We have implemented our approach using *Smodels* and *DLV* . We present several experimental comparisons between these systems and the *SHOP* planning system.

- We demonstrate that our method outperforms the transformation of a classical (i.e., action-based) planning problem into ASP proposed in [Son *et al.*, 2001] by a factor of 40-100. The particular relevance of that transformation method to our work is that, in their work, [Son *et al.*, 2001] proposed to use a form of control knowledge to speed up the classical planning process. In this chapter, we show that HTN control knowledge provides more time-efficient transformations compared to the control strategies presented in [Son *et al.*, 2001].

- We investigate on how grounding affects the performance. It seems that systems allowing for unbound variables (without grounding) are better suited and would come closer in performance to *SHOP* than current ASP systems.

We have created a website where all our formalisations can be downloaded in a form ready to run on *DLV* and *Smodels* : `<http://www.in.tu-clausthal.de/index.php?id=aspplan>`. This site will be maintained and new examples will be added as we progress in our research.

This chapter is organised as follows. In the next section, we present the approaches in the literature which, we believe, are directly related to our efforts. In Section 3, we describe the HTN planning paradigm and the *SHOP* planning system. In Section 4, we present our causal theory for HTN planning and our translation method for transforming HTN planning problems into logic programs with negation. Section 5 contains our results. Our main theorem is that our translation method is correct and complete with respect to *SHOP*. We also present a variety of experimental results along with some discussions on the sources of complexity. In particular, we compare the performance of *DLV* and *Smodels* on planning benchmarks. Finally, we conclude with Section 6 and provide our future research directions.

2 Related Work

The published literature includes many efforts at formulating actions in logic programs and solving planning problems by using formulations such as [Gelfond and Lifschitz, 1998; Turner, 1997; Lifschitz, 1999; Lifschitz, 2002]. [Gelfond and Lifschitz, 1998] describes three different action description languages that formalise theories of actions. These languages provide means to implement logic programs to solve planning problems effectively and efficiently [Lifschitz, 1999; Giunchiglia and Lifschitz, 1998]. The $\mathcal{C}$ language consists of general templates to define actions that have preconditions and effects. [McCain and Turner, 1997] presents a language for causal theories. They have also developed a system called *Ccalc* , which is a model checker for the language of causal theories translated from propositions in the $\mathcal{C}$ action language using rewrite rules [McCain, 1999]. The idea in all these works is to represent a given computational problem by a logic program whose models correspond to the solutions for the original problem. This idea was the main inspiration for the work presented here.

[Eiter *et al.*, 2002] proposes a declarative language, called the K language, for planning with incomplete information. The K language makes it possible to describe transitions between knowledge states that describe the agent's knowledge about the world. Knowledge states may be incomplete, compared to the actual states of the world. This language is implemented as a front-end to the *DLV* logic programming system. [Eiter *et al.*, 2003] describes a language K^c , which extends the language K for dealing with the action costs during planning. In particular, the language K^c can express planning problems with optimality criteria, such as computing the shortest or the least-cost plans.

[Baral et al., 2002] presents a language about actions using causal laws to reason in probabilistic settings and solves the planning problems in such settings. The language resembles those described above, but the action theory incorporates probabilities and probabilistic reasoning techniques—as described in [Pearl, 1988]—to solve the planning problems with uncertainty.

[Dimopoulos et al., 1997] presents a framework for encoding planning problems in logic programs with negation-as-failure. In this work, the idea is almost the same as ours, that is, the models of the logic program correspond to the plans. However, this work incorporates ideas from planners such as GraphPlan and SATplan, and it does not consider any sort of search-control knowledge in the logic-program encodings. In this respect, our approach is completely different.

[Son et al., 2001] discusses solving planning programs by logic programs. The difference between this work and the one described in [Dimopoulos et al., 1997] is that [Son et al., 2001] incorporates domain-dependent control knowledge to improve the performance of the planning. In this respect the work is similar to HTN planning. However, the encoding is conceptually different from HTN planning: it exploits domain constraints to define the ordering relationships between the actions, and uses these constraints to prune the search for correct sequence of actions to solve a planning problem. This technique does not eliminate an action in a state if it is applicable in that state and it satisfies the input constraints, although that action is not part of any solution for the input planning problem. In HTN planning, on the other hand, the search-control knowledge eliminates actions from consideration for the states that the planner visits during its search, which provides better search control.

3   Hierarchical Task Network (HTN) Planning

HTN planning is like classical (i.e., action-based) planning in that each state of the world is represented by a set of atoms, and each action corresponds to a deterministic state transition. However, HTN planners differ from classical planners in what they plan for, and how they plan for it.

The purpose of an HTN planner is to produce a sequence of actions to perform some activity or *task*. The description of a planning domain includes a set of operators similar to those of action-based planning, and also a set of *methods*, each of which is a prescription for how to decompose a task into its *subtasks* (smaller tasks). Within a domain, the description of a planning problem contains an initial state. Instead of a goal formula, however, there is a partially ordered set of tasks to accomplish.

Planning proceeds by decomposing tasks recursively into smaller and smaller subtasks, until *primitive tasks*, which can be performed directly using the planning operators, are reached. For each task, the planner chooses an applicable method, instantiates it to decompose the task into subtasks, and then chooses and instantiates other methods to decompose the subtasks even further. If the constraints on the subtasks or the interactions among them prevent the plan from being feasible, the planning system will backtrack and try other methods.

HTN planning has been proved to be more expressive than action-based planning [Erol *et al.*, 1996]. Moreover, HTN planning algorithms have been experimentally proved to be more efficient than their action-based counterparts. This is because the domain knowledge and the notion of decomposing a task network while satisfying the given constraints enable the planner to focus on a much smaller portion of the search space than is typically searched by action-based planning procedures. Due to their ability to generate plans very efficiently, HTN planners are used in a large variety of real-world applications [Wilkins, 1990; Currie and Tate, 1991; Nau *et al.*, 2003; Nau *et al.*, 2005].

3.1   HTN planning using Ordered Task Decomposition (OTD)

In this chapter, we are interested in a special case of HTN planning, namely HTN planning with Ordered Task Decomposition (OTD). This special case was first introduced in the *SHOP* system [Nau *et al.*, 1999; Nau *et al.*, 2000]. The difference between *SHOP* and most other HTN-planning algorithms is that *SHOP* plans for tasks in the same order that they will later be executed. Planning for tasks in the order that those tasks will be performed makes it possible to know the current state of the world at each step in the planning process, which reduces the complexity of reasoning by eliminating a great deal of uncertainty about the world. This makes it easy to incorporate substantial inferencing and reasoning power into the planning system, including the ability to call external programs and the ability to perform numeric computations.

In order to do planning in a given planning domain, *SHOP* needs to be given knowledge about that domain. *SHOP*'s knowledge base contains *operators* and *methods*. Each operator is a description of what needs to be done to accomplish some primitive task, and each method is a prescription for how to decompose some compound (abstract) task into a totally ordered sequence of subtasks, along with various restrictions that must be satisfied in order for the method to be applicable. More than one method may be applicable to the same task, in which case there will be more than one possible way to decompose that task. Given the next task to accomplish, the *SHOP*

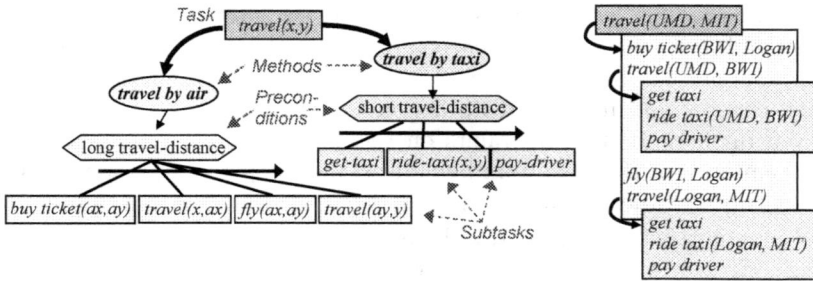

Figure 1. Travel planning example.

algorithm nondeterministically chooses an applicable method, instantiates it to decompose the task into subtasks, and then chooses and instantiates other methods to decompose the subtasks even further. The deterministic implementation of the *SHOP* algorithm uses depth-first backtracking: if the constraints on the subtasks prevent the plan from being feasible, then the implementation will backtrack and try other methods.

As an example, Figure 1 shows two methods for the task of travelling from one location to another: *travelling by air*, and *travelling by taxi*. Travelling by air involves the subtasks of purchasing a plane ticket, travelling to the local airport, flying to an airport close to our destination, and travelling from there to our destination. Travelling by taxi involves the subtasks of *calling a taxi, riding in it to the final destination,* and *paying the driver*. Note that each method's preconditions are not used to create subgoals (as would be done in action-based planning). Rather, they are used to determine whether or not the method is applicable: thus in Figure 1, the *travel by air* method is only applicable for long distances, and the *travel by taxi* method is only applicable for short distances. Now, consider the task of travelling from the University of Maryland to MIT. Since this is a long distance, the *travel by taxi* method is not applicable, so we must choose the *travel by air* method. As shown in Figure 1, this decomposes the task into the following subtasks: *(1)* purchase a ticket from Baltimore-Washington International (BWI) airport to Logan airport, *(2)* travel from the University of Maryland to BWI, *(3)* fly from BWI airport to Logan airport, and *(4)* travel from Logan airport to MIT. For the subtasks of travelling from the University of Maryland to BWI and travelling from Logan to MIT, we can use the *travel by taxi* method to produce additional subtasks as shown in Figure 1.

Here are some of the complications that can arise during the planning process:

- The planner may need to recognise and resolve interactions among the subtasks. For example, in planning how to travel to the airport, one needs to make sure one will arrive at the airport in time to catch the plane. To make the example in Figure 1 more realistic, such information would need to be specified as part of *SHOP*'s methods and operators.

- In the example in Figure 1, it was always obvious which method to use. But in general, more than one method may be applicable to a task. If it is not possible to solve the subtasks produced by one method, *SHOP* will backtrack and try another method instead.

3.2   HTN-planning with OTD: Syntax and Semantics

We use the same definitions for variable and constant symbols, predicate symbols, and terms, as in the *SHOP* planning system [Nau *et al.*, 1999; Nau *et al.*, 2000]. Our definitions for logical atoms, literals, states, tasks, task networks, axioms, operators, and methods are adapted from *SHOP*.

Following the notation used in *SHOP*, we will write logical atoms using the format (**name** $t_1 t_2 \ldots t_n$), where **name** is a predicate symbol, and $t_1, t_2, \ldots, t_n$ are terms. In *SHOP* we can classify the atoms into three kinds:

- *Rigid Atoms:* These are atoms whose truth values never change during planning. These atoms appear in states, but do not appear in the effects of planning operators nor in the heads of Horn clauses.

- *Primary Atoms:* These atoms can appear in states and in the effects of planning operators, but cannot appear in the heads of Horn clauses.

- *Secondary Atoms:* These are the ones whose truth values are inferred rather than being stated explicitly. They can appear in the heads of Horn clauses, but cannot appear in states nor in the effects of planning operators.

Now, we define the states and the axioms as in *SHOP*:

DEFINITION 1 (States ($\mathcal{S}$), Axioms ($\mathcal{AX}$)). A *state* $\mathcal{S}$ is a set of ground primary atoms. An *axiom* is an expression of the form

$$a \leftarrow l_1, \ldots, l_n,$$

where a is a secondary atom and the $l_1, \ldots, l_n$ are literals that constitute either primary or secondary atoms.

Axioms need not be ground. We assume that the set of axioms does not contain cycles through negation.[1]

---

[1]This is just to ensure that a unique stable model always exist and thus the state is

SHOP starts with a state $\mathcal{S}$ and modifies this state by taking into account the delete and add lists of the operators in the plan. Axioms are used only to check whether the preconditions of methods are satisfied. A precondition might not be explicitly satisfied (in the sense that the atom in question is contained in $\mathcal{S}$), but might be *caused by* $\mathcal{S}$ and the axioms $\mathcal{AX}$. The precise definition of this relation *"caused by"* is given as follows and extended in Subsection 4.

DEFINITION 2 (Literal caused by $(\mathcal{S}, \mathcal{AX})$). A literal l is caused by $(\mathcal{S}, \mathcal{AX})$ if l is true in all answer sets of $\mathcal{S} \cup \mathcal{AX}$.

Because of our assumption on $\mathcal{AX}$, the set of axioms constitutes a stratified logic program which has exactly one answer set. This ensures that any state described by the stable model of $\mathcal{S} \cup \mathcal{AX}$ is complete: any literal is either caused or its negation is caused.

In order to check which literals follow from $(\mathcal{S}, \mathcal{AX})$, *SHOP* uses an axiomatic inference procedure. To discuss this procedure, we need to make a distinction between the abstract *SHOP* algorithm and the *SHOP* implementation. On one hand, the abstract *SHOP* algorithm is nondeterministic and it makes no commitment to what inference procedure is used for checking whether literals follow from $(\mathcal{S}, \mathcal{AX})$. The completeness proof for *SHOP* [Nau *et al.*, 2000] says that if the inference procedure is complete, then *SHOP* is complete (i.e., if a planning problem has a solution, then at least one of *SHOP*'s execution traces will find a solution).

On the other hand, the *SHOP* implementation uses an inference procedure that does a depth-first search similar to the one in Prolog. This inference procedure is complete only if the axioms satisfy some restrictions similar to those needed in Prolog (no positive cycles, no cycles through negation).[2] However, all the axioms $\mathcal{AX}$ we are dealing with in this chapter are of this sort. In fact, checking causality for these simple instances can be done in linear time.

A *task* is an expression of the form (name $t_1 t_2 \ldots t_n$), where name (the task's name) is a task symbol, and $t_1, t_2, \ldots, t_n$ (the task's arguments) are terms. A *ground task* is a task that has no variables in its arguments. A task can be either *primitive* (if it is to be accomplished directly in the world) or *compound* (if it is to be decomposed into other tasks). We use a prefix ? to denote a variable (such as ?x and ?y) and ! to denote the name of a

---

always complete (see the next definition). Without this condition, our approach is still complete but no more correct wrt. *SHOP*: *SHOP* does not terminate while our translation still gets meaningful results.

[2]In addition, the *SHOP* implementation also computes its task decompositions using a depth-first search. Thus, in order to achieve completeness, the HTN methods also need to satisfy a similar acyclicity restriction.

primitive task. For example, to tell the planner that getting a taxi, riding in it, and paying the driver are primitive tasks, we would give them names like !get-taxi, !ride-taxi, and !pay-driver. Tasks using these names are (!get-taxi ?x), (!ride-taxi ?x ?y), or (!pay-driver ?x ?y).

A *task list* is a list of tasks, like the following:

((!get-taxi ?x) (!ride-taxi ?x ?y) (!pay-driver ?x ?y)))

A *ground task list* is a task list that consists of only ground tasks, like the following:

((!get-taxi umd) (!ride-taxi umd mit) (!pay-driver umd mit)))

An operator specifies how to accomplish a primitive task by modifying the current state of the world by removing every atom in its delete list and by adding every atom in its add list, as defined below.

DEFINITION 3 (Operator: $(\mathbf{Op}\ h\ \epsilon_{del}\ \epsilon_{add})$). An *operator* is an expression of the form $(\mathbf{Op}\ h\ \epsilon_{del}\ \epsilon_{add})$, where h (the *head*) is a primitive task and ϵ_{add} and ϵ_{del} are lists of primary atoms (called the *add-* and *delete-lists*, respectively). The set of variables in the atoms in ϵ_{add} and ϵ_{del} must be a subset of the set of variables in h.[3]

As an example, here is a possible implementation of the get-taxi operator from Figure 1:

```
(:Op (!get-taxi ?x)
    ((taxi-called-to ?x))
    ((taxi-standing-at ?x)))
```

Operators are used in decomposition of primitive tasks during planning:

DEFINITION 4 (Decomposition of Primitive Tasks). Let t be a primitive task, and let $\mathsf{Op} = (\mathbf{Op}\ h\ \epsilon_{del}\ \epsilon_{add})$ be an operator. Suppose that θ is a unifier for h and t. Then the ground operator instance $(\mathsf{Op})\theta$ is *applicable* to t, in which case we define the *decomposition* of t by Op to be $(\mathsf{Op})\theta$.

The decomposition of a primitive task by an operator results in a ground instance of that operator – i.e., it results in an action that can be applied in a state of world. We now define the result of such an application:

---

[3] Unlike the operators used in action-based planning, ours have no preconditions. Preconditions are not needed for operators in our formulation, because they occur in the methods that invoke the operators.

DEFINITION 5 (Plans, result($\mathcal{S}$,π)). A *plan* is a list of heads of ground operator instances.[4] A plan π is called a *simple plan* if it consists of the head of just one ground operator instance.

Given a simple plan $\pi = (h)$, we define result($\mathcal{S}, \pi$) to be the set

$$\mathcal{S} \setminus \epsilon_{del} \cup \epsilon_{add}$$

obtained by deleting from $\mathcal{S}$ all atoms in ϵ_{del} and by adding all ground instances of atoms in ϵ_{add}.

If $\pi = (h_1, h_2, \ldots, h_n)$ is a plan and $\mathcal{S}$ is a state, then the *result* of applying π to $\mathcal{S}$ is the state result($\mathcal{S}, \pi$) = result(result($\ldots$(result $(\mathcal{S}, h_1), h_2), \ldots$), h_n).

In *SHOP*, a method specifies a possible way to accomplish a compound task. The set of methods relevant for a particular compound task can be seen as a recursive definition of that task.

DEFINITION 6 (Method: **(Meth** h ρ **t)**). A *method* is an expression of the form **(Meth** h ρ **t)** where h (the method's *head*) is a compound task, ρ (the method's *preconditions*) is a conjunction of literals and **t** is a totally-ordered list of subtasks, called the *decomposition list* of the method. The set of variables that appear in the decomposition list of a method must be a subset of the variables in h (the head of the method) and ρ (the preconditions of the method). [5]

Here is a possible implementation of the travel-by-taxi method from the same figure:

```
(:Meth (travel ?x ?y)
    ((smaller-distance ?x ?y))
    ((!get-taxi ?x) (!ride-taxi ?x ?y) (!pay-driver ?x ?y)))
```

Let $m = $ **(Meth** h ρ **t)** be a method. Note that there may be variables in ρ that do not appear in the head h of the method m. These variables are called the *unbound variables* of m. During planning, these variables are grounded when the method is used for the decomposition of a compound task, as described below.

---

[4] In Definition 8, we will require that in any planning domain, every planning operator must have a unique name. This is sufficient to guarantee that every plan specifies an unambiguous sequence of operator instances.

[5] This restriction is needed to ensure that our programs do not violate the safeness restrictions of the ASP systems we are using. However, the restriction has no effect on the expressivity of our formalism. Any method that does not satisfy the restriction can easily be translated into an equivalent method that does satisfy the restriction, by introducing a dummy precondition that can always be satisfied.

DEFINITION 7 (Decomposition of Compound Tasks). Let t be a compound task, $\mathcal{S}$ be the current state, $Meth = (\textbf{Meth } h \; \rho \; \textbf{t})$ be a method, and $\mathcal{AX}$ be an axiom set. Suppose that θ is a unifier for h and t, and that θ' is a unifier such that all literals in $(\rho)\theta\theta'$ are caused wrt. $\mathcal{S}$ and $\mathcal{AX}$ (see Definition 2).

Then, the ground method instance $(Meth)\theta\theta'$ is *applicable* to t in $\mathcal{S}$, and the result of applying it to t is the ground task list $\textbf{r} = (\textbf{t})\theta\theta'$. The task list $\textbf{r}$ is the *decomposition* of t by $Meth$ in $\mathcal{S}$.

Note that the decomposition of a compound task by a method does not change the state of the world. The result of such a decomposition is a ground task list that needs to be further decomposed until we get a list of only ground operator instances — i.e., a plan.

DEFINITION 8 (Planning Domain Descriptions and Problems). A *planning domain description* $\mathcal{D}$ is a triple consisting of (1) a set of axioms, (2) a set of operators such that no two operators have the same head, and (3) a set of methods.

A *planning problem* is a triple $(\mathcal{S}, \textbf{t}, \mathcal{D})$, where $\mathcal{S}$ is a state, $\textbf{t} = (t_1, t_2, \ldots, t_k)$ is a ground task list, and $\mathcal{D}$ is a planning domain description.

We now define a *solution* of a planning problem.

DEFINITION 9 (Solutions). Let $P = (\mathcal{S}, \textbf{t}, \mathcal{D})$ be a planning problem and $\pi = (h_1, h_2, \ldots, h_n)$ be a plan. Then, π is a *solution* for P,[6] if any of the following is true:

- Case 1: $\textbf{t}$ and π are both empty, (i.e., $k = 0$ and $n = 0$);
- Case 2: $\textbf{t} = (t_1, t_2, \ldots, t_k)$, t_1 is a ground primitive task, (h_1) is the decomposition of t_1, and $(h_2 \ldots h_n)$ solves $(\text{result}(\mathcal{S}, (h_1)), (t_2, \ldots, t_k), \mathcal{D})$;
- Case 3: $\textbf{t} = (t_1, t_2, \ldots, t_k)$, t_1 is a ground compound task, and there is a decomposition $(r_1 \ldots r_j)$ of t_1 in $\mathcal{S}$ such that π solves $(\mathcal{S}, (r_1, \ldots, r_j, t_2, \ldots, t_k), \mathcal{D})$.

The planning problem $(\mathcal{S}, \textbf{t}, \mathcal{D})$ is *solvable* if there is a plan that solves it.

One important issue that we want to point out about this definition is that the *SHOP* formalism does not require the tasks to be ground. This and the restriction in Definition 6 are both necessary in the formalism of our translation method, simply because, otherwise, the logic programs that are generated by our translation would contain rules that violate the safeness conditions that are imposed by current ASP systems. However, this is a mild restriction and can always be ensured by adding dummy predicates.

---

[6]Or equivalently, we say that π *solves* P, or π *achieves* $\textbf{t}$ from $\mathcal{S}$ in $\mathcal{D}$ (we will omit the phrase "in $\mathcal{D}$" if the identity of $\mathcal{D}$ is obvious).

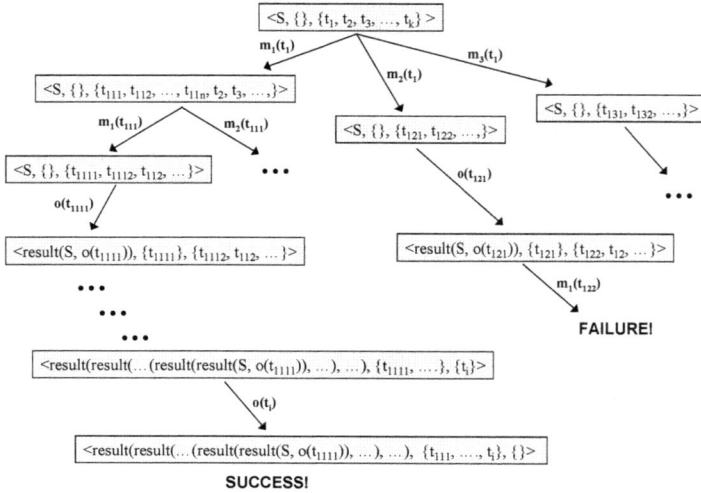

Figure 2. Search Tree for $(\mathcal{S}, \mathbf{t}, \mathcal{D})$. Edge labellings $m_i(t)$ (resp. $o(t)$) represent a method (resp. an operator) application to a task t, which is compound (resp. primitive).

It will be very helpful for the main proof of Theorem 30 to introduce the notion of a *search tree*. The successful paths of this tree correspond to the solutions of the planning problem.

DEFINITION 10 (Search Tree for $\mathfrak{Trans}(\cdot)$).
Given a planning problem $(\mathcal{S}, \mathbf{t}, \mathcal{D})$, we define the search tree for $(\mathcal{S}, \mathbf{t}, \mathcal{D})$ as follows. Nodes of the tree are triples of the form $\langle \mathcal{S}', \mathbf{t}_{caused}, \mathbf{t}' \rangle$, where $\mathcal{S}'$ is a state, $\mathbf{t}_{caused}$ is an ordered list of ground primitive tasks, and $\mathbf{t}'$ is a (possibly empty) ordered list of ground (compound or primitive) tasks.

The start node consists of the triple $\langle \mathcal{S}, \emptyset, \mathbf{t} \rangle$. Leaf nodes are those of the form $\langle \mathcal{S}, \mathbf{t}_{caused}, \emptyset \rangle$. Branches ending in such leaves are called *successful*. Given a node $\langle \mathcal{S}', \mathbf{t}_{caused}, \mathbf{t}' \rangle$ with $\mathbf{t}' \neq \emptyset$, its children are defined as follows:

- If the first task in $\mathbf{t}'$ is primitive and there is an operator in $\mathcal{D}$ for it, then there is exactly one child $\langle \mathcal{S}^\star, \mathbf{t}^\star_{caused}, \mathbf{t}^\star \rangle$. $\mathbf{t}^\star_{caused}$ is the old list $\mathbf{t}_{caused}$ plus this first task appended. $\mathcal{S}^\star$ is obtained by modifying $\mathcal{S}'$ according to the add and delete lists of the operator. $\mathbf{t}^\star$ is $\mathbf{t}'$ with the first element deleted. The edge to this child is labelled with the name of this first task. If the first task in $\mathbf{t}'$ is primitive and there is no operator, then there is no child and the branch is marked with *Failure*.

- If the first task in $\mathbf{t}'$ is compound, and there exist method instances applicable to it (according to Definition 7), then each such method instance m_i leads to a child node $\langle \mathcal{S}, \mathbf{t}_{caused}, \mathbf{t}^i \rangle$ the edge to which is labelled with m_i. $\mathbf{t}^i$ is obtained from $\mathbf{t}'$ by replacing the first task by the subtasks according to m_i. If the first task in $\mathbf{t}'$ is compound, and there are no methods applicable to it, then the branch is marked *Failure*.

We define the *task-depth* of a node and its edges as follows. The start node gets task-depth 0. Whenever a method is used to extend a node, the children nodes keep the same task-depth. When an operator is applied, and thus a task is moved from $\mathbf{t}'$ into $\mathbf{t}_{caused}$ then the task depth of the child node is incremented by one. Obviously, the task depth of a node is the size of the list of tasks in $\mathbf{t}_{caused}$.

Such a tree (or a part thereof) is depicted in Figure 2. Note that there can be different paths (corresponding to the application of different methods) that finally lead to the same plans (as a list of the heads of ground instances of operators).

DEFINITION 11 (Solution Set of a Planning Problem: $\mathbf{Sol}(\mathcal{S}, \mathbf{t}, \mathcal{D})$). Let $P = (\mathcal{S}, \mathbf{t}, \mathcal{D})$ be a planning problem, and suppose $\mathcal{T}$ is the search tree for P. Then, $\mathbf{Sol}(\mathcal{S}, \mathbf{t}, \mathcal{D})$ is a *multi set*: it contains exactly the ordered lists $\mathbf{t}_{caused}$ in the leaf nodes that are reached by the successful paths of $\mathcal{T}$.

We also say $\mathcal{T}$ represents $\mathbf{Sol}(\mathcal{S}, \mathbf{t}, \mathcal{D})$. Note that $\mathbf{Sol}(\mathcal{S}, \mathbf{t}, \mathcal{D})$ may contain more than one copy of the same plan.

4 Encoding HTN planning in Nonmonotonic Logic Programming

Our approach of encoding HTN planning problems as logic programs is based on *SHOP*'s representation of a planning problem as described in the last section. We now present the first steps of a causal theory of HTN planning based on that formalism. This theory serves as an intermediate step and a motivation for our translation methodology, which is given in the next subsection. We conclude this section with the formalisation of a particular example.

4.1 Causal Theory for HTN Planning

In this section we prepare the ground for our translation in the next subsection. We give some definitions of a causal theory for HTN planning in a

SHOP-like ordered task decomposition.

DEFINITION 12 (Causable Literals). Let $\mathcal{S}$ be a state, and let $\mathcal{D}$ be a planning domain description. A literal l is causable wrt. $(\mathcal{S},\mathcal{D})$ if it is caused by $(\mathcal{S},\mathcal{AX})$ (according to Definition 2), where $\mathcal{AX}$ is the set of axioms in $\mathcal{D}$.

A conjunction of literals is causable wrt. $(\mathcal{S},\mathcal{D})$ if every literal in the conjunct is causable wrt. $(\mathcal{S},\mathcal{D})$ (according to the Definition 2).

DEFINITION 13 (Causable Tasks). Let $\mathcal{S}$ be a state, and let $\mathcal{D}$ be a planning domain description. The definition of an ordered list of ground tasks to be causable wrt. $(\mathcal{S},\mathcal{D})$ comes in three steps.

1. The empty list $[\,]$ is causable wrt. $(\mathcal{S},\mathcal{D})$.

2. An ordered list of ground primitive tasks $t_1,\ldots,t_n$ is causable wrt. $(\mathcal{S},\mathcal{D})$ if for each t_i, there exists an operator $(\mathbf{Op}\ h\ \epsilon_{del}\ \epsilon_{add}) \in \mathcal{D}$ and there is a unifier θ such that $t_i = (h)\theta$.

3. An ordered list of ground tasks $t_1,\ldots,t_j,\ldots,t_n$, where t_j is a ground compound task and all tasks $t_1,\ldots,t_{j-1}$ are ground primitive tasks, is causable wrt. $(\mathcal{S},\mathcal{D})$ if the following holds:

   • There exists a method $(\mathbf{Meth}\ h\ \rho\ \{t_{j_1},\ldots,t_{j_m}\}) \in \mathcal{D}$ for t_j, and there is a unifier θ such that $t_j = (h)\theta$; and

   • There exists a unifier θ' such that the preconditions list $(\rho)\theta\theta'$ is causable wrt. $(\mathrm{result}(\mathcal{S},(t_1,\ldots t_{j-1})),\mathcal{D})$; and

   • The ordered decomposition list

$$(t_1,\ldots,t_{j-1},(t_{j_1})\theta\theta',\ldots,(t_{j_m})\theta\theta',t_{j+1},\ldots t_n)$$

   is causable wrt. $(\mathcal{S},\mathcal{D})$.

Note that this is a recursive definition. The condition in the last part (*compound tasks*) eventually ends when there are only primitive tasks left, and thus the second part (*primitive tasks*) can be applied. The notion of *literals being causable* is used to make sure that the appropriate methods (used to decompose the task t_j) can be applied in the current state.

Using this causal theory as an intermediate step, we developed a systematic translation method for mapping planning problems to logic programs with negation which we illustrate in the subsequent section. The next theorem states the equivalence of the original *SHOP* planning framework as presented in the last section with the notion of causable tasks just introduced.

THEOREM 14. *Let $P = (\mathcal{S}, \mathbf{t}, \mathcal{D})$ be a planning problem. Then, there is a solution to P if and only if* the task list $\mathbf{t}$ is causable wrt. $(\mathcal{S}, \mathcal{D})$.

Proof. Rather than giving a full proof using structural induction, we give a detailed proof sketch from which the full proof can be easily worked out.

The proof starts by recursively constructing the solution of an HTN planning problem $(\mathcal{S}, \mathbf{t}, \mathcal{D})$ and showing the causal relationships based on our causal theory at the same time.

Suppose there exists a solution to $(\mathcal{S}, \mathbf{t}, \mathcal{D})$. If $\mathbf{t}$ is empty, then $(\mathcal{S}, \mathbf{t}, \mathcal{D})$ has exactly one solution, namely the empty plan. This is because of the fact that there will be no tasks to be accomplished—thus, no task is causable. If $\mathbf{t}$ is not empty, and consists of primitive tasks only, then there must be operators for all these tasks and thus, by Definition 13 (2nd step), $\mathbf{t}$ is causable wrt. $(\mathcal{S}, \mathcal{D})$.

We now reduce the general case to the case where only primitive tasks occur. Let $\mathbf{t}$ be non-empty, and assume it contains compound tasks. Then we recursively carry out the task decompositions (see Definition 7) until we reach a list of ground primitive tasks. This is possible because a solution exists (this solution is given by the final list of primitive tasks). Note that there might exist different such lists, corresponding to different choices of decompositions via methods. But this list is causable according to the second part of Definition 13 (*primitive tasks*).

Now, according to the third part of Definition 13, we can recursively replace the primitive tasks with the compound tasks they were induced from (via methods) and we get the result that all of the intermediate ordered lists obtained in that way themselves are causable wrt. $(\mathcal{S}, \mathcal{D})$. Note that the conditions in the third part of Definition 13 correspond exactly to the notion of the decomposition of a compound task (Definition 7).

Therefore, it follows from the recursive construction above that if a list of ground tasks $\mathbf{t}$ is achieved according to our planning theory, it must be causable as well.

The proof of the converse is similar. Once a list of ground tasks $\mathbf{t}$ is causable wrt. $(\mathcal{S}, \mathcal{D})$, we can find a list of primitive tasks that is causable. This list is obtained by certain decompositions, and these decompositions constitute a solution of the planning problem. ∎

4.2   Encoding Planning Problems as Logic Programs

In this section, we present our translation method for encoding planning problems as logic programs with ASP semantics. Our translation method is a general technique that is independent from the implementation details and syntactic requirements of the any underlying ASP system. Note that there are several differences between the syntactic requirements of the ASP systems. In this respect, the presentation in this section is given in a more conceptual level in general; however, where necessary, we used the syntax of *Smodels* .[7]

Translating a planning problem $(\mathcal{S}, \mathbf{t}, \mathcal{D})$ to its logic program counterpart requires encoding the initial state and the state transition characteristics of *SHOP*, the goal tasks and the ordered task decomposition technique in *SHOP*, and the domain description including the axioms, the operators, and the methods, which are given in the description of a planning problem. For this reason, we describe our translation method in several steps such that each step encodes a part of the complete translation corresponding to the components of a planning problem as described above. Combining these steps yield a complete logic program in ASP semantics that is capable of solving planning problems in the way that *SHOP* does.

We now present our main definition:

DEFINITION 15 ($\mathfrak{Trans}((\mathcal{S}, \mathbf{t}, \mathcal{D}))$: Translation for the Planning Problem). Let $P = (\mathcal{S}, \mathbf{t}, \mathcal{D})$ be a planning problem. The logic program $\mathfrak{Trans}((\mathcal{S}, \mathbf{t}, \mathcal{D}))$ that solves P is defined as

$$\mathfrak{Trans}((\mathcal{S}, \mathbf{t}, \mathcal{D})) = \mathfrak{Trans}(\bot) \cup \mathfrak{Trans}(\mathcal{S}) \cup \mathfrak{Trans}(\mathbf{t})$$
$$\cup\, \mathfrak{Trans}(\mathcal{AX}) \cup \mathfrak{Trans}(\mathcal{OP}) \cup \mathfrak{Trans}(\mathbf{F}) \cup \mathfrak{Trans}(\mathcal{METH}),$$

where

- $\mathfrak{Trans}(\bot)$ is the logic program segment that marks the successful termination of the planning process,

- $\mathfrak{Trans}(\mathcal{S})$ is the logic program segment that encodes the initial state $\mathcal{S}$,

- $\mathfrak{Trans}(\mathbf{t})$ is the logic program segment that encodes the goal task list $\mathbf{t}$,

- $\mathfrak{Trans}(\mathcal{AX})$ is the logic program segment that encodes the axioms given in the domain description $\mathcal{D}$,

- $\mathfrak{Trans}(\mathcal{OP})$ is the logic program segment that encodes the operator descriptions given in $\mathcal{D}$, and

- $\mathfrak{Trans}(\mathbf{F})$ is the logic program segment that encodes the state-transition characteristics of *SHOP* , and

- $\mathfrak{Trans}(\mathcal{METH})$ is the logic program segment that encodes the method descriptions given in $\mathcal{D}$.

---

[7]However, note that when implementing our translation methodology, one must address the syntax requirements of the underlying system that is being used. In this chapter, we concentrated on the two ASP systems *Smodels* and *DLV*, so we made system-specific syntactic changes to the conceptual description of our translation method during the implementation in these systems. Our complete implementation of planning examples for both *DLV* and *Smodels* are available at <http://www.in.tu-clausthal.de/index.php?id=aspplan>.

In the following subsections, we give the definitions for the logic program segments mentioned above.

The Time Line

In order to be able to keep track of different states of the world in an answer set of a logic program, we attached a *time variable* that will occur in many predicates in our translation. The domain of a time variable is a *time line*, which is a set of integers $\{0, 1, \ldots, \tau\}$, where τ is called the end point of the time line.

Given a planning problem $P = (\mathcal{S}, \mathbf{t}, \mathcal{D})$, we need to know τ in advance in order for our translation methodology to work. P does not specify this information since *SHOP* does not need a notion of time line during planning — the search process naturally differentiates the states of the world. One way to determine a correct value for τ is an incremental approach. That is, we start with $\tau = 0$, use our translation to produce the logic program $\mathfrak{Trans}((\mathcal{S}, \mathbf{t}, \mathcal{D}))$, and generate the answer sets of $\mathfrak{Trans}((\mathcal{S}, \mathbf{t}, \mathcal{D}))$. Then, we increment τ, and repeat the whole process until no new answer sets are generated by $\mathfrak{Trans}((\mathcal{S}, \mathbf{t}, \mathcal{D}))$. Note that, using this approach, we eventually generate all of the possible answer sets of $\mathfrak{Trans}((\mathcal{S}, \mathbf{t}, \mathcal{D}))$, as shown in Theorem 30.

An alternative approach would be as follows. Let $\mathbf{Sol}(\mathcal{S}, \mathbf{t}, \mathcal{D})$ be the set of solutions of P, as in Definition 11. Then, we define τ as

$$\tau = max\{|\pi| \; : \; \pi \in \mathbf{Sol}(\mathcal{S}, \mathbf{t}, \mathcal{D})\},$$

where $|\pi|$ denotes the size of the solution π. Note that we can find the set $\mathbf{Sol}(\mathcal{S}, \mathbf{t}, \mathcal{D})$ by solving P using *SHOP* .

We use time variables in various rules in our translation, so before going into the details of the translation, we believe that the following points are worth noting:

1. As described above, we assume that planning starts at time point 0 (see Definitions 16 and 19).

2. Planning proceeds by selecting a task to be accomplished next (see Definition 19. The rules about *taskTBA* and *causable* are given in Definition 17). Note that the task to be decomposed may be either primitive or compound.

3. The time variable T is incremented only when the task to be decomposed is a primitive task and there is an operator for it in the domain description provided as a part of the planning problem (see Definition 18).

This means that, in general, there may be more than one task that is selected and decomposed at a particular time point T. However, among these tasks, there is only one primitive task at any particular point in time. For example, consider Figure 2. Task t_1 is a compound task and so are t_{111} and t_{1111} (obtained by respective methods). So $taskTBA(t_1, 0), taskTBA(t_{111}, 0), taskTBA(t_{1111}, 0)$ are all true, i.e., hold in a stable model. Only after the primitive task t_{1111} has been accomplished by an operator is the time incremented by 1.

4. As a result of this formulation, the task depth in the search tree corresponds to the value of the time variable T.

Encoding the Initial State

SHOP's initial state is a set of ground atoms. In this respect, given a planning problem $(\mathcal{S}, \mathbf{t}, \mathcal{D})$, the logic program encoding for the initial state $\mathcal{S}$ is defined as follows:

DEFINITION 16 ($\mathfrak{Trans}(\mathcal{S})$: Translation for Initial State). Given a planning problem $(\mathcal{S}, \mathbf{t}, \mathcal{D})$, for each ground atom $a \in \mathcal{S}$, the logic program $\mathfrak{Trans}(\mathcal{S})$ contains the rule

$$in\_state(a, 0) : -$$

where 0 indicates the initial time.

Encoding the Goal Task(s)

Given a planning problem $(\mathcal{S}, \mathbf{t}, \mathcal{D})$, where $\mathcal{S}$ is the initial state, $\mathbf{t}$ is a ground task list, and $\mathcal{D}$ is a domain description for this planning problem, the aim of the planning process is to find a plan that accomplishes all of the (goal) tasks in $\mathbf{t}$ from the initial state $\mathcal{S}$ in the order they are given (according to Definition 8). A task is accomplished if and only if it is *causable* with respect to the initial state and the domain description given in the planning problem, and this is due to the Definition 13 and a direct consequence of Theorem 14.

In this respect, planning proceeds by selecting a task as the "current task" – i.e., the task that the planner will try to accomplish next. In the logic programs produced by our translation, this is encoded by using a special predicate defined as follows:

DEFINITION 17 (Tasks To Be Accomplished). Given a planning problem $(\mathcal{S}, \mathbf{t}, \mathcal{D})$, we define a special predicate $taskTBA\_n$ for each possible task (e.g. primitive or compound) such that if the task that needs to be decomposed at time T is $h \equiv (name_h \; arg_1 \; arg_2 \; \ldots \; arg_N)$ then $taskTBA\_n(name_h, arg_1, arg_2, \ldots, arg_N, T)$ denotes this fact and n is a natural number which equals $N + 2$ (n is the number of arguments of this predicate).

As an example, if the task to be accomplished is travelling from UMD to MIT denoted as (`travel` umd mit), then we use the predicate $taskTBA\_4$ $(travel, umd, mit, T)$ to denote this fact. For the sake of clarity, we will use the shorthand notation $taskTBA(h, T)$ in the rest of the chapter.

We define the fact that whether a task is causable as follows:

DEFINITION 18 (CAUSABLE).   Given a ground task t, we define $CAUSABLE(t, T_s, T_a)$ as one of the following:

$$
\begin{cases}
false & \text{if } t \text{ is a primitive task and} \\
 & \text{there is no operator for it in } \mathcal{D}, \\
false & \text{if } t \text{ is a compound task and} \\
 & \text{there is no method for it in } \mathcal{D}, \\
taskTBA(t, T_s) \text{ and } T_a = T_s + 1 & \text{if } t \text{ is a primitive task and} \\
 & \text{there is an operator for it in } \mathcal{D}, \\
causable(t, T_s, T_a) & \text{if } t \text{ is a compound task and} \\
 & \text{there is a method for it in } \mathcal{D}.
\end{cases}
$$

where T_s denotes the time when the task t was selected to be decomposed and T_a denotes the time when t is actually accomplished (i.e., T_a is the time when t is caused).

Intuitively, $CAUSABLE(t, T_s, T_a)$ is a placeholder for the purposes of clarity in our formalism. Given a task t, if the HTN planning domain description specifies no planning operators or HTN methods for t, then $CAUSABLE(t, T_s, T_a)$ refers to the truth value $false$. If t is primitive and there is an operator for it, then t is causable by Definition 13. Therefore, whether t is causable or not is specified by whether the predicate $taskTBA(t, T_s)$ is $true$ or $false$ in a stable model in our formalism. In this case, $CAUSABLE(t, T_s, T_a)$ refers to the predicate $taskTBA(t, T_s)$ and $T_a = T_s + 1$ since after the primitive task t is accomplished by an operator, the time is incremented by 1 as described above. Similarly, if t is a compound task, then whether it is causable is specified by whether the predicate $causable(t, T_s, T_a)$ is $true$ or not in a stable model. Therefore, in this case, $CAUSABLE(t, T_s, T_a)$ refers to the predicate $causable(t, T_s, T_a)$.

In the definition above, $causable(t, T_s, T_a)$ is a shorthand notation for the predicate $causable\_n(name_t, arg_1, arg_2, \ldots, arg_N, T_s, T_a)$ in which the symbol $n = N + 3$ denotes the number of arguments of the predicate $causable\_n$. For the sake of clarity, we will use $causable(t, T_s, T_a)$ in the rest of the chapter.

We are now ready to define the logic program segment that encodes the goal task list of a given planning problem.

DEFINITION 19 ($\mathfrak{Trans}(\mathbf{t})$: Translation for Goal Tasks). Given a planning problem $(\mathcal{S}, \mathbf{t}, \mathcal{D})$, let $\mathbf{t} = h_1, h_2, \ldots, h_n$ be the ordered sequence of ground tasks. Then, $\mathfrak{Trans}(\mathbf{t})$ is the logic program that contains one rule for each ground task h_i, where $i = 1, 2, \ldots, n$, as follows:

1. Case 1: $i = 1$,

    $taskTBA(h_1, 0) \quad : -$

2. Case 2: Otherwise,

    $taskTBA(h_i, T_i) \quad : - \quad CAUSABLE(h_{i-1}, T_{i-1}, T_i),\ T_i \geq T_{i-1}.$

Note that if there exists only one goal task to be accomplished for the problem in hand, then only defining the first rule will suffice. Definition 19 enforces the fact that a goal task h_i is designated as the current task to be accomplished if the previous goal task h_{i-1} in $\mathbf{t}$ is causable. This is a direct consequence of our Theorem 14.

The planning process terminates successfully when all of the goal tasks are accomplished (i.e., caused) in the order they are given in the planning problem. The following definition is given to encode the successful termination of the planning process.

DEFINITION 20 ($\mathfrak{Trans}(\perp)$: Successful Termination). Given a planning problem $(\mathcal{S}, \mathbf{t}, \mathcal{D})$, the logic program segment $\mathfrak{Trans}(\perp)$ that encodes the successful termination of the planning process (i.e., the fact that a solution to the given planning problem is found) is defined as follows:

$$plan\_found \quad : - \quad CAUSABLE(h_n, T_n, T_{n+1}), T_{n+1} \geq T_n.$$
$$plan\_found \quad : - \quad not\ plan\_found.$$

where h_n is the last goal task in $\mathbf{t}$, T_n denotes the time that h_n is decomposed by a simple reduction for it (see Definition 7), and T_{n+1} is the time at which h_n is causable (accomplished).

These two rules together state that if the last goal task is causable then there is a plan (solution) for the planning problem $(\mathcal{S}, \mathbf{t}, \mathcal{D})$ as a result of Definition 19. Otherwise, there is none.

Encoding the Axioms

We now define the logic program segment that encodes the axioms of a domain description. We start with a notion of *translation* for a literal.

DEFINITION 21 (Translation for Literals). Given a literal, l, we define $C(l, T)$, the *translation* of l at time T, as

$$C(l, T) := \begin{cases} in\_state(a, T) & \text{if } l = a \text{ is a positive literal,} \\ not\ in\_state(a, T) & \text{otherwise.} \end{cases}$$

DEFINITION 22 ($\mathfrak{Trans}(\mathcal{AX})$: Translation for Axioms). Given a planning problem $(\mathcal{S}, \mathbf{t}, \mathcal{D})$, $\mathfrak{Trans}(\mathcal{AX})$ is the logic program segment that contains the following rules: for all " $a \leftarrow l_1, \ldots, l_n$ " $\in \mathcal{AX}$,

$$in\_state(a, T) \ :- \ C(l_1, T), C(l_2, T), \ldots, C(l_n, T),$$

where $C(l_i, T)$ is the translation of the literal l_i, as defined in Definition 21 above.

Encoding the Operators and the State Transitions

SHOP uses the operator descriptions in $\mathcal{D}$ in decomposition of primitive tasks that needs to be accomplished during planning. In the translation, the logic program segment that encodes the operators in $\mathcal{D}$ is given as follows:

DEFINITION 23 ($\mathfrak{Trans}(\mathcal{OP})$: Translation for Operators). Given a planning problem $(\mathcal{S}, \mathbf{t}, \mathcal{D})$, for all $\mathsf{Op} \in \mathcal{OP}$, $\mathfrak{Trans}(\mathsf{Op})$ is the logic program that contains the following rules:

$$out\_state(a, T + 1) \ :- \ taskTBA(h, T), \quad \forall a \in \mathrm{Del}(\mathsf{Op})$$
$$in\_state(a, T + 1) \ :- \ taskTBA(h, T), \quad \forall a \in \mathrm{Add}(\mathsf{Op}),$$

where h is a primitive task – i.e., the ground head of the operator that is used in the decomposition of h.

Note that these two rules encode the delete- and the add-lists of the operator respectively. The result of a decomposition of a primitive task by an operator application is a new state of the world, which is generated by deleting all of the atoms that are in the delete-list of that operator from the current state and by adding all of the atoms that are in the add-list of that operator to the current state.

An operator only describes the *change* it causes to occur in the current state. The planner is still responsible for keeping track of the other facts that remain *unchanged* after an operator application. This is known as the famous *Frame Problem* in AI Planning. In the translation, the logic program segment that addresses the frame problem is defined as follows:

DEFINITION 24 ($\mathfrak{Trans}(\mathbf{F})$: Keeping Track of the State $\mathcal{S}$). The logic program segment $\mathfrak{Trans}(\mathbf{F})$ that encodes the frame axiom is defined as follows:

$$in\_state(A, T+1) \quad :- \quad in\_state(A, T), \, not \, out\_state(A, T+1).$$

Note that the state of the world in *SHOP* consists of only positive ground primary atoms. Basically the above rule states that if a positive ground atom, a, is initially true, then it should be true in the next state unless it has been marked as to be deleted during the transition from the current state to the next state.

Encoding the Methods

SHOP uses the method descriptions in $\mathcal{D}$ in decompositions of compound tasks that need to be accomplished during planning. Before proceeding with the definition of the logic program segment that encodes the methods in $\mathcal{D}$, we give the following definition:

DEFINITION 25 (Methods for a Compound Task). Given a planning problem $(\mathcal{S}, \mathbf{t}, \mathcal{D})$ and a compound task $h \equiv (name_h \ arg_1 \ arg_2 \ \ldots \ arg_N)$, we define a special predicate $method\_n\_i(name_h, arg_1, arg_2, \ldots, arg_N, T)$ for each method $m_i \in \mathcal{D}$ whose head unifies with h. The symbol n in the predicate name denotes the number of arguments of the predicate, i.e. $n = N+2$.

For purposes of clarity, we will use the following shorthand notations in the rest of this chapter: Given a planning problem $(\mathcal{S}, \mathbf{t}, \mathcal{D})$ and a compound task h, if there is only one method m whose head unifies with h in $\mathcal{D}$ then we will use $method(h, T)$ to refer to m, instead of the $method\_n\_1(name_h, arg_1, arg_2, \ldots, arg_N, T)$ predicate as defined above. If there are more than one method m_i for h then we will use the notation $method_i(h, T)$ for each such method m_i.

We present the translation for methods of *SHOP* in four steps: (1) the translation for the nondeterministic choice of applying alternative methods to a particular compound task; (2) the translation for evaluating the preconditions of a method in order to decide whether it is applicable to a particular compound task; (3) the translation for the task decomposition specified by a method; and (4) the translation for the accomplishment of a particular task by a method. Given a *SHOP* method, if the translations in these four steps are performed, then we produce a logic program segment, $\mathfrak{Trans}(\mathcal{METH})$, which is the ASP encoding of that method.

DEFINITION 26 (Translation for Encoding Alternative Methods). Given a planning problem $P = (\mathcal{S}, \mathbf{t}, \mathcal{D})$, let h be a compound task that needs to be accomplished in the solution of P. Suppose $\mathcal{D}$ contains N methods whose heads unify with h; namely, m_1, m_2, $\ldots$, m_N. Then, the logic program segment that encodes the nondeterministic choice of which method to apply to the task h is as follows: for $i, j = 1, \ldots, N$,

$$
\begin{aligned}
method_i(h, T) \quad :- \quad & \bigwedge_{i \neq j} not \; method_j(h, T), \\
& taskTBA(h, T).
\end{aligned}
$$

Intuitively, the rules in Definition 26 enforce the translation to create N different answer sets for N possible method applications to the task h. This is due to the fact that each such method specifies a different way to decompose the task h, and therefore, the planner can find different solutions due to each such method.

Next, we present the translation for evaluating the preconditions of each such method. Note that if the precondition of the method is not satisfied in the state of the world, then the method cannot be applied to the task h.

DEFINITION 27 (Translation for Precondition Evaluations). Let h be a compound task that needs to be accomplished, m_i be a method whose head unifies with h, and ρ be the preconditions of m_i. Then, we have two steps:

1. Let $p \in \rho$ be a precondition of m and $\chi_1, \chi_2, \ldots, \chi_f$ be the unbound variables in p such that if $f = 0$ then p has no such variables. Suppose that R_j denotes the *range* of χ_j in p – i.e., R_j is the set of all possible instantiations of χ_j in the world. Then, for each unbound variable χ_j of p, we create new variable symbols $\chi_{j,k}$ such that $j = 1, \ldots, f$ and $k = 1, \ldots, |R_j|$.

2. For each precondition $p \in \rho$,

   - if p contains no unbound variables or it is a negative literal, then

   $$
   checked\_state(p, T) \quad :- \quad C(p, T), method_i(h, T).
   $$

   where $C(p, T)$ is as defined in Definition 21.

- Otherwise, we have

$$
\begin{aligned}
checked\_state(p(\chi_{1,1}, \chi_{2,1}, \ldots, \chi_{f,1}), T) \ : \ - \\
method_i(h, T), \ in\_state(p(\chi_{1,1}, \chi_{2,1}, \ldots, \chi_{f,1}), T), \\
not\, checked\_state(p(\chi_{1,1}, \chi_{2,1}, \ldots, \chi_{f,2}), T), \\
\vdots \\
not\, checked\_state(p(\chi_{1,1}, \chi_{2,1}, \ldots, \chi_{f,|R_f|}), T), \\
\vdots \\
not\, checked\_state(p(\chi_{1,1}, \chi_{2,|R_2|}, \ldots, \chi_{f,|R_f|}), T), \\
not\, checked\_state(p(\chi_{1,2}, \chi_{2,1}, \ldots, \chi_{f,1}), T), \\
\vdots \\
not\, checked\_state(p(\chi_{1,|R_1|}, \chi_{2,|R_2|}, \ldots, \chi_{f,|R_f|}), T), \\
\bigwedge_{j=1}^{f} \chi_{j,1} \ ! = \chi_{j,2} \ ! = \ldots \ ! = \chi_{j,|R_j|}.
\end{aligned}
$$

where $\chi_{j,k}$ are the new variable symbols created for the unbound variables $\chi_1, \chi_2, \ldots, \chi_f$ in p, and $p(\cdot)$ denote the atoms that are the same as p except that, each $p(\cdot)$ contains the new variable symbols $\chi_{j,k}$ such that $j = 1, \ldots, f$ and $k = 1, \ldots, |R_j|$.

Intuitively, the rules of Definition 27 create an answer set for each possible instantiation of the unbound variables of m_i in the world. This is due to the fact that *SHOP* creates an instance of m_i for each instantiation of the unbound variables in m_i, and decomposes the task h with each such method instance. In order for our translation to be correct, we need to simulate this behavior of *SHOP* in our translation since ASP systems do not provide such semantics, to the best of our knowledge.

Note that for each precondition $p \in \rho$, there must be at least one answer set in which both $method_i(h, T)$ and $checked\_state(p, T)$ are true for the method m_i to be applicable to the task h in the current state of the world (denoted by the time variable T). If there is no such answer set then it means that m_i is not applicable to h.

Now that we have established the rules for checking the applicability of m_i to h, we are ready to give the definition for the decomposition of the task h by m_i.

DEFINITION 28 (Translation for Method Decomposition). Let h be a compound task that needs to be accomplished, and let m_i be a method that is applicable to h. Then, the decomposition of h by m_i is encoded by the following rules:

$$taskTBA(t_1, T) \quad :- \quad method_i(h, T),$$
$$\bigwedge_{p \in \rho} checked\_state(p(\chi_1, \chi_2, \ldots, \chi_f), T).$$
$$taskTBA(t_2, T_2) \quad :- \quad method_i(h, T),$$
$$\bigwedge_{p \in \rho} checked\_state(p(\chi_1, \chi_2, \ldots, \chi_f), T),$$
$$CAUSABLE(t_1, T, T_2),$$
$$T_2 \geq T.$$

$$\vdots \qquad \vdots \qquad\qquad \vdots$$

$$taskTBA(t_n, T_n) \quad :- \quad method_i(h, T),$$
$$\bigwedge_{p \in \rho} checked\_state(p(\chi_1, \chi_2, \ldots, \chi_f), T),$$
$$CAUSABLE(t_{n-1}, T_{n-1}, T_n),$$
$$T_n \geq T_{n-1}.$$

where $\chi_1, \ldots, \chi_f$ are the unbound variables of the precondition p, and $t_1, \ldots, t_n$ are the subtasks of m –i.e., the ordered list of tasks in the decomposition list of m.

The translation in Definition 28 encodes the fact that the decomposition of each subtask t_k can only be done if the previous subtask t_{k-1} has been already accomplished – i.e. t_{k-1} has been already $CAUSABLE$. The only exception is the first task, which is decomposed *if and only if* the particular method, h, has been chosen to be applied to the current task in the planning process. This property is encoded by using the $CAUSABLE(t_k, T_k, T_{k+1})$ construct for each subtask t_k (see Definition 18). The time point T_k denotes the time when the current task is decomposed and T_{k+1} denotes the time when it is accomplished – i.e. *causable*. In this way, we can identify the exact place where a specific task is causable in the entire search tree.

Finally, we have the definition of the accomplishment of a task by a particular method.

DEFINITION 29 (Translation for Accomplishment of Compound Tasks). Let h be a compound task that needs to be accomplished, let m_i be a method that is applicable to h, and suppose that h has been already decomposed into its subtasks t_k specified by the decomposition list of m_i. Then, the accomplishment (i.e., causation) of h by the method m_i is encoded as follows:

$$causable(h, T, T_{n+1}) \quad :- \quad method_i(h, T),$$
$$\bigwedge_{p \in \rho} checked\_state(p(\chi_1, \chi_2, \ldots, \chi_f), T),$$
$$CAUSABLE(t_n, T_n, T_{n+1}),$$
$$T_{n+1} \geq T_n.$$

where $\chi_1, \ldots, \chi_f$ are the unbound variables of the precondition p of m.

Intuitively, h is accomplished (i.e., caused) when/if the last subtask in its decomposition by m_i is accomplished (i.e., caused.) This is a direct consequence of Theorem 14.

Note that some of the rules given Definitions 26 and 27 could also be encoded into disjunctive rules, which seems to be conceptually simpler. However, not all ASP systems do handle disjunctions. Therefore, we decided to use non-disjunctive rules. At this point, we refer to the discussion in subsection 5.2.

4.3 A Translation Example: An Elevator Domain

One of the planning domains in the AIPS-2000 planning competition was the Miconic-10 Elevator domain. In order to accommodate the representational power of different planning systems, several different versions of this domain were used in the competition. The simplest version, which we will call Miconic-10-simtest,[8] has the following specifications: (1) the planner simply has to generate plans to serve a group of passengers of whom the origin and destination floors are given, and (2) there are no constraints such as satisfying space requirements of passengers or achieving optimal elevator controls.

Below, we describe how to use the techniques in the previous section to translate an HTN version of the domain into an ASP encoding. For the sake of simplicity and clarity, we present our translation methodology on a simplified and modified version of this problem domain in *Smodels* syntax. Some of the *SHOP* axioms, operators, and methods for this problem domain are shown in Figures 3, 4, and 5, respectively. Our complete encodings of the original Miconic-10-simtest domain for both *DLV* and *Smodels* are available at <http://www.in.tu-clausthal.de/index.php?id=aspplan>.

In our modified elevator example, there is only one person to be transported in a five floor building. The elevator starts its operation at the ground floor. Our passenger is at the top floor and wants to go down to the ground floor. The elevator can move between any two floors in one step; however, this movement can be either slow or fast, depending on the distance between those floors. The fast movement of the elevator depends on the amount of energy available to it. The elevator has initially enough energy for such movements. However, a fast movement decreases the total energy of the elevator by a specific amount. More specifically, a fast movement between two adjacent floors consumes one unit of energy. Unlike fast movements, slow movements do not require energy. The elevator always makes slow movements when it is empty, in order to conserve energy.

---

[8]We use this name because the domain is available at <http://www.informatik. uni-freiburg.de/~koehler/elev/simtests.tar.gz>.

Now, we will describe the basics of the translation process step by step as described in the previous section.

Prelude

We have to formulate all of the possible atoms that can ever be used during the planning process. Due to the fact that each variable in *Smodels* semantics must have a range of values, we must also define type predicates in our translation as described in the previous section. In a *SHOP* domain description, we do not need to make these definitions about the set of all possible atoms and tasks, nor about the type predicates.

In our elevator example, we have the following rules for specifying the possible atoms:

$$
\begin{aligned}
atom(boarded(P)) \quad &: -\quad person(P). \\
atom(goal(P)) \quad &: -\quad person(P). \\
atom(lift\_at(F)) \quad &: -\quad floor(F). \\
atom(destination(P, F)) \quad &: -\quad person(P), floor(F). \\
atom(on\_floor(P, F)) \quad &: -\quad person(P), floor(F). \\
atom(on\_lift(P)) \quad &: -\quad person(P). \\
atom(total\_energy(E)) \quad &: -\quad energy\_levels(E).
\end{aligned}
$$

And also the following rules for the type predicates such as:

$$
\begin{aligned}
time(0..10) \quad &: - \\
person(p0) \quad &: - \\
floor(0..4) \quad &: - \\
energy\_levels(0..10) \quad &: -
\end{aligned}
$$

Note that we define the time line of our logic program to be the set $\{0, 1, \ldots, 10\}$. This definition is just for illustrative purposes in this example. Normally, we use one of the two techniques for defining the time lines, as described in the previous section.

Encoding the Initial State

The logic program segment $\mathfrak{Trans}(\mathcal{S})$ consists of the following rules to specify the initial state in our encoding of the elevator example:

$$
\begin{aligned}
in\_state(lift\_at(0), 0) \quad &: - \\
in\_state(goal(p0), 0) \quad &: - \\
in\_state(on\_floor(p0, 4), 0) \quad &: - \\
in\_state(destination(p0, 0), 0) \quad &: - \\
in\_state(total\_energy(10), 0) \quad &: -
\end{aligned}
$$

As it can be seen from these rules, they specify certain ground atoms to be in the initial state of the planner (Definition 16 in the previous section).

```
;; SHOP Axioms for the simplified Miconic-10-simtest Domain.
(:- (can-move-fast ?floor1 ?floor2)
    ((> ?floor1 ?floor2)(total_energy ?energy)
     (≥ (- ?floor1 ?floor2) ?energy))
(:- (can-move-fast ?floor1 ?floor2)
    ((≤ ?floor1 ?floor2)(total_energy ?energy)
     (≥ (- ?floor2 ?floor1) ?energy)))
(:- (floorDiff ?floor1 ?floor2 ?d)
    ((> ?floor1 ?floor2)(assign ?d (- ?floor1 ?floor2)))
(:- (floorDiff ?floor1 ?floor2 ?d)
    ((≤ ?floor1 ?floor2)(assign ?d (- ?floor2 ?floor1))))
```

Figure 3. Examples of the *SHOP* axioms for the simplified version of Miconic-10-simtest planning domain.

The last argument for each $in\_state(A, T)$ predicate is the time T at which the atom A holds. As mentioned before, we define the starting time of the planning process to be 0.

The frame axiom $\mathfrak{Trans}(\mathbf{F})$ is as follows:

$$in\_state(A, T+1) \quad :- \quad time(T), atom(A), in\_state(A, T),$$
$$not\ out\_state(A, T+1).$$

Encoding the Goal Task(s).

Suppose we have a single goal task to accomplish, namely the task of transporting our passenger. Then, the logic program segment $\mathfrak{Trans}(\mathbf{t})$ will encode this task via the following rule:

$$taskTBA(transport\_person, p0, 0) \quad :-$$

This rule specifies that our goal task to be accomplished at the beginning of planning process is the task $(transport\_person\ p0)$ — i.e., the task of transporting the person $p0$.

Encoding the Axioms

Suppose that we have the axioms shown in Figure 3. The intended meaning of these axioms is to decide whether the elevator can make a fast movement between the specified two floors. The criteria for this decision is that if the distance between the two floors is greater than or equal to the total energy of the elevator, then it can move fast between these two floors. The encoding

```
;; SHOP Operators for the simplified Miconic-10-simtest Domain.
  (:operator (!markServed ?person)
      ((goal ?person))
      ())
  (:operator (!moveSlow ?floor1 ?floor2)
      ((lift_at ?floor1))
      ((lift_at ?floor2)))
  (:operator (!moveFast ?floor1 ?floor2 ?old ?new)
      ((lift_at ?floor1)(total_energy ?old))
      ((lift_at ?floor2)(total_energy ?new))
  (:operator (!board ?person ?floor)
      ((on ?person ?floor))
      ((on_lift ?person)))
  (:operator (!debark ?person ?floor)
      ((on_lift ?person))
      ((on ?person ?floor)))
```

Figure 4. Examples of the *SHOP* operators for the simplified version of Miconic-10-simtest planning domain.

of this axiom as the logic program segment $\mathfrak{Trans}(\mathcal{AX})$ is straightforward:

$$in\_state(can\_move\_fast(F1, F2), T) \quad :- \quad time(T), floor(F1),$$
$$floor(F2), energy\_level(E),$$
$$in\_state(total\_energy(E), T),$$
$$F1 > F2, F1 - F2 \geq E.$$

$$in\_state(can\_move\_fast(F1, F2), T) \quad :- \quad time(T), floor(F1),$$
$$floor(F2), energy\_level(E),$$
$$in\_state(total\_energy(E), T),$$
$$F1 \leq F2, F2 - F1 \geq E.$$

$$in\_state(floorDiff(F1, F2, F1 - F2), T) : - \quad time(T), floor(F1),$$
$$floor(F2), F1 > F2.$$

$$in\_state(floorDiff(F1, F2, F2 - F1), T) : - \quad time(T), floor(F1),$$
$$floor(F2), F1 \leq F2.$$

Encoding the Operators

Suppose that in the domain description of our elevator example, we have the planning operators shown in Figure 4.

The first operator in Figure 4 is for the primitive task of marking a person served. This operator basically removes the (*goal ?person*) atom from the state of the world, which means that the goal of transporting the person *?person* has been achieved. Note that this operator does not add any atoms to the state of the world. The second operator is for the primitive task of moving the elevator slowly from one floor to another. It simply deletes the $(lift\_at \ ?floor1)$ atom from the state, which describes the location of the elevator before it started its move, and adds the atom $(lift\_at \ ?floor2)$, which describes the location of the elevator after it completed its move. The third operator is for the primitive task of moving the elevator fast. Note that this operator also changes the total amount of energy that the elevator has. The fourth and the fifth operators are for boarding a person to the elevator and for debarking a person from the elevator, respectively.

In our translation, the markServed operator is encoded by a single rule:

$$out\_state(goal(P), T+1) \quad : - \quad time(T), person(P),$$
$$taskTBA\_3(markServed, P, T).$$

The encoding of the moveSlow operator corresponds the following:

$$out\_state(lift\_at(F1), T+1) \quad : - \quad time(T), floor(F1), floor(F2),$$
$$taskTBA\_4(moveSlow, F1, F2, T).$$
$$in\_state(lift\_at(F2), T+1) \quad : - \quad time(T), floor(F1), floor(F2),$$
$$taskTBA\_4(moveSlow, F1, F2, T).$$

The moveFast, board, and debark operators are encoded similarly.

Encoding the Methods

We have the following four methods in our domain description: We have two methods for fast transporting the person from his/her original floor to his/her destination floor. The first method is for the case in which the elevator and the person are on the same floor, so the person can be immediately boarded to the elevator and transported to his/her destination. The second method is for the case in which the elevator and the person are not on the same floor; the elevator must be first moved to the floor of the person so that the person can be transported. Figure 5 shows the *SHOP* specification of these two methods.

We have also two methods for slow transportation of the person, similar to the ones described above. The two groups of methods – i.e., the first two and the second two – correspond to a branching (i.e, backtracking choice) point in the planner's search space, in which different branches may possibly lead to different solution plans. However, the first two methods cannot both yield to a solution due to the way their preconditions are defined. The same is true for the second two as well.

According to Definition 26, the translation of the nondeterministic choice among these four methods is the following set of rules, which correspond to the same branching point in the search space. Here, we only give the rule(s)

```
;; SHOP Methods for the simplified Miconic-10-simtest Domain.
;; Method for FAST transporting a person when the lift is at the same floor
;; as him/her.
 (:method (transport_person ?person)
   ;; preconditions
    ((lift_at ?floor1)(on ?person ?floor1)(destination ?person ?floor2)
     (total_energy ?old)(can_move_fast ?floor1 ?floor2)
     (floorDiff ?floor1 ?floor2 ?d)(assign ?new (- ?old ?d)))
   ;; decomposition task list
    (((!board ?person ?floor1) (!moveFast ?floor1 ?floor2 ?old ?new)
      (!debark ?person ?floor2) (!markServed ?person)))

 ;; Method for FAST transporting a person when the lift is not at the same floor
 ;; as the person.
  (:method (transport_person ?person)
   ;; preconditions
    ((lift_at ?floorX)(on ?person ?floor1)(destination ?person ?floor2)
     (total_energy ?old)(can_move_fast ?floor1 ?floor2)
     (floorDiff ?floor1 ?floor2 ?d)(assign ?new (- ?old ?d)))
   ;; decomposition task list
    (((!moveSlow ?floorX ?floor1)(!board ?person ?floor1)
      (!moveFast ?floor1 ?floor2 ?old ?new)(!debark ?person ?floor2)
      (!markServed ?person)))
```

Figure 5. Examples of the *SHOP* methods for the simplified version of Miconic-10-simtest planning domain.

for the first method; the others are almost identical.

$$
\begin{aligned}
method\_3\_1(transport\_person, P, T) \ : - \\
time(T), person(P), taskTBA\_3(transport\_person, P, T), \\
not \, method\_3\_2(transport\_person, P, T), \\
not \, method\_3\_3(transport\_person, P, T), \\
not \, method\_3\_4(transport\_person, P, T).
\end{aligned}
$$

We now describe the encoding for evaluating the preconditions of this method. For that matter, we need to encode the rules given by Definition 27 for every precondition of this method. As an example, consider the first precondition, which is $(lift\_at \ ?floor1)$ (see Figure 5). This precondition has only one unbound variable; namely, $?floor1$. Then according to Definition 27, we have the following rule:

$$
\begin{aligned}
checked\_state(lift\_at(F1, T) \ : - \ \ method\_3\_1(transport\_person, P, T), \\
in\_state(lift\_at(F1), T).
\end{aligned}
$$

Note that, although $?floor1$ is an unbound variable, we did not create
new variable symbols for it and did not use the first rule given in Definition 27. The reason for this is that the elevator can be at one and only one
floor at any time in this domain.

The rules for the other preconditions are almost identical to the one
above, so we do not give them here due to space limitations (for a complete
encoding of this domain, see <http://www.in.tu-clausthal.de/index.
php?id=aspplan>. This finishes our encoding of the precondition evaluation
for the first method of Figure 5.

The following rules encode the decomposition list of this method:

$taskTBA\_4(board, P, F1, T) \ : -$
    $floor(F1), floor(F2), person(P), energy\_level(Old), number(D),$
    $time(T), \ checked\_state(lift\_at(F1), T), \ checked\_state(on(P, F1), T),$
    $checked\_state(destination(P, F2), T),$
    $checked\_state(total\_energy(Old), T),$
    $checked\_state(can\_move\_fast(F1, F2), T),$
    $checked\_state(floorDiff(F1, F2, D), T),$
    $method\_3\_1(transport\_person, P, T).$

$taskTBA\_6(moveFast, F1, F2, Old, Old - D, T2) \ : -$
    $floor(F1), floor(F2), person(P), energy\_level(Old), number(D),$
    $time(T), time(T2), checked\_state(lift\_at(F1), T),$
    $checked\_state(on(P, F1), T), \ checked\_state(destination(P, F2), T),$
    $checked\_state(total\_energy(Old), T),$
    $checked\_state(can\_move\_fast(F1, F2), T),$
    $checked\_state(floorDiff(F1, F2, D), T),$
    $method\_3\_1(transport\_person, P, T),$
    $causable\_5(board, P, F1, T, T2), \ T2 \geq T.$

$taskTBA\_4(debark, P, F2, T3) \ : -$
    $floor(F1), floor(F2), person(P), energy\_level(Old), number(D),$
    $time(T), time(T2), time(T3), energy\_level(New),$
    $checked\_state(lift\_at(F1), T), \ checked\_state(on(P, F1), T),$
    $checked\_state(destination(P, F2), T),$
    $checked\_state(total\_energy(Old), T),$
    $checked\_state(can\_move\_fast(F1, F2), T),$
    $checked\_state(floorDiff(F1, F2, D), T),$
    $method\_3\_1(transport\_person, P, T),$
    $causable\_7(moveFast, F1, F2, Old, New, T2, T3), \ T3 \geq T2.$

$taskTBA\_3(markedServed, P, T4)$: −
  $time(T), time(T3), time(T4), person(P), floor(F1), floor(F2),$
  $energy\_level(Old), number(D), method\_3\_1(transport\_person, P, T),$
  $checked\_state(lift\_at(F1), T),$ $checked\_state(on(P, F1), T),$
  $checked\_state(destination(P, F2), T),$
  $checked\_state(total\_energy(Old), T),$
  $checked\_state(can\_move\_fast(F1, F2), T),$
  $checked\_state(floorDiff(F1, F2, D), T),$
  $causable\_5(debark, P, F2, T3, T4),$ $T4 \geq T3.$

The rules for the decomposition list of the method define the successor subtasks with the order they were specified in that method. Note that in the formalism for HTN planning with Ordered Task Decomposition, the ordering of the subtasks enforces the fact that a subtask t can be selected as the current task for decomposition only if all of the subtasks preceding t are accomplished successfully. This is achieved in our translation by the causable properties of the tasks (see Definition 13).

Note also that although the first method of Figure 5 has an unbound variable ?*new*, we have not encoded this variable as an unbound variable in our rules. This is due to the fact that this variable is used in the method for storing the energy that is left after the elevator moves fast between two levels, and we can encode this as we did in the head of the second rule above. In fact, *SHOP*'s assign statement (see Figure 5) serves the same purpose; it was a design choice in *SHOP* to handle these cases in the preconditions of a method using an assign statement, rather than handling them in the arguments of the subtasks as we did in our encodings.

At this point, we want to emphasise again that the translation method presented in the previous section is a general technique; one can make several optimisations and modifications during actual implementation.

Now, we are ready to give the rule for accomplishment of the task of transporting the person via the method encoded above:

$causable\_4(transport\_person, P, T, T5)$: −
  $time(T), time(T4), time(T5), person(P), floor(F1), floor(F2),$
  $energy\_level(Old), number(D), method\_3\_1(transport\_person, P, T),$
  $checked\_state(lift\_at(F1), T),$ $checked\_state(on(P, F1), T),$
  $checked\_state(destination(P, F2), T),$
  $checked\_state(total\_energy(Old), T),$
  $checked\_state(can\_move\_fast(F1, F2), T),$
  $checked\_state(floorDiff(F1, F2, D), T),$
  $causable\_4(markedServed, P, T4, T5),$ $T5 \geq T4.$

5　Results: Theory and Practice

In this section, we present our theoretical results on the correctness and completeness of our translation method. These results in soundness and completeness theorems are for the resulting logic programs under the answer set semantics for planning problems. We also describe in detail the experiments we have undertaken. All the detailed formalisations as well as more implementation related information can be obtained from <http://www.in.tu-clausthal.de/index.php?id=aspplan>.

5.1　Soundness and Completeness

Our first theorem states that our translation indeed corresponds to HTN planning as done in *SHOP*. Soundness and completeness are the two important requirements for any planning system. *Soundness* means that all of the plans that are generated by the planner are actually true solutions to the given planning problem; that is, no plan, which is not solution to the problem, should be generated. *Completeness* means that the planning system must be able to generate all of the possible plans (solutions) for the given planning problem. A more formal treatment and the fact that *SHOP* is sound and complete is contained in [Nau *et al.*, 2000; Nau *et al.*, 1999].

Let $\mathfrak{Trans}(\cdot)$ be the translation method described in the previous section. Given any HTN planning problem described in *SHOP*'s formalism, we are interested in the relationship between the solutions to the problem and the models (or answer sets) of $\mathfrak{Trans}(\cdot)$.

THEOREM 30 ($\mathfrak{Trans}(\cdot)$ and HTN planning using OTD). *Given a planning problem* $(\mathcal{S}, \mathbf{t}, \mathcal{D})$, *where* $\mathcal{S}$ *is the initial state,* $\mathbf{t}$ *is the list of ground tasks to be accomplished and* $\mathcal{D}$ *is the domain description, let* $\mathfrak{Trans}((\mathcal{S}, \mathbf{t}, \mathcal{D}))$ *be the corresponding logic program with negation. We assume that the set of axioms in* $\mathcal{D}$ *does not contain any cycles through negation. Furthermore, let* $\mathbf{Sol}(\mathcal{S}, \mathbf{t}, \mathcal{D})$ *be the set of solutions as defined in Definition 11. Then,*

1. $\mathbf{Sol}(\mathcal{S}, \mathbf{t}, \mathcal{D}) = \emptyset$ *if and only if* $\mathfrak{Trans}((\mathcal{S}, \mathbf{t}, \mathcal{D}))$ *has no answer sets.*

2. *If* $\mathbf{Sol}(\mathcal{S}, \mathbf{t}, \mathcal{D}) \neq \emptyset$, *then the following holds:*

    (a) *For every plan* $P \in \mathbf{Sol}(\mathcal{S}, \mathbf{t}, \mathcal{D})$, *there is an answer set of* $\mathfrak{Trans}((\mathcal{S}, \mathbf{t}, \mathcal{D}))$ *and a sequence of primitive tasks* $t_0, t_1, \ldots, t_n$, *such that the predicates* $taskTBA(t_i, i)$ *that are true in this answer set and the* t_i *correspond exactly to the steps* p_i *in* P.

    (b) *For every answer set of* $\mathfrak{Trans}((\mathcal{S}, \mathbf{t}, \mathcal{D}))$ *there is a sequence of primitive tasks* $t_0, \ldots, t_n$, *such that the predicates* $taskTBA(t_i, i)$

are true in this answer set, and this sequence constitutes a plan
$[t_0, \ldots, t_n] \in \mathbf{Sol}(\mathcal{S}, \mathbf{t}, \mathcal{D})$.

Proof. The proof is in three steps:

Step 1: We take a close look at the *search tree*, introduced in Definition 10, which gives a formal and handy description of the causal theory introduced in Section 4. Through Theorem 14, this search tree is linked with our planning problem as introduced in Definition 8.

Step 2: Here we list some facts about the rules used in the translation $\mathfrak{Trans}(\cdot)$. They will be used in the next step.

Step 3: We show by induction the precise relationship between branches in the search tree and the existence of stable models and the predicates true in them.

We now give the details of the steps.

Step 1

It is immediate that the 2nd entry in the triple $\langle \mathcal{S}', \mathbf{t}_{caused}, \mathbf{t}' \rangle$ represents the caused tasks (as defined in Definition 13): they are all ground and primitive. The operators that are applied to these tasks are the markings on the edges above that node $\langle \mathcal{S}', \mathbf{t}_{caused}, \mathbf{t}' \rangle$. There are also edges marked by methods: they just denote which method has been used in the process to decompose the first compound task in the 3rd entry of the node immediately above it.

As long as the 3rd entry $\mathbf{t}'$ is not yet empty, the tree is expanded until a successful branch is built up. Of course, it can also lead to (1) a dead end: the final node might be unsuccessful, or (2) it might never end and the same methods are applied and lead to longer and longer lists of tasks in $\mathbf{t}'$.

By the very construction of the tree and by the Definition 13, we have the following

$$\mathbf{t} \text{ is causable wrt. } (\mathcal{S}, \mathcal{D})$$
$$\textit{if and only if}$$
$$\text{there is a successful branch for the tree for } \langle \mathcal{S}, \emptyset, \mathbf{t} \rangle.$$

Moreover, comparing the definition of the tree with the original definition of a planning problem, we get that all the plans $\pi \in \mathbf{Sol}(\mathcal{S}, \mathbf{t}, \mathcal{D})$ are obtained by traversing the successful branches and putting together the edge labellings that correspond to the operator applications (note that these are the ground instances of primitive tasks, i.e., the actions in our plan).

Step 2

We now show formally the relation of the tree to the stable models of $\mathfrak{Trans}((\mathcal{S}, \mathbf{t}, \mathcal{D}))$. Several things are worth noticing before we give the formal proof.

1. All the predicates used in $\mathfrak{Trans}((\mathcal{S}, \mathbf{t}, \mathcal{D}))$ carry as last argument a time variable T. This is used, informally, to denote the time when this predicate is active. In a stable model where $taskTBA(t_1, 5)$ holds, this is interpreted as *task t_1 is to be accomplished at time* 5. Similarly, $method_3(h, 7)$ means that the third method for the task h is selected to decompose h at time 7.

2. The first task at time 0 is h_1 in $\mathbf{t}'$ (forced by Definition 19) in all stable models of $\mathfrak{Trans}((\mathcal{S}, \mathbf{t}, \mathcal{D}))$.

3. The state predicates represent the current state at any point in time: $in\_state(a, 5)$ means that atom a is true at time 5, $out\_state(a, 5)$ means that a is not true at time 5. Note that the initial state is encoded in Definition 16, and all changes (when operators are applied) are formalised in Definition 23: $out\_state(.)$ is responsible for the delete lists. The $checked\_state$ predicate uses the state predicates to check whether the preconditions of a method are satisfied in the state at time T.

4. The first part of $\mathfrak{Trans}((\mathcal{S}, \mathbf{t}, \mathcal{D}))$ defined in Definition 26 ensures the following: in all stable models of $\mathfrak{Trans}((\mathcal{S}, \mathbf{t}, \mathcal{D}))$, for all time points T, exactly one of the methods $m_1, \ldots, m_N$ that are applicable to the current task is selected at time T to decompose it. This is because stable models are minimal.

5. We note that the $taskTBA$ predicate can be true for several tasks at a particular time point. Suppose we have a stable model of $\mathfrak{Trans}((\mathcal{S}, \mathbf{t}, \mathcal{D}))$ where $taskTBA(t, 5)$ holds and t is a compound task. Suppose further that a method m is selected to decompose t (i.e. $method(t, 5)$ holds in that stable model), and that the preconditions of m are true in the current state so t can be decomposed into t_1, t_2. Then, because of the first rule in Definition 28, $taskTBA(t_1, 5)$ is true as well. If t_1 were a compound task and a method decomposes it into other tasks, then the first of these tasks would again be the current task at time 5. This goes on as long as a primitive task is found (which would also be true at time 5). In that case, there are no rules of the form in Definition 26 available (since these rules are only stated for compound tasks). Then the time T is incremented by one, due to the 2nd case in the definition of $CAUSABLE$.

6. $causable(h, T_1, T_{n+1})$ informally means that the task h is caused at time point T_{n+1} using a method, which was decomposed at time T_1. Such a predicate is true in a stable model only if there is a path on the search tree in which h has been decomposed and its subtasks have all been caused.

7. Note that the two rules mentioning the predicate *plan_found* (see Step 5. above) ensure that either there is at least one stable model (in which the causable predicate is true) or there are no stable models at all (because there is a negative cycle). This is the only place where there is a potential cycle through negation which could lead to the nonexistence of stable models.

Step 3

Let us consider the translation $\mathfrak{Trans}^*((\mathcal{S}, \mathbf{t}, \mathcal{D}))$, which is exactly like $\mathfrak{Trans}((\mathcal{S}, \mathbf{t}, \mathcal{D}))$, except that the clause *plan_found : − not plan_found* is not included. Note that this is the only cycle through negation in $\mathfrak{Trans}((\mathcal{S}, \mathbf{t}, \mathcal{D}))$ which can be the cause of nonexistence of stable models. The other two places where there are potential conflicts are in $\mathfrak{Trans}(\mathcal{AX})$ and in $\mathfrak{Trans}(\mathcal{METH})$ (via the *checked_state* predicates and the *method* predicates in 1.). However, we explicitly assumed that $\mathcal{AX}$ is free of cycles through negation, and it can be easily seen that the complete instantiation of the *checked_state* predicates is also stratified. Also the rules for the methods ensure that if the *taskTBA* predicate is true, there are always stable models (there are only even cycles through negation).

It is trivial (but tedious) to show formally by induction on the length of a path in the search tree that the following holds:

1. If *path* is a path in the search tree for $(\mathcal{S}, \mathbf{t}, \mathcal{D})$, then there is an answer set *Ans* of $\mathfrak{Trans}^*((\mathcal{S}, \mathbf{t}, \mathcal{D}))$ such that the following holds:

    (a) For all tasks t and time point i: $taskTBA(t, i)$ holds in *Ans* if and only if task t (compound or not) occurs as the first task in $\mathbf{t}'$ at a node of task-depth i.

    (b) For all tasks t and time points i: $method_j(t, i)$ holds in *Ans* if and only if $method_j$ is the labelling of an edge at task-depth i of the path, for every $j = 1 \ldots N$.

    (c) For all tasks h and time points i, e, f: $causable(h, e, f)$ holds in *Ans* if and only if the following holds:

        • the task h has been decomposed at the path at task-depth e, and

- all h's successor tasks t_j are causable at time points g such that $e \leq g \leq f$ (i.e. $causable(t_j, \cdot, g_j)$ holds in Ans for all successor tasks t_j of h, where we have $e \leq g_j \leq f$).

    (d) For all atoms a and time point i: $in\_state(a, i)$ holds in Ans if and only if a is true in the state represented by the node of task-depth i in the path.

2. If Ans is an answer set of $\mathfrak{Trans}^*((\mathcal{S}, \mathbf{t}, \mathcal{D}))$, then there is a path in the search tree for $(\mathcal{S}, \mathbf{t}, \mathcal{D})$ such that the above properties 1(a)–1(d) hold.

Note that the above formulation includes the situation where the planning problem has no solution. In that case, there is no successful path in the search tree. But there still might be infinite paths. These are generated because tasks are decomposed without being replaced, eventually, by primitive tasks. So even for those paths, there are corresponding stable models (in which the time variable T is unbounded). Of course, the predicate $plan\_found$ does not hold in these stable models. Thus, if we include the rule $plan\_found : - \, not \, plan\_found$, then we get the desired result: successful branches exist if and only if there exist stable models. ∎

COROLLARY 31 (Soundness and Completeness of $\mathfrak{Trans}(\cdot)$). *The answer sets of $\mathfrak{Trans}((\mathcal{S}, \mathbf{t}, \mathcal{D}))$ correspond exactly to the plans in* **Sol** *($\mathcal{S}$,t,$\mathcal{D}$). There is a bijection between these two sets and each plan in* **Sol** *($\mathcal{S}$,t,$\mathcal{D}$) can be reconstructed from its corresponding answer set in $\mathfrak{Trans}((\mathcal{S}, \mathbf{t}, \mathcal{D}))$ and vice versa.*

COROLLARY 32 (Soundness and Completeness of $\mathfrak{Trans}(\cdot)$ wrt SHOP). *If the axioms $\mathcal{AX}$ in $\mathcal{D}$ do not contain any (positive or negative) cycles, then the answer sets of $\mathfrak{Trans}((\mathcal{S}, \mathbf{t}, \mathcal{D}))$ correspond exactly to the plans computed by SHOP .*

The corollary follows easily from the theorem and the fact that SHOP itself has been shown to be a sound and complete planner.

Note that the only reason why we assume the axiomatic theory in $\mathcal{D}$ is free of cycles is because the SHOP implementation cannot handle such axioms. In principle, such axioms would cause no problem for the abstract SHOP procedure, which makes no commitment about what kind of inference procedure to use. The soundness and completeness result for the abstract SHOP procedure [Nau *et al.*, 1999; Nau *et al.*, 2000] says something like "if the inference procedure is sound and complete, then so is the planning procedure." However, in the inference procedure used in the implementation of SHOP, SHOP would go into an infinite loop even for simple axioms like

$a \leftarrow a$. So our overall result (all plans returned by *SHOP* are also obtained using our ASP framework) still holds, even without this assumption. In fact, our method computes plans that an ideal version of *SHOP* would compute as well. It is thus complete for such a version of *SHOP*.

5.2 Experimental Study

In our experiments, we used the following three different planning domains:

The Simple-Travel Domain This domain is one of the domains included in the distribution of the *SHOP* planning system. The scenario for the domain, as described in [Nau *et al.*, 1999; Nau *et al.*, 2000], is that we want to travel from one location to another in a city. We have three locations: downtown, uptown, and park. There are two possible means of transportation: by taxi and by bus. Taxi travel involves hailing the taxi, riding to the destination and paying the driver $1.50 plus $1.00 for each mile travelled. Bus travel involves hailing the bus, paying the driver $1.00, and riding to the destination. Thus, different plans are possible depending on the weather conditions, the distance between our current location and the one we want to go, and how much money we have.

The Miconic-10-simtest Domain This is the domain as described in Section 4.3. It is contained in a series of benchmarks <http://www.informatik.uni-freiburg.de/~koehler/elev/elev.html> and it was recently used not only to measure the performance of various planners but also for other translation methods from planning problems into ASP (see http://www.fcs.nmsu.edu/~tson/asp_planner/>.

The Zeno-Travel Domain The Zeno-Travel problem was one of the domains that were introduced as recent benchmarks in International Planning Competition (IPC-2002).[9] This domain involves transporting people around in planes, using different modes of movement: fast and slow. There were four versions of this domain in the competition; namely STRIPS, NUMERIC, SIMPLE-TIME, and TIME. In the NUMERIC version, aircrafts consume fuel at different rates according to the mode of travel. Using a small set of symbolic fuel levels, the SIMPLE-TIME version manages to combine the benefits of fast travel (shorter journey times) with the associated costs (higher fuel consumption) that must be balanced with the cost of refuelling to arrive at time-efficient plans. In the TIME version, the domain uses

---

[9]IPC-2002 was organised within the Sixth International Conference on AI Planning and Scheduling 2002 (AIPS-2002). For more information on AIPS-2002 and IPC-2002, please see <http://www.dur.ac.uk/d.p.long/competition.html>.

Table 1. Comparison between our *Smodels* encoding of Miconic-10-simtest and the encoding described in [Son *et al.*, 2001]. In this table, "no solution" means that the computer used in these experiments ran out of memory.

Problem	Trans(·)	[Son *et al.*, 2001]
S1-0	0.160	8.29
S2-0	1.160	73.41
S3-0	4.450	162.24
S4-0	12.790	964.01
S5-0s1	44.090	no solution
S5-0s2	44.490	no solution
S6-0	46.300	no solution

numbers to encode fuel consumptions, dependent on distances, and speeds to calculate travel times in each travel mode. This version is essentially the domain used to illustrate the Zeno planning system developed by Penberthy and Weld [<http://www.cs.washington.edu/ai/zeno.html>]. In our experiments we used the STRIPS version of the domain.

We describe our experiments in the following subsections. In these experiments, we used the *Smodels* system v2.27 (which is available at <http://www.tcs.hut.fi/Software/smodels/>) and the *DLV* system (available at <http://www.dbai.tuwien.ac.at/proj/dlv/>). For the experiments in the *Smodels* system, we used *lparse* v1.0.11 as a grounding front-end. We ran our experiments on an HP Notebook PC with an AMD 900Mhz Processor and 256MB RAM running Linux RedHat v7.2 operating system. In all of our experiments we were finding all of the solutions to each planning problem.

Note that we also redid the experiments of [Son *et al.*, 2001] on our machine so that fair comparisons could be done. All of our source codes are available at <http://www.in.tu-clausthal.de/index.php?id=aspplan>. In our experiments on the Simple-Travel Domain using our method together with *DLV*, we got a speed-up of two orders of magnitude compared to *Smodels*.

Comparing our method with [Son et al., 2001]

This section describes our comparison of the time performance of the logic programs produced by using our translation methodology with that of the logic-program encodings presented in [Son *et al.*, 2001].

Note that the encodings proposed in [Son *et al.*, 2001] do not produce actual HTN encodings. Instead, they make use of only a few properties of HTNs, in order to implement some control knowledge in logic programs that perform action-based planning. In their paper, Son *et al.* showed that employing that control knowledge has improved the time performance of the logic program that encodes an action-based planner.

For our experimental comparison, we had to be very careful about how we wrote our HTN formulation of the planning domain. If two different formulations of a planning problem perform differently, then there are several different ways in which this can occur:

- The two formulations may be based on different ways of conceptualising the problem. For example, one formulation might involve reasoning about the movement of the elevator and where it needs to go next, and another formulation might involve reasoning about the movement of the people and where they need to go next. These two problem formulations would produce very different search spaces.

- The two different formulations may use basically the same tasks, and use them to mean basically the same thing. However, one formulation may take less time because it has lower overhead, or because it does a better job of deciding which tasks should actually be generated and explored.

For our experiments, we did not want to use a different conceptual representation than the one used by [Son *et al.*, 2001], because we wanted our experiments to test the performance of the two different approaches, not our ability to devise a clever conceptual representation! Thus, we were careful to write our HTN formulation of Miconic-10-simtest so that we used basically the same conceptual representation that they did.

The problems that we used in these experiments are from http://www. cs.nmsu.edu/~tson/asp_planner>. Table 1 shows both our results and the results from [Son *et al.*, 2001], which were also obtained on the *Smodels* system. Our results show that the logic programs produced by our translation methodology were about 1.5 to 2 orders of magnitude faster than the logic programs produced by the methodology described in [Son *et al.*, 2001]. In addition, our encoding was able to solve problems for which the encoding in [Son *et al.*, 2001] ran out of memory. In this respect, these results confirm the fact that a *SHOP*-like HTN planning approach is a much more effective way for solving planning problems if a good set of HTN methods is available, because the HTN methods constrain the size of the search space. They also illustrate that our translation method provides a way to produce efficient HTN-logic programs with ASP semantics to solve planning problems compared to action-based encoding methodologies in the style of [Son

Table 2. Comparing *Smodels* and *DLV* on the Simple-Travel Domain

Problem	*Smodels*	*DLV*	*DLV* grounding+*Smodels*
P1	3.430	0.050	0.040+0.020
P2	3.330	0.050	0.050+0.020
P3	3.190	0.030	0.030+0.000
P4	3.340	0.070	0.060+0.010
P5	3.410	0.060	0.050+0.030
P6	3.230	0.030	0.030+0.010
P7	3.340	0.050	0.050+0.010
P8	3.260	0.040	0.040+0.010
P9	3.230	0.040	0.040+0.010
P10	3.410	0.070	0.050+0.000
P11	3.340	0.050	0.040+0.000
P12	3.250	0.020	0.030+0.010
P13	3.410	0.070	0.060+0.010
P14	3.350	0.060	0.050+0.020
P15	3.270	0.030	0.030+0.000
P16	3.380	0.060	0.050+0.010
P17	3.300	0.050	0.050+0.010
P18	3.260	0.030	0.030+0.000

et al., 2001].

Comparing *Smodels* and *DLV* using planning benchmarks

We believe that our translation methodology provides more efficient logic programs with ASP semantics if the system on which those programs are implemented allows the usage of unbound variables in the programs, or, at least, produces an intelligent grounding. Otherwise, the system tries to make every rule ground in the input program, which decreases the efficiency of planning by causing a combinatorial explosion in the size of the search space.

As we described earlier, neither *Smodels* nor *DLV* is designed to work on the logic programs with unbound variables. However, *DLV*'s grounding differs from that of *Smodels*, and is generally believed to be *more intelligent* and smaller in size. Also, *DLV* is based on deductive database techniques and we expected a better handling of unbound variables. To test this hypothesis, we applied our translation methodology to our elevator and travelling examples. While testing this hypothesis, one must note the way the *Smodels* system solves a problem. *Smodels* uses a front-end called *lparse*

Table 3. Comparison of *Smodels* and *DLV* using 𝔗𝔯𝔞𝔫𝔰(·) on the Miconic-10-simtest Domain

Problem	*Smodels*	*DLV*	*DLV* grounding+*Smodels*
S1-0	0.160	0.040	0.030+0.010
S2-0	1.160	0.060	0.050+0.010
S3-0	4.450	0.080	0.010+0.090
S4-0	12.790	0.260	0.100+0.530
S5-0s1	44.090	0.640	0.080+1.540
S5-0s2	44.490	0.680	0.090+1.840
S6-0	46.300	0.980	0.170+3.560

to preprocess the programs in order to get them grounded. After the programs are grounded, the *Smodels* ASP solver solves the ground program itself. Therefore, the problems solved in this system may not reflect the actual performance of the ASP solver. To accommodate this issue, we also designed a set of experiments in which we used *DLV* to produce groundings of the programs. These groundings were then given to *Smodels*. The aim in these experiments was to determine the effect of grounding done by the front-end *lparse* on the overall performance of the *Smodels* system.

Tables 2 and 3 show our results on the Simple-Travel and Miconic-10-simtest problems. We compared our encodings using *Smodels* with *lparse* for grounding, *DLV*, and *Smodels* with *DLV* for grounding. These results suggest that our programs are much faster on *DLV* than on *Smodels* using *lparse* for grounding. One possible reason for this behavior is as follows. In *Smodels* we have to define type predicates for each variable in the problem domain description as well as all possible ground instances of the atoms that can ever be used in the planning process. The result is that as the number of variables and the number of their possible instantiations increase, the time performance of the logic program decreases. However, we do not have such constraints in *DLV*.

In order to test the effect of grounding in two systems, we have also designed the following experiments. We used *DLV* to produce the ground programs, and fed these into *Smodels*, instead of using *lparse* for grounding. The result is that the overall performance of *Smodels* is increased to almost that of *DLV* on the Simple-Travel problems. Note that the last column in Table 2 contains the sum of (1) the time for producing the grounded version by *DLV*, and (2) the time it takes for *Smodels* to produce the solution based on this grounding.

We observed similar behaviour between the two systems with our elevator problems. As can be seen in Table 3, *DLV* performed better than *Smodels* not only in direct comparison, but also when *Smodels* used the grounding obtained by *DLV*. The reasons behind the differences in the performance of both *DLV* and *Smodels* on the problems in Miconic-10-simtest and Simple-Travel domains are twofold: (1) the number of solutions for a problem, and (2) the task depth – i.e., the length of a solution. In Simple-Travel domain, the hardest problems has only 3 solutions, whereas the hardest problem in Miconic-10-simtest domain has 120 solutions. Furthermore, the length of the solutions in Simple-Travel domain is at 4 — i.e., there are 4 steps in the longest plan found in this domain. However, in Miconic-10-simtest domain, we have solutions with 20 actions in them. This means that the search trees that correspond to Miconic-10-simtest problems are much larger than the ones that correspond to Simple-Travel problems. As a result, our programs required much more time to find all of the solutions for the problems in Miconic-10-simtest domain than they required for Simpe-Travel problems.

Note that, in Miconic-10-simtest problems, the performance of *Smodels* using *DLV*'s groundings is not as good as that of *DLV* on the same problems. One possible explanation for *DLV*'s dominance in performance over *Smodels* on these problems is that the implemented search heuristics for the *guess-and-check* method in *DLV* are better suited for planning problem such as the ones in these experiments.

The above results led us to investigate more about the performances of the ASP solvers (without the grounding part) of the two systems. To do this, we designed a new set of experiments using the Zeno-Travel domain, which was introduced as a benchmark problem in the recent AIPS-2002 planning competition. These experiments involved harder problems than the previous ones: the hardest one (e.g. p25 in Table 4) has over 20000 solutions (i.e stable models), whereas no problem in Tables 2 and 3 has more than 120 models. In these experiments, we compared the performance of *Smodels* using *lparse* for grounding with that of *Smodels* using ground programs produced by *DLV*. Therefore, we were able to investigate the performances of the ASP solvers implemented in the two systems.

The results for these experiments are shown in Table 4. *Smodels* using *lparse* for grounding was only able to solve the first two problems, and on those problems its performance was about an order of magnitude worse than that of *DLV* . *Smodels* with *lparse* was unable to solve the rest of the problems because *lparse* ran out of memory. Table 4 also shows the time it takes for *DLV* to ground the problems. As it can be seen, *DLV* performs better than *Smodels*, even when *Smodels* is using the grounding produced by *DLV*. These results clearly indicate the difference in model generation algorithms implemented in the two systems. According to these results, we

Table 4. Comparison of *Smodels* and *DLV* on the Zeno-Travel Domain. In this table, "no solution" means that the computer used in these experiments ran out of memory.

Problem	DLV	DLV grounding+Smodels
P1	0.590	0.510+0.330
P2	0.670	0.590+0.330
P3	0.410	0.380+0.060
P4	0.320	0.290+0.040
P5	0.490	0.490+0.080
P6	0.360	0.350+0.040
P7	16.440	14.340+35.210
P8	26.180	22.630+85.390
P9	38.390	36.160+76.860
P10	27.220	24.840+52.730
P11	30.370	28.150+55.550
P12	22.930	20.930+21.310
P13	16.560	14.920+22.650
P14	18.230	16.240+66.310
P15	17.020	14.960+38.190
P16	78.060	70.880+152.190
P17	66.300	62.450+62.75
P18	85.000	81.370+194.940
P19	146.030	139.700+138.240
P20	168.630	163.660+329.940
P21	120.080	117.160+106.330
P22	2025.16	1578.69+no solution
P23	4275.25	4236.60+no solution
P24	3612.96	3462.32+no solution
P25	4619.24	4585.35+no solution

can conclude that *DLV* is better suited for the encodings of planning with ordered task decomposition.

Note that on the hardest problems, namely P22-P25, of Zeno Travel, *Smodels* could not find a solution (Table 4). This was not because the ASP solver itself could not solve the problems, but the ground programs generated by *DLV* were too big for *Smodels*'s front-end *lparse* to parse and convert them into the input syntax required by *Smodels*. On these problems, *lparse* ran out of memory and the operating system killed its process.

Table 5. Comparison of *Smodels* and *DLV* on the Simple-Travel Domain using disjunctions

Problem	*Smodels*	*DLV*	*DLV* grounding+*Smodels*
P1	3.640	0.050	0.050+0.010
P2	3.560	0.050	0.030+0.000
P3	3.360	0.030	0.030+0.000
P4	3.590	0.080	0.060+0.010
P5	3.610	0.050	0.050+0.000
P6	3.500	0.030	0.030+0.000
P7	3.580	0.050	0.050+0.020
P8	3.480	0.040	0.030+0.010
P9	3.430	0.030	0.030+0.000
P10	3.560	0.050	0.050+0.010
P11	3.550	0.040	0.030+0.010
P12	3.450	0.020	0.010+0.000
P13	3.570	0.060	0.050+0.020
P14	3.550	0.050	0.050+0.020
P15	3.470	0.030	0.030+0.000
P16	3.580	0.050	0.050+0.010
P17	3.460	0.040	0.040+0.010
P18	3.510	0.030	0.030+0.000

Influence of using disjunctions

Finally, we did another set of experiments to find out the influence of using disjunctions in our transformation. In Definition 26 we used a set of rules to represent nondeterministic choice of methods. Conceptually, these rules can be much more easily formulated by just one disjunctive rule:

$$method_1(h, T) \vee method_2(h, T) \vee \ldots \vee method_N(h, T)$$
$$\leftarrow taskTBA(h, T).$$

Tables 5 and 6 show the results. While disjunctions pay off for the simpler problem (times are slightly better), they do not pay off for the harder Zeno-Travel problems. We do not have an explanation for this yet. We had expected that since *DLV* is especially designed to deal with disjunctions, it would perform better when using disjunctions than when we coded them into equivalent non-disjunctive versions.

Table 6. Comparison of *Smodels* and *DLV* on the Zeno-Travel Domain with disjunctions

Problem	DLV	DLV grounding+*Smodels*
P1	0.600	0.510+0.340
P2	0.650	0.620+0.310
P3	0.410	0.410+0.070
P4	0.330	0.310+0.040
P5	0.520	0.490+0.080
P6	0.380	0.370+0.050
P7	19.390	14.420+34.790
P8	29.990	22.720+84.950
P9	43.150	35.630+75.690
P10	30.960	24.630+53.670
P11	34.170	27.920+56.080
P12	25.220	20.890+20.590
P13	18.530	14.620+22.550
P14	20.540	16.060+70.750
P15	19.460	14.880+39.430
P16	89.050	70.640+150.000
P17	73.030	62.540+63.600
P18	90.510	81.300+194.900
P19	162.110	142.210+133.450
P20	171.910	164.510+335.090
P21	121.900	120.060+106.680

Comparison with SHOP

Encouraged by the performances of the logic programs produced by our translation, we prepared a set of experiments to compare the time performances of our logic-program encodings on both the Simple-Travel and Zeno-Travel examples with those of the SHOP planning system itself.

Tables 7 and 8 augment our results from Tables 2 and 3 to include comparisons with *SHOP* on the Simple-Travel and Miconic-10-simtest domains. For the Simple-Travel domain, the comparison is inconclusive, because the amount of time taken by *SHOP* was too small for us to measure accurately. For the Miconic-10-simtest domain, the time needed by our program using *DLV* was about 1 to 2 orders of magnitude more than the time needed by *SHOP*.

The results of our experiments on the Zeno-Travel Domain can be seen

Table 7. Comparison of *Smodels* and *DLV* with *SHOP* on the Simple-Travel Domain

Problem	*Smodels*	*DLV*	*DLV* grounding+*Smodels*	*SHOP*
P1	3.430	0.050	0.040+0.020	0.000
P2	3.330	0.050	0.050+0.020	0.000
P3	3.190	0.030	0.030+0.000	0.000
P4	3.340	0.070	0.060+0.010	0.000
P5	3.410	0.060	0.050+0.030	0.000
P6	3.230	0.030	0.030+0.010	0.000
P7	3.340	0.050	0.050+0.010	0.000
P8	3.260	0.040	0.040+0.010	0.000
P9	3.230	0.040	0.040+0.010	0.000
P10	3.410	0.070	0.050+0.000	0.000
P11	3.340	0.050	0.040+0.000	0.000
P12	3.250	0.020	0.030+0.010	0.000
P13	3.410	0.070	0.060+0.010	0.000
P14	3.350	0.060	0.050+0.020	0.000
P15	3.270	0.030	0.030+0.000	0.000
P16	3.380	0.060	0.050+0.010	0.000
P17	3.300	0.050	0.050+0.010	0.000
P18	3.260	0.030	0.030+0.000	0.000

at Table 9. In most cases, the time needed by our program using *DLV* was 2.5 to 3.5 orders of magnitude more than *SHOP*.

Our experimental results are encouraging in several ways:

- To the best of our knowledge, this is the first time that an ASP-based approach has been able to do this well on planning problems of this calibre. Our ASP programs were slower than *SHOP*, but this is to be expected since *SHOP*, and also its successor *SHOP2*, are known to be efficient planning systems. For example, *SHOP2* was one of the fastest planning systems in the International Planning Competition in 2002 (`<http://www.dur.ac.uk/d.p.long/competition.html>`).

- Consider the "performance ratio" for our programs, i.e., the ratio of the amount of time they require to the time required by *SHOP*. If the average-case time complexity of our programs were worse than that of *SHOP*, then we would expect their performance ratio to get worse with increasing problem size. In our experiments, the performance ratio did not seem to get worse with increasing problem size.

Table 8. Comparison of *Smodels* and *DLV* using 𝔗𝔯𝔞𝔫𝔰(·) with *SHOP* on the Miconic-10-simtest Domain

Problem	*Smodels*	*DLV*	*DLV* grounding +*Smodels*	*SHOP*	(*DLV* / *SHOP*) Ratio
S1-0	0.160	0.040	0.030+0.010	0.000	-
S2-0	1.160	0.060	0.050+0.010	0.010	6
S3-0	4.450	0.080	0.010+0.090	0.000	-
S4-0	12.790	0.260	0.100+0.530	0.020	13
S5-0s1	44.090	0.640	0.080+1.540	0.060	10.67
S5-0s2	44.490	0.680	0.090+1.840	0.060	11.33
S6-0	46.300	0.980	0.170+3.560	0.090	10.89

Although there is not enough data to say so conclusively, this suggests that the average-case time complexity of our programs may be roughly the same as that of *SHOP*. This gives reason to hope that future improvements in our programs and in ASP solvers may make it possible to get performance competitive with planning systems such as *SHOP*.

- In the HTN formulation of the Miconic-10-simtest domain, the average branching factor (the average number of subtasks of each task) is smaller than in the Zeno Travel domain. The performance ratio for our programs is about two orders of magnitude better in the Miconic-10-simtest domain than in the Zeno Travel domain. As explained later in Section 6, a likely explanation is that the ASP systems are creating ground instances of clauses that are irrelevant for the planning process, because the number of such irrelevant ground instances grows combinatorially with the branching factor.

If this hypothesis is correct, then it may be possible to improve the performance of our ASP systems—possibly by several orders of magnitude—if we can avoid creating the ground instances. In the near future, we will test our system on more planning domains and compare our approach with other well-known planning systems. We are also planning to implement our approach on two systems, namely the *XSB* system ([Rao et al., 1997]) and the front-end software developed by P. Bonatti for *Smodels* ([Bonatti, 2001b; Bonatti, 2001a]), both of which can handle unbound variables, unlike *DLV* and *Smodels* .

Table 9. Comparison of *DLV* with *SHOP* on the Zeno-Travel Domain

Problem	DLV	SHOP	Performance Ratio (DLV / SHOP)
P1	0.590	0.000	-
P2	0.670	0.010	67.00
P3	0.410	0.000	-
P4	0.320	0.010	32.00
P5	0.490	0.000	-
P6	0.360	0.000	-
P7	16.440	0.020	822.00
P8	26.180	0.030	872.67
P9	38.390	0.070	548.43
P10	27.220	0.040	680.50
P11	30.370	0.030	1012.34
P12	22.930	0.020	1146.50
P13	16.560	0.060	276.00
P14	18.230	0.020	911.50
P15	17.020	0.020	851.00
P16	78.060	0.090	867.34
P17	66.300	0.060	1105.00
P18	85.000	0.070	1214.29
P19	146.030	0.120	1216.92
P20	168.630	0.130	1297.15
P21	120.080	0.100	1200.80
P22	2025.16	3.050	663.99
P23	4275.25	12.250	349.00
P24	3612.96	7.980	452.75
P25	4619.24	12.860	359.19

6 Conclusions and Future Research Directions

In this chapter, we have described a way to encode HTN planning problems into logic programs under the answer set semantics. This transformation is not only sound and complete, but it also corresponds closely to HTN planning systems which generate plans by using ordered task decompositions. Previous encodings (as first introduced in [Dimopoulos *et al.*, 1997]) either consider only action-based planning or they take a special view of HTN planning (as constraint-based planning in [Son *et al.*, 2001]).

In general, classical preconditions do not have enough expressive power to provide the amount of control that is needed for efficient planning. It is not hard to formulate examples of planning problems for which a classical representation of the problem will have a much larger search space than the one that can be generated using a more expressive representation such as *SHOP*'s HTNs, or the temporal-logic control rules used in *TLplan* [Bacchus and Kabanza, 2000] and *TALplanner* [Kvarnström and Doherty, 2001]. The practical result of this is that *SHOP* and *TLplan* and *TALplanner* have been able to solve far more planning problems than classical (i.e., action-based) planning systems, and they have been able to solve them in several orders of magnitude less time in extensive empirical comparisons across dozens of planning domains [Nau *et al.*, 1999; Bacchus and Kabanza, 2000; Bacchus, 2001; Fox and Long, 2002].

To test our approach, we have used it to create both *Smodels* and *DLV* logic programs, for three different AI planning domains: the Miconic-10-simtest domain, the Simple-Travel Domain, and the Zeno-Travel domain. Here is a summary of our experimental results:

HTN vs. action-based In our experiments, our *Smodels* logic programs clearly outperformed the corresponding ones described in [Son *et al.*, 2001]. Our programs took several orders of magnitude less time to solve planning problems, and they solved several problems that were inaccessible to the action-based planning based ASP systems. This is due largely to the HTN-style control knowledge that our translation methodology encodes into the logic programs. HTN planning is more expressive than action-based planning [Erol *et al.*, 1996], and in particular, the domain description for an HTN formulation of a domain can include domain-specific knowledge about how to carry out the planning process—knowledge that cannot be expressed in an action-based formalism. To develop this domain-specific knowledge can take a significant amount of human effort, but it can enable the planning system to search much smaller search spaces than the ones explored by action-based formalisms.

DLV **vs.** *Smodels* In our experiments on the Simple-Travel Domain using our method together with *DLV*, we got a speed-up of two orders of magnitude compared to *Smodels*. We believe one of the reasons for this is that *Smodels* requires type predicates to be defined as an input, which creates combinatorially many ground instances of the clauses in the logic program. For any given problem instance, most of these clauses are likely to be irrelevant. But our experiments also revealed that the grounding of *Smodels* is not the only source responsible for this behaviour. By using the grounding obtained from *DLV* and then running *Smodels*, we still got a performance of *Smodels* that does not compete with *DLV*.

SHOP vs ASP   In our experiments, our logic-program encodings were not competitive with *SHOP*. That is not particularly surprising, since *SHOP* is a state-of-the-art HTN planning system. However, we find it quite interesting and encouraging that in our experiments, the time requirement for our logic-program encodings did not seem to grow any faster than proportional to *SHOP*'s time requirement.

We believe that most of the difference in performance is due to the grounding mechanism underlying ASP systems. In order to run our logic-program encoding of an HTN domain, both *DLV* and *Smodels* must first ground the program. This can generate many ground instances for the rules that correspond to the methods. For example, suppose that a method m corresponds to r different rules, that the number of unground variables in each rule is c, and that the number of possible values for each variable is k. Then there will be rk^c ground instances for the rules. Now, suppose we are trying to accomplish a task t that is unified with a method m. Then in *DLV* and *Smodels*, the branching factor of the search space may potentially be as high as rk^c in the worst case. At the same node of the search space, the branching factor for *SHOP* will typically be much less, because *SHOP* will be able to use method instances that are only partially ground. Furthermore, if there are p different predicate symbols in the domain description, then there will be pk^c ground instances of the frame axiom for the domain. *SHOP* will not have any of these instances because it does not need the frame axiom.

If the difference in performance is because of the reasons described above, then we should be able to get a great improvement in performance by using systems like XSB. We have just started investigating this aspect.

We emphasise the fact that our method does not use any particular features of the engine for computing answer sets. Obviously, taking advantage of the particular search method of *Smodels*, or the bottom-up evaluation of *DLV*, it would be possible to write even more efficient translations. But our aim is to develop a translation that is independent of the underlying nonmonotonic engine.

As a byproduct, we believe our method can be easily used as a way to transfer benchmarks from the planning community to benchmarks for comparing nonmonotonic systems based on computing answer sets. This is because our method does not rely on the features of a particular ASP system. In the near future, there are several additional investigations that we would like to perform:

- We want to test the benchmarks on the *XSB* system, a Prolog system which not only allows function symbols but also unbound variables at the same time. These are features that neither *Smodels* nor *DLV* provide. We believe that we can get a competitive planning system once we can apply our translation into a nonmonotonic system with these two features.

- In all of our experiments, we were finding all of solutions to each planning problem. However, in most planning domains, one just wants to find a single solution (hopefully a near-optimal one). In *SHOP*, finding a single solution takes exponentially less time than finding all solutions. Whether it takes exponentially less time for our logic-program encodings is an open question, and we would like to run additional experiments to find out.

- We are also planning to compare our method with *Smodels* equipped with a front-end to allow for (restricted use of) unbound variables ([Bonatti, 2001b; Bonatti, 2001a]). The latter system has been developed by Piero Bonatti and is a front-end system that can be added to any system computing answer sets and based on grounding. This would also allow for comparisons of systems with built-in grounding to those who do not require this (but are, in general, slower). Again, we believe that serious benchmarks from the planning community can help a lot to evaluate nonmonotonic systems.

Our overall aim is to investigate to what extent state-of-the-art nonmonotonic theorem provers can compete with dedicated planners (in particular those based on HTN) and what lessons we can learn from the different translation methods. We expect that optimal translations (if they exist) depend on the particular application area. Developing a methodology to determine or classify such domains seems to us to be worthwhile.

Acknowledgments

This work was supported in part by the following grants, contracts, and awards: Air Force Research Laboratory F30602-00-2-0505, Army Research Laboratory DAAL0197K0135, and Naval Research Laboratory N00173021G005. Opinions expressed in this chapter are those of authors and do not necessarily reflect opinion of the funders.

BIBLIOGRAPHY

[Aarup et al., 1994] M. Aarup, M. M. Arentoft, Y. Parrod, J. Stader, and I. Stokes. OPTIMUM-AIV: A Knowledge-based Planning and Scheduling System for Spacecraft AIV. In M. Zweben and M. S. Fox, editors, *Intelligent Scheduling*, pages 451–469. Morgan Kaufmann, 1994.

[Agosta, 1995] J. M. Agosta. Formulation and Implementation of an Equipment Configuration Problem with the SIPE-2 Generative Planner. In *Proceedings of AAAI-95 Spring Symposium on Integrated Planning Applications*, pages 1–10, 1995.

[Apt *et al.*, 1999] K. R. Apt, V. Marek, M. Truszczynski, and D. S. Warren, editors. *The Logic Programming Paradigm: Current Trends and Future Directions*, Berlin, 1999. Springer.

[Bacchus and Kabanza, 2000] F. Bacchus and F. Kabanza. Using Temporal Logics to Express Search Control Knowledge for Planning. *Artificial Intelligence*, 116(1-2):123–191, 2000.

[Bacchus, 2001] F. Bacchus. AIPS'00 Planning Competition. *AI Magazine*, 22(3):47–56, 2001.

[Baral *et al.*, 2002] C. Baral, N. Tran, and L. Tuan. Reasoning about Actions in a Probabilistic Setting. In *AAAI/IAAI 2002*, pages 507–512. AAAI Press, 2002.

[Biundo and Schattenberg, 2001] S. Biundo and B. Schattenberg. From Abstract Crisis to Concrete Relief. A Preliminary Report on Flexible Integration on Nonlinear and Hierarchical Planning. In A. Cesta and D. Borrajo, editors, *Proceedings of the 6th European Conference on Planning (ECP-01)*, LNAI, pages 157–168, Toledo, Spain, 2001. Springer-Verlag.

[Bonatti, 2001a] P.A. Bonatti. Prototypes for Reasoning with Infinite Stable Models and Function Symbols. In Th. Eiter, M. Truszczyński, and W. Faber, editors, *Logic Programming and Non-Monotonic Reasoning, Proceedings of the Sixth International Conference*, LNCS 2173, pages 416–419, Berlin, September 2001. Springer.

[Bonatti, 2001b] P.A. Bonatti. Reasoning with Infinite Stable Models. In B. Nebel, editor, *Proceedings of IJCAI-01*, pages 603–608, Seattle, Washington, August 2001. Morgan Kaufmann.

[Chen and Warren, 1996] Weidong Chen and David S. Warren. Tabled Evaluation with Delaying for General Logic Programs. *Journal of the ACM*, 43(1):20–74, January 1996.

[Currie and Tate, 1991] K. Currie and A. Tate. O-Plan: The Open Planning Architecture. *Artificial Intelligence*, 52(1):49–86, 1991.

[Dimopoulos *et al.*, 1997] Y. Dimopoulos, B. Nebel, and J. Koehler. Encoding Planning Problems in Non-monotonic Logic Programs. In S. Steel and R. Alami, editors, *Proceedings of the Fourth European Conference on Planning*, pages 169–181, Toulouse, France, September 1997. Springer-Verlag.

[Dix *et al.*, 2001] Jürgen Dix, Ulrich Furbach, and Ilkka Niemelä. Nonmonotonic Reasoning: Towards Efficient Calculi and Implementations. In Andrei Voronkov and Alan Robinson, editors, *Handbook of Automated Reasoning*, pages 1121–1234. Elsevier-Science-Press, 2001.

[Dix *et al.*, 2002] Jürgen Dix, Hector Munoz-Avila, and Dana Nau an Lingling Zhang. Theoretical and Empirical Aspects of a Planner in a Multi-Agent Environment. In Giovambattista Ianni and Sergio Flesca, editors, *Proceedings of Journees Europeens de la Logique en Intelligence Artificielle (JELIA '02)*, LNCS 2424, pages 173–185. Springer, 2002.

[Dix *et al.*, 2003] Jürgen Dix, Hector Munoz-Avila, Dana Nau, and Lingling Zhang. IMPACTing SHOP: Putting an AI planner into a Multi-Agent Environment. *Annals of Mathematics and AI*, 37(4):381–407, 2003.

[Eiter *et al.*, 1998] Thomas Eiter, Nicola Leone, Cristinel Mateis, Gerald Pfeifer, and Francesco Scarcello. The KR System dlv: Progress Report, Comparisons and Benchmarks. In A. G. Cohn, L. Schubert, and S. C. Shapiro, editors, *Proceedings Sixth International Conference on Principles of Knowledge Representation and Reasoning (KR'98)*, pages 406–417. Morgan Kaufmann, 1998.

[Eiter *et al.*, 2002] T. Eiter, W. Faber, N. Leone, G. Pfeifer, and A. Polleres. The DLV-K Planning System: Progress Report. In Giovambattista Ianni and Sergio Flesca, editors, *Proceedings of Journees Europeens de la Logique en Intelligence Artificielle (JELIA '02)*, LNCS 2424, pages 541–544. Springer, 2002.

[Eiter *et al.*, 2003] T. Eiter, W. Faber, N. Leone, G. Pfeifer, and A. Polleres. Answer Set Planning Under Action Costs. *Journal of Artificial Intelligence Research*, 19:25–71, 2003.

[Erol *et al.*, 1994] K. Erol, J. Hendler, and D.S. Nau. UMCP: A Sound and Complete Procedure for Hierarchical Task-Network Planning. In K. Hammond, editor, *Proceedings of AIPS-94*, pages 249–254, Chicago, IL, 1994. AAAI Press.

[Erol *et al.*, 1996] K. Erol, J. Hendler, and D. Nau. Complexity Results for Hierarchical Task-Network Planning. *Annals of Mathematics and Artificial Intelligence*, 18:69–93, 1996.

[Estlin *et al.*, 1997] T. A. Estlin, S. A. Chien, and X. Wang. An Argument for Hybrid HTN/Operator-Based Approach to Planning. In S. Steel and R. Alami, editors, *Proc. Fourth European Conference on Planning (ECP-97)*, pages 184–196, Toulouse, France, September 1997. Springer-Verlag.

[Fox and Long, 2002] M. Fox and D. Long. International planning competition. `http://www.dur.ac.uk/d.p.long/competition.html`, 2002.

[Gelfond and Lifschitz, 1991] M. Gelfond and V. Lifschitz. Classical Negation in Logic Programs and Disjunctive Databases. *New Generation Computing*, 9:365–385, 1991.

[Gelfond and Lifschitz, 1998] M. Gelfond and V. Lifschitz. Action Languages. *Electronic Transactions on AI*, 2(3–4):193–210, 1998.

[Giunchiglia and Lifschitz, 1998] E. Giunchiglia and V. Lifschitz. An Action Language Based on Causal Explanation: Preliminary Report. In *Proc. AAAI-98*, pages 623–630, Madison, Wisconsin, 1998. AAAI Press.

[Hebbar *et al.*, 1996] K. Hebbar, S.J. Smith, I. Minis, and D. Nau. Plan-Based Evaluation of Designs for Microwave Modules. In *Proc. ASME Design Technical Conference and Computers in Engineering Conference*, Irving, California, August 1996.

[Kvarnström and Doherty, 2001] J. Kvarnström and P. Doherty. TALplanner: A temporal logic based forward chaining planner. *Annals of Mathematics and Articial Intelligence*, 30:119–169, 2001.

[Lifschitz, 1999] V. Lifschitz. Action Languages, Answer Sets and Planning. In K. R. Apt, V. W. Marek, M. Truszczynski, and D. S. Warren, editors, *The Logic Programming Paradigm: A 25-Year Perspective*, pages 357–373. Springer-Verlag, 1999.

[Lifschitz, 2002] V. Lifschitz. Answer Set Programming and Plan Generation. *Artificial Intelligence*, 138(1–2), 2002.

[Marek and Truszczyński, 1999] W. Marek and M. Truszczyński. Stable Models and an Alternative Logic Programming Paradigm. In Apt et al. [1999], pages 375–398.

[McCain and Turner, 1997] N. McCain and H. Turner. Causal Theories of Action and Change. In *Proceedings of the 14th National Conference on Artificial Intelligence (AAAI-97)*, pages 460–465, Menlo Park, CA, 1997. AAAI Press.

[McCain, 1999] N. McCain. Using Causal Calculator with the C Input Language. Technical report, University of Texas at Austin, 1999.

[Muñoz-Avila *et al.*, 2001] H. Muñoz-Avila, D.W. Aha, D.S. Nau, R. Weber, L. Breslow, and F. Yaman. SiN: Integrating Case-based Reasoning with Task Decomposition. In B. Nebel, editor, *Proceedings of IJCAI-2001*, Seattle, Washington, August 2001. Morgan Kaufmann.

[N. Leone and Scarcello, 2001] S. Perri N. Leone and F. Scarcello. Improving ASP instantiators by join-ordering methods. In Th. Eiter, M. Truszczyński, and W. Faber, editors, *Logic Programming and Non-Monotonic Reasoning, Proceedings of the Sixth International Conference*, LNCS 2173, pages 280–294, Berlin, September 2001. Springer.

[Nau *et al.*, 1999] D.S. Nau, Y. Cao, A. Lotem, and H. Muñoz-Avila. SHOP: Simple Hierarchical Ordered Planner. In T. Dean, editor, *Proceedings of IJCAI-99*, pages 968–973. Morgan Kaufmann, 1999.

[Nau *et al.*, 2000] D.S. Nau, Y. Cao, A. Lotem, and H. Muñoz-Avila. SHOP and M-SHOP: Planning with Ordered Task Decomposition. Technical Report CS TR 4157, University of Maryland, 2000.

[Nau et al., 2001] D.S. Nau, Y. Cao, A. Lotem, H. Muñoz-Avila, and S. Mitchell. Total-Order Planning with Partially Ordered Subtasks. In B. Nebel, editor, *Proceedings of IJCAI-01*, pages 425–430, Seattle, Washington, August 2001. Morgan Kaufmann.

[Nau et al., 2003] Dana Nau, Tsz-Chiu Au, Okhtay Ilghami, Ugur Kuter, William Murdock, Dan Wu, and Fusun Yaman. SHOP2: An HTN planning system. *Journal of Artificial Intelligence Research*, 20:379–404, December 2003.

[Nau et al., 2005] Dana Nau, Tsz-Chiu Au, Okhtay Ilghami, Ugur Kuter, Hector Muñoz-Avila, William Murdock, Dan Wu, and Fusun Yaman. Applications of SHOP and SHOP2. *IEEE Intelligent Systems*, 20(2):34–41, March–April 2005.

[Niemelä and Simons, 1996] Ilkka Niemelä and Patrik Simons. Efficient Implementation of the Well-founded and Stable Model Semantics. In M. Maher, editor, *Proceedings of the Joint International Conference and Symposium on Logic Programming*, pages 289–303, Bonn, Germany, September 1996. The MIT Press.

[Niemelä, 1999] Ilkka Niemelä. Logic Programs with Stable Model Semantics as a Constraint Programming Paradigm. volume 25(3–4) of *Special Issue of the Annals in Mathematics and Artificial Intelligence*, pages 241–273. Baltzer Science Publishers, 1999.

[Pearl, 1988] Judea Pearl. *Probabilistic Reasoning in Intelligent Systems - Networks of Plausible Inference*. Morgan Kaufmann, San Mateo, 1988.

[Rao et al., 1997] Prasad Rao, K. Sagonas, T. Swift, D. S. Warren, and J. Freire. XSB: A System for Efficiently Computing Well-Founded Semantics. In J. Dix, U. Furbach, and A. Nerode, editors, *Logic Programming and Non-Monotonic Reasoning, Proceedings of the Fourth International Conference*, LNAI 1265, pages 430–440, Berlin, June 1997. Springer.

[Sacerdoti, 1990] Earl D. Sacerdoti. The Nonlinear Nature of Plans. In James Allen, James Hendler, and Austin Tate, editors, *Readings in Planning*, pages 162–170. Morgan Kaufmann, 1990. Originally appeared in *Proc. IJCAI-75*, pp. 206-214.

[Smith et al., 1997] S. J. Smith, K. Hebbar, D. Nau, and I. Minis. Integrating Electrical and Mechanical Design and Process Planning. In M. Mantyla, S. Finger, and T. Tomiyama, editors, *Knowledge Intensive CAD*, volume 2, pages 269–288. Kluwer Academic Publishers, Boston, 1997.

[Smith et al., 1998a] S. J. Smith, D. S. Nau, and T. Throop. Success in Spades: Using AI Planning Techniques to Win the World Championship of Computer Bridge. In *AAAI-98/IAAI-98*, pages 1079–1086. AAAI Press, 1998.

[Smith et al., 1998b] Stephen J. J. Smith, Dana S. Nau, and Thomas Throop. Computer Bridge: A Big Win for AI Planning. *AI Magazine*, 19(2):93–105, 1998.

[Son et al., 2001] T.C. Son, C. Baral, and S. McIlraith. Planning with domain-dependent knowledge of different kinds – an answer set programming approach. In Th. Eiter, M. Truszczyński, and W. Faber, editors, *Logic Programming and Non-Monotonic Reasoning, Proceedings of the Sixth International Conference*, LNCS 2173, pages 226–239, Berlin, September 2001. Springer.

[Tate et al., 1994] A. Tate, B. Drabble, and R. Kirby. O-Plan2: An Architecture for Command, Planning and Control. In M. Zweben and M. S. Fox, editors, *Intelligent Scheduling*, volume 1, pages 213–239, San Francisco, CA, 1994. Morgan-Kaufmann.

[Tate, 1977] A. Tate. Generating project networks. In R. Reddy, editor, *Proc. IJCAI-77*, pages 888–893, Boston, MA, USA, 1977. Morgan Kaufmann.

[Turner, 1997] H. Turner. Representing Actions in Logic Programs and Default Theories: A Situation Calculus Approach. *The Journal of Logic Programming*, 31(1-3):245–298, 1997.

[Wilkins, 1988] D.E. Wilkins. *Practical Planning - Extending the Classical AI Planning Paradigm*. Morgan Kaufmann, 1988.

[Wilkins, 1990] D. Wilkins. Can AI Planners Solve Practical Problems? *Computational Intelligence*, 6(4):232–246, 1990.

Negation and Involutive Adjunctions

KOSTA DOŠEN AND ZORAN PETRIĆ

Dedicated to Dov Gabbay on the occasion of his 60th birthday

1 Introduction

Adjunction is one of the central notions of category theory, and according to leading categorists a central notion of mathematics in general. The goal of this note is to present a phenomenon of adjunction found in assumptions about an involutive negation connective, like classical negation. Proof-theoretical assumptions concerning such a negation make an adjoint situation that we call an *involutive adjunction*. The notion of involutive adjunction amounts, in a sense to be made precise, to adjunction where an endofunctor is adjoint to itself, which in [Došen and Petrić, 2003a] is called *self-adjunction*.

In a series of papers, which starts with [Gabbay, 1988] (see [Gabbay, 1991], [Gabbay and Wansing, 1996] and [Gabbay and Hunter, 1999]), Dov Gabbay has been working on characterizations of negation in terms of assumptions about a consequence relation. Sometimes, as in this note, Gabbay concentrates on negation in the absence of any other connective. The context of the present note replaces Gabbay's logical framework of a consequence relation by a consequence *graph*, as this is done in categorial proof theory. We do not have any more just a relation between premises and conclusions, but we have arrows between them, and there may be more than one such arrow. We are interested in equalities between these arrows. Often these equalities, which are proof-theoretically motivated, exemplify important notions of category theory. This note shows that with an involutive negation we fall on a particular notion of adjunction. This is yet another corroboration of Lawvere's thesis that all logical constants are tied to adjoint situations (see [Lawvere, 1969]), and of Mac Lane's slogan that adjunction arises everywhere (see [,], Preface).

2   Self-adjunctions

To fix notation and terminology, we will rely on the following definition of
the notion of adjunction (cf. [,], Section IV.1, and [Došen, 1999], Section
4.1.3).

An *adjunction* is a sextuple $\langle \mathcal{A}, \mathcal{B}, F, G, \varphi, \gamma \rangle$ where

$\mathcal{A}$ and $\mathcal{B}$ are categories,

F from $\mathcal{B}$ to $\mathcal{A}$ and G from $\mathcal{A}$ to $\mathcal{B}$ are functors,

φ is a natural transformation of $\mathcal{A}$ from the composite functor
FG to the identity functor of $\mathcal{A}$, which means that the following
equation holds in $\mathcal{A}$ for every arrow $f\colon A_1 \to A_2$ of $\mathcal{A}$:

$$(\varphi\ nat) \qquad f \circ \varphi_{A_1} = \varphi_{A_2} \circ FGf,$$

γ is a natural transformation of $\mathcal{B}$ from the identity functor of
$\mathcal{B}$ to the composite functor GF, which means that the following
equation holds in $\mathcal{B}$ for every arrow $g\colon B_1 \to B_2$ of $\mathcal{B}$:

$$(\gamma\ nat) \qquad GFg \circ \gamma_{B_1} = \gamma_{B_2} \circ g,$$

the following *triangular* equations hold in $\mathcal{A}$ and $\mathcal{B}$ respectively:

$$(\varphi\gamma F) \qquad \varphi_{FB} \circ F\gamma_B = \mathbf{1}_{FB},$$
$$(\varphi\gamma G) \qquad G\varphi_A \circ \gamma_{GA} = \mathbf{1}_{GA}.$$

A *self-adjunction* is a quadruple $\langle \mathcal{S}, L, \varphi, \gamma \rangle$ where $\langle \mathcal{S}, \mathcal{S}, L, L, \varphi, \gamma \rangle$ is an
adjunction (this notion is taken over from [Došen and Petrić, 2003a], Section
10). So, in a self-adjunction, L is an endofunctor, and the equations $(\varphi\ nat)$
and $(\gamma\ nat)$ become

$$f \circ \varphi_{A_1} = \varphi_{A_2} \circ LLf,$$
$$LLf \circ \gamma_{A_1} = \gamma_{A_2} \circ f,$$

while the triangular equations become

$$(\varphi\gamma L) \qquad \varphi_{LA} \circ L\gamma_A = L\varphi_A \circ \gamma_{LA} = \mathbf{1}_{LA}.$$

A $\mathcal{K}$-*self-adjunction* is a self-adjunction that satisfies the additional equation

$$(\varphi\gamma\mathcal{K}) \qquad L(\varphi_A \circ \gamma_A) = \varphi_{LA} \circ \gamma_{LA},$$

and a *$\mathcal{J}$-self-adjunction* is a self-adjunction that satisfies the additional equation

$$(\varphi\gamma\mathcal{J}) \qquad \varphi_A \circ \gamma_A = \mathbf{1}_A$$

(these notions are also from [Došen and Petrić, 2003a], Section 10). It is easy to see that every $\mathcal{J}$-self-adjunction is a $\mathcal{K}$-self-adjunction (the converse need not hold).

A $\mathcal{J}$-self-adjunction that satisfies

$$(\gamma\varphi) \qquad \gamma_A \circ \varphi_A = \mathbf{1}_{LLA}$$

is called a *trivial* self-adjunction. Note that for trivial self-adjunctions it is superfluous to assume the equations (γ *nat*) and ($\varphi\gamma G$), or alternatively (φ *nat*) and ($\varphi\gamma F$); these equations can be derived from the remaining ones.

The *free self-adjunction* $\langle \mathcal{S}, L, \varphi, \gamma \rangle$ generated by $\{p\}$, where we call p a *letter*, is defined as follows. The category $\mathcal{S}$ has as objects the formulae of the propositional language generated by $\{p\}$ with a unary connective L. We may identify the formulae p, Lp, $LLp,\ldots$ of this language with the natural numbers $0, 1, 2,\ldots$

The arrow terms of $\mathcal{S}$ are defined inductively out of the primitive arrow terms

$$\mathbf{1}_A\colon A \to A, \qquad \varphi_A\colon LLA \to A, \qquad \gamma_A\colon A \to LLA,$$

for every object A of $\mathcal{S}$, with the help of the operations of composition $\circ$ and the unary operation that assigns to the arrow term $f\colon A \to B$ the arrow term $Lf\colon LA \to LB$. On these arrow terms we impose the equations of self-adjunctions. In the set of these equations we have of course all the equations $f = f$, and this set is closed under symmetry and transitivity of equality, and under the rules

$$(\text{cong } \circ) \qquad \frac{f = f_1 \qquad g = g_1}{f \circ g = f_1 \circ g_1}$$

$$(\text{cong } L) \qquad \frac{f = g}{Lf = Lg}$$

We assume for f and g in (*cong* $\circ$) that they have composable types, such that $f \circ g$ is defined; the same assumption is made for f_1 and g_1.

We define analogously the free $\mathcal{K}$-self-adjunction, the free $\mathcal{J}$-self-adjunction and the free trivial self-adjunction generated by $\{p\}$, just by imposing additional equations.

3   Involutive adjunctions

Consider a category $\mathcal{A}$ and a contravariant functor $\neg$ from $\mathcal{A}$ to $\mathcal{A}$, which means that for $f \colon A \to B$ in $\mathcal{A}$ we have $\neg f \colon \neg B \to \neg A$ in $\mathcal{A}$, and the following equations are satisfied:

$(\neg 1)$ $\qquad \neg \mathbf{1}_A = \mathbf{1}_{\neg A}$

$(\neg 2)$ $\qquad \neg(f \circ g) = \neg g \circ \neg f, \quad$ for $f \colon A \to B$ and $g \colon C \to A$.

The contravariant functor $\neg$ may be conceived either as a functor from the category $\mathcal{A}^{op}$ to $\mathcal{A}$, which we denote by $\neg$ too, or as a functor from $\mathcal{A}$ to $\mathcal{A}^{op}$, which we denote by $\neg^{op}$.

Suppose that for every object A of $\mathcal{A}$ we have an arrow $n_A^{\rightarrow} \colon \neg\neg A \to A$ of $\mathcal{A}$. The arrow $n_A^{\rightarrow}$ becomes the arrow $n_A^{\rightarrow\,op} \colon A \to \neg\neg A$ in $\mathcal{A}^{op}$.

We say that $\langle \mathcal{A}, \neg, n^{\rightarrow} \rangle$ is an $n^{\rightarrow}$-adjunction when

$$\langle \mathcal{A}, \mathcal{A}^{op}, \neg, \neg^{op}, n^{\rightarrow}, n^{\rightarrow\,op} \rangle$$

is an adjunction. This means that in $\mathcal{A}$ we have for every $f \colon A_1 \to A_2$ the equation

$(n^{\rightarrow}\ nat)$ $\qquad f \circ n_{A_1}^{\rightarrow} = n_{A_2}^{\rightarrow} \circ \neg\neg f,$

alternatively written $f \circ n_{A_1}^{\rightarrow} = n_{A_2}^{\rightarrow} \circ \neg\neg^{op} f$, which also delivers $(n^{\rightarrow\,op}\ nat)$ in $\mathcal{A}^{op}$, and the equation

$(n^{\rightarrow}\ triang)$ $\qquad n_{\neg A}^{\rightarrow} \circ \neg n_A^{\rightarrow} = \mathbf{1}_{\neg A},$

which delivers both the equation $(\varphi\gamma F)$, i.e. $(n^{\rightarrow} n^{\rightarrow\,op} \neg)$, in $\mathcal{A}$, and the equation $(\varphi\gamma G)$, i.e. $(n^{\rightarrow} n^{\rightarrow\,op} \neg^{op})$, in $\mathcal{A}^{op}$.

Suppose now that we have as before a category $\mathcal{A}$ and a contravariant functor $\neg$ from $\mathcal{A}$ to $\mathcal{A}$, and that for every object A of $\mathcal{A}$ we have an arrow $n_A^{\leftarrow} \colon A \to \neg\neg A$ of $\mathcal{A}$. The arrow $n_A^{\leftarrow}$ becomes the arrow $n_A^{\leftarrow\,op} \colon \neg\neg A \to A$ in $\mathcal{A}^{op}$.

We say that $\langle \mathcal{A}, \neg, n^{\leftarrow} \rangle$ is an $n^{\leftarrow}$-adjunction when

$$\langle \mathcal{A}^{op}, \mathcal{A}, \neg^{op}, \neg, n^{\leftarrow\,op}, n^{\leftarrow} \rangle$$

is an adjunction. This means that in $\mathcal{A}$ we have for every $f \colon A_1 \to A_2$ the equation

$(n^{\leftarrow}\ nat)$ $\qquad \neg\neg f \circ n_{A_1}^{\leftarrow} = n_{A_2}^{\leftarrow} \circ f,$

which also delivers $(n^{\leftarrow\,op}\ nat)$ in $\mathcal{A}^{op}$, and the equation

$(n^{\leftarrow} \ triang) \qquad \neg n_A^{\leftarrow} \circ n_{\neg A}^{\leftarrow} = 1_{\neg A},$

which delivers both the equation $(\varphi \gamma F)$, i.e. $(n^{\leftarrow \ op} n^{\leftarrow} \neg^{op})$, in $\mathcal{A}^{op}$, and the equation $(\varphi \gamma G)$, i.e. $(n^{\leftarrow \ op} n^{\leftarrow} \neg)$, in $\mathcal{A}$. Note that what we call $n^{\leftarrow}$-adjunction is called *self-adjunction* in [Pavlović, 1997] (Section 3.1; cf. also [Mac Lane and Moerdijk, 1992], Section I.8), which should not be confused with our notion of self-adjunction in the preceding section.

We say that $\langle \mathcal{A}, \neg, n^{\rightarrow}, n^{\leftarrow} \rangle$ is an *involutive* adjunction when $\langle \mathcal{A}, \neg, n^{\rightarrow} \rangle$ is an $n^{\rightarrow}$-adjunction and $\langle \mathcal{A}, \neg, n^{\leftarrow} \rangle$ is an $n^{\leftarrow}$-adjunction.

A $\mathcal{K}$-*involutive* adjunction is an involutive adjunction that satisfies the additional equation

$(n^{\rightarrow} n^{\leftarrow} \mathcal{K}) \qquad \neg(n_A^{\rightarrow} \circ n_A^{\leftarrow}) = n_{\neg A}^{\rightarrow} \circ n_{\neg A}^{\leftarrow},$

and a $\mathcal{J}$-*involutive* adjunction is an involutive adjunction that satisfies the additional equation

$(n^{\rightarrow} n^{\leftarrow} \mathcal{J}) \qquad n_A^{\rightarrow} \circ n_A^{\leftarrow} = 1_A.$

It is easy to see that every $\mathcal{J}$-involutive adjunction is a $\mathcal{K}$-involutive adjunction (the converse need not hold).

A $\mathcal{J}$-involutive adjunction that satisfies

$(n^{\leftarrow} n^{\rightarrow}) \qquad n_A^{\leftarrow} \circ n_A^{\rightarrow} = 1_{\neg \neg A}$

is called a *trivial* involutive adjunction.

Note that for trivial involutive adjunctions it is superfluous to assume the equations $(n^{\leftarrow} \ nat)$ and $(n^{\leftarrow} \ triang)$, or alternatively $(n^{\rightarrow} \ nat)$ and $(n^{\rightarrow} \ triang)$; these equations can be derived from the remaining ones. In trivial involutive adjunctions we have the equations

$$n_{\neg A}^{\leftarrow} = \neg n_A^{\rightarrow},$$
$$n_{\neg A}^{\rightarrow} = \neg n_A^{\leftarrow}.$$

The *free involutive adjunction* $\langle \mathcal{A}, \neg, n^{\rightarrow}, n^{\leftarrow} \rangle$ generated by $\{p\}$ is defined as follows. The category $\mathcal{A}$ has as objects the formulae of the propositional language generated by $\{p\}$ with a unary connective $\neg$. We may identify these formulae with the natural numbers.

The arrow terms of $\mathcal{A}$ are defined inductively out of the primitive arrow terms

$$1_A : A \rightarrow A, \qquad n_A^{\rightarrow} : \neg \neg A \rightarrow A, \qquad n_A^{\leftarrow} : A \rightarrow \neg \neg A,$$

for every object A of $\mathcal{A}$, with the help of the operations of composition $\circ$ and the unary operation that assigns to the arrow term $f \colon A \to B$ the arrow term $\neg f \colon \neg B \to \neg A$. On these arrow terms we impose the equations of involutive adjunctions. In the set of these equations we have of course all the equations $f = f$, and this set is closed under symmetry and transitivity of equality, under the rule ($cong \circ$), and also under the rule

$$(cong \,\neg) \qquad \frac{f = g}{\neg f = \neg g}$$

We define analogously the free $\mathcal{K}$-involutive adjunction, the free $\mathcal{J}$-involutive adjunction and the free trivial involutive adjunction generated by $\{p\}$, just by imposing additional equations.

Note that the category of the free involutive adjunction generated by an arbitrary set having more than one letter would be the disjoint union of isomorphic copies of the category $\mathcal{A}$ of the free involutive adjunction generated by $\{p\}$. An analogous remark applies to the category of the free self-adjunction generated by an arbitrary set having more than one member: it would be the disjoint union of isomorphic copies of the category $\mathcal{S}$ of the free self-adjunction generated by $\{p\}$.

4   Self-adjunctions and involutive adjunctions

We are now going to prove that in the free self-adjunction $\langle \mathcal{S}, L, \varphi, \gamma \rangle$ and the free involutive adjunction $\langle \mathcal{A}, \neg, n^{\rightarrow}, n^{\leftarrow} \rangle$, both generated by $\{p\}$, the categories $\mathcal{S}$ and $\mathcal{A}$ are isomorphic categories.

First, we define $\neg$, $n^{\rightarrow}$ and $n^{\leftarrow}$ in $\mathcal{S}$ in the following manner. On objects we have that $\neg$ is L, while for the arrow term $f \colon A \to B$ of $\mathcal{S}$ we define the arrow term $\neg f \colon \neg B \to \neg A$ of $\mathcal{S}$ inductively as follows:

$$\neg 1_A = L1_A = 1_{LA} = 1_{\neg A},$$
$$\neg \varphi_A = L\gamma_A,$$
$$\neg \gamma_A = L\varphi_A,$$
$$\neg(f \circ g) = \neg g \circ \neg f,$$
$$\neg Lf = L\neg f.$$

That this defines an operation $\neg$ on the arrows of $\mathcal{S}$ is shown by verifying that if $f = g$ in $\mathcal{S}$, then $\neg f = \neg g$ in $\mathcal{S}$; we verify, namely, that the equations of $\mathcal{S}$ are closed under the rule ($cong \,\neg$) of the preceding section. This is done by a straightforward induction on the length of the derivation of $f = g$ in $\mathcal{S}$. For that we use the fact that for every arrow term f of $\mathcal{S}$ the arrow term $\neg f$ is equal in $\mathcal{S}$ to an arrow term of the form Lf'.

Finally, we have

$$n\overrightarrow{A} =_{df} \varphi_A, \qquad n\overleftarrow{A} =_{df} \gamma_A.$$

Next, we define L, φ and γ in $\mathcal{A}$ in the following manner. On objects we have that L is $\neg$, while for the arrow term $f \colon A \to B$ of $\mathcal{A}$ we define the arrow term $Lf \colon LA \to LB$ of $\mathcal{A}$ inductively as follows:

$$L\mathbf{1}_A = \neg\mathbf{1}_A = \mathbf{1}_{\neg A} = \mathbf{1}_{LA},$$
$$Ln\overrightarrow{A} = \neg n\overleftarrow{A},$$
$$Ln\overleftarrow{A} = \neg n\overrightarrow{A},$$
$$L(f \circ g) = Lf \circ Lg,$$
$$L\neg f = \neg Lf.$$

That this defines an operation L on the arrows of $\mathcal{A}$ is shown by verifying that if $f = g$ in $\mathcal{A}$, then $Lf = Lg$ in $\mathcal{A}$; we verify, namely, that the equations of $\mathcal{A}$ are closed under the rule $(cong\ L)$ of §2 above. This is done by a straightforward induction on the length of the derivation of $f = g$ in $\mathcal{A}$. For that we use the fact that for every arrow term f of $\mathcal{A}$ the arrow term Lf is equal in $\mathcal{A}$ to an arrow term of the form $\neg f'$.

Finally, we have

$$\varphi_A =_{df} n\overrightarrow{A}, \qquad \gamma_A =_{df} n\overleftarrow{A}.$$

We verify easily by induction on the complexity of the arrow term f that both in $\mathcal{S}$ and in $\mathcal{A}$ we have the equation

$$(LL\neg\neg) \qquad LLf = \neg\neg f.$$

Next we verify that the equations of involutive adjunctions hold for the defined $\neg$, $n^{\to}$ and $n^{\leftarrow}$ in $\mathcal{S}$. This is done in a straightforward manner by induction on the length of derivation. In the basis of this induction, we use $(LL\neg\neg)$, $(\varphi\ nat)$ and $(\gamma\ nat)$ to verify $(n^{\to}\ nat)$ and $(n^{\leftarrow}\ nat)$, while the equations $(n^{\to}\ triang)$ and $(n^{\leftarrow}\ triang)$ reduce to $(\varphi\gamma L)$. In the induction step, we rely on the closure of $\mathcal{S}$ under $(cong\ \neg)$, which we established above.

We verify also that the equations of self-adjunctions hold for the defined L, φ and γ in $\mathcal{A}$. This is done again in a straightforward manner by induction on the length of derivation. In the basis of this induction, we use $(LL\neg\neg)$, $(n^{\to}\ nat)$ and $(n^{\leftarrow}\ nat)$ to verify $(\varphi\ nat)$ and $(\gamma\ nat)$, while the equations $(\varphi\gamma L)$ reduce to $(n^{\to}\ triang)$ and $(n^{\leftarrow}\ triang)$. In the induction step, we rely on the closure of $\mathcal{A}$ under $(cong\ L)$, which we established above.

We have a functor $F_{\mathcal{A}}$ from $\mathcal{S}$ to $\mathcal{A}$ that maps the object of $\mathcal{S}$ corresponding to the natural number n to the object of $\mathcal{A}$ corresponding to n, and that maps every arrow of $\mathcal{S}$ to the homonymous arrow in the defined $\mathcal{S}$ structure of $\mathcal{A}$. For example,

$$F_{\mathcal{A}}\,\varphi_{LLp} = \varphi_{\neg\neg p} = n^{\rightarrow}_{\neg\neg p}\,.$$

We define analogously a functor $F_{\mathcal{S}}$ from $\mathcal{A}$ to $\mathcal{S}$. That $F_{\mathcal{A}}$ and $F_{\mathcal{S}}$ are indeed functors follows from what we established above.

It is trivial that on objects we have that $F_{\mathcal{S}}F_{\mathcal{A}}A$ is A, and that $F_{\mathcal{A}}F_{\mathcal{S}}B$ is B. We show next by induction on the complexity of f that in $\mathcal{S}$ we have

$$F_{\mathcal{S}}F_{\mathcal{A}}\,f = f.$$

When f is of the form Lf', we make an auxiliary induction on the complexity of f', in which we use $(LL\neg\neg)$. We show analogously that in $\mathcal{A}$ we have

$$F_{\mathcal{A}}F_{\mathcal{S}}\,g = g.$$

This concludes the proof that $\mathcal{S}$ and $\mathcal{A}$ are isomorphic categories.

We demonstrate analogously that the categories of, respectively,

      the free $\mathcal{K}$-self-adjunction and the free $\mathcal{K}$-involutive adjunction,

      the free $\mathcal{J}$-self-adjunction and the free $\mathcal{J}$-involutive adjunction,

      the free trivial self-adjunction and the free trivial involutive adjunction,

all generated by $\{p\}$, are isomorphic categories.

The interest of considering $\mathcal{K}$ and $\mathcal{J}$ versions of self-adjunctions and involutive adjunctions comes from connections with Temperley-Lieb algebras and the associated geometrical interpretation (see [Došen and Petrić, 2003a] and references therein). Roughly speaking, $\mathcal{K}$ is what we find in Temperley-Lieb algebras, where only the number of circles (which correspond to $\varphi_A \circ \gamma_A$ or $n^{\rightarrow}_A \circ n^{\leftarrow}_A$) counts, while in $\mathcal{J}$ circles are disregarded.

The free trivial self-adjunction, and hence also the free trivial involutive adjunction, are preorders; namely, all arrows with the same source and target are equal. This follows from the results of [Došen and Petrić, 2003a] (unabridged version) or [Došen and Petrić, 2003b].

With that we conclude our note about categorial assumptions underlying the proof theory of negation. We believe this points towards a new dimension in investigations of negation through consequence, such as has been undertaken in papers of Dov Gabbay (see the introduction).

BIBLIOGRAPHY

[Došen, 1999] K. Došen, *Cut Elimination in Categories*, Kluwer, Dordrecht, 1999

[Došen and Petrić, 2003a] K. Došen and Z. Petrić, Self-adjunctions and matrices, *Journal of Pure and Applied Algebra*, vol. 184 (2003), pp. 7-39 (unabridged version available at: http:// arXiv.org/math.GT/0111058)

[Došen and Petrić, 2003b] K. Došen and Z. Petrić. The geometry of self-adjunction, *Publications de l'Institut Mathématique (N.S.)*, vol. 73(87) (2003), pp. 1-29 (incorporated in the unabridged version of [Došen and Petrić, 2003a])

[Gabbay, 1988] D. M. Gabbay, What is negation in a system?, *Logic Colloquium '86* (F. R. Drake and J. K. Truss, editors), North-Holland, Amsterdam, 1988, pp. 95-112

[Gabbay, 1991] D. M. Gabbay. Modal provability foundations for negation by failure, *Extensions of Logic Programming* (P. Schroeder-Heister, editor), Springer, Berlin, 1991, pp. 179-222

[Gabbay and Hunter, 1999] D. M. Gabbay and A. Hunter, Negation and contradiction, *What is Negation?* (D.M. Gabbay and H. Wansing, editors), Kluwer, Dordrecht, 1999, pp. 89-100

[Gabbay and Wansing, 1996] D. M. Gabbay and H. Wansing. What is negation in a system? Part II: Negation in structured consequence relations, *Logic, Action and Information* (A. Fuhrmann and H. Rott, editors), de Gruyter, Berlin, 1996, pp. 328-350

[Lawvere, 1969] F. W. Lawvere, Adjointness in foundations, *Dialectica*, vol. 23 (1969), pp. 281–296.

[,] S. MAC LANE, *Categories for the Working Mathematician*, Springer, Berlin, 1971 (expanded second edition, 1998).

[Mac Lane and Moerdijk, 1992] S. Mac Lane and I. MOERDIJK. *Sheaves in Geometry and Logic: A First Introduction to Topos Theory*, Springer, Berlin, 1992.

[Pavlović, 1997] D. Pavlović. Chu I: cofree equivalences, dualities and ∗-autonomous categories, *Mathematical Structures in Computer Science*, vol. 7 (1997), pp. 49–73.

Modal Tableaux: Completeness vs. Termination

LUIS FARIÑAS DEL CERRO, OLIVIER GASQUET,
ANDREAS HERZIG AND MOHAMAD SAHADE

Introduction

Basic modal logics are normal modal logics defined by any combination of axioms D, T, B, 4, 5. For all these logics the complexity of satisfiability is in PSPACE, and research (in particular in the field of description logics) has focussed on tableaux decision procedures for them that stay within that boundary. For all these logics one can find sound, complete and terminating tableau algorithms in the standard textbooks [Fitting, 1983; Goré, 1998], and implemented tableau theorem provers such as the LogicWorkBench (LWB) [[Abate and Goré, 2003] are readily available.

This contrasts with the situation for more complex modal logics that are used to reason about mental states of artificial agents, about knowledge and communication, or norms and regulations. In these domains we have to deal with BDI logics [Wooldridge, 2002], logics of knowledge, common knowledge and time [Fagin *et al.*, 1995], logics of action and obligation [Carmo and Pacheco, 2001], etc. Such logics are not as 'stable' as the traditional modal logics we find in textbooks: for example, it often happens that one would like to slightly modify the interplay between the modalities (action-knowledge, action-obligation, belief-knowledge, time-belief, etc.), or to add a new modality to an existing logic. The variety and complexity of these logics make that in practice we do not always have at our disposal tableau algorithms that enjoy soundness, completeness and termination.

The contributions of our paper are the following:

1. We argue that it is sufficient in practice to prove soundness and termination, and to conjecture completeness (at least as long as the right calculus together with a satisfactory terminating algorithm have not been found)

2. We provide a general formal framework for tableaux systems, within which virtually all tableaux algorithms for modal logics can be imple-

mented. This framework is based on rooted directed acyclic graphs (RDAGs) in order to stay close to possible worlds models, and its basic mechanism is graph rewriting. It provides a simple language to define tableau rules and a simple language to define strategies. We formally define the semantics of the usually rather implicit or purely operational notions of rules and strategy.

3. This allows us to state general termination theorems for a class of strategies that contains strategies for the modal logics K, T, S4, S5, KB4, and even LTL and PDL.

4. We finally give two specific terminating strategies for the more complex logics of density and confluence.

This paper is organized as follows: after a brief introduction focussing on our contribution, we proceed in section 1 by making some remarks about tableaux-based approaches for modal logics, espectially concerning some of their limits, and we propose some historical explanations of why more general approaches (like ours) have not yet been thoroughly investigated. Then, in section 2, we present some basics about modal logics and graphs, in order to proceed, in section 3 and 4, by presenting our framework for monomodal logics and for multimodal ones: we show how to obtain complete tableau calculi. In section 5, we discuss the problem of termination and its intrication with completeness, and finally argue for giving priority to termination and consider that a conjecture of completeness may be satisfactory. In section 6, we give two termination theorems for tableaux algorithms that cover most standard modal logics like K or S4, up to some less standard ones like LTL and PDL. We finish in section 7 by a succint presentation of the generic tableau prover LoTREC that we develop in the LILaC group of IRIT[1], and which theoretical elements are those contained in previous sections. Concluding remarks follows in the last section.

1   Methodology: model building vs. theorem proving

Tableaux systems can be viewed as algorithms that search for countermodels of a given formula. As a consequence the objects they work on are more or less close to models of the logic under concern.

Somewhat contrarily to that perspective, the objects of standard tableaux systems for modal logics have a much flatter structure than that: while pos-

---

[1] A first basic version of LoTREC was presented in [Fariñas et al., 2001], and the very new version is presented in the system description [Gasquet et al., 2005]. Materials are available at the LoTREC webpage at
http://www.irit.fr/LILAC/Lotrec,
together with a version of LoTREC that is accessible via Internet.

sible worlds models have accessibility relations, traditional tableaux systems for modal logics work on sets or lists of formulas, trees of formulas, or lists of labelled formulas. It seems to us that there are several reasons to this mismatch, which are mainly historical.

The first one is due to a perspective on tableaux systems that is closer to Gentzen's sequent calculi and views tableaux construction not as model building, but as search for a refutational proof. And this has been consolidated by the close connection between cut-free sequents and tableaux for propositional logic. Such sequent-like systems don't fit well modal logics. For example, it is part of the sequent calculus 'philosophy' that the introduction and elimination rules concern one connective at a time. Such a principle can only be obeyed in the modal case at the price of complex devices such as higher order sequents [Došen, 1985] or display techniques [Wansing, 1998], but they only work for a limited number of modal systems. Well, and after all it is well-known that proofs are trees.

The second reason is due to the fact that the very first modal logics investigated from the point of view of tableaux can be characterized semantically by classes of trees (trees, reflexive trees, transitive trees for respectively K, KT, K4). In this setting, each tree represents a tentative model of the input formula. (Due to disjunction there are several trees for a given formula.) Naturally, most further tableaux methods dealt with new logics in this framework by adding supplementary mechanisms to the underlying tree framework in order to handle new semantical properties. For instance Massacci has added backward rules to take into account symmetric accessibility relations [Massacci, 1994; Massacci, 2000].

Another reason, connected to the second one, is that the success and rigor of the use of labels (thoroughly investigated in [Gabbay, 1998]) in tableaux has led people to consider them as a rather universal tool. But still, most of the time the underlying structure of a tableau is a tree and the accessibility relation is encoded via the prefix relation between labels: label x is implicitly related to label $x.n$, where n is a fresh integer. Thus labels can cope with properties like confluence only by adding some additionnal extra reasoning over them (called the label algebra in [Gabbay, 1998].)

Last, but not least, the tree approach to tableaux has also a computational motivation: most of the basic modal logics have a satisfiability problem which has been proved to be complete for the class PSPACE. In order to obtain decision procedures that are optimal w.r.t. this complexity class it is crucial to avoid storing the whole tableau (whose size is most of the time exponential in the size of the input formula) but rather to *explore* it branch by branch. (We are talking here of the branches of the tree of tentative models.) This is clearly an advantage of the tree approach w.r.t.

the model building approach, but on the one hand for some experimentations one may wish to compute the whole model disregarding the (ab)use of memory, and on the other hand this argument fails when dealing with modal logics beyong PSPACE (like K+confluence) where the required memory is inherently exponential.

Of course, the model building perspective does not suffer from any of these restrictions. We believe that in some sense, it reflects a compromise between the tree approach to tableaux and the translation approach (into first-order logic). More precisely, we believe that the model building approach is nothing but the careful construction of a Herbrand universe, either a finite one (in the case of a logic as K) or a finitely generated one (in the case of a logic like S4 with a loop test).

Another advantage of the model building approach is that the task of designing tableau rules for a given modal logic is rather straightforward: in the case of the basic modal logic K, the tableau rules exactly correspond to the truth conditions; in the case of more complex modal logics the accessibility relation properties can be 'compiled' into the tableau rule in a way that is straightforward in many cases.

Our position while being close to model building is different at least because, as much as possible, we do not completely compute the accessibility relation: for example, we do not compute its transitive closure but simulate it by propagation of formulas. For handling properties like symmetry, we do not need to make the structure symmetric, but rather we use a propagation rule like "if $\Box A \in y$ and xRy then add A to x": such a rule *simulates* symmetry. In fact, only existential properties require to explicitly compute the accessibility relation.

We argue that rooted directed acyclic graphs (RDAG for short), which are DAGs having a distinguished node called the root, are better suited than trees. They allow to naturally handle some properties that do not marry easily with tree structures (like confluence, density, and also, permutation in the multi-modal case), while other properties (like transitivity, symmetry, ...) can still be handled by the propagation of formulas.

This leads us to identify two kinds of modal tableaux rules:

1. propagation rules

2. structural rules.

The former correspond to closure properties of the accessibility relation (e.g. transitivity, symmetry,...), and the latter to existential properties (e.g. confluence, density,...). Soundness of the obtained calculi is straightforward (as usual) and completeness proofs are modular and neatly separated into three components:

1. the Relational Closure Lemma (lemma 8) where the properties of the closure of an RDAG under some relational properties are expressed in terms of the initial RDAG,

2. the Structural Lemma (lemma 10) where we check that the closure of an RDAG under some relational properties preserves some of its initial features (e.g. the transitive closure of a confluent RDAG yields a confluent relation, but not necessarily an RDAG),

3. the Box Lemma (lemma 25) where we check that whenever x and y are related in the closure, and $\Box A \in x$ then the set of associated rules ensures that A was transported into y.

For backgrounds about tableaux for modal logic, the reader may look at [de Swart, 1980], [Fitting, 1983], [Fitting, 1993], [Goré, 1998].

2 Preliminaries

2.1 Modal logics and relational properties

The modal logics we consider are all obtained by extending the basic modal logic K by one or several of the well-known axioms T, B, 4, 5, D, De (axiom of density: $\Diamond p \to \Diamond\Diamond p$) and C (axiom of confluence: $\Diamond\Box p \to \Box\Diamond p$). Thus KDC4 denotes the modal logic obtained by adding the axioms D, C and 4 to the basic system K.

A relational property of the accessibility relation of the Kripke models can be associated with each of these axioms :

Axiom	Property	Notation
T $= \Box p \to p$	reflexivity	Ref
4 $= \Box p \to \Box\Box p$	transitivity	Tr
B $= \Diamond\Box p \to p$	symmetry	Sym
5 $= \Diamond\Box p \to \Box p$	euclideanness	$Eucl$

Group 1: Properties handled by propagation rules

Axiom	Property	Notation
D $= \Box p \to \Diamond p$	seriality	Ser
De $= \Diamond p \to \Diamond\Diamond p$	density	$Dens$
C $= \Diamond\Box p \to \Box\Diamond p$	confluence	$Conf$

Group 2: Properties handled by structural rules

As a consequence of Sahlqvist's theorem [Sahlqvist, 1975], a system based on K plus any combination of these axioms is characterized by the Kripke models whose accessibility relation satisfies the corresponding properties. Thus, KD4 is characterized by Kripke models where the accessibility relation is both serial and transitive; for KT5 reflexivity and euclideanness are required (and, as a consequence, transitivity, seriality and symmetry).

From now on we will indistinctly denote a modal system by $KA_1 \ldots A_n$, where each A_i belongs to group 1 or 2, or by a set ρ of its accessibility relation properties; we will write $\rho = \rho_1 \cup \rho_2$ where ρ_1 is a maximal subset of properties of group 1 (maximal here means "including all those of group 1 which are a consequence of it": thus, symmetry and transitivity imply euclideanness: any set ρ_1 that contain Sym and Tr must also contains $Eucl$), and ρ_2 is a subset of properties of group 2. E.g. KCD4 will be denoted by $\{Ser, Tr, Conf\}$, KDeB4 by $\{Sym, Tr, Eucl, Dens\}$ (since euclideanness is a consequence of transitivity and symmetry).

DEFINITION 1. Given a set ρ of relational properties among group 1 and 2, a ρ-model is a Kripke model whose accessibility relation satisfies ρ. A formula is ρ-satisfiable iff it is satisfiable in a ρ-model. It is ρ-valid iff it is valid in the class of all ρ-models, this will be denoted $\models_\rho A$. Thus A is a theorem of a system denoted by a set ρ of properties iff it is ρ-valid.

2.2   Graphs

DEFINITION 2. Let R and S be two relations over N. As usual, $R(x)$ will denote the set of nodes accessible from x by R: $R(x) = \{y \in N : (x, y) \in R\}$, we will indistinctly denote that y is an R-successor of x by either xRy, $y \in R(x)$ or $(x, y) \in R$. The inverse relation $\overline{R}$ is $x\overline{R}y$ iff yRx, the composition of R and S is $(x, y) \in R \circ S$ iff $\exists z : (x, z) \in R$ & $(z, y) \in S$. The diagonal relation: $\{(x, x) : x \in N\}$ will be denoted by I and is a neutral element for composition. Exponentiation R^n is defined inductively by: $R^0 = I$ and $R^{n+1} = R \circ R^n$. As usual, the transitive closure of R denoted by R^+ is defined by $R^+ = \bigcup_{i \geq 0} R^i$ and the reflexive-transitive closure of R denoted by R^* is $R^+ \cup I$.

The tableau calculi we are going to present are based on RDAG (rooted directed acyclic graphs) having additional properties; let ρ be the set of these additional properties, we define:

DEFINITION 3. (RGRAPH)
A ρ-directed graph (or graph for short, ρ being implicit) is a triple $G = \langle N, \mathcal{R}, F \rangle$ where

- N is a set of labels (names of nodes);

- $\mathcal{R} \subseteq N \times I_{\mathcal{R}} \times N$ is a ternary relation where $I_{\mathcal{R}}$ is some index set, and such that $(N, \mathcal{R})$ satisfies properties of ρ;

- F is a binary relation between N and the set of formulas; (this concerns monomodal logics).

G is said to be monorelational if $\mathcal{R}$ is a singleton.

The elements $\langle n, R, n' \rangle$ of $\mathcal{R}$ are also written nRn' or $n' \in R(n)$. nRn' means that n' is accessible from n via the relation R. The elements $\langle n, A \rangle \in F$ are sometimes also written $A \in F(n)$. $A \in F(n)$ means that the formula A is associated with node n in the graph. Graphs are denoted by $G, g, g', \ldots$. We will use G for the graph under computation at some iteration, and g for subgraphs of G to which a rule applies.

The union of the accessibility relations $\bigcup \mathcal{R}$ is the set of all tuples $\langle n, n' \rangle$ such that there is $R \in I_{\mathcal{R}}$ with nRn'. $(\bigcup \mathcal{R})^+$ is the transitive closure of $\bigcup \mathcal{R}$, and $(\bigcup \mathcal{R})^*$ is the reflexive and transitive closure of $\bigcup \mathcal{R}$.

DEFINITION 4. (RDAG)

A ρ-*rooted directed acyclic graph* (RDAG) is a couple $\langle G, n_0 \rangle$ where $G = \langle N, \mathcal{R}, F \rangle$ is a ρ-directed graph without $\bigcup \mathcal{R}$-cycles, and $n_0 \in N$ is a root of G.[2]

REMARK 5. By abuse of notation, we will often denote a graph $g = \langle N, \mathcal{R}, F \rangle$ by $\mathcal{R}$.

DEFINITION 6. (union of graphs, difference, inclusion)

Given two graphs $g = \langle N, \mathcal{R}, F \rangle$ and $g' = \langle N', \mathcal{R}', F' \rangle$, we extend the standard set-theoretic operations as expected:

- $g \cup g' = \langle N \cup N', \mathcal{R} \cup \mathcal{R}', F \cup F' \rangle$;

- $g \setminus g' = \langle N \setminus N', \mathcal{R} \setminus \mathcal{R}', F \setminus F' \rangle$;

- $g \subseteq g'$ iff $N \subseteq N'$, $\mathcal{R} \subseteq \mathcal{R}'$ and $F \subseteq F'$.

We first investigate the monomodal case, and will come back to the multimodal one in section 4.

2.3   Closure of RGRAPH

We define the following closure operation on monorelational RGRAPH:

DEFINITION 7. Let R be an RGRAPH over a set N and ρ a set of relational properties of group 1; the ρ-closure of R (denoted by R^ρ) is the least RGRAPH that contains R and which satisfies every property of ρ.

---

[2]for every $n' \in N$ we have $\langle n_0, n' \rangle \in (\bigcup \mathcal{R})^*$.

This ρ-closure always exists if the properties are among $\{Ref, Tr, Sym, Eucl\}$. A very important point is that for properties of group 1, the closure can be expressed in terms of the initial RGRAPH. E.g. the transitive closure of an RGRAPH R is defined by: $(x, y) \in R^{Tr}$ iff $\exists n \geq 1$ such that $(x, y) \in R^n$. Note that we do not consider here properties of group 2: it makes no sense to talk about closure under a property of group 2. This is the reason why they are handled in a different way: no propagation rule can simulate them.

LEMMA 8 (Relational Closure Lemma).
Let R be an RDAG *over a set N of nodes:*

- $(x, y) \in R^{Ref}$ *iff* $(x, y) \in R$ *or* $x = y$, *or simply* $R^{Ref} = I$.

- $(x, y) \in R^{Sym}$ *iff* $(x, y) \in R$ *or* $(y, x) \in R$, *or simply* $R^{Sym} = R \cup \overline{R}$.

- $(x, y) \in R^{Tr}$ *iff* $\exists n \geq 1$ *such that* $(x, y) \in R^n$; *it is usually denoted by* R^+.

- $(x, y) \in R^{Eucl}$ *iff* $(x, y) \in R$ *or* $\exists u \in N$ $\exists n \geq 1$ $\exists m \geq 1$ *such that* $(u, x) \in R^n$ *and* $(u, y) \in R^m$, *or equivalently* , $R^{Eucl} = R \cup \overline{R}^+ \circ R^+$.

- $(x, y) \in R^{Ref,Sym}$ *iff* $(x, y) \in R$ *or* $x = y$ *or* $(y, x) \in R$, *or* $R^{Sym,Ref} = R \cup \overline{R} \cup I$.

- $(x, y) \in R^{Ref,Tr}$ *iff* $\exists n \geq 0$ *such that* $(x, y) \in R^n$, *usually denoted by* R^*.

- $(x, y) \in R^{Ref,Eucl}$ *iff* $\exists n \geq 0$ $\exists x_0 = x, \ldots x_i, x_{i+1}, \ldots, x_n = y : (x_i, x_{i+1}) \in R$ *or* $(x_{i+1}, x_i) \in R$, *or equivalently* $R^{Eucl,Ref} = \overline{R}^* \circ R^*$.

- $(x, y) \in R^{Sym,Tr}$ *iff* $\exists n \geq 1$ $\exists x_0 = x, \ldots x_i, x_{i+1}, \ldots, x_n : (x_i, x_{i+1}) \in R$ *or* $(x_{i+1}, x_i) \in R$.

- $(x, y) \in R^{Tr,Eucl}$ *iff* $\exists u \in N$ $\exists n \geq 0$ $\exists m \geq 1$ *such that* $(u, x) \in R^n$ *and* $(u, y) \in R^m$.

Proof. See [Castilho *et al.*, 1997].   ∎

LEMMA 9. *The remaining cases are reducible to those of the previous lemma:* • $R^{Sym,Eucl} = R^{Sym,Tr,Eucl} = R^{Sym,Tr}$
• $R^{Ref,Sym,Tr} = R^{Ref,Tr,Eucl} = R^{Ref,Sym,Eucl} = R^{Ref,Sym,Tr,Eucl} = R^{Ref,Eucl}$

Proof. Straightforward.   ∎

The above lemma is a useful tool for proving completeness: it will allow to define a model for a formula from an open tableau. But this is not the whole story. As we have said previously, some properties are handled structurally; roughly speaking seriality, density and confluence are treated by the underlying "kind" of RDAG of the tableaux. When in the completeness proof we must close the RDAG under one or several properties of group 1 (after this closure operation, the initial RDAG is no longer an RDAG but an RGRAPH), we must also check that its structural properties are preserved after this closure (i.e. that it is still of the same "kind"). E.g. we must prove that the transitive closure of a confluent RDAG is still confluent. This is the aim of the lemma below:

LEMMA 10 (Structural Lemma). *Let ρ_2 be a subset of group 2, ρ_1 a subset of group 1 and let R be a ρ_2-RGRAPH over a set N of nodes. Then R^{ρ_1} is also a ρ_2-RGRAPH and hence is a $(\rho_1 \cup \rho_2)$-RGRAPH.*

Proof. See [Castilho *et al.*, 1997].   ∎

3   Tableaux as sets of graphs

In the paper [Castilho *et al.*, 1997] we have advocated that tableaux systems should work on rooted directed acyclic graphs (RDAGs). Such a choice not only has the advantage of generality: as RDAGs are closer to Kripke models than sequents and trees, it also makes the way from an open tableau to a countermodel shorter. Here we pursue that approach.

3.1   Rules

DEFINITION 11. (tableau rule)
A *tableau rule* ρ is a couple $\rho = \langle g, g' \rangle$ where g and g' are graphs such that $g \subseteq g'$. The left part of a rule is called a *pattern* for the rule. ρ is *analytic* if every formula appearing in g' is a subformula of some formula appearing in g. ρ is *strictly analytic* if every formula appearing in $g \setminus g'$ is a strict subformula of some formula appearing in g.

REMARK 12. g and g' may contain formula variables (that are instantiated when the rule is applied). We can think of such rules as the (infinite) set of its instances.

REMARK 13. Of course, $\rho(g)$ may contain new nodes and/or edges that do not belong to g. Such new nodes will be denoted by a ground first-order term with elements of g as arguments. As an example, if some node n contains a formula $\Diamond A$ then applying the rule for $\Diamond$ will introduce the new node $\Diamond(n, A)$ (see below).

We view rule $\rho = \langle g_1, g_2 \rangle$ as a partial function that returns $\rho(g_1') = g_2'$

if g_1' matches g_1 (by instantiating the variables in g_1), and is undefined otherwise.

We can extend rules from partial to total functions as follows.

DEFINITION 14. (semantics of a rule)
Let ρ be a rule and G a graph. $\rho(G) = G \cup \bigcup_{\{g \subseteq G : \rho(g) \text{defined}\}} \rho(g)$.

REMARK 15. Note that the usual informal statement that rules apply only once is taken into account here by the facts that ρ is a function and that application is defined by set union. Mathematically speaking, if a rule ρ has already been applied on some subgraph g of some graph G then $G \cup \rho(g) = G$.

DEFINITION 16. Here are the rules we need (Since $g \subseteq g'$ we only give the new expressions of g').

- Classical rules:

    - Rule ($\perp$): $g = \langle \{n\}, \emptyset, \{\langle n, A \rangle, \langle n, \neg A \rangle\} \rangle$
      and $\wedge(g) = \langle \{n\}, \emptyset, \{\langle n, \perp \rangle\} \rangle$
      (Adds formulas $\perp$ to node n if it contains contradictory formulas.)

    - Rule ($\neg$): $g = \langle \{n\}, \emptyset, \{\langle n, \neg\neg A \rangle\} \rangle$
      and $\wedge(g) = \langle \{n\}, \emptyset, \{\langle n, A \rangle\} \rangle$
      (Simplifes double negation.)

    - Rule ($\wedge$): $g = \langle \{n\}, \emptyset, \{\langle n, A \wedge B \rangle\} \rangle$
      and $\wedge(g) = \langle \{n\}, \emptyset, \{\langle n, A \rangle\}, \langle n, B \rangle\} \rangle$
      (Adds formulas A and B to node n.)

    - Rule ($\vee$): $g = \langle \{n\}, \emptyset, \{\langle n, A \vee B \rangle\} \rangle$
      and $\vee(g) = \langle \{n\}, \emptyset, \{\langle n, C \rangle\} \rangle$
      (Nondeterministically chooses C among A and B, and adds it to node n. In practice, one may choose to manage several graphs at the same time —this is what our implemented system LoTREC does— and in this case to simply duplicate $\vee(g)$ into two $\vee_r(g)$ and $\vee_l(g)$, respectively corresponding to the choice of A and B.)

- Diamond rule:

    - Rule ($\Diamond$): $g = \langle \{n\}, \emptyset, \{\langle n, \Diamond A \rangle\} \rangle$
      and $\Diamond(g) = \langle \{(\Diamond(n, A))\}, \{nR\Diamond(n, A)\}, \{\langle \Diamond(n, A), A \rangle\} \rangle$
      (Creates the new node $\Diamond(n, A)$ as a R-successor of n and associates to it the formula A.)

- Propagation rules:

- Rule ($\square$): $g = \langle \{n, n'\}, \{nRn'\}, \{\langle n, \square A\rangle\}\rangle$
  and $\square(g) = \langle \emptyset, \emptyset, \{\langle n', A\rangle\}\rangle$
  (Adds the formula A to the node n' if it is an R-successor node of n.)

- Rule (T): $g = \langle \{n\}, \emptyset, \{\langle n, \square A\rangle\}\rangle$
  and $T(g) = \langle \emptyset, \emptyset, \{\langle n, A\rangle\}\rangle$
  (Adds the formula A to the node n.)

- Rule (4): $g = \langle \{n, n'\}, \{nRn'\}, \{\langle n, \square A\rangle\}\rangle$
  and $4(g) = \langle \emptyset, \emptyset, \{\langle n', \square A\rangle\}\rangle$
  (Adds the formula $\square A$ to the node n' if it is an R-successor node of n.)

- Rule (B): $g = \langle \{n, n'\}, \{nRn'\}, \{\langle n', \square A\rangle\}\rangle$
  and $B(g) = \langle \emptyset, \emptyset, \{\langle n, A\rangle\}\rangle$
  (Adds the formula A to the node n if it is an R-predecessor node of n.)

- Rule ($5_\rightarrow$): $g = \langle \{n, n', n''\}, \{nRn', nRn''\}, \{\langle n', \square A\rangle\}\rangle$
  and $5_\rightarrow(g) = \langle \emptyset, \emptyset, \{\langle n'', \square A\rangle\}\rangle$
  (Moves boxed formulas to siblings.)

- Rule ($5_\uparrow$): $g = \langle \{n, n'\}, \{nRn'\}, \{\langle n', \square A\rangle\}\rangle$
  and $5_\uparrow(g) = \langle \emptyset, \emptyset, \{\langle n, \square A\rangle\}\rangle$
  (Move boxed formulas upwards.)

- Rule ($5_\downarrow$): $g = \langle \{n, n', n''\}, \{nRn', n'Rn''\}, \{\langle n', \square A\rangle\}\rangle$
  and $5_\downarrow(g) = \langle \emptyset, \emptyset, \{\langle n'', \square A\rangle\}\rangle$
  (Move boxed formulas downwards except for the root - otherwise this will simulate transitivity.)

- Structural rules:

  - Rule D: $g = \langle \{n\}, \emptyset, \emptyset\rangle$
    and $D(g) = \langle \{D(n)\}, \langle nR\ D(n)\rangle, \emptyset\rangle$
    (Creates the new node $D(n)$ as a R-successor of n).

  - Rule C: $g = \langle \{n, n', n''\}, \{nRn', nRn''\}, \emptyset\rangle$
    and $C(g) = \langle C(n, n', n''), \{n'R\ C(n, n', n''), n''R\ C(n, n', n'')\}, \emptyset\rangle$
    (Creates a new common successor to such n' and n'', including the special case when $n' = n''$.)

  - Rule De: $g = \langle \{n, n'\}, \{nRn', nRn''\}, \emptyset\rangle$
    and $De(g) = \langle De(n, n'), \{nR\ De(n, n'), De(n, n')Rn''\}, \emptyset\rangle$
    (Creates a new node between such n and n'.)

REMARK 17. Using a function like C above, instead of simply introducing a new node, ensures that the confluent node of two nodes n and n' is unique (and similarly for density and seriality). This make precise the saying that a rule is applied only once. Also, it allows to keep track of what rule did create such or such node.

Because of this, a corollary of completeness will be that in existential properties (of group 2), the existential quantifier may be replaced by $\exists!$ (there exists a unique...), giving finer semantical characterization to these modal logics.

In the sequel *RULES* will denote the set of tableau rules for a given logic $\mathcal{L}$.

3.2   Tableau calculi

In order to define a tableau calculus for a system denoted by a set ρ of properties, we must associate a set of rules with it. All the tableaux calculi we are going to define contain: the classical rules and the rule $\Diamond$ plus the rule $\Box$ (as these rules are common to all tableaux calculi, we will henceforth omit them) plus zero or more structural and propagation rules.

A tableau calculus for a system denoted by $(\rho_1 \cup \rho_2)$ is obtained by taking (in addition to classical, $\Diamond$ and K rules) the rules corresponding to properties of $(\rho_1 \cup \rho_2)$; this correspondance is given in the figure below.

	Properties	Rules	
Group 1	*Ref*	T	Propagation
	Sym	B	Rules
	Tr	4	
	Eucl	$5_\uparrow$ $5_\downarrow$ $5_\rightarrow$	
Group 2	*Ser*	D	Structural
	Dens	De	Rules
	Conf	C	

We define what we call *naive tableaux*, i.e. tableaux computed with no strategy, particularly strategies ensuring termination.

DEFINITION 18. A (naive) $(\rho_1 \cup \rho_2)$-tableau for a formula A is the least fixpoint G of a sequence $G_0, \ldots, G_i, G_{i+1}, \ldots$ where $G_i = (N_i, R_i, F_i)$, i.e. $G = \bigcup_{i \geq 0} G_i$.
In addition we must have:

- G_0 is an RDAG consisting of only one node whose associated set of formulas is $\{A\}$,

- G_{i+1} is obtained from G_i by applying either a classical rule, or the $\Diamond$ rule, or the rule $\Box$, or a rule of $(\rho_1 \cup \rho_2)$

- and in which every applicable rule has been applied once.

DEFINITION 19. A tableau is closed if some node in it contains $\bot$; it is open otherwise. A formula is $\rho_1 \cup \rho_2$-closed iff all its $(\rho_1 \cup \rho_2)$-tableaux are closed [3].

As given, definition 18 is ambiguous: the rules application order is left imprecise, as well as the important notion of "all applicable rules have been applied once" also known as *fairness*: it is usually taken into account by marking formulas as active making algorithms even more procedural. Another point is: are the subscripts in the sequence $G_0, \ldots, G_i, \ldots$ integers or may they be cardinal numbers? From a mathematical point of view, since all rules are monotonic, the least fixpoint is unique, and these ambiguities are harmless, but from a computational point of view they cannot be left underdetermined. To explicit these, we need to introduce the notion of algorithm.

Our position is 1) that applying a rule to a set of graphs (representing all the current tableaux under computation) consists in applying it (cf. rules above) to every pattern of each of these graphs, and 2) that with this view, algorithms may simply be described by a language based on function composition and least fixpoint iteration (Kleene star).

3.3   Algorithms

DEFINITION 20. Algorithms over a set of rules *RULES* are inductively defined as follows:

- Any rule alone is an algorithm;

- If a_1 is an algorithm and r a rule then $r; a$ is an algorithm (where ; denotes reverse function composition: $f; g(x) = g(f(x))$);

- If a is an algorithm then so is a^*.

DEFINITION 21. Let α be an algorithm, and G a graph, the result of applying α to G is defined as follows:

- If α is a rule then see definition 16;

- $r; \alpha(G) = \alpha(r(G))$ (; is associative);

---

[3]Remember that due to the rule $\vee$, a formula may have several distinct tableaux.

- $\alpha^*(G) = \alpha^*(\alpha(G))$.

DEFINITION 22. Now a tableau G under algorithm α is the least fix-point of a sequence $G_0, \ldots, G_i, G_{i+1}, \ldots$ where $G_{i+1} = r(G_i)$ for some r determined by the algorithm, we define *terminating* and *fair* algorithms as follows:

- α terminates iff there exists an integer n such that $\alpha(G_0) = G_n$;

- α is fair iff the indexes are integers (i.e. iff $\alpha(G_0) = \bigcup_{i \in \mathbb{N}} G_i$.) This can be possible only if any subterm of α of the form β^* terminates.

DEFINITION 23. The basic fair algorithms over a set $RULES = \{r_1, \ldots, r_n\}$ of rules are the following: $(r_\sigma(1); r_\sigma(2); \ldots; r_\sigma(n))^*$, where σ is any permutation over $\{1, \ldots, n\}$.

3.4 Completeness

In this subsection we prove the completeness of our tableaux calculi. We show how, from a given open $(\rho_1 \cup \rho_2)$-tableau for A we can construct a $(\rho_1 \cup \rho_2)$-model for A.

Let G be an open $(\rho_1 \cup \rho_2)$-tableau for A. G is a ρ_2-RDAG where $G = (N, R, F)$ with root r, since structural rules corresponding to ρ_2 ensure that G satisfies ρ_2.

Now let $\mu = (W, R, \tau)$ be the Kripke model defined as follows:

DEFINITION 24.

- $W = N$

- R is the ρ_1-closure of R, i.e. $R = R^{\rho_1}$

- for all $w \in W$, $w \in \tau(p)$ iff $p \in w$ (in fact iff $p \in F(w)$).

By construction, μ satisfies properties of ρ_1 and, by the Structural Lemma (lemma 10), it also satisfies the properties of ρ_2; hence it is a $(\rho_1 \cup \rho_2)$-model. What remains is to prove that it satisfies the formula A. We first establish the following important lemma:

LEMMA 25 (Box Lemma). *Let* $G = (N, R, F)$ *be a* $(\rho_1 \cup \rho_2)$-*tableau with root* r. *Let* x, y *be such that* $(x, y) \in R^{\rho_1}$ *and* $\Box A \in x$; *then* $A \in y$.

Proof. In [Castilho *et al.*, 1997].   ■

The following fundamental lemma brings us to the desired conclusion:

LEMMA 26 (Truth Lemma). *Let G be an open $(\rho_1 \cup \rho_2)$-tableau for A, μ be the $(\rho_1 \cup \rho_2)$-model defined as in definition 24 w.r.t. G and let $B \in$ Subformulas(A) then: (i) if $B \in x$ then $\mu, x \models B$.*

Proof. (By induction on the structure of B: W.l.o.g we can suppose that B is written with only $\neg$, $\wedge$, $\perp$ and $\square$). See [Castilho *et al.*, 1997]. ∎

As a direct consequence of the previous lemma, we have:

COROLLARY 27. *If A has a fair open $(\rho_1 \cup \rho_2)$-tableau then A is $(\rho_1 \cup \rho_2)$-satisfiable. Hence our tableaux calculi are complete under the fairness assumption.*

3.5 Soundness

In this subsection, we prove the soundness of our tableaux calculi: if a formula A is $(\rho_1 \cup \rho_2)$-closed then A is $(\rho_1 \cup \rho_2)$-unsatisfiable. The technique we use for proving the soundness of our tableaux is simple. We prove that all rules preserve the "satisfiability" of the pattern involved in its application. In our sense, a pattern is $(\rho_1 \cup \rho_2)$-satisfiable iff there exists a $(\rho_1 \cup \rho_2)$-model that contains it and satisfies its formulas. We formally develop this below.

DEFINITION 28. Let $G = (N, R, F)$ be a labelled $(\rho_1 \cup \rho_2)$-RDAG and $\mu = (W, R, \tau)$ be a $(\rho_1 \cup \rho_2)$-model; let h be a function such that $h(N) \subseteq W$ and $\forall n_1, n_2 \in N \colon (n_1, n_2) \in R \Rightarrow (h(n_1), h(n_2)) \in R$.

- h is called an *embedding* from G to μ (or h matches G to μ);

- μ satisfies G via h iff $\forall n \in N \colon A \in F(n) \Rightarrow \mu, h(n) \models A$;

- μ satisfies G iff there exists an embedding h from G to μ such that μ satisfies G via h.

LEMMA 29. *Let $\langle g, g' \rangle$ be a rule; then if some ρ-model μ satisfies g then it satisfies g'.*

Proof. See [Castilho *et al.*, 1997]. ∎

COROLLARY 30. *If A is $(\rho_1 \cup \rho_2)$-satisfiable then it has an open $(\rho_1 \cup \rho_2)$-tableau. Hence our tableaux calculi are sound.*

Proof. If A is $(\rho_1 \cup \rho_2)$-satisfiable by some world x of some $(\rho_1 \cup \rho_2)$-model μ, then its starting labelled RDAG: $(\{n_0\}, \emptyset, \{(n_0, A)\})$ is satisfied by μ (via the embedding $h\colon n_0 \mapsto x$). Hence, at least one of its $(\rho_1 \cup \rho_2)$-tableaux must be open since no closed tableau is satisfiable by μ.   ∎

It is worth noticing that soundness proofs may be totally formalized by the use of provers like Isabelle that allows higher order reasoning. This is subject of ongoing work.

4   Complete tableaux for some bimodal logics with permutation and/or confluence

4.1   Modal logics and relational properties

The language of our bimodal logic have $\Box_a$, $\Box_b$,... and $\Diamond_a$, $\Diamond_b$,... as additionnal connectives w.r.t. classical logic. As usual, $\Diamond_a$ and $\Diamond_b$ abbreviate $\neg\Box_a\neg$ and $\neg\Box_b\neg$. Thus we will use the same axioms as above but with indexes a or b.

We investigate some systems based on K(a,b) plus some of the axioms described below. Among them, axioms involving both modalities are known as *interaction axioms*.

As in the monomodal case, each of these axioms can be associated with a relational property of the accessibility relations of the Kripke models (which are of the form (W, R_a, R_b, m) where R_a and R_b are binary relations over W):

Axiom	Property	Notation
$T_a = \Box_a p \to p$	reflexivity	Ref_a
$4_a = \Box_a p \to \Box_a \Box_a p$	transitivity	Tr_a
$T_b = \Box_b p \to p$	reflexivity	Ref_b
$4_b = \Box_b p \to \Box_b \Box_b p$	transitivity	Tr_b

Group 3: Properties handled by propagation rules

Axiom	Property	Notation
$D_a = \Box_a p \to \Diamond_a p$	seriality	Ser_a
$De_a = \Diamond_a p \to \Diamond_a \Diamond_a p$	density	$Dens_a$
$C_a = \Diamond_a \Box_a p \to \Box_a \Diamond_a p$	confluence	$Conf_a$
$D_b = \Box_b p \to \Diamond_b p$	seriality	Ser_b
$De_b = \Diamond_b p \to \Diamond_b \Diamond_b p$	density	$Dens_b$
$C_b = \Diamond_b \Box_b p \to \Box_b \Diamond_b p$	confluence	$Conf_b$
$Per = \Box_a \Box_b p \leftrightarrow \Box_b \Box_a p$	permutation	P
$C = \Diamond_b \Box_a p \to \Box_a \Diamond_b p$	confluence	C

Group 4: Properties handled by structural rules

Where properties P and C are defined by:

- P:
  $$\forall x, y, z : ((x,y) \in R_b \& (y,z) \in R_a) \to (\exists u: (x,u) \in R_a \& (u,z) \in R_b)),$$
  and
  $$\forall x, y, z : ((x,y) \in R_a \& (y,z) \in R_b) \to (\exists u: (x,u) \in R_b \& (u,z) \in R_a))$$

  This conjunction is equivalent to $R_b \circ R_a = R_a \circ R_b$

- C:
  $$\forall x, y, z : ((x,y) \in R_b \& (x,z) \in R_a) \to (\exists u: (y,u) \in R_a \& (z,u) \in R_b)),$$

  Equivalent to $\overline{R_b} \circ R_a = R_a \circ \overline{R_b}$

As a consequence of Sahlqvist's theorem [Sahlqvist, 1975] (or a multi-modal version of it, see [Catach, 1989]), a system based on K(a,b) plus any combination of these axioms is characterized by the Kripke models (W, R_a, R_b) whose accessibility relation satisfies the corresponding properties.

Decision procedures for systems based on K(a,b) plus *non-interaction* axioms are treated in [Catach, 1991], and those based on K(a,b) plus *interaction* axioms of the form $\Diamond_b p \to \Diamond_a p$ in [Demri, 2000].
The system K(a,b)+P+C is also known as the weakest product modal logic K× K investigated in [Gabbay and Shehtman, 1998].

4.2 Preliminaries and notations

DEFINITION 31. A labelled ρ-RDAG is a triple (N, R, F) as in the monomodal case but where R is partitionned into R^a and R^b.

4.3 Rules

Classical, $\diamond$ and propagation rules are the immediate copies of their monomodal counterpart, but in addition, in order to handle the permutation properties, we need the following structural rules:

- Rule P: $g = \langle \{n, n', n''\}, \{nR_b n', n'R_a n''\}, \emptyset \rangle$
  and $P(g) = \langle P(n, n', n''), \{nR_a P(n, n', n''), P(n, n', n'')R_b n''\}, \emptyset \rangle$
  or $g = \langle \{n, n', n''\}, \{nR_a n', n'R_b n''\}, \emptyset \rangle$
  (Given two nodes related by R_b followed by R_a, this rule make them related by R_a followed by R_b by creating a unique intermediary node.)
  and $P(g) = \langle P(n, n', n''), \{nR_b P(n, n', n''), P(n, n', n'')R_a n''\}, \emptyset \rangle$
  (Similar to the above case, but inverting R_a and R_b roles.)

- Rule C $g = \langle \{n, n', n''\}, \{nR_b n', nR_a n''\}, \emptyset \rangle$
  and $C(g) = \langle C(n, n', n''), \{n'R_a C(n, n', n''), n''R_b C(n, n', n'')\}, \emptyset \rangle$
  (Given two nodes having a common predecessor via respectively R_a and R_b, this rule creates a unique common successor via respectively R_b and R_a.)

In order to define a tableau calculus for a logical system, we must associate a set of rules with it. Tableaux calculi we are going to define contain:

- Classical rules

- Rules $\diamond_\mathsf{a}$ and $\diamond_\mathsf{b}$

- Rule $\square_\mathsf{a}$ and $\square_\mathsf{b}$

- Rules corresponding to the axioms of the system.

4.4 Complete tableaux

It is easy to check that in this bimodal case the Relational Closure lemma (8) still holds, and the Truth lemma (26) holds as long as the Box lemma (25) holds too. Thus completeness of tableaux calculi will hold if both the Box lemma and the Structural lemma (10) hold. It is straightforward with these properties (as well as for symmetry and/or euclideanness) to check that the Box lemma indeed holds ($(R)_a$ will denote the restriction of $R = R^\mathsf{a} \cup R^\mathsf{b}$ to a-edges, i.e. $(R)_a = R^\mathsf{a}$):

LEMMA 32. *Let $G = (N, R = R^\mathsf{a} \cup R^\mathsf{b}, F)$ be a $(\rho_1 \cup \rho_2)$-tableau with root r. Let x, y be such that $(x, y) \in ((R)^{\rho_1})_a$ and $\square_\mathsf{a} A \in x$; then $A \in y$. The same holds for $(x, y) \in (R^\mathsf{b})^{\rho_1}$ and $\square_\mathsf{b} A \in x$.*

Proof. See [Fariñas del Cerro and Gasquet, 2002]. ■

What about the Structural lemma (3)? It holds in fact as it is in the case of closure under reflexivity and transitivity (and the proof is the same), but needs some additionnal requirements to hold for closure under symmetry or euclideanness: for example, in the case of closure under symmetry of R_b, if $R = R^a \cup R^b$ satisfies $R^b \circ R^a = R^a \circ R^b$ then R^{ρ_1} does not and some structural rules need to be added !

LEMMA 33. *Let ρ_2 be a subset of properties of group 4, let ρ_1 be a subset of group 3 and let $R = R^a \cup R^b$ be a ρ_2-RGRAPH over a set N of nodes. Then R^{ρ_1} is also a ρ_2-RGRAPH and hence is a $(\rho_1 \cup \rho_2)$-RGRAPH.*

Proof. See [Fariñas del Cerro and Gasquet, 2002]. ■

Soundness is for free since new rules (w.r.t. the monomodal case) are only structural rules which are trivially sound). Thus, for bimodal logics of permutation containing in addition any axiom listed in groups 3 and 4 above, we obtain very easily naive tableaux that are sound and complete (for fair algorithms).

5 Methodology: completeness vs. termination

Standard presentations of tableaux systems for modal logics such as Fitting's [Fitting, 1983], Massacci's [Massacci, 1994; Massacci, 2000], or Goré's [Goré, 1998] start by defining a set of tableau rules *RULES* for a given logic, and then proceed by proving its soundness and completeness w.r.t. the underlying logic $\mathcal{L}$.

While soundness proofs are generally straightforward, completeness proofs are strictly more complex. The underlying algorithms being usually very procedural, the proofs are not so formal. But there exists a fairly standard methodology as exposed e.g. in [Fitting, 1983]. It is usually based on a fair strategy, which gives equal rights to all tableau rules (and corresponds more or less to our basic fair algorithm): it is supposed that every applicable rule $\rho \in RULES$ will eventually be applied. This strategy ensures that an $\mathcal{L}$-model can be associated to an open tableau.

Such an algorithm does not always terminate for modal logics $\mathcal{L}$ beyond the simple ones; for example the fair application of the standard S4 tableau rules to the formula $\Box \Diamond p$ runs forever. This is because the application of the π-rules (in the Smullyan-Fitting terminology) to formulas of the form $\Diamond p$ in node n creates new nodes that are linked to the RDAG root.

Therefore the last part of standard presentations contains an algorithm that combines rule applications in a way such that termination is guaranteed. Typically:

- The application of 'many' rules is blocked if they have already been applied. This is the case for the α-, β-, and π-rules, but not for the ν-rules.

- The application of the π-rules to node n is blocked if n is included in (or identical to) some ancestor node.

Then a combinatorial argument limits the number of possible nodes that can be created in a graph, and the number of possible graphs.

It is important to note that *completeness has to be re-proved for such a terminating algorithm.* Usually informal arguments are employed here. They basically say that the restriction on rule applicability imposed by the terminating algorithm is harmless in what concerns completeness.

Clearly, while proving completeness for the fair strategy is already a hard task, to prove completeness for the terminating strategy is even harder. As said above, when working with complex modal logics one often would like to modify the interplay between the modalities. Moreover, even for a given logic one would often like to fine-tune the strategy in order to improve performance of the algorithm, shorten proofs, decrease the size of the resulting model, etc. For all these reasons it seems to be too hard a task to prove both completeness and termination of the tableau algorithms for these logics.

We believe that for such complex logics formal termination proofs are more important in practice than formal completeness proofs, and that it is sufficient to only *conjecture completeness.*

To support our claim, suppose we have at our disposal a tableau algorithm that is both terminating and sound for logic $\mathcal{L}$, together with an algorithm that allows us to build a model from an open tableau (usually consisting in closing the accessibility relation under some property). Suppose we want to know whether a given formula A is $\mathcal{L}$-satisfiable or not. Let us proceed as follows:

- If the tableaux algorithm returns a set of graphs all of which are closed, soundness of the algorithm ensures that A is $\mathcal{L}$-unsatisfiable.

- If the tableau algorithm returns at least one open G graph then our completeness conjecture says that it can be turned into an $\mathcal{L}$-model of A. Now this can be verified in the following way:

  1. Apply the model building algorithm to build a model M, with actual world w_0.

2. Check whether M is a $\mathcal{L}$-model.

3. Model check whether $M, w_0 \models A$.

4. If $M, w_0 \models A$ then A is indeed $\mathcal{L}$-satisfiable; else we have discovered that our conjecture was erroneous, and that the tableau algorithm has to be modified to 'get more complete' completeness w.r.t. $\mathcal{L}$-unsatisfiability.

Thus we propose to postpone the work on completeness until the good logic has been found together with a satisfactory (terminating) strategy.

6 Proving termination

Let us forget for a moment the specific systems we were investigating above, and let us concentrate on arbitrary sets of rules and algorithms (henceforth called strategies). The setting we are concerned with is a well-delimited one (labelled graphs and monotonic rules, regular expression-like strategies), and it would be desirable to have general termination results. Unfortunately there seems to be no results in the literature that apply to such a case. For example techniques from rewriting theory such as Knuth-Bendix completion cannot be applied here (in particular because rewriting focusses on rule application in any order, Church-Rosser property etc, while we insist on the contrary). Suppose we have defined a set of rewriting rules $RULES$ for our logic $\mathcal{L}$, and we now want to design a strategy and prove it terminates.

As they are, our rules with no constraints and our strategies with no control, we may obtain (especially in the case of transitivity-like properties) non terminating calculi.

We propose to add constraints on rules, in order to prevent them from being applied at some point. Of course, completeness is lost, but as we said above, it is just postponed. Now, to ensure termination we must add some control to rule application. This will be realized by adding side conditions (or constraints). While infinite branching is roughly avoided by the use of functions to generate new nodes, we also need to avoid infinitely long paths in the graph. With this aim, we introduce the following definitions:

DEFINITION 34. A function d which associates an integer with each node of N is said to be a ranking function iff $d(x) < d(y)$ whenever x is an ancestor of y (i.e. $x(\bigcup \mathcal{R})y$). Typically, for monomodal logics, d will be the distance from the root.

We here give two kinds of rule constraints, which are enough to handle many logics ranging from K to PDL.

DEFINITION 35. Let $G = \langle N, \mathcal{R}, F \rangle$ be a graph and n a node of G;

- empty(n) is true if $F(n) = \emptyset$

- loop(n) is true if there exists an ancestor n' of n such that $F(n) \subseteq F(n')$.

6.1 Adding strict subformulas only

Our first kind of strategies only adds strict subformulas in the action part. It might also be called strictly analytic.

THEOREM 36. *Let RULES be a set of rules such that for every $\rho \in$ RULES, with $r = \langle g, g' \rangle$, $n_0, \ldots, n_k$ being the nodes appearing in g:*

- ρ *is strictly analytic (only* strict subformulas *of the original formula appear in nodes of g');*

- *if g' contains new nodes then*

    - *at least one of empty(n_i) is false*

    - $d(n) > \max(d(n_0), \ldots, d(n_k))$, *i.e. new nodes are strictly farther from the root (this will ensure strict decrease of their contents),*

Let α be any strategy on RULES. Let G be a finite graph. Then the application of α to input G terminates, and $\alpha(G)$ is a finite graph.

Proof. We prove that

- application of α can never lead to RDAGs with infinite branching factor;

- application of α can never lead to RDAGs of infinite depth.

The proof is part of the proof of the next theorem.   ■

This theorem covers the case of logics like K, KD, KT, K+confluence (but non optimally), K.2 (linearity) and star-free PDL, but the condition on d excludes K+density. More generally, such a strategy can account for many modal logics without transitive-like accessibility relations, in particular those with symmetric accessibility relations such as KB, KDB, and KTB.

6.2 Testing loops

Our second criterion applies in particular to logics with transitive accessibility relations. In this case, the strict decrease of node content is no more ensured, since we need rules which are not strictly analytic (like rule 4, $5_{\to}, \ldots$).

THEOREM 37. *Let RULES be a set of rules such that for every $\rho \in$ RULES,, with $r = \langle g, g' \rangle$, $n_0, \ldots, n_k$ being the nodes appearing in g:*

- *ρ is analytic (only* subformulas *of the original formula appear in nodes of g');*

- *if g' contains new nodes then g is $\langle \{n\}, \emptyset, F \rangle$, ρ is one of the $\Diamond$ rule of RULES, and loop(n) is false.*

Let α be any strategy on RULES. Let G be a finite graph. Then the application of α to input G terminates, and $\alpha(G)$ is a finite graph.

Proof. First, observe that due to the subformula condition:

- every node only contains a finite number of formulas, and

- there can only be a finite number of nodes with differing associated set of formulas.

We prove that (1) application of α can never lead to RDAGs of infinite depth, and (2) application of α can never lead to RDAGs with infinite branching factor.

For theorem 36, the argument is that creation of new nodes is subject to non-emptiness, but with our constraints, a branch cannot be of length more than the modal degree of the input formula: because of the strict subformula condition, nodes situated farther would be empty, contrary to the side condition. Infinite branching in node n could only be produced by introducing infinitely many new nodes of rank r strictly greater than those of the pattern and of the form $H(\ldots, n, \ldots)$ (where H is the Skolem function associated with the existential property corresponding to some rule). This would induce that there are infinitely many pattern involving n and thus infinitely many nodes of rank $r-1 \ldots$ leading to infinitely many roots. Which is contradictory. Thus the branching factor is bounded.

For theorem 37, the argument is that creation of new nodes is subject to non-inclusion in an ancestor node, which due to the condition on α, is always tested before node creation. Therefore a branch of infinite length would contain an infinite number of nodes having different associated formula sets.

This cannot be the case because our rules are monotonic: no formulas are erased. This even gives us the classical upper bound: the length of branches is bounded by an exponential in the length of the input formula.

Infinite branching in node n could only be produced by introducing infinitely many new nodes of the form $\Diamond(n, A)$. But given a node n, the number of $\Diamond$ formulas being bounded by the (linear) number of subformulas of the input formula, each node may only have a linear number of successors. Thus the branching factor is bounded. This ends the proof. ∎

This theorem covers the case of logics build over the axioms K, T, 4, 5, B, 2 (linearity).

To sum it up, the above theorems give us some general termination criteria for our strategies. These strategies cover all standard modal logics, including Linear Temporal Logic LTL and Propositional Dynamic Logic PDL (see the LoTREC webpage for the rules). It remains to prove their completeness, this cannot be done without considering a specific logic.

7 Lotrec: a generic theorem prover

The above generic approach has been a theoretical basis for the development of the generic modal theorem prover Lotrec at the *Institut de Recherche en Informatique de Toulouse*. It is described in [Fariñas del Cerro et al., 2001] and in [Gasquet et al., 2005], and may be downloaded at //www.irit. fr/LILAC/Lotrec. It has been designed in order to answer the need for designing provers for new modal logics. In general, most implementors of provers for modal logics put the emphasis on performance, and thus have restricted their prover to few *fixed* logics, hacking sophisticated strategies in their systems.

The choice of one logic over another among possibly infinitely many modal logics is driven by modeling needs and computational constraints of one's applications. A logic about actions and time is likely to have different semantical and computational properties from a logic about database schemata. Even with the same logic, different search strategies may be needed for different applications.

To answer the needs of users wishing to *experiment and model with different logics or strategies* there is a need for a *generic theorem prover for modal logics* playing the same role as Isabelle [Paulson, 1994] or PVS [Owre et al., 1992] for higher order logics, while being less complex. If the user is not the same person as the programmer of the prover, one needs flexibility and portability of the implementation, high-level languages for tableau rules and strategy definition and user-friendly interfaces.

Lotrec is such a generic tableau prover. It aims at covering all logics having possible worlds semantics. Lotrec has been implemented by D. Fauthoux [Fauthoux, 2000]. It is written in Java. All entities are modeled as objects, in particular the tableaux, the nodes and links of a tableau, the tableau rules, and the strategy. Within such an object-based programming language, Lotrec raises Java's event-based architecture to a declarative approach, in order to be able to manipulate and manage the computation in an easy but strict way.

Discussion and conclusion

We have defined a graph-based formal language in terms of rules and strategies that enables programming tableaux systems for modal logics. We have given a semantics to it, and proved termination results for two important classes of strategies, as well as terminating strategies for the modal logics of density and confluence. We have stressed the importance of termination in comparison with completeness.

All these strategies can be tested in our generic tableaux theorem prover LoTREC that is accessible via Internet.

LoTREC is similar in its aims to the Tableaux Workbench (TWB), which also aims at a generic tableaux theorem prover. They differ in perspective: while LoTREC adopts the more semantic model building view, TWB opts for the more syntactic sequent calculus view.

It remains to investigate whether our termination results can be extended to larger classes of strategies.

BIBLIOGRAPHY

[[Abate and Goré, 2003] P. Abate and R. Goré. System Description: The Tableaux Work Bench. In: *Proc. TABLEAUX 2003*, LNAI 2796, Springer Verlag, pp. 164–175, 2003.

[Baldoni, 1998] M. Baldoni. Normal Multimodal Logics: Automatic Deduction and Logic Programming Extension. PhD thesis, Univ. Torino, Italy, 1998.

[Baldoni et al., 1998] M. Baldoni, L. Giordano, and A. Martelli. A Tableau Calculus for Multimodal Logics and some (Un)Decidability Results. In H. de Swart (Ed) *TABLEAUX'98*, LNAI 1397,Springer-Verlag, 1998.

[Carmo and Pacheco, 2001] J. Carmo, O. Pacheco. Deontic and Action Logics for Organized Collective Agency, Modeled through Institutionalized Agents and Roles. Fundamenta Informaticae 48,2001.

[Castilho et al., 1997] M. A. Castilho, L. Fariñas del Cerro, O. Gasquet, and A. Herzig. Modal tableaux with propagation rules and structural rules. *Fundamenta Informaticae*, 32(3/4):281-297, 1997.

[Catach, 1989] L. Catach, Les logiques multi-modales. Ph.D. thesis, Université Paris VI, France, 1989

[Catach, 1991] L. Catach, TABLEAUX: A General Theorem Prover for Modal Logics. Journal of Automated Reasoning, 7:489-510, 1991.

[De Giacomo and Massacci, 1996] G. De Giacomo, F. Massacci. Tableau and Algorithm for Propositional Dynamic Logic with Converse. In M. A. McRobbie, J. K. Slaney editor, *Proc. CADE-13, LNAI 1104, Springer, 1996.*

[Demri, 2000] S. Demri Complexity of simple dependent bimodal logics. In *Proc. Int. Conf. Automated Reasoning with Analytic Tableaux and Related Methods* (TABLEAUX'2000), LNCS 1847, Springer, 2000.

[de Swart, 1980] H. C. M. de Swart. Gentzen-type systems for C, K and several extensions of C and K; constructive completeness proofs and effective decision procedure for these systems. *Logique et Analyse*, 1980.

[Došen, 1985] K. Došen. Sequent-systems for modal logic *The Journal of Symbolic Logic* 50, pp. 149-168, 1985.

[Fagin *et al.*, 1995] R. Fagin, J. Y. Halpern, Y. Moses, M. Y. Vardi. *Reasoning about knowledge.* MIT Press, 1995

[Fariñas *et al.*, 2001] L. Fariñas del Cerro, D. Fauthoux, O. Gasquet, A. Herzig, D. Longin, F. Massacci. Lotrec: a generic tableau prover for modal and description logics. In *Proc. Int. Joint Conference on Automated Reasoning* (IJCAR'01), LNCS 2083, Springer, 2001.

[Fariñas del Cerro and Gasquet, 2000] L. Fariñas del Cerro, and O. Gasquet. Tableaux Based Decision Procedures for Modal Logics of Confluence and Density. *Fundamenta Informaticae*, 41(1):1-17, 2000.

[Fariñas del Cerro and Gasquet, 2002] L. Fariñas del Cerro, and O. Gasquet. A General Framework for Pattern-Driven Modal Tableaux. In: *Logic Journal of the IGPL*, V. 10 N. 1, p. 51-83, 2002.

[Fariñas del Cerro *et al.*, 2001] L. Fariñas del Cerro, D. Fauthoux, O. Gasquet, A. Herzig, D. Longin and F. Massacci. Lotrec: The Generic Tableau Prover for Modal and Description Logics. In: *International Joint Conference on Automated Reasoning, System Description*, Siena, Italy, 2001.

[Fauthoux, 2000] D. Fauthoux. Lotrec, un outil javanais de traitement formel sur les graphes. Tech. rep., IRIT, June 2000. Master Thesis (rapport de D.E.A).

[Fitting, 1983] M. Fitting. *Proof methods for modal and intuitionistic logics.* D. Reidel, Dordrecht, 1983.

[Fitting, 1993] M. Fitting. Basic modal logic. In D. Gabbay et al., editor, *Handbook of Logic in Artificial Intelligence and Logic Programming: Logical Foundations*, vol. 1, Oxford University Press, 1993.

[Gabbay, 1998] D. M. Gabbay. *Labelled Deductive Systems I.* Oxford Science Publications, Clarendon Press, Oxford, 1996.

[Gabbay and Shehtman, 1998] D. M. Gabbay, V. Shehtman. Products of modal logics, Part 1. *Logic Journal of the IGPL*, 6:73-146, 1998.

[Gasquet *et al.*, 2005] O. Gasquet, A. Herzig, D. Longin, M. Sahade. LoTREC: Logical Tableaux Research Engineering Companion. In *Proc. Int. Conf. on Automated Reasoning with Analytic Tableaux and Related Methods* (TABLEAUX 2005). LNAI, Springer Verlag (to appear).

[Goré, 1998] R. Goré. Methods for Modal and Temporal Logics. In D. Gabbay et al., editor, *Handbook of Tableau Methods*, Kluwer Dordrecht, 1998.

[Kripke, 1959] S. Kripke. A completeness theorem in modal logic. *Journal of Symbolic Logic (24)*, 1959.

[Kripke, 1963] S. Kripke. Semantical analysis of modal logic I: Normal modal propositional calculi. *Zeitschrift fr Mathematische Logik und Grundlagen der Mathematik (9)*, 1963.

[Ladner, 1977] R. Ladner. The computational complexity of provability in systems of modal logic. *SIAM Journal on Computing*, 6:466-480, 1977.

[Marx and Mikulas, 2002] M. Marx,S. Mikulas An elementary constrution for a non-elementary procedure *Studia Logica*,72(2), 2002 (to appear).

[Massacci, 1994] F. Massacci. Strongly analytic tableaux for normal modal logics. In Alan Bundy editor, *Proc. CADE-12, LNAI 814, Springer, 1994.*

[Massacci, 2000] F. Massacci. Single Step Tableaux for Modal Logics: Computational properties, complexity and methodology. *Journal of Automated Reasoning*, 24(3): 319-364, April 2000.

[Owre *et al.*, 1992] S. Owre, J. M. Rushby, and N. Shankar. PVS: A prototype verification system. In *Proc. of CADE'92*, LNAI, pp. 748–752, 1992.

[Paulson, 1994] L. C. Paulson. Isabelle: A Generic Theorem Prover. LNCS, Springer-Verlag, 1994.

[Sahlqvist, 1975] H. Sahlqvist. Completeness and correspondence in the first and second order semantics for modal logics. In S.Kanger editor, *Proc. 3rd Scandinavian Logic Symposium 1973, Studies in Logic 82* (1975), North-Holland, 1973

[Wansing, 1998] H. Wansing. Displaying Modal Logic. Trends in Logic Series vol. 3, Springer, 1998.

[Wooldridge, 2002] M. Wooldridge. *An Introduction to Multiagent Systems*. John Wiley & Sons, 2002.

Mathematical Foundations of Answer Set Programming

PAOLO FERRARIS AND VLADIMIR LIFSCHITZ

1 Introduction

Answer set programming (ASP) is a form of declarative logic programming oriented towards difficult combinatorial search problems. ASP has been applied, for instance, to developing a decision support system for the Space Shuttle [Nogueira *et al.*, 2001] and to graph-theoretic problems arising in zoology and linguistics [Brooks *et al.*, 2005]. This paper is about the design of provably correct ASP programs and about the mathematical theory it is based on. Our description of ASP may be useful as a complement to the monograph [Baral, 2003] and to the manuals on the software systems SMODELS[1] and DLV[2].

Syntactically, ASP programs look like Prolog programs, but the computational mechanisms used in ASP are different: they are based on the ideas that have led to the creation of fast satisfiability solvers for propositional logic.

ASP has emerged from interaction between two lines of research—on the semantics of negation in logic programming [Gelfond and Lifschitz, 1988] and on applications of satisfiability solvers to search problems [Kautz and Selman, 1992]. It was identified as a new programming paradigm in [Lifschitz, 1999; Marek and Truszczyński, 1999; Niemelä, 1999].

The main definition of ASP, discussed in Section 2 below, tells us under what conditions a model of a propositional formula F (that is, a truth assignment satisfying F) is called "stable." The idea of ASP is to represent the search problem we are interested in as the problem of finding a stable model of a formula, and then find a solution using an *answer set solver*— a system for generating stable models, such as the systems SMODELS and DLV mentioned above. Information on these and other available answer set solvers can be obtained from the library of logic programming systems

---

[1] http://www.tcs.hut.fi/Software/smodels/lparse.ps .

[2] http://www.dbai.tuwien.ac.at/proj/dlv/man/ .

maintained at the University of Koblenz and Landau.[3] Such systems are not applicable, at least directly, to arbitrary propositional formulas; what they expect as input is a conjunction, or list, of formulas of several special types that are particularly useful from the programmer's perspective; these formulas are assumed to be written in "logic programming notation"—as rules, often similar to Prolog rules.

Properties of stable models are discussed here in Section 2, and designing provably correct ASP programs is the topic of Section 3. Proofs of theorems are relegated to Section 4. These sections are followed by Appendix A, which provides the necessary background information about propositional logic, and Appendix B, which reviews the original 1988 definition of a stable model and relates it to the definition used in the main part of this paper.

As we will see, two equivalent formulas do not necessarily have the same stable models. For instance, the only stable model of the implication

(1) $\neg p \rightarrow q$

is the truth assignment that makes p false and q true; the only stable model of its contrapositive

(2) $\neg q \rightarrow p$

makes p true and q false. (In ASP, it is customary to identify a truth assignment with the set of atoms that get the value t. So we can say that the only stable model of (1) is $\{q\}$, and the only stable model of (2) is $\{p\}$.) The fact that there is an essential difference between (1) and (2) will not surprise a Prolog programmer: in the world of programs with negation as failure, placing q in the head of a rule and the negation of p in the body is very different than placing p in the head and the negation of q in the body. A logician, on the other hand, will note that formulas (1) and (2) are equivalent to each other classically, but not intuitionistically. We will see, in fact, that intuitionistically equivalent formulas always have the same stable models. Hence intuitionistically equivalent transformations play an important role in ASP.

A conjunction $F \wedge G$ may have a stable model that is not a stable model of F. For instance, the formula

(3) $(\neg p \rightarrow q) \wedge p$

has one stable model $\{p\}$, which is not a stable model of its first conjunctive term (1). Thus appending an additional conjunctive term to a formula may give it a new stable model. In this sense, the concept of a

---

[3]http://www.uni-koblenz.de/ag-ki/LP/lp_systems.html .

stable model is nonmonotonic. Early work on stable models was an out-growth of research on formal nonmonotonic reasoning [McCarthy, 1980; McDermott and Doyle, 1980; Reiter, 1980], and, more specifically, of the study in [Gelfond, 1987] of the relationship between autoepistemic logic [Moore, 1985] and the semantics of negation in logic programming.

2 Stable Models

After defining the concept of a stable model in Section 2.1, we apply this definition to three special cases that are often encountered in the practice of answer set programming: Horn formulas, choice formulas and constraints (Sections 2.2–2.4). Section 2.5 is a brief introduction to the use of the answer set solver SMODELS. Then we discuss two mathematical ideas that play an important role in the design of provably correct ASP programs: strong equivalence (Section 2.6) and splitting (Section 2.7).

2.1 Definition and Examples

Recall that we identify truth-valued functions on the set of atoms with subsets of that set (Section A.1).

The *reduct* F^X of a formula F relative to a set X of atoms is the formula obtained from F by replacing each maximal subformula that is not satisfied by X with $\perp$ [Ferraris, 2005].[4] We say that X is a *stable model* (or an *answer set*) of F if X is minimal among the sets satisfying F^X. The minimality of X is understood here in the sense of set inclusion: no proper subset of X satisfies F^X.

Clearly, every set that is a stable model of F according to this definition is a model of F. Indeed, if X does not satisfy F then F^X is $\perp$.

According to the definition, we can verify that X is a stable model of F as follows:

(i) mark in F the maximal subformulas that are not satisfied by X;

(ii) replace each of these subformulas with $\perp$ (after that, equivalent trans-formations of classical propositional logic can be used to simplify the result);

(iii) check that the resulting formula is satisfied by X;

(iv) check that it is not satisfied by any proper subset of X.

For instance, to check that $\{q\}$ is an answer set of (1), we do the following:

---

[4]This is somewhat different from the definition of the reduct proposed in [Gelfond and Lifschitz, 1988]. The relationship between the two definitions is discussed in Appendix B.

(i) mark the only subformula of (1) that is not satisfied by $\{q\}$:

$$\neg \underline{p} \to q;$$

(ii) replace that subformula with $\bot$:

$$\neg \bot \to q;$$

simplify:

$$q;$$

(iii) check that the last formula is satisfied by $\{q\}$;

(iv) check that it is not satisfied by $\emptyset$.

The other two models of (1)

$$\{p\}, \ \{p, q\}$$

are not stable. We check, for instance, that $\{p\}$ is not stable as follows. First we mark the subformulas of (1) that are not satisfied by $\{p\}$:

$$\underline{\neg p} \to \underline{q}.$$

After replacing these subformulas with $\bot$, we get $\bot \to \bot$, or $\top$. Clearly $\{p\}$ is not minimal among the sets satisfying this reduct: the empty set satisfies it as well.

Alternatively, we can conclude that $\{p\}$ and $\{p, q\}$ cannot be stable models of (1) from the general property of stable models stated below. An atom A is a *head atom* of a formula F if at least one occurrence of A in F is strictly positive (see Section A.1 for the definition). This terminology is related to the fact that in logic programming it is customary to write implications $F \to G$ as "rules" $G \leftarrow F$ and to call G the "head" of the rule and H its "body." Clearly, every head atom of a rule $G \leftarrow F$ occurs in its head G.

THEOREM 1 ([Ferraris, 2005]). *Any stable model of F is a subset of the set of head atoms of F.*

Since the only head atom of (1) is q, Theorem 1 shows that stable models of that formula can be found only among the subsets of $\{q\}$.

Many formulas have several stable models. For instance, the conjunction of (1) and (2)

(4) $(\neg p \to q) \land (\neg q \to p)$

has two stable models $\{p\}$ and $\{q\}$, and so does the formula $p \vee q$. On the other hand, $\neg\neg p$ and $\neg p \rightarrow p$ are examples of satisfiable formulas that have no stable models.

It is easy to see that for any X

$$\bot^X = \bot;$$

$$A^X = \begin{cases} A, & \text{if } X \models A, \\ \bot, & \text{otherwise} \end{cases}$$

$(A$ is an atom$)$;

$$(F \odot G)^X = \begin{cases} F^X \odot G^X, & \text{if } X \models F \odot G, \\ \bot, & \text{otherwise} \end{cases}$$

$(\odot$ is $\wedge$, $\vee$ or $\rightarrow)$.

These equalities provide an alternative, recursive definition of the reduct. There is no clause for negation here, because we treat it as an abbreviation (Section A.1). It is easy to check that

$$(\neg F)^X = \begin{cases} \bot, & \text{if } X \models F, \\ \top, & \text{otherwise.} \end{cases}$$

2.2   Horn Formulas

A *Horn formula* is a conjunction of several (0 or more) implications of the form $F \rightarrow A$, where F is a conjunction of several (0 or more) atoms, and A is an atom. The theorem below shows that any Horn formula has exactly one stable model.

For any Horn formula F, the intersection of all models of F is a model of F also; it is called the *minimal model* of F.

THEOREM 2. *For any Horn formula F, the minimal model of F is the only stable model of F.*

For instance, the formula

(5) $p \wedge (p \rightarrow q) \wedge (q \wedge r \rightarrow s)$

has one stable model—its minimal model $\{p, q\}$. The only model of the empty conjunction $\top$ is the empty set.

2.3 Choice Formulas

From formulas with a unique stable model discussed above we turn now to an example of formulas that have exponentially many stable models.

For any finite set Z of atoms, by Z^c we denote the formula

$$\bigwedge_{A \in Z} (A \vee \neg A).$$

PROPOSITION 3. *For any finite set Z of atoms, a set X of atoms is a stable model of Z^c iff $X \subseteq Z$.*

Proof. For any subset X of Z, the reduct of Z^c relative to X is

$$\bigwedge_{A \in X} (A \vee \bot) \wedge \bigwedge_{A \in Z \backslash X} (\bot \vee \neg \bot),$$

which is equivalent to $\bigwedge_{A \in X} A$. This formula is satisfied by X, but is not satisfied by any proper subset of X. We have proved that if X is a subset of Z then X is a stable model of Z^c. The converse is immediate from Theorem 1. ∎

For instance, $\{p, q\}^c$ is

$$(p \vee \neg p) \wedge (q \vee \neg q).$$

This formula has 4 answer sets—arbitrary subsets of $\{p, q\}$. Generally, if Z consists of n atoms then Z^c has 2^n stable models. To form one of them, we choose for every element of Z arbitrarily whether to include it in the model. We will call formulas of the form Z^c *choice formulas*. (The superscript ^{c} is used in this notation because it is the first letter of the word "choice.")

2.4 Constraints

The art of answer set programming is based on the possibility of representing the collection of sets that we are interested in as the collection of stable models of a formula. This is often achieved by conjoining a choice formula, which provides an approximation from above for the collection of sets that we want to describe, with formulas of a special syntactic form, called constraints, that eliminate the unsuitable stable models.

As discussed in the introduction, conjoining a formula F with another formula generally affects the collection of stable models of F nonmonotonically. But this does not happen if the second conjunctive term begins with negation. According to the proposition below, conjoining a formula F

with $\neg G$ has a simple effect on the set of stable models of F: it eliminates the stable models that do not satisfy the additional conjunctive term.

PROPOSITION 4. *A set of atoms is a stable model of $F \wedge \neg G$ iff it is a stable model of F that satisfies $\neg G$.*

Proof. *Case 1:* X satisfies $F \wedge \neg G$. Then X does not satisfy G, and $(F \wedge \neg G)^X$ is $F^X \wedge \neg\bot$, which is equivalent to F^X. Consequently X is minimal among the sets satisfying F^X iff it is minimal among the sets satisfying $(F \wedge \neg G)^X$. *Case 2:* X does not satisfy $F \wedge \neg G$. Then X cannot be a model of F that satisfies $\neg G$. ∎

In the terminology of ASP, a *constraint* is simply a formula beginning with negation. To illustrate the special role of constraints as additional conjunctive terms, let us go back to example (3), where adding the conjunctive term p to formula (1) changed its collection of stable models nonmonotonically. If we conjoin (1) with the constraint $\neg p$ instead then we will get the formula

$$(\neg p \rightarrow q) \wedge \neg p.$$

Since the only stable model $\{q\}$ of (1) satisfies the constraint, the conjunction has $\{q\}$ as the only stable model as well. If we conjoin (1) with the constraint $\neg\neg p$ then we will get the formula

$$(\neg p \rightarrow q) \wedge \neg\neg p.$$

Since the only stable model $\{q\}$ of (1) does not satisfy this constraint, the conjunction has no stable models.

2.5 LPARSE and SMODELS

We briefly interrupt now the discussion of the theory of stable models to talk about the capabilities of one of the widely used answer set solvers, SMODELS.[5] Its frontend LPARSE serves also as the frontend of three other systems for computing stable models: GNT[6], ASSAT[7] and CMODELS[8]. This frontend requires that the input formula be represented in a special format, as a list (conjunction) of "rules," somewhat similar to Prolog rules. In this paper the reader will find many examples of representing formulas in the input language of LPARSE. A detailed description of that language can be found in the LPARSE manual, available online (see Footnote 1).

Our first example illustrates representing Horn formulas. Formula (5) would be represented in an LPARSE input file as follows:

---

[5]http://www.tcs.hut.fi/Software/smodels/ .
[6]http://www.tcs.hut.fi/Software/gnt/ .
[7]http://assat.cs.ust.hk/ .
[8]http://www.cs.utexas.edu/users/tag/cmodels/ .

```
p.
q :- p.
s :- q, r.
```

Note that each conjunctive term here is written as a separate rule, followed by a period. In each rule $A \leftarrow F$, the left arrow is written as :- and conjunction as a comma.

To instruct SMODELS to find the stable model of (5), we save the three lines shown above in a file, called, say, `file5`, and invoke LPARSE and SMODELS as follows:

```
% lparse file5 | smodels
```

The main part of the output generated in response to this command line is the program's stable model:

```
Answer: 1
Stable Model: q p
```

Negation in front of an atom is represented in the language of LPARSE as `not`. For instance, formula (4) would be written in a file, say, `file4`, as

```
q :- not p.
p :- not q.
```

The command line

```
% lparse file4 | smodels 0
```

instructs SMODELS to find the stable models of this formula. The zero at the end indicates that we want to compute all stable models; a positive number k in this position would tell SMODELS to terminate after computing k stable models. SMODELS will produce the following output:

```
Answer: 1
Stable Model: q
Answer: 2
Stable Model: p
```

To represent a choice formula $\{A_1, \ldots, A_n\}^c$ in the language of LPARSE, we simply drop the superscript ^{c}. A constraint $\neg F$, where F is a conjunction of literals, is written as :- F. For instance, the formula

$$\{p, q, r\}^c \wedge \neg\neg p \wedge \neg(q \wedge \neg r)$$

can be written in the syntax of LPARSE as[9]

---

[9]The choice construct was originally defined as an addition to the language of LPARSE in [Simons *et al.*, 2002], and it was treated there as a primitive, rather than abbreviation. The equivalence of our treatment of choice formulas to that definition follows from results of [Ferraris and Lifschitz, 2005] and [Ferraris, 2005].

```
{p,q,r}.
:- not p.
:- q, not r.
```

The search process employed in SMODELS is quite sophisticated, and it guarantees, in principle, that every stable model of the given input will be found. It can be viewed as a modification of the Davis-Putnam-Logemann-Loveland procedure for the propositional satisfiability problem (SAT) [Davis *et al.*, 1962]. We should note that finding a stable model of a formula is more difficult than SAT: the existence of a stable model is a Σ_2^P-complete property [Eiter and Gottlob, 1993]. But most uses of ASP involve formulas of special syntactic forms for which this property is known to be in class NP.[10]

Systems ASSAT and CMODELS operate in a different way: they reduce the problem of computing stable models of a given formula to an instance (or a series of instances) of SAT and then invoke SAT solvers to do search.

2.6 Strong Equivalence

As discussed in the introduction, two formulas have the same stable models if they are intuitionistically equivalent (see Section A.3 for the definition). The theorem below is stronger than this claim in several ways.

We say that a formula F is *strongly equivalent* to a formula G if any formula F' that contains an occurrence of F has the same stable models as the formula G' obtained from F' by replacing that occurrence with G. (As the term "strongly equivalent" suggests, this relation turns out to be stronger than equivalence in the sense of classical logic.) For instance, $p \to q$ has the same stable model as $p \to r$ (the empty set), but these two formulas are not strongly equivalent. Indeed, take F' to be $(p \to q) \wedge p$. Then G' is $(p \to r) \wedge p$. These two formulas have different stable models: $\{p, q\}$ and $\{p, r\}$ respectively. The theorem below shows, however, that intuitionistically equivalent formulas are strongly equivalent. This is not surprising in view of the replacement property of intuitionistic logic (Section A.3): if F is intuitionistically equivalent to G then F' is intuitionistically equivalent to G'.

The role of strong equivalence in the practice of answer set programming is determined by the fact that it allows us to simplify a part of a program without looking at the rest of it. For instance, we can observe that the formula $p \wedge (p \to q)$ is intuitionistically equivalent to $p \wedge q$. It follows that

---

[10]There are exceptions, however, and the answer set solvers GNT and CMODELS are among the systems that are not limited to "NP cases" of answer set programming. System DLV (http://www.dbai.tuwien.ac.at/proj/dlv/) is the earliest answer set solver of this kind.

in any program containing the rules

p.

q :- p.

replacing the second rule by

q.

will have no effect on the set of stable models.

If formulas F and G are not strongly equivalent to each other then this can be always demonstrated using a counterexample F' that is not much more complicated than F. We can always take F' to be a formula of the form $F \wedge H$, where H is a Horn formula (see Section 2.2 for the definition). One can say even more. A formula H is *unary* if it is a conjunction of several (0 or more) atoms and implications of the form $A_1 \to A_2$, where A_1 and A_2 are atoms. The theorem below shows that formulas F and G are strongly equivalent iff, for every unary H, $F \wedge H$ and $G \wedge H$ have the same stable models.

Furthermore, in the statement of the theorem intuitionistic logic is replaced with a stronger subsystem of classical logic, called "the logic of here-and-there." This logic is reviewed in Section A.2. Its role in the theory of stable models was first recognized in [Pearce, 1997], where it was used to define a nonmonotonic "equilibrium logic"; the definition of a stable model in Section 2 is equivalent to the semantics of equilibrium logic [Ferraris, 2005, Theorem 1].

Finally, the theorem below asserts not only that equivalence in the logic of here-and-there implies strong equivalence, but that the converse holds also. Thus the logic of here-and-there provides a complete characterization of strong equivalence. This fact, as well as the property of unary formulas pointed out above, was first established in [Lifschitz *et al.*, 2001].

THEOREM 5. *For any formulas F and G, the following conditions are equivalent:*

(i) *F is strongly equivalent to G,*

(ii) *for every unary formula H, $F \wedge H$ and $G \wedge H$ have the same stable models,*

(iii) *F is equivalent to G in the logic of here-and-there.*

Here are some examples of the use of the most important part of this theorem, the implication from (iii) to (i). As mentioned in Section 2.1, formulas $\neg\neg p$ and $\neg p \to p$ have no stable models. We can now say more:

since these formulas are intuitionistically equivalent (Section A.3), the result of replacing the subformula $\neg p \rightarrow p$ in any formula with $\neg\neg p$ does not change its stable models. In particular, any program containing the rule

```
p :- not p.
```

can be simplified by replacing that rule with

```
:- not p.
```

Another example: the formulas $p \vee \neg p$ and $\neg\neg p \rightarrow p$ are equivalent to each other in the logic of here-and-there (Section A.3); consequently, they are strongly equivalent. We know that the first of these formulas can be written in the input language of LPARSE as

```
{p}.
```

That language allows us to represent the double negation of an atom in the body of a rule as well: $\neg\neg A$ can be written as $\{\text{not } A\}0$. (This is a special case of "cardinality" notation discussed in Section 3.1 below.) In particular, we can write $\neg\neg p \rightarrow p$ as the rule

```
p :- {not p} 0.
```

One more example of the use of Theorem 5 is given by the proof of the following proposition:

PROPOSITION 6. *Let Z be the set of atoms occurring in a formula F. A subset X of Z satisfies F iff X is a stable model of $Z^c \wedge F$.*

Proof. From Propositions 3 and 4, a subset X of Z satisfies F iff X is a stable model of $Z^c \wedge \neg\neg F$. It remains to observe that $\neg\neg F \leftrightarrow F$ can be intuitionistically derived from Z^c, because Z^c is the conjunction of the excluded middle formulas $A \vee \neg A$ for all atoms A occurring in this equivalence. ∎

Proposition 6 provides a reduction of the propositional satisfiability problem to ASP: to find a model of F, look for a stable model of the conjunction of F with the excluded middle formulas $A \vee \neg A$ for all atoms A occurring in F.

An alternative approach to proving the strong equivalence of propositional formulas, based on [Lifschitz et al., 1999, Section 4] and [Turner, 2003], does not require the knowledge of intuitionistic logic or the logic of here-and-there. We can show that F is strongly equivalent to G by checking that, for every set X of atoms, the reducts F^X and G^X are equivalent to each other in classical logic:

THEOREM 7 ([Ferraris, 2005]). *For any formulas F and G, F is strongly equivalent to G iff, for every set X of atoms, F^X is equivalent to G^X in classical logic.*

For instance, the fact that $\neg p \to p$ is strongly equivalent to $\neg\neg p$ can be established by the following computation:

$$\begin{aligned}
(\neg p \to p)^{\{p\}} &= \bot \to p &\leftrightarrow \top, \\
(\neg\neg p)^{\{p\}} &= \neg\bot &= \top;
\end{aligned}$$

$$\begin{aligned}
(\neg p \to p)^\emptyset &= \bot, \\
(\neg\neg p)^\emptyset &= \bot.
\end{aligned}$$

2.7 Splitting

Theorem 2 describes stable models of Horn formulas; Proposition 3 describes stable models of choice formulas. Many ASP programs involve conjunctions of formulas of these two types. To design such programs, we need to understand the structure of their stable models.

Consider the following example:

(6) $\{p, q\}^c \wedge (p \to r) \wedge (q \wedge r \to s).$

The first conjunctive term of this formula has 4 stable models:

(7) $\emptyset, \{p\}, \{q\}, \{p, q\}.$

The rest of the conjunction can be viewed as a "definition," characterizing r and s in terms of p and q. Appending this definition to the choice formula $\{p, q\}^c$ does not affect the total number of its stable models, but it can change each of the models (7) by adding to it some of the atoms r, s. In view of the implication $p \to r$, atom r is added to each model containing p. In view of the implication $q \wedge r \to s$, atom s is added to each model containing both q and r. Thus we can expect that (6) will have the following stable models:

(8) $\emptyset, \{p, r\}, \{q\}, \{p, q, r, s\}.$

The validity of this claim can be justified using the theorem below, which shows that in some cases we can compute the stable models of a conjunction by "splitting" it into conjunctive terms and computing first the stable models of one of these terms. Splitting was proposed in [Lifschitz and Turner, 1994] and generalized and simplified in [Erdoğan and Lifschitz, 2004] and [Ferraris, 2005].

THEOREM 8. *Let F and G be formulas such that F does not contain any head atoms of G. A set X of atoms is a stable model of $F \wedge G$ iff there* *exists a stable model* $\{A_1, \ldots, A_n\}$ *of* F *such that* X *is a stable model of* $A_1 \wedge \cdots \wedge A_n \wedge G$.

(For the definition of a head atom see Section 2.1.)

In application to (6), we take $\{p, q\}^c$ to be F and $(p \to r) \wedge (q \wedge r \to s)$ to be G. Since F does not contain any of the head atoms r, s of G, the stable models of $F \wedge G$ can be generated by taking each of the stable models (7) of F, conjoining its elements with G, and listing all stable models of each of the resulting formulas

$$(p \to r) \wedge (q \wedge r \to s),$$
$$p \wedge (p \to r) \wedge (q \wedge r \to s),$$
$$q \wedge (p \to r) \wedge (q \wedge r \to s),$$
$$p \wedge q \wedge (p \to r) \wedge (q \wedge r \to s).$$

Since these are Horn formulas, each of them has one stable model—its minimal model. As we have conjectured, these stable models are the sets (8) shown above.

Here is an example of the use of Theorem 8 in the case when F is not a choice formula. We want to find the stable models of the conjunction

(9) $(\neg p \to q) \wedge (q \to r)$.

The only stable model of the first conjunctive term is $\{q\}$ (Section 2.1). According to Theorem 8, it follows that (9) has the same stable models as

$$q \wedge (q \to r).$$

This is a Horn formula, and its minimal model $\{q, r\}$ is its only stable model.

Theorem 8 can be useful also when G is not a Horn formula. But in such cases $A_1 \wedge \cdots \wedge A_n \wedge G$ will not be a Horn formula either, and computing its stable models may require additional work. The following two propositions can often help at this stage.

Notation: F_G^A stands for the formula obtained from a formula F by substituting a formula G for all occurrences of an atom A.

PROPOSITION 9. *For any atom* A *that is not a head atom of* F, F *has the same stable models as* $F_\perp^A$.

Proof. Since A is not a head atom of F, A does not belong to any of the stable models of F (Theorem 1). Consequently, F has the same stable models as $F \wedge \neg A$ (Proposition 4). Similarly, $F_\perp^A$ has the same stable model as $F_\perp^A \wedge \neg A$. It remains to observe that $F \wedge \neg A$ and $F_\perp^A \wedge \neg A$ are intuitionistically equivalent to each other by the replacement property of intuitionistic logic (Section A.3). ∎

Example: Using Proposition 9, we can find the stable model of $\neg p \rightarrow q$ without directly referring to the definition of a stable model as in Section 2.1. Since p is not a head atom of $\neg p \rightarrow q$, this formula has the same stable models as $\neg \bot \rightarrow q$, which is intuitionistically equivalent to the Horn formula q. Consequently, the only stable model of $\neg p \rightarrow q$ is q.

PROPOSITION 10. *For any atom A, a set X of atoms is a stable model of $F \wedge A$ iff there exists a stable model Y of $F_{\top}^{A}$ such that $X = Y \cup \{A\}$.*

Proof.   By the replacement property of intuitionistic logic, $F \wedge A$ is intuitionistically equivalent to $F_{\top}^{A} \wedge A$, so that the two formulas have the same stable models. By Theorem 8, X is a stable model of $F_{\top}^{A} \wedge A$ iff there exists a stable model $\{A_1, \ldots, A_n\}$ of $F_{\top}^{A}$ such that X is a stable model of $A_1 \wedge \cdots \wedge A_n \wedge A$. The only stable model of this Horn formula is $\{A_1, \ldots, A_n, A\}$, which can be written as $\{A_1, \ldots, A_n\} \cup \{A\}$.   ■

Example: Proposition 10 can be used to verify the claim that the only stable model of (3) is $\{p\}$. To find the stable models of $(\neg p \rightarrow q) \wedge p$, we need to add p to each stable model of $\neg \top \rightarrow q$. Since this formula is intuitionistically equivalent to $\top$, its only stable model is the empty set.

The following example shows how Propositions 9 and 10 can be used in combination with splitting. We want to find the stable models of

(10) $\{p, q\}^c \wedge (\neg p \rightarrow r)$.

By Theorem 8, this can be done by computing the stable models of each of the conjunctions

$$\neg p \rightarrow r,$$
$$p \wedge (\neg p \rightarrow r),$$
$$q \wedge (\neg p \rightarrow r),$$
$$p \wedge q \wedge (\neg p \rightarrow r).$$

Proposition 9 shows that the only stable model of the first of these formulas is $\{r\}$. Proposition 10 shows that the only stable model of the second formula is $\{p\}$. Proposition 9 shows that the only stable model of the third formula is $\{q, r\}$. Proposition 10 shows that the only stable model of the last formula is $\{p, q\}$. Consequently, (10) has 4 stable models:

$$\{p\}, \ \{r\}, \ \{q, r\}, \ \{p, q\}.$$

Using Theorem 8 twice, we can derive the following useful fact:

PROPOSITION 11. *Let F and G be formulas such that F does not contain head atoms of G, and G does not contain head atoms of F. A set of atoms is a stable model of $F \wedge G$ iff it can be represented as the union of a stable model of F and a stable model of G.*

Proof. By Theorem 8, X is a stable model of $F \wedge G$ iff there exists a stable model $\{A_1, \ldots, A_n\}$ of F such that X is a stable model of

(11) $G \wedge (A_1 \wedge \cdots \wedge A_n)$.

By Theorem 1, for any stable model $\{A_1, \ldots, A_n\}$ of F, atoms $A_1, \ldots, A_n$ are head atoms of F. Consequently they are different from the head atoms of G, so that the head atoms of G do not occur in the second conjunctive term of (11). By Theorem 8, X is a stable model of (11) iff there exists a stable model $\{B_1, \ldots, B_m\}$ of G such that X is a stable model of

$$B_1 \wedge \cdots \wedge B_m \wedge A_1 \wedge \cdots \wedge A_n,$$

that is to say, such that

$$X = \{A_1, \ldots, A_n\} \cup \{B_1, \ldots, B_m\}.$$

∎

3 Programming

By an ASP program we understand a propositional formula that can be easily communicated to an answer set solver. We want to learn how to represent a given search problem as the problem of computing a stable model of such a formula.

After discussing in Sections 3.1 and 3.2 a few features of the language of LPARSE that have not been mentioned earlier, we give several examples of computational problems that can be solved using ASP (Sections 3.3–3.8). Then we talk about answer set programming with strong negation (Section 3.9) and about its application to representing actions (Section 3.10).

3.1 Cardinality Expressions

In answer set programming we often need formulas expressing conditions on cardinalities of sets. The following notation is useful. For any nonnegative integer l ("lower bound") and formulas $F_1, \ldots, F_n$,

(12) $l \leq \{F_1, \ldots, F_n\}$

stands for the disjunction

$$\bigvee_{I \subseteq \{1,\ldots,n\}, \, |I|=l} \bigwedge_{i \in I} F_i.$$

For instance,

$$2 \leq \{F_1, F_2, F_3\}$$

stands for

$$(F_1 \wedge F_2) \vee (F_1 \wedge F_3) \vee (F_2 \wedge F_3).$$

By

(13) $\{F_1, \ldots, F_n\} \leq u$

where u is a nonnegative integer ("upper bound") we denote the formula

$$\neg(u + 1 \leq \{F_1, \ldots, F_n\}).$$

Finally,

(14) $l \leq \{F_1, \ldots, F_n\} \leq u$

stands for

$$(l \leq \{F_1, \ldots, F_n\}) \wedge (\{F_1, \ldots, F_n\} \leq u).$$

It is clear that any set of atoms

- satisfies (12) iff it satisfies at least l of the formulas $F_1, \ldots, F_n$;

- satisfies (13) iff it satisfies at most u of the formulas $F_1, \ldots, F_n$;

- satisfies (14) iff it satisfies at least l and at most u of the formulas $F_1, \ldots, F_n$.

The input language of LPARSE allows us to use expressions (12)–(14) in the bodies of rules, with the symbol $\leq$ dropped, if all formulas $F_1, \ldots, F_n$ are literals.[11] We saw an example in Section 2.3: the implication $\neg\neg p \rightarrow p$ can be represented in an LPARSE input file as

```
p :- {not p} 0.
```

because

$$\{\neg p\} \leq 0 \;=\; \neg(1 \leq \{\neg p\}) \;=\; \neg\neg p.$$

If $A_1, \ldots, A_n$ are pairwise distinct atoms then we will write

$$
\begin{array}{lll}
l \leq \{A_1, \ldots, A_n\}^c & \text{for} & \{A_1, \ldots, A_n\}^c \wedge (l \leq \{A_1, \ldots, A_n\}), \\
\{A_1, \ldots, A_n\}^c \leq u & \text{for} & \{A_1, \ldots, A_n\}^c \wedge (\{A_1, \ldots, A_n\} \leq u), \\
l \leq \{A_1, \ldots, A_n\}^c \leq u & \text{for} & \{A_1, \ldots, A_n\}^c \wedge (l \leq \{A_1, \ldots, A_n\} \leq u).
\end{array}
$$

The following proposition explains why these are useful abbreviations.

PROPOSITION 12. *For any pairwise distinct atoms $A_1, \ldots, A_n$, nonnegative integers l and u, and a set X of atoms,*

---

[11]What we said in Footnote 9 about the invention of choice formulas applies also to "cardinality formulas" (12)–(14) and to their combinations with choice formulas introduced below. These expressions were originally introduced in [Simons *et al.*, 2002] as primitives. The equivalence of our presentation to that definition follows from results of [Ferraris and Lifschitz, 2005] and [Ferraris, 2005].

(i) X is a stable model of $l \leq \{A_1, \ldots, A_n\}^c$ iff $X \subseteq \{A_1, \ldots, A_n\}$ and $l \leq |X|$;

(ii) X is a stable model of $\{A_1, \ldots, A_n\}^c \leq u$ iff $X \subseteq \{A_1, \ldots, A_n\}$ and $|X| \leq u$;

(iii) X is a stable model of $l \leq \{A_1, \ldots, A_n\}^c \leq u$ iff $X \subseteq \{A_1, \ldots, A_n\}$ and $l \leq |X| \leq u$.

Proof: Immediate from Proposition 6.   ■

For instance, the stable models of

$$2 \leq \{p, q, r\}^c \leq 2$$

are

$$\{p, q\}, \ \{p, r\}, \ \{q, r\}.$$

Expressions of the forms

$$l \leq \{\cdots\}^c, \ \{\cdots\}^c \leq u, \ l \leq \{\cdots\}^c \leq u$$

can be used in LPARSE code in the head of a rule, with both $\leq$ and the superscript c dropped:

$$1 \ \{\ldots\}, \ \ \{\ldots\} \ \mathrm{u}, \ \ 1 \ \{\ldots\} \ \mathrm{u}.$$

Note that LPARSE understands expressions of these types in different ways depending on whether they occur in the body or in the head of a rule: a choice formula is included in the second case, but not in the first. For instance, the LPARSE rules

```
r :- 1 {p,q}.
1 {p,q} :- r.
```

stand for

$$1 \leq \{p, q\} \rightarrow r$$

and

$$r \rightarrow 1 \leq \{p, q\}^c$$

respectively.

3.2   Variables in the Language of **LPARSE**

A group of rules that follow a pattern can be often described concisely in the input language of LPARSE using schematic variables. As in Prolog, variables must be capitalized. Here is an example:

```
p(1..4).
#domain p(I).
q(I) :- not q(I-1).
```

Assume that these 3 lines are saved as file **var**. The first line of **var** is an LPARSE abbreviation for a group of 4 rules:

```
p(1).  p(2).  p(3).  p(4).
```

It defines the auxiliary "domain" predicate[12] **p**, which is used in the second line to declare **I** to be a variable with the domain $\{1, \ldots, 4\}$. The last line of **var** is interpreted then as a schematic representation of 4 rules:

```
q(1) :- not q(0).
q(2) :- not q(1).
q(3) :- not q(2).
q(4) :- not q(3).
```

Generating these rules from the schematic expression at the end of file **var** is an example of "grounding," which is the main computational task performed by LPARSE.

To sum up, LPARSE interprets **var** as the conjunction of the formulas

$$(15) \quad \begin{array}{l} p(i), \\ \neg q(i-1) \rightarrow q(i) \end{array}$$

$(1 \leq i \leq 4)$. In response to the command

```
% lparse var | smodels 0
```

SMODELS will compute the only stable model of this conjunction:

```
Stable Model: q(1) q(3) p(1) p(2) p(3) p(4)
```

The auxiliary atoms **p(1)**,...,**p(4)** in the output can be suppressed by including the declaration

```
hide p(_).
```

---

[12]The general definition of a domain predicate in the LPARSE manual is somewhat complicated, and it has been changing from one version of the system to another.

("do not display atoms of the form p(_)") in file var. Alternatively, displaying information about domain predicates can be suppressed using the -d option of LPARSE, as follows:

```
% lparse -d none var | smodels 0
```

Besides #domain declarations, there is another mechanism for telling LPARSE how to ground schematic rules: domain predicates can be included directly in the bodies of these rules. For instance, file var can be rewritten in the following way:

```
p(1..4).
q(I) :- p(I), not q(I-1).
```

These two lines represent the conjunction of the formulas

$$(16) \quad \begin{array}{l} p(i), \\ p(i) \land \neg q(i-1) \to q(i) \end{array}$$

($1 \leq i \leq 4$). Since the conjunction of formulas (16) is intuitionistically equivalent to the conjunction of formulas (15), these two conjunctions have the same stable models.

In the language of LPARSE, variables can be used also to describe a list of literals that is formed according to a pattern. For instance, LPARSE understands

```
p(1..4).
2 {q(I) : p(I)} 3.
```

as shorthand for

```
p(1).  p(2).  p(3).  p(4).
2 {q(1), q(2), q(3), q(4)} 3.
```

3.3 Graph Coloring

As we are turning to actual programming examples, we would like to facilitate representing propositional formulas by input files of answer set solvers. To this end, we will usually write formulas in "logic programming notation." A conjunction will be written as a list of its conjunctive terms; if a conjunctive term is an implication $F \to G$ then it will be written as $G \leftarrow F$; if a conjunctive term is a constraint $\neg F$ then it will be written as $\leftarrow F$. If the body F in such an expression is a conjunction then we will separate its conjunctive terms by commas; if the body or one of its conjunctive terms is a literal $\neg A$ then we will write it as *not A*.

An *n-coloring* of a graph G is a function f from its set of vertices to $\{1, \ldots, n\}$ such that $f(x) \neq f(y)$ for every pair of adjacent vertices x, y.

We would like to use ASP to find an n-coloring of a given graph or to determine that it does not exist. To this end, we will write a program whose answer sets are in a 1–1 correspondence with the n-colorings of G.

Let V be the set of vertices of the graph, and E the set of its edges. The program consists of the rules

(17) $1 \leq \{color(x,1), \dots, color(x,n)\}^c \leq 1$ $(x \in V)$,

(18) $\leftarrow color(x,i), color(y,i)$ $(\{x,y\} \in E; \ 1 \leq i \leq n)$.

It has the desired property:

PROPOSITION 13. *A set X of atoms is a stable model of the conjunction of (17) and (18) iff X is*

(19) $\{color(x, f(x)) : x \in V\}$

for some n-coloring f of $\langle V, E \rangle$.

Proof. By Proposition 12(iii), each of the formulas (17) has n stable models $\{color(x,i)\}$ $(i = 1, \dots, n)$. By Proposition 11, it follows that arbitrary stable models of the conjunction of these formulas are unions of such singletons, one per each $x \in V$. In other words, stable models of (17) can be characterized as sets of the form (19), where f is a function from V to $\{1, \dots, n\}$.

By Proposition 4, it follows that the stable models of the conjunction of (17) with the constraints (18) can be characterized as the sets of the form (19) that do not satisfy the bodies of the constraints. The last condition can be expressed by saying that the equalities $f(x) = i$ and $f(y) = i$ cannot hold simultaneously when $\{x,y\} \in E$, which means that $f(x) \neq f(y)$ whenever $\{x,y\} \in E$. ∎

Program (17), (18) can be encoded in the language of LPARSE as the following file `color`:

```
c(1..n).
1 {color(X,I) : c(I)} 1 :- v(X).
:- color(X,I), color(Y,I), e(X,Y), c(I).
```

The domain predicates v and e, characterizing the vertices and edges of G, are assumed to be defined in a separate file, called, say, **graph**. For instance, if G is the 3-dimensional cube then that file may look like this:

```
v(0..7).

e(0,1).  e(1,2).  e(2,3).  e(3,0).
e(4,5).  e(5,6).  e(6,7).  e(7,4).
e(0,4).  e(1,5).  e(2,6).  e(3,7).
```

(There is no need to include atoms with the opposite order of arguments, such as e(1,0); it is only essential that the adjacency relation of G be the symmetric closure of e.) The command line uses the -c option of LPARSE to specify the value of the symbolic constant n, and it instructs LPARSE to concatenate the files graph and color:

```
% lparse -c n=2 -d none graph color | smodels
```

In response, SMODELS produces the set of atoms describing a 2-coloring of the cube:

```
Stable Model: color(0,1) color(1,2) color(2,1) color(3,2)
color(4,2) color(5,1) color(6,2) color(7,1)
```

As can be seen from the proof of Proposition 13, the first part (17) of our coloring program describes a "simple" superset of the set of n-colorings of G that we are trying to capture; the second half (18) consists of the constraints that weed out the "bad" elements of that superset. This "generate-and-test" organization is typical for simple ASP programs. But it would be a mistake to think that answer set solvers operate by generating the elements of the superset described in the generate part and verifying which of these elements satisfy the test conditions. As briefly discussed at the end of Section 2.5, the search algorithms implemented in these systems are based on very different ideas.

3.4 Cliques

A *clique* in a graph G is a set of pairwise adjacent vertices of G. We would like to use ASP to find a clique of a cardinality $\geq n$ in a given graph or to determine that it does not exist. To this end, we will write a program whose answer sets are in a 1–1 correspondence with cliques of cardinalities $\geq n$.

As before, V is the set of vertices of the graph, and E the set of its edges. The program consists of the rules

(20) $n \leq \{in(x) : x \in V\}^c$,

(21) $\leftarrow in(x), in(y)$ $(x, y \in V;\ x \neq y;\ \{x, y\} \notin E).$

PROPOSITION 14. *A set X of atoms is a stable model of the conjunction of (20) and (21) iff X is*

(22) $\{in(x) : x \in C\}$

for some clique C in G such that $|C| \geq n$.

Proof. By Proposition 12(i), the stable models of (20) can be characterized as sets of the form (22), where C is a set of vertices of a cardinality $\geq n$. By Proposition 4, it follows that the stable models of the conjunction of (20) with the constraints (21) can be characterized as the sets of the form (22) that do not satisfy the bodies of the constraints. The last condition can be expressed by saying that the conditions $x \in C$ and $y \in C$ cannot hold simultaneously for two different non-adjacent vertices x, y, which means that C is a clique.   ■

Here is an LPARSE encoding of (20), (21):

```
n {in(X) : v(X)}.
:- in(X), in(Y), v(X;Y), X!=Y, not e(X,Y), not e(Y,X).
```

The domain predicates v and e are assumed to characterize the vertices and edges of G, as in Section 3.3. In the body of the second rule, v(X;Y) is an LPARSE abbreviation for v(X),v(Y), and != represents $\neq$. The pair of conditions not e(X,Y), not e(Y,X) expresses that X and Y are non-adjacent.

3.5   Schur Numbers

A set S of integers is called *sum-free* if there are no numbers x, y in S such that $x + y$ is in S. For instance, $\{1, 3, 5\}$ is sum-free, and $\{2, 3, 5\}$ and $\{2, 4\}$ are not. We would like to use ASP to find, for given k and n, a partition of the interval $\{1, \ldots, n\}$ into at most k sum-free sets or to determine that such a partition does not exist. (The largest n such that $\{1, \ldots, n\}$ can be partitioned into k sum-free set is called the k-th *Schur number* and denoted by $S(k)$.)

In the following program the atoms $s_i(x)$ $(1 \leq i \leq k, 1 \leq x \leq n)$ are used to express that x belongs to the i-th set S_i in a partition of $\{1, \ldots, n\}$ into sum-free sets $S_1, \ldots, S_k$:

(23) $1 \leq \{s_1(x), \ldots, s_k(x)\}^c \leq 1$ $\qquad (1 \leq x \leq n)$,

(24) $\leftarrow s_i(x), s_i(y), s_i(x + y)$ $\qquad (1 \leq i \leq k;\ x, y \geq 1;\ x + y \leq n)$.

The proposition below expresses the correctness of this program. Note that the conditions on the sets S_i in the statement of the proposition allow these sets to be empty, so that the list $S_1, \ldots, S_k$ represents a partition into *at most* k sets, not exactly k.

PROPOSITION 15.   *A set X of atoms is a stable model of the conjunction of (23) and (24) iff X is*

(25) $\{s_i(x) : 1 \leq i \leq k;\ x \in S_i\}$

for sum-free pairwise disjoint sets $S_1, \ldots, S_k$ *such that*

(26) $S_1 \cup \cdots \cup S_k = \{1, \ldots, n\}.$

Proof. By Proposition 12(iii), each of the formulas (23) has k stable models $\{s_i(x)\}$ ($i = 1, \ldots, k$). By Proposition 11, it follows that arbitrary stable models of the conjunction of these formulas are unions of such singletons, one per each $x \in \{1, \ldots, n\}$. In other words, the stable models of (23) can be characterized as sets of the form (25), where the sets S_i are pairwise disjoint and satisfy (26).

By Proposition 4, it follows that the stable models of the conjunction of (23) with the constraints (24) can be characterized as sets of the form (25), where S_i are pairwise disjoint, satisfy (26), and do not satisfy the bodies of the constraints. The last condition can be expressed by saying that each S_i is sum-free. ∎

Here is how rules (23), (24) can be written in the language of LPARSE:

```
subset(1..k).
number(1..n).
#domain number(X;Y).

1 {s(I,X) : subset(I)} 1.
:- s(I,X), s(I,Y), s(I,X+Y), subset(I), X+Y<=n.
```

(In the last rule, `<=` is the LPARSE symbol for $\leq$.) In response to the command

```
% lparse -c k=3 -c n=13 -d none schur | smodels
```

SMODELS produces the output

```
Stable Model: s(3,1) s(1,2) s(1,3) s(3,4) s(2,5) s(2,6) s(2,7)
s(2,8) s(2,9) s(3,10) s(1,11) s(1,12) s(3,13)
```

which represents a partition of the interval $\{1, \ldots, 13\}$ into 3 sum-free sets:

$$\{2, 3, 11, 12\} \cup \{5, 6, 7, 8, 9\} \cup \{1, 4, 10, 13\}.$$

If we replace 13 by 14 in the command line then SMODELS will report that the program has no stable models; thus $S(3) = 13$.

3.6 Tiling

We would like to use ASP to find a way to cover an 8×8 chessboard by twenty-one 3×1 tiles and one 1×1 tile.

The idea of the solution below is due to Ashish Gupta (personal communication). The problem can be reformulated as follows: place twenty-one 3×1 tiles on an 8×8 chessboard without overlaps. If a tile is placed on the chessboard horizontally then we will describe its position by the atom $h(x, y)$ ($0 \le x \le 5$, $0 \le y \le 7$), where x, y are the coordinates of the tile's southwest corner. If a tile is placed on the chessboard vertically then we will describe its position by the atom $v(x, y)$ ($0 \le x \le 7$, $0 \le y \le 5$); x and y have the same meaning. Call these 96 atoms $A_1, \ldots, A_{96}$. The stable models of the rule

(27) $21 \le \{A_1, \ldots, A_{96}\}^c \le 21$

correspond to all possible ways to place 21 tiles on the chessboard. To this "generate" part we now add the constraints testing an arrangement for overlaps. Overlaps between two horizontal tiles are eliminated by the rules

(28) $\leftarrow h(x, y), h(x + i, y)$ ($0 \le x, y \le 7$; $i = 1, 2$).

For overlaps between the vertical tiles, we include

(29) $\leftarrow v(x, y), v(x, y + i)$ ($0 \le x, y \le 7$; $i = 1, 2$).

Finally, we eliminate overlaps between a horizontal tile and a vertical tile:

(30) $\leftarrow h(x, y), v(x + i, y - j)$ ($0 \le x, y \le 7$; $0 \le i, j \le 2$).

The stable models of program (27)–(30) correspond to the solutions to the tiling problem we are interested in.

The program above can be represented in the language of LPARSE as follows:

```
number(0..7).
#domain number(X;Y;I;J).

hpos(X,Y)  :- X<=5.
vpos(X,Y)  :- Y<=5.

21 {h(XX,YY) : hpos(XX,YY), v(XX,YY) : vpos(XX,YY)} 21.

:- h(X,Y), h(X+I,Y), 0<I, I<=2.
:- v(X,Y), v(X,Y+I), 0<I, I<=2.
:- h(X,Y), v(X+I,Y-J), I<=2, J<=2.
```

The domain predicates hpos and vpos represent the possible positions of horizontal and vertical tiles. In the output of SMODELS we read:

```
Stable Model: h(5,1) h(5,0) h(3,7) h(3,6) h(3,5) h(3,4) h(3,3)
h(3,2) h(2,1) h(2,0) h(0,7) h(0,6) v(7,5) v(7,2) v(6,5) v(6,2)
v(2,3) v(1,3) v(1,0) v(0,3) v(0,0)
```

3.7 Hamiltonian Cycles

Each of the programs in Sections 3.3–3.6 is a conjunction of choice formulas and constraints. In the next example we will have a chance to use Horn formulas as well.

A *Hamiltonian cycle* in a directed graph G is a closed path that passes through each vertex of G exactly once. We would like to use ASP to find a Hamiltonian cycle in a given directed graph or to determine that it does not exist.

The program below uses the atoms $in(x, y)$ for all edges $\langle x, y \rangle$ of G to express that $\langle x, y \rangle$ belongs to the path. The generate part of the program consists of the choice rules

$$(31) \quad \{in(x, y)\}^c \qquad (\langle x, y \rangle \in E)$$

(E stands for the set of edges of G). We need to conjoin them with constraints that eliminate all subsets of E other than Hamiltonian cycles.

Two useful constraints are

$$(32) \quad \leftarrow 2 \le \{in(x, y) : y \in A_x\} \qquad (x \in V),$$

where A_x stands for $\{y : \langle x, y \rangle \in E\}$, and

$$(33) \quad \leftarrow 2 \le \{in(x, y) : x \in B_y\} \qquad (y \in V),$$

where B_y stands for $\{x : \langle x, y \rangle \in E\}$. They ensure that two *in*-edges neither start nor end at the same vertex, so that the set of *in*-edges is a path or a union of disjoint paths. In addition, we want to require that every vertex of G be reachable by a sequence of *in*-edges from some fixed vertex x_0. We will do this using the auxiliary atoms $r(x)$ ("x is reachable from x_0") for all vertices x of G. The following two rules provide a "recursive definition" of r:

$$(34) \quad r(x) \leftarrow in(x_0, x) \qquad (x \in V),$$

$$(35) \quad r(y) \leftarrow r(x), in(x, y) \qquad (\langle x, y \rangle \in E)$$

(V stands for the set of vertices of G). Now we are ready to impose the reachability constraints:

(36) $\leftarrow not\ r(x)$ $(x \in V)$.

Besides the generate part (31) and the test part (32), (33), (36), this program contains the rules (34) and (35), which define the auxiliary atoms used in one of the test rules. This "generate-define-test" structure is typical for more advanced ASP programs. As in the example above, the definitions of auxiliary atoms are often recursive.

The following proposition expresses the correctness of program (31)–(36). In its statement, the *essential part* of a set X of atoms is the set of atoms in X that have the form $in(x, y)$.

PROPOSITION 16. *A set X of atoms is the essential part of a stable model of (31)–(36) iff X has the form*

(37) $\{in(x, y)) : \langle x, y \rangle \in H\}$

where H is the set of edges of a Hamiltonian cycle in G. Furthermore, different stable models of this program have different essential parts.

The last sentence shows that if we "hide" the atoms of the form $r(x)$ in the list of stable models of this program produced by an answer set solver then the output will contain each Hamiltonian cycle of G exactly once.

For any set $H \subseteq E$, by R_H we denote the set of atoms $r(x)$ for all vertices x to which there is a path of nonzero length from x_0 over edges in H.

LEMMA 17. *A set X of atoms is a stable model of the conjunction of formulas (31), (34) and (35) iff X is*

(38) $\{in(x, y)) : \langle x, y \rangle \in H\} \cup R_H$

for some subset H of E.

Proof. Denote the conjunction of formulas (31) by F, and the conjunction of formulas (34), (35) by G. By Theorem 8, X is a stable model of $F \wedge G$ iff there exists a stable model $\{A_1, \ldots, A_n\}$ of F such that X is a stable model of $A_1 \wedge \cdots \wedge A_n \wedge G$. By Proposition 3, it follows that the stable models of $F \wedge G$ can be characterized as the stable models of formulas of the form

(39) $\bigwedge_{\langle x,y \rangle \in H} in(x, y) \wedge G$

for arbitrary subsets H of E. Formula (39) is a Horn formula, and its minimal model is its only stable model (Theorem 2). It remains to observe that the minimal model of (39) is (38). ∎

Proof of Proposition 16. A set $H \subseteq E$ is the set of edges of a Hamiltonian cycle in G iff it satisfies the following conditions:

(i) H does not contain two different edges leaving the same vertex.

(ii) H does not contain two different edges ending at the same vertex.

(iii) For every vertex x of G, there exists a path of nonzero length from x_0 to x over edges in H.

By Lemma 17 and Proposition 4, a set X of atoms is a stable model of program (31)–(36) iff X has the form (38), where $H \subseteq E$, and does not satisfy the bodies of the constraints (32), (33), (36). It is clear that

- (i) holds iff (38) does not satisfy the bodies of constraints (32);

- (ii) holds iff (38) does not satisfy the bodies of constraints (33);

- (iii) holds iff (38) does not satisfy the bodies of constraints (36).

Consequently X is a stable model of (31)–(36) iff X has the form (38) for a subset H of E satisfying conditions (i)–(iii). Both parts of the statement of Proposition 16 now follow, because the essential part of (38) is (37). ∎

The discussion of the Hamiltonian cycles example above is based on [Erdoğan and Lifschitz, 2004, Section 5].

Here is a representation of program (31)–(36) in the language of LPARSE, assuming that x_0 is 0:

```
{in(X,Y)} :- e(X,Y).

:- 2 {in(X,Y) : e(X,Y)}, v(X).
:- 2 {in(X,Y) : e(X,Y)}, v(Y).

r(X) :- in(0,X), v(X).
r(Y) :- r(X), in(X,Y), e(X,Y).

:- not r(X), v(X).

hide r(_).
```

3.8 The Blocks World

The blocks world consists of several blocks $1, \ldots, n$, placed on the table so that they form a tower or several towers. For instance, if $n = 2$ then the blocks world can be in 3 states:

```
    1              2              ·
    2              1          1   2
------       -------      -------
```

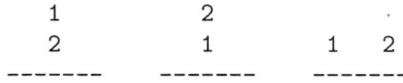

If $n = 3$ then 13 states are possible: 6 configurations in which the blocks form one tower; 6 configurations in which 2 blocks form a tower and the third is on the table; one configuration in which all blocks are on the table.

Blocks can be moved around, and in Section 3.10 we show how ASP can be used to find a sequence of actions that takes the blocks world from a given initial state to a given goal state (or, more generally, to a state satisfying a given goal condition). As a preliminary step, in this section we write an ASP program that represents the set of all possible configurations of n blocks.

Positions of blocks are described in this program by the atoms $on(x, y)$, where $x \in \{1, \ldots, n\}$, $y \in \{1, \ldots, n, table\}$, $x \neq y$. The first rule of the program is the choice rule allowing us to choose arbitrarily, for each block x, a unique location:

(40) $1 \leq \{on(x, y) : y \in \{1, \ldots, n, table\} \setminus \{x\}\}^c \leq 1$

$(1 \leq x \leq n)$. Furthermore, we do not allow two blocks to be on top of the same block:

(41) $\leftarrow 2 \leq \{on(x, y) : x \in \{1, \ldots, n\} \setminus \{y\}\}$

$(1 \leq y \leq n)$. These constraints are not sufficient, however, for eliminating all "bad" stable models of (40), because they allow subsets of blocks to form circular configurations "floating in space," such as $on(1, 2)$ and $on(2, 1)$.

The absence of such configurations can be expressed using an auxiliary recursively defined predicate, similar to the predicate r used in Section 3.7 to describe Hamiltonian cycles. The atoms $s(x)$, where $1 \leq x \leq n$, will express that x is supported by the table, that is to say, belongs to a tower of blocks that rests on the table. They are defined by the rules

(42) $s(x) \leftarrow on(x, table)$ $(1 \leq x \leq n)$,

(43) $s(x) \leftarrow s(y), on(x, y)$ $(1 \leq x, y \leq n; x \neq y)$.

The absence of blocks floating in space is expressed by the constraints

(44) $\leftarrow not \ s(x)$ $(1 \leq x \leq n)$.

In the language of LPARSE:

```
block(1..n).

1 {on(X,Y) : block(Y) : X!=Y, on(X,table)} 1 :- block(X).

:- 2 {on(X,Y) : block(X) : X!=Y}, block(Y).

s(X) :- on(X,table), block(X).
s(X) :- s(Y), on(X,Y), block(X;Y), X!=Y.

:- not s(X), block(X).

hide s(_).
```

3.9   Strong Negation

Some applications of ASP, including those related to actions and planning, are facilitated by the use of a second kind of negation, called "strong" (or "classical," or "true"), proposed in [Gelfond and Lifschitz, 1991].

Recall that propositional formulas are formed from atoms and the 0-place connective $\perp$ using the binary connectives $\wedge$, $\vee$ and $\rightarrow$ (Section A.1). Assume that we distinguish between atoms of two kinds, *positive* and *negative*, and that each negative atom is an expression of the form $\sim A$, where A is a positive atom. The symbol $\sim$ is called *strong negation*.

Note that syntactically strong negation is not really a connective, according to this definition: it is allowed to occur in front of positive atoms only. For example, expressions $\sim \sim p$ and $\sim (p \wedge q)$ are not formulas.[13]

A set of atoms is *coherent* if it does not contain "complementary" pairs of atoms A, $\sim A$.

Consider, for instance, the program

$$
\begin{array}{l}
\{p\}^c, \\
(45) \quad q, \\
\sim q \leftarrow \neg p.
\end{array}
$$

It contains two positive atoms p, q and one negative atom $\sim q$. It is easy to check using the method of Section 2.7 that the stable models of this program are $\{p,q\}$ and $\{q, \sim q\}$. The first of them is coherent, and the second is not.

The problem of computing the coherent stable models of a formula can be easily reduced to the problem of computing arbitrarily stable models:

PROPOSITION 18. *A set X of atoms is a coherent stable model of a formula F iff X is a stable model of the formula*

---

[13]Alternatively, strong negation can be treated as an additional connective, in the spirit of [Nelson, 1949].

$$(46) \quad F \wedge \bigwedge_A \neg (A \wedge \sim A),$$

where the big conjunction extends over all positive atoms A such that both A and $\sim A$ are head atoms of F.

Proof. By Proposition 4, X is a stable model of (46) iff X is a stable model of F which does not have subsets of the form $\{A, \sim A\}$ such that A, $\sim A$ are head atoms of F. By Theorem 1, this condition on X is equivalent to saying that X is a coherent stable model of F. ∎

In the input language of LPARSE, strong negation is written as $-$. When the input program contains strong negation, LPARSE should be called with the option `--true-negation` . The answer set solvers that accept input programs with strong negation, such as SMODELS, generate coherent answer sets only. For instance, if we save the rules

```
{p}.
q.
-q :- not p.
```

as `file45` and give the command

```
% lparse --true-negation file45 | smodels 0
```

then the output will contain only one model:

```
Stable Model: p q
```

Strong negation allows us to distinguish between the assertions "A is false" and "A is not known to be true" in ASP programs. The former is expressed by the presence of the negative atom $\sim A$ in a coherent stable model; the latter, by the absence of the positive atom A, which is obviously a weaker condition. The rule

$$(47) \quad \sim A \leftarrow \mathit{not}\ A$$

("A is false if there is no evidence to the contrary") is an ASP representation of the closed world assumption [Reiter, 1978] for the positive atom A. The following proposition describes the effect of adding this rule on the stable models of a program.

PROPOSITION 19. *Let F be a formula and A a positive atom such that $\sim A$ does not occur in F. For any set X of atoms, X is a coherent stable model of*

$$(48) \quad F \wedge (\neg A \rightarrow \sim A)$$

iff

   (i) *X is a stable model of F and $A \in X$, or*

   (ii) *$X = Y \cup \{\sim A\}$, where Y is a stable model of F such that $A \notin Y$.*

Case (ii) is the case when there is no evidence that A is true, and the closed world assumption leads us to the conclusion that A is false.

Proof. By Theorem 8, X is a stable model of (48) iff X is a stable model of a formula of the form

$$(49)\quad A_1 \wedge \cdots \wedge A_n \wedge (\neg A \rightarrow \sim A),$$

where $\{A_1, \ldots, A_n\}$ is a stable model of F. *Case 1:* A equals one of the atoms A_i. Then (49) is intuitionistically equivalent to $A_1 \wedge \cdots \wedge A_n$, and X is a stable model of (49) iff $X = \{A_1, \ldots, A_n\}$. *Case 2:* A is different from all atoms A_i. Then A is not a head atom of (49). By Proposition 9, it follows that X is a stable model of (49) iff X is a stable model of the formula

$$A_1 \wedge \cdots \wedge A_n \wedge (\neg \bot \rightarrow \sim A),$$

which is intuitionistically equivalent to

$$A_1 \wedge \cdots \wedge A_n \wedge \sim A.$$

So X is a stable model of (48) iff $X = Y \cup \{\sim A\}$, where Y stands for $\{A_1, \ldots, A_n\}$.   ■

The rule

$$(50)\quad A \leftarrow not \sim A$$

expresses the inverse closed world assumption: A is true if there is no evidence to the contrary.

3.10 Planning

We would like to use ASP to find a sequence of actions that takes the blocks world (Section 3.8) from a given initial state to a state satisfying a given goal condition. To be more precise, we will be looking for a sequence of *sets* of actions, because some actions can be executed concurrently. There are n^2 possible actions, where n is the number of blocks: any block $x \in \{1, \ldots, n\}$ can be moved to any location $l \in \{1, \ldots, n, table\}$ different from x.

We assume that

- a block can be moved only when there are no blocks on top of it, and

- at most k actions can be executed concurrently

(think of a robot with k grippers that can only grasp a block from above). We also assume that

- a block x can be moved onto a block y only if y is not being moved at the same time

(the robot's ability to coordinate the movements of the grippers is not good enough for that).

A *history* is a finite sequence

$$s_0, e_0, s_1, e_1, \ldots, e_{m-1}, s_m$$

where $s_0, s_1, \ldots, s_m$ are states of the blocks world, and each e_i ($0 \leq i < m$) is a set of actions (an "event"), which, when executed concurrently in state s_i, lead to state s_{i+1}. We will write a program whose stable models represent the histories with a given initial state s_0 and a given length m such that their final state s_m satisfies a given goal condition.

Histories will be described by

- the atoms $on(x, l, i)$ ($x \in \{1, \ldots, n\}$, $l \in \{1, \ldots, n, table\}$, $x \neq l$, $i \in \{0, \ldots, m\}$), expressing that x is on l in state s_i, and

- the atoms $move(x, l, i)$ ($x \in \{1, \ldots, n\}$, $l \in \{1, \ldots, n, table\}$, $x \neq l$, $i \in \{0, \ldots, m - 1\}$), expressing that x is moved onto l as part of event e_i.

One of the rules of the program (rule (55) below) uses these atoms to describe the effect of moving a block: if x is moved onto l as part of event e_i then x is on l in state s_{i+1}.

The program contains strong negation (Section 3.9), which is applied to the atoms $on(x, l, i)$. The usefulness of strong negation in ASP programs describing effects of actions is related to the frame problem [Shanahan, 1997]—the problem of describing what does *not* change when actions are executed. If $x_1, x_2, \ldots$ are the blocks that are moved in the course of event e_i then rule (55) tells us where these blocks are going to be afterwards. But what about locations of the blocks *other* than $x_1, x_2, \ldots$? An adequate formalization should allow us to conclude that the locations of all the other blocks will not change; in state s_{i+1} each of them will stay where it was in state s_i.

An elegant way to ensure this is to postulate the default that Leibniz stated in his *Introduction to a Secret Encyclopedia* and that is now called the *commonsense law of inertia*: "Everything is presumed to remain in the state in which it is" [Leibniz, 1995, p. 9]. In particular, the location of a block after event e_i is presumed, in the absence of evidence to the contrary, to remain the same as it was before the event. Blocks $x_1, x_2, \ldots$ are exceptions: since they are moved, rule (55) provides evidence that their locations may not remain the same. We have seen in Section 3.9 that strong negation helps us formalize another default—the closed world assumption; here strong negation will be used to solve the frame problem.

The generate part of the program expresses that each event e_i can be composed of up to k actions, chosen arbitrarily:

(51) $\{move(x, l, i) : 1 \leq x \leq n,\ l \in \{1, \ldots, n, table\},\ x \neq l\}^c \leq k$

$(1 \leq i < m)$. This rule is followed by constraints expressing that a block can be moved only if it is clear

(52) $\leftarrow move(x, l, i), on(y, x, i)$

$(1 \leq x, y \leq n,\ l \in \{1, \ldots, n, table\},\ x \neq l,\ x \neq y,\ 0 \leq i < m)$ and if the destination is not a block that is being moved also:

(53) $\leftarrow move(x, y, i), move(y, l, i)$

$(1 \leq x, y \leq n,\ l \in \{1, \ldots, n, table\},\ x \neq y,\ y \neq l,\ 0 \leq i < m)$.

The next part of the program defines the locations of blocks in state s_i in terms of their initial locations and the events $e_0, \ldots, e_{i-1}$. It begins with the rules

(54) $on(x, init(x), 0)$

$(1 \leq x \leq n)$, where $init(x)$ stands for the initial location of x, and

(55) $on(x, l, i + 1) \leftarrow move(x, l, i)$

$(1 \leq x \leq n,\ l \in \{1, \ldots, n, table\},\ x \neq l,\ 0 \leq i < m)$. The next rule expresses the uniqueness of the location of a block using strong negation:

(56) $\sim on(x, l, i) \leftarrow on(x, l', i)$

$(1 \leq x \leq n,\ l, l' \in \{1, \ldots, n, table\},\ x \neq l,\ x \neq l',\ l \neq l',\ 0 \leq i \leq m)$. The last rule in this group expresses the commonsense law of inertia for the blocks world:

(57) $on(x, l, i + 1) \leftarrow on(x, l, i), not \sim on(x, l, i + 1)$

$(1 \leq x \leq n, l \in \{1, \ldots, n, table\}, x \neq l, 0 \leq i < m)$. It says that

> x is on l in state s_{i+1} (the head of the rule) if
>
> x is on l in state s_i (the first term of the body) and
>
> the rules of the program provide no evidence to the contrary (the second term of the body).

Note that except for the presence of the term $on(x, l, i)$ in the body, (57) has the same syntactic form as (50).

Finally, we need to include constraints expressing that $s_1, \ldots, s_m$ are valid states of the blocks world, and that s_m satisfies the goal condition G:

(58) $\leftarrow 2 \leq \{on(x, y) : x \in \{1, \ldots, n\} \setminus \{y\}\}$

$(0 \leq y \leq n, 0 \leq i < m)$;

(59) $\leftarrow not\ G$.

Rule (58) says that two blocks cannot be on top of the same block; this is a counterpart of rule (41). Counterparts of the other properties of valid states discussed in Section 3.8 are not needed in the new framework. Indeed, the existence of the location of every block and the absence of circular configurations are assumed to hold in the initial state described by the function *init*, and these properties are preserved when blocks are moved; the uniqueness of the location of a block is expressed by (56).

Here is program (51)–(59) in the language of LPARSE:

```
step(0..m).
block(1..n).
location(1..n;table).

#domain step(I).
#domain block(X;Y;Z).
#domain location(L;L1).

{move(XX,LL,I) : block(XX) : location(LL) : XX!=LL} k :- I<m.

:- move(X,L,I), on(Y,X,I), X!=L, X!=Y, I<m.
:- move(X,Y,I), move(Y,L,I), X!=Y, Y!=L, I<m.

on(X,L,0)   :- init(X,L).
on(X,L,I+1) :- move(X,L,I), X!=L, I<m.
-on(X,L,I)  :- on(X,L1,I), X!=L, X!=L1, L!=L1.
on(X,L,I+1) :- on(X,L,I), not -on(X,L,I+1), X!=L, I<m.
```

```
:- 2 {on(XX,Y,I) : block(XX) : XX!=Y}.
:- not goal.
```

```
hide.
show move(_,_,_).
```

The last two lines instruct SMODELS to display no atoms except for the actions `move(...)`. The initial state and the goal condition are assumed to be defined in a separate file, for instance:

```
init(1,2).  init(2,table).  init(3,4).
init(4,table).  init(5,6).  init(6,table).
```

```
goal :- on(2,1,m), on(3,2,m), on(6,5,m), on(5,4,m).
```

The idea of the solution to the frame problem given by rule (57) goes back to [Reiter, 1980, Section 1.1.4], but implementing that idea was not straightforward [Hanks and McDermott, 1987], and it was achieved years later [Turner, 1997]. The fact that planning can be reduced to finding a stable model was noted in [Subrahmanian and Zaniolo, 1995], and first experiments on generating plans using SMODELS were reported in [Dimopoulos et al., 1997]. The discussion of blocks world planning in this section is based on [Lifschitz, 2002, Section 5].

4   Proofs of Theorems

4.1   Proof of Theorem 1

LEMMA 20. *If* $X \models F$ *and a set* Y *contains all head atoms of* F *then* $X \cap Y \models F^X$.

Proof. The proof is by structural induction on F. Assume that $X \models F$. Clearly F is not $\bot$. *Case 1:* F is an atom A. Since $X \models F$, F^X is A and $A \in X$. Since A is a head atom, we can further conclude that $A \in X \cap Y$. *Case 2:* F is $G \wedge H$. Since $X \models F$, we know that F^X is $G^X \wedge H^X$, $X \models G$ and $X \models H$. Since all head atoms of G and H belong to Y, from the induction hypothesis we conclude that $X \cap Y \models G^X$ and $X \cap Y \models H^X$. Consequently $X \cap Y \models F^X$. *Case 3:* F is $G \vee H$. Similar to Case 2. *Case 4:* F is $G \to H$. Since $X \models F$, F^X is $G^X \to H^X$. *Case 4.1:* $X \models G$. Then $X \models H$. Since all head atoms of H belong to Y, from the induction hypothesis we conclude that $X \cap Y \models H^X$. Consequently $X \cap Y \models F^X$. *Case 4.2:* $X \not\models G$. Then G^X is $\bot$, so that F^X is a tautology. ∎

THEOREM 1. *Any stable model of* F *is a subset of the set of head atoms of* F.

Proof. Let X be a stable model of F, and Y the set of head atoms of F. By Lemma 20, $X \cap Y \models F^X$. Since X is minimal among the sets satisfying F^X, it follows that $X \cap Y = X$, and consequently $X \subseteq Y$. ∎

4.2 Proof of Theorem 2

LEMMA 21. *For any Horn formula F and any two sets X and Y of atoms, if $X \subseteq Y$ and $Y \models F$ then $X \models F$ iff $X \models F^Y$.*

Proof. Assume first that F is a single implication

(60) $A_1 \wedge \cdots \wedge A_n \to A.$

Case 1: $A_1, \ldots, A_n$ belong to Y. Under the assumption $Y \models F$ the consequent A of F belongs to Y also, so that $F^Y = F$. *Case 2:* for some i, $A_i \notin Y$. Under the assumption $X \subseteq Y$, $A_i \notin X$, so that X satisfies F. On the other hand, F^Y is the tautology $\bot \to A^Y$, so that X satisfies F^Y as well.

If F is a conjunction $F_1 \wedge \cdots \wedge F_m$ of several implications of the form (60) then X satisfies F iff X satisfies each F_j. Under the assumption $Y \models F$, F^Y is $F_1^Y \wedge \cdots \wedge F_m^Y$; consequently X satisfies F^Y iff X satisfies each of the conjunctive terms F_j^Y. The assertion of the lemma follows from the special case proved above. ∎

THEOREM 2. *For any Horn formula F, the minimal model of F is the only stable model of F.*

Proof. Let M be the minimal model of a Horn formula F. Lemma 21, applied to M as Y, shows that F^M is satisfied by M but is not satisfied by any proper subset of M. Consequently M is a stable model of F. Now take any stable model Y of F. By the choice of M, $M \subseteq Y$. Lemma 21, applied to M as X, shows that $M \models F^Y$. By the definition of a stable model, Y is minimal among the sets satisfying F^Y. Consequently $Y \subseteq M$. We have proved that $Y = M$. ∎

4.3 Proof of Theorems 5 and 7

LEMMA 22. *For any formula F and any set X of atoms, $X \models F^X$ iff $X \models F$.*

Proof. Reduct F^X is obtained from F by replacing some subformulas that are not satisfied by X with $\bot$. ∎

LEMMA 23. *For any two formulas F and G and any set X of atoms,*

  (a) $(F \wedge G)^X$ *is equivalent to* $F^X \wedge G^X$ *in classical logic, and*

(b) $(F \vee G)^X$ is equivalent to $F^X \vee G^X$ in classical logic.

Proof. Part (a): consider two cases, depending on whether X satisfies $F \wedge G$. If it does then the two formulas are equal to each other; if not then each of them is equivalent to $\perp$. For part (b), the proof is similar. ∎

LEMMA 24. *For any formula F and any two sets X and Y of atoms, $X \models F^Y$ iff $\langle X \cap Y, Y \rangle \models F$.*

Proof. The proof is by structural induction on F. If F is $\perp$ then the assertion of the lemma is trivial. If F is an atom A,

$$
\begin{aligned}
X \models A^Y \quad &\text{iff} \quad A \in Y \text{ and } A \in X \\
&\text{iff} \quad A \in X \cap Y \\
&\text{iff} \quad \langle X \cap Y, Y \rangle \models A.
\end{aligned}
$$

If F is $G \wedge H$ then, using Lemma 23(a),

$$
\begin{aligned}
X \models (G \wedge H)^Y \quad &\text{iff} \quad X \models G^Y \wedge H^Y \\
&\text{iff} \quad X \models G^Y \text{ and } X \models H^Y \\
&\text{iff} \quad \langle X \cap Y, Y \rangle \models G \text{ and } \langle X \cap Y, Y \rangle \models H \\
&\text{iff} \quad \langle X \cap Y, Y \rangle \models G \wedge H.
\end{aligned}
$$

If F is $G \vee H$ then the reasoning is similar, using Lemma 23(b). Finally, if F is $G \rightarrow H$,

$$
\begin{aligned}
X \models (G \rightarrow H)^Y \quad &\text{iff} \quad Y \models G \rightarrow H \text{ and } X \models G^Y \rightarrow H^Y \\
&\text{iff} \quad Y \models G \rightarrow H \text{ and} \\
&\qquad\qquad\qquad\qquad X \not\models G^Y \text{ or } X \models H^Y \\
&\text{iff} \quad Y \models G \rightarrow H \text{ and} \\
&\qquad\qquad\qquad \langle X \cap Y, Y \rangle \not\models G \text{ or } \langle X \cap Y, Y \rangle \models H \\
&\text{iff} \quad \langle X \cap Y, Y \rangle \models G \rightarrow H.
\end{aligned}
$$

∎

LEMMA 25. *Let F, G, F', G' be formulas such that G' is obtained from F' by replacing some (zero or more) occurrences of F with G. For any set X of atoms, if F^X is equivalent to G^X then $(F')^X$ is equivalent to $(G')^X$.*

Proof. Assume that F^X is equivalent to G^X. By Lemma 22, it follows that

(61) $X \models F \leftrightarrow G$.

We will prove that $(F')^X$ is equivalent to $(G')^X$ by structural induction on F'. This assertion is trivial when F' equals F and also when the number of occurrences of F in F' that are being replaced is 0; in particular, the cases when F' is $\bot$ or an atom are trivial. Assume that F' has the form $F'_1 \odot F'_2$, and G' is $G'_1 \odot G'_2$, where G'_i is obtained from F'_i by replacing some occurrences of F with G. *Case 1:* $X \not\models F'$. In view of (61), $X \not\models G'$, so that $(F')^X = \bot$ and $(G')^X = \bot$. *Case 2:* $X \models F'$. In view of (61), $X \models G'$, so that $(F')^X = (F'_1)^X \odot (F'_2)^X$ and $(G')^X = (G'_1)^X \odot (G'_2)^X$, and the claim follows by the induction hypothesis.   ∎

COMBINED STATEMENT OF THEOREMS 5 AND 7. *For any formulas F and G, the following conditions are equivalent:*

(i) *F is strongly equivalent to G,*

(ii) *for every unary formula H, $F \wedge H$ and $G \wedge H$ have the same stable models,*

(iii) *F is equivalent to G in the logic of here-and-there,*

(iv) *for any set X of atoms, F^X is equivalent to G^X in classical logic.*

Proof. From (i) to (ii): obvious.

From (ii) to (iii): assume that F is not equivalent to G in the logic of here-and-there, and let $\langle X, Y \rangle$ be an HT-interpretation that satisfies, say, F but not G. Then $X \subseteq Y$ and, by Lemma 24, $X \models F^Y$, $X \not\models G^Y$. Since $X \models F^Y$, F^Y is not $\bot$, which implies that $Y \models F$. By Lemma 22, it follows that $Y \models F^Y$. *Case 1:* $Y \not\models G^Y$. By Lemma 22, $Y \not\models G$, so that Y is not a stable model of $G \wedge H$ for any H. But if we take H to be $\bigwedge_{A \in Y} A$ then Y is a stable model of $F \wedge H$. Indeed, by Lemma 23(a), $(F \wedge H)^Y$ is equivalent to $F^Y \wedge H^Y$, which is the same as $F^Y \wedge H$; both conjunctive terms of this formula are satisfied by Y, but the second term is not satisfied by any proper subset of Y. *Case 2:* $Y \models G^Y$. Since $X \not\models G^Y$, X is different from Y; consequently X is a proper subset of Y. Let H be the unary formula

$$\bigwedge_{A \in X} A \wedge \bigwedge_{A, A' \in Y \setminus X} (A \to A').$$

Set Y is not a stable model of $F \wedge H$. Indeed, just as in Case 1, $(F \wedge H)^Y$ is equivalent to $F^Y \wedge H$; X is a proper subset of Y that satisfies both conjunctive terms. We will show, on the other hand, that Y is a stable model of $G \wedge H$, which contradicts condition (ii). In view of Lemma 23(a),

$(G \wedge H)^Y$ is equivalent to $G^Y \wedge H$. Clearly Y satisfies both conjunctive terms; the only proper subset of Y that satisfies H is X, and X does not satisfy G^Y.

From (iii) to (iv): if F and G are satisfied by the same HT-interpretations then, by Lemma 24, for any set Y of atoms, F^Y and G^Y are satisfied by the same sets of atoms.

From (iv) to (i): immediate from Lemma 25. ∎

4.4 Proof of Theorem 8

LEMMA 26. *If X is a stable model of F then F^X is equivalent to $\bigwedge_{A \in X} A$.*

Proof. Since all atoms occurring in these two formulas belong to X, it is sufficient to show that the formulas are satisfied by the same subsets of X. By the definition of a stable model, the only subset of X satisfying F^X is X. ∎

LEMMA 27. *Let S be a set of atoms that contains all atoms occurring in a formula F but does not contain any head atoms of a formula G. For any set X of atoms, if X is a stable model of $F \wedge G$ then $X \cap S$ is a stable model of F.*

Proof. Since X is a stable model of $F \wedge G$, $X \models F$, so that $X \cap S \models F$, and, by Lemma 22, $X \cap S \models F^{X \cap S}$. It remains to show that no proper subset Y of $X \cap S$ satisfies $F^{X \cap S}$. Let S' be the set of head atoms of G, and let Z be $X \cap (S' \cup Y)$. Set Z has the following properties:

(i) $Z \cap S = Y$;

(ii) $Z \subset X$;

(iii) $Z \models G^X$.

To prove (i), note that since S' is disjoint from S, and Y is a subset of $X \cap S$,

$$Z \cap S = X \cap (S' \cup Y) \cap S = X \cap Y \cap S = (X \cap S) \cap Y = Y.$$

To prove (ii), note that set Z is clearly a subset of X. It cannot be equal to X, because otherwise we would have, by (i),

$$Y = Z \cap S = X \cap S;$$

this is impossible, because Y is a proper subset of $X \cap S$. Property (iii) follows from Lemma 20, because $X \models G$, and $S' \cup Y$ contains all head atoms of G.

Since X is a stable model of $F \wedge G$, from property (ii) we can conclude that $Z \not\models (F \wedge G)^X$. Consequently, by Lemma 23(a) and property (iii), $Z \not\models F^X$. Since all atoms occurring in F belong to S, $F^X = F^{X \cap S}$, so that we can rewrite this formula as $Z \not\models F^{X \cap S}$. Since all atoms occurring in $F^{X \cap S}$ belong to S, it follows that $Z \cap S \not\models F^{X \cap S}$. By property (i), we conclude that $Y \not\models F^{X \cap S}$. ∎

THEOREM 8. *Let F and G be formulas such that F does not contain any head atoms of G. A set X of atoms is a stable model of $F \wedge G$ iff there exists a stable model $\{A_1, \ldots, A_n\}$ of F such that X is a stable model of*

(62) $A_1 \wedge \cdots \wedge A_n \wedge G.$

Proof. Take formulas F and G such that F does not contain any head atoms of G, and let S the set of atoms occurring in F. Observe first that if a set X of atoms is a stable model of a formula of the form (62), where $A_1, \ldots, A_n \in S$, then $X \cap S = \{A_1, \ldots, A_n\}$. Indeed, by Lemma 27 with $A_1 \wedge \cdots \wedge A_n$ as F, $X \cap S$ is a stable model of $A_1 \wedge \cdots \wedge A_n$, and the only stable model of this formula is $\{A_1, \ldots, A_n\}$. Consequently, the assertion to be proved can be reformulated as follows: a set X of atoms is a stable model of $F \wedge G$ iff

(i) $X \cap S$ is a stable model of F, and

(ii) X is a stable model of $\bigwedge_{A \in X \cap S} A \wedge G$.

If $X \cap S$ is not a stable model of F then X is not a stable model of $F \wedge G$ by Lemma 27. Now suppose that $X \cap S$ is a stable model of F. Then, by Lemma 26, $F^{X \cap S}$ is equivalent to $\bigwedge_{A \in X \cap S} A$. Consequently, by Lemma 23(a),

$$(F \wedge G)^X \leftrightarrow F^X \wedge G^X = F^{X \cap S} \wedge G^X \leftrightarrow \bigwedge_{A \in X \cap S} A \wedge G^X$$

$$= \left(\bigwedge_{A \in X \cap S} A \right)^X \wedge G^X \leftrightarrow \left(\bigwedge_{A \in X \cap S} A \wedge G \right)^X.$$

We can conclude that X is a stable model of $F \wedge G$ iff X is a stable model of $\bigwedge_{A \in X \cap S} A \wedge G$. ∎

5 Conclusion

Many publications in the area of answer set programming are directed towards practical applications, and the titles of several papers of this kind included in the bibliography[14] show the remarkable diversity of the areas

---

[14][Soininen and Niemelä, 1998], [Erdem *et al.*, 2000], [Nogueira *et al.*, 2001], [Heljanko and Niemelä, 2003], [Baral *et al.*, 2004], [Brooks *et al.*, 2005], [Leone *et al.*, 2005], [Hermansson *et al.*, 2005].

of science and technology where ASP may be useful. Success in this work would have been impossible without efficient, reliable, carefully crafted answer set solvers.

The main topic of this paper, however, is theoretical. We have seen that ASP is based on interesting mathematics, including some ideas developed in the early days of modern logic. The senior author (VL) is particularly pleased to contribute a paper on mathematical foundations of answer set programming to a volume in honor of Dov Gabbay in view of the important role that intuitionistic logic plays in this theory. Intuitionistic logic is what both of us were interested in as beginning researchers many years ago, when we first learned about each other's work.

A   Propositional Logic

A.1   Syntax and Semantics

(Propositional) formulas are formed from propositional atoms and the 0-place connective $\perp$ using the binary connectives $\wedge$, $\vee$ and $\rightarrow$. We use

$$\begin{array}{rl} \top & \text{as shorthand for} \quad \perp \rightarrow \perp, \\ \neg F & \text{as shorthand for} \quad F \rightarrow \perp, \\ F \leftrightarrow G & \text{as shorthand for} \quad (F \rightarrow G) \wedge (G \rightarrow F). \end{array}$$

Atoms and negated atoms are called *literals*.

The relation $X \models F$ between a set X of atoms and a formula F is defined recursively:

- for an atom A, $X \models A$ if $A \in X$;

- $X \not\models \perp$;

- $X \models F \wedge G$ if $X \models F$ and $X \models G$;

- $X \models F \vee G$ if $X \models F$ or $X \models G$;

- $X \models F \rightarrow G$ if $X \not\models F$ or $X \models G$.

If $X \models F$ then we say that X *satisfies* F, or is a *model* of F. A formula is a *tautology* if it is satisfied by every set of atoms. A formula F is *equivalent* to a formula G if $F \leftrightarrow G$ is a tautology (or, equivalently, if F and G have the same models).

An occurrence of an atom A in a formula F is *positive* if the number of implications containing that occurrence in the antecedent is even, and *negative* otherwise. For instance, both occurrences of p in the formula

(63) $((p \rightarrow q) \wedge r) \rightarrow p$

are positive, and q, r are negative.   An occurrence of an atom A in a formula F is *strictly positive* if it does not belong to the antecedent of any implication in F. For instance, the second occurrence of p in (63) is strictly positive, and the first is not. Since $\neg F$ is shorthand for $F \rightarrow \bot$, no occurrence of an atom in a formula of the form $\neg F$ can be strictly positive.

A.2   Logic of Here-and-There

The logic of here-and-there is a 3-valued logic that was originally proposed by the inventor of intuitionistic logic Arend Heyting as a technical tool for the purpose of proving that intuitionistic logic is weaker than classical [Heyting, 1930]. (He remarks that the truth values in his truth tables "can be interpreted as follows: 0 denotes a correct proposition, 1 denotes a false proposition, and 2 denotes a proposition that cannot be false but whose correctness is not proved.") We will identify a function from the set of atoms to the extended set of truth values $\{0, 1, 2\}$ with the ordered pair consisting of the set X of atoms that are mapped to 0 and the set Y of atoms that are mapped to 0 or 2. (If an atom belongs to X then it is true "here"; if an atom belongs to Y then it is true "there".)

An *HT-interpretation* is an ordered pair $\langle X, Y \rangle$ of sets of atoms such that $X \subseteq Y$. The *satisfaction* relation $\models$ between an HT-interpretation $\langle X, Y \rangle$ and a formula F is defined recursively:

- for an atom A, $\langle X, Y \rangle \models A$ if $A \in X$;

- $\langle X, Y \rangle \not\models \bot$;

- $\langle X, Y \rangle \models F \wedge G$ if $\langle X, Y \rangle \models F$ and $\langle X, Y \rangle \models G$;

- $\langle X, Y \rangle \models F \vee G$ if $\langle X, Y \rangle \models F$ or $\langle X, Y \rangle \models G$;

- $\langle X, Y \rangle \models F \rightarrow G$ if

  (i) $\langle X, Y \rangle \not\models F$ or $\langle X, Y \rangle \models G$, and
  (ii) $Y \models F \rightarrow G$.

(The symbol $\models$ in the last line refers to the satisfaction relation of classical logic defined in Section A.1.)

A formula is *valid in the logic of here-and-there* if it is satisfied by every HT-interpretation. A formula F is *equivalent* to a formula G *in the logic of here-and-there* if $F \leftrightarrow G$ is valid in the logic of here-and-there (or, equivalently, if F and G are satisfied by the same HT-interpretations).

The following facts relate the satisfaction relation of the logic of here-and-there to the satisfaction relation of classical logic:

(64) $\langle X, X \rangle \models F$ iff $X \models F$.

(65) If $\langle X, Y \rangle \models F$ then $Y \models F$.

(66) $\langle X, Y \rangle \models \neg F$ iff $Y \models \neg F$.

From property (64) we see that a formula can be valid in the logic of here-and-there only if it is a tautology. It follows that two formulas can be equivalent to each other in the logic of here-and-there only if they are classically equivalent. To see where the two equivalence relations differ from each other, note that $\neg\neg p$ is not equivalent to p in the logic of here-and-there. Indeed, by (66), the HT-interpretation $\langle \emptyset, \{p\} \rangle$ satisfies $\neg\neg p$, but it clearly does not satisfy p.

A.3 Natural Deduction

In the natural deduction system for propositional logic, the derivable objects are *sequents*—expressions of the form $\Gamma \Rightarrow F$ ("F under the assumptions Γ"), where F is a formula and Γ is a finite set of formulas. Notationally, we will identify the set of assumption in a sequent with the list of its elements. For instance, we will write $\Gamma, F \Rightarrow G$ for $\Gamma \cup \{F\} \Rightarrow G$.

The axiom schemas are

(67) $F \Rightarrow F$

and

(68) $\Rightarrow F \vee \neg F$.

The latter is called *the law of excluded middle*. The inference rules are

$$(\wedge I)\ \frac{\Gamma \Rightarrow F \quad \Delta \Rightarrow G}{\Gamma, \Delta \Rightarrow F \wedge G} \qquad (\wedge E)\ \frac{\Gamma \Rightarrow F \wedge G}{\Gamma \Rightarrow F} \quad \frac{\Gamma \Rightarrow F \wedge G}{\Gamma \Rightarrow G}$$

$$(\vee I)\ \frac{\Gamma \Rightarrow F}{\Gamma \Rightarrow F \vee G} \quad \frac{\Gamma \Rightarrow G}{\Gamma \Rightarrow F \vee G} \qquad (\vee E)\ \frac{\Gamma \Rightarrow F \vee G \quad \Delta_1, F \Rightarrow H \quad \Delta_2, G \Rightarrow H}{\Gamma, \Delta_1, \Delta_2 \Rightarrow H}$$

$$(\to I)\ \frac{\Gamma, F \Rightarrow G}{\Gamma \Rightarrow F \to G} \qquad (\to E)\ \frac{\Gamma \Rightarrow F \quad \Delta \Rightarrow F \to G}{\Gamma, \Delta \Rightarrow G}$$

$$(C)\ \frac{\Gamma \Rightarrow \bot}{\Gamma \Rightarrow F}$$

$$(W)\ \frac{\Gamma \Rightarrow F}{\Gamma' \Rightarrow F}$$

$$\text{if } \Gamma \subseteq \Gamma'$$

Among the first six inference rules, the rules in the left column are *introduction rules*, and the rules in the right column are *elimination rules*. Rule (C) is the *contradiction rule*, and (W) is *weakening*.

Since we defined $\neg F$ as an abbreviation for $F \to \bot$ (Section A.1), "negation introduction"

$$\frac{\Gamma, F \Rightarrow \bot}{\Gamma \Rightarrow \neg F}$$

is a special case of $(\to I)$, and "negation elimination"

$$\frac{\Gamma \Rightarrow F \quad \Delta \Rightarrow \neg F}{\Gamma, \Delta \Rightarrow \bot}$$

is a special case of $(\to E)$. Similarly, the introduction and elimination rules for equivalence

$$\frac{\Gamma \Rightarrow F \to G \quad \Delta \Rightarrow G \to F}{\Gamma, \Delta \Rightarrow F \leftrightarrow G} \qquad \frac{\Gamma \Rightarrow F \leftrightarrow G}{\Gamma \Rightarrow F \to G} \qquad \frac{\Gamma \Rightarrow F \leftrightarrow G}{\Gamma \Rightarrow G \to F}$$

are special cases of $(\wedge I)$ and $(\wedge E)$.

To prove a formula F in this system means to prove the sequent $\Rightarrow F$. For instance, here is a proof of the equivalence

(69) $(\neg p \to p) \leftrightarrow \neg\neg p$.

1.	$\neg p \to p \Rightarrow \neg p \to p$	— axiom
2.	$\neg p \Rightarrow \neg p$	— axiom
3.	$\neg p, \neg p \to p \Rightarrow p$	— by $(\to E)$ from 2, 1
4.	$\neg p, \neg p \to p \Rightarrow \bot$	— by $(\to E)$ from 3, 2
5.	$\neg p \to p \Rightarrow \neg\neg p$	— by $(\to I)$ from 4
6.	$\Rightarrow (\neg p \to p) \to \neg\neg p$	— by $(\to I)$ from 5
7.	$\neg\neg p \Rightarrow \neg\neg p$	— axiom
8.	$\neg p, \neg\neg p \Rightarrow \bot$	— by $(\to E)$ from 2, 7
9.	$\neg p, \neg\neg p \Rightarrow p$	— by (C) from 8
10.	$\neg\neg p \Rightarrow \neg p \to p$	— by $(\to I)$ from 9
11.	$\Rightarrow \neg\neg p \to (\neg p \to p)$	— by $(\to I)$ from 10
12.	$\Rightarrow (\neg p \to p) \leftrightarrow \neg\neg p$	— by $(\wedge I)$ from 6, 11

The deductive system described above is sound and complete: a formula F is provable in this system iff F is a tautology.

A formula is *intuitionistically provable* if it can be proved in this deductive system without references to axiom schema (68). A formula F is *intuitionistically equivalent* to a formula G if $F \leftrightarrow G$ is intuitionistically provable.

For instance, the implication $\neg p \rightarrow p$ is intuitionistically equivalent to the formula $\neg\neg p$, because the proof of (69) above contains no references to the law of excluded middle. On the other hand, this implication is not intuitionistically equivalent to p: the equivalence obtained from (69) by dropping the double negation in the right-hand side cannot be proved without (68).

According to the replacement property of intuitionistic logic, if F is a subformula of a formula F', and G' is obtained from F' by replacing an occurrence of F with another formula G, then $F' \leftrightarrow G'$ is intuitionistically derivable from $F \leftrightarrow G$. For instance, from the fact that $\neg p \rightarrow p$ is intuitionistically equivalent to $\neg\neg p$ we can conclude that $(\neg p \rightarrow p) \wedge q$ is intuitionistically equivalent to $\neg\neg p \wedge q$.

Every intuitionistically provable formula is valid in the logic of here-and-there; if two formulas are intuitionistically equivalent then they are equivalent in the logic of here-and-there. Moreover, these assertions remain true if, instead of intuitionistic logic, we talk about the stronger deductive system, obtained from classical by replacing (68) with the axiom schema expressing *the weak law of excluded middle*:

$$(70) \quad \Rightarrow \neg F \vee \neg\neg F.$$

We can use this fact, for example, to check that the formulas $p \vee \neg p$ and $\neg\neg p \rightarrow p$ are equivalent to each other in the logic of here-and-there, as follows:

1.	$p \vee \neg p \Rightarrow p \vee \neg p$	— axiom
2.	$p \Rightarrow p$	— axiom
3.	$\neg p \Rightarrow \neg p$	— axiom
4.	$\neg\neg p \Rightarrow \neg\neg p$	— axiom
5.	$\neg p, \neg\neg p \Rightarrow \bot$	— by $(\rightarrow E)$ from 3, 4
6.	$\neg p, \neg\neg p \Rightarrow p$	— by (C) from 5
7.	$p \vee \neg p, \neg\neg p \Rightarrow p$	— by $(\vee E)$ from 1, 2, 6
8.	$p \vee \neg p \Rightarrow \neg\neg p \rightarrow p$	— by $(\rightarrow I)$ from 7
9.	$\Rightarrow (p \vee \neg p) \rightarrow (\neg\neg p \rightarrow p)$	— by $(\rightarrow I)$ from 8
10.	$\neg\neg p \rightarrow p \Rightarrow \neg\neg p \rightarrow p$	— axiom
11.	$\Rightarrow \neg p \vee \neg\neg p$	— axiom
12.	$\neg p \Rightarrow p \vee \neg p$	— by $(\vee I)$ from 3
13.	$\neg\neg p, \neg\neg p \rightarrow p \Rightarrow p$	— by $(\rightarrow E)$ from 4, 10
14.	$\neg\neg p, \neg\neg p \rightarrow p \Rightarrow p \vee \neg p$	— by $(\vee I)$ from 13
15.	$\neg\neg p \rightarrow p \Rightarrow p \vee \neg p$	— by $(\vee E)$ from 11, 12, 14
16.	$\Rightarrow (\neg\neg p \rightarrow p) \rightarrow (p \vee \neg p)$	— by $(\rightarrow I)$ from 15
17.	$\Rightarrow (p \vee \neg p) \leftrightarrow (\neg\neg p \rightarrow p)$	— by $(\wedge I)$ from 9, 16

Here is an axiom schema that is even stronger than (70) and that can be used for establishing the validity of formulas in the logic of here-and-there as well:

$(71) \Rightarrow F \vee (F \rightarrow G) \vee \neg G.$

Nothing stronger would be acceptable: A propositional formula is valid in the logic of here-and-there iff it is provable in the deductive system obtained from intuitionistic logic by adding axiom schema (71). This theorem is due to Lex Hendriks [Lifschitz *et al.*, 2001, Section 2.2].

B   Traditional Definition of a Stable Model

In [Gelfond and Lifschitz, 1988], a logic program is assumed to consist of rules of the form

$(72) \ A_0 \leftarrow A_1, \ldots, A_m, not \ A_{m+1}, \ldots, not \ A_n$

where $n \geq m \geq 0$ and $A_0, \ldots, A_n$ are atoms; we will call such expressions *traditional rules*. A finite set of traditional rules with $m = n$, that is, rules of the form

$(73) \ A_0 \leftarrow A_1, \ldots, A_m$

is essentially a Horn formula in the sense of Section 2.2.

The *traditional reduct* of a traditional program Π relative to a set X of atoms is the set of rules (73) for all rules (72) in Π such that

$$A_{m+1}, \ldots, A_n \notin X.$$

According to the 1988 definition, the stable model of a traditional program Π is a set X of atoms with the following property: X *is the minimal model of the traditional reduct of* Π *relative to* X. This is equivalent to our definition of a stable model (Section 2.1) limited to traditional programs:

PROPOSITION 28. *For any traditional program* Π, *a set* X *of atoms is the minimal model of the traditional reduct of* Π *relative to* X *iff* X *is a stable model of* Π.

Proof. Let $\Pi^{\underline{X}}$ denote the traditional reduct of Π relative to X.

Case 1: $X \not\models \Pi$. Set X is not a stable model of Π. On the other hand, Π contains a rule (72) such that $A_1, \ldots, A_m \in X$ and $A_0, A_{m+1}, \ldots, A_n \notin X$. The corresponding rule (73) in $\Pi^{\underline{X}}$ is not satisfied by X, so that X is not the minimal model of $\Pi^{\underline{X}}$.

Case 2: $X \models \Pi$. We will show that Π^X and $\Pi^{\underline{X}}$ are satisfied by the same subsets of X. Since Π^X is the conjunction of the formulas R^X for all rules R

of Π, and $\Pi^{\underline{X}}$ is the union of the programs $\{R\}^{\underline{X}}$ for all rules R of Π, it is sufficient to verify this claim for the case when Π is a single rule (72). If X contains at least one of the atoms $A_{m+1}, \ldots, A_n$ then $\Pi^{\underline{X}}$ is empty and Π^X is the tautology $\bot \to A_0^X$. Otherwise $\Pi^{\underline{X}}$ is (73). If $A_1, \ldots, A_m \in X$ then $A_0 \in X$, because $X \models \Pi$; consequently Π^X is the result of replacing $A_{m+1}, \ldots, A_n$ in (72) with $\bot$, which is equivalent to (73). It remains to consider the case when $A_{m+1}, \ldots, A_n \notin X$ and at least one of the atoms $A_1, \ldots, A_m$, say A_1, does not belong to X. In this case Π^X is the tautology $\bot \to A_0^X$. On the other hand, $\Pi^{\underline{X}}$ is the rule (73) whose body contains A_1 and consequently is not satisfied by any subset of X. It follows that every subset of X satisfies $\Pi^{\underline{X}}$. ∎

Intuitively, a rule (73) can be viewed as a rule for generating atoms: we are allowed to generate its head A_0 as soon as all atoms $A_1, \ldots, A_m$ in the body have been generated. The minimal model of a set of rules of the form (73) is the set of all atoms that can be generated by this process, starting from the empty set. The traditional definition of a stable model can be thought of as an extension of this idea to rules containing negative literals in the body. A rule (72) allows us to generate A_0 as soon as we generated the atoms $A_1, \ldots, A_m$ *provided that none of the atoms $A_{m+1}, \ldots, A_n$ can be generated using the rules of the program*. There is a vicious circle in this sentence: to decide whether a rule of Π can be used to generate a new atom, we need to know which atoms can be generated using the rules of Π. The traditional definition of a stable model overcomes this difficulty using a "fixpoint construction." Take a set X that you suspect may be exactly the set of atoms that can be generated using the rules of Π. Under this assumption, Π has the same meaning as the traditional reduct of Π relative to X, which is a set of rules of the form (73). Consider the minimal model of the traditional reduct. If this model is exactly identical to the set X that we started with then X was a "good guess"; it is indeed a stable model of Π.

The definition of a stable model for traditional programs can be viewed as a possible definition of a "correct" answer to a query in Prolog. Let Π be a Prolog program without variables (or the set of ground rules obtained from a Prolog program with variables by replacing each rule with all its ground instances). If Π is a traditional program with a unique stable model then the correct answer to a ground query A is *yes* or *no* depending on whether A belongs to that model.

From this perspective, a program with several stable models is "bad"—it does not provide an unambiguous specification for the behavior of a Prolog system. Programs without answer sets are "bad" also. In answer set programming, on the other hand, programs without a unique answer set are quite useful: they correspond to computational problems with many solutions, or with no solutions.

The concept of a stable model is only one of several available definitions of the semantics of negation as failure. Two other definitions frequently referred to in the literature are based on program completion [Clark, 1978] and the well-founded model [Van Gelder et al., 1991]. These three definitions are not completely equivalent to each other, but each of them provides an adequate description of the behavior of Prolog.

Acknowledgements

This paper grew out of a graduate seminar that one of the authors has taught at the University of Texas and his course included in the program of the 2004 European Summer School on Logic, Language and Information. We are grateful to the students who took those classes for their feedback. We would like to thank also Esra Erdem, Artur Garcez, Joohyung Lee, Grigori Mints, Jayadev Misra and Hudson Turner for comments on a draft of this paper. This work was partially supported by the National Science Foundation under Grant IIS-0412907.

BIBLIOGRAPHY

[Baral et al., 2004] Chitta Baral, Karen Chancellor, Nam Tran, Nhan Tran, Anna Joy, and Michael Berens. A knowledge based approach for representing and reasoning about cell signaling networks. In Proceedings of European Conference on Computational Biology (ECCB), Supplement on Bioinformatics, pages 15–22, 2004.

[Baral, 2003] Chitta Baral. Knowledge Representation, Reasoning and Declarative Problem Solving. Cambridge University Press, 2003.

[Brooks et al., 2005] Daniel R. Brooks, Esra Erdem, James W. Minett, and Donald Ringe. Character-based cladistics and answer set programming. In Proceedings of International Symposium on Practical Aspects of Declarative Languages (PADL), pages 37–51, 2005.

[Clark, 1978] Keith Clark. Negation as failure. In Herve Gallaire and Jack Minker, editors, Logic and Data Bases, pages 293–322. Plenum Press, New York, 1978.

[Davis et al., 1962] Martin Davis, George Logemann, and Donald Loveland. A machine program for theorem proving. Communications of ACM, 5(7):394–397, 1962.

[Dimopoulos et al., 1997] Yannis Dimopoulos, Bernhard Nebel, and Jana Koehler. Encoding planning problems in non-monotonic logic programs. In Sam Steel and Rachid Alami, editors, Proceedings of European Conference on Planning, pages 169–181. Springer-Verlag, 1997.

[Eiter and Gottlob, 1993] Thomas Eiter and Georg Gottlob. Complexity results for disjunctive logic programming and application to nonmonotonic logics. In Dale Miller, editor, Proceedings of International Logic Programming Symposium (ILPS), pages 266–278, 1993.

[Erdem et al., 2000] Esra Erdem, Vladimir Lifschitz, and Martin Wong. Wire routing and satisfiability planning. In Proceedings of International Conference on Computational Logic, pages 822–836, 2000.

[Erdoğan and Lifschitz, 2004] Selim T. Erdoğan and Vladimir Lifschitz. Definitions in answer set programming. In Vladimir Lifschitz and Ilkka Niemelä, editors, Proceedings of *International Conference on Logic Programming and Nonmonotonic Reasoning (LPNMR)*, pages 114–126, 2004.

[Ferraris and Lifschitz, 2005] Paolo Ferraris and Vladimir Lifschitz. Weight constraints as nested expressions. *Theory and Practice of Logic Programming*, 5:45–74, 2005.

[Ferraris, 2005] Paolo Ferraris. Answer sets for propositional theories. In *Proceedings of International Conference on Logic Programming and Nonmonotonic Reasoning (LPNMR)*, 2005. To appear.

[Gelfond and Lifschitz, 1988] Michael Gelfond and Vladimir Lifschitz. The stable model semantics for logic programming. In Robert Kowalski and Kenneth Bowen, editors, *Proceedings of International Logic Programming Conference and Symposium*, pages 1070–1080, 1988.

[Gelfond and Lifschitz, 1991] Michael Gelfond and Vladimir Lifschitz. Classical negation in logic programs and disjunctive databases. *New Generation Computing*, 9:365–385, 1991.

[Gelfond, 1987] Michael Gelfond. On stratified autoepistemic theories. In *Proceedings of National Conference on Artificial Intelligence (AAAI)*, pages 207–211, 1987.

[Hanks and McDermott, 1987] Steve Hanks and Drew McDermott. Nonmonotonic logic and temporal projection. *Artificial Intelligence*, 33(3):379–412, 1987.

[Heljanko and Niemelä, 2003] Keijo Heljanko and Ilkka Niemelä. Bounded LTL model checking with stable models. *Theory and Practice of Logic Programming*, 3:519–550, 2003.

[Hermansson et al., 2005] Martin Hermansson, Andreas Uphoff, Reijo Käkelä, and Pentti Somerharju. Automated quantitative analysis of complex lipidomes by liquid chromatography/mass spectrometry. *Analytical Chemistry*, 77:2166–2175, 2005.

[Heyting, 1930] Arend Heyting. Die formalen Regeln der intuitionistischen Logik. *Sitzunsberichte der Preussischen Akademie von Wissenschaften. Physikalisch-mathematische Klasse*, pages 42–56, 1930.

[Kautz and Selman, 1992] Henry Kautz and Bart Selman. Planning as satisfiability. In *Proceedings of European Conference on Artificial Intelligence (ECAI)*, pages 359–363, 1992.

[Leibniz, 1995] Gottfried Wilhelm Leibniz. *Philosophical Writings*. Everyman, 1995.

[Leone et al., 2005] Nicola Leone, Thomas Eiter, Wolfgang Faber, Michael Fink, Georg Gottlob, Gianluigi Greco, Giovambattista Ianni, Edyta Kalka, Domenico Lembo, Maurizio Lenzerini, Vincenzino Lio, Bartosz Nowicki, Riccardo Rosati, Marco Ruzzi, Witold Staniszkis, and Giorgio Terracina. The INFOMIX system for advanced integration of incomplete and inconsistent data. In *Proceedings of ACM Symposium on Principles of Database Systems (PODS)*, pages 915–917. ACM, 2005. Demo paper.

[Lifschitz and Turner, 1994] Vladimir Lifschitz and Hudson Turner. Splitting a logic program. In Pascal Van Hentenryck, editor, *Proceedings of International Conference on Logic Programming (ICLP)*, pages 23–37, 1994.

[Lifschitz et al., 1999] Vladimir Lifschitz, Lappoon R. Tang, and Hudson Turner. Nested expressions in logic programs. *Annals of Mathematics and Artificial Intelligence*, 25:369–389, 1999.

[Lifschitz et al., 2001] Vladimir Lifschitz, David Pearce, and Agustin Valverde. Strongly equivalent logic programs. *ACM Transactions on Computational Logic*, 2:526–541, 2001.

[Lifschitz, 1999] Vladimir Lifschitz. Action languages, answer sets and planning. In *The Logic Programming Paradigm: a 25-Year Perspective*, pages 357–373. Springer Verlag, 1999.

[Lifschitz, 2002] Vladimir Lifschitz. Answer set programming and plan generation. *Artificial Intelligence*, 138:39–54, 2002.

[Marek and Truszczyński, 1999] Victor Marek and Mirosław Truszczyński. Stable models and an alternative logic programming paradigm. In *The Logic Programming Paradigm: a 25-Year Perspective*, pages 375–398. Springer Verlag, 1999.

[McCarthy, 1980] John McCarthy. Circumscription—a form of non-monotonic reasoning. *Artificial Intelligence*, 13:27–39,171–172, 1980. Reproduced in [McCarthy, 1990].

[McCarthy, 1990] John McCarthy. *Formalizing Common Sense: Papers by John McCarthy*. Ablex, Norwood, NJ, 1990.

[McDermott and Doyle, 1980] Drew McDermott and Jon Doyle. Nonmonotonic logic I. *Artificial Intelligence*, 13:41–72, 1980.

[Moore, 1985] Robert Moore. Semantical considerations on nonmonotonic logic. *Artificial Intelligence*, 25(1):75–94, 1985.

[Nelson, 1949] David Nelson. Constructible falsity. *Journal of Symbolic Logic*, 14:16–26, 1949.

[Niemelä, 1999] Ilkka Niemelä. Logic programs with stable model semantics as a constraint programming paradigm. *Annals of Mathematics and Artificial Intelligence*, 25:241–273, 1999.

[Nogueira et al., 2001] Monica Nogueira, Marcello Balduccini, Michael Gelfond, Richard Watson, and Matthew Barry. An A-Prolog decision support system for the Space Shuttle. In *Proceedings of International Symposium on Practical Aspects of Declarative Languages (PADL)*, pages 169–183, 2001.

[Pearce, 1997] David Pearce. A new logical characterization of stable models and answer sets. In Jürgen Dix, Luis Pereira, and Teodor Przymusinski, editors, *Non-Monotonic Extensions of Logic Programming (Lecture Notes in Artificial Intelligence 1216)*, pages 57–70. Springer-Verlag, 1997.

[Reiter, 1978] Raymond Reiter. On closed world data bases. In Herve Gallaire and Jack Minker, editors, *Logic and Data Bases*, pages 119–140. Plenum Press, New York, 1978.

[Reiter, 1980] Raymond Reiter. A logic for default reasoning. *Artificial Intelligence*, 13:81–132, 1980.

[Shanahan, 1997] Murray Shanahan. *Solving the Frame Problem: A Mathematical Investigation of the Common Sense Law of Inertia*. MIT Press, 1997.

[Simons et al., 2002] Patrik Simons, Ilkka Niemelä, and Timo Soininen. Extending and implementing the stable model semantics. *Artificial Intelligence*, 138:181–234, 2002.

[Soininen and Niemelä, 1998] Timo Soininen and Ilkka Niemelä. Developing a declarative rule language for applications in product configuration. In Gopal Gupta, editor, *Proceedings of International Symposium on Practical Aspects of Declarative Languages (PADL)*, pages 305–319. Springer-Verlag, 1998.

[Subrahmanian and Zaniolo, 1995] V.S. Subrahmanian and Carlo Zaniolo. Relating stable models and AI planning domains. In *Proceedings of International Conference on Logic Programming (ICLP)*, 1995.

[Turner, 1997] Hudson Turner. Representing actions in logic programs and default theories: a situation calculus approach. *Journal of Logic Programming*, 31:245–298, 1997.

[Turner, 2003] Hudson Turner. Strong equivalence made easy: nested expressions and weight constraints. *Theory and Practice of Logic Programming*, 3(4,5):609–622, 2003.

[Van Gelder et al., 1991] Allen Van Gelder, Kenneth Ross, and John Schlipf. The well-founded semantics for general logic programs. *Journal of ACM*, 38(3):620–650, 1991.

A Mistake on My Part

MELVIN FITTING

1 The Background

I first met Dov at a logic conference in Manchester, in August 1969, though we had begun a mathematical correspondence the previous year. Here are a few photos from the conference. As I recall, Dov planned to get married shortly after the conference. I found a letter in my files mentioning that I sent Dov a copy of the pictures in 1969, so this is for everybody else.

Figure 1 shows Dov at Jodrell Bank, the huge radio telescope complex run by the University of Manchester.

Figure 1.

Figure 2 shows, from left to right, Saul Kripke, Dov, Michael Rabin, and someone I can't identify.

Figure 2.

Figure 3 shows the following. Back row: David Pincus, Miriam Lucian, Dov, Saul Kripke; front row: George Rousseau, and me.

It was also at this conference that a mistake I made got straightened out. I don't know if it had an effect on Dov's work, but it may have. But let's discuss that in a section of its own.

2   Where I Went Wrong

Kripke's semantics for intuitionistic logic came along in [Kripke, 1965], and prompted a considerable amount of research. It was a possible worlds semantics, and in it domains for quantifiers varied from world to world, though subject to a monotonicity condition. It was a natural question: what would a restriction to constant domain intuitionistic models impose. In my dissertation, [Fitting, 1969], I had shown in passing that, for formulas without universal quantifiers, constant domain and monotonic domain semantics validated the same formulas (unlike in classical logic, the two quantifiers are not interdefinable intuitionistically). This is not the case once universal quantifiers are present, so just what is the logic of constant domain Kripke intuitionistic models. For modal logics, Kripke had addressed an analogous problem by showing constant domains correspond to the Barcan formula, but the question was still open for intuitionistic logic.

Dov asked me, in a letter, about a conjectured axiomatization for the

Figure 3.

constant domain version of intuitionistic logic. I replied on March 13, 1969, and my response was quite decisive. Here is an excerpt from my letter to Dov.

> ... However, I can settle one of the problems you raised in your letter.
>
> You asked does the following axiomatize constant domains:
>
> (1) $(\forall x)[P(x) \vee Q(x)] \supset [(\exists x)P(x) \vee (\forall x)Q(x)]$.
>
> No. About a month ago I began working with intuitionistic logic plus the axiom
>
> (2) $(\forall x)[A \vee B(x)] \supset [A \vee (\forall x)B(x)]$,
>
> where x does not occur free in A, and it has turned out to be a most interesting system. Schemas (1) and (2) are equivalent. If, in (1) we let $P(x)$ be A, and $Q(x)$ be $B(x)$, (1) is (2). If, in (2) we let A be $(\exists y)P(y)$, and $B(x)$ be $Q(x)$, (2) implies (1).
>
> Call intuitionistic logic **I**, and call **I** plus (2) (or (1)), **SI**. A simple way to characterize **SI** is by Beth tableaus. Use the system of *The Foundations of Mathematics* section 145, but replace rule vi^b by the corresponding classical rule, vi^b of section 92 (then all quantifier rules are classical). This is **SI**.
>
> **I** can be embedded in **SI** as follows. Let *true* be a truth constant (e.g. let *true* be $X \supset X$). For any formula F, define a translate,

F^* to be like F except that subformulas of the form $(\forall x)A(x)$ are replaced by $true \supset (\forall x)A(x)$. Then

(3) $\vdash_{\mathbf{I}} F$ if and only if $\vdash_{\mathbf{SI}} F^*$.

To return to your original question, the translate of (2) is not provable in **SI**, but is valid in all constant domain models.

The Beth tableau system hybrid referred to above is perhaps more conveniently seen as the combination of an intuitionistic propositional system using signed formulas, described in [Fitting, 1983; Fitting, 1998], with the classical rules of Smullyan, from [Smullyan, 1968].

3   What Was Wrong

I wonder if my negative March response stopped Dov from following up on the conjectured axiomatization. At any rate, he did not do so. But Sabine Görneman was present at the August Manchester conference, and she had just proved that (2) was precisely the axiom one needed to add to intuitionistic logic to axiomatize constant domain models. Her dissertation containing this result was written later—I received my copy in February of 1970, and a paper based on it appears as [Görneman, 1971]. I well remember Dov devoting much of a conference bus trip, to a nearby site of interest, to walking up and down the aisle and trying to reconstruct Görneman's proof. Apparently he succeeded, because in her Journal of Symbolic Logic paper, Görnemann mentions "another proof has been given by D. Gabbay." The citation is [Gabbay, 1969], but I have never seen this.

Equivalence (3) is correct, and so the translate of (2) is, indeed, not provable in the tableau system, since (2) is not an intuitionistic theorem. What is not true is my assertion that intuitionistic propositional plus classical quantifier tableau rules give a proof procedure equivalent to axiomatic intuitionistic first-order logic plus (2). In fact, all the axioms are provable in the tableau system. What we have here is a seemingly natural set of tableau rules for which cut elimination does not hold. Indeed, this should have been seen even without Görneman's crusher. Formula (2) is provable in the tableau system, but its translate is not. However, a formula and its translate differ by having some subformulas Z replaced with subformulas $true \supset Z$, and the equivalence of these can be proved using just intuitionistic propositional tableau rules. If cut elimination held, one could prove that (2) and its translate were provably equivalent, so both or neither should have been provable.

Well, I was young and careless. I'm older now. But even if Dov did not get this result first, there have certainly been many other firsts over a long career. Happy birthday, Dov.

BIBLIOGRAPHY

[Fitting, 1969] Melvin C. Fitting. *Intuitionistic Logic Model Theory and Forcing*. North-Holland Publishing Co., Amsterdam, 1969.

[Fitting, 1983] Melvin C. Fitting. *Proof Methods for Modal and Intuitionistic Logics*. D. Reidel Publishing Co., Dordrecht, 1983.

[Fitting, 1998] Melvin C. Fitting. Introduction. In Marcello D'Agostino, Dov Gabbay, Reiner Hähnle, and Joachim Posegga, editors, *Handbook of Tableau Methods*. Oxford University Press, 1998.

[Gabbay, 1969] Dov Gabbay. Montague type semantics for nonclassical logics. I. Technical report, The Hebrew University of Jerusalem, 1969.

[Görneman, 1971] Sabine Görneman. A logic stronger than intuitionism. *Journal of Symbolic Logic*, 36:249–261, 1971.

[Kripke, 1965] Saul Kripke. Semantical analysis of intuitionistic logic I. In J. N. Crossley and M. Dummett, editors, *Formal Systems and Recursive Functions, Proc. of the Eight Logic Colloquium, Oxford 1963*, pages 92–130, Amsterdam, 1965. North-Holland.

[Smullyan, 1968] Raymond M. Smullyan. *First-Order Logic*. Springer-Verlag, Berlin, 1968. Revised Edition, Dover Press, New York, 1994.

DAG Sequent Proofs with a Substitution Rule

MARCELO FINGER

1 Introduction

In this paper, we investigate a sequent proof method known to have short proofs even for the hardest known propositional formulas. We explore some of the proof-theoretical properties of this inference system and investigate how it can be transformed in a tableau-like decision procedure.

This work is in the spirit of recent work on efficient propositional inference systems, by Dov Gabbay and the author, in which we studied several families of tractable subclassical logics that are less complex than propositional classical logic [Finger and Gabbay, 2005]. Each element of those families has a polynomial time decision procedure. That investigation restricted the use of the cut rule in a non cut-eliminable formulation of propositional classical logic.

In this work, we investigate classical proof theory, especially the role of admissible rules, in another direction. Here add the admissible *substitution rule* (or s-rule) to the set of inference rules in a Gentzen sequent system. Let a *substitution* σ be a formula transformation that maps atoms into formulas, and is extended to all formulas in a homomorphic way. If A is a formula let $A\sigma$ be the application of σ on A. Similarly, if Γ is a set of formulas, then $\Gamma\sigma$ is the result of applying σ to every formula in Γ. In this setting, the s-rule states that if $\Gamma \vdash \Delta$ is a derivable sequent and σ is a substitution, then we can infer $\Gamma\sigma \vdash \Delta\sigma$. Furthermore, this rule can be *defocusing*, that is, the same source sequent $\Gamma \vdash \Delta$ can receive several substitutions, transforming the usual tree structure of a sequent proof into a direct acyclic graph (DAG), as illustrated in Figure 1.

In a usual tree-like sequent proofs, a rule may contain one premiss or several premisses; in the latter case, the rule is called *focusing* according to the terminology of [Carbone and Semmes, 2000] illustrated in Figure 2. Usual rules (see Figures 3 and 4) are linear or *focusing*, that is, rules are viewed as providing directed edges from the premisses to the conclusion, such that there is only one conclusion but possibly one or more premisses;

$$\Gamma \vdash \Delta$$

$$\sigma_1 \qquad \sigma_n$$

$$\Gamma\sigma_1 \vdash \Delta\sigma_1 \quad \cdots \quad \Gamma\sigma_n \vdash \Delta\sigma_n$$

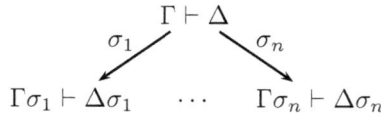

Figure 1. The Defocusing Substitution Rule

premiss$_1$ $\cdots$ premiss$_k$ premiss

conclusion   conclusion$_1$ $\cdots$ conclusion$_k$

(a) Focusing Rule   (b) Defocusing Rule

Figure 2. Focusing and Defocusing Rules

the former is considered a linear rule, the latter a focusing one. With usual rules, no defocusing is possible, so the proof necessarily has a tree-like structure with a single directed path between any node and the deduced sequent at the root of the tree.

In a DAG proof, due to the presence of defocusing substitution nodes (or s-nodes) there may be more than one directed path from any sequent in the proof to the root sequent. In this way, a DAG proof avoids repetition of isomorphic branches and is thus more compact than a tree proof, so proofs of the same sequent may be shorter with a DAG structure. As the substitution σ may be the identity substitution, there is no need to have any other defocusing rule.

Notation: The propositional connectives we consider here are $\wedge$, $\vee$, $\rightarrow$ and $\neg$. We measure the size of a formula A, $|A|$, as the number of symbols it contains. There are basically two ways of measuring the size of a proof:

(a) The *number of lines*: this is the number of sequents (usually called in the literature the number of lines) occurring in a proof Π, represented by $|\Pi|$.

(b) The *number of symbols*: this is the sum of the sizes of all formulas occurring in a proof Π, represented by $\|\Pi\|$.

We say that a proof system S_1 p-simulates a proof system S_2 if there exists a polynomial $p(x)$ such that for every proof Π_2 of a theorem A in S_2 there is a proof Π_1 of A in S_1 such that $|\Pi_1| \leq p(|\Pi_2|)$. We say that two proof systems are equivalent if each one p-simulates the other.

1.1 Related Works

In their seminal work, Cook and Reckhow [1979] defined a generalisation of the usual propositional proof system called *Frege systems*. A Frege system is based on a finite and complete set of propositional connectives, and has a finite set of schematically defined rules of inference with one or more premisses, and one conclusion, such that the set of rules is sound and complete. A *proof* is a direct acyclic graph, where each node is labelled with a formula or a sequent. The class of Frege systems include inference systems like Hilbert-style axiomatisations, Natural Deduction and Gentzen sequent systems. Furthermore, Cook and Reckhow showed that any Frege system $\mathcal{F}_1$ can p-simulate any other Frege system $\mathcal{F}_2$.

Cook and Reckhow [1979] also defined the notion of an *extended Frege system*, which is a Frege proof system augmented with the introduction of inferences of the form:

$$\vdash p \leftrightarrow A$$

where p is a propositional symbol that does not occur in A, nor in any previous formula in that branch of the proof, nor in the final final formula at the root of the proof. This inference allows the atom p to be an abbreviation of the formula A, which has the potential of reducing the number of symbols in the proof. Extended Frege systems were also shown to p-simulate each other. It remains an open problem if Frege systems can p-simulate extended Frege systems.

In a similar way, the notion of *substitution Frege system* consists of a Frege system augmented with the substitution rule for formulas (or sequents), as in Figure 1. It was shown that substitution Frege systems and extended Frege systems are equivalent, that is, any substitution Frege system $s\mathcal{F}$ can p-simulate any extended Frege proof system $e\mathcal{F}$ and vice-versa [Cook and Reckhow, 1979; Krajíček and Pudlák, 1989]. It was also shown that this result holds even if the substitution is restricted to the mere renaming of propositional symbols [Buss, 1995].

Thus, if we extend a sequent system with a defocusing renaming rule as a weaker version of the substitution rule, we are still guaranteed to have a proof system that can p-simulate any extended Frege system.

One thing that comes to mind when one is discussing extended Frege systems, or any of its equivalent formulations, is if they can ever be used in practice in a real prover. This paper tries to contribute to this question.

1.2 Organisation of this Paper

Most of the literature on substitution and renaming Frege system concentrated basically on the length of proofs. Here we take one particular substitution system and study its intrinsic logic properties.

We start by defining the DAG-sequent proof system and show its soundness and completeness, and show how the substitution rule can be eliminated (Section 2). We then study the complexity of cut elimination and show that, unlike traditional sequent system, in the presence of the substitution rule cut elimination does not provoke an exponential blow up in the size of proofs.

This motivates us to examine how this technique can be brought to semantic tableaux (Section 3). We thus present s-tableaux by extending semantic tableaux with a substitution closure rule, and we prove that it actually corresponds to the DAG-sequent proof system.

As an example of s-tableaux, we apply it to the family of formulas that encode the pigeon hole principle (PHP_n), and show that we have linear s-tableau proofs for PHP_n (Section 4).

2   DAG proofs

We start by formally defining the sequent proof system we will be studying. A sequent is a pair of the form $\Gamma \vdash \Delta$, where Γ and Δ are *multisets*; Γ is the antecedent and Δ is the consequent of the sequent. We then have schematic rules that apply to sequents in a rule, divided in the usual two groups of logical and structural rules. The logical (connective) rules are shown in Figure 3. The structural rules are shown in Figure 4.

Due to the definition of the antecedent and consequent as multisets, the structural rules of *associativity* and *commutativity* are implicit in this formulation. The structural rule of *monotonicity* or *weakening* is taken care of by the *Axiom Rule*. The rules in Figures 3 and 4 are focusing rules and thus generate only tree-like proofs.

The generalisation to DAGs comes when we introduce the defocusing *substitution Rule* (s-rule); see Figure 1. A *substitution* σ is a set of pairs of propositional atoms and formulas, that we represent as $\sigma = [p_1 := A_1, \ldots, p_k := A_k]$. If the pair $\langle p_i, A_i \rangle$ is in σ we write $(p_i := A_i) \in \sigma$. For sets, multisets or sequents, the application of σ means the application of substitution to each of its elements.

There is nothing implicitly "defocusing" in the use of the s-rule, and in principle any node that is used more than once in a proof could be defocusing. We have decided to concentrate the defocusing effect only on the s-rule to obtain a true extension of the usual tree-like sequent proofs.

We can then formally define a *DAG-sequent proof* Π as a direct acyclic graph constructed inductively from the application of only the Axiom, Logical, Structural and Substitution rules. A sequent $\Gamma \vdash \Delta$ is *derivable* or *provable* if there is a DAG sequent proof Π having $\Gamma \vdash \Delta$ as the only node without leaving arrows (a *drain* in the graph). Note that axioms are the only

$$\frac{A, \Gamma \vdash \Delta}{A \wedge B, \Gamma \vdash \Delta} \ (\wedge \vdash)$$

$$\frac{\Gamma \vdash \Delta, A \quad \Gamma \vdash \Delta, B}{\Gamma \vdash \Delta, A \wedge B} \ (\vdash \wedge)$$

$$\frac{A, \Gamma \vdash \Delta \quad B, \Gamma \vdash \Delta}{A \vee B, \Gamma \vdash \Delta} \ (\vee \vdash)$$

$$\frac{\Gamma \vdash \Delta, A}{\Gamma \vdash \Delta, A \vee B} \ (\wedge \vdash)$$

$$\frac{\Gamma_1 \vdash \Delta_1, A \quad B, \Gamma_2 \vdash \Delta_2}{A \rightarrow B, \Gamma_1, \Gamma_2 \vdash \Delta_1, \Delta_2} \ (\rightarrow \vdash)$$

$$\frac{\Gamma, A \vdash \Delta, B}{\Gamma \vdash \Delta, A \rightarrow B} \ (\vdash \rightarrow)$$

$$\frac{\Gamma \vdash \Delta, A}{\neg A, \Gamma \vdash \Delta} \ (\neg \vdash)$$

$$\frac{A, \Gamma \vdash \Delta}{\Gamma \vdash \Delta, \neg A} \ (\vdash \neg)$$

Figure 3. Usual Logical Rules for the Sequent Calculus

$$\frac{}{\Gamma, A \vdash A, \Delta} \ (\text{Axiom})$$

$$\frac{\Gamma_1 \vdash \Delta_1, A \quad A, \Gamma_2 \vdash \Delta_2}{\Gamma_1, \Gamma_2 \vdash \Delta_1, \Delta_2} \ (\text{Cut})$$

$$\frac{\Gamma_1, A, A, \Gamma_2 \vdash \Delta}{\Gamma_1, A, \Gamma_2 \vdash \Delta} \ (\text{Contraction} \vdash)$$

$$\frac{\Gamma \vdash \Delta_1, A, A, \Delta_2}{\Gamma \vdash \Delta_1, A, \Delta_2} \ (\vdash \text{Contraction})$$

Figure 4. Usual Structural Rules

source nodes in a proof, that is, the only nodes with no incoming arrows.

A *prefix* Π' of a DAG Π at a node n is the subgraph containing n, such that if all nodes and arcs pointing to to some node in Π' are also in Π'.

LEMMA 1. *Let S be a sequent in a proof Π. Let Π' be the prefix of Π at S. Then Π' is a proof of S.*

Proof. Directly from the definitions of DAG-sequent proofs and of prefix. Just note that S must be a drain in Π' since, by acyclicity of Π, no arc coming out from S may point into Π'.   ■

The result above shows that any intermediate sequent generated in a proof is indeed a derivable sequent, as in usual sequent proofs. We say that a sequent proof is *usual* if it is a proof without the use of the substitution rule, and hence it has a tree-like structure.

Consider the classical semantics for propositional formulas based on propositional valuations. We write $\Gamma \models A$ if any valuation that satisfies all formulas in Γ also satisfies A. *Soundness* means that $\Gamma \vdash A$ implies $\Gamma \models A$ and *completeness* means that $\Gamma \models A$ and $\Gamma \vdash A$.

THEOREM 2. *The DAG-sequent calculus is sound and complete.*

Proof. Completeness is trivial, since the usual sequent calculus is contained in the DAG-sequent calculus. For soundness, given the soundness of the usual calculus, all we have to do is show that the substitution rule takes a valid sequent into a valid sequent, which follows directly from the fact that the substitution has the effect of reducing the number of valuations available to the valid sequent in the premiss of a substitution rule.   ■

We now consider the admissibility of the substitution and cut rules from DAG proof, by showing their elimination. It must be obvious from the soundness and completeness results that the substitution rule does not add or remove any derivable sequents from classical ones. However, what we are concerned here with the exponential explosion that occurs when we eliminate it from the proof.

LEMMA 3. *If there is a DAG-sequent proof Π of sequent S then there is a DAG-sequent proof Π' of S without the use of the substitution rule, such that $|\Pi'|$ is bounded by an exponential function on $|\Pi|$.*

Proof. (Sketch) The exponential explosion occurs when we eliminate a convergence point generated by two or more *distinct* applications of the substitution rule, to a single node, a illustrated below.

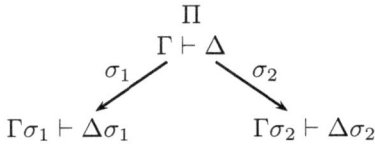

$$
\begin{array}{c}
\Pi \\
\Gamma \vdash \Delta
\end{array}
$$

$$
\sigma_1 \swarrow \qquad \searrow \sigma_2
$$

$$
\Gamma\sigma_1 \vdash \Delta\sigma_1 \qquad\qquad \Gamma\sigma_2 \vdash \Delta\sigma_2
$$

Here we see that a proof Π leads to a sequent $\Gamma \vdash \Delta$ to which the substitution rule is applied twice (or n-times, for $n \geq 2$). When we eliminate this application of the rule, two instances of the proof proof Π are created: $\Pi\sigma_1$, and $\Pi\sigma_2$, where by $\Pi\sigma$ we mean the application of the substitution σ to all formulas in Π, thus generating:

$$
\begin{array}{cc}
\Pi\sigma_1 & \Pi\sigma_2 \\
\Gamma\sigma_1 \vdash \Delta\sigma_1 & \Gamma\sigma_2 \vdash \Delta\sigma_2
\end{array}
$$

The duplication of proof Π leads to the exponential growth both in number of lines and in number of symbols when there is a chain of eliminations on the same path in the proof   ∎

The fact that the two substitutions σ_1 and σ_2 are distinct is fundamental for the exponential growth, otherwise a simple use of the contraction rule would have done the job without the duplication of Π.

This phenomenon of exponential growth is also known for usual cut elimination, for there are known sequents whose cut-free traditional proofs can only be exponentially larger than some versions with cuts [Statman, 1979; Orevkov, 1982; Boolos, 1984]. Proofs can, of course, be free of substitution and of cut, by first eliminating substitution as above and then eliminating the cuts by some traditional cut elimination process [Girard, 1987b; Takeuti, 1987].

We now show that cut can be eliminated without exponential explosion if the substitution rule is used.

THEOREM 4. *If there is a usual tree-like sequent proof Π of sequent S, possibly with the use of cuts, then there is a DAG-sequent proof Π' of S without the use of the cut-rule, such that $|\Pi'|$ is linear with respect to $|\Pi|$ and $\|\Pi'\|$ is linear with respect to $\|\Pi\|$.*

Proof. The work of Carbone [Carbone, 1997; Carbone and Semmes, 2000] has shown that the duplication of a chunk of a branch during cut elimination occurs when that branch contains a contraction of a formula A followed by a cut over A. The exponential explosion then occurs when a branch contains several occurrences of contraction-cut sequences.

In fact, according to the usual cut elimination processes [Girard, 1987b; Takeuti, 1987], when cut is eliminated from an axiom, the size of Π' actually decreases. When we eliminate a cut where the cut-formula was introduced by a connective rule, the cut is pushed up towards the leaves. At each such step, the number of lines may increase only in a constant way and the number of symbols may increase only linearly, guaranteeing the final linear bound on number of lines and quadratic bound in the number of symbols.

So all we have to do is focus in the case where cut is eliminated from a contracted formula

$$
\dfrac{\dfrac{\Pi_1}{\dfrac{\Gamma_1, A, A \vdash \Delta_1}{\Gamma_1, A \vdash \Delta_1}} \qquad \dfrac{\Pi_2}{\Gamma_2 \vdash A, \Delta_2}}{\Gamma_1, \Gamma_2 \vdash \Delta_1, \Delta_2} \ (\text{Cut})
$$

generating the following configuration, with the duplication of branch Π_2:

$$
\dfrac{\dfrac{\dfrac{\Pi_1}{\Gamma_1, A, A \vdash \Delta_1} \quad \dfrac{\Pi_2}{\Gamma_2 \vdash A, \Delta_2}}{\Gamma_1, \Gamma_2, A \vdash \Delta_1, \Delta_2} \ (\text{Cut}) \quad \dfrac{\Pi_2}{\Gamma_2 \vdash A, \Delta_2}}{\dfrac{\Gamma_1, \Gamma_2, \Gamma_2 \vdash \Delta_1, \Delta_2, \Delta_2}{\Gamma_1, \Gamma_2 \vdash \Delta_1, \Delta_2}} \begin{matrix} (\text{Cut}) \\ \\ (\text{Contractions}) \end{matrix}
$$

However, with the application of a single substitution rule to $\Gamma_2 \vdash A, \Delta_2,$, with two defocusing applications of identity substitution ι, we can avoid the proof duplication that leads to the exponential explosion

$$
\dfrac{\dfrac{\dfrac{\Pi_1}{\Gamma_1, A, A \vdash \Delta_1} \quad \Gamma_2 \vdash A, \Delta_2}{\Gamma_1, \Gamma_2, A \vdash \Delta_1, \Delta_2} \ (\text{Cut}) \quad \Gamma_2 \vdash A, \Delta_2}{\dfrac{\Gamma_1, \Gamma_2, \Gamma_2 \vdash \Delta_1, \Delta_2, \Delta_2}{\Gamma_1, \Gamma_2 \vdash \Delta_1, \Delta_2}} \begin{matrix} (\text{Cut}) \\ \\ (\text{Contractions}) \end{matrix}
$$

The proof chunk Π_2 is not repeated, the number of added lines is a constant and the number of added formulas is linearly bounded, just as in the connective cases. ∎

Note that to replace cuts with substitution rules all we needed was the employment of identity substitutions. It is not known how to eliminate the use of substitution with the use of cuts. If this were possible, then it would follow that Frege systems can p-simulate extended Frege systems.

3 Tableaux with Substitution

Theorem 4 motivates us to explore a possible adaptation of the substitution rule to decision algorithms such as analytic tableaux. We follow Smullyan's presentation of tableaux dealing with *signed formulas*, in which formulas are prefixed with an F or T sign [Smullyan, 1968]. The signed formulas $F\ A$ and $T\ A$ are called *conjugates*. Any propositional valuation f is simply extended to signed formulas by making $f(T\ A) = 1$ iff $f(A) = 1$ and $f(F\ A) = 1$ iff $f(A) = 0$.

A tableau for a sequent $A_1, \ldots A_n \vdash B_1, \ldots, B_m$ is an attempt to refute it by asserting the antecedent with T-signed formulas $T\ A_1, \ldots T\ A_n$ while denying the consequent with F-signed formulas $F\ B_1, \ldots, F\ B_m$. The tableau's *expansion rules* will then expand the tableau into a tree of signed formulas. If every branch of that tree *closes*, than the initial sequent has been shown; otherwise, a *falsifying valuation* is obtained, which validates the sequent's antecedent and falsifies its consequent.

Signed formulas are classified into α, *neg* and β formulas, as indicated in Figure 5; it must be noted that Smullyan splits *neg*-formulas arbitrarily into α and β. The *expansion of a branch* consists in choosing a signed formula in that branch and then proceeding as follows. If it is an α formula, then add both conclusions α_1 and α_2 to the end of the branch. If it is a *neg* formula, add the *pos* formula to the end of the branch. If it is a β formula, split the branch by adding β_1 to one branch and β_2 to the other. The expansion rules are illustrated in Figure 6.

α	α_1	α_2
$T\ A \wedge B$	$T\ A$	$T\ B$
$F\ A \vee B$	$F\ A$	$F\ B$
$F\ A \rightarrow B$	$T\ A$	$F\ B$

neg	*pos*
$T\ \neg A$	$F\ A$
$F\ \neg A$	$T\ A$

β	β_1	β_2
$F\ A \wedge B$	$F\ A$	$F\ B$
$T\ A \vee B$	$T\ A$	$T\ B$
$T\ A \rightarrow B$	$F\ A$	$T\ B$

Figure 5. Smullyan's Notation for Signed Formulas

We consider a branch Θ as a set of formulas. A branch is *partially expanded* if some of its signed formulas, but not necessarily all, have been expanded. branch. In usual analytic tableaux, a branch *closes* if it contains a pair of conjugate formulas, meaning that a contradiction was reached on that branch. If all formulas in Θ have been expanded and the branch is not

$$
\frac{\alpha}{\begin{array}{c}\alpha_1\\\alpha_2\end{array}} \qquad \frac{neg}{pos} \qquad \begin{array}{c}\beta\\ {/}\quad{\backslash}\\ \beta_1 \qquad \beta_2\end{array}
$$

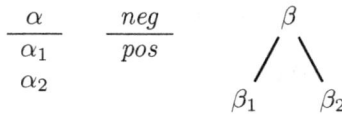

Figure 6. Tableau Expansion Rules

closed, than it is an *open*. The tableau closes if all its branches are closed; if a single branch is open the tableau is open.

Traditional analytic tableaux are always tree-like. By extending tableaux with substitution rules, one could expect us to transform its tree-like structure into a DAG. However, we do not do it. Instead, we add a new branch closure rule.

Substitution Closure Rule (s-closure): If a tableau has partially expanded branches Θ_1 and Θ_2 such that there exists a substitution σ satisfying

$$
\Theta_1 \subseteq \Theta_2\sigma
$$

where by $\Theta_2\sigma$ we mean the set obtained by the application of σ to every signed formula in Θ_2. Then Θ_2 is closed.

The usual closure of a branch with $T\ A$ and $F\ A$ is a *simple closure*. We call a tableau extended with the substitution closure rule a *substitution tableau* (or an *s-tableau*).

In order to facilitate the application of the condition $\Theta_1 \subseteq \Theta_2\sigma$, it would be nice if we could increase the number of equivalent formulas that are also identical in their representation. For example, the formula $p_1 \wedge (p_2 \wedge p_3)$ is equivalent to $(p_1 \wedge p_2) \wedge p_3$ but not syntactically identical. The idea is thus to introduce a connective $\bigwedge$ that operate over a set of formulas, such that both formulas can be represented as $\bigwedge\{p_1, p_2, p_3\}$; similarly, we introduce $\bigvee$. By convention, $\bigwedge\{A\} = \bigvee\{A\} = A$, $\bigwedge\emptyset = \top$, $\bigvee\emptyset = \bot$. The transformation of maximal conjunctions and disjunctions into, respectively, $\bigwedge$-conjunctions and $\bigvee$-disjunctions is immediate. And the tableau rules for $F\bigwedge$ and $T\bigvee$ are, obviously, n-ary branching rules, and the tableau rules for $T\bigwedge$ and $F\bigvee$ are linear with multiple consequences. In the following we will assume that formulas are transformed into this set notation for large conjunctions and disjunctions without mentioning it.

An example of s-tableaux for a family of "hard" propositional formulas is given in Section 4. We now analyse some properties of s-tableaux. First, we note that completeness is trivial for s-tableaux are an extension of usual semantic tableaux and inherit its completeness. Soundness deserves more care.

LEMMA 5. *S-tableaux are sound and complete.*

Proof. Completeness follows immediately from the completeness of usual tableaux, as all usual tableau rules and closure conditions are present in s-tableaux.

For soundness, the tableaux rules maintain their usual property, namely if there is a valuation that satisfies α it also satisfies α_1 and α_2, if there is a valuation that satisfies *neg* it also satisfies *pos* and if there is a valuation that satisfies β it satisfies β_1 of β_2. In this case, if there is an open saturated branch, a valuation can be constructed that falsifies the original sequent; if there is a closed branch in the usual way containing $T\ A$ and $F\ A$ for some A, then no valuation can satisfy all of the branches signed formulas, meaning that the expansion taken on that branch do not lead to a counter-valuation. It remains to be proved that closing a tableau with the substitution closure rule keeps the soundness of the process, that is, we are not closing a branch that has the possibility of becoming saturated open.

In fact, suppose there are partially expanded branches Θ_1 and Θ_2 such that there exists a substitution σ satisfying $\Theta_1 \subseteq \Theta_2\sigma$, so that Θ_2 is closed. It is easy to see that, inductively, every expansion of a signed formula $X\phi$ in Θ_1 can be mimicked by an expansion of $X\phi\sigma$ in Θ_2. So if Θ_1 closes due to $T\ A, F\ A\Theta_1$ for some A, Θ_2 will also close due to $T\ A\sigma, F\ A\sigma \in \Theta_2$. On the other hand, if Θ_1 becomes saturated open, the sequent is not provable anyway. Thus, in both cases, the closure of Θ_2 preserves the soundness of the inference.   ∎

We also have the direct correspondence between closed s-tableaux and DAG sequent proofs with substitution. For a tableau $\mathcal{T}$, we define $|\mathcal{T}|$ as the total number of signed formulas (lines) occurring in it, and $\|\mathcal{T}\|$ as the total number symbols occurring in it.

LEMMA 6. *For every closed s-tableau $\mathcal{T}$ there corresponds a DAG s-sequent proof Π such that $|\Pi|$ is linear with $|\mathcal{T}|$ and $\|\Pi\|$ is quadratic with $\|\mathcal{T}\|$.*

Proof. (Sketch) We apply the usual transformation of tableau proofs into sequent proofs, that is, we transform analytic tableaux into *block tableau*; see [Smullyan, 1968]. In case a branch Θ_2 is closed due to the substitution closing rule, there is a branch Θ'_1 whose prefix Θ_1 is a partially expanded branch verifying $\Theta_1 \subseteq \Theta_2\sigma$. Consider three sub branches: the partially expanded "joint" branch $\Theta_1 \cap \Theta_2$, the "left" expansion $\Theta_1 \setminus \Theta_2$ and the "right" expansion that closes $\Theta_2 \setminus \Theta_1$. The corresponding last elements of the left and right branches correspond to sequents S and $S\sigma$, which become the receiving ends of an application of an s-rule:

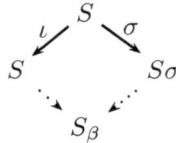

The formula over which a β-expansion was applied to generate Θ_1 and Θ_2 becomes the sequent S_β, which is the point where the two branches rejoin. The rest of the details can be easily filled in, so it is omitted. ∎

4   Example: The Pigeon Hole Problem

The Pigeon Hole Problem (PHP) is a notoriously famous hard case for theorem provers. An initial polynomial-size proof for extended Frege systems for PHP_n was given by [Cook and Reckhow, 1979], and the existence of a polynomial size Frege proof for PHP_n was shown by [Buss, 1987].

The Pigeon Hole Principle of size n (PHP_n) states that if there are $n+1$ pigeons to be placed at n holes, at least one hole will get more than one pigeon. Pigeons are numbered from 1 to $n+1$, holes are numbered from 1 to n, and the fact that pigeon i is placed in hole j is coded by the atomic symbol p_{ij}. These are the only atomic symbols employed, hence there are $n(n+1)$ atomic symbols.

This situation is encoded with a sequent $\Gamma_n \vdash \Delta_n$, where Γ_n expresses that every pigeon goes to a hole, and Δ_n expresses that there is a hole with at least two pigeons. In this way, Γ_n encodes that, for each of the $n+1$ pigeons, it is placed in one of the n holes, that is:

$$\Gamma_n = \left\{ \bigvee_{j=1}^{n} p_{ij} | 1 \leq i \leq n+1 \right\}$$

and Δ_n encodes that for some of the n holes, there are two distinct pigeons placed at it, that is:

$$\Delta_n = \{p_{kj} \wedge p_{ij} | 1 \leq j \leq n, 1 \leq k < i \leq n+1\}.$$

An initial tableau for PHP_n is constructed by T-signing all formulas in Γ_n and F-signing all formulas in Δ_n. The size of Γ_n is $O(n^2)$ and the size of Δ_n is $O(n^3)$, so the size of a PHP_n sequent is $O(n^3)$.

The big symmetries fond in the PHP problems have been pointed as the cause of its high complexity, for all pigeons and all holes "look the same". It is this very symmetry that is exploited to generate a small s-tableau proof. Note that formulas in Γ_n look like a $(n+1) \times n$ matrix, where each line correspond to a pigeon i and each column correspond to a hole j:

$$\Gamma_n = \quad p_{11} \quad \vee \quad p_{12} \quad \vee \quad \ldots \quad \vee \quad p_{1n},$$

$$\vdots \qquad\qquad\qquad\qquad\qquad \vdots$$

$$p_{n+1,1} \quad \vee \quad p_{n+1,2} \quad \vee \quad \ldots \quad \vee \quad p_{n+1,n}$$

which evidences that if we swap lines i' and i'' (that is, if we apply the substitution $\sigma = [i' := i'', i'' := i']$) Γ_n remains the same, and similarly, if we swap columns j' and j'' Γ_n also remains the same. It is perhaps harder to see, but no less true, that the same holds for the formulas of Δ_n, that is, if we swap i' with i'', or j' with j'', in all formulas of Δ_n, Δ_n remains the same. Thus, the symmetry of PHP can be expressed by the following.

LEMMA 7. Let $\sigma_i = [i' := i'', i'' := i']$ and $\sigma_j = [j' := j'', j'' := j']$. Then $\Gamma_n \sigma_i = \Gamma_n$, $\Gamma_n \sigma_j = \Gamma_n$, $\Delta_n \sigma_i = \Delta_n$ and $\Delta_n \sigma_j = \Delta_n$.

We can then start expanding the tableau with an n-branch over the $(n+1)$st (ie, the last) line of Γ_n, as illustrated below.

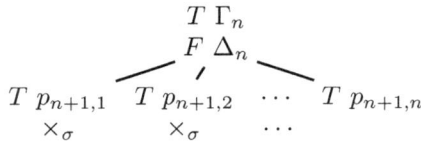

$$T\ \Gamma_n$$
$$F\ \Delta_n$$

$$T\ p_{n+1,1} \quad T\ p_{n+1,2} \quad \cdots \quad T\ p_{n+1,n}$$
$$\times_\sigma \qquad\quad \times_\sigma \qquad \cdots$$

By Lemma 7 there is a substitution that, for any pair of partially expanded branches, transforms one branch into the other. We choose to map every branch into the last one, so that all the first $n-1$ branches are s-closed, which is indicated above by the symbol $\times_\sigma$. Let the branch containing $T\ p_{n+1,n}$ be the *main* branch.

We then concentrate on all the n formulas in Δ_n of the form $F\ p_{i,n} \wedge p_{n+1,n}$, $1 \le i \le n$. We branch each of these formulas, so that the branch containing $F\ p_{n+1,n}$ will simply close due to the presence of $T\ p_{n+1,n}$, and $F\ p_{i,n}$ is added to the main branch, for $1 \le i \le n$. We next consider the first n lines of Γ_n with a branch that generates $T\ p_{i1} \vee p_{i2} \vee \ldots \vee p_{i,n-1}$ and $T\ p_{in}$. Clearly, the branch containing $T\ p_{i,n}$ will simply close due to the presence of $F\ p_{i,n}$ in the main branch, so that we have added to the main branch n formulas of the form, $T\ p_{i1} \vee p_{i2} \vee \ldots \vee p_{i,n-1}$, $1 \le i \le n$, which consists of Γ_{n-1}. By noting that $\Delta_{n-1} \subset \Delta_n$, we have shown how to reduce

$\Gamma_n \vdash \Delta_n$ into $\Gamma_{n-1} \vdash \Delta_{n-1}$. We illustrated below the main branch.

$$T \; \Gamma_n$$
$$F \; \Delta_n$$
$$T \; p_{n+1,n}$$
$$F \; p_{1,n}$$
$$\vdots$$
$$F \; p_{n,n}$$
$$T \; \Gamma_{n-1}$$
$$F \; \Delta_{n-1} \text{ by copying from above}$$

Note that in this process of reducing PHP_n to PHP_{n-1}, we have used $O(n)$ formulas (lines) and $O(n^2)$ symbols. If we repeat this process n times we end up with $\Gamma_1 \vdash \Delta_1$, which clearly closes. We have thus shown the following.

THEOREM 8. *There is an s-tableau $\mathcal{T}$ for PHP_n such that $|\mathcal{T}| = O(n^2)$ and $\|\mathcal{T}\| = O(n^3)$.*

The number of atoms in PHP_n is $O(n^2)$ and the number of symbols in PHP_n is $O(n^3)$, so we have a proof for PHP_n whose size in number of formulas is linear in the number of atoms of the input sequent, and whose size in number of symbols is linear in the number of symbols of the input sequent.[1]

There are two interesting points from the proof above we would like to highlight. First, that all the substitutions used in the s-closure of the branches are actually variable renamings. Second, that all the application of s-closure starts with the identification of a set of substitutions *in the original problem* that make the problem invariant, that is, that map the problem into itself. It is the presence of this substitution invariance that allows one to look for substitutions for the application of the s-closure rule. This seems to indicate a way of applying s-closure in practice, that is, the identification of the invariant substitution and the search for adequate substitutions after branching. The problem is that identifying the existence of an initial substitution seems to be as hard as theorem proving itself.

5 Conclusion

We have defined in this paper an extension of classical sequent proofs with a substitution rule and a DAG proof structure, and we have shown how this technique can be transposed to semantic tableaux. This technique can be generalised in several directions.

---

[1]Without the use of the s-closure rule, the corresponding semantic tableau $\mathcal{T}'$ would be such that $|\mathcal{T}'| = O(n!)$ and $\|\mathcal{T}'\| = O(n!)$.

First, with respect to tableaux, there is nothing particular to semantic tableaux that we have used, and other forms of tableaux can be extended with an s-closure rule, such as KE tableaux.

Second, with respect to sequent proofs, the techniques explored in this paper are not restricted to classical propositional logic and can be directly applied to extensions of propositional logic such as: modal logics, temporal logics, description logics and first-order logic. It may be even possible to apply those techniques to non-classical logics that possess the uniform substitution property, such as most substructural logics [Restall, 2000; Bull and Segerberg, 1984; Dalen, 1984; Girard, 1987a].

It remains an open problem whether Frege proof systems can p-simulate extended Frege proof systems. In the current setting, this problem can be formulated as the search for a systematic way in which the use of the substitution rule can be simulated in ordinary sequent calculus by means of the cut rule.

BIBLIOGRAPHY

[Boolos, 1984] George Boolos. Don't eliminate cut. *Journal of Philosophical Logic*, 13:373–378, 1984.

[Bull and Segerberg, 1984] R. Bull and K. Segerberg. Basic Modal Logic. In D. Gabbay and F. Guenthner, editors, *Handbook of Philosophical Logic*, volume II, pages 1–88. D. Reidel Publishing Company, 1984.

[Buss, 1987] Samuel Buss. Polynomial size proofs of the propositional pigeonhole principle. *Journal of Symbolic Logic*, 52:916–927, 1987.

[Buss, 1995] Samuel R. Buss. Some remarks on lengths of propositional proofs. *Archive for Mathematic Logic*, 34:377–394, 1995.

[Carbone and Semmes, 2000] Alessandra Carbone and Stephen Semmes. *A Graphic Apology for Symmetry and Implicitness*. Oxford Mathematical Monographs. Oxford University Press, 2000.

[Carbone, 1997] A. Carbone. Interpolants, cut elimination and flow graphs for the propositional calculus. *Annals of Pure and Applied Logic*, 83:249–299, 1997.

[Cook and Reckhow, 1979] S. A. Cook and R. A. Reckhow. The relative efficiency of propositional proof systems. *Journal of Symbolic Logic*, 44:36–50, 1979.

[Dalen, 1984] Dirk Van Dalen. Intuitionistic logic. In D. Gabbay and F. Guenthner, editors, *Handbook of Philosophical Logic*, volume III, 1984.

[Finger and Gabbay, 2005] Marcelo Finger and Dov Gabbay. Cut and pay. Manuscript, 2005.

[Girard, 1987a] J.-Y. Girard. Linear logic. *Theoretical Computer Science*, 50:1–102, 1987.

[Girard, 1987b] J.-Y. Girard. *Proof Theory and Logical Complexity*, volume volume 1 of *Studies in Proof Theory*. Monographs. Bibliopolis, 1987.

[Krajíček and Pudlák, 1989] Jan Krajíček and Pavel Pudlák. Propositional proof systems, the consistency of first order theories and the complexity of computations. *Journal of Symbolic Logic*, 54(3):1063–1079, 1989.

[Orevkov, 1982] V. Orevkov. Lower bounds for increasing complexity of derivations after cut elimination. *Journal of Soviet Mathematics*, 20(4):2337–2350, 1982.

[Restall, 2000] G. Restall. *An Introduction to Substructural Logics*. Routledge, 2000.

[Smullyan, 1968] Raymond M. Smullyan. *First-Order Logic*. Springer-Verlag, 1968.

[Statman, 1979] R. Statman. Lower bounds on herbrand's theorem. *Proceedings of the American Mathematical Society*, 75(1):104–107, June 1979.

[Takeuti, 1987] G. Takeuti. *Proof Theory*. North Holland, second edition edition, 1987.

Polymorphic Quantifiers and Underspecification in Natural Language

CHRIS FOX AND SHALOM LAPPIN

1 Introduction

Consider the following sentences.

1. Every art dealer introduced a critic to an artist.

2. Someone believes all that Mary believes.

3. The algorithm assigns an integer to each theorem in the logic.

These examples illustrate the well-known problem of quantifier scope ambiguity; the appropriate ordering of the quantifiers in the formal representation is ambiguous, which motivates the various treatments of underspecification. In addition, they also illustrate the polymorphic nature of natural language quantification. In these examples in particular, *introduced* denotes a relation among three (basic) individuals, *believes* a relation between an individual and a proposition, and *assigns* a relation that holds among an abstract individual (an algorithm), a number, and a proposition.

Existing treatments of underspecification and scope ambiguity do not explicitly address the issue of polymorphism in natural language quantifiers. Ideally in these cases, we might wish for the type of the semantic representation of a noun phrase to reflect the nature of the objects which are being quantified over (i.e. individuals, propositions, numbers etc.), and for the type of the representation of the core verb to be able to capture constraints on the appropriate types for its arguments.

There are existing treatments and discussions of polymorphism in linguistic theories, for example [van Benthem, 1991; Chierchia, 1982; Chierchia and Turner, 1988; Emms, 1993; Kamareddine and Klein, 1993; Parsons, 1979]. In the realm of formal semantics, one idea is to adopt a universal type [Chierchia, 1982; Chierchia and Turner, 1988]. All types are then subtypes of this universal type. Any more specific types are then subsumed by occurrences of this universal type.[1]

Unfortunately, when considering polymorphism that arises with coordination, it can be argued that this approach is too unrestricted, and a more fine-grained notion of polymorphism is required. Such an approach is explored by [Fox and Lappin, 2005b; Fox and Lappin, 2005a]. That work also develops an approach to underspecification. The treatment of underspecification there is couched in terms that make it amenable to inhomogeneous quantifiers, which range over different types of objects. However, this assumes that the determiners are polymorphic in nature, and the way in which this is achieved is not explained. This paper addresses this shortcoming.

First we reprise some common ways of formulating polymorphic types (section 2) and the polymorphic nature of natural language (section 3) and then we show how we can use implicit polymorphism for generalised quantifiers (section 4). Finally we indicate how this account can be exploited in a polymorphic theory of underspecification (section 5), and then we state our conclusions (section 6).

2 Polymorphic Types

A *polymorphic type* is, in broad terms, a type expression T that contains a type variable X which may be instantiated by any type.[2]

With *schematic* polymorphism, type variables are merely a meta-theoretic device. They abbreviate all the possible types that are generated through substitution of a concrete type for a variable. This is convenient for stating certain kinds of inference rules, for example, which have the same form for all possible instantiations of the type variables.

In contrast, *non-schematic* polymorphism forms a genuine extension to a type theory where the type variables are new additions to the theory. These type variables cannot be trivially eliminated through simple substitution.

There are two flavours of non-schematic polymorphism.[3] In *implicit* polymorphism, the polymorphic types are of the form $\Pi X.T$ where T is a complex type in which X appears, and Π acts like a universal quantifier over

---

[1] A natural generalisation of this approach is given by [Kamareddine and Klein, 1993], where types other than the universal type can have subtypes.

[2] Much of the work on theories of polymorphism has been done in the framework of theoretical computer science (for example, [Milner, 1978; Abadi *et al.*, 1991; Harper and Lillibridge, 1993]). There is relatively little discussion of formal theories of polymorphism in the literature on formal semantics of natural language.

[3] There are other classes of polymorphism in the computer science literature, such as *structural polymorphism* [Ruehr, 1998] but these are not relevant to the current discussion.

types. Membership of this type is then governed by a rule of the form

4. $e \in \Pi X.T$ iff $\forall X (e \in T)$

where quantification over types is governed by the expected rules.

The polymorphism is "implicit" because appropriate values of the type variable are not made explicit within a type derivation. For example, if we consider $x \in T$ and $f \in \Pi X.(X \Rightarrow \mathsf{Prop})$, then we can infer that $fx \in \mathsf{Prop}$. The information that the relevant instantiation of the type variable X is T is not made explicit.

By contrast, in *explicit* polymorphism the appropriate value of a type variable in a polymorphic type must be made explicit. With this kind of polymorphism, in place of $e \in \Pi X.T$ we might write

5. $e[X] \in T$

where X is to be understood as a parametric type that occurs in T. When e is applied to a term t , that term has to supply an appropriate type with which to instantiate X. This would normally be the type of t. Both schematic polymorphism and explicit polymorphism could be thought of as forms of *parametric* polymorphism, the difference between the two being whether the type parameter is part of the language of types or not.[4]

Implicit polymorphism carries a complication. If we allow type variables to range over polymorphic types themselves, then the polymorphism becomes impredicative; that is, the set of polymorphic types cannot be considered to be defined by the rule in (4) as the type $\Pi X.T$ on the left falls within the domain of quantified statement $\forall X(e \in T)$ on the right. Fortunately, it appears that weak polymorphism is sufficient for natural language [van Benthem, 1991], where quantified type variables range only over non-polymorphic types.

The approach advocated by [Chierchia, 1982; Chierchia and Turner, 1988] and [Kamareddine and Klein, 1993] can be seen as variants of implicit polymorphism. In the former case a universal type Δ is allowed to dominate other types, so that an expression of the form $\Delta \Rightarrow T$ will take an argument of any type and produce an expression of type T. In [Kamareddine and Klein, 1993], a more general notion of subtype is adopted.

3   Polymorphism and Natural Language

There is strong evidence that natural language requires flexibility in its semantic type system. [Chierchia, 1982] points out how certain type general

---

[4]Note however that [Cardelli and Wegner, 1985] take the term *parametric polymorphism* to include the cases where the type "parameter" is implicit as well as explicit.

verbal expressions, such as *is fun*, can combine with a range of categories, such as gerunds, infinitives and nouns:

6. Tennis is fun.

7. Playing tennis is fun.

8. To play tennis is fun.

This nominalisation data provides motivation for some intensional theories, such as those by [Bealer, 1982; Chierchia and Turner, 1988], with less rigid notions of types than conventional Montague Semantics [Montague, 1973].

Moreover natural language conjunctions, such as *and* and *or* generate coordinate structures over a range of diverse syntactic categories.

Chierchia argues that this motivates the introduction of a universal type Δ, so *is fun* is of the type $(\Delta \Rightarrow \mathsf{Prop})$ and *and* is of the type $(\Delta \Rightarrow (\Delta \Rightarrow \Delta))$. This move is excessively permissive. In the case of conjunction, it does not capture the typical requirement that the arguments of *and*, and the resultant coordinate structure are all of the same category (type). In fact, these constraints are captured if in place of a universal type, and the sub-type polymorphism that results, we adopt implicit polymorphism, and assign conjunction the type $\Pi X.(X \Rightarrow (X \Rightarrow X))$. It is this form of polymorphism that is incorporated into PTCT [Fox and Lappin, 2005a].[5]

The verbal expression *is fun* can be given the type $\Pi X.(X \Rightarrow \mathsf{Prop})$.[6] It seems, then, that polymorphic types are appropriate for capturing the type flexibility of natural language. We might worry that using polymorphic types in place of universal types is still too permissive, particularly in this case, where the type $\Pi X.(X \Rightarrow \mathsf{Prop})$ allows *is fun* to form propositions with any type of expression.

Even granting this concern, polymorphic types do seem to offer the right sort of formal device for dealing with the type flexibility of natural language expressions. The promiscuity that they introduce can be restricted by formulating constraints on type variables through subtyping.

4 Polymorphic Generalised Quantifiers

To deal with examples such as (1)–(3) we could consider endowing determiners with a type parameter whose values are supplied by the noun. A determiner would then have the type

$$(X \Rightarrow \mathsf{Prop}) \Rightarrow ((X \Rightarrow \mathsf{Prop}) \Rightarrow \mathsf{Prop})$$

---

[5] Indeed, the type system of PTCT excludes a universal type, so Chierchia's proposal [Chierchia, 1982] cannot be adopted directly.

[6] Note that this typing is perhaps excessively inclusive; we may have to add constraints on the felicitous values of the type variable X.

where X is the type of the argument required for the noun to form a proposition. If the noun's type was $(A \Rightarrow \mathsf{Prop}))$, then the type of the resulting noun phrase would be

$$((A \Rightarrow \mathsf{Prop}) \Rightarrow \mathsf{Prop})$$

We could achieve this effect by using schematic polymorphism. As we have already observed, this is not a particularly interesting approach if we wish to explore the nature of polymorphism in natural language in more than a superficial, notational manner.

A possible alternative is to to adopt explicit polymorphism, as discussed in [Fox and Lappin, 2005a]. For the above example, this requires some means of instantiating the type parameter/variable X by A within the theory. Sketching this in a little more detail, the representation d of a determiner (or any other polymorphic expression) would be augmented by a type parameter X, giving it the general form

$$d[X]$$

When combining with the representation of a noun n of type $(A \Rightarrow \mathsf{Prop})$, we want X to be instantiated by A, that is, we want the representation of the generalised quantifier formed by d and n to be $d[A](n)$. The expression $d[A]$ explicitly instantiates the type of the determiner to

$$((A \Rightarrow \mathsf{Prop}) \Rightarrow ((A \Rightarrow \mathsf{Prop}) \Rightarrow \mathsf{Prop}))$$

allowing it to felicitously combine with the representation of the noun. The type of the resultant generalised quantifier is thus

$$((A \Rightarrow \mathsf{Prop}) \Rightarrow \mathsf{Prop})$$

as required.[7]

To implement the details of this proposal, we need various components: (i) a way of indicating which type is "supplied", e.g. by a noun; (ii) a form of abstraction for types, to indicate which type variable is to be substituted for, e.g. in the type of the determiner; and (iii) some form of type application, which in this case will be used to perform the substitution of the type supplied by the noun for the appropriate variable in the type of the determiner. In effect, the noun's representation has to make available both its logical representation, and the expected type of its argument. In turn,

---

[7]Note that following the way such theories are typically presented, we might expect the polymorphic type of the determiner could be given as $(X \Rightarrow (X \Rightarrow \mathsf{Prop}))$, where X would then be instantiated by the type of the noun $(A \Rightarrow \mathsf{Prop})$, rather than the type of the argument of the noun. However, this would require an additional constraint to indicate that X is intended to be of the form $(Y \Rightarrow \mathsf{Prop})$.

the determiner then has to provide an additional argument place. The value of that argument will be substituted for the relevant variable in the type of the determiner.

Working with this approach has made it clear to us that this is not a trivial undertaking, especially when dealing with the details of the parametric substitution of types. It can dramatically increase the combinatorial complexity of the semantics and make it difficult to manage underspecified representations. The introduction of type abstraction and type application also change the character of the underlying theory.

There may be more elegant ways of implementing explicit polymorphism for these examples, but actually there is no need to adopt it. We propose to use implicit polymorphism instead and to give up on the direct use of logical quantifiers (or their intensions) as representations of natural language quantifiers. We assume that all natural language determiners are expressed by non-logical constants interpreted as implicitly polymorphic functions from property terms to generalised quantifiers (GQs), rather than the usual logical quantifiers (or their intensions). This liberates us from the need to impose any constraints on the typing of logical quantifiers to express our semantic theory. Such typing is distracting when it comes to the typing of natural language determiners.

We introduce polymorphic non-logical analogues all$'$ and some$'$ of the logical quantifiers $\forall, \exists$. These are typed as

9. $\Pi X.(X \Rightarrow \mathsf{Prop}) \Rightarrow ((X \Rightarrow \mathsf{Prop}) \Rightarrow \mathsf{Prop})$

Given this typing, the determiners can combine with any expression whose type is of the form $(A \Rightarrow \mathsf{Prop})$, to produce a generalised quantifier of the type $((A \Rightarrow \mathsf{Prop}) \Rightarrow \mathsf{Prop})$. This gives rise to the following theorem.

10. If $p, q \in (A \Rightarrow \mathsf{Prop})$, for some non-polymorphic type A, then all$'(p)(q) \in \mathsf{Prop}$ and some$'(p)(q) \in \mathsf{Prop}$.

The proof is straightforward.

This simple addition allows for polymorphic determiners and quantifiers without incorporating additional compositional machinery to deal with type parameters.

Next we specify the truth conditions for the quantifiers.

11. If all$'(p)(q) \in \mathsf{Prop}$ and $p \in (A \Rightarrow \mathsf{Prop})$, then all$'(p)(q)$ is true iff for all $x \in A$, if px is true, then qx is true.

12. If some$'(p)(q) \in \mathsf{Prop}$ and $p \in (A \Rightarrow \mathsf{Prop})$, then some$'(p)(q)$ is true iff for some $x \in A$, px and qx are both true.

In (11) and (12), $(A \Rightarrow \mathsf{Prop})$ has the character of a schematic polymorphic type. This is a consequence of the usual schematic character of inference rules. It is important to note that by using implicit polymorphism we are incorporating polymorphic types into our representation language rather than using schematic meta-variables that abbreviate a disjunction of possible type values.

5   Underspecification in PTCT

We now integrate our implicit polymorphic treatment of typing for generalised quantifiers into a PTCT account of underspecified scope interpretations.

PTCT is a theory of fine-grained intensionality with Curry typing. The theory has a first-order presentation, with a rich language of terms. The types of the theory is also representable within the language of terms. Here we sketch a part of the theory. The interested reader is referred to [Fox and Lappin, 2005a] for more details.

13. The language of PTCT

   Logical constants l ::= $\hat{\sim} \mid \hat{\wedge} \mid \hat{\vee} \mid \hat{\rightarrow} \mid \hat{\leftrightarrow} \mid \hat{\forall} \mid \hat{\exists} \mid \doteq_T \mid \hat{\cong}_T \mid \epsilon T$

   Terms $\qquad\qquad\quad$ t ::= $x \mid c \mid l \mid T \mid \lambda x(t) \mid (t)t$

   Types $\qquad\qquad\quad$ T ::= $B \mid \mathsf{Prop} \mid T \Rightarrow S \mid X \mid \{x \in T.\varphi'\} \mid \{x.\varphi'\}$
   $\qquad\qquad\qquad\qquad\quad \mid \Pi X.T \mid S \otimes T$

   Atomic wffs $\qquad$ α ::= $(t =_T s) \mid t \in T \mid t \cong_T s \mid {}^\mathsf{T}t$

   Wffs $\qquad\qquad\quad$ φ ::= $\alpha \mid \sim \varphi \mid (\varphi \wedge \psi) \mid (\varphi \vee \psi) \mid (\varphi \rightarrow \psi) \mid (\varphi \leftrightarrow \psi)$
   $\qquad\qquad\qquad\qquad\quad \mid (\forall x \varphi) \mid (\exists x \varphi) \mid (\forall X \varphi) \mid (\exists X \varphi)$

where type variables X are intended to range only over non-polymorphic types, and φ' are term representable wffs, defined by the following.

14. Term representable wffs

   φ' ::= $\alpha' \mid (\sim \varphi') \mid (\varphi' \wedge \psi') \mid (\varphi' \vee \psi') \mid (\varphi' \rightarrow \psi') \mid (\varphi' \leftrightarrow \psi')$
   $\qquad \mid (\forall_T x \varphi') \mid (\exists_T x \varphi')$

   Atomic representable wffs α' ::= $(t =_T s) \mid t \cong_T s \mid {}^\mathsf{T}t$

Here $\forall_T x \varphi'$ and $\exists_T x \varphi'$ correspond to the wffs $\forall x (x \in T \rightarrow \varphi')$ and $\exists x (x \in T \wedge \varphi')$ respectively.

15. Sample Proof Rules. Here we exemplify some of these kinds of rules as they apply to conjunction, both as it appears in the language of wff ($\wedge$), and in the language of terms ($\hat{\wedge}$)

(a) The basic connectives of the wff

$$\frac{\varphi \quad \psi}{\varphi \wedge \psi} \wedge i \quad \frac{\varphi \wedge \psi}{\varphi} \wedge e \quad \frac{\varphi \wedge \psi}{\psi} \wedge e$$

(b) Typing rules for λ-terms

$$t \in \mathsf{Prop} \wedge t' \in \mathsf{Prop} \rightarrow (t \hat{\wedge} t') \in \mathsf{Prop}$$

(c) Truth conditions for Propositions

$$t \in \mathsf{Prop} \wedge t' \in \mathsf{Prop} \rightarrow ({}^\mathsf{T}(t \hat{\wedge} t') \leftrightarrow {}^\mathsf{T}t \wedge {}^\mathsf{T}t')$$

Here is an axiom that governs the polymorphic types $\Pi X.T$ in PTCT.

16. $f \in \Pi X.T \leftrightarrow \forall X(f \in T)$

Because PTCT already includes implicit polymorphism, the proposal for polymorphic generalised quantifiers given above can be adopted without modifying the theory.

The fact that the language of terms in PTCT is the untyped λ-calculus gives rise to some interesting possibilities. in particular, it allows us to encode various data-structures and algorithms within the theory. In particular, we can use product types to represent heterogeneous lists which can then be used to mimic Cooper storage [Cooper, 1983], but where quantifiers may be of different types. Product types allow us to type pairs of elements using the following axiom.

17. $\langle x, y \rangle \in (S \otimes T) \leftrightarrow x \in S \wedge y \in T$

We can then form typed heterogeneous lists $\langle x_1, \langle x_2, \langle x_{n-1}, x_n \rangle \ldots \rangle \rangle$, which we can write $\langle x_1, x_2, \ldots, x_{n-1}, x_n \rangle$.

In order to construct underspecified scope representations we need a procedure that generates the $k!$-tuple of possible scope readings that corresponds to the possible permutations of the surface sequence of GQ arguments in a core propositional relation. Using the approach of [Fox and Lappin, 2005b], we specify a function *perms_scope* that generates all $k!$ indexed permutation products of a k-ary product term $\langle t_1, \ldots, t_k \rangle$ as part of the procedure for generating the set of possible scope readings of a sentence.

In this treatment of underspecification, *perms_scope* needs to take a k-ary product of scope taking elements (in the order in which they appear in the surface syntax) and a k-ary relation, expressed as a λ-term representing the core proposition, as arguments. The scope taking elements and the core relation can be combined into a single product, e.g. as a pair consisting of the

k-tuples of GQs as its first element and the core relation as its second. The permutation function *perms_scope* produces the $k!$-ary product of scoped readings. When a k-tuple of GQs is permuted, the λ-operators that bind the quantified argument positions in the core relation are permuted in the same order as the quantifiers in the k-tuple. This correspondence is necessary to preserve the connection between each GQ and its argument position in the core relation across scope permutations.

A scope reading is generated by applying the elements of the k-tuple of quantifiers in sequence to the core relation, reducing its arity with each such operation until a proposition results. The ith scope reading is identified by projecting the ith element of the indexed product of propositions that is the output of our *perms_scope* function. It is important to recognise that it is not necessary to compute the full $k!$-ary product of permutations for this function. The term consisting of the application of *perms_scope* to its argument pair can be used in computing interpretations and inferences without computing its value until a later point in subsequent discourse. Filters on possible scope readings can be acquired through additional discourse information, where these filters reduce the search space of $k!$ scope readings that is the full value of the *perms_scope* term.

Because the underspecified representations are expressed within the logical theory, it is possible to exploit the full power of PTCT to state filters and constraints on the legitimate scopings of the underspecified representations. This means that unlike other proposals, PTCT can achieve expressive completeness in the sense of[Ebert, 2005] for the possible scope readings generated by an underspecified representation.

5.1   Underspecification with Polymorphic Quantifiers

Here we see how the approach of [Fox and Lappin, 2005a] can be applied naturally to all′ and some′ of Section 4. This analysis relies on the fact that the treatment of underspecified representations in PTCT are based on heterogeneous lists, which can accommodate the polymorphism of all′ and some′.

Consider the example *Every man loves a woman*, with the GQs interpreting the subject and object NPs, the core relation, and the term expressing the underspecified representation of the sentence given, respectively, as follows

18. $Q_1 = \mathsf{all}'(\mathsf{man}')$

19. $Q_2 = \mathsf{some}'(\mathsf{woman}')$

20. $\lambda uv.\mathsf{loves}'(u, v)$

21. $perms\_scope(\langle\langle Q_1, Q_2\rangle, \lambda uv.\mathsf{loves}'(u, v)\rangle)$

The permutations of the quantifiers relative to the core relation that *perms_scope* produces are

22. $\langle\langle\langle Q_1, Q_2\rangle, \lambda uv.\mathsf{loves}'(u, v)\rangle, \langle\langle Q_2, Q_1\rangle, \lambda vu.\mathsf{loves}'(u, v)\rangle\rangle$

Applying relation reduction [Keenan, 1992; van Eijck, 2003] to computing the final propositions gives us a product containing the two readings.

23. $perms\_scope(\langle\langle Q_1, Q_2\rangle, \lambda uv.\mathsf{loves}'(u, v)\rangle) =$
       $\langle\mathsf{all}'(\mathsf{man}')(\lambda u.\mathsf{some}'(\mathsf{woman}')(\lambda v.\mathsf{loves}'(u, v))),$
       $\mathsf{some}'(\mathsf{woman}')(\lambda v.\mathsf{all}'(\mathsf{man}')(\lambda u.\mathsf{loves}'(u, v)))\rangle$

To obtain resolved scope readings from an underspecified representation, we define a function $project\_scope^i_k$ that computes the ith permutation of a k-ary product of propositions. Specifically, it returns the i-th proposition in the product of scope readings that *perms_scope* gives as its value.

In this example all' and some' are implicitly polymorphic functions that take as arguments property terms typed as functions from individuals to propositions in order to generate GQ functions from property terms of the same type to propositions.

Analogous underspecified representations are available for (1)–(3) where appropriate polymorphic functions interpret each of the determiners in these sentences. We illustrate with examples (1) and (2).

In the case of (1), repeated here,

(1) Every art dealer introduced a critic to an artist.

for simplicity we will introduce the predicate $\mathsf{introduced\_to}'$ of type $B \Rightarrow B \Rightarrow B \Rightarrow \mathsf{Prop}$. The tuple in the underspecified representation would then be

24. $\langle\lambda d(\mathsf{all}'(\mathsf{art\_dealer}')(d)),$
     $\lambda d(\mathsf{exists}'(\mathsf{critic}')(d)),$
     $\lambda d(\mathsf{exists}'(\mathsf{artist}')(d)),$
     $\lambda uvw(\mathsf{introduced\_to}'(uvw))\rangle$

which, assuming the atomic predicates are of type $B \Rightarrow \mathsf{Prop}$, would have the type

25. $((B \Rightarrow \mathsf{Prop}) \Rightarrow \mathsf{Prop}) \otimes ((B \Rightarrow \mathsf{Prop}) \Rightarrow \mathsf{Prop}) \otimes ((B \Rightarrow \mathsf{Prop}) \Rightarrow \mathsf{Prop}) \otimes (B \Rightarrow B \Rightarrow B \Rightarrow \mathsf{Prop})$

The application of relation reduction to the tuple (24), or a *perms_scope* permutation of it, will yield a proposition.

For (2), repeated here,

(2) Someone believes all that Mary believes.

we need to provide an analysis of *someone*. We shall assume that *someone* always quantifies over basic individuals, sortally constrained to people. That is, it can be given the representation

26. $\lambda d(\mathsf{some}'(\mathsf{person}')(d)$

where $\mathsf{person}' \in B \Rightarrow \mathsf{Prop}$.

Assuming some appropriate analysis of the relative clause and proper name, the tuple in the underspecified representation of the sentence would then be something like

27. $\langle \lambda d(\mathsf{some}'(\mathsf{person}')(d), \lambda d(\mathsf{all}'(\lambda x(\mathsf{believes}'(\mathsf{Mary}', x)))(d),$
    $\lambda uv(\mathsf{believes}'(u, v)) \rangle$

which would be of type

28. $((B \Rightarrow \mathsf{Prop}) \Rightarrow \mathsf{Prop}) \otimes ((\mathsf{Prop} \Rightarrow \mathsf{Prop}) \Rightarrow \mathsf{Prop}) \otimes (B \Rightarrow (\mathsf{Prop} \Rightarrow \mathsf{Prop})))$

Again, the application of relation reduction to the tuple (27), or a *perms_scope* permutation of it, will yield a proposition.

Here the requirement that only propositions can be the objects of belief is enforced by the representation. In the alternative analysis of such polymorphism provided by [Chierchia, 1982; Chierchia and Turner, 1988], there would have to be an external constraint that only propositions are the objects of believe; PTCT can express such type constraints internally, while that analysis cannot.

If the sentence were

29. Someone believes everything that Mary believes.

we would need some polymorphic analysis of *everything*, where its domain of quantification is determined by its argument, much like all' and some'. So its semantic representation, $\mathsf{everything}'$, can be of type

30. $\Pi X.(X \Rightarrow \mathsf{Prop}) \Rightarrow \mathsf{Prop}$

However, the analysis of the relative clause might cause some difficulty. We could instead seek to analysis this as *every thing*, but then we would need to determine an appropriate type for the representation of *thing*. If we gave it the type $\Pi X.X \Rightarrow \mathsf{Prop}$, then that would cause difficulties with the typing of quantifiers.[8] This suggests that there may be some further subtle

---

[8]Note that we are trying to avoid impredicative polymorphism, so type variables cannot range over polymorphic types.

issues to consider in deciding the appropriate way to deal with polymorphic quantifiers.

In a similar vein, example (3) serves as an illustration of where care has to be taken in deciding whether it is appropriate to adopt type constraints rather than sortal constraints.

(3) The algorithm assigns an integer to each theorem in the logic.

The question is whether it is appropriate to restrict one or more of *algorithms*, *integers*, *theorems* and *logic* to appropriate types within the theory. Although it appears plausible to restrict the type of objects of *believe* to the propositions, we might question whether it is always appropriate to interpret natural language sortal expressions as giving rise to a type constraint in the logic; for example, do we wish to always interpret discussion of numbers as discussion of elements of the type Num in the logic? We shall leave this issue for future work.

6   Conclusions and Future Work

This paper has discussed polymorphic types and their relevance to natural language semantics. We have proposed an account of implicit polymorphic typing of determiners, and we have applied it to underspecified semantic representations.

An adequate type-theoretic treatment of quantified NPs must accommodate both the polymorphic properties of determiners and verbs, and scope ambiguity. Our account provides a straightforward and integrated type-theoretic account of both.

Our results suggest that a type system that includes general function spaces, weak polymorphism, and subtypes is sufficiently expressive to allow an adequate typed theory of natural language semantics.

Some details remain to be worked out. For example, here we offer no type for the *perms_scope* operator. It is possible to state constraints on the arguments of this operator, to require that they will form propositions following relation reduction. The 'output' of *perms_scope* will then be a tuple of propositions, whose arity depends upon the number of legitimate permutations of the input. The type theory of PTCT would have to be adapted to express such types.[9]

A further issue that remains to be explored is whether the type system that we have sketched is excessively unconstrained. For example, it may be the case that only a small subset of all functional types are instantiated

---

[9]There are alternatives, for example where a particular reading is projected, depending upon the value of an integer input. The type of underspecified readings would then be Num $\Rightarrow$ Prop [Fox and Lappin, 2005b].

in natural language semantics. There should also be additional restrictions on the types which can occur as arguments of verbs. In this regard, it is possible that the types of natural language expressions should take into account whether or not a verbal expression has tense [Kamareddine and Klein, 1993].

Although the implicit polymorphism of PTCT seems broadly suited for analysing coordination and type general verbs, we might legitimately question whether this is the most appropriate way of dealing with apparent polymorphism in natural language quantification, or whether we should avoid tying sortal expressions in natural language so strongly to the types of the semantic theory. In some cases, this might just reduce to a question of taste.

In future work we intend to explore the formulation of appropriate conditions to define a constrained type system that corresponds more closely to the observed semantic properties of natural language. We will also explore the connections between these typed logics and type-theoretic grammars, such as those of [Steedman, 2000] and [Morrill, 1994].

BIBLIOGRAPHY

[Abadi et al., 1991] Martín Abadi, Luca Cardelli, Benjamin Pierce, and Gordon Plotkin. Dynamic typing in a statically typed language. *ACM Transactions on Programming Languages and Systems*, 13(2):237–268, April 1991.

[Bealer, 1982] G. Bealer. *Quality and Concept*. Clarendon Press, Oxford, 1982.

[Cardelli and Wegner, 1985] Luca Cardelli and Peter Wegner. On understanding types, data abstraction, and polymorphism. *ACM Computing Surveys*, 17(4):471–522, 1985.

[Chierchia and Turner, 1988] G. Chierchia and R. Turner. Semantics and property theory. *Linguistics and Philosophy*, 11:261–302, 1988.

[Chierchia, 1982] G. Chierchia. Nominalisation and Montague grammar: a semantics without types for natural languages. *Linguistics and Philosophy*, 5:303–354, 1982.

[Cooper, 1983] Robin Cooper. *Quantification and Syntactic Theory*. Synthese Language Library. D. Reidel, Dordrecht, 1983.

[Ebert, 2005] C. Ebert. *Formal Investigation of Underspecified Representations*. PhD thesis, Department of Computer Science, King's College London, 2005. Unpublished.

[Emms, 1993] Martin Emms. Parsing with polymorphism. In *Sixth Conference of the European Asociation for Computational Linguistics*, pages 120–129, Utrecht, 1993.

[Fox and Lappin, 2005a] C. Fox and S. Lappin. *Formal Foundations of Intensional Semantics*. Blackwell, Oxford, 2005.

[Fox and Lappin, 2005b] C. Fox and S. Lappin. Underspecified interpretations in a curry-typed representation language. *The Journal of Logic and Computation*, 15:131–143, 2005.

[Harper and Lillibridge, 1993] Robert Harper and Mark Lillibridge. Explicit polymorphism and CPS conversion. In *Conf. Record 20th Ann. ACM SIGPLAN-SIGACT Symp. on Principles of Programming Languages, POPL'93, Charleston, SC, USA, Jan. 1993*, pages 206–219. ACM Press, New York, 1993.

[Kamareddine and Klein, 1993] Fairouz Kamareddine and Ewan Klein. Nominalization, predication and type containment. *Journal of Logic, Language and Information*, 2(3):171–215, 1993.

[Keenan, 1992] E. Keenan. Beyond the fregean boundary. *Linguistics and Philosophy*, 15:199–221, 1992.

[Milner, 1978] R. Milner. A theory of type polymorphism in programming. *Journal of Computer and Systems Sciences*, 17:348–375, 1978.

[Montague, 1973] R. Montague. The proper treatment of quantification in ordinary English. In K. J. J. Hintikka, J. M. E. Moravcsik, and P. Suppes, editors, *Approaches to Natural Language*. D. Reidel, Dordrecht, 1973. Synthese Library.

[Morrill, 1994] Glyn Morrill. *Type Logical Grammar: Categorial Logic of Signs*. Kluwer Academic Publishers, Dordrecht, 1994.

[Parsons, 1979] T. Parsons. The theory of types and ordinary language. In S. Davis and M. Mithun, editors, *Linguistics, Philosophy and Montague Grammar*. University of Texas Press, Austin, Texas, 1979.

[Ruehr, 1998] F. Ruehr. Structural polymorphism, 1998.

[Steedman, 2000] Mark Steedman. *The Syntactic Process*. Linguistic Inquiry Monographs. MIT Press, Cambridge, 2000.

[van Benthem, 1991] J. van Benthem. *Language in Action*. Studies in Logic. North-Holland, Amsterdam, 1991.

[van Eijck, 2003] J. van Eijck. *Computational Semantics and Type Theory*. unpublished ms., CWI, Amsterdam, 2003.

Some Formal Considerations on Gabbay's Restart Rule in Natural Deduction and Goal-Directed Reasoning

MICHAEL GABBAY AND MURDOCH GABBAY

1 Introduction

In this paper we make some observations about Natural Deduction derivations [Prawitz, 1965; van Dalen, 1986; Bell and Machover, 1977]. We assume the reader is familiar with it and with proof-theory in general. Our development will be simple, even simple-minded, and concrete. However, it will also be evident that general ideas motivate our examples, and we think both our specific examples and the ideas behind them are interesting and may be useful to some readers.

In a sentence, the bare technical content of this paper is: *Extending natural deduction with global well-formedness conditions can neatly and cheaply capture classical and intermediate logics.*

The interest here is in the 'neatly' and 'cheaply'. By 'neatly' we mean 'preserving proof-normalisation',[1] and 'maintaining the subformula property', and by 'cheaply' we mean 'preserving the formal structure of deductions' (so that a deduction in the original system is still, formally, a deduction in the extended system, and in particular it requires no extra effort to write just because it is in the extended system).

To illustrate what we have in mind consider intuitionistic first-order logic (**FOL**) [van Dalen, 1986] as a paradigmatic example of a formal notion of deduction. A natural deduction derivation (or **deduction**) is an inductively defined tree structure where each node contains an instance of a formula. A deduction is **valid** when each successive node follows from its predecessors in accordance with some predetermined inference rules.

A particular attraction of Natural Deduction is its clean and economical presentation. Here for example are deduction (fragments) proving $A \wedge B$ from A and B, and $\forall x.\ (P(x) \wedge Q(x))$ from $\forall x.\ P(x)$ and $\forall x.\ Q(x)$:

$$\frac{A \quad B}{A \wedge B}\ (\wedge I) \qquad \frac{\dfrac{\forall x.\ P(x)}{P(x)}\ (\forall E) \quad \dfrac{\forall x.\ Q(x)}{Q(x)}\ (\forall E)}{\dfrac{P(x) \wedge Q(x)}{\forall x.\ (P(x) \wedge Q(x))}\ (\forall I)}\ (\wedge I)$$

Of interest is also the Curry-Howard correspondence [M. H. B. Sorensen, 1998; Barendregt, 2000]: these deductions have a natural notion of proof-normalisation. As is well-

---

[1] Perhaps in this paper we should call it 'deduction-normalisation' or 'derivation-normalisation' but, as the phrase is a common one, we shall use the term 'proof-normalsation' to refer to a property of deductions (the property of reducing to normal forms).

known, studying proof-normalisation is an important first step for giving deductions semantics, for example arrows in Cartesian-Closed Categories [Taylor, 1999], sets-and-functions-between-them, or normal forms in functional programs [M. H. B. Sorensen, 1998].

Now consider classical FOL. We can express it in Natural Deduction style by adding the law of excluded middle or double-negation elimination to intuitionistic FOL:

$$\frac{}{A \vee \neg A} \, (EM) \qquad\qquad \frac{\neg\neg A}{A} \, (DNE)$$

However, these rules compromise proof-normalisation and are in that sense unsatisfactory (computational semantics of deductions in classical logic are a research area in their own right, see [Coquand, 1996] for a solid survey; we do *not* discuss this huge field here).

So how to recover normal forms? Gentzen's elegant solution [Gentzen, 1934] was a multiple-conclusioned logic. We are so familiar with this idea nowadays, we may forget that this is an expensive option in the sense that we have to add explicit context and co-context *everywhere*. For example, the deduction-fragments above become:

$$\frac{\Gamma \vdash A, \Delta \quad \Gamma \vdash B, \Delta}{\Gamma \vdash A \wedge B, \Delta} \, (\wedge I) \qquad \frac{\dfrac{\dfrac{\Gamma \vdash \forall x.\, P(x), \Delta}{\Gamma \vdash P(x), \Delta} \, (\forall E) \quad \dfrac{\Gamma \vdash \forall x.\, Q(x), \Delta}{\Gamma \vdash Q(x), \Delta} \, (\forall E)}{\Gamma \vdash P(x) \wedge Q(x), \Delta} \, (\wedge I)}{\Gamma \vdash \forall x.\, (P(x) \wedge Q(x)), \Delta} \, (\forall I)$$

(In practice we might prefer left- and right-introduction rules, but that is not important here.) So to do classical FOL we have to pay at every stage of the deduction by threading Γ and Δ through every part of the deduction, even the purely intuitionistic parts inherited from the intuitionistic core of the logic. (This is analogous to emulating global state in a purely functional programming language by threading it through function calls [Peyton-Jones, 2001].)

Now for our proposal. We extend intuitionistic FOL with **classical restart**, which is the following Natural Deduction rule:

$$(Restart) \qquad \frac{A}{B}$$

This is not a misprint. From A we may proceed to B.

The *side-condition*, of course there is one, is that below every occurrence of restart from A to B, there is (at least) one occurrence of A. For example $\dfrac{\top}{B}$ is not a valid deduction, but the following deduction is valid:

$$\frac{\dfrac{\dfrac{[A]}{B} \, (Restart)^*}{A \rightarrow B} \, (\rightarrow I) \qquad [(A \rightarrow B) \rightarrow A]}{\dfrac{A^\dagger}{((A \rightarrow B) \rightarrow A) \rightarrow A} \, (\rightarrow I)} \, (\rightarrow E)$$

The restart at * is justified at †. The conclusion of this deduction is *Peirce's Law*, a famous classical tautology, and we have just proved one half of the next theorem.

Natural deduction has a notion of state given by the undischarged assumptions. Although Restart insists we return to A, on our return we may find ourselves in a more clement state with respect to the undischarged assumptions. This turns out to give precisely the extra deductive power of classical logic:

THEOREM 1. *Propositional Intuitionistic Logic plus (Restart) has the same entailment relation as Classical Logic.*

The proof is elementary; one half follows from the derivation of Peirce's Law above. We give the other half in the next section (see also Theorem 2 on page 725).

So *if* we impose global well-formedness conditions on Natural Deduction derivations, *then* we have a way of strengthening a logic 'cheaply'. There is a bit more: it is 'neat' too and proof-normalisation is inherited in a natural way, and we shall explore that briefly in the next section.

In conclusion: There are (at least) two techniques to imposing side-conditions on the form of a deduction. One is to build them in as sequents at every stage, another is to impose them globally on a (natural) deduction. Sometimes these side-conditions are easy to express in both techniques. For example:

- *Natural Deduction:* In ∀-introduction $\dfrac{P}{\forall x.\,P}$, x may not occur in assumptions on which P depends.

- *Sequents:* In ∀-left-introduction $\dfrac{\Gamma \vdash P}{\Gamma \vdash \forall x.\,P}$, x may not occur in Γ.

- *Natural Deduction:* In ⇒-introduction $\dfrac{\overset{\displaystyle [A]}{\overset{\vdots}{B}}}{A \Rightarrow B}$, A must occur precisely once (to obtain a 'linearity' condition on logical implication).

- *Sequent Calculus:* In ⇒-right-introduction $\dfrac{\Gamma, A \vdash B}{\Gamma \vdash A \Rightarrow B}$, Γ is a multiset and comma denotes multiset union.

An interesting historical example where Natural Deduction seemed to do better than sequent systems was Prawitz's natural deduction system for the modal logic S5 [Prawitz, 1965] for which the obvious sequent system does not have cut-elimination (an issue addressed later by [Braüner, 2000]). Conversely Natural Deduction could not neatly (in the sense we define) capture classical logic, until now:

- *Sequent Calculus:* Sequents are multiple-conclusion $\Gamma \vdash \Delta$ (for classical logic).

- *Natural Deduction:* We propose: Restart.[2]

A little history: A rule corresponding to Classical Restart was invented by Dov Gabbay [Gabbay and Reyle, 1984; Gabbay, 1998; Gabbay and Olivetti, 2000b] in the framework of **goal directed deductions**; these are for proof-search and the rules are read

---

[2]Methods of obtaining a calculus for classical logic that satisfies a normal form theorem are known.

bottom up (we discuss this further below). As a bottom-up goal-directed rule, restart is: at a goal B we can 'restart' at any previous goal A we have considered. The second author then re-presented the same rule top-down in Prawitz natural deduction [Gabbay, 2004].

2   Classical Restart

We begin by studying in detail Intuitionistic Logic (IL) augmented with our restart natural deduction rule. We show it has the same derivability relation as Classical Logic (CL); We prove that the standard proof-normalisation procedure for Natural Deduction is preserved; We then consider consider restart in other deduction formats than natural deduction.

2.1   Proof that IL+(Restart)=CL

For simplicity we consider just the propositional case. Fix a set of **atomic formulae** $p, q, r, \ldots$. **Formulae** $A, B \ldots$ are defined by the grammar[3]

$$A ::= p \mid A \to A$$

The Natural Deduction rules of IL are standard:

$$\frac{\substack{[A] \\ \vdots \\ B}}{A \to B} (\to I) \qquad \frac{A \quad A \to B}{B} (\to E) \qquad \frac{\bot}{C} (\bot E)$$

Other connectives such as $\wedge$ and $\vee$ are possible.

A **deduction** of IL is formed from $(\to I)$, $(\to E)$, $(\bot E)$. Call a **proto-deduction** a deduction formed as just described, but also with $(Restart)$. A proto-deduction is **valid** when it satisfies the condition that under every restart $\dfrac{A}{B}$ there is a later instance of A. We represent this in the following diagram:

$$\frac{A}{B} (Restart)$$
$$\vdots$$
$$A$$

For example in [Stalmarck, 1991; Prawitz, 1965] the rule

$$\frac{\substack{\neg A \\ \vdots \\ \bot}}{A} (PIP)$$

is used. This rule may be considered as an elimination rule for $\bot$, it is also known as the **principle of indirect proof** (PIP). We confine discussion of this rule to a footnote because it is not purely structural, it cannot be formulated unless at least one of $\bot$ or $\neg$ is in the language. So for example we cannot use it to formulate the implication-only fragment of classical propositional logic which should have $((A \to B) \to A) \to A$ as a theorem independently of whether we have negation or $\bot$ in the language as well. We are interested here in the properties of Restart as a purely structural rule that makes the difference between intuitionistic, classical and intermediate logics.

[3] Our results easily generalise to $\wedge$ and $\vee$.

Let **Classical Logic (CL)** be IL augmented with Peirce's Law $((A \to B) \to A) \to A$ for all A and B. We have already seen how to use (*Restart*) to prove Peirce's Law, now we consider the converse.

Any application of (*Restart*) as above may be replaced by

$$
\cfrac{\cfrac{\cfrac{A \quad [A \to B]^1}{B}\,(\to E)}{\vdots\\ \cfrac{A}{(A \to B) \to A}\,(\to I)^1} \qquad \cfrac{}{((A \to B) \to A) \to A}\,\text{Assumed}}{A}\,(\to E).
$$

This proves Theorem 1, and in view of this we may call (*Restart*) **Classical Restart**.

2.2 Proof-normalisation

Unlike adding Peirce's Law, adding (*Restart*) to IL does not compromise proof-normalisation. The essential cases are as before, for example:

$$
\cfrac{\cfrac{\begin{array}{c}[A]^1\\ \vdots\\ B\end{array}}{A \to B}\,(\to I)^1 \quad \begin{array}{c}\vdots\ \Pi\\ A\end{array}}{B}\,(\to E) \quad\Longrightarrow\quad \begin{array}{c}\vdots\ \Pi\\ A\\ \vdots\\ B\end{array}
$$

The only possible problem is when $A \to B$ above justifies one or more instances of (*Restart*) in the deduction above; then restructure them as follows:

$$
\cfrac{A \to B}{C}\,(Restart) \quad\Longrightarrow\quad \cfrac{\cfrac{A \to B \quad \begin{array}{c}\vdots\ \Pi\\ A\end{array}}{B}\,(\to E)}{C}\,(Restart)
$$

Now each of the problematic instances of (*Restart*) is justified one line further down, by B rather than $A \to B$, and we can proceed with elimination of the essential case. Note how Π is **teleported** deep inside the deduction.

More generally, in the presence of conjunction and disjunction, we can always restructure a deduction so that no formula both completes a restart rule and is the major premise of an elimination rule. Schematically, we can replace

$$
\cfrac{A}{B}\,(Restart)
$$
$$
\vdots
$$
$$
\cfrac{A}{C}\,(?E)
$$

with this

$$\frac{\dfrac{\dfrac{A}{C}\ (?E)}{B}\ (Restart)}{\vdots}$$

$$\frac{\dot{A}}{C}\ (?E)$$

and now the premise of the restart is C and its side condition is met by a formula one step closer to the conclusion (so we may conclude that ultimately, we can restructure the deduction so that no formula both is the major premise of an elimination rule and validates a previous application of restart).

Notice also that we can always restructure a deduction so that the conclusion of an application of $\vee E$ (or $\exists E$ if we extend the results to the quantified case) is also not the major premise of any elimination rule.[4]

We can now complete a normalisation argument using familiar methods, we sketch the argument here. Say a *segment* is a sequence of occurrences of a formulae $A_1 \ldots A_n$, where each A_i is an instance of the same formula A, and A_i is a minor premise of a rule the conclusion of which is A_{i+1}. Say a segment is *maximal* when A_1 is the conclusion of an introduction rule, and A_n is the major premise of an elimination rule. Notice that since we can always restructure a deduction so that no major premise of an elimination rule is also the conclusion of $\vee E$ (or $\exists E$) then we can assume that any maximal segment is a sequence of only two formulae. But then since we may always restructure a deduction so that no major premise of an elimination rule is necessary to validate a previous application of restart, we may remove the maximal segment altogether (as shown by the essential cases above).

Additionally, we may neaten up a deduction by thinking of $(Restart)$ as an introduction rule and simplify $(Restart)$ followed by an elimination rule as we would any essential case:

$$\frac{\dfrac{A}{B \to C}\ (Restart) \qquad \dfrac{\Pi}{B}}{C}\ (\to E) \qquad \Longrightarrow \qquad \frac{A}{C}\ (Restart).$$

We do not simplify restart in the minor premise ($\dfrac{\Pi}{B}$ above); this destroys confluence of the reductions, we discuss this later.

---

[4]So for example in the case of the existential quantifier (A_t^x is the formula obtained by simultaneously replacing all free occurrences of the variable x in A by the term t):

$$\frac{\dfrac{\exists x A \qquad \overset{[A_a^x]}{\underset{\vdots}{B}}}{\dfrac{B}{C}\ (?E)}}{C}\ (\exists E) \qquad \Longrightarrow \qquad \frac{\exists x A \qquad \dfrac{\overset{[A_a^x]}{\underset{\vdots}{B}}}{C}\ (?E)}{C}\ (\vee E)$$

In such a reduction it may be necessary to replace a (in the deduction of B from A_a^x) by some other constant if a occurs in C. This is because of the side condition on $(\exists E)$, see section 8.1.

Disjunction. Our syntax in (2.1) does not have disjunction $\vee$ but later in this paper it will be convenient to suppose we do. The natural reduction is as follows:

$$\frac{A}{B \vee C}\;(Restart) \qquad \frac{[B]}{D} \quad \frac{[C]}{D} \atop {\displaystyle\rule{3cm}{0.4pt}}\;(\vee E) \qquad \Longrightarrow \qquad \frac{A}{D}\;(Restart).$$

Again, a similar simplification is possible if a restart terminates either of the two minor premises to D, but adding it would destroy confluence.

2.3   From restart to multiple conclusions

Suppose we have a deduction $\begin{array}{c}\Gamma \\ \vdots \\ A\end{array}$ where the side conditions of some restarts, say from members of Δ, above A have not (yet) been met. We can write this as a sequent $\Gamma \vdash A; \Delta$ (the predicate A is 'active'; a minor tweak of standard sequents, see below). We apply the restart rule from A to B to obtain this deduction tree

$$\begin{array}{c}\Gamma \\ \vdots \\ A \\ \hline B\end{array}\;(Restart)$$

Now we describe this new deduction tree as $\Gamma \vdash B; \Delta \cup \{A\}$ or just $\Gamma \vdash B; \Delta, A$. We can construct a new sequent system (with the semicolon as an additional structural entity) with rules

$$\frac{\Gamma \vdash A; \Delta}{\Gamma \vdash B; \Delta, A}\;(Restart1) \qquad \frac{\Gamma \vdash A; A, \Delta}{\Gamma \vdash A; \Delta}\;(Restart2)$$

The second of these two rules captures the side condition 'provided A is deduced again later on'. If we add rules for the other connectives we obtain a sequent system for classical logic. For example:

$$\frac{\Gamma, A \vdash B; \Delta}{\Gamma \vdash A \to B; \Delta}\;(\to R) \qquad \frac{\Gamma \vdash A; \Delta \quad \Gamma, B \vdash C; \Delta}{\Gamma, A \to B \vdash C; \Delta}\;(\to L)$$

Here is a (well-known) deduction of Peirce's law:

$$\frac{\dfrac{\dfrac{A \vdash A}{A \vdash B; A}\;(Restart1)}{\vdash (A \to B); A}\;(\to R) \qquad A \vdash A; A}{\dfrac{\dfrac{(A \to B) \to A \vdash A; A}{(A \to B) \to A \vdash A;}\;(Restart2)}{\vdash ((A \to B) \to A) \to A;}\;(\to L)}\;(\to R)$$

In the usual multiple conclusion sequent calculus the conclusion is not split as above into an 'active formula' and a set of 'outstanding formulae'. This extra structure imposes no restrictions since we can easily change the active formula:

$$\frac{\Gamma \vdash A; \Delta}{\Gamma \vdash B; \Delta, A} \ (Restart1)$$
$$\frac{}{\Gamma \vdash A; \Delta, A, B} \ (Restart1)$$
$$\frac{}{\Gamma \vdash A; \Delta, B} \ (Restart2).$$

Thus we may interpret the sequent $\Gamma \vdash A; \Delta$ semantically just as we would an ordinary multiple conclusion logic:

If the Γ are true and the Δ are false then A is true.

or even

If the Γ are true then so is at least one of A, Δ.

Similarly we may understand a deduction tree $\begin{smallmatrix}\Gamma\\\vdots\\A\end{smallmatrix}$ as a deduction of A from the assumptions of Γ and further assumptions of the negations of the premises of restarts that do not have their side conditions met.[5]

The reader may recognise that the observations of this subsection and the last correspond, via the Curry-Howard correspondence, to the $\lambda\mu$-calculus [Parigot, 1992]. In fact (using the terminology we have developed) the $\lambda\mu$-calculus only allows restarts from any A to $\bot$. Also our proof-normalisation procedure is not restricted at all to following that specified by $\lambda\mu$; a restart from A to B need not necessarily be tied to a particular later instance of A, and we can choose arbitrarily from which to teleport; so the system here is a little more general.

Of course, all this is not limited to first-order logic. Indeed restart as described is valid in the presence of any abstract machine with state, and perhaps investigating 'stateful automata with restart' belongs to interesting future work. We return to this in the Conclusion.

3 Restart in frameworks other than Natural Deduction

3.1 Restart in linear natural deduction

We have formulated (Restart) for a Prawitz style natural deduction system. Its tree structure makes dependency between two formulae easy to track, being a matter of whether there is a path in the tree between them. (Restart) as discussed so far exploits this. For example in this deduction

$$\begin{array}{c}A\\\overline{B}\ (Restart)\\\vdots\\A\end{array}$$

---

[5]Perhaps this subsection is too humbly titled. 'No', we can say: 'a multiple conclusion sequent $\Gamma \vdash \Delta$ expresses the existence of a/any deduction from assumptions Γ with a conclusion $A \in \Delta$ and incomplete applications of restart to formulae $B \in \Delta - \{A\}$'. This is philosophically more satisfying than an interpretation of multiple conclusions which assumes the logic has an explicit disjunction $\vee$.

formulae occurring between B and A may be said to **depend** on that particular application of restart. We do not have a full deduction until this dependency is discharged (until A occurs again).

In a linear natural deduction system, dependency is harder to track. When we extend with restart we must extend whatever mechanism we have for handling dependencies to take the new rule into account.

Consider adding restart to a Lemmon style natural deduction system. Each formula in the deduction has three sets of labels (this is standard): a line number; the line numbers of the premises if it is the conclusion of a rule; and the line numbers of the assumptions it depends on. Here is a deduction of $\neg\neg(A \lor \neg A)$:

Dependency	Line	Formula	Wherefrom
1	(1)	$\neg(A \lor \neg A)$	
2	(2)	A	
2	(3)	$A \lor \neg A$	$2(\lor I)$
1, 2	(4)	$\bot$	$1, 3(\neg E)$
1	(5)	$\neg A$	$2, 4(\neg I)$
1	(6)	$A \lor \neg A$	$5(\lor I)$
1	(7)	$\bot$	$6, 1(\neg E)$
	(8)	$\neg\neg(A \lor \neg A)$	$1, 7(\neg I)$

The 'deduction rule' $(\neg I)$ is the following schema:

$$
\begin{array}{lll}
n & (n) \quad A & \\
 & \vdots & \\
\{n_1 \ldots n_r, n\} & (m) \quad \bot & \ldots \\
 & \vdots & \\
\{n_1 \ldots n_r\} & (k) \quad \neg A & n, m(\neg I)
\end{array}
$$

We say 'schema' because the premises (lines n and m) can occur at some distance above the conclusion (line k). The assumption A is discharged by removing its line number from the set of dependency labels at line k, and this is how dependencies are tracked.

We say that B at line m depends on A at line n when the sets of dependency labels are $\{n\}$ at line n and $\{m_1 \ldots m_k, n\}$ at line m (i.e. A occurs as an assumption and B has A's line number in its dependency column). So now we have $\Gamma \vdash A$ when there is a deduction in which there is a line containing A that depends only on elements of Γ.

To add restart, extend labels with a dependency label $R(n)$ where R is for 'restart' and n is the line number of the premise. We then add two rules, one for the restart:

$$
\begin{array}{lll}
\{n_1 \ldots n_r\} & (n) \quad A & \ldots \\
 & \vdots & \\
\{n_1 \ldots n_r, R(n)\} & (k) \quad B & n(Restart1)
\end{array}
$$

... and one for the side condition that the premise of the restart reoccur:

$$\vdots$$

$$\{m_1 \ldots m_r, R(n)\} \quad (m) \quad A \quad \ldots \qquad \text{Provided the formula}$$
$$\text{at line } n \text{ is } A$$

$$\vdots$$

$$\{m_1 \ldots m_r\} \qquad (k) \quad A \quad m(Restart2)$$

So for example we can now deduce Peirce's law:

Dependency	Line	Formula	Wherefrom
1	(1)	A	
2	(2)	$(A \to B) \to A$	
$1, R(1)$	(3)	B	$1(Restart1)$
$R(1)$	(4)	$A \to B$	$1, 3(\to I)$
$2, R(1)$	(5)	A	$2, 4(\to E)$
2	(6)	A	$5(Restart2)$
	(7)	$((A \to B) \to A) \to A$	$2, 6(\to I)$

The restart rule looks less magical now, since these rules are so like a rule for indirect proof; $R(n)$ resembles a label for $\neg A$, $(Restart1)$ resembles a rule for deducing B from A and $\neg A$, and $(Restart2)$ resembles a rule for deducing that if A follows from $\neg A$ then we can conclude A without a dependency on $\neg A$. It is not hard to translate between deductions using the two rules above, and deductions using Indirect Proof or Peirce's Law.

In Prawitz form, restart is interesting because we get a normalising deduction system for classical logic from an intuitionistic one with no structural changes (labels, multiple conclusions). In linear natural deduction systems, where dependency is tracked by labelling anyway, this advantage is, if not exactly lost, certainly less apparent.

3.2 Goal-directed deduction with restart

A striking formulation of the restart rule is the original one from [Gabbay, 1990]. This is in terms of Goal-Directed reasoning which we can understand here as a Fitch style natural deduction system such that all subdeductions have a goal which must be met. That is, a subdeduction is not open-ended, but can be closed only when the goal is reached. For example, here is a Goal-Directed deduction that $[A \to (B \to C)] \to [(A \to B) \to (A \to C)]$:

1	$A \to (B \to C)$ $?(A \to B) \to (A \to C)$		
2	$A \to B$ $?(A \to C)$		
3	A $?C$		
4	$B \to C$	$(\to E), 1, 3$	
5	B	$(\to E), 2, 3$	
6	C	$(\to E), 4, 5$	
7	$A \to C$	$(\to I), 3, 6$	
8	$(A \to B) \to (A \to C)$	$(\to I), 2, 7$	
9	$[A \to (B \to C)] \to [(A \to B) \to (A \to C)]$	$(\to I), 1, 8$	

Notice that an individual subdeduction is completed only when its goal is reached, at which point we can terminate the subdeduction and introduce an implication. The restart rule then becomes an additional rule on when a subdeduction is completed.

A subdeduction with a hypothesis A and goal $?B$ may be terminated (introducing an implication $A \to B$) by the deduction of not only its goal, but by the deduction of any previous goal.

For example here is a Goal-Directed deduction of Peirce's Law:

1	$(A \to B) \to A)$ $?A$	
2	$A \to B$ $?A$	
3	A $?B$	
4	A	Recalling , 3
5	$A \to B$	*(Restart)* to $?A$, 4, 3, 2
6	A	$(\to E)$, 2, 5
7	$[(A \to B) \to A)] \to A$	$(\to E)$, 1, 6

We get line 5 by restarting the goal $?B$ to a goal $?A$ (and declaring ourselves satisfied since that is what we have at line 4); instead of terminating the subdeduction with B, we terminate it with the previous goal A.

It is not hard to see that this restart is sound: we observe that we can obtain a valid deduction in a system with an indirect proof rule by replacing any goal $?A$ with an additional assumption $\neg A$ and then using the principle of indirect proof. Any previous 'goal' is really an assumption that the goal is false. So note in the example above, reading $?A$ as $\neg A$, that the introduction of $A \to B$ is valid, as we can deduce B from A and the previous assumption that $\neg A$. Because $?A$ was a *previous* goal, when A is deduced again later in the deduction (line 6), we may consider the assumption of $\neg A$ (i.e. $?A$) as being discharged by the Principle of Indirect Proof:

n		
m	$\neg A$	
$m+1$	A	(PIP), n, m

The system of Goal-Directed logic [Gabbay, 1990; Gabbay and Olivetti, 2000a] is mainly for implication- and quantifier-fragments of various logics.

4 n-depth restart

Say a Kripke model has **height bounded by** n when for all series of worlds $w_1, \ldots, w_{n+1}$, if $w_i \, R \, w_{i+1}$ for $1 \le i \le n$ then an i exists with $w_i = w_j$ ("the model has depth at most n"). Here R denotes the Kripke accessibility relation.

A characteristic axiom scheme for models of depth 1 is $P^1(A,B) \equiv ((A \to B) \to A) \to A$. A characteristic axiom scheme for models of depth $i+1$ is taken from [Gabbay, 1981] and given as follows:

$$P^{i+1}(A_1, \ldots, A_{i+1}) \overset{\text{def}}{=} P^1(A_{i+1}, P^i(A_1, \ldots, A_i)).$$

For example $P^2(A,B,C) = ((A \to (((B \to C) \to B) \to B)) \to A) \to A$.

4.1 A natural deduction restart for bounded height

An appealing *intuition* (we do not make it formal) behind n-depth restart is that each A_i is proved at world w_i, where w_n is the final world. Whence the condition on discharge after handles, which corresponds to a restriction not to descend below w_i until we have justified the restart at that world. The restart at the final world is an ordinary 1-depth restart, jumping from A_n to an arbitrary B is reasonable because there is no future world and so any B holds in any future world.

$$
\begin{array}{ccc}
\vdots \Pi_1 \quad \vdots \Pi_n & & \\
\dfrac{A_1 \quad \cdots \quad A_n}{\begin{array}{c} B \\ \vdots \Pi'_n \\ A_n \\ \vdots \Pi'_2 \\ A_2 \\ \vdots \Pi'_1 \\ A_1 \end{array}} & &
\end{array}
$$

$$
\dfrac{\dfrac{\dfrac{\dfrac{\dfrac{[B]^1 \qquad \dfrac{[A]^2}{A \vee \neg A^a}}{\bot}}{\neg A^2}}{A \vee \neg A^a}}{B \to (A \vee \neg A)^1}}{B \vee (B \to (A \vee \neg A))^b}
$$

with $B \vee (B \to (A \vee \neg A))^b$ at top.

$$
\dfrac{\dfrac{\dfrac{[A]^1}{A \vee \neg A^a} \qquad \dfrac{[A]^2}{A \vee \neg A^b}}{\bot}}{\neg A^{1,2}}
$$
$$
A \vee \neg A^{a,b}
$$

Invalid (two 1-depth restarts)

n-depth Restart   **Proof using 2-depth restart**

There is an extra side-condition on n-depth restart. Say that the restart is *handled* by each re-deduction of $A_n \ldots A_n$ (so each application of an n-depth restart rule gets handled n times). The side condition is this

> In n-depth restart no assumption or application of restart in the deduction Π_i of A_i may be discharged or handled under B before A_{i+1} has been handled

We can prove $B \vee (B \to (A \vee \neg A))$ using 2-depth restart, as illustrated. The 'deduction' on the right of $A \vee \neg A$ using the 2-depth restart rule is *invalid* because it discharges A at ^{2} before $A \vee \neg A$ at ^{b}.

Proof-normalisation is fairly evident and similar to the 1-depth case (ordinary restart); an elimination rule following A_i in the lower part of the deduction is teleported to the A_i in the upper part of the deduction. Normal forms are unique. We omit the proof of this, which is not hard.

5 Soundness of n-depth restart

We shall argue that any deduction containing applications of n-depth restart may be replaced by a deduction, of the same conclusion, using the n-depth axiom instead. We shall do this by showing how to replace any application of n-depth restart with a use of the n-depth restart rule.

The construction, although simple to the mind, is unfortunately not so simple to the eye. Take any application of n-depth restart:

$$
\begin{array}{c}
\vdots\ \Pi_1 \qquad\qquad \vdots\ \Pi_n \\
\dfrac{A_1 \quad \cdots \quad A_n}{B} \\
\vdots\ \Pi'_n \\
A_n \\
\vdots\ \Pi'_2 \\
A_2 \\
\vdots\ \Pi'_1 \\
A_1
\end{array}
$$

and we can begin to rewrite it as follows:

$$
\begin{array}{c}
\vdots\ \Pi_n \\
\dfrac{A_n \qquad [A_n \to B]^{i_n}}{B} \\
\vdots\ \Pi'_n \\
\dfrac{\dfrac{A_1 \quad \cdots \quad A_{n-1}}{[(A_n \to B) \to A_n] \to A_n} \qquad \dfrac{A_n}{(A_n \to B) \to A_n}\ (\to I)^{i_n}}{A_n}\ (\to E) \\
\vdots\ \Pi'_2 \ldots \Pi'_{n-1} \\
A_2 \\
\vdots\ \Pi'_1 \\
A_1
\end{array}
$$

with $\Pi_1 \ldots \Pi_{n-1}$ above $A_1 \ldots A_{n-1}$.

We can routinely replace all applications of n-depth restart with this 'rule':

$$
\begin{array}{c}
\vdots \qquad\quad \vdots \qquad\qquad\qquad \vdots \\
\dfrac{\dfrac{A_1 \quad \cdots \quad A_{n-1}}{P^1(A_n, B)} \qquad (A_n \to B) \to A_n}{A_n}\ (\to E) \\
\vdots \\
A_1
\end{array}
$$

If there is no A_{n-1} (i.e. if $n = 1$) then A_{n-1} and Π_{n-1} are empty and we have replaced the n-depth restart with an appeal to Peirce's Law.

$$
\overline{P^1(A_1, B)}
$$

However, if $n > 1$ then we must make further replacements. To save space, let P^m be short for $P^m(A_{(n-(m-1))} \ldots A_n, B)$

We may replace any application of

$$
\cfrac{
\cfrac{
\begin{array}{ccc}
\vdots\ \Pi_1 & & \vdots\ \Pi_{n-m} \\
A_1 & \cdots & A_{n-m}
\end{array}
}{P^m}
\qquad
\cfrac{
\vdots\ \Pi_{n-(m-1)}, \Pi'_{n-(m-1)} \cdots \Pi_n, \Pi'_n
}{(A_{(n-(m-1))} \to P^{m-1}) \to A_{(n-(m-1))}}\ (\to E)
}{A_{n-(m-1)}}
$$
$$
\vdots\ \Pi'_1 \ldots \Pi'_{n-m}
$$
$$
A_1
$$

so that we use instead an application of

$$
\cfrac{
\cfrac{
\begin{array}{ccc}
\vdots\ \Pi_1 & & \vdots\ \Pi_{n-(m+1)} \\
A_1 & \cdots & A_{n-(m+1)}
\end{array}
}{P^{m+1}}
\qquad
\cfrac{
\vdots\ \Pi_{n-m}, \Pi'_{n-m} \cdots \Pi_n, \Pi'_n
}{(A_{n-m} \to P^m) \to A_{n-m}}\ (\to E)
}{A_{n-m}}
$$
$$
\vdots\ \Pi'_1 \ldots \Pi'_{n-(m+1)}
$$
$$
A_1
$$

in a similar way:

Notice that the side condition ensures that no assumption in $\Pi_1 \ldots \Pi_{n-(m+1)}$ is discharged in Π'_{n-m} so we can bring Π'_{n-m} out and across from underneath $\Pi_1 \ldots \Pi_{n-(m+1)}$ without affecting the deduction.

Ultimately, after repeated applications of this we will have replaced all applications of n-depth restart by this:

$$
\cfrac{
\cfrac{
\begin{array}{ccc}
\vdots & & \vdots \\
A_1 & \cdots & A_{n-n}
\end{array}
}{P^n}
\qquad
(A_{n-(n-1)} \to P^{n-1}) \to A_{n-(n-1)}
}{A_{n-(n-1)}}\ (\to E)
$$
$$
\vdots
$$
$$
A_1
$$

and since there is no A_0 this is simply

$$
\cfrac{(A_1 \to P^{n-1}) \to A_1 \quad \overline{P^n}}{A_1}\ (\to E)
$$

which is just an appeal to the n-depth Kripke model axiom.

6   Completeness of n-depth restart

Notice that within an application of an $n+1$-depth restart rule we can find an n-depth restart rule (considering only $A_n \ldots A_2$ and ignoring A_1). We have already shown that any instance of Peirce's law P^1 can be derived using 1-depth restart. We can now argue that P^n can be derived using the n-depth restart rule.

The argument is by induction on n. Suppose for the induction hypothesis that the r-depth restart rule can be used to derive the r-depth axiom, where $r = n - 1$ and $r \geq 1$.

$$\frac{A_2 \ldots A_n}{B}$$
$$\vdots$$
$$A_n$$
$$\vdots$$
$$\dot{A}_2$$

where A_2 is $P^{n-1}(A_2 \ldots A_n, B)$. Now we can edit it in the following way, first we add an extra premise:

$$\frac{\dfrac{A_1}{P^n(A_1 \ldots A_n, B)} \; (\to I) \quad A_2 \ldots A_n}{B}$$
$$\vdots$$
$$A_n$$
$$\vdots$$
$$\dot{A}_2$$

and now we continue on from A_2 as follows

$$\cfrac{\cfrac{\cfrac{[A_1]^i}{P^n(A_1 \ldots A_n, B)} \; (\to I) \quad A_2 \ldots A_n}{B}}{\vdots \; A_n \; \vdots \; \dot{A}_2}$$

$$\frac{[(A_1 \to A_2) \to A_1]^l \qquad \dfrac{A_2}{A_1 \to A_2} \; (\to I)^i \; (\to E)}{\dfrac{A_1}{((A_1 \to A_2) \to A_1) \to A_1}}$$

and since A_2 is $P^{n-1}(A_2 \ldots A_{n+1}, B)$ then $((A_1 \to A_2) \to A_1) \to A_1$ is really $P^n(A_1 \ldots A_n, B)$ which is the n-depth axiom. Furthermore we have converted the r-depth restart into an instance of $r + 1$-depth restart. This completes the induction.

6.1   Goal-directed n-depth restart

n-depth restart can be formulated in Goal Directed reasoning as well. Say that a goal $?A$ is *embedded* in another goal $?B$ when $?B$ is a goal previous to $?A$,[6] and B occurs,

---

[6]i.e. the subdeduction for which $?A$ is a goal is a subdeduction (of a subdeduction of. . .) the subdeduction for which $?B$ is a goal.

not as a goal, somewhere between $?B$ and $?A$.

Schematically, $?A$ is embedded in $?B$ when this happens:

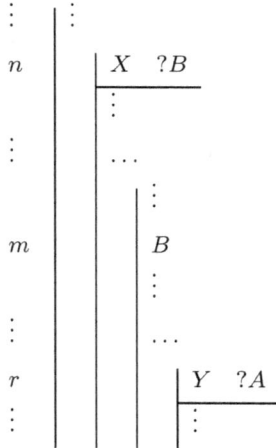

The n-depth goal directed restart rule is this:

> A subdeduction with a hypothesis A and goal $?B$ may be terminated (introducing an implication $A \to B$) by the deduction of not only its goal, but of a previous goal $?C$ provided that there are goals $?C_1 \ldots ?C_{n-1}$ such that $?C$ is embedded in $C_n - 1$ and each C_i is embedded in C_{i-1}.

So for example, here is a deduction using 2-depth restart of $((A \to (\neg\neg B \to B)) \to A) \to A$:

1	$(A \to (\neg\neg B \to B)) \to A$ $?A$	
2	A $?\neg\neg B \to B$	
3	$\neg\neg B$ $?B$	
4	B $?\bot$	
5	B	Recalling , 4
6	$\neg B$	2-restart, $?B$ is embedded in $?A$
7	$\bot$	$(\to E)$, 2, 6
8	B	$(\bot E)$, 7
9	$\neg\neg B \to B$	$(\to I)$, 3, 8
10	$A \to (\neg\neg B \to B)$	$(\to I)$, 2, 9
11	A	$(\to E)$, 1, 10
12	$((A \to (\neg\neg B \to B)) \to A) \to A$	$(\to I)$, 1, 11

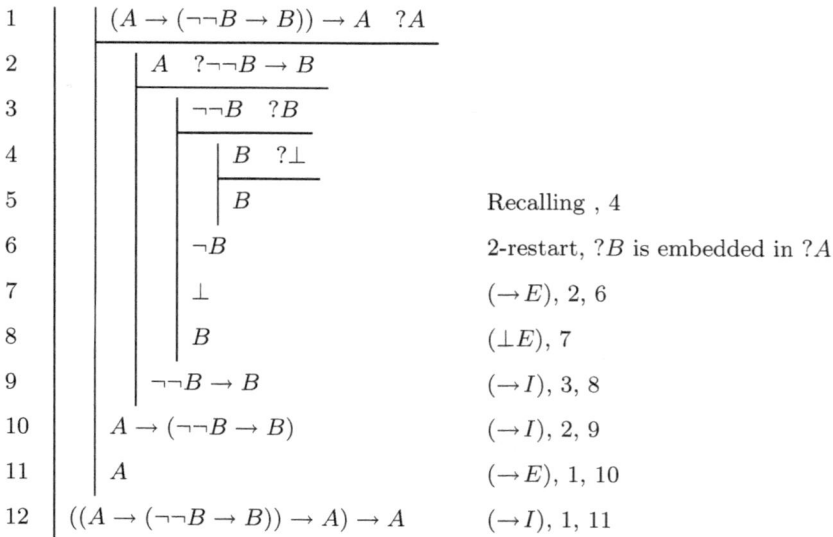

The application of restart was to complete the Goal $?\bot$ by completing instead the previous goal $?B$. Notice that above the goal $?B$ is an occurrence of A and above that is a goal $?A$. So $?B$ is embedded in at least one other goal. This legitimises they use of 2-depth restart.

6.2 Soundness of goal-directed n-depth restart

It is a somewhat tedious matter to show how to convert a deduction using the goal-directed n-depth restart rule into a deduction using the n-depth axioms. We shall not describe the construction in detail here, mainly because it is similar to the construction proving soundness of the natural deduction n-depth restart rule.

But a sketch-proof of why the rule is sound may be found by observing that the goal-directed n-depth restart rule meets the conditions of the natural deduction n-depth restart rule.

To apply the goal directed restart rule, we act as if we have achieved a goal $?B$ when in fact we have deduced only A. Effectively we do this:

$$\frac{A}{B}$$

But A was not just any formula, it was a previous goal, which means it will have to be met at a later point for the deduction to be complete:

$$\frac{A}{B}$$
$$\vdots$$
$$A$$

But A is embedded in a goal $?C_1$, which means that we have C_1 to complete the goal $?C_1$ after we have achieved goal $?A$:

$$\frac{A}{B}$$
$$\vdots$$
$$A$$
$$\vdots$$
$$C_1$$

and also we have already deduced C_1 when deduced A:

$$\frac{A \quad C_1}{B}$$
$$\vdots$$
$$A$$
$$\vdots$$
$$C_1$$

but discharge in goal-directed reasoning is done only by completing goals, it follows (since C_1 was deduced in between the goal $?A$ and the prior goal $?C_1$) that no assumption on which C_1 depends is discharged (has its goal met) before A is deduced again (that is, before the goal $?A$ is achieved). The same holds if $?A$ must be embedded in n

further goals:

$$\frac{A, C_1 \ldots C_n}{B}$$
$$\vdots$$
$$A$$
$$\vdots$$
$$C_1$$
$$\vdots$$
$$C_n$$

This is precisely the side condition on the $n+1$-depth natural deduction restart rule.

7   Hash logic

7.1   Natural deduction for hash logic

We can internalise the restart rule and use it as an elimination rule for a connective. Here are the rules for a unary connective #.

$$\frac{A}{\#A}\ (\#I) \qquad \frac{\#B \quad A}{B}\ (\#E) \qquad \frac{\#A_1 \quad \cdots \quad \#A_n \quad \overset{\displaystyle [A_1]^i \ldots [A_n]^i}{\underset{\displaystyle B}{\vdots}}}{\#B}\ (\#P)^i$$

Say an application of $(\#E)$ is *completed* when A is deduced again below it. Until A is deduced again the application of $(\#E)$ is incomplete.

The side condition on the rule $(\#E)$ is this:

(*i*) A must be deduced again after the application of the rule and
(*ii*) no (occurrence of an) assumption on which $\#B$ depends may be discharged until A is re-deduced and
(*iii*) no $(\#E)$ rule on which $\#B$ depends is completed until A is re-deduced.

So the rule is this

$$\frac{\#B \quad A}{B}\ (\#E)$$
$$\vdots$$
$$A$$

where nothing on which $\#B$ depends may be discharged until the $(\#E)$ rule is completed and no $(\#E)$ rule that is incomplete at $\#B$ is completed until A re-occurs.

We should be liberal with our interpretation of these rules. We should not regard the rules as requiring that an application of the $(\#E)$ rule is completed by the *first* re-occurrence of its minor premise that is deduced again. The rule demands only that the minor premise is deduced at some later point, it may appear many times before we can regard the rule as being completed.[7]

---

[7]So suppose we encounter something like this in the deduction

$$\frac{\dfrac{\#\#A \quad B}{\#A}\ (\#E) \quad C}{A}\ (\#E)$$
$$\vdots$$
$$B$$

To see what the Hash rules do, note the following deduction of $A \vee (A \to B)$:[8]

$$\frac{\#B \quad \dfrac{[A]^1}{A \vee (A \to B)} \ \vee I}{\dfrac{\dfrac{B}{A \to B} \ (\to I)(1)}{A \vee (A \to B)} \ \vee I} \ (\#E)$$

Furthermore, we can replace instances of $(\#E)$ with instances of this $\#B \to (A \vee (A \to B))$:

$$\frac{\overset{\vdots \ \Pi_1}{\#B} \quad \overset{\vdots \ \Pi_2}{A}}{\underset{\vdots \ \Pi}{\underset{A}{B}}} \ (\#E)$$

may be replaced by

$$\frac{\dfrac{\#B \to (A \vee (A \to B)) \quad \overset{\vdots \ \Pi_1}{\#B}}{A \vee (A \to B)} \ (\to E) \qquad [A]^i \qquad \dfrac{\dfrac{[A \to B]^i \quad \overset{\vdots \ \Pi_2}{A}}{B} \ (\to E)}{\underset{\vdots \ \Pi}{\underset{A}{}}}}{A} \ \vee E(i)$$

And the side condition ensures that moving Π out from underneath Π_1 does not invalidate the deduction.

We have shown that for every deduction using $(\#E)$ there is a deduction, of the same conclusion, using an appeal to an axiom schema $\#A \to (A \vee (A \to B))$ instead. Furthermore, it is not hard to see that we can replace the $(\#I)$ rule throughout a deduction by an appeal to the axiom schema $A \to \#A$ and $(\to E)$. Finally, we can remove all occurrences of the $\#P$ rule and replace them with appeals to this axiom $\#(A \to B) \to (\#A \to \#B)$.[9]

It is shown in [Gabbay, 1981] that adding the axioms

$$\#B \to (A \vee (A \to B))$$
$$A \to \#A$$
$$\#(A \to B) \to (\#A \to \#B)$$

to intuitionistic logic gives (the new connective) $\#A$ the following truth conditions:

$\#A$ is true at world w (in a Kripke model) when A is true in all w' s.t. wRw' and $w \neq w'$.

---

and the first application of $(\#E)$ cannot be completed by the re-occurrence of B (suppose it is the first) unless C has already re-occurred. If the completed the second application of $(\#E)$ then, even though B has been deduced again, both application of $(\#E)$ remain incomplete.

[8] Here we use disjunction to harmonise our results with [Gabbay, 1981], but from the next section it can be seen disjunction is not essential to make the point made here.

[9] The minor premise of $(\#P)$ gives us that $A_1 \to \ldots A_n \to B$ and so we have $\#(A_1 \to \ldots A_n \to B)$ using the axiom $A \to \#A$ and then the $\#$ operator may be distributed across all implications using the axiom above.

Given this, $\#^n\bot$ means that the Kripke model has a depth of at most n worlds.[10]

I shall not discuss the hash logic much here. We shall note that the additional rules for $\#$ preserve normalisation.

7.2   Normalisation for hash logic

The essential case for $\#$ is only slightly more complex than for the other connectives. Suppose $\#$ is introduced, subjected to a number of applications of $\#P$, and then eliminated. First note that two (and hence any number) of successive applications of $(\#P)$

$$
\cfrac{
\#A_1 \quad \cdots \quad \#A_n \quad \cfrac{\begin{array}{c}[A_1]^i \ldots [A_n]^i \\ \vdots \\ B\end{array}}{}\ (\#P)(i) \quad B_1 \quad \cdots \quad \#B_m \quad \cfrac{\begin{array}{c}[B]^j\ \ [B_1]^j \ldots [B_m]^j \\ \vdots \\ C\end{array}}{}
}{\#C}\ (\#P)(j)
$$

may be reduced to just one:

$$
\cfrac{
\#A_1 \quad \cdots \quad \#A_n \quad \#B_1 \quad \cdots \quad \#B_m \quad \cfrac{\begin{array}{c}[A_1]^i \ldots [A_n]^i \\ \vdots \\ B \quad \begin{array}{c}[B_1]^i \ldots [B_m]^i \\ \vdots \\ C\end{array}\end{array}}{}
}{\#C}\ (\#P)(i)
$$

and now an introduction, permutation $((\#P)$ rule) and subsequent elimination of $\#$

$$
\cfrac{
\cfrac{\cfrac{\begin{array}{c}\vdots\ \Pi \\ A\end{array}}{\#A}\ (\#I) \quad \#A_1 \quad \cdots \quad \#A_n \quad \cfrac{\begin{array}{c}[A]^i\ \ [A_1]^i \ldots [A_n]^i \\ \vdots \\ B\end{array}}{}}{\#B}\ (\#P)(i) \quad C
}{B}\ (\#E)
$$
$$
\begin{array}{c}\vdots \\ C\end{array}
$$

---

[10] Actually this is not true. The axioms for $\#$ considered in [Gabbay, 1981] are these:

$\#B \to (A \vee (A \to B))$
$A \to \#A$
$(A \to B) \to (\#A \to \#B)$
$\#A \to \neg\neg A$

and the semantics for $\#$ is that $\#A$ is true at world w (in a Kripke model) when either

  (i) w is not an endpoint and A is true in all w' s.t. wRw' and $w \neq w'$.

  (ii) w is an endpoint and A is true in w.

But it is not hard to convert Gabbay's proof into one that the axioms we present are complete for our simpler semantics. We find our simpler semantics much more convenient. For one thing we can express the hash operator of [Gabbay, 1981] using conjunction: $\#A \wedge \neg\neg A$. Furthermore it is useful not to have $\#\bot$ as necessarily false, for we can use it (on our semantics for $\#$) to express the property of being an endpoint (according to our semantics, $\#A$ is true at w iff w has no worlds accessible to it other than itself).

may be reduced to:

$$
\cfrac{
\#A_1 \quad \cdots \quad \#A_n \qquad
\cfrac{\begin{matrix} \vdots\ \Pi \\ A \end{matrix} \quad [A_1]^i \ldots [A_n]^i \\ \vdots \\ B}{\#B}
}{B} \ (\#P)(i) \quad C \ (\#E)
$$

$$
\begin{matrix} \vdots \\ C \end{matrix}
$$

and if any of the $\#A_i$ are introduced by $(\#I)$ then they may be dispatched similarly.

The normalisation argument then proceeds in almost identical fashion to the argument for restart. We must argue that we can always restructure the deduction so that the major premise of an elimination rule never be required to complete an application of restart. In such a case we must teleport the elimination rule away, so this

$$
\cfrac{\cfrac{\#B \quad \begin{matrix} \vdots\ \Pi \\ A \end{matrix}}{B} \ (\#E)}{\cfrac{\begin{matrix} \vdots\ \Pi' \\ A \end{matrix}}{C}} \ (?E)
$$

gets replaced by this:

$$
\cfrac{\#B \quad \cfrac{\cfrac{\begin{matrix} \vdots\ \Pi \\ A \end{matrix}}{\vdots}}{C} \ (?E)}{\cfrac{\begin{matrix} B \\ \vdots\ \Pi' \\ A \end{matrix}}{C}} \ (\#E)
$$
$$
(?E)
$$

and we must treat any incomplete restarts in the minor premises of Π as being completed only *after* Π' (even if the appropriate formulae happen, by chance, to re-occur in Π').[11]

7.3 Goal directed Hash logic

The Goal-directed reasoning of [Gabbay and Reyle, 1984; Gabbay, 1998; Gabbay and Olivetti, 2000b] does not have rules for disjunction. However, we can formulate the axioms for $\#$ in terms of implication. It is not hard to show that we obtain equivalent logics when using the axiom $\#B \rightarrow [((A \rightarrow B) \rightarrow A) \rightarrow A]$ instead of $\#B \rightarrow (A \vee (A \rightarrow B))$. For example, here is an argument that there is a deduction of $(A \vee (A \rightarrow B))$ if we add $((A \rightarrow B) \rightarrow A) \rightarrow A$ as an axiom, for brevity let X be shorthand for

---

[11]Note that for a hash logic normalisation theorem we must work with a slightly different notion of a segment: a segment is either (i) a sequence in a deduction $A_1 \ldots A_n$ where each A_i is an occurrence of A or (ii) a sequence $\#B_1 \ldots \#B_n$ where each $\#B_{i+1}$ is deduced from B_i by the $\#P$ rule.

$[A \vee (A \to B)]$.

1	$\vdash ((X \to B) \to X) \to X$	by the alternative axiom
2	$A \vdash X$	by the disjunction axioms
3	$A, X \to B \vdash B$	from 2, by Cut and Modus Ponens
4	$X \to B \vdash A \to B$	from 3 using the Deduction theorem
5	$X \to B \vdash X$	from 4 using the disjunction axioms and Cut
6	$\vdash (X \to B) \to X$	from 5 by the Deduction theorem
7	$\vdash X$	Cut from 1 and 6

Here are the rules for the $\#$ operator in goal directed logic. Firstly is the obvious introduction rule:

> We may infer $\#A$ if we have previously inferred A.

and now the less obvious elimination rule:

> We may infer A from an instance of B if we have a goal $?B$ prior to that instance of B and we have deduced $\#A$ prior to that goal $?B$.

Schematically this rule looks like this

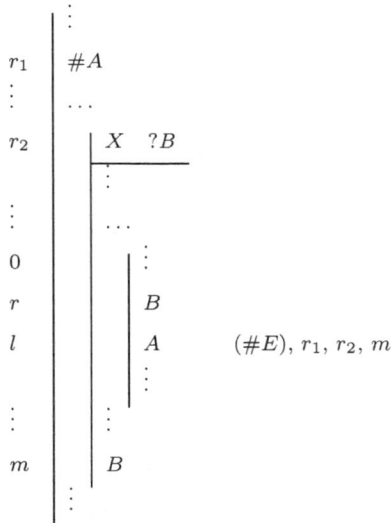

$$
\begin{array}{ll}
& \vdots \\
r_1 & \#A \\
\vdots & \\
& \cdots \\
r_2 & \quad X \quad ?B \\
& \qquad \vdots \\
\vdots & \\
& \cdots \\
0 & \qquad \vdots \\
r & \qquad B \\
l & \qquad A \qquad (\#E),\, r_1, r_2, m \\
& \qquad \vdots \\
\vdots & \qquad \vdots \\
m & \qquad B \\
& \vdots
\end{array}
$$

   Line l is inferred from r by the rule for $\#$, prior to B is a goal $?B$ and prior that is $\#A$.

To see the rule is sound observe that we can convert it into a deduction that uses an instance of the axiom $\#A \to [((B \to A) \to B) \to B]$.

$$
\begin{array}{ll}
r_1 & \#A \\
 & \#A \to [((B \to A) \to B) \to B] \qquad \text{Axiom} \\
r_1' & ((B \to A) \to B) \to B \\
 & \ldots \\
r_2 & \quad X \ \ ?B \\
r' & \qquad B \to A \ \ ?B \\
 & \\
0 & \\
r & \qquad\qquad B \\
l & \qquad\qquad A \qquad\qquad\qquad \text{Modus Ponens, } r', r \\
 & \\
m & \qquad\qquad B \\
m' & \qquad (B \to A) \to B \\
 & \qquad B \qquad\qquad\qquad \text{Modus Ponens, } r_1', m'
\end{array}
$$

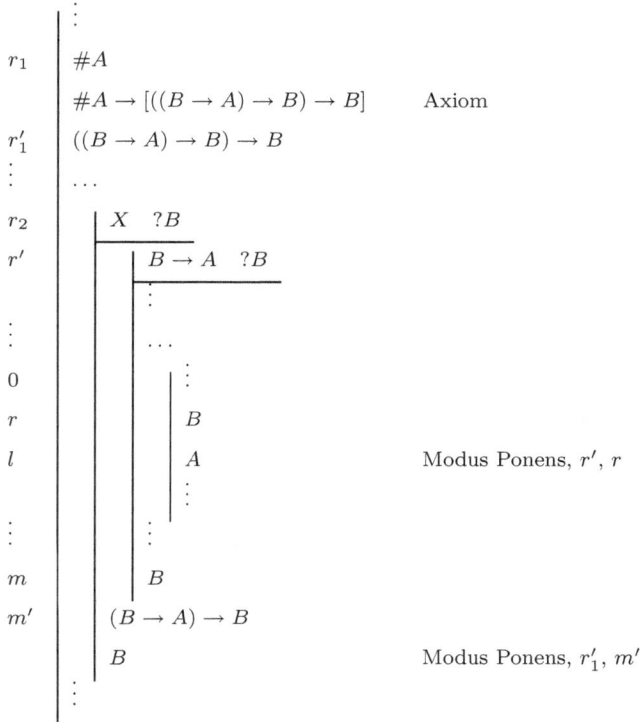

Looking within this construction we can see an alternative proof that the simple goal-directed restart rule is sound that proceeds by replacing applications of restart to an appeal to Peirce's law rather than the principle of indirect proof.

To ensure that the goal-directed hash calculus is complete we need only add the following rule (the equivalent of the $\#P$ rule) which we represent here schematically:

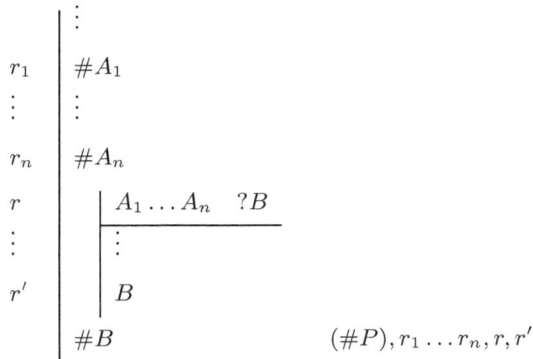

$$
\begin{array}{ll}
r_1 & \#A_1 \\
\vdots & \vdots \\
r_n & \#A_n \\
r & \quad A_1 \ldots A_n \ \ ?B \\
\vdots & \vdots \\
r' & \quad B \\
 & \#B \qquad\qquad (\#P), r_1 \ldots r_n, r, r'
\end{array}
$$

To show completeness we must use the new rules to derive the three axioms $\#A \to [((B \to A) \to B) \to B]$, $A \to \#A$ and $\#(A \to B) \to \#A \to \#B$, this is left for the reader.

8   Restart in first order logic

8.1   Quantifier rules

We have concentrated on restart for propositional logic, but there is no reason not to do exactly the same with predicate logic.

In first order logic, the restart rule remains unchanged (to extend intuitionistic to classical logic), but the side conditions on $\forall I$ and $\exists E$ rules must be modified a little.[12]

$$\frac{A}{\forall x A} \ (\forall I)$$
Provided x does not occur free in any assumptions on which A depends *nor* (free) in any formula that is required to handle any restarts that are incomplete at A.

$$\frac{\exists x A \quad C}{C} \ (\exists E)^i$$
with
$$[A_c^x]^i$$
$$\vdots$$

Where c is a constant which does not occur in any assumptions on which C depends, except A_c^x, nor in $\exists x A$, nor in any formula that is required to handle any restarts that are incomplete at C.

A formula B is required to handle a restart at (a node containing an instance of) A in a deduction when there is a (prior) application of restart that requires B to be deduced again but (an appropriate instance of) B has not yet been deduced again.

We obtain these side conditions systematically as follows: in the soundness proof of restart (page 705) we convert the instances of restart into an appeal to Peirce's law. That is, we show that for every deduction using the restart rule there is a direct way of replacing the appeal to restart with an appeal to Peirce's law. Actually, it is not hard to see that we can replace the applications of restart with appeals to a particular instance of Peirce's law: $((A \to \bot) \to A) \to A$.

The deduction (segment) with which we replace the application of restart contains the assumption, and subsequent discharge, of $(A \to \bot) \to A$. This assumption of $(A \to \bot) \to A$ (which formerly did not appear in the deduction at all) may interfere with the side conditions of any applications of $(\forall I)$ or $(\exists E)$ below it. Schematically, if we replace an application of restart

$$\frac{A}{B} \ (Restart)$$
$$\vdots \ \Pi$$
$$A$$

by this

$$\frac{\cfrac{A \quad [A \to \bot]^1}{\bot} \ (\to E)}{\cfrac{\cfrac{B}{\vdots \ \Pi}{A}}{(A \to B) \to A} \ (\to I)^1 \qquad \cfrac{}{((A \to \bot) \to A) \to A} \ \text{Assumed}}{A} \ (\to E)$$

then any applications of $(\forall I)$ or $(\exists E)$ in Π may become invalidated if x or c is free in A. The new side condition ensures that this does not happen as the application of restart

---

[12] A_t^x is the formula obtained by simultaneously replacing all free occurrences of the variable x in A by the term t

in question is incomplete throughout Π and so no application of $(\forall I)$ or $(\exists E)$ makes critical use of any x or c that is free in A.

It is now not hard to derive the following theorem:

THEOREM 2. *Intuitionistic FOL plus (Restart) has the same entailment relation as Classical FOL.*

We have just shown the left right direction of this theorem, and the right left direction is shown in the same way as with Theorem 1

Normalisation follows by an easy extension of the arguments of Section 2.2. For example here is the reduction case for the existential quantifier (given the familiar side condition on $\exists I$ we may suppose that no variable in t is bound in A_t^x).

$$
\begin{array}{ccc}
\begin{array}{c}
\vdots\ \Pi_1 \\
A_t^x \\
\hline
\exists x A
\end{array} (\exists I)
&
\begin{array}{c}
[A_c^x]^i \\
\vdots\ \Pi_2 \\
B
\end{array}
&
\\
\hline
\multicolumn{2}{c}{B} (\exists E)^i
&
\end{array}
\quad\Longrightarrow\quad
\begin{array}{c}
\vdots\ \Pi_1 \\
A_t^x \\
\vdots\ \Pi_2' \\
B
\end{array}
$$

Where Π_2' is obtained from Π_1 by making the following replacements:

1. First replace any constant or variable symbol in Π_2 that is not free in any assumption of Π_2 but is free in assumptions of Π_1 by some other variable or constant symbol that is not free in any assumptions of Π_1.

2. Finally, replace c by t throughout Π_2

It is left to the reader to verify, bearing in mind the rules for the quantifiers, that Π_2' may be appended to the end of Π_1 to yield a valid deduction.

There is much interesting research to be done in the relation between more complex restart rules, hash logic and first order intermediate logics. Restart seems quite robust and suggests in a natural way the side-conditions and reduction rules necessary to make things work. We now continue with another example.

8.2   Endpoint Restart

Consider this restart rule, call it *endpoint restart*:

$$
\begin{array}{c}
A \\
\hline
B \\
\vdots \\
A \\
\vdots \\
\bot
\end{array}
$$

which allows us to infer B from A provided we infer A again later and then $\bot$ still later. In relation to the side conditions on the universal quantifier, if x is free in A, then we cannot introduce $\forall x D$ by $\forall I$ until after A is deduced again (but we can do so before $\bot$ is deduced again as x is not free in $\bot$).

This more complex restart rule allows us to deduce the following formula

$$NS = \forall x \neg\neg A \rightarrow \neg\neg \forall x A$$

$$\cfrac{\cfrac{[\forall x\neg\neg A]^3}{\neg\neg A}\ \forall E \qquad \cfrac{\cfrac{[A]^1}{\bot}}{\neg A}\ (\to I)(1)}{\cfrac{\cfrac{\bot}{A}\ (\bot E)}{\forall x A}\ \forall I}\ (\to E)$$

$$\cfrac{[\neg\forall x A]^2 \qquad \cfrac{\cfrac{\bot}{A}\ (\bot E)}{\forall x A}\ \forall I}{\cfrac{\bot}{\cfrac{\neg\neg\forall x A}{\forall x\neg\neg A \to \neg\neg\forall x A}\ (\to I)(2)}\ (\to I)(3)}\ (\to E)$$

and the application of $\forall I$ is legitimate because the requirement to re-deduce A has been met (although the application of restart is not fully complete as $\bot$ must be deduced). Furthermore observe that for any deduction using the new restart there is a deduction of the same conclusion using the axiom $\neg\neg\forall x(((A \to B) \to A) \to A)$. We can convert any application of endpoint restart thusly

$$\cfrac{\neg\neg\forall x(((A \to B) \to A) \to A) \qquad \cfrac{\cfrac{[\forall x(((A \to B) \to A) \to A)]^2}{((A \to B) \to A) \to A} \qquad \cfrac{\cfrac{\vdots}{\cfrac{\dot{A}\ [A \to B]^1}{B}\ (\to E)}\ \vdots}{\cfrac{\dot{A}}{(A \to B) \to A}\ (\to I)^1}}{\cfrac{A}{\cfrac{\vdots}{\cfrac{\bot}{\neg\forall x(((A \to B) \to A) \to A)}\ (\to I)^2}}\ \forall E}}{\bot}$$

But it is a fact of intuisionistic logic that

$$\forall x\neg\neg A \to \neg\neg\forall x A \vdash \neg\neg\forall x(((A \to B) \to A) \to A)^{13}$$

and so we can convert any applications of endpoint restart into an appeals to NS.

Adding NS as an axiom to intuitionistic logic yields an intermediate logic complete for Kripke frames where each world has an endpoint accessible to it (see [van Dalen, 1986; Gabbay, 1981]), that is:

$$\forall w \exists w'(wRw' \& \forall w''(w'Rw'' \to w' = w''))$$

Notice that we can already express such a condition using the hash operator $\neg\neg\#\bot$. And indeed, we can deduce NS from $\neg\neg\#\bot$ (see 8.3).

Interestingly however, we cannot deduce $\neg\neg\#\bot$ if we add NS as an axiom. To see this, consider a Kripke frame with a set of worlds W and an accessibility relation such that wRw' for every $w, w' \in W$, i.e. the accessibility relation is an equivalence class on W. In such a frame NS is clearly true at all worlds, but $\#\bot$ is not true at any world.

What has happened is that NS does not force a Kripke frame to have endpoints, but it does force a frame into something indistinguishable from one with endpoints

---

[13]$((A \to B) \to A) \to A$ is a propositional classical theorem and so $\neg\neg(((A \to B) \to A) \to A)$ is an intuitionistic theorem, and therefore so is its universal generalisation etc.

(a frame where there are always worlds where all propositions are determined). But with the additional expressive power of the $\#$ operator we can make such distinctions and so, in the presence of $\#$, the intermediate logic with NS as an axiom (named MH in [Gabbay, 1981]) is not complete for frames where every world has an accessible endpoint. It follows that NS and endpoint restart are complete for Kripke frames where every world has an accessible endpoint only if the language does not contain $\#$.

Normalisation of endpoint restart follows almost exactly as with ordinary restart: because $\bot$ has no introduction rule, the extra condition that $\bot$ be deduced after A has no effect on the normalisation argument.

The goal-directed formulation of endpoint restart is this

> We may replace a goal $?B$ with a previous goal $?A$ (i.e. act as if we have deduced B when we have really deduced A) provided that there is a goal $?\bot$ previous to the goal $?A$.

To see this work it is convenient to have the following introduction rule for the universal quantifier:

$$
\begin{array}{ll}
r & \quad\quad ?A_y^x \quad\quad y \text{ does not occur above} \\
\vdots & \quad\quad \vdots \\
r' & \quad\quad A_y^x \\
& \forall x A \quad\quad\quad\quad \forall I,\, r,\, r'
\end{array}
$$

That is, we assume nothing and set $A[x/y]$ as our goal where y is new to the deduction, then when we reach that goal we may conclude $\forall x A$. Now, here is a deduction of NS using goal-directed endpoint restart:

1	$\forall x \neg\neg A \quad ?\neg\neg\forall x A$	
2	$\neg\forall x A \quad ?\bot$	
3	$?A_y^x$	
4	$\neg\neg A_y^x$	$\forall E,\, 1$
5	$A_y^x \quad ?\bot$	
6	A_y^x	recalling, 5
7	$\neg A_y^x$	endpoint restart, 3, 5, 6
8	$\bot$	$(\rightarrow E),\, 4,\, 7$
9	A_y^x	$(\bot E),\, 8$
10	$\forall x A$	$\forall I,\, 3,\, 9$
11	$\bot$	$(\rightarrow E),\, 2,\, 10$
12	$\neg\neg\forall x A$	$(\rightarrow I),\, 2,\, 11$
13	$\forall x \neg\neg A \rightarrow \neg\neg\forall x A$	$(\rightarrow I),\, 1,\, 12$

What is interesting about this particular restart rule is that it does not mention quantifiers at all. Looking carefully at the rule we can see that it has no effect on the propositional fragment of intuitionistic logic. Without quantifiers it corresponds to an appeal to the double negation of Peirce's law which is an intuitionistic theorem.[14] With quantifiers endpoint restart corresponds to an appeal to the double negation of a universal generalisation of Peirce's law (and that is not an intuitionistic theorem).

8.3 More fun with hash

Endpoint restart was not discovered by accident, it was noticed by externalising the effect of adding $\neg\neg\#\bot$ as an axiom. That is, $\#$ internalises into intuitionistic logic the deduction structure of restart. And so more complex formulae involving hash internalise more complex deduction structures. It is interesting to see what restarts we get if we attempt to put these structures back into the external structure of natural deduction (in the form of more complex restart rules). It is not hard to see how the axiom $\neg\neg\#\bot$ justifies endpoint restart:[15]

$$\frac{\dfrac{[\#\bot]^1 \quad A}{\dfrac{\bot}{B}\,(\#E)}\,(\bot E)}{\vdots} $$

$$\frac{\neg\neg\#\bot \qquad \dfrac{\dfrac{\vdots}{A}}{\dfrac{\vdots}{\bot}} \atop \qquad\qquad \dfrac{}{\neg\#\bot}\,(\to I)(1)}{\bot}$$

and we obtained endpoint restart from considering that deduction. We suspect many more restart rules can be obtained in this manner.

9   Conclusions and future work

It is always nice when we can extend a system without having to reengineer our existing material (say it is **inherited**). For example it is nice to be able to extend a logic in such a way that deductions of the unextended logic are still deductions of the extended logic, and valid proof-reductions of the unextended logic are still valid proof-reductions of the extended logic.

Examples of this principle in action are not hard to find wherever people have invested effort in a particular system, and then find they want to extend it. Besides saved effort, the chances are that if formal structure is preserved, so may the good properties (e.g. proof-normalisation!).

This paper presents a robust and elementary method by which logical derivability may be extended, without affecting the structure of deductions or their normalisation. We have applied it here both to intuitionistic Natural Deduction and to Gabbay's Goal-Directed Deduction. We consider consider restart-like methods of obtaining interesting and sometimes exotic logics. This also gives some new perspectives on their computational content.

On a philosophical note we are following work of Gentzen, Prawitz and Dummett (to name a few) that a deduction theoretic account of the meaning of a logical connective can be found in its introduction and elimination rules (intro/elim pair). In this view, the

---

[14]Also, propositional intuitionistic logic is complete for finite Kripke models. So the condition that every world has an accessible endpoint is trivially met in the propositional case.
[15]And since endpoint restart can be used to deduce NS it follows that $\neg\neg\#\bot$ deduces NS.

intro/elim pair of, say implication, define its meaning. A *useful* definition of meaning is an intro/elim pair which normalises in the context of the rest of the logic. As we have seen, natural deduction plus restart is convenient for defining classical logics in a way that, for example, natural deduction plus double negation elimination is not.

We made it obvious that Classical Restart corresponds to multiple conclusions, but it is not *identical* to multiple conclusions, any more than Natural Deduction is *identical* to sequent systems. Indeed, we can argue that Restart is the true meaning of multiple conclusions!

There is a much work on 'exotic' deduction systems for logics a little off the beaten track. We have dabbled in this in the paper with the hash logic and with *n*-depth restart. **Hypersequents**, introduced by Arnon Avron [Avron, 1996], are another example in the literature. Amusingly hypersequents go in entirely the opposite direction to this paper; instead of doing away with sequent structure, they enrich it. And yet they seem to arrive at similar applications to us:

Ciabattoni and Ferrari [Ciabattoni and Ferrari, 2000] present deduction-systems for logics complete for Kripke models with restrictions on their geometry, including bounded depth like our *n*-depth restart but also bounded cardinality, bounded width, and so on — this is much more general than what we managed.

The connection between our determinedly Natural Deduction restart rules, and the superficially quite different determinedly Sequent Calculus hypersequent systems, is in the connection between the *proof-teleportation* which we have seen in this paper, which arises naturally from restart proof-normalisation, in which fragments of deduction are copied (teleported) deep inside intermediate parts of other deductions; on the other hand hypersequents on the other hand derive their power by allowing us to *rearrange and copy assumptions* inside intermediate steps in the sequent deduction. In the first, we move a fragment of deduction to a new context, and in the second we move the new context to be around a fragment of deduction — in the end, it seems to come to almost the same thing.

Hypersequents seem to be more general and expressive than restart rules in that they express a wider range of intermediate logics. Yet Restart rules are nice when they work, they retain the simple Natural Deduction presentation and so are 'cheap' in the sense of the Introduction. This can be seen in the ease by which we can internalise the restart rule into the hash logic which is a conservative extension of classical logic (this follows from the normal form theorem on hash logic). The hash logic itself is interesting and further work into the extra expressive power and computational content of #, and connectives axiomatisable using #, seems promising. Indeed, as far as we are aware, the intermediate logic MH (see section 8.2) has not been formulated in terms of hypersequents.

Viewed computationally, Classical Restart leads immediately to a λ-calculus very similar to Parigot's $\lambda\mu$-calculus (restarts-yet-to-be-justified, and the current goal, correspond to multiple conclusions). Parigot's great accomplishment was to 'pull out of the air' the $\lambda\mu$-calculus in 1992, and in retrospect we now see that Dov Gabbay had pulled it out of the air too, in 1984 and in a different field as goal-directed restart. This should serve as a taster to interesting calculi that can be produced for the predicate case, for hash logic, and for any other restart rule.

Perhaps the most striking presentation of restart rules is still in the original presentation in goal directed formulations. The restart rule is somewhat obvious in the rigorous and highly annotated structure of (Lemmon style) linear natural deduction systems. The effects of restart and hash rules in their natural deduction formulations are unexpected at first, but then intelligible. But in their goal-directed formulations the restart and hash rules seem, at least to the authors of this article, almost magical. The discoverer of the restart rule, who spotted it directly in its goal directed formulation, is surely to be commended on such a leap of the imagination.

Acknowledgements

We are grateful to Agata Ciabattoni, Alexis Saurin, and David Richter, for useful comments, and to two anonymous referees.

BIBLIOGRAPHY

[Avron, 1996] A. Avron. The method of hypersequents in the proof theory of propositional non-classical logics. In W. Hodges, M. Hyland, C. Steinhorn, and J. Truss, editors, *Logic: from foundations to applications. Proc. Logic Colloquium, Keele, UK, 1993*, pp. 1–32. oup, New York, 1996.
[Barendregt, 2000] H. P. Barendregt. Lambda calculi with types. In S. Abramsky, D. M. Gabbay, and T. S. E. Maibaum, editors, *Handbook of Logic in Computer Science, Volume 2*, pp. 117–309. OUP, 2000.
[Bell and Machover, 1977] J. Bell and M. Machover. *A course in mathematical logic*. North-Holland, 1977.
[Braüner, 2000] T. Braüner. A cut-free gentzen formulation of the modal logic s5. *Logic Journal of IGPL*, 8(5):629–643, 2000.
[Ciabattoni and Ferrari, 2000] A. Ciabattoni and M. Ferrari. Hypertableau and path-hypertableau calculi for some families of intermediate logics. In *Automated Reasoning with Tableaux and Related Methods (Tableaux'2000)*, volume 1847 of *LNAI*, pp. 160–175, 2000.
[Coquand, 1996] T. Coquand. Computational content of classical logic. In P. Dybjer and A. Pitts, editors, *CLICS Cambridge Summer School.*, 1996.
[de Groote, 1995] P. de Groote. A simple calculus of exception handling. In *LNCS*, volume 902, pp. 201–215. Springer, 1995. TLCA'95.
[Gabbay and Olivetti, 2000a] D. Gabbay and N. Olivetti. *Goal Directed Algorithmic Proof Theory*. Kluwer Academic, 2000.
[Gabbay and Olivetti, 2000b] D. Gabbay and N. Olivetti. *Goal-directed Proof Theory*. Kluwer Academic Publishers, 2000.
[Gabbay and Reyle, 1984] D. Gabbay and U. Reyle. N-prolog: An extension of prolog with hypothetical implications i. *Journal of Logic Programming*, (1):391–355, 1984.
[Gabbay, 1981] D. Gabbay. *Semantical investigations in Heyting's Intuitionistic Logic*, volume 148 of *Synthèse*. D. Reidel, 1981.
[Gabbay, 1990] D. Gabbay. Algorithmic proof with diminishing resources part i. In *CSL'90*, volume 533 of *LNCS*, pp. 156–173. Springer, 1990.
[Gabbay, 1998] D. Gabbay. *Elementary Logic: A Procedural Perspective*. Prentice Hall, 1998.
[Gabbay, 2004] Michael Gabbay. *Proof-Normalisation and Truth by Definition*. PhD thesis, King's College, London, 2004.
[Gentzen, 1934] G. Gentzen. Untersuchungen über das logische Schliessen. *Mathematische Zeitschrift*, 39:176–210 and 405–431, 1934.
[M. H. B. Sorensen, 1998] P. Urzyczyn M. H. B. Sorensen. Lectures on the curry-howard isomorphism. Technical Report 98/14, DIKU, 1998.
[Parigot, 1992] M. Parigot. Lambda-mu-calculus: An algorithmic interpretation of classical natural deduction. In *LPAR'92*, pp. 190–201, 1992.
[Peyton-Jones, 2001] S. Peyton-Jones. Tackling the awkward squad: monadic input/output, concurrency, exceptions, and foreign-language calls in haskell. In *Engineering theories of software construction, Marktoberdorf Summer School*, NATO ASI, pp. 47–96. 105 Press, 2001.
[Prawitz, 1965] D. Prawitz. *Natural Deduction - A proof theoretical study*. Almquist and Wiksell, 1965.
[Stalmarck, 1991] G. Stalmarck. Normalization theorems for full first order classical natural deduction. *Journal of Symbolic Logic*, 56:129–149, 1991.
[Taylor, 1999] P. Taylor. *Practical Foundations of Mathematics*. Number 59 in Cambridge Studies in Advanced Mathematics. 1999.
[van Dalen, 1986] D. van Dalen. Intuitionistic logic. In Gabbay and Günthner, editors, *Handbook of Philosophical Logic vol III*, volume 166 of *Synthèse*, chapter 4. D. Reidel, 1986.

a-logic

MURDOCH J. GABBAY AND MICHAEL J. GABBAY

1 Introduction

Mathematics and computer science pervasively use logics formal languages
for reasoning in. One of the most common is classical First-Order Logic with
equality (FOL) [Bell and Machover, 1977; Gabbay and Günthner, 1986].

For example FOL is routinely used as a foundational tool to express
Zermelo-Fraenkel set theory (**ZF sets**) [Johnstone, 1987; Machover, 1996]
and thus as a (language to express) a foundation for classical mathematics.
The implication is that first-order logic is a minimal and convenient logical
framework for doing, well, triangles and numbers and other mathematical
objects.

But is it so good for the mathematics of computation, i.e. for computer
science? Perhaps not, since it cannot express substitution as an object-level
operation on its own terms. The λ-calculus [Barendregt, 1984] can and does,
in the sense that we can write $(\lambda a.s)t$, which β-reduces [Barendregt, 1984]:

$$(\lambda a.s)t \rightarrow s[a{\mapsto}t].$$

This is a pervasive phenomenon and theories of computation are frequently
based on notions of substitution. (For this reason, some researchers refer
informally to a general notion of '*movement*', which in the examples here is
implemented by substitution.)

Another example is channel communication as appears for example in
the π-calculus [Milner *et al.*, 1992]:

$$\overline{a}b \mid a\langle c\rangle.P \rightarrow P[c{\mapsto}b].$$

Now note that first-order logic cannot express this *directly*; consider a
ternary **explicit substitution** term-former which we write $s[t{\mapsto}u]$. The
intuition is that if t *is* a variable symbol, then $s[t{\mapsto}u]$ denotes the term
obtained by replacing every instance of t in s by u.

This is ridiculous because even if t *were* a variable symbol, there is no
way we could guarantee it would *still be* a variable symbol if that variable
symbol is instantiated, as arises from the following consideration:

We want our logic to satisfy cut-elimination and thus we want to simplify a proof of the form

$$\frac{\dfrac{\Gamma \vdash P, \Delta}{\Gamma \vdash \forall x.\, P, \Delta} \qquad \dfrac{\Gamma, P[x \mapsto s] \vdash \Delta}{\Gamma, \forall x.\, P \vdash \Delta}}{\Gamma \vdash \Delta} \text{(Cut)}$$

to one without the cut. It suffices to demonstrate the **Substitution Lemma**:

LEMMA. *If $\Gamma \vdash \Delta$ is derivable in our logic without cut, then $\Gamma[x \mapsto s] \vdash \Delta[x \mapsto s]$ is also derivable in our logic without cut, and with a no larger proof.*

(— 'no larger' according to whatever measure is useful for making an inductive argument work given the other transformations we make on the proof while eliminating cuts. The substitution action on Γ and Δ is element-wise. See Lemma 5.)

Likewise it is not possible to hypothesise binary term-formers λ and $\circ$ with axioms in first-order logic such that if a is a variable symbol then $(\lambda a.s) \circ u = s[a \mapsto u]$.

Yet there is a need to express the λ-calculus in logic, for any of several reasons:

- The untyped λ-calculus is a foundational system to rival ZF sets. Why should we be able to write down axioms for one in FOL, but not for the other?

- FOL has a well-developed and tractable meta-theory.

- Functions *are* useful.

Theoretical computer science as a field has reacted to these issues by working in higher-order frameworks, but this commits to certain very particular choices, e.g. committing to β-reduction as the only notion of substitution (*'movement'*), a particular discipline of types (which may prevent us from quantifying over all elements of the universe), and it may be expensive for the meta-theory (models are more complex, Interpolation need not hold, compactness fails, and so on).

So a first-order logic which can express explicit substitution as a term-former as described above, offers the prospect of combining the power of a functional theory, with the simplicity and flexibility of a first-order system.

In this paper we define *a*-logic, investigate its relation to First-Order Logic, prove some meta-theory for the logic as a formal system, and then axiomatise the λ-calculus as a finite first-order theory. Finally, we consider axiomatisations of other theories.

The '*a*' in '*a*-logic' refers to the fact that the logic is for making statements about its variable symbols a, b, c. Any resemblance to a pronoun is coincidental.

2 *a*-logic

2.1 Terms, directions, predicates

We assume a countably infinite set x, y, z, a, b, n of **(term) variable symbols**, and a finite set of **term-formers** f to each of which is associated an **arity** $ar(f)$ which is a nonnegative number which may be 0.[1] If $ar(f) = 0$ we may call f a **constant**.

Terms are inductively generated as follows:

$$s, t ::= x \mid f(s, \ldots, s)$$

where f has $ar(f)$ arguments. If f is a constant we may write $f()$ as just f.

For reasons which will become clear later, we arbitrarily call some of the f **inevitably term-formers**. (For future reference, λ and application will be inevitably term-formers, but explicit substitution is not. We come back to this later.)

We write $s \equiv t$ for 's and t are syntactically identical terms'. We write $s \not\equiv t$ for 's and t are syntactically different terms'.

Assume some set of **predicate constant symbols** $p, q, r \ldots \in \mathbb{P}$, each with its arity $ar(p)$, $ar(q)$, $ar(r)$,... which is a nonnegative number which may be 0.

We assume distinguished predicate constants:

1. We call **at** the predicate **is an atom** (or **atom** for short), of arity 1.

2. We call $\bot$ **false** or **contradiction** of arity 0.

3. We call $=$ **equality**.

Write $p(t_1, \ldots, t_n)$ for a predicate constant applied to a list of terms of length n where we shall always assume $n = ar(p)$. Call this a **primitive proposition**.

Propositions are generated by the grammar

$$P ::= p(ts) \mid P \supset P \mid \forall x.\, P.$$

---

[1] Our reasons for using a as a variable symbol, not a constant, will become clear.

$\forall x.\, P$ is binding, nothing else is. Consistent with our convention for terms we write $P \equiv Q$ for 'P and Q are identical predicates' (up to α-equivalence).

Write VP for **the variables occurring in P**, defined by: $Vp(t_1, \ldots, t_n) = \bigcup Vt_i$, $V(P \supset Q) = VP \cup VQ$, and $V(\forall n.\, P) = VP \backslash \{n\}$.

Here, if S and T are sets then $S \backslash T$ is set of elements of S which are not elements of T. In particular $VP \backslash \{n\}$ denotes the set of elements of VP not equal to n. We use this notation without comment henceforth.

We may write $x \in P$ for $x \in VP$, we read it 'x **occurs in** P'. Then $x \notin P$ means 'x does not occur in P'. (We strictly equate predicates up to α-equivalence, we will *not* talk about 'occurring free' or 'occurring bound'.) Write $P[x \mapsto s]$ for P with every instance of the variable x replaced by s in the usual, capture-avoiding, manner.

Later, when we develop our notions of derivability and model, if we happen to know that **at** holds of a term, then we may choose a name like a, b, or n to represent it.

2.2   Contexts and judgements

A **(logical) context** Γ, Δ is a set of propositions. We way write $V\{P, Q\}$ for $VP \cup VQ$, and $V\Gamma$ for the natural meaning as a union; we may even mix, as in $V(\Gamma, P, \Gamma')$. We shall write $x \notin \Gamma, P, \Gamma'$ for $x \notin V(\Gamma, P, \Gamma')$, and so on.

A **judgment** is a pair of contexts which we write $\Gamma \vdash \Delta$. When a context is on the right-hand side of a judgement we call it a **cocontext**. The **valid** or **derivable** judgments are inductively defined by the following derivation rules (notation is defined below):

$$\frac{}{\Gamma, P \vdash P, \Delta}\ (\mathbf{Ax}) \qquad \frac{}{\Gamma, \bot \vdash \Delta}\ (\bot\mathbf{L})$$

$$\frac{\Gamma, P \vdash Q, \Delta}{\Gamma \vdash P \supset Q, \Delta}\ (\supset \mathbf{R}) \qquad \frac{\Gamma \vdash P, \Delta \quad \Gamma, Q \vdash \Delta}{\Gamma, P \supset Q \vdash \Delta}\ (\supset \mathbf{L})$$

$$\frac{\Gamma \vdash P, \Delta}{\Gamma \vdash \forall x.\, P, \Delta}\ (\forall \mathbf{R})^{(*)} \qquad \frac{\Gamma, P[x \mapsto s] \vdash \Delta}{\Gamma, \forall x.\, P \vdash \Delta}\ (\forall \mathbf{L}) \qquad \frac{\Gamma \vdash P, \Delta \quad \Gamma, P \vdash Q, \Delta}{\Gamma \vdash Q, \Delta}\ (\mathbf{Cut})$$

$$\frac{(s \text{ inevitably a term})}{\Gamma, \mathbf{at}\, s \vdash \Delta}\ (\mathbf{at\, L}) \qquad \frac{\Gamma, \mathbf{at}\, a \vdash \Delta}{\Gamma \vdash \Delta}\ (\mathbf{Fresh})^{(*)}$$

$$\frac{}{\Gamma \vdash t = t, \Delta}\ (=\mathbf{R}) \qquad \frac{\Gamma, P[x \mapsto s'] \vdash \Delta}{\Gamma, s' = s, P[x \mapsto s] \vdash \Delta}\ (=\mathbf{L})$$

Here x and a are variables; bound by $\forall$, free otherwise. Conditions in brackets are side-conditions whose validity can be decided just by examining syntax.

In particular the bracketed asterisks annotating ($\forall$**R**) and (**Fresh**) indicate that the rules are subject to the side-condition $x \notin \Gamma, \Delta$ or $a \notin \Gamma, \Delta$. (**Fresh**) is not mis-typed. It *eliminates* a variable. Comma, as in 'Γ, P', indicates set union. Note that this means that for example in ($\supset$ **R**), it may still be that $P \in \Gamma$ or indeed that $P \supset Q \in \Delta$.

We say that s is **inevitably a term** when it is of the form $f(t_1, \ldots, t_n)$ and f is an inevitable term-former. An example of a term which is not inevitably a term, is a term without any top-level term-former at all, namely y a variable symbol.

(Later on, we will encounter explicit substitution $s\langle t \mapsto u \rangle$ which is not inevitably a term, and λ-abstraction and application which are.)

Say a judgement $\Gamma \vdash \Delta$ is a **theorem** when a derivation exists concluding in $\Gamma \vdash \Delta$. Note that Δ may be empty in which case we write $\Gamma \vdash$; we may call Γ **inconsistent** when $\Gamma \vdash$ is derivable.

A subset of the logic is classical propositional logic, so we use standard sugar such as writing $\neg P$ for $P \supset \bot$, $P \vee Q$ for $(\neg P) \supset Q$, $\exists x. P$ for $\neg(\forall x. \neg P)$, $P \wedge Q$ for $\neg(P \supset (\neg Q))$, and in general we shall use other well-known shorthands for classical equivalences. We may also use 'derivation rules' such as ($\exists$**L**) and ($\neg$**L**), to represent several rules of the core system appropriately pasted together.

2.3 The first-order theory corresponding to *a*-logic

We now set up some notation. Uses of this notation will be scattered throughout the text; **all definitions relevant to theories are here.**

Let **FOL** be classical first-order logic with equality.

Say a triple of

- a finite set of **theory term-formers**, which may already be in the base logic (write a typical term-former f),

- a finite set of **theory atomic predicate symbols**, which may already be in the base logic (write a typical atomic predicate symbol ϕ),

- a possibly infinite set of **theory sentences** or **axioms**, which are predicates in the language of the base logic extended with any of the new term-formers and atomic predicate symbols,

is a **theory**. If the set of sentences is finite, call it a **finite theory**, otherwise call it an **infinite theory**.

When a theory has no, or a trivial, signature, we may equate it with the set of its axioms and name it using calligraphic letters, for example $\mathcal{T}$. Otherwise, we tend to give theories names in sans-serif font, for example Q or lambda. Theories named in sans-serif font will *always* be finite.

As a matter of notation we may write Unc+Cle for the theory obtained by taking the unions of the theories' term-formers, predicate symbols, and sentences. We write ¬Unc for the theory obtained by negating all the theory sentences. Finally, if both Unc and Cle are finite, we may identify them with the predicate which is the logical conjunction of their senteces, and write things like Unc ⊃ Cle or Unc ⊃ ¬Cle.

The rest of this subsection proves that a-logic corresponds to the following FOL theory alogic in a formal sense we make precise in a moment:

- There are some term-formers (e.g. f with arity n).

- There is one unary predicate symbol **at** .

- There is an axiom
$$\exists a.\, \mathbf{at}\, a \qquad \textbf{(folfresh)}$$

- For each inevitable term-former f of arity n, there is an axiom:

$$\forall x_1, \ldots, x_n.\, \neg \mathbf{at}\, (f(x_1, \ldots, x_n)) \qquad \textbf{(folat Lf)}$$

Here is the characteristic property of inevitable term-formers:

LEMMA 1. alogic, $\mathbf{at}\, a$, $f(s_1, \ldots, s_n) = a$ *is inconsistent.*
(Formally: alogic, $\mathbf{at}\, a, f(s_1, \ldots, s_n) = a \vdash$ *.)*

Proof. We use **(folat Lf)**.   ∎

That is, 'in a-logic, terms which are inevitably terms, cannot be atoms'.

The intuition of an atom is a term which the logic identifies as being a variable symbol; the only way this can happen is if $\mathbf{at}\, t$ is derivable. A term is inevitably a term when the logic *always* identifies it as *not* being a variable symbol (and this happens when $\neg \mathbf{at}\, t$ is derivable).

LEMMA 2. $\Gamma \vdash \Delta$ *in a-logic if and only if* alogic, $\Gamma \vdash \Delta$ *in FOL (with its signature extended with* **at** *and so on; **we shall not mention this again**).*

Proof. We can translate from a derivation in a-logic to one in FOL with alogic by transforming instances of **(Fresh)** and **(at L)** as shown (where for simplicity we suppose f is unary):

$$\frac{\mathbf{at}\, a,\, \exists a.\, \mathbf{at}\, a,\, \Gamma \vdash \Delta}{\exists a.\, \mathbf{at}\, a,\, \Gamma \vdash \Delta}\ (\exists \mathbf{L}) \quad (a \notin \Gamma, \Delta)$$

$$\frac{\dfrac{}{\mathbf{at}\, f(s),\, \forall x.\, \neg\mathbf{at}\, f(x),\, \Gamma \vdash \Delta,\, \mathbf{at}\, f(s)}\ (\mathbf{Ax})}{\dfrac{\neg\mathbf{at}\, f(s),\, \mathbf{at}\, f(s),\, \forall x.\, \neg\mathbf{at}\, f(x),\, \Gamma \vdash \Delta}{\mathbf{at}\, f(s),\, \forall x.\, \neg\mathbf{at}\, f(x),\, \Gamma \vdash \Delta}\ (\neg\mathbf{L})}\ (\forall\mathbf{L})$$

∎

So roughly speaking: (**Fresh**) just says that

> *an atom exists,*

and (**at L**) says that

> *they cannot be directly described by (some) term-formers.*

alogic is parametarised over the choice of which f are inevitable terms-formers, and we will always make that choice explicit.

COROLLARY 3. *a-logic is consistent (⊢ is not derivable).*

Proof. FOL is sound and complete with respect to a standard notion of model [van Dalen, 1985; Machover, 1996], so $\Gamma \vdash \Delta$ in a FOL theory it suffices to show that the FOL theory has a model.

This is given by a two-element set $\{anatom, aterm\}$ (here *anatom* and *aterm* are merely suggestive names for two distinct elements, though in view of their uniqueness we might also call them *theatom* and *theterm*). Write $[\![\text{-}]\!]$ for the interpretation of a term or predicate.

$[\![f]\!]$ is the constant map to *aterm* no matter what the arity of f. $[\![\mathbf{at}]\!]$ is $\{anatom\}$. $[\![=]\!] = \{\langle anatom, anatom \rangle, \langle aterm, aterm \rangle\}$. Predicate constants can be interpreted arbitrarily. ∎

Recall from the Introduction that we may call a variable symbol a of which we know **at** a an **atom**.

So is *a*-logic trivial, if it has such a simple model? Of course a model of pure FOL is the one-element set. The interest comes in provability and its properties, and in models of more complex predicates than 'T'. The same holds for *a*-logic. What we have demonstrated is that an *a*-logic theory is a special case of a first-order theory, so we can be optimistic that *a*-logic should behave meta-theoretically very much like FOL (since that is what it is).

2.4 Comments on the rules

This is a simple logic consisting of First-Order Logic with equality, augmented with a unary predicate **at** with two rules (**Fresh**) and (**at L**) as usual, though instead of left- and right-introduction rules we have a left-introduction and -elimination rule.

Recall from the last section that the distinguishing feature of an 'inevitable' term-former f is that (**at L**) may be used for a term of which it is the top-level term-former, or equivalently one for which the corresponding alogic has an axiom (**folat Lf**). The *need* for this becomes clear in §4.1. But why do we say 'inevitably a term'?

One of the important basic lemmas of First-Order Logic is the *Substitution Lemma* (see Lemma 5 for the a-logic version):

If $\Gamma \vdash \Delta$ is derivable, then so is $\Gamma[a \mapsto s] \vdash \Delta[a \mapsto s]$.

This makes formal the idea that 'variables stand for unknown terms' and gives them a slightly dynamic character in the logic. Suppose **at** $f(t_1, \ldots, t_n)$ gives rise to contradiction for certain f. Then we may also obtain a contradiction from **at** $f(t_1[a \mapsto s], \ldots, t_n[a \mapsto s])$. The property of 'having top-level term-former f' is invariant under substitution. Therefore, (**at L**) is not incompatible with the Substitution Lemma after all.

We say 'inevitably' because substitutions do not affect the top-level term-former. We say 'a term' because the property of being a variable symbol is the canonical example of a property which is *not* inevitable (i.e. variables get substituted; that is what they are there for!).

Accordingly, there is no (**at R**) rule such as

$$\Gamma \vdash \mathbf{at}\, a, \Delta$$

if a *is* a variable symbol.

at $\langle a, a \rangle \vdash$ is a theorem with a trivial proof by (**at L**). (We assume a pair term-former $\langle \text{-}, \text{-} \rangle$.)

In summary, **at** a in Γ represents a promise that a will never be instantiated, or at least, not to the wrong kind of term. We may call a variable symbol a of which we know **at** a an **atom**.

3 Meta-properties of a-logic

3.1 Cut-elimination

The proof-theory of a-logic is hardly more complex than that of FOL itself.

LEMMA 4 (Weakening). *If $\Gamma \vdash \Delta$ then $\Gamma, \Gamma' \vdash \Delta', \Delta$.*

Proof. It suffices to consider the derivation rules one-by-one and check that if we weaken the conclusion of an instance with Γ' and Δ', then we may weaken the hypotheses in the same way, and still have an instance of the derivation rule. We shall see that we also have to remember a systematic 'freshening' of the variables, but this is no problem.

This is quite obvious for $(\mathbf{Ax})$, $(\supset \mathbf{R})$, $(\supset \mathbf{L})$, and $(\bot\mathbf{L})$, because these rules make no comment on Γ and Δ so they 'might as well be as big as we like'.

$(\forall\mathbf{R})$ has a side-condition that $x \notin \Gamma, \Delta$ (we say a **is fresh**). If we weaken context and cocontext to Γ, Γ' and Δ, Δ' the instance of the rule will be invalid precisely when $x \in \Gamma'$ or $x \in \Delta'$.

However, we can use *any* fresh x in $(\forall\mathbf{R})$, and we simply choose some x' *not* in Γ, Γ', Δ, or Δ' and rename x to x' systematically from then on as we move upwards, before weakening.[2]

The rest of the rules are just as easy: $(\mathbf{Fresh})$ has a condition similar to $(\forall\mathbf{R})$, which can be dealt with similarly. ∎

Call a derivation **cut-free** when it does not use $(\mathbf{Cut})$. Say it has **depth** n when the greatest number of deduction steps from the conclusion to an assumption is n. If S is a set of predicates write $S[y \mapsto t]$ for the set obtained by applying the substitution $[y \mapsto t]$ pointwise to S. We now prove the **substitution lemma**:

LEMMA 5. *If $\Gamma \vdash \Delta$ has a cut-free derivation of length n then $\Gamma[y \mapsto t] \vdash \Delta[y \mapsto t]$ has a cut-free derivation of length at most n.*

Proof. It suffices to consider the derivation rules one-by-one and check that if we apply $[y \mapsto t]$ to the conclusion, we may do the same to the hypotheses, and still have an instance of the derivation rule. As for Weakening above we may need to systematically accumulate some other transformations, but they are not a problem.

Rules $(\mathbf{Ax})$, $(\supset \mathbf{R})$, $(\supset \mathbf{L})$, and $(\bot\mathbf{L})$ are not a problem because they are propositional and do not care what terms the predicates mention.

$(\forall\mathbf{R})$ is problematic — what if $x \equiv y$ so that $[y \mapsto t]$ affects P in the hypotheses, and what if $x \in s$ and $y \in \Gamma, \Delta$ so that $x \in \Gamma[y \mapsto t], \Delta[y \mapsto t]$ (violating the side-condition)? As for weakening, these problems are solved simply by choosing a fresher x'.

---

[2]What if we choose an x' which is introduced by some other $(\forall\mathbf{R})$ further up in the proof? Valid answers are: *you have the proof in your hand; check, and make sure you don't* or *don't look ahead, but freshen them up if and when you come to them.* This is a form of α-equivalence at the level of proofs which is common to the theory of proofs of any logic with binding.

In ($\forall$**L**) it suffices to observe that by α-equivalence on formulae we may assume $x \not\equiv y$ and $x \notin t$ so that

$$(\forall x.\, P)[y{\mapsto}t] \equiv \forall x.\, (P[y{\mapsto}t]) \qquad P[x{\mapsto}s][y{\mapsto}t] \equiv P[y{\mapsto}t][x{\mapsto}s[y{\mapsto}t]]$$

and after the substitution we still have a valid instance of ($\forall$**L**) where s is replaced by $s[y{\mapsto}t]$.

In (**at L**) it suffices to observe that if s is not a variable, neither is $s[y{\mapsto}t]$.

(**Fresh**) is not a problem unless ('by accident') $a \equiv y$; if we apply $[y{\mapsto}t]$ naïvely then after the substitution we no longer have an instance of the rule. The answer is, as for ($\forall$**R**), to rename a to some fresher a' and then apply the substitution. ∎

LEMMA 6. (**Fresh**) *followed by any other rule* ($*$), *may be commuted to* ($*$) *followed by* (**Fresh**).

As a corollary, all instances of (**Fresh**) *in a derivation may be commuted downwards to the head of the proof.*

Proof. We verify that derivations may be transformed as follows:

$$\cfrac{\Gamma \vdash P, \Delta \qquad \cfrac{\Gamma, P, \mathbf{at}\, a \vdash \Delta}{\Gamma, P \vdash \Delta}\ (\mathbf{Fresh})}{\Gamma \vdash \Delta}\ (\mathbf{Cut}) \implies \cfrac{\cfrac{\Gamma, \mathbf{at}\, a \vdash P, \Delta \qquad \Gamma, \mathbf{at}\, a, P \vdash \Delta}{\Gamma, \mathbf{at}\, a \vdash \Delta}\ (\mathbf{Cut})}{\Gamma \vdash \Delta}\ (\mathbf{Fresh})$$

This is easy, we just need Weakening to add **at** a to the derivation in the top left. Similarly:

$$\cfrac{\cfrac{\Gamma, P[x{\mapsto}s], \mathbf{at}\, b \vdash \Delta}{\Gamma, P[x{\mapsto}s] \vdash \Delta}\ (\mathbf{Fresh})}{\Gamma, \forall x.\, s \vdash \Delta}\ (\forall\mathbf{L}) \implies \cfrac{\cfrac{\Gamma, \mathbf{at}\, b, P[x{\mapsto}s] \vdash \Delta}{\Gamma, \mathbf{at}\, b, \forall x.\, P \vdash \Delta}\ (\forall\mathbf{L})}{\Gamma \vdash \Delta}\ (\mathbf{Fresh})$$

This is easy, since if $b \notin \Gamma, P[x{\mapsto}s], \Delta$ then certainly $b \notin \Gamma, \forall x.\, P, \Delta$.

$$\cfrac{\cfrac{\Gamma,\ \mathbf{at}\, a \vdash P,\ \Delta}{\Gamma \vdash P,\ \Delta}\ (\mathbf{Fresh})}{\Gamma \vdash \forall y.\, P,\ \Delta}\ (\forall\mathbf{R}) \implies \cfrac{\cfrac{\Gamma,\ \mathbf{at}\, a \vdash P,\ \Delta}{\Gamma,\ \mathbf{at}\, a \vdash \forall y.\, P,\ \Delta}\ (\forall\mathbf{R})}{\Gamma \vdash \forall y.\, P,\ \Delta}\ (\mathbf{Fresh})$$

By the side-condition on (**Fresh**) we can assume $a \not\equiv y$ and $a \notin \Gamma, \Delta, P$. By the side-condition on ($\forall$**R**) we also know $y \notin \Gamma, \Delta$. Therefore, $y \notin \Gamma, \Delta, \mathbf{at}\, a, P$ and $a \notin \Gamma, \Delta, \forall y.\, P$. Therefore in the transformed derivation above, the instances of (**Fresh**) and ($\forall$**R**) are still valid.

We consider only one more case:

$$\dfrac{\dfrac{\Gamma, \ \mathbf{at}\, a, \ P[x{\mapsto}s'] \vdash \Delta}{\Gamma, \ P[x{\mapsto}s'] \vdash \Delta} \ (\mathbf{Fresh})}{\Gamma, \ P[x{\mapsto}s], \ s' = s \vdash \Delta} \ (=\!\mathbf{L}) \ \Longrightarrow \ \dfrac{\dfrac{\Gamma, \ \mathbf{at}\, a, \ P[x{\mapsto}s'] \vdash \Delta}{\Gamma, \ \mathbf{at}\, a, \ s' = s, \ P[x{\mapsto}s] \vdash \Delta} \ (=\!\mathbf{L})}{\Gamma, \ s' = s, \ P[x{\mapsto}s] \vdash \Delta} \ (\mathbf{Fresh})$$

We need only verify that $a \notin s$. If this is not the case, we rename a in the derivation of $\Gamma, \ \mathbf{at}\, a, \ P[x{\mapsto}s'] \vdash \Delta$ first.   ∎

THEOREM 7 (Cut elimination). *If $\Gamma \vdash \Delta$ has a derivation then it has a cut-free derivation.*

Proof. We work by induction on the pair of numbers (*cuts, depth*) where *cuts* is the number of instances of (**Cut**) in the derivation, and *depth* is its depth — lexicographically ordered. The essential cases are just as for first-order logic (since **at** has no right rule, there can be no essential case for it); we use the previous lemmas in standard ways [van Dalen, 1985; Machover, 1996].

The commutation cases are also as for first-order logic except we worry about commuting (**Fresh**) from between a left- and a right-introduction. It suffices to verify that (**Fresh**) may be commuted *down* through all the other rules. This follows by the previous lemma.   ∎

3.2 Interpolation

The following result is known as Craig's Interpolation Theorem for First-Order Logic ([D.M.Gabbay, 2005, Theorem 5.65, page 169], [G.Boolos, 1989, Section 23]):

THEOREM 8. *In classical first-order logic with equality, if $P \supset Q$ is provable then there is some I mentioning only predicate symbols and term-formers mentioned in both P and Q, and possibly also =, such that $P \supset I$ and $I \supset Q$ are provable.*

We call I the **the interpolant** and say it is in the **common sublanguage** of P and Q. Because = may appear in the interpolant even if it does not occur in P and Q, we call such an atomic predicate a **logical symbol**.

An easy corollary is:

THEOREM 9. *In classical a-logic with equality, if $P \supset Q$ then there is an interpolant I, such that $P \supset I$ and $I \supset Q$ are provable; = and **at** are logical symbols.*

*More specifically, if P or Q mention **at** (but not necessarily both) it may still be that I mentions **at**, but if neither P nor Q mention **at** then the interpolant will not.*

Proof. In §2.3 we characterised a-logic as a theory alogic in FOL. There-fore $P \supset Q$ is equivalent to $P \wedge$ alogic $\supset Q$ and $P \supset$ (alogic $\supset Q$). By Interpolation, interpolants exist.

We now observe that alogic mentions **at** and an unspecified collection of term-formers. By partitioning these axioms judiciously between the left- and right-hand sides of the implication to minimise the term-formers which appear on both sides, we can obtain the result of the theorem.   ∎

3.3   Models

Since alogic is a FOL theory and FOL has Soundness and Completeness, to give a theory of models it suffices to recall the theory of models for FOL, which is first-order classical logic with equality.

Here we give a *very brief* recap:

A **model Q** of a theory Q is:

- An **underlying set** $|Q|$, which we just write **Q**.

- For each n-ary function-symbol f, a function $[\![f]\!] : \mathbf{Q}^n \to \mathbf{Q}$.

- For each atomic predicate symbol ϕ, a function $[\![\phi]\!] : \mathbf{Q}^n \to Bool$, where $Bool = \{True, False\}$ is a two-element set representing truth values. We may treat $[\![\phi]\!]$ as a set, writing $[\![\phi]\!] \subseteq \mathbf{Q}^n$.

(Recall that we are working, for simplicity, with an unsorted logic, so there is only one underlying set to consider.)

We can either insist that equality be identity in the model, or treat it as just another one of the atomic predicate symbols, in which case the model should also satisfy

> If $(p, q) \in [\![=]\!]$ and $\langle p_1, \ldots, p, \ldots, p_n \rangle \in [\![\phi]\!]$, then $\langle p_1, \ldots, q, \ldots, p_n \rangle \in [\![\phi]\!]$.

We use this latter approach in this paper.

The interpretation of atomic predicates can be extended to predicates in a standard way. Write σ for functions from variable symbols to elements of **Q**, and call these **valuations**.

- $[\![\perp]\!]\sigma$ is false.

- $[\![P \supset Q]\!]\sigma$ is true when *if* $[\![Q]\!]\sigma$ is true, then $[\![P]\!]\sigma$ is true.

- $[\![\forall x.\, P]\!]\sigma$ is true when $[\![P]\!]([x \mapsto p]\sigma)$ is true, for every $p \in \mathbf{Q}$. Here $[x \mapsto p]\sigma$ represents the valuation which acts like σ, only x maps to p.

Say $[\![\mathbf{Q}]\!] \vDash P$ when for all valuations, $[\![P]\!]\sigma$ is true. Say $\Gamma \vDash_{[\![\mathbf{Q}]\!]} \Delta$ when $[\![\mathbf{Q}]\!] \vDash \bigwedge_i G_i \supset \bigvee_j D_j$, where $\Gamma = \{G_1, \ldots\}$ and $\Delta = \{D_1, \ldots\}$ and we interpret empty conjunction as $\top \equiv \bot \supset \bot$ and empty disjunction as $\bot$.

Then a standard theorem is **Soundness and Completeness** for FOL:

THEOREM 10. $\Gamma \vdash \Delta$ *if and only if* $\Gamma \vDash_{[\![\mathbf{Q}]\!]} \Delta$ *for all models* $[\![\mathbf{Q}]\!]$.

4 Substitution

We now give a *theory of substitution*.

4.1 Explicit substitution

Let sub have a ternary term-former *sub* which we sugar to $s\langle u \mapsto t \rangle$. Being a ternary term-former, $s\langle u \mapsto t \rangle$ is valid syntax whether or not u is a variable symbol — however, our axioms allow us to prove nothing about $s\langle u \mapsto t \rangle$ unless **at** u is known.

There are no new predicate symbols.

As a matter of notation, write

$$a\#u \quad \text{is sugar for} \quad \mathbf{at}\, a \,\wedge\, \forall x.\, u\langle a \mapsto x \rangle = u.$$

Intuitively we can read $a\#u$ as 'a is fresh for u' or 'a does not occur (free) in u'. Note that this is not a syntactic judgement; for example a does not occur in the syntax of x, but if we know nothing about x, intuitively $a\#x$ may or may not hold.

Then, sentences are:

$$\mathbf{at}\, a \supset u\langle a \mapsto a \rangle = u \qquad \mathbf{at}\, a \supset a\langle a \mapsto x \rangle = x \qquad \mathbf{at}\, a \wedge \mathbf{at}\, b \supset (a \neq b \Leftrightarrow a\#b)$$

$$\mathbf{at}\, a \wedge b\#u \supset u\langle a \mapsto b \rangle \langle b \mapsto y \rangle = u\langle a \mapsto y \rangle \qquad \mathbf{at}\, a \wedge a\#x \supset a\#u\langle a \mapsto x \rangle$$

$$\mathbf{at}\, b \wedge a\#b \wedge a\#y \supset u\langle a \mapsto x \rangle \langle b \mapsto y \rangle = u\langle b \mapsto y \rangle \langle a \mapsto x \langle b \mapsto y \rangle \rangle$$

(Note that $a\#u$ implies $\mathbf{at}\, a$.) Here a, b, x, y, z, and u are all variable symbols which are universally quantified. For example, the first axiom is in fact $\forall a, u.\, \mathbf{at}\, a \supset u\langle a \mapsto a \rangle = u$.

We do *not* introduce a sentence

$$\forall x, y, z.\, \neg\mathbf{at}\, (x\langle y \mapsto z \rangle) \qquad (\mathbf{folat\ L})$$

nor the corresponding deduction rule ($\mathbf{at\ L}$) for *sub*. This is to avoid the following derivation being valid:

$$\frac{\dfrac{\dfrac{\rule{2cm}{0.4pt}}{\mathbf{at}\, a\langle a \mapsto a \rangle \vdash} (\mathbf{at\ L})}{\mathbf{at}\, a \vdash} (\text{the first axiom})}{\vdash} (\mathbf{Fresh})$$

LEMMA 11. sub+alogic *is consistent.*

Proof. As in Lemma 2 it suffices to give a FOL model, call it $\mathcal{N}$. We recall the model from Lemma 2, but here $\mathcal{N}$ is (intuitively) all atoms with no terms.

The underlying set is $\mathbb{N}$ where $\mathbb{N}$ is the **natural numbers** $0, 1, \ldots$. $[\![=]\!] = \bigcup_{n \in \mathbb{N}} \{\langle n, n \rangle\}$ is the usual graph of equality on the underlying set. $[\![\mathbf{at}]\!] = \mathbb{N}$. Clearly $[\![\exists a.\, \mathbf{at}\, a]\!]$ is 'true'; choose $[\![a]\!] = 1$.

Write p, q, r for arbitrary elements of our model (**we tend to do the same later** when we consider other models). By abuse of notation, henceforth we write $p\langle q \mapsto r \rangle$ instead of $[\![sub]\!](p, q, r)$, and $\mathbf{at}\, p$ instead of $p \in [\![\mathbf{at}]\!]$, and so on, without comment. **We also use this kind of shorthand later.** $[\![sub]\!]$ is defined by:

$$n\langle n \mapsto m \rangle = m \qquad n\langle n' \mapsto m \rangle = n \quad (n' \neq n)$$

It remains to verify that the derivation rules are valid of these interpretations and that they make the interpretations of the axioms valid. This is easy:

- We check in the definition above that $p\langle n \mapsto n \rangle = p$ for all p and all n ($\mathbf{at}\, p$ holds always).

- We similarly check that $n\langle n \mapsto p \rangle = p$ always.

- We check that $n\#m$ holds precisely when $n \neq m$.

- The other equalities follow by checking of cases.

■

We can easily extend the model $\mathcal{N}$ with 'terms'; for example, introduce an element $aterm$ such that $aterm \notin \mathbf{at}$, $aterm\langle n \mapsto p \rangle = aterm$, and $p\langle aterm \mapsto q \rangle = p$. It is not hard to check that the axioms are still satisfied.

To see more complex models, the reader can either verify that 'ordinary syntax' with 'ordinary substitution' (e.g. the syntax of first-order logic with capture-avoiding substitution of terms for variables) is also a model, or they can wait for later when we use models of ECA to construct models of lambda (see below for terminology).

We conclude with some simple lemmas:

LEMMA 12.

1. $a\#u$ if and only if for some $b\#u$, $u\langle a \mapsto b \rangle = u$. *(So to test freshness, it suffices to try freshening the variable and checking for equality.)*

2. If $\mathbf{at}\, a$ and $b\#x$ and $b\#y$, then $b\#x\langle a \mapsto y \rangle$. *(A useful case.)*

Proof.

1. Suppose $a\#u$. Then $u\langle a{\mapsto}b\rangle = u$ by definition. Conversely suppose $u\langle a{\mapsto}b\rangle$ for some $b\#u$. Then by an axiom, $u\langle a{\mapsto}b\rangle\langle b{\mapsto}y\rangle = u\langle a{\mapsto}y\rangle$ for any y.

2. By an axiom $x\langle a{\mapsto}y\rangle = x\langle a{\mapsto}b\rangle\langle b{\mapsto}y\rangle$. Then $b\#x\langle a{\mapsto}b\rangle\langle b{\mapsto}y\rangle$ follows by another axiom.

∎

LEMMA 13. sub + alogic, **at** a, **at** $b \nvdash a \neq b$. *(Two atoms with different names are not necessarily distinct.)*

Proof. Consider a valuation on a and b in the model above, making them equal. The result follows by soundness and completeness of FOL. ∎

LEMMA 14. *Any model of* sub *must have at least two elements.*

Proof. By (**Fresh**), the underlying set is non-empty, since it contains an element a such that **at** a. Now suppose this is the only element. It is easy to check that $a\#a$, which contradicts the right-to-left direction of the third axiom of sub (**at** $a \wedge$ **at** $b \wedge a\#b \supset a \neq b$). ∎

$\exists x, y.\, x \neq y$ alone is sufficient, in the presence of the other axioms, to give the right-to-left direction of the third axiom of sub.

The reader familiar with Nominal Techniques [Gabbay and Pitts, 2001] might expect an axiom along the lines of

$$(\textbf{Fresh}\#) \qquad \forall x.\, \exists a.\, a\#x$$

('no x can mention all atoms'), or even something with the meaning of 'no x can mention more than finitely many atoms' [Gabbay and Pitts, 2001, (Fresh), page 8] (call this **finite support**). This would certainly be useful.

The problem is, (**Fresh**#) only allows us to choose a fresh for x, not necessarily $x_1, \ldots, x_n$ for arbitrarily large n. Unless the language of terms has something like pairing — e.g. a binary term-former of which we can deduce $a\#(x, y) \Leftrightarrow a\#x \wedge a\#y$ — then the only option is to introduce an axiom scheme over all n. Rather than do that (and make our theories infinite), we wait for a more specific theory (lambda for example) and then add made-to-measure axioms with the effect of (**Fresh**#). As for insisting on finite support, we do not bother; it would make some things easier, but for our purposes we have enough.

4.2 'Provably closed'

Meta-level substitution and explicit substitution are quite different. For example $x[y \mapsto z] \equiv x$ but $x\langle y \mapsto z \rangle = x$ is not derivable in general. This fact is clearly brought out by the following definition, which will be useful in a moment:

In sub, write

$$\bullet s \quad \text{for} \quad \forall a.\, \text{at}\, a \supset a \# s$$

where $a \notin s$. Read $\bullet s$ as 's is **(provably) closed**'. By abuse of terminology, we may use 's **is open**' to mean 's is not provably closed'.

It is important to note that $\bullet s$ does not at all imply that the term s is *actually* closed (viewed as syntax). For example, $\bullet x \vdash \bullet x$ but x *is* a variable. Similarly, if $\bullet x$ then $\forall a.\, \text{at}\, a \supset \bullet(a\langle a \mapsto x\rangle)$, but $a\langle a \mapsto x\rangle$ clearly has free variables a and x.

However, in an underlying model $\bullet p$ indicates precisely the notion of closure given by 'is invariant under the explicit substitution'. For example recalling $\mathcal{N}$ the model of sub used in the proof of Lemma 11, it is easy to prove

LEMMA 15. *$\mathcal{N}$ has no closed elements. Recalling also the extension of $\mathcal{N}$ with aterm, the element aterm is closed.*

Proof. By routine concrete calculations which we omit. ∎

Note also that there are no binders in terms, in the sense of a term-former which quotients the syntax of terms by α-equivalence. The *only* binder anywhere in this paper, is $\forall$ which is inherited from FOL and exists at the level of predicates (and the few times we mention 'real' λ-abstraction if we want to build a 'real' function).

(Recall from the end of §2.2 that when Γ entails the empty set we write $\Gamma \vdash$ and say Γ is inconsistent.)

So if a term can contain variable symbols and *still* be 'closed', what does 'closed' mean? Here are two examples of things we can derive:

LEMMA 16. *In* sub...

- **at** $a, \bullet a \vdash$ *is derivable (atoms cannot be closed).*

- **at** $a, \bullet x \vdash \bullet a\langle a \mapsto x\rangle$ *is derivable (if x is closed then, well, x is closed).*

Proof. In the following derivation, we omit sub.

$$\frac{\text{at}\, a,\ a \# a \vdash \qquad \text{at}\, a \vdash \text{at}\, a}{\forall b.\, \text{at}\, b \supset b \# a,\ \text{at}\, a \vdash} (\forall \mathbf{L}), (\supset \mathbf{L})$$

at a, $a\#a \vdash \perp$ follows easily from **sub**.

For the second part, we can use either

$$\textbf{at } a \supset a\langle a \mapsto x \rangle = x \qquad \text{or} \qquad \textbf{at } a \wedge a\#x \supset a\#u\langle a \mapsto x \rangle$$

— we omit the derivations. ∎

Note that $aterm_i$ from the model used in Lemma 11 are closed, and they the only closed element of the model (all the other elements are atoms).

Say a model of **sub** is **non-trivial** when there exist p and q such that $\bullet p$ and $\bullet q$ and $p \neq q$. The model used in Lemma 11 demonstrates that **sub** has at least one nontrivial model. Using Soundness and Completeness of FOL we have:

LEMMA 17. **sub** + **alogic**, $\bullet x$, $\bullet y \not\vdash x = y$.

So let us take stock. We have constructed a logic with a theory **sub** which 'internalises the meta-level' enough to express substitution as an explicit term-former. We can now exploit this to directly and finitely axiomatise the untyped λ-calculus as a first-order theory, and explore in what sense that axiomatisation is correct.

5 λ-calculus

5.1   The theory lambda

Assume **sub**, the theory of explicit substitution above.

We now construct **lambda**, an axiomatisation of the extensional λ-calculus. The signature has binary term-formers **application** and **lambda**, write them st and $\lambda s.t$, and axioms:

$$
\begin{array}{ll}
(\alpha) & \forall a, b, x, y.\, \textbf{at } a \wedge b\#x \supset \lambda a.x = \lambda b.x\langle a \mapsto b \rangle \\
(\beta) & \forall a, x, y.\, \textbf{at } a \wedge \bullet y \supset (\lambda a.x)y = x\langle a \mapsto y \rangle \\
(\xi) & \forall x, y.\, \bullet x \wedge \bullet y \supset (\forall z.\, \bullet z \supset xz = yz) \supset x = y \\
(\sigma\lambda) & \forall a, b, x, y.\, \textbf{at } b \wedge a\#y \supset (\lambda a.x)\langle b \mapsto y \rangle = \lambda a.(x\langle b \mapsto y \rangle) \\
(\sigma\textbf{app}) & \forall a, x, y, z.\, \textbf{at } a \supset (xy)\langle a \mapsto z \rangle = (x\langle a \mapsto z \rangle)(y\langle a \mapsto z \rangle) \\
(\#\textbf{app}) & \forall x, y.\, \exists a.\, a\#x \wedge a\#y \wedge \forall b.\, (b\#axy \Leftrightarrow (b\#a \wedge b\#x \wedge b\#y))
\end{array}
$$

We explore in what precise sense this can be said to be an 'axiomatisation of the extensional λ-calculus' in the rest of this section.

Note that **lambda** is built on top of **sub**. So let us suppose that **sub** *is* 'substitution' and that $a\#x$ *is* 'a does not occur (free) in x' — whatever those sentences mean. In that case we can say:

- (α) is α-conversion.

- (β) is β-conversion. The condition that it be applied to a provably closed term means that a term like $(\lambda a.a)b$ (for a and b atoms) will *not* reduce; as if it were waiting for b to 'become' something, which it can do if an explicit substitution arrives from a surrounding term. Compare this with [Fernández *et al.*, 2005], where a reduction strategy is considered which only allows reductions for *closed* terms; in their case the motivation is efficient reduction.

- (ξ) is a form of extensionality, restricted to the world of provably closed terms.

- ($\sigma\lambda$) and (σ**app**) are the usual rules for moving a substitution inside a term (the action of substitution on atoms is specified in sub).

- (#**app**) says 'there are infinitely many atoms and no x can mention them all'. See Lemma 18.

Interesting points about this axiomatisation are:

- λ is a binary term-former — *just like application*. The special status of the first argument is mediated through **at** .

- This is a finitely axiomatised FOL theory; quantification over all elements is $\forall x.blah$; quantification over all *atoms* (i.e. object-level unknowns) is $\forall x.$**at** $x \supset blah$.

- The underlying domain need not be an inductive type. It is some set with interpretations for the signature, that is, λ, explicit substitution, and **at** .

- The theory includes an *explicit* term-former for substitution, substitution is *not* emulated by applying a λ-abstraction to a term.

- The condition $\bullet y$ in (β), and similar conditions in (ξ), give terms a two-level structure: provably closed terms are the 'real' λ-terms, because we can β-reduce on them.

  This gives open terms something of a flavour of an *internal metalanguage* for talking about terms.

- Accordingly, we can think intuitively as follows:

  1. Unknown λ-terms are variables.

  2. Unknown λ-term *variables* are atoms (terms of which we can prove **at**).

3. Known λ-terms (the 'real' λ-terms) are provably closed (terms of which we can prove $\bullet$).

- Explicit substitution acts on u whether or not u is provably closed (and so is part of the internal meta-language).

- The axioms of lambda represent just one choice, out of many possible systems.

For example, rules

$$(\beta') \quad \forall a, x, y. \, \text{at} \, a \wedge \neg\text{at} \, y \supset (\lambda a.x)y = x\langle a\mapsto y\rangle$$
$$(\xi') \quad \forall x, y. \, \neg\text{at} \, x \wedge \neg\text{at} \, y \supset (\forall z. \bullet z \supset xz = yz) \supset x = y$$

are quite reasonable; they just equate more terms of the internal meta-language.

- The following axiom is *wrong* and *inconsistent*:

$$(\beta_{\textbf{FALSE}}) \qquad \forall a, x, y. \, \text{at} \, a \supset (\lambda a.x)y = x\langle a\mapsto y\rangle$$

Using it, it is not hard to show that $\text{at} \, a \vdash a = (\lambda a.a)a$. However, $\neg\text{at} \, ((\lambda a.a)a)$ (because the top-level term-former is application), so $\text{at} \, a \vdash \bot$. Using (**Fresh**) it is easy to derive $\vdash$, i.e. the system becomes inconsistent and has no models.

Write lambda as shorthand for lambda + sub + alogic. We conclude this subsection with a question, whose answer also throws light on the significance of ($\#\textbf{app}$):

How many atoms are there, and how fresh can they be?

LEMMA 18.

- lambda $\vdash \exists a, b. \, \text{at} \, a \wedge \text{at} \, b \wedge a\#b$. *('There exist two distinct atoms'.)*

- lambda $\vdash \exists a, b, c. \, \text{at} \, a \wedge \text{at} \, b \wedge \text{at} \, c \wedge a\#b \wedge b\#c \wedge a\#c$. *('There exist three distinct atoms'.)*

- *'There exist n distinct atoms' (we do not write out the predicate in full).*

- lambda $\vdash \exists a. \, a\#x_1 \wedge \cdots \wedge a\#x_n$. *('for any x_1 to x_n, there is an a fresh for them all'.)*

Proof.

- By (**Fresh**) there exists some atom a. We then use (#**app**), taking $x = a = y$. (We cannot use (#**app**) directly, since without (**Fresh**) the empty model is a model of lambda.)

- We use (#**app**) again, taking $x = ba = y$ and using the previous part.

- We just consider the case $n = 4$. We use (#**app**), taking $x = cba = y$.

- For the case $n = 1$ we use (#**app**) with $x = x_1 = y$, to obtain $a\#x_1$. For $n = 2$ we use (#**app**) again with $x = ax_1$ and $y = x_2$ to obtain $b\#a, x_1, x_2$. For $n = 3$ we use (#**app**) with $x = bx_1x_2$ and $y = x_3$ to obtain some $c\#b, x_1, x_2, x_3$, and so on.

$\blacksquare$

5.2   Moving from theories of lambda to λ-models

We now show how provably closed terms in lambda correspond to elements of models of the extensional λ-calculus in a more traditional sense (thus giving our axioms some justification).

Let **ECA** (Extensional Combinatory Algebras) be the FOL theory with signature:

- A binary term-former **application** (we write it infix and invisible, as the name suggests).

- Constants (0-ary term-formers) **s** and **k**. Write **i** for **sk**.

— and axioms:

$$\forall x, y. \, \mathsf{k}xy = x \qquad \mathsf{s}xyz = (xy)(xz) \qquad \forall x, y. \, (\forall z. \, xz = yz) \supset x = y.$$

(We do not need extensionality, but the extensional theory is marginally easier to work with.)

An **extensional λ-model** is a FOL model of ECA. This is a traditional notion of model for the λ-calculus [Salibra, 2003b; Selinger, 1997].

We now show how to build a λ-model out of a theory in lambda+sub+alogic. Take any model **Q** of lambda+sub+alogic.

Let Λ be the set of all provably closed elements of (the underlying set of) **Q**. That is, $\Lambda = \{p \in \mathbf{Q} \mid \bullet p\}$.

THEOREM 19. Λ *is an extensional λ-model.*

Proof. We know by Lemma 18 that infinitely many atoms exist in **Q**, so pick three different ones *a*, *b*, *c* (**different** here means that $a\#b$, $b\#c$, and $a\#c$ are all derivable). Set $\mathsf{s} \equiv \lambda abc.(ab)(ac)$ and $\mathsf{k} \equiv \lambda ab.a$. Here we abuse notation and use the same notation for elements which are atoms in the model, as for terms of which we can prove **at** — **we do this without comment henceforth**.

We check that:

- $\bullet\lambda abc.(ab)(ac)$ and $\bullet\lambda ab.a$ are derivable.

- $\lambda abc.(ab)(ac) = \lambda a'b'c'.(a'b')(a'c')$ is derivable, for any other choice of three different atoms.

- Likewise $\lambda ab.a = \lambda a'b'.a'$ is derivable.

The first part is from Lemma 20 which follows. The second two parts are by using (α), and the other axioms.

We must verify that $\mathsf{s}pqr = (pq)(pr)$. We verify this using (β) and the rules of **sub**. Similarly, we can verify $\mathsf{k}pq = p$.

Finally we must verify that with these definitions,

$$\forall p, q. (\forall r. pr = qr) \supset p = q.$$

This follows directly from (ξ). ∎

LEMMA 20. *Assume* lambda + sub + alogic *is in the context. Then:*

$$\mathbf{at}\, a \vdash a\#\lambda a.x \qquad and \qquad b\#x,\ \mathbf{at}\, a \vdash b\#\lambda a.x.$$

Proof.

- Fix some *u* and use Lemma 18 to choose $b\#a, x, u$. By axioms, $\lambda a.x = \lambda b.(x\langle a{\mapsto}b\rangle)$ and

$$(\lambda a.x)\langle a{\mapsto}u\rangle = \lambda b.(x\langle a{\mapsto}b\rangle\langle a{\mapsto}u\rangle) = \lambda b.(x\langle a{\mapsto}b\rangle) = \lambda a.x.$$

- Suppose **at** *a* and $b\#x$. If $b = a$ we use the previous part. Otherwise, $b\#a$ by an axiom, we use the last part of Lemma 18 to obtain a fresh *c* and reason as follows:

$$(\lambda a.x)\langle b{\mapsto}u\rangle = (\lambda c.(x\langle a{\mapsto}c\rangle))\langle b{\mapsto}u\rangle = \lambda c.(x\langle a{\mapsto}c\rangle\langle b{\mapsto}u\rangle) =$$
$$\lambda c.(x\langle a{\mapsto}c\rangle) = \lambda a.x$$

For the last step, we use the second part of Lemma 12.

■

So informally described, we have observed that elements corresponding to s and k live inside a model of lambda and are provably closed. Application is already in lambda, so we can just take the provably closed elements and quotient by provable equality to get an extensional λ-model.

5.3   Moving back

In the previous subsection we showed how to obtain a model of ECA from a model of lambda. Now we consider going back.

Given a model $\mathbf{Q}$ of ECA there are two issues with obtaining a model of lambda:

- How to add atoms and explicit substitution.

- How to interpret λ.

These problems can be handled in a relatively standard way using a suitable quotient of a free term model over the elements of $\mathbf{Q}$. Extend $\mathbf{Q}$ formally with a disjoint countably infinite set $a, b \in \mathbb{A}$ and close formally under application. We obtain a language $\mathcal{L}$ generated by the grammar

$$l ::= (a \in \mathbb{A}) \mid (q \in \mathbf{Q}) \mid l \cdot l.$$

We may drop $\cdot$ and the bracket; $\cdot$ associates to the left.

Quotient $\mathcal{L}$ by

$$(\mathsf{s})l_1 l_2 l_3 = (l_1 l_3)(l_2 l_3), \quad (\mathsf{k})l_1 l_3 = l_1, \quad \text{and} \quad (p)(q) = (pq)$$

provided that l_3 does not mention any atoms.

(The final rule says that if p and q are from $\mathbf{Q}$ then (p) applied to (q) in $\mathcal{L}$ is related to the result of applying p to q in $\mathbf{Q}$, injected into $\mathcal{L}$.)

Call this quotiented set $\mathbf{Q}'$. We shall be **lax** about the difference between l, an element of $\mathcal{L}$, and its equivalence class, an element of $\mathbf{Q}'$.

Here is a simple result:

LEMMA 21.

1. *If l does not mention atoms, then $l = (p)$ for some $p \in \mathbf{Q}$.*

2. *$p \neq q$ in $\mathbf{Q}$ if and only if $(p) \neq (q)$ in $\mathbf{Q}'$.*

3. *As a corollary of the last part, $(\mathsf{s}) \neq (\mathsf{k})$ in $\mathbf{Q}'$.*

Proof. For the first part, we notice that in the absence of atoms we may use the final rule above to concatenate the elements of l.

For the second part, we notice that the rules for deducing $l = l'$ simulate the equalities satisfied in $\mathbf{Q}$ anyway.   ■

Consider $\mathbb{A}$ as a subset of $\mathbf{Q}'$ by mapping a to (a). We may be **lax** and write a when we mean (a), and more generally we may write p for (p). So for example s in the equation below should be (s).

Let $[\![\mathbf{at}\,]\!] = \mathbb{A} \subseteq \mathbf{Q}'$.

Define an interpretation of explicit substitution as follows:

- $l\langle a{\mapsto}u\rangle$ is (the equivalence class of) the string obtained by formally replacing every a in l by u.

  It is not hard to show that this is well-defined, i.e. that:

  $$(\mathsf{s}l_1 l_2 l_3)\langle a{\mapsto}u\rangle = ((l_1 l_2)(l_1 l_3))\langle a{\mapsto}u\rangle \qquad (\mathsf{k}l_1 l_2)\langle a{\mapsto}u\rangle = l_1\langle a{\mapsto}u\rangle$$
  $$((p)(q))\langle a{\mapsto}u\rangle = (pq)\langle a{\mapsto}u\rangle.$$

  For example, for the last part we simply observe that both sides are equal to (pq).

- $p\langle q{\mapsto}u\rangle = p$ for $q \notin \mathbb{A}$ (this is just a convenient default value).

LEMMA 22.

- $a\#ll'$ if and only if $a\#l$ and $a\#l'$.

- $a\#(b)$ if $a \neq b$.

- $a\#(a)$ is not the case.

Therefore, $a\#l$ if and only if a occurs in l,

As a corollary, $\bullet l$ if and only if l does not mention atoms, and $\bullet l$ if and only if $l = (p)$ for some $p \in \mathbf{Q}$.

Proof.

- Direct from the definition of explicit substitution, observing that

  $$(ll')\langle a{\mapsto}l''\rangle = (l\langle a{\mapsto}l''\rangle)(l'\langle a{\mapsto}l''\rangle).$$

- Direct from the definition.

- Direct from the observation that $a\langle a{\mapsto}(\mathsf{s})\rangle = (\mathsf{s})$ and $a\langle a{\mapsto}(\mathsf{k})\rangle = (\mathsf{k})$, and $(\mathsf{s}) \neq (\mathsf{k})$ by a previous part of this result.

For the last part, observe that elements of $\mathbf{Q}'$ have a strict inductive structure as lists, and by induction on structure we conclude that $a\#l$ if and only if a occurs in l.

For the corollary, we use the results just proved and the first part of Lemma 21. ∎

LEMMA 23. *Using the interpretations above, $\mathbf{Q}'$ is a model of* sub.

Proof. We just verify that the axioms of sub are valid for the interpretations given. This is quite long and uses the previous parts of this lemma and Lemma 21, but it is also routine, since the interpretation of substition *really is* substitution. ∎

We now interpret λ in a standard way for combinatory algebra:

- $\lambda a.a = $ skk.

- $\lambda a.l = $ kl if $a\#l$.

- Otherwise, $\lambda a.(ll') = $ s$(\lambda a.l)(\lambda a.l')$.

See [Barendregt, 1984, Definition 7.1.5] or [Selinger, 1997, Subsection 2.2.2] for a futher treatment.

LEMMA 24. $a\#\lambda a.l$, *and* $a\#\lambda b.l$ *if* $a\#l$.

Proof. By a simple induction on the structure of l. For example, $a\#$skk so $a\#\lambda a.a$. ∎

We must now verify that this interpretation satisfies the axioms of lambda:

THEOREM 25. *With the interpretations above, $\mathbf{Q}'$ is a model of* alogic $+$ sub $+$ lambda. *That is:*

1. $\lambda a.l = \lambda b.(l\langle a\mapsto b\rangle)$ *if* $b\#l$.

2. $(\lambda a.l)l' = l\langle a\mapsto l'\rangle$ *if* $\bullet l'$.

3. *If* $\bullet l$ *and* $\bullet l'$ *then if for all* l'' *such that* $\bullet l''$, $ll'' = l'l''$, *then* $l = l'$.

4. $(\lambda a.l)\langle b\mapsto m\rangle = \lambda a.(l\langle b\mapsto m\rangle)$ *if* $a\#m$.

5. $(ll')\langle b\mapsto m\rangle = (l\langle b\mapsto m\rangle)(l'\langle b\mapsto m\rangle)$.

6. *For any* l *and* l', *there exists some* $a\#l, l'$ *such that for all* b, $b\#all'$ *if and only if* $b\#a, l, l'$.

Here we use $a\#l, l'$ as shorthand for $a\#l$ and $a\#l'$, and similarly for $b\#a, l, l'$.

Proof.

1. If $a\#l$ then a does not occur in l so $l\langle a\mapsto b\rangle$ *is* l with b replaced by a. The result follows by induction on l.

2. Suppose l' mentions no atoms. We work by induction on l.

   - $(\lambda a.a)l' = \mathsf{skk}l' = (\mathsf{k}l')(\mathsf{k}l') = l'$ (since l' mentions no atoms; we do not mention this again).
   - If $a\#l$ then $(\lambda a.l)l' = \mathsf{k}ll' = l'$.
   - $(\lambda a.l_1 l_2)l' = ((\lambda a.l_1)l')((\lambda a.l_2)l')$. We now use the inductive hypothesis.

   ∎

Recall that we may write lambda when we mean alogic+sub+lambda.

So, lambda really does axiomatise the λ-calculus. We conclude with two more simple observations:

Say a model of lambda is **non-trivial** when there exist two distinct provably closed elements.

THEOREM 26. lambda *has a non-trivial model.*

Proof. ECA has non-trivial models by standard results of the λ-calculus [Barendregt, 1984]. We use the theorem above and Lemma 21. ∎

THEOREM 27. *If we travel from* **Q** *an extensional λ-model to* **Q**$'$ *a model of* lambda *and back again as described above, we return to* Q.

Proof. By construction Q maps into the model of lambda, with $p\mapsto(p)$. By Lemma 22 this map surjects onto the provably closed elements, and by Lemma 21 the map is an injection. The result follows, since mapping back from **Q**$'$ to **Q** is just taking closed elements. ∎

The converse does not hold: if we travel from **Q**$'$ to **Q** and back to obtain **Q**$''$, it need not be the case that **Q**$''$ is identical to **Q**$'$; fewer equalities may hold in **Q**$''$. For example, consider **Q**$'$ equal to terms up to provable equality with

$$\text{lambda}, \ \forall a, b. \, \text{at } a \wedge \text{at } b \supset (\lambda a.a)b = b$$

(and the rest of lambda). This identity does not hold in **Q**$''$.

We can say that "the map from **Q**$'$ to **Q** throws away the *internal metalanguage*" (mentioned in §5).

So models of lambda are, in a suitably formal sense, models of the λ-calculus, of perhaps a slightly new kind, since open terms are given a direct semantics too.

6 Other theories

We now return to our example theories.

6.1 Meta-level substitution

Until now we have dealt with two notions of substitution: the **meta-level** substitution $[b \mapsto t]$ and the **object-level** (explicit) substitution $\langle a \mapsto s \rangle$.

How do they interact? As follows:[3]

$$u\langle a \mapsto s \rangle [b \mapsto t] \equiv u[b \mapsto t]\langle a \mapsto s[b \mapsto t] \rangle.$$

Note that meta-level substitution is an operation on terms, whereas the explicit substitution is a term-former with a theory of equality. This is why $[b \mapsto t]$ can distribute down under $\langle a \mapsto s \rangle$, without avoiding capture. (The first author builds a λ-calculus of meta-variables based on this idea in [Gabbay, 2005].)

We can internalise this. Extend sub with a binary predicate $\leq$ and axioms:

$$\textbf{at } a \wedge \textbf{at } b \wedge a \leq b \wedge b \leq a \ \supset \ a = b$$

$$\textbf{at } a \wedge \textbf{at } b \wedge \textbf{at } c \wedge a \leq b \wedge b \leq c \ \supset \ a \leq c \qquad \textbf{at } a \supset a \leq a.$$

Write $a < b$ for $a \leq b \wedge a \neq b$. Modify $(\sigma\lambda)$ as follows:

$$(\sigma\lambda) \qquad \begin{aligned} &\textbf{at } b \wedge a\#b \wedge a\#y \wedge b \leq a \ \supset \\ &\quad u\langle a \mapsto x \rangle \langle b \mapsto y \rangle = s\langle b \mapsto y \rangle \langle a \mapsto x\langle b \mapsto y \rangle \rangle \end{aligned}$$

$$(\sigma\lambda') \qquad \begin{aligned} &\textbf{at } b \wedge \textbf{at } b \wedge a < b \ \supset \\ &\quad u\langle a \mapsto x \rangle \langle b \mapsto y \rangle = s\langle b \mapsto y \rangle \langle a \mapsto x\langle b \mapsto y \rangle \rangle \end{aligned}$$

This endows $a < b$ with the intuition 'b is a meta-variable with respect to a', in the sense that *the characteristic of a meta-variable* is that it may be replaced by any term, possibly capturing names in the term under surrounding bindings.

6.2 'Substitution instance of'

Return to sub as originally presented. Substitution $\langle t \mapsto u \rangle$ is not a term, only substitution *applied* to a term, that is $s\langle t \mapsto u \rangle$, is a term (and even then, only meaningfully so if $\textbf{at } t$.

Assume term-formers for pairs and lists and use standard notations (s, t), **Nil**, and $hd :: tl$. Introduce a binary term-former $\cdot$ and axioms such that:

$$s \cdot [] = s \qquad s \cdot (a, u) :: tl = s\langle a \mapsto u \rangle \cdot tl$$

(here a must be an atom).

---

[3]In the equation we slipped in a simplification: $a[b \mapsto t] \equiv a$.

Then a list of pairs whose first elements are atoms behaves like an internal notion of substitution. It is easy to detect when a term is a substitution; let σ, σ',... vary over them.

We can then write predicates such as:

1. 'σ and σ' are extensionally equal':

$$Eq(\sigma, \sigma') = \forall x.\, x \cdot \sigma = x \cdot \sigma'$$

   (since *a*-logic is untyped, we may wish in practice to restrict equality to some subset of all terms, e.g. those not of the form σ'').

   We can then write '*s* is a substitution instance of *t*' as $\exists \sigma.\, s = t \cdot \sigma$. Similarly we can write $\sigma \leq \sigma'$ when $\exists \sigma''.\, Eq(\sigma, \sigma'\sigma'')$ (where $\sigma'\sigma''$ denotes list concatenation), and thus 'σ is a (most general) unifier of *s* and *t*', and so on.

2. The discussion above extends to predicates in a weaker sense. We can write $a\#P$ to mean $\forall u.\, P\langle a \mapsto u \rangle \Leftrightarrow P$ where here $P\langle a \mapsto u \rangle$ is that predicate obtained by replacing every term *s* in *P* by $s\langle a \mapsto u \rangle$. We can express $\bullet P$ as $\forall a.\, \mathbf{at}\, a \supset a\#P$, and so on.

6.3   Logic Programming.

As *a*-logic gives **at** proof-rules, and can express properties of unification using explicit substitutions, it is interesting to consider logic programming in it. We give a sketch of how this might proceed. Define *a*-**hereditarily harrop formulae** by

$$G ::= \top \mid p(ts) \mid G{\wedge}G \mid \mathbf{at}\, t \mid \exists x.G \mid \forall x.G \mid G{\vee}G \mid D{\supset}G$$
$$D ::= \top \mid p(ts) \mid D{\wedge}D \mid G{\supset}D \mid \mathbf{at}\, t \mid \forall x.D$$

Write $\mathcal{G}$ and $\mathcal{D}$ for the sets of all *G* and *D* respectively. Say a derivation is **uniform** when every subderivation concluding a non-atomic goal ends in a right rule or (**FreshL**). Say $(\mathcal{G}, \mathcal{D})$ is a **logic programming language** when for any $G \in \mathcal{G}$ and $\Delta \subseteq \mathcal{D}$, if $\Delta \vdash G$ is derivable then a uniform derivation exists.

THEOREM 28. *a-hereditarily harrop formulae are an abstract logic programming language.*

Proof. We establish commutation rules for those right rules down through those left rules, which can appear in a proof of $\Delta \vdash G$. Most of the proof is as for FOL [Miller *et al.*, 1991]; then by Lemma 6 (**Fresh**) commutes down through any left rule, and it is easily verified that (**at L**) commutes up through any right rule.   ∎

7   Conclusions and related work

Think of an atom as a 'generic element'; certainly, it is distinguished from other kinds of elements (just like the reader can distinguish an 'arbitrary triangle' from other kinds of triangles) — but only if we use **at**. This brings out a motivation for this work, from Kit Fine's book 'reasoning with arbitrary elements' [Fine, 1985]. Fine's 'generic elements' are not quite the atoms in a-logic; for a start he does not insist that generic elements exist 'naked' in the semantics, only that there be some notions of dependence and specialisation. However, we were inspired by the idea of a semantics *in which generic elements are first-class members of the denotation*. To us, 'generic element' suggests 'variable', and 'denotation' suggests 'underlying set'.

We were influenced by the logic of generic judgements by Dale Miller and Alwen Tiu [Miller and Tiu, 2003]. This uses higher-order techniques in logic-programming (with a 'definitional' equality [Schroeder-Heister, 1993]) suited to logic-programming but not as readily to model theory; we suspect it is unsuited for our applications. Their system exhibits a quantifier $\nabla x.Px$ which, roughly, reads 'generate a new variable symbol and assert P applied to x, where P has higher type'. While Miller and Tiu seem to insist on there being no semantic content to this (being logic-programmers, they treat the logic as a computational system and it is perfectly reasonable to request it supply a fresh variable symbol; the meaning of this operation is given by the resulting notions of derivability and proof-search, semantics in the sense of maps to sets is irrelevant). The first author was curious to know what ∇ actually *means*. This paper does not study the mathematical relation of a-logic with the logic of generic judgements (but we will come back to ∇ in more detail in a later paper) — however, the underlying idea suggested by what a semantics for ∇ *could* be, is clearly visible in the intuition we suggest for **at** that it identifies an element as being a variable symbol.

We have explored the semantics of a-logic, but not treated them in this paper (except via the FOL theory alogic and standard FOL model theory). We would like to point out that a nice way of *obtaining* a semantics is via FOL model theory, by doing to a-logic itself what we did to the λ-calculus with lambda: build a signature for the syntax of a-logic in a suitable extension of sub, so that the syntax of a-logic can be interpreted in that theory. Write down axioms for implication and universal quantification (which we can do because we have explicit substitution); much as we did for lambda. Prove that formal derivability in a-logic, coincides with the entailment between interpretations in the theory. Then a model of the theory in a-logic, is a model of a-logic. The peculiarity of this notion of model is that variable symbols exist in the underlying set and are represented in the underlying model by atoms.

This paper has considered the λ-calculus as a finitely-axiomatised object-level theory in first-order logic with a finite signature. We were aware of Salibra's work on Lambda Abstraction Algebras,[4] [Lusin and Salibra, 2004; Salibra, 2003a] which are an axiomatisation of the λ-calculus as an *algebraic* theory (our axiomatisation is not algebraic because of all the hypotheses such as **at** a, $\bullet x$, and $b\#x$), albeit with a infinite signature (roughly, Salibra assumes a unary term-former λa, one for each of a countably infinite set of names a) and infinitely many algebraic equalities (indexed over the a). The parallel with our binary term-former λ (using **at** in axioms to make sure that the first element is an atom) is obvious. However, Salibra can state that names are infinite as a fact of the model, he uses the infinite signature and an appropriate infinite axiom scheme, whereas we have internalised names into the object level and so must introduce a single but non-algebraic axiom (#**app**) to do the same job. In addition, we must introduce conditions $\bullet y$ to avoid inconsistencies (see *wrong* and *inconsistent* on page 749), giving, as we commented there, our open λ-terms the flavour of an *internal meta-language*.

Work by van Benthem and others on Dynamic Logic [van Benthem, 1997] is related to ours. The idea is to treat valuations σ (§3.3) as possible worlds, with an accessibility relation for each variable corresponding to changing the value of σ at x. Substitution of terms for variables becomes movement in a Kripke structure, and universal and existential quantifiers become modalities.

Similar ideas appear in work by Venema [Venema, 1996] and others. Kripke structures are a very general and powerful framework, and *a*-logic slots nicely into this picture as a logic with a predicate **at** which internalises whether the current world assigns a term to a variable (this is to our knowledge a new idea).

Taking the broader view, there are roughly two ways of enriching FOL with structure — add a new predicate with deduction rules directly, or view the structure as a Kripke frame and add modalities and suitable axioms. In this view, *a*-logic and Dynamic logic pursue similar goals by different means. This does raise the question of what other aspects of syntax can be internalised in *a*-logic style.

Michael Beeson created Otter2 [Beeson, 2004; Beeson, 2001] a first-order theorem-prover which extends the term-language of Otter (a first-order

---

[4]If the reader has not heard of these, but knows about Cylindric Algebras [L. Henkin, ; Burris and Sankappanavar, 1981], then they can think of Lambda Abstraction Algebras as infinite-dimensional cylindric algebras for the λ-calculus, and not be too far off the mark.

theorem-prover [Kalman, 2001]) with λ-terms. He calls the underlying logic 'Lambda-Logic'; which we can think of as a single-sorted FOL with λ-abstraction in the language of terms. His motivation was good proof-search principles (from first-order logic) combined with expressivity (from the stronger term-language). Variables cannot be positively identified as such within the logic (which is what **at** does), but a scheme of infinitely many axioms asserts equalities such as $Ap(\lambda x.t, s) = t[x \mapsto s]$ where here Ap is a distinguished binary term-former for application and $t[x \mapsto s]$ is the result of *actually* performing a capture-avoiding substitution of x for s in t.

We can think of the term-language of Lambda-Logic as being λ-terms up to α-equivalence, but not up to β-equivalence. This treatment of higher-orders brings many advantages. The argument (as we interpret it) is that β-reduction is just *too powerful* to put into the language of terms (it is Turing-complete, after all, so unification and matching are undecidable), but still we want functions to practice computer science, so we put in functions but leave β-reduction as an equality to be asserted, rather than being a literal quotient on terms. (Compare this with, say, Isabelle/HOL [Paulson, 2001], which practices higher-order unification.)

In this terminology the theory lambda (§5.1) does the same as Lambda-Logic, only we do not even quotient by α-equivalence. Consequently, λ can be treated as a binary term-former and the whole affair becomes just one theory in FOL enriched with **at** . It may or may not be the case that lambda is well-suited to theorem-proving (it does not delegate α-equivalence to the meta-level, and this might well be a great nuisance), but it would be possible to express the axioms of lambda in *any* theorem-prover with first-order logic, as we have demonstrated.

a-logic draws on Nominal Logic [Pitts, 2003; Gabbay and Cheney, 2004] and to some extent FreshML [FreshML, ; Shinwell *et al.*, 2003]. The terminology 'atom' is directly drawn from that work, as is the notion of 'fresh' #. However, in Nominal Logic atoms populate a distinct type, and # is defined in terms of permutations using the NEW quantifier N [Gabbay and Pitts, 2001], rather than substitutions using $\forall$. The 'nominal way' gives atoms the character of a collection of atomic constants representing precisely themselves up to identity, rather than a subset of the universe which are atoms representing generic elements. Consistently with the 'atomic' nature of atoms, Nominal Logic and FreshML are designed for *deep embeddings* of syntax in an ambient logic, so that the syntax is an inductive datatype [Azurat and Prasetya, 2002; Wildmoser and Nipkow, 2004]. Thus, terms like $(1 + 1)$ are equal in the logic when they are provably equal as syntax, and not when they are equal in some denotational sense (so $1 + 1$ is not equal to 2). Any *other* notions of equality, such as arithmetic equality or

β-convertibility, are expressed (probably inductively) as relations. In our implementation, equality is a mix of the two; denotational on closed elements, and syntactic (except for the action of the explicit substitution) on open ones.

It is important to realise that Nominal Unification [Urban *et al.*, 2004] and Nominal Rewriting [Maribel Fernández, 2004] introduces a two-level hierarchy of variables, 'unknowns' X and 'atoms' a, which is *not* present in Nominal Logic [Pitts, 2003]. This corresponds to the 'meta-level substitution' we considered in this paper in §6.1, though there are technical differences (only substitution, no permutations) and the axiomatisation corresponds more closely to the New Calculus of Contexts [Gabbay, 2005].

We have shown how the λ-calculus can be expressed in first-order logic. Compare this with logic expressed in the λ-calculus. For example Beeson (having extended terms of FOL with λ-abstraction to obtain Lambda Logic) uses the λ-terms to express an object-level logic [Beeson, 1998] (we could do the same here, starting from lambda). Barendregt and others have pursued the idea of taking λ-calculus as *the* foundation, and expressing logic in it (and avoiding inconsistency while doing so, which turns out to be hard) [H. Barendregt, 1998b; H. Barendregt, 1998a] and [Barendregt, 1984, p.560]. Finally, many systems are based directly on higher-order logic [Leivant, 1994; Paulson, 1989; Prawitz, 1964], which simply quotients terms by β-reduction and gains much power as a result (we cannot discuss the trade-offs here [Leivant, 1994; Beeson, 2004]).

To avoid inconsistency in lambda, we restrict β-reduction to st where t is provably closed. We note that this idea is similar to work by Fernández et el [Fernández *et al.*, 2005] (they use a similar condition for a different purpose, to develop efficient reduction strategies and eliminate the need for α-conversion).

This paper is one of very many which try to bring the advantages of the λ-calculus to FOL, and more generally to reconcile them (in new and interesting ways!). We do this in a general way, which we demonstrate would be applicable also to other systems with binders, and we do this working directly in FOL using standard tools.

BIBLIOGRAPHY

[Azurat and Prasetya, 2002] A. Azurat and I.S.W.B. Prasetya. A survey on embedding programming logics in a theorem prover. Technical Report 7, Institute of information and computing sciences, Utrecht University, 2002.

[Barendregt, 1984] H. P. Barendregt. *The Lambda Calculus: its Syntax and Semantics (revised ed.)*, volume 103 of *Studies in Logic and the Foundations of Mathematics*. North-Holland, 1984.

[Beeson, 1998] M. Beeson. Unification in lambda-calculi with if-then-else. In *CADE-*

15: Proceedings of the 15th International Conference on Automated Deduction, pages 103–118, 1998.

[Beeson, 2001] M. Beeson. A second-order theorem prover applied to circumscription. *IJCAR*, pages 318–324, 2001.

[Beeson, 2004] M. Beeson. Lambda logic. *IJCAR*, pages 460–474, 2004.

[Bell and Machover, 1977] J. Bell and M. Machover. *A course in mathematical logic.* North-Holland, 1977.

[Burris and Sankappanavar, 1981] S. Burris and H. Sankappanavar. *A Course in Universal Algebra.* Springer, 1981. Also available online.

[D.M.Gabbay, 2005] L.Maksimova D.M.Gabbay. *Interpolation and Definability: Modal and Intuitionistic Logics.* Clarendon Press, 2005.

[Fernández et al., 2005] Fernández, Mackie, and Sinot. Closed reductions: Explicit substitutions without α-conversion. *Mathematical Structures in Computer Science*, 15(2):343–381, 2005.

[Fine, 1985] K. Fine. *Reasoning with Arbitrary Objects.* Blackwell, 1985.

[FreshML,] FreshML. FreshML homepage, . http://www.freshml.org.

[Gabbay and Cheney, 2004] M. J. Gabbay and J. Cheney. A sequent calculus for nominal logic. In *Proc. 19th IEEE Symposium on Logic in Computer Science (LICS 2004)*, pages 139–148, 2004.

[Gabbay and Günthner, 1986] D. Gabbay and F. Günthner, editors. *Handbook of philosophical logic.* Number 166 in Synthese Library. D.Reidel Publishing company, 1986.

[Gabbay and Pitts, 2001] M. J. Gabbay and A. M. Pitts. A new approach to abstract syntax with variable binding. *Formal Aspects of Computing*, 13:341–363, 2001.

[Gabbay, 2005] M. J. Gabbay. A new calculus of contexts. In *PPDP*, 2005.

[G.Boolos, 1989] R. Jeffrey G. Boolos. *Computability and Logic.* Cambridge University Press, 1989.

[H. Barendregt, 1998a] M. Bunder H. Barendregt, W. Dekkers. Completeness of the propositions-as-types interpretation of intuitionistic logic into illative combinatory logic. *J. Symbolic Logic*, 3:869–890, 1998.

[H. Barendregt, 1998b] M. Bunder H. Barendregt, W. Dekkers. Completeness of two systems of illative combinatory logic for first-order propositional and predicate calculus. *Archive für Mathematische Logik*, 37:327–341, 1998.

[Johnstone, 1987] P. T. Johnstone. *Notes on logic and set theory.* Cambridge University Press, 1987.

[Kalman, 2001] J. A. Kalman. *Automated Reasoning with Otter.* Rinton Press, 2001.

[L. Henkin,] A. Tarski L. Henkin, J. D. Monk. *Cylindric algebras.* North Holland. Part I (1971), Part II (1985).

[Leivant, 1994] D. Leivant. Higher order logic. In *Handbook of Logic in Artificial Intelligence and Logic Programming*, volume 2, pages 229–322. 1994.

[Lusin and Salibra, 2004] S. Lusin and A. Salibra. The lattice of lambda theories. *Journal of Logic and Computation*, 14 n.3:373–394, 2004.

[Machover, 1996] M. Machover. *Set Theory, Logic and their Limitations.* Cambridge University Press, 1996.

[Maribel Fernández, 2004] I. Mackie M. Fernández, Murdoch J. Gabbay. Nominal rewriting. Submitted, January 2004.

[Miller and Tiu, 2003] D. Miller and A. Tiu. A proof theory for generic judgments: An extended abstract. In *Proceedings of LICS 2003*, pages 118–127. IEEE, June 2003.

[Miller et al., 1991] D. Miller, G. Nadathur, F. Pfenning, and A. Scedrov. Uniform proofs as a foundation for logic programming. *Annals of Pure and Applied Logic*, 51:125–157, 1991.

[Milner et al., 1992] R. Milner, J. Parrow, and D. Walker. A calculus of mobile processes, II. *Information and Computation*, 100(1):41–77, September 1992.

[Paulson, 1989] L. C. Paulson. The foundation of a generic theorem prover. *Journal of Automated Reasoning*, 5(3):363–397, 1989.

[Paulson, 2001] L. Paulson. *The Isabelle reference manual.* Cambridge University Computer Laboratory, February 2001.

[Pitts, 2003] A. M. Pitts. Nominal logic, a first order theory of names and binding. *Information and Computation,* 186:165–193, 2003.

[Prawitz, 1964] D. Prawitz. Completeness and hauptsatz for second-order logic. *Theoria,* 33:246–258, 1964.

[Salibra, 2003a] A. Salibra. Lambda calculus: models and theories. In G. Scollo F. Spoto and A. Nijhol, editors, *Proceedings of the Third AMAST Workshop on Algebraic Methods in Language Processing (AMiLP-2003),* number 21 in TWLT Proceedings, pages 39–54, University of Twente, 2003. Invited Lecture.

[Salibra, 2003b] A. Salibra. Topological incompleteness and order incompleteness of the lambda calculus. *ACM Trans. Comput. Logic,* 4(3):379–401, 2003.

[Schroeder-Heister, 1993] P. Schroeder-Heister. Definitional reflection and the completion. In *Extensions of Logic Programming,* volume 798 of *Springer Lecture Notes in Artificial Intelligence,* pages 333–347, 1993.

[Selinger, 1997] P. Selinger. *Functionality, polymorphism, and concurrency: a mathematical investigation of programming paradigms.* PhD thesis, University of Pennsylvania, June 1997.

[Shinwell et al., 2003] M. R. Shinwell, A. M. Pitts, and M. J. Gabbay. FreshML: Programming with binders made simple. In *Eighth ACM SIGPLAN International Conference on Functional Programming (ICFP 2003), Uppsala, Sweden,* pages 263–274. ACM Press, August 2003.

[Urban et al., 2004] C. Urban, A. M. Pitts, and M. J. Gabbay. Nominal unification. *Theoretical Computer Science,* 323:473–497, 2004.

[van Benthem, 1997] J. van Benthem. Modal foundations for predicate logic. *Logic Journal of the IGPL,* 5(2):259–286, 1997.

[van Dalen, 1985] D. van Dalen. Intuitionistic logic. In Gabbay and Günthner, editors, *Handbook of Philosophical Logic vol III,* volume 166 of *Synthèse,* chapter 4. D. Reidel, 1985.

[Venema, 1996] Y. Venema. A modal logic of substitution and quantification. *Logic Colloquium '92,* pages 293–309, 1996.

[Wildmoser and Nipkow, 2004] M. Wildmoser and T. Nipkow. Certifying machine code safety: Shallow versus deep embedding. In *TPHOLs,* volume 3223 of *LNCS,* pages 305–320. Springer, 2004.

INDEX

www.ingramcontent.com/pod-product-compliance
Lightning Source LLC
Chambersburg PA
CBHW060414220326
41598CB00021BA/2179